MW00837635

Management of Contaminated Site Problems

Management of Contaminated Site Problems

Second Edition

Kofi Asante-Duah
Scientific Advisor/Environmental Consultant
Human Health Risk Assessment Practice
Washington, DC, USA

CRC Press
Taylor & Francis Group
Boca Raton London New York

CRC Press is an imprint of the
Taylor & Francis Group, an **informa** business

CRC Press
Taylor & Francis Group
6000 Broken Sound Parkway NW, Suite 300
Boca Raton, FL 33487-2742

© 2019 by Taylor & Francis Group, LLC
CRC Press is an imprint of Taylor & Francis Group, an Informa business

No claim to original U.S. Government works

Printed on acid-free paper

International Standard Book Number-13: 978-1-4987-6156-7 (Hardback)

This book contains information obtained from authentic and highly regarded sources. Reasonable efforts have been made to publish reliable data and information, but the author and publisher cannot assume responsibility for the validity of all materials or the consequences of their use. The authors and publishers have attempted to trace the copyright holders of all material reproduced in this publication and apologize to copyright holders if permission to publish in this form has not been obtained. If any copyright material has not been acknowledged, please write and let us know so we may rectify in any future reprint.

Except as permitted under U.S. Copyright Law, no part of this book may be reprinted, reproduced, transmitted, or utilized in any form by any electronic, mechanical, or other means, now known or hereafter invented, including photocopying, microfilming, and recording, or in any information storage or retrieval system, without written permission from the publishers.

For permission to photocopy or use material electronically from this work, please access www.copyright.com (http://www.copyright.com/) or contact the Copyright Clearance Center, Inc. (CCC), 222 Rosewood Drive, Danvers, MA 01923, 978-750-8400. CCC is a not-for-profit organization that provides licenses and registration for a variety of users. For organizations that have been granted a photocopy license by the CCC, a separate system of payment has been arranged.

Trademark Notice: Product or corporate names may be trademarks or registered trademarks, and are used only for identification and explanation without intent to infringe.

Library of Congress Cataloging-in-Publication Data

Names: Asante-Duah, D. Kofi, author.
Title: Management of contaminated site problems / by Kofi Asante-Duah.
Description: 2nd edition. | Boca Raton : Taylor & Francis, 2019. | Includes
bibliographical references and index.
Identifiers: LCCN 2018052582| ISBN 9781498761567 (hardback : alk. paper) |
ISBN 9780429198021 (ebook)
Subjects: LCSH: Hazardous waste sites—Management.
Classification: LCC TD1052 .A83 2019 | DDC 628.5/5—dc23
LC record available at https://lccn.loc.gov/2018052582

Visit the Taylor & Francis Web site at
http://www.taylorandfrancis.com

and the CRC Press Web site at
http://www.crcpress.com

To: Dad—George Kwabena Duah

To: Mom—Alice Adwoa Twumwaa

To: Kojo Asante-Duah (a.k.a., K.J.)

To: Kwabena Asante Duah (a.k.a., K.B./Daddy-K./Koby)

To: Adwoa Twumwaa Asante-Duah (a.k.a., A.T./
Obaa-Sima/Maame-T./Naana)

To: My Extraordinary Families (of Abaam, Kade and Nkwantanan)
—and all the offsprings and offshoots
and
To the Everlasting and Ever-Loving Memories of:
Daddy, George Kwabena Duah (a.k.a., GKD/Agya Duah)
Grandma, Nana Martha Adwoa Oforiwaa and Grandpa
Grandma, Nana Abena Nketia Owusua and Grandpa
Atta Kakra (a.k.a., Emmanuel K. Asare Duah)
Atta Panin (a.k.a., Ebenezer K. Asare Duah)
Osagyefo Dr. Kwame Nkrumah @GH

Contents

PART I   A General Overview of Contaminated Site Problems

PART II   Contaminated Site Problem Diagnosis

PART III   *Contaminated Site Risk Assessment: Concepts, Principles, Techniques, and Methods of Approach*

PART IV   Development, Design, and Implementation of Risk-Based Solutions and Restoration Plans for Contaminated Site Problems

PART V   Appendices

Preface

Wide-ranging environmental media that have been contaminated to various degrees by numerous chemicals or pollutants exist in several locations globally, and these constitute huge numbers of derelict or contaminated site problems with distinct complexities. Without a doubt, environmental contamination and related contaminated site issues represent a rather complex problem with worldwide implications. The greatest concern from such impacted sites relates to the fact that the properties of interest tend to pose significant risks to numerous stakeholders (e.g., to the general public because of the potential health and environmental effects of the contaminants, to property owners and financiers due to possible financial losses and reduced property values, and to other potentially responsible parties because of possible financial liabilities that could arise from their consequential effects). Risks to both human and ecological health that could arise from environmental contamination and related contaminated site problems are indeed a matter of particularly grave concern. The effective management of contaminated sites has therefore become a very important environmental policy and risk management issue that will remain a growing social challenge for years to come. In efforts to tackle this, risk assessment promises a systematic way for developing appropriate strategies to aid risk management and policy decisions associated with contaminated site problems. In fact, risk assessment represents one of the fastest evolving tools available for developing appropriate and cost-effective strategies to aid contaminated site restoration and risk management decisions—as reflected in the elaborate presentation offered by this book.

In practice, risk assessment generally serves as a tool that can be used to organize, structure, and compile scientific information to help identify existing hazardous situations or problems, anticipate potential problems, establish priorities, and provide a basis for regulatory controls and/ or corrective actions. It may also be used to help gauge the effectiveness of corrective measures or remedial actions. By and large, risk assessment has become an important contemporary foundational tool in the development of effectual contaminated site risk management strategies and policies. In fact, to ensure public health and environmental sustainability, decisions relating to contaminated site management should generally be based on a systematic and scientifically valid process, such as via the use of risk assessment. In the interest of usually limited resources available for environmental restoration programs, it is always prudent to identify credible scientific methods that allow technically sound decisions to be formulated even in the absence of active cleanup actions undertaken at contaminated sites. With that said, it is noteworthy that not all contaminated sites necessarily require cleanup, and neither do equally contaminated sites necessarily require the same degree of cleanup or restoration efforts, and this is where risk-based decision-making becomes the tool of choice.

This book outlines risk-based strategies that are useful in the investigation, characterization, management, and restoration of environmental contamination and related contaminated site problems. It draws on many real-world problem scenarios from across the globe to illustrate several crucial points. It also contains a review of several useful remediation techniques and their common applicability scenarios. Overall, this book provides a wealth of information that can be applied to the planning, appraisal, assessment, evaluation, and implementation of contaminated site investigation and restoration programs by bringing together the common themes and concepts necessary for making credible decisions about the reclamation of contaminated sites for redevelopment and/or protection of the health of populations in the vicinity. Specifically, the book presents some very important tools and methodologies that can be used to address contaminated site risk management problems in a consistent, efficient, and cost-effective manner. To this end, this book focuses on the key principles necessary for an effective management of contaminated site problems that are achievable through an application of the risk assessment paradigm in the overall implementation process. Major attributes of the book include the following:

- Discusses the site assessment processes involved in mapping out the types and extents of contamination at potentially contaminated and derelict lands
- Addresses issues relevant to the investigation and management of potentially contaminated and derelict land problems that span problem diagnosis and site characterization to the development and implementation of site restoration programs
- Elaborates the types of activities generally required of environmental site assessments and often conducted for contaminated site problems, including a discussion of the steps involved in the design of comprehensive site investigation workplans
- Presents concepts and techniques/methodologies in risk assessment that may be applied to contaminated site problems, and then provides a guidance framework for the formulation of contaminated site risk assessments as well as the follow-on risk management and site restoration decisions
- Describes specific methods of approach for conducting human health and ecological risk assessments, consisting of an elaboration of the requisite protocols and tasks involved in its application to contaminated site problems
- Offers prescriptions and decision protocols to be utilized or followed when making key decisions in corrective action assessment and response programs for contaminated site problems
- Identifies the key elements necessary for the design of effectual corrective action programs, comprised of the use of appropriate diagnostic tools to determine the degree of contamination and/or the concomitant risk as well as the subsequent development of an effectual risk management and/or site restoration program
- Elucidates the scientific and technical requirements and strategies for the diagnostic assessments of potentially contaminated and derelict lands, and also for developing site restoration goals and corrective action plans for such problems
- Synthesizes and summarizes the several relevant generic principles that may be applied to the characterization and management of the variety of contaminated site situations often encountered in practice
- Presents a comprehensive review of several useful remediation techniques and their common applicability scenarios
- Serves as a practical guide to the effective application of risk assessment and risk-based strategies to contaminated site restoration and management problems, offering an understanding of the scientific basis of risk assessment and its applicability to contaminated site management decisions

All in all, the above-noted features fundamentally encompass a collection and synthesis of the principal elements of the risk assessment process in a format that can be applied to contaminated site problems. Several illustrative example problems (consisting of a variety of analyses relevant to the design of effectual site characterization, risk management, and corrective action programs for potentially contaminated site problems) are interspersed throughout the book to help present the book in an easy-to-follow, pragmatic manner.

Meanwhile, it must be acknowledged that the nature and extent of contaminated site problems tend to be rather diverse, so much as to make it almost impractical to cover every aspect of it in one book. Nevertheless, the concepts presented in this book will generally be applicable to the broad spectrum of contaminated site situations often encountered in practice. In fact, despite the variability in the nature of contaminated site problems typically encountered at different locations, there seems to be growing consensus on the general technical principles and procedures that should be used in the management of such sites. Along those lines, this book presents a synthesis of technically sound principles and concepts that should find unbiased global application to contaminated site problems. Indeed, the basic technical principles and concepts involved in the management of contaminated site problems will generally not differ in any significant way from one geographical region to another; basic differences in strategies tend to lie in regional or local policies and legislation. Consequently, this book will serve as a useful resource for any sector of the global community that is faced with contaminated site risk management and site restoration decisions.

On the whole, the subject matter of this book should be of interest to many professionals encountering environmental contamination problems in the broader sense but particularly as they relate to contaminated site risk management problems alongside risk-based site restoration decision-making. The specific intended audience includes environmental consulting professionals (environmental consultants and scientists/engineers); hazardous waste site managers; national, regional, and local government agencies; public policy analysts; environmental attorneys; environmental and public health regulatory agencies; manufacturing and process industries; real-estate developers and investors involved in property transfers; utility companies; chemical and pharmaceutical industries; petroleum and petrochemical industries; environmental NGOs; consumer service industries; environmental firms and construction companies; banks and lending institutions; and a miscellany of health, environmental, and consumer advocacy interest groups. The book also serves as a useful *educational and training resource for both students and professionals in the environmental and allied fields*, particularly those who have to deal with contaminated or derelict land risk management as well as land reclamation or site restoration issues. Written for both the novice and the experienced, this book is an attempt to offer a simplified and systematic presentation of contaminated site characterization and risk management strategies through a design/layout that will carefully navigate the user through the major processes involved.

Finally, it is noteworthy that a key objective in preparing this revised and expanded edition of the book has been to, insofar as practicable, incorporate new key developments and/or updates in the field since the previous version was last published. Also included in this revised edition are more elaborate discussions relating to the application of risk-based solutions to contaminated site problems and associated site restoration decision-making modalities or processes. Another notable feature of the revised edition is the sectional reorganization that has been carried out for some topics, all meant to help with the overall flow of the presentations but especially to facilitate a more holistic learning process/experience afforded by this new book. All in all, the book is organized into five parts, consisting of 14 chapters, together with a bibliographical listing and a set of four appendices. It is the author's hope that the five-part presentation will provide adequate guidance and direction for the successful completion of corrective action assessment and response programs that are to be designed for any type of contaminated site problem at any geographical location. The structured layout should also help with any efforts to develop effective classroom curricula for teaching purposes. Ultimately, the systematic risk-based protocols presented in this book should indeed aid many environmental scientists and related environmental professionals to formulate and manage contaminated/derelict land issues and associated problems more efficiently.

Acknowledgments

I am indebted to several people for both the direct and indirect support afforded me during the period I worked on this book project, especially my proximate family for providing a loving environment and roadway to embark on this journey. Many thanks to the extensive Duah family (Abaam, Kade, Nkwantanan, and beyond) and offshoots as well as to some personal role models and motivational/inspirational personalities who provided much-needed moral and enthusiastic support throughout the preparation of this book.

Special thanks, once again, does indeed belong to my immediate/core family (Adwoa, Kwabena/ Koby, Kojo, and Tish) for, among several other things, putting up with all the time I had to spend away from them to successfully complete this book. Thanks also to several colleagues associated with Premier [Occupational and Environmental] Health Services Group (an occupational and environmental health firm offering consulting and primary healthcare services) and The Environmental Risk Solutions Group (an environmental consulting firm with practice focusing on human health risk assessments) for providing miscellaneous assistance.

The support of the publishing, editorial, and production staff at CRC Press/Taylor & Francis Group Publishers in helping to bring this book project to a successful conclusion is very much appreciated. I also wish to thank every author whose work is cited in this book for having provided some pioneering work to build on.

Finally, it should be acknowledged that this book benefited greatly from the review comments of several anonymous individuals as well as from discussions with a number of professional colleagues. Any shortcomings that remain are, however, the sole responsibility of the author.

Kofi Asante-Duah
Washington, DC, USA
(08-08-2018)

Author

Kofi Asante-Duah presently serves as chief scientific advisor to the DC Department of Energy and Environment. In this capacity, he provides technical support/oversight to numerous real-life multipathway environmental risk management projects comprised of multimedia environmental investigations and characterization, development of data quality objectives and sampling design, statistical analysis of environmental data and multimedia environmental modeling, ecological and human health risk assessments, probabilistic risk assessments, corrective action planning, development of risk-based cleanup goals for corrective action and site restoration decisions, remedial technologies screening and selection, 'no-further-action' documentation and site closure/redevelopment, and risk management programs.

Part I

A General Overview of
Contaminated Site Problems

This part of the book is comprised of three sections—consisting of the following specific chapters:

- Chapter 1, *Introduction*, presents archetypical background information on the sources/ origins of environmental contamination and chemical exposure problems in general and contaminated site problems in particular. This chapter also provides a broad overview on the general types of issues that need to be addressed in order to establish an effective risk management or corrective action program for contaminated site problems.
- Chapter 2, *The Nature of Contaminated Site Problems*, presents a discussion on the general nature and types of contaminated site problems often encountered in practice; it also elaborates on the general implications of such types of environmental contamination problems, as well as offers guidance to developing proper classification nomenclatures for the comparative evaluation of diverse sites.
- Chapter 3, *Legislative-Regulatory Considerations in Contaminated Site Decision-Making*, presents a summary review of selected environmental legislation potentially relevant to the management of contaminated site problems. Some of the well-established environmental laws found in the different regions of the world are introduced here.

1 Introduction

Wide-ranging environmental media that have been contaminated to various degrees by numerous chemicals or pollutants exist in several locations globally—and these constitute huge numbers of derelict or contaminated site problems with distinct complexities. Indeed, environmental contamination and related contaminated site issues represent a rather complex problem with worldwide implications. The greatest concern from such impacted sites and facilities relates to the fact that the properties of interest tend to pose significant risks to numerous stakeholders—as, for example, to the general public because of the potential health and environmental effects of the contaminants, to property owners and financiers due to possible financial losses and reduced property values, and to other potentially responsible parties (PRPs) because of possible financial liabilities that could arise from their consequential effects. Risks to both human and ecological health that could potentially arise from environmental contamination and related contaminated site problems are indeed a matter of particularly grave concern and significant interest. The effective management of contaminated sites has therefore become a very important environmental policy and risk management issue that will remain a growing social challenge for years to come—and the key to achieving this lies in the use of an informed strategy to develop and implement case-specific corrective action programs; in efforts to tackle this, risk assessment promises a systematic way for developing appropriate strategies to aid the contaminated site risk management and risk policy decisions. In fact, risk assessment notably represents one of the fastest evolving tools available for developing appropriate and cost-effective strategies to aid contaminated site restoration and risk management decisions.

This book, as a whole, outlines risk-based and related or complementing strategies that are useful in the investigation, characterization, management, and restoration of contaminated site problems. It draws on many real-world problem scenarios from across the globe to illustrate several crucial points. It also contains a review of several useful remediation techniques and their applicability scenarios. Indeed, to ensure public health and environmental sustainability, decisions relating to contaminated site management should generally be based on systematic and scientifically valid processes—such as via the use of risk assessment. Thus, this book highlights the application of risk assessment concepts and principles in the development of effective contaminated site management programs—including site identification, characterization, and restoration efforts.

To start off, this introductory chapter presents archetypical background information on the sources/origins of environmental contamination and chemical exposure problems in general and contaminated site problems in particular. This chapter also provides a broad overview on the general types of issues that need to be addressed in order to establish an effective risk management or corrective action program for contaminated site problems.

1.1 TRACKING THE CAUSES AND POTENTIAL IMPACTS OF ENVIRONMENTAL CONTAMINATION AND CONTAMINATED SITE PROBLEMS

A look into unfortunate lessons from the past—such as the Love Canal incident in New York (in the United States)—clearly demonstrates the dangers that could arise from the presence of contaminated sites within or near residential communities. Love Canal was a disposal site for chemical wastes for about 25–30 years (see, e.g., Gibbs, 1982; Levine, 1982), and subsequent use of this site culminated in residents of a township in the area suffering from various health impairments; it is believed that several children in the neighborhood apparently were born with serious birth defects. Analogous incidents are recorded in other localities in the United States, in Europe, in Japan, and indeed in numerous other locations in Asia and elsewhere around the world. The presence of potentially contaminated sites can therefore create potentially hazardous situations and pose significant risks of concern to society.

Indeed, it is apparent that the mere existence or presence of contaminated sites within a community or a human population habitat zone can invariably lead to contaminant releases and potential receptor exposures—possibly resulting in both short- and long-term effects on a diversity of populations within the "zone of influence" of the contaminated site. By and large, any consequential chemical intake or uptake by exposed organisms can cause severe health impairments or even death, if taken in sufficiently large amounts. Also, there are those chemicals of primary concern that can cause adverse impacts even from limited exposures. Still, human and other ecological populations are continuously in contact with varying amounts of chemicals present in air, water, soil, food, and other consumer products—among several other possible sources. Such human and ecological exposures to chemical constituents can indeed produce several adverse health effects in the target receptors, as well as potentially impart significant socioeconomic woes to the affected communities. What is more, historical records (as exemplified in Table 1.1) clearly demonstrate the dangers that may result from the presence of environmental contamination and chemical exposure situations (including those attributable to contaminated site problems) within or near residential communities and human work environments or habitats (Alloway and Ayres, 1993; Ashford and Miller, 1998; BMA, 1991; Brooks et al., 1995; Canter et al., 1988; Gibbs, 1982; Grisham, 1986; Hathaway et al., 1991; Kletz, 1994; Levine, 1982; Long and Schweitzer, 1982; Meyer et al., 1995; Petts et al., 1997; Rousselle et al., 2014; Williams et al., 2008).

1.1.1 Coming to Terms with Environmental Contamination and Chemical Exposure Problems?

Broadly speaking, the key environmental chemicals of greatest concern in much of modern societies are believed to be anthropogenic organic compounds. These typically include pesticides—for example, lindane, chlordane, endrin, dieldrin, toxaphene, and dichlorodiphenyl trichloroethane [DDT]; industrial compounds—for example, solvents such as trichloroethylene (or, trichloroethene) [TCE] and fuel products derived from petroleum hydrocarbons; and by-products of various industrial processes—for example, hexachlorobenzene [HCB], polychlorinated biphenyls [PCBs], polychlorinated dibenzodioxins (or, polychlorodibenzo-p-dioxins) [PCDDs], and polychlorinated dibenzofurans (or, polychlorodibenzofurans) [PCDFs] (see, e.g., Dewailly et al., 1993, 1996; Walker, 2008). Many industries also produce huge quantities of highly toxic waste by-products that include cyanide ions, acids, bases, heavy metals, oils, dyes, and organic solvents. Further yet, other rather unsuspecting sources of environmental contaminants are beginning to add to the multitude of chemical exposure problems that contemporary societies face (Asante-Duah, 2017); for instance, a number of scientists and regulatory agencies around the world have come to recognize/acknowledge pharmaceuticals to be an emerging environmental problem of significant concern.

Anyhow, due in part to the types of observations and occurrences identified in Table 1.1, a cultural mind-set in which issues relating to public safety are completely ignored is apparently a thing of the past. Indeed, all people are becoming increasingly aware and sensitive to the dangers of inadequacies in environmental management programs and related contaminated site management practices. The notion of *what you don't know doesn't hurt* is no longer tolerated, since even the least educated of populations have some conscious awareness of the potential dangers associated with hazardous materials. Consequently, in addition to managing the actual risks associated with environmental pollution, professionals may also have to deal with the psychosocial effects that the mere presence of toxic materials and/or contaminated sites may have on communities (Peck, 1989). Indeed, this is a picture probably well digested by several individuals, even by lay persons. Armed with this kind of information and knowledge, there is increasing public concern about the several problems and potentially dangerous situations associated with contaminated sites. Such concerns, together with the legal provisions of various legislative instruments and regulatory programs,

TABLE 1.1

Selected/Typical Examples of Potential Environmental Contamination and Resultant Human Exposures to Hazardous Chemicals/Materials—as Observed around the World

Chemical Hazard Location	Source/Nature of Problem	Contaminants of Concern	Nature of Exposure Settings and Scenarios, and Observed Effects
Love Canal, Niagara Falls, New York, United States	Section of an abandoned excavation for a canal was used as industrial waste landfill	Various carcinogenic and volatile organic chemicals—including hydrocarbon residues from pesticide manufacture.	Section of an abandoned excavation for a canal that lies within suburban residential setting had been used as industrial waste landfill. Problem first uncovered in 1976.
	Site received over 20,000 metric tons of chemical wastes containing more than 80 different chemicals.		Industrial waste dumping occurred from the 1940s through the 1950s; this subsequently caused entire blocks of houses to be rendered uninhabitable.
			Potential human exposure routes included direct contact and also various water pathways.
			Several apparent health impairments—including birth defects and chromosomal abnormalities—observed in residents living in vicinity of the contaminated site.
Chemical Control, Elizabeth, New Jersey, United States	Fire damage to drums of chemicals—resulting in leakage and chemical releases.	Various hazardous wastes from local industries.	The Chemical Control site was adjacent to an urban receptor community; site located at the confluence of two rivers.
			Leaked chemicals from fire-damaged drums contaminated water (used for fire-fighting)—that subsequently entered adjacent rivers.
			Plume of smoke from fire-deposited ash on homes, cars, and playgrounds.
			Potential exposures mostly via inhalation of airborne contaminants in the plume of smoke from the fire that blew over surrounding communities.
Bloomington, Indiana, United States	Industrial wastes entering municipal sewage system.	PCBs	PCB-contaminated sewage sludge used as fertilizer—resulting in crop uptakes.
	Sewage material was used for garden manure/fertilizer.		Also discharges and runoff into rivers resulted in potential fish contamination.
			Direct human contacts and also exposures via the food chain (as a result of human ingestion of contaminated food).
Times Beach, Missouri, United States	Dioxins (tetrachlorinated dibenzo(p) dioxin, TCDD) in waste oils sprayed on public access areas for dust control.	Dioxins (TCDD)	Waste oils contaminated with dioxins (TCDD) were sprayed in several public areas (residential, recreational, and work areas) for dust control of dirt roads, and so on. in the late 1960s and early 1970s.
			Problem was deemed to present extreme danger in 1982—that is, from direct contacts, inhalation, and probable ingestion of contaminated dust and soils.

(Continued)

TABLE 1.1 (Continued)
Selected/Typical Examples of Potential Environmental Contamination and Resultant Human Exposures to Hazardous Chemicals/Materials—as Observed around the World

Chemical Hazard Location	Source/Nature of Problem	Contaminants of Concern	Nature of Exposure Settings and Scenarios, and Observed Effects
Triana, Alabama, United States	Industrial wastes dumped in local stream by a pesticide plant.	DDT and other compounds.	High DDT metabolite residues detected in fish consumed by community residents. Potential for human exposure via food chain—that is, resulting from consumption of fish.
Woburn, Massachusetts, United States	Abandoned waste lagoon with several dumps.	Arsenic compounds, various heavy metals, and organic compounds.	Problem came to light in 1979 when construction workers discovered more than 180 large barrels of waste materials in an abandoned lot alongside a local river. Potential for leachate to contaminate groundwater resources and also for surface runoff to carry contamination to surface water bodies. High levels of carcinogens found in several local wells—which were then ordered closed. Potential human receptors and ecosystem exposure via direct contacts and water pathways indicated. Inordinately high degree of childhood leukemia observed. This apparent excess of childhood leukemia was linked to contaminated well water in the area. In general, leukemia and kidney cancers in the area were found to be higher than normal.
Santa Clarita, California, United States	Runoff from an electronics manufacturing industry resulted in contamination of drinking water.	Trichloroethylene (TCE) and various other volatile organic compounds (VOCs).	TCE and other VOCs contaminated drinking water in this community (due to runoff from industrial facility). Excess of adverse reproductive outcomes, and excess of major cardiac anomalies among infants suspected.
Three Mile Island, Pennsylvania, United States	Overheating of nuclear power station in March 1979.	Radioactive materials	Small amount of radioactive materials escaped into atmosphere. Emission of radioactive gases—and potential for radioactivity exposures. Unlikely that anyone was harmed by radioactivity from incident. Apparently, the discharge of radioactive materials was too small to cause any measurable harm.

(Continued)

TABLE 1.1 (*Continued*)

Selected/Typical Examples of Potential Environmental Contamination and Resultant Human Exposures to Hazardous Chemicals/Materials—as Observed around the World

Chemical Hazard Location	Source/Nature of Problem	Contaminants of Concern	Nature of Exposure Settings and Scenarios, and Observed Effects
European Union Member States, EU	Furniture treated with dimethylfumarate (DMFu)—together with possible (persisting) cross-contamination from the primary sources. Numerous patients in Europe were reported to suffer from DMFu-induced dermatitis. Dermatological symptoms attributed to contact with DMFu-treated consumer products—mostly shoes and sofas/furniture.	DMFu	Furniture identified as possible cause/source of numerous cases of dermatitis [induced by DMFu] in several European Union Member States. Apparently, DMFu had been used to prevent mold development in various items—including furniture; these DMFu-contaminated items in dwellings ostensibly posed substantial threats to the health of the occupants. Thousands of patients were diagnosed with severe dermatitis, with a few cases even requiring hospitalization; studies concluded that the likely cause of this furniture dermatitis epidemic was contact allergy due to DMFu. DMFu had typically been used as a biocide for preventing mold development that can deteriorate furniture or shoes during storage or transport—thus serving as an anti-mold agent for various polyurethane, polyvinyl chloride, leather and similar products.
Flixborough, England, United Kingdom	Explosion in nylon manufacturing factory in June 1974.	Mostly hydrocarbons	Hydrocarbons processed in reaction vessels/reactors (consisting of oxidation units, etc.). Destruction of plant in explosion, causing death of 28 men on site and extensive damage and injuries in surrounding villages. Explosive situation—that is, vapor cloud explosion.
Chernobyl, Ukraine (then part of the former USSR)	Overheating of a water-cooled nuclear reactor in April 1986.	Radioactive materials	Nuclear reactor blew out and burned, spewing radioactive debris over much of Europe. General concern relates to exposure to radioactivity. About 30 people reported killed immediately or died within a few months that may be linked to the accident. It has further been estimated that several thousands more may/could die from cancer during the next 40 years or so as a result of incident.
Seveso (near Milan), Italy	Discharge containing dioxin contaminated a neighboring village over a period of approximately 20 minutes in July 1976.	Dioxin and caustic soda.	Large areas of land contaminated—with part of it being declared uninhabitable. Mostly dermal contact exposures (resulting from vapor-phase/gas-phase deposition on the skin)—especially from smoke particles containing dioxin falling onto skins, and so on. About 250 people developed the skin disease, chloracne, and about 450 were burned by caustic soda.

(Continued)

TABLE 1.1 (Continued)

Selected/Typical Examples of Potential Environmental Contamination and Resultant Human Exposures to Hazardous Chemicals/Materials—as Observed around the World

Chemical Hazard Location	Source/Nature of Problem	Contaminants of Concern	Nature of Exposure Settings and Scenarios, and Observed Effects
Lekkerkirk (near Rotterdam), The Netherlands	Residential development built on land atop layer of household demolition waste and covered with relatively thin layer of sand. Housing project spanned 1972–1975. Problem of severe soil contamination was discovered in 1978. Evacuation of residents commenced in the summer of 1980.	Various chemicals—comprised mainly of paint solvents and resins (containing toluene, lower boiling point solvents, antimony, cadmium, lead, mercury, and zinc).	Rising groundwater carried pollutants upward from underlying wastes into the foundations of houses. This caused deterioration of plastic drinking water pipes, contamination of the water, noxious odors inside the houses, and toxicity symptoms in garden crops. Several houses had to be abandoned, while the waste materials were removed and transported by barges to Rotterdam for destruction by incineration. Polluted water was treated in a physicochemical purification plant.
Union Carbide Plant, Bhopal, India	Leak of methyl isocyanate (MIC) from storage tank in December 1984.	MIC	Leak of over 25 tonnes of MIC from storage tank occurred at Bhopal, India. In general, exposure to high concentrations of MIC can cause blindness, damage to lungs, emphysema, and ultimately death. MIC vapor discharged into the atmosphere—and then spread beyond plant boundary, killing well over 2,000 people and injuring several tens of thousands more.
Kamioka Zinc Mine, Japan	Contaminated surface waters.	Cadmium	Water containing large amounts of cadmium discharged from the Kamioka Zinc Mine into river used for drinking water and also for irrigating paddy rice. Ingestion of contaminated water and consumption of rice contaminated by crop uptake of contaminated irrigation water. Long-term exposures resulted in kidney problems for population.
Minamata Bay and Agano River at Niigata, Japan	Effluents from wastewater treatment plants entering coastal waters near a plastics manufacturing factory.	Mercury—giving rise to the presence of the highly toxic methylmercury.	Accumulation of methylmercury in fish and shellfish. Human consumption of contaminated seafood—resulting in health impairments, particularly severe neurological symptoms.

have compelled both the industrial and governmental authorities in several jurisdictions to carefully formulate effective management plans for potentially contaminated sites. These plans include techniques and strategies needed to provide good preliminary assessments, site characterizations, impact assessments, and the development of cost-effective corrective action programs.

1.2 MANAGING CONTAMINATED SITE PROBLEMS

It is apparent that *not all* contaminated lands necessarily require cleanup and neither do equally contaminated sites necessarily all require the same degree of cleanup or restoration efforts—and this is where risk-based decision-making becomes the tool of choice. Indeed, at least in the interest of the usually limited resources available for environmental restoration programs, it is always prudent to identify credible scientific methods that allow technically sound decisions to be formulated even in the absence of active cleanup actions being undertaken at contaminated sites.

In general, when it is suspected that a potential hazard exists at a particular locale, then it becomes necessary to further investigate the situation—and to fully characterize the prevailing or anticipated hazards. This activity may be accomplished by the use of a well-designed site investigation program. To this end, contaminated site investigations typically consist of the planned and managed sequence of activities carried out to determine the nature and distribution of hazards associated with potential contaminated site and/or chemical exposure problems. The activities involved usually are comprised of several specific tasks—broadly listed to include the following (BSI, 1988):

- Problem definition/formulation (including identifying project objectives and data needs)
- Identification of the principal hazards or site contaminants
- Design of sampling and analysis programs
- Collection and analysis of appropriate samples
- Recording or reporting of laboratory results for further evaluation
- Logical analysis of sampling data and laboratory analytical results
- Interpretation of study results (consisting of enumeration of the implications of and decisions on corrective action)

On the whole, a thorough investigation—usually culminating in a risk assessment—that establishes the nature and extent of receptor exposures usually becomes necessary, in order to arrive at appropriate and realistic corrective action and/or risk management decisions.

Meanwhile, it is worth the mention here that, in order to get the most out of the environmental contamination and/or chemical exposure characterization for a contaminated site problem, this activity must be conducted in a systematic manner. Indeed, systematic methods help focus the purpose, the required level of detail, and the several topics of interest—such as physical characteristics of the potential receptors, contacted chemicals, extent and severity of possible exposures, effects of chemicals on populations potentially at risk, probability of harm to human and/or ecological health, and possible residual hazards following implementation of risk management and corrective action activities (Cairney, 1993). Ultimately, the data derived from the site investigation may be used to perform a risk assessment that becomes a very important element in the contaminated site risk management decision.

1.2.1 Risk Assessment as a Tool in the Management of Contaminated Site Problems

A key underlying principle of contaminated site risk assessment is that some risks are tolerable—a reasonable and even sensible view, considering the fact that nothing is wholly safe per se. In fact, whereas human and ecological exposures to large amounts of a toxic substance may be of major concern, exposures of rather limited extent may be trivial and therefore should not necessarily be a cause for alarm. In order to be able to make a credible decision on the cutoff between what really constitutes a "dangerous dose" and a "safe dose," systematic scientific tools—such as those afforded by risk assessment—may be utilized. In this regard, therefore, risk assessment seems to have become an important contemporary foundational tool in the development of effectual environmental and public health risk management strategies and policies in relation to contaminated site problems. The risk assessment generally serves as a tool that can be used to organize, structure, and compile scientific information—in order to help identify existing hazardous situations or problems, anticipate potential problems, establish priorities, and provide a basis for regulatory controls and/or corrective actions. It may also be used to help gage the effectiveness of corrective measures or remedial actions.

It is noteworthy that, as part of the efforts aimed at designing an effectual risk assessment paradigm or framework in the application of the various risk assessment tools (meant to help resolve a given problem on hand), one should be cognizant of the fact that developments in other fields of study—such as data management systems—are likely to greatly benefit the risk analyst. In fact, an important aspect of contaminated site risk management with growing interest relates to the coupling of a contaminated site's environmental or risk data with Geographic Information System (GIS)—in order to allow for an effectual risk mapping of a study area with respect to the location and proximity of risk to identified or selected populations. In a nutshell, the GIS can process geo-referenced data and provide answers to questions such as the distribution of selected phenomena and their temporal changes, the impact of a specific event on populations, or the relationships and systematic patterns of site locations and chemical exposures vis-à-vis observed health trends in a region. Indeed, it has been suggested that, as a planning and policy tool, the GIS technology could be used to "regionalize" a risk analysis process. Once risks have been mapped using GIS, it may then be possible to match estimated risks to risk reduction strategies and also to delineate spatially the regions where resources should be invested, as well as the appropriate risk management strategies to adopt for various geographical dichotomies. Meanwhile, it should also be recognized that there are several direct and indirect legislative issues that affect contaminated site risk assessment programs in different regions of the world (see Chapter 3). Differences in legislation among different nations (or even within a nation) tend to result in varying types of contaminated site risk management strategies being adopted or implemented. Undeniably, legislation remains the basis for the administrative and management processes in the implementation of most contaminated site policy agendas. Despite the good intents of most regulatory controls, however, it should be acknowledged that, in some cases, the risk assessment seems to be carried out simply to comply with the prevailing legislation—and thus may not necessarily result in any significant hazard or risk reduction.

On the whole, the benefits of risk assessment outweigh any possible disadvantages, but it must be recognized that this process will not be without tribulations. Indeed, risk assessment is by no means a panacea. Its use, however, is an attempt to widen and extend the decision maker's knowledge base and thus improve the decision-making capability. Accordingly, the method deserves the effort required for its continual refinement as an environmental management tool.

1.2.2 Contaminated Site Risk Management: Establishing the Need for Corrective Actions at Contaminated Sites

In practice, the general objectives of a contaminated site risk management program usually will encompass the following typical tasks:

- Determine if a hazardous substance exists at a contaminated site and/or may be contacted by human and ecological receptors.
- Estimate the potential threat to ecological and public health, as posed by the chemical substances of concern.
- Determine if immediate response action is required to abate potential problems.

- Identify possible remedy or corrective action strategy(s) for the situation.
- Provide for environmental and public health informational needs of the population-at-risk in the potentially affected community.

Overall, risk assessment provides one of the best mechanisms for completing the prototypical tasks involved here. Undeniably, a systematic and accurate assessment of risks associated with a given contaminated site and chemical exposure problem is crucial to the development and implementation of a cost-effective corrective action plan. Consequently, risk assessment should generally be considered as an integral part of most contaminated site risk management programs that are directed at controlling the potential effects of chemical exposure problems. The application of risk assessment can indeed provide for prudent and technically feasible and scientifically justifiable decisions about corrective actions that will help protect public health and the environment in a most cost-effective manner. Meanwhile, it is noteworthy that, in any attempts to shape contaminated site risk management and restoration programs, contemporary risk assessment methods of approach may have to be used to facilitate the design of more reliable risk management and remedial strategies/schemes.

Finally, corrective action decisions can be formulated to address the specific problem on hand; corrective actions for contaminated sites are characteristically developed and implemented with the principal objective being to protect public health and the environment. Classically, to design an effectual corrective action program, contaminated sites have to be extensively studied to determine the areal extent of contamination, the quantities of contaminants that human and ecological receptors could potentially be exposed to, the human health and ecological risks associated with the site, and the types of corrective or remedial actions necessary to abate risks from such sites. Oftentimes in these types of situations, soils tend to become the principal focus of attention in the investigation of the contaminated sites. This is because soils at such sites do not only serve as a medium of exposure to potential receptors but also as a long-term reservoir for contaminants that may be released to other media. At any rate, the primary focus of a risk appraisal for such cases is the assessment of whether existing or potential receptors are presently, or in the future may be, at risk of adverse effects as a result of exposure to conditions from potentially hazardous situations. This evaluation then serves as a basis for developing mitigation measures in risk management and risk prevention programs. Typically, the risk assessment will help define the level of risk as well as set performance goals for various response alternatives. The application of risk assessment can indeed provide for prudent and technically feasible and scientifically justifiable decisions about corrective actions that will help protect public health and the environment in a most cost-effective manner. Meanwhile, it is worth the mention here that, in order to arrive at a cost-effective corrective action or risk management decision, answers will typically have to be generated for several pertinent questions when one is confronted with a potential environmental contamination or contaminated site problem (Box 1.1); in general, the answers to these questions will help define the corrective action needs for the subject site.

BOX 1.1   QUESTIONS RELEVANT TO DEFINING CORRECTIVE ACTION NEEDS: MAJOR ISSUES IMPORTANT TO MAKING COST-EFFECTIVE ENVIRONMENTAL MANAGEMENT DECISIONS FOR AN ENVIRONMENTAL CONTAMINATION PROBLEM

- What is the nature of contamination?
- What are the sources of, and the "sinks" for, the site contamination?
- What is the current extent of contamination?
- What population groups are potentially at risk?
- What are the likely and significant exposure pathways and scenarios representative of the site—that connect contaminant source(s) to potential receptors?

(Continued)

> **BOX 1.1 (*Continued*)   QUESTIONS RELEVANT TO DEFINING CORRECTIVE ACTION NEEDS: MAJOR ISSUES IMPORTANT TO MAKING COST-EFFECTIVE ENVIRONMENTAL MANAGEMENT DECISIONS FOR AN ENVIRONMENTAL CONTAMINATION PROBLEM**
>
> - What is the likelihood of health and environmental effects resulting from the contamination?
> - What interim measures, if any, are required as part of a risk management and/or risk prevention program?
> - What corrective action(s) may be appropriate to remedy the prevailing situation?
> - What level of residual contamination will be tolerable or acceptable for the situation—or for site restoration, if the site needs to be cleaned up?

1.2.3   RESTORING THE INTEGRITY OF CONTAMINATED SITES VIA A COST-EFFECTIVE CONTAMINATED SITE DECISION-MAKING PROCESS

To avert possible detrimental effects to human health and the environment, an attempt is often made to restore the integrity of contaminated sites. Incidentally, it is worth emphasizing the fact that, in a number of situations for the processes involved, it becomes necessary to implement interim corrective measures prior to the development and full implementation of a comprehensive site restoration program (viz., remedial action and/or risk management program). The types of preliminary corrective actions may consist of a variety of site control activities—such as the installation of security fences (to restrict access to the site), the construction of physical barriers (to restrict site access and also to minimize potential runoff from the site), the application of dust suppressants (to reduce airborne migration of contaminated soils to offsite locations), the removal of "hot-spots" and drums (where fire, explosion, or acute toxic exposure could possibly occur). In any case, a thorough site investigation that establishes the nature and extent of contamination, followed by an appropriate level of a risk assessment, usually may be necessary—in order to be able to arrive at an appropriate and realistic corrective action decision. Ultimately, a systematic process involving the development and full implementation of a comprehensive remedial action and/or risk management program may be undertaken for the project site—and possibly, to be followed by a site closure.

Indeed, there seems to be a number of potential environmental, public health, political, and socioeconomic implications associated with most environmental contamination problems—including those related to contaminated site issues. It is, therefore, important to generally use systematic and technically sound methods of approach in the relevant risk management programs. Risk assessment provides one of the best mechanisms for completing the tasks involved. In fact, a systematic and accurate assessment of current and future risks associated with a given contaminated site problem is crucial to the development and implementation of a cost-effective corrective action plan. Consequently, risk assessment is commonly considered as an integral part of the corrective action assessment processes used in the study of potentially contaminated site problems.

At the end of the day, developing credible tools with hallmarks of clarity and understandability—such as may be afforded by a well-designed risk assessment program—becomes important in facilitating effective risk prevention or minimization, risk management or control, and risk communication for the miscellany of chemical exposure problems (including those directly or indirectly attributable to contaminated sites) that have become ubiquitous/prevalent, perhaps even inescapable, for much of the contemporary societies. Indeed, if risk assessment is carried out properly, then any concomitant risk perception may hopefully lean more towards pragmatic reality—and thus take away some of the unwarranted fears that at times force risk managers to "misallocate" resources to deal with relatively low-risk issues, while potentially high- or significant-risk problems sit unattended.

Ultimately, understanding and/or knowing the true dimension of the prevailing risks would help to mitigate or control any potential threats in a more prudent/meaningful way.

1.3   KEY ATTRIBUTES, SCOPE/COVERAGE, AND ORGANIZATION OF THIS BOOK

The focus of this volume is on the effective management of contaminated site problems—generally achievable through the design of effectual corrective action programs. Indeed, many sites contaminated to various degrees by several chemicals or pollutants exist in a number of geographical locations that must be tackled; the management of these contaminated sites is of great concern in view of the risks associated with such sites. For instance, apart from their immediate and direct health and environmental threats, contaminated sites can contribute to the long-term contamination of the ambient air, soils, surface waters, groundwater resources, and the food chain. Such detrimental outcomes are unavoidable if the impacted sites are not managed efficaciously and/or if remedial actions are not taken in an effective manner; this book sets out to provide a road map to reaching a desirable goal on this matter.

Broadly speaking, the book outlines risk-based strategies that are useful in the investigation, characterization, management, and restoration of environmental contamination and related contaminated site problems. It draws on many real-world problem scenarios from across the globe to illustrate several crucial points. It also contains a review of several useful remediation techniques and their common applicability scenarios. On the whole, this book provides a wealth of information that can be applied to the planning, appraisal, assessment, evaluation, and implementation of contaminated site investigation and restoration programs—by bringing together the common themes and concepts necessary for making credible decisions about the reclamation of contaminated sites for redevelopment and/or protection of the health of populations in the vicinity. In particular, the book presents some very important tools and methodologies that can be used to address contaminated site risk management problems in a consistent, efficient, and cost-effective manner. To this end, this book focuses on the key principles necessary for an effective management of contaminated site problems, achievable through an application of the risk assessment paradigm in the overall implementation process. Major attributes of the book include the following:

- Discusses the site assessment processes involved in mapping out the types and extents of contamination at potentially contaminated and derelict lands.
- Addresses issues relevant to the investigation and management of potentially contaminated and derelict land problems—spanning problem diagnosis and site characterization, to the development and implementation of site restoration programs.
- Elaborates the types of activities generally required of environmental site assessments, and often conducted for contaminated site problems—including a discussion of the steps involved in the design of comprehensive site investigation work plans.
- Presents concepts and techniques/methodologies in risk assessment that may be applied to contaminated site problems—and then provides guidance framework for the formulation of contaminated site risk assessments, as well as the follow-on risk management and site restoration decisions.
- Describes methods of approach for conducting human health and ecological risk assessments—consisting of an elaboration of the requisite protocols and tasks involved in its application to contaminated site problems.
- Offers prescriptions and decision protocols to be utilized/followed when making key decisions in corrective action assessment and response programs for contaminated site problems.

- Identifies the key elements necessary for the design of effectual corrective action programs—comprised of the use of appropriate diagnostic tools to determine the degree of contamination and/or the concomitant risk and the subsequent development of an effectual risk management and/or site restoration program.
- Elucidates the scientific and technical requirements and strategies for the diagnostic assessments of potentially contaminated and derelict lands and also for developing site restoration goals and corrective action plans for such problems.
- Synthesizes and summarizes the several relevant generic principles that may be applied to the characterization and management of the variety of contaminated site situations often encountered in practice.
- Presents a comprehensive review of several useful remediation techniques and their common applicability scenarios.
- Serves as practical guide to the effective application of risk assessment and risk-based strategies to contaminated site restoration and management problems—offering an understanding of the scientific basis of risk assessment and its applicability to contaminated site management decisions.

All in all, the above-noted features fundamentally include a collection and synthesis of the principal elements of the risk assessment process in a format that can be efficaciously applied to contaminated site problems. A number of illustrative example problems (consisting of a variety of analyses relevant to the design of effectual site characterization, risk management, and corrective action programs for potentially contaminated site problems) are interspersed throughout the book, in order to help present the book in an easy-to-follow, pragmatic manner.

Meanwhile, it must be acknowledged here that the nature and extent of contaminated site problems tend to be rather diverse—and indeed, in such a manner as to make it almost impractical to cover every aspect of it in one volume; notwithstanding the concepts presented in this book will generally be applicable to the broad spectrum of contaminated site situations often encountered in practice. In fact, despite the variability in the nature of contaminated site problems typically encountered at different locations, there seems to be growing consensus on the general technical principles and procedures that should be used in the management of such sites. Along those lines, this book presents a synthesis of technically sound principles and concepts that should find unbiased global application to contaminated site problems. Indeed, the basic technical principles and concepts involved in the management of contaminated site problems will generally not differ in any significant way from one geographical region to another; the basic differences in strategies tend to lie in the regional or local policies and legislation. Consequently, this book will serve as a useful resource for any sector of the global community that is faced with contaminated site risk management and site restoration decisions.

On the whole, the subject matter of this book should be of interest to many professionals encountering environmental contamination problems in general and contaminated site risk management problems alongside risk-based site restoration decision-making in particular. The specific intended audience include environmental consulting professionals (viz., environmental consultants and scientists/engineers); national, regional, and local government environmental agencies; public policy analysts; environmental attorneys; environmental and public health regulatory agencies; manufacturing and process industries; real-estate developers and investors involved in property transfers; waste management companies; utility companies; chemical and pharmaceutical production industries; petroleum and petrochemical industries; environmental nongovernmental organization (NGOs); consumer service industries; environmental firms and construction companies; banks and lending institutions; and a miscellany of health, environmental, and consumer advocacy interest groups. The book is also expected to serve as a useful *educational and training resource for both students and professionals in the environmental and allied fields*—particularly those who have to deal with contaminated or derelict land risk management, as well as land reclamation or site restoration issues. Written for both the novice and the experienced, the subject matter of this book is an attempt at offering a simplified and systematic presentation of contaminated site characterization and risk management strategies—all these facilitated by a design/layout that will carefully navigate the user through the major processes involved.

Finally, it is noteworthy that a key objective in preparing this revised and expanded edition to this book has been to, insofar as practicable, incorporate new key developments and/or updates in the field since the previous version was last published; also included in this revised edition are the elaborate discussions relating to the application of risk-based solutions to contaminated site problems and associated site restoration decision-making modalities or processes. Another notable feature of the revised edition is the sectional reorganization that has been carried out for some topics—all meant to help with the overall flow of the presentations but especially to facilitate a more holistic learning process/experience afforded by this new book. All in all, the book is organized into five parts—consisting of 14 chapters, together with a bibliographical listing and a set of four appendices. It is the hope of the author that, the five-part presentation offered by this title will provide adequate guidance and direction for the successful completion of corrective action assessment and response programs that are to be designed for any type of contaminated site problem and at any geographical location for that matter. The structured presentation should also help with any efforts to develop effectual classroom curricula for teaching purposes. Ultimately, the systematic risk-based protocols presented in this volume should indeed aid many an environmental scientist and related environmental professional to formulate and manage contaminated/derelict land issues and associated problems more efficiently.

2 The Nature of Contaminated Site Problems

Site contamination is generally said to have occurred when chemicals are detected at locations where such constituents are not naturally expected and/or are not desired. Contaminated site problems typically are the result of soil contamination due to placement of wastes on or in the ground; as a result of accidental spills, lagoon failures or contaminated runoff; and/or from leachate generation and migration. Contaminants released to the environment are indeed controlled by a complex set of processes that include various forms of transport and cross-media transfers, transformation, and biological uptake. For instance, atmospheric contamination may result from emissions of contaminated fugitive dusts and volatilization of chemicals present in soils; surface water contamination may result from contaminated runoff and overland flow of chemicals (from leaks, spills, and so on) and chemicals adsorbed onto mobile sediments; and groundwater contamination may result from the leaching of toxic chemicals from contaminated soils or the downward migration of chemicals from lagoons and ponds. Consequently, contaminated soils can potentially impact several other environmental matrices.

In general, environmental pollutants will tend to move from the source matrix into other receiving media—and which could in turn become secondary release sources for such pollutants or their transformation products. In this changing dynamics, it is notable that several different physical and chemical processes can affect contaminant migration from a contaminated site, as well as the cross-media transfer of contaminants at any given site. This chapter presents a brief discussion on the general nature and types of contaminated site problems often born out of these processes and thus likely to be encountered in practice; it also elaborates on the general implications of such types of environmental contamination problems.

2.1 THE MAKING AND ANATOMY OF CONTAMINATED SITES: WHAT ARE THE ORIGINS OF CONTAMINATED SITES?

Contaminated sites may arise in a number of ways—many of which are the result of manufacturing and other industrial activities or operations. It is apparent that industrial activities and related waste management facilities or programs may indeed be responsible for most contaminated site problems. The key contributing waste management activities may relate to industrial wastewater impoundments, land disposal sites for solid wastes, land spreading of sludge materials, accidental chemical spills, leaks from chemical storage tanks and piping systems, septic tanks and cesspools, disposal of mine wastes, or indeed a variety of waste treatment, storage and disposal facilities (TSDFs). In any case, once a chemical constituent is released into the environment, it may be transported and/or transformed in a variety of complex ways—and these types of situations have, in several ways, contributed to the widespread occurrence of contaminated site problems globally.

Undeniably, many of the contaminated site problems encountered in a number of places around the world are the result of waste generation associated with various forms of industrial activities. Wastes are generated from several operations associated with industrial (e.g., manufacturing and mining), agricultural, military, commercial (e.g., automotive repair shops, utility companies, fueling stations, dry-cleaning facilities, transportation centers, and food processing industries), and domestic activities. In particular, the chemicals and allied product manufacturers are generally seen as the major sources of industrial hazardous waste generation that might culminate in contaminated site problems. Box 2.1 presents a listing of some major industrial sectors that are potential sources of waste generation; these industries generate several waste types—such as organic waste sludge and still bottoms (containing chlorinated solvents, metals, oils, and so on); oil and grease (contaminated with polychlorinated biphenyls [PCBs], polyaromatic hydrocarbons [PAHs], metals, and so on); heavy metal solutions (of arsenic, cadmium, chromium, lead, mercury, and so on); pesticide and herbicide wastes; anion complexes (containing cadmium, copper, nickel, zinc, and so on); paint and organic residuals; and several other chemicals and byproducts that need special handling/management (and generally have the potential to contaminate lands or facilities). Table 2.1 summarizes the typical hazardous waste streams potentially generated by selected major industrial operations. Indeed, all of these waste streams have the potential to enter the environment by several mechanisms. Ultimately, this may lead to the birth of contaminated sites and related environmental or chemical exposure problems—conceptually depicted in Figure 2.1 and displaying some of the typical or potential consequences related to the migration and chemical exposures from contaminated sites.

TABLE 2.1
Examples of Typical Potentially Hazardous Waste Streams from Selected Industrial Sectors

Sector/Source	Typical Hazardous Waste Stream or Contaminants
Agricultural and food production	Acids and alkalis; fertilizers (e.g., nitrates); herbicides (e.g., dioxins); insecticides; unused pesticides (e.g., aldicarb, aldrin, dichlorodiphenyl trichloroethane [DDT], dieldrin, parathion, toxaphene); fungicides; arsenic; copper; grain fumigants
Airports	Hydraulic fluids; oils
Auto/vehicle servicing and equipment repair	Acids and alkalis; heavy metals; lead-acid batteries (e.g., cadmium, lead, nickel); degreasing agents; solvents; paint and paint sludge; scrap metals; waste oils; various ignitable wastes
Battery recycling and disposal	Arsenic; cadmium; chromium; copper; lead; nickel; zinc; acids
Chemicals/pharmaceuticals production	Acids and alkalis; biocide wastes; cyanide wastes; heavy metals (e.g., arsenic, mercury); infectious and laboratory wastes; organic residues; PCBs; solvents
Domestic	Acids and alkalis; dry-cell batteries (e.g., cadmium, mercury, zinc); heavy metals; insecticides; solvents (e.g., ethanol, kerosene)
Dry-cleaning/laundries	Detergents (e.g., boron, phosphates); dry-cleaning filtration residues; halogenated solvents
Educational/research institutions	Acids and alkalis; ignitable wastes; reactives (e.g., chromic acid, cyanides; hypochlorites, organic peroxides; perchlorates, sulfides); solvents; perchloroethylene
Electrical transformers	PCBs
Electroplating operations	Various metals—such as cadmium; chromium; copper; cyanide; nickel
Glass manufacture	Arsenic; lead
Leather tanning	Inorganic chemicals (e.g., chromium, lead); solvents
Machinery manufacturing	Acids and alkalis; cyanide wastes; heavy metals (e.g., cadmium, lead); oils; solvents
Medical/health services	Laboratory wastes; pathogenic/infectious wastes; formaldehyde; radionuclides; photographic chemicals; solvents; mercury
Metal treating/manufacture	Acids and alkalis; cyanide wastes; heavy metals (e.g., antimony, arsenic, cadmium, cobalt); ignitable wastes; reactives; solvents (e.g., toluene, xylenes)
Military training grounds	Heavy metals
Mineral processing/extraction	High-volume/low-hazard wastes (e.g., mine tailings); red muds

(Continued)

TABLE 2.1 (*Continued*)

Examples of Typical Potentially Hazardous Waste Streams from Selected Industrial Sectors

Sector/Source	Typical Hazardous Waste Stream or Contaminants
Motor freight/railroad terminals	Acids and alkalis; heavy metals; ignitable wastes (e.g., acetone; benzene; methanol); lead-acid batteries; solvents
Paint manufacture	Heavy metals (e.g., antimony, cadmium, chromium); PCBs; solvents; toxic pigments (e.g., chromium oxide)
Paper manufacture/printing	Acids and alkalis; dyes; heavy metals (e.g., chromium, lead); inks; paints and resins; solvents; dioxins; furans
Petrochemical industry/gasoline stations	Benzo-a-pyrene (BaP); hydrocarbons; oily wastes; lead; phenols; spent catalysts; fuels; oil and grease
Photofinishing/photographic industry	Acids; silver; solvents
Plastic materials and synthetics	Heavy metals (e.g., antimony, cadmium, copper, mercury); organic solvents
Shipyards and repair shops	Heavy metals (e.g., arsenic, mercury, tin); solvents; paints; acids
Textile processing	Dyestuff heavy metals and compounds (e.g., antimony, arsenic, cadmium, chromium, mercury, lead, nickel); halogenated solvents; mineral acids; PCBs
Timber/wood preserving industry	Heavy metals (e.g., arsenic); nonhalogenated solvents; oily wastes; preserving agents (e.g., creosote, chromated copper arsenate, pentachlorophenol [PCP]); PCBs; PAHs; dioxin

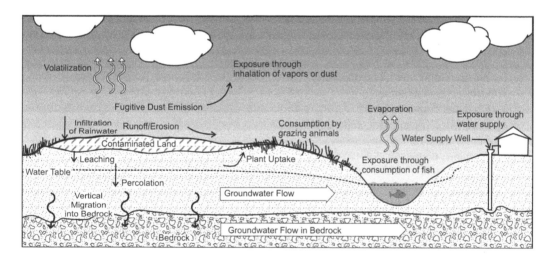

FIGURE 2.1   A diagrammatic representation of an archetypal contaminated site and possible implications: Conceptual illustration of key potential consequences associated with a contaminated site problem.

Site contamination by a variety of toxic chemicals has indeed become a major environmental issue in several locations around the world—with such contamination scenarios generally arising from the presence or release of hazardous materials or wastes at any given locale, site, or TSDF. [In a broad sense, a hazardous material is that product which is capable of producing some adverse effects and/or reactions in potential biological receptors; toxic substances generally present "unreasonable" risk of harm to human health and/or the environment, and need to be regulated. Specifically, hazardous wastes include those by-products with the potential to cause detrimental effects on human health and/or the natural environment; such wastes may be toxic, bioaccumulative, persistent (i.e., nondegradable), radioactive, carcinogenic (i.e., cancer-causing), mutagenic (i.e., causing gene alterations), and/or teratogenic (i.e., capable of damaging a developing fetus). Among the most persistent of the organic-based contaminants arising from a hazardous waste stream, for instance, are the organo-halogens; organohalogens are rarely found in nature and therefore relatively few biological systems can break them down, in contrast to compounds such as petroleum hydrocarbons which are comparatively more easily degraded (Fredrickson et al., 1993).] Meanwhile, it is worth the mention here that the nature of the hazardous constituents of interest will typically determine its potential to cause detrimental impacts to human health and the environment. The proper identification and/or classification of the types of potentially hazardous contaminants present in the environment is therefore important in the determination of potential risks attributable to the site contaminants of interest; subsequently, the outcome here should provide important clues on the nature of best corrective actions and/or risk management strategies to adopt for a case-specific problem.

BOX 2.1   TYPICAL MAJOR INDUSTRIES AND MANUFACTURERS POTENTIALLY CONTRIBUTING TO ENVIRONMENTAL CONTAMINATION AND/OR CONTAMINATED SITE PROBLEMS

- Aerospace
- Agriculture
- Ammunitions
- Automobile
- Batteries
- Beverages
- Chemical production
- Computer manufacture
- Dry cleaning
- Electronics and electrical
- Electroplating and metal finishing
- Explosives manufacture
- Food and dairy products
- Herbicides, insecticides, and pesticides
- Ink formulation
- Inorganic pigments
- Iron and steel
- Leather tanning and finishing
- Medical care facilities
- Metal smelting and refining
- Mineral exploration and mining
- Paint products
- Perfumes and cosmetics
- Petroleum products
- Pharmaceutical products
- Photographic equipment and supplies
- Printing and publishing
- Pulp and paper mills
- Rubber products, plastic materials, and synthetics
- Scrap metals operations
- Shipbuilding
- Soap and detergent manufacture
- Textile products
- Wood processing and preservation

2.2 HEALTH, ENVIRONMENTAL, AND SOCIOECONOMIC IMPLICATIONS OF CONTAMINATED SITE PROBLEMS

The mere existence of contaminated sites at any locale or region can result in contaminant releases and possible receptor exposures (Figure 2.1). This can in turn result in both short- and long-term effects on various populations potentially at risk. Exposure to chemical constituents present at contaminated sites can indeed produce adverse effects in both human and nonhuman ecological receptors. For example, human exposures to certain hazardous chemicals present at contaminated sites may result in such diseases as allergic reaction, anemia, anxiety, asthma, blindness, bronchitis, various cancers, contact dermatitis, convulsions, embryotoxicity, emphysema, heart disease, hepatitis, obstructive lung disease, memory impairment, nephritis, neuropathy, and pneumonoconiosis; ecological exposures to certain chemicals may result in such toxic manifestations as bioaccumulation and/or biomagnification in aquatic organisms, with potential for lethal or sublethal consequences in some situations (Asante-Duah, 1998). In general, any chemical present at a contaminated site can cause severe health impairment or even cause death if taken by an organism in sufficiently large amounts. On the other hand, there are those chemicals of primary concern that, even in small doses, can cause adverse health impacts. In the end, several health effects may arise if/when human and/or ecological receptors are exposed to some agent or stressor present in the environment.

The following represent the major broad categories of human health and ecologic effects that could result from exposure to chemicals typically found in contemporary societies or present at a contaminated site (Andelman and Underhill, 1988; Asante-Duah, 1996, 1998, 2017; Ashford and Miller, 1998; Bertollini et al., 1996; Brooks et al., 1995; Grisham, 1986; Hathaway et al., 1991; Lippmann, 1992):

Human Health
- Carcinogenicity (i.e., capable of causing cancer in humans and/or laboratory animals)
- Heritable genetic defects and chromosomal mutation [mutagenesis] (i.e., capable of causing defects or mutations in genes and chromosomes that will be passed on to the next generation)
- Developmental toxicity/abnormalities and teratogenesis (i.e., capable of causing birth defects or miscarriages, or damage to developing fetus)
- Reproductive toxicity (i.e., capable of damaging the ability to reproduce)
- Acute toxicity (i.e., capable of causing death from even short-term exposures)
- Chronic toxicity (i.e., capable of causing long-term damage other than cancer)
- Neurotoxicity (i.e., capable of causing harm to the nervous system—viz., central nervous system [CNS] disorders)
- Alterations of immunobiological homeostasis
- Congenital anomalies

Ecologic
- Environmental toxicity (i.e., capable of causing harm to wildlife and vegetation)
- Persistence (i.e., does not breakdown easily, thus persisting and accumulating in portions of the environment)
- Bioaccumulation (i.e., can enter the bodies of plants and animals but is not easily expelled, thus accumulating over time through repeated exposures).

Several different symptoms, health effects, and other biological responses may indeed be produced from exposure to various specific toxic chemicals commonly encountered at contaminated sites and/or in the environment.

Broadly speaking, a number of chemicals encountered at contaminated sites are known or suspected to cause cancer; several others may not have carcinogenic properties but are nonetheless of significant concern due to their systemic toxicity effects. Table 2.2 lists typical symptoms, health

TABLE 2.2

Some Typical Human Health and Ecological Effects/Manifestations from Chemical Exposures: A Listing for Selected Toxic Chemicals Commonly Found in the Environment or Contaminated Sites

Chemical	Typical Health Effects/Symptoms and Toxic Manifestations/Responses
Arsenic and compounds	Acute hepatocellular injury, anemia, angiosarcoma, cirrhosis, developmental disabilities, embryotoxicity, heart disease, hyperpigmentation, peripheral neuropathies
Antimony	Heart disease
Asbestos	Asbestosis (scarring of lung tissue)/fibrosis (lung and respiratory tract)/lung cancer, mesothelioma, emphysema, irritations, pneumonia/pneumoconioses
Benzene	Aplastic anemia, CNS depression, embryotoxicity, leukemia and lymphoma, skin irritant
Beryllium	Granuloma (lungs and respiratory tract)
Cadmium	Developmental disabilities, kidney damage, neoplasia (lung and respiratory tract), neonatal death/fetal death, pulmonary edema
Carbon tetrachloride	Narcosis, hepatitis, renal damage, liver tumors
Chromium and compounds	Asthma, cholestasis (of liver), neoplasia (lung and respiratory tract), skin irritant
Copper	Gastrointestinal irritant, liver damage
Cyanide	Asthma, asphyxiation, hypersensitivity, pneumonitis, skin irritant
DDT	Ataxic gait, convulsions, human infertility/reproductive effects, kidney damage, neurotoxicity, peripheral neuropathies, tremors
Dieldrin	Convulsions, kidney damage, tremors
Dimethylfumarate	Dermatological symptoms/effects (contact dermatitis)—viz., skin irritation and skin sensitization, cutaneous allergic reactions; possible respiratory allergic symptoms or diseases
Dioxins and furans (polychlorinated dibenzodioxins/polychlorinated dibenzofurans)	Hepatitis, neoplasia, spontaneous abortion/fetal death; bioaccumulative
Formaldehyde	Allergic reactions; gastrointestinal upsets; tissue irritation
Lead and compounds	Anemia, bone marrow suppression, CNS symptoms, convulsions, embryotoxicity, neoplasia, neuropathies, kidney damage, seizures; biomagnifies in food chain
Lindane	Convulsions, coma and death, disorientation, headache, nausea and vomiting, neurotoxicity, paresthesias
Lithium	Gastroenteritis, hyperpyrexia, nephrogenic diabetes, Parkinson's disease
Manganese	Bronchitis, cirrhosis (liver), influenza (metal-fume fever), pneumonia, neurotoxicity
Mercury and compounds	Ataxic gait, contact allergen, CNS symptoms; developmental disabilities, neurasthenia, kidney and liver damage, Minamata disease; biomagnification of methyl mercury
Methylene chloride	Anesthesia, respiratory distress, death
Naphthalene	Anemia
Nickel and compounds	Asthma, CNS effects, gastrointestinal effects, headache, neoplasia (lung and respiratory tract)
Nitrate	Methemoglobinemia (in infants)
Organochlorine pesticides	Hepatic necrosis, hypertrophy of endoplasmic reticulum, mild fatty metamorphosis
PCP	Malignant hyperthermia

(Continued)

TABLE 2.2 *(Continued)*

Some Typical Human Health and Ecological Effects/Manifestations from Chemical Exposures: A Listing for Selected Toxic Chemicals Commonly Found in the Environment or Contaminated Sites

Chemical	Typical Health Effects/Symptoms and Toxic Manifestations/ Responses
Phenol	Asthma, skin irritant
PCBs	Embryotoxicity/infertility/fetal death, dermatoses/chloracne, hepatic necrosis, hepatitis, immune suppression, endocrine effects, neurologic effects, cardiovascular effects, musculoskeletal issues, gastrointestinal systems effects
Silver	Blindness, skin lesions, pneumonoconiosis
Toluene	Acute renal failure, ataxic gait, neurotoxicity/CNS depression, memory impairment
Trichloroethylene	CNS depression, deafness, liver damage, paralysis, respiratory and cardiac arrest, visual effects
Vinyl chloride	Leukemia and lymphoma, neoplasia, spontaneous abortion/fetal death, tumors, death
Xylene	CNS depression, memory impairment
Zinc	Corneal ulceration, esophagus damage, pulmonary edema

Source: Compiled from various sources—including, Blumenthal, 1985; Chouaniere et al., 2002; Grisham, 1986; Hughes, 1996; Lave and Upton, 1987; Rousselle et al., 2014; Rowland and Cooper, 1983; Williams et al., 2008; and personal communication with Dr. Kwabena Duah, Australia (2002).

effects, and other biological responses of primary concern that could be produced from exposures to some representative toxic chemicals commonly encountered at contaminated sites and in the general environment. Meanwhile, it is notable that the potential for adverse health effects on populations contacting hazardous chemicals present at a contaminated site can involve any organ system; the target and/or affected organ(s) tend to be dependent on several factors—but most significantly, the specific chemicals contacted; the extent of exposure (i.e., dose or intake); the characteristics of the exposed individual (e.g., age, gender, body weight, psychological status, genetic makeup, immunological status, susceptibility to toxins, hypersensitivities); the metabolism of the chemicals involved; weather conditions (e.g., temperature, humidity, barometric pressure, season); and the presence or absence of confounding variables such as other diseases (Asante-Duah, 2017; Brooks et al., 1995; Grisham, 1986).

Invariably, exposures to chemicals escaping into the environment can result in significant impacts to the populations potentially at risk. Ultimately, potential receptor exposures to chemicals can result in a reduction of life expectancy—and possibly a period of reduced quality of life (e.g., as caused by anxiety from exposures, and diseases). The presence of toxic chemicals and/or existence of unregulated contaminated site problems in society can therefore be perceived as a potential source of several health, environmental, and additionally possible socioeconomic problems to society at large. In fact, this situation can actually reduce the future development potentials of a community. Conversely, it has been noted that sensible policies and actions that protect the environment can at the same time contribute to long-term economic progress (Schramm and Warford, 1989). In general, potential health, ecological and socioeconomic problems are averted by carefully implementing substantive corrective action and/or risk management programs appropriate for the specific chemical exposure or contaminated site problems; a variety of methods for identifying and linking all the multiple chemical sources to the potential receptor exposures are often used to facilitate the development of a sound risk management program (Asante-Duah, 2017). Several analysts have indeed concluded that environmental problems generally are inseparable from socio-economic development problems, and that long-term economic growth depends on protecting the environment (The World Bank, 1989). Economic development and environmental protection should therefore complement each other, since improving one has the potential to enhance the other.

2.3 CONTAMINATED SITE CLASSIFICATION AND DESIGNATION SYSTEMS

It is important to recognize the fact that there may be varying degrees of hazards associated with different contaminated site problems, and that there are good technical and economic advantages for ranking potentially contaminated sites according to the level of hazard they present. A typical site risk categorization scheme for contaminated site problems may involve a grouping or clustering of the "candidate" sites on the basis of the potential risks attributable to various plausible conditions—such as high- versus intermediate/medium- versus low-risk problems, conceptually depicted by Figure 2.2. Such a classification can indeed facilitate the development and implementation of a more efficient risk management or corrective action program. In the investigation of potentially contaminated site problems, a broad categorization scheme may be used for a variety of contaminated area designations—and which will ultimately facilitate the planning, development, and implementation of appropriate site management or restoration programs. Sites may indeed be clustered into different groups in accordance with the level of effort required to implement the appropriate or necessary response actions. Archetypical designations, which are by no means exhaustive, are enumerated in the following:

- *Low-risk areas*, represented by areas with no confirmed contamination; this may consist of areas/locations (e.g., suspected sources) where the results of records search and site investigations show that no hazardous substances were stored for any substantial period of time, released into the environment or site structures, or disposed of on the site property—or areas where the occurrence of such storage, release, or disposal is not considered to have

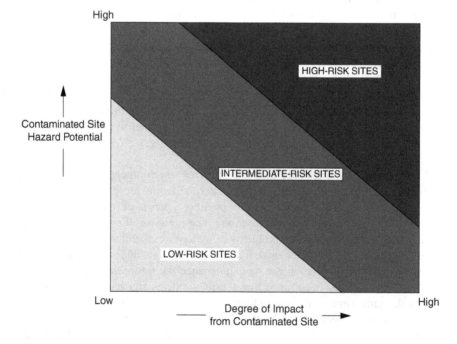

FIGURE 2.2 A conceptual representation of classical risk categories for an archetypal contaminated sites classification system.

been probable. The determination of a low-risk area can be made at any of several decision stages during a site characterization activity.

- *Intermediate/medium-risk areas*, represented by areas with limited contamination; this may consist of impacted areas/locations where no response or remedial action is necessarily required to ensure protection of human health and the environment. Intermediate-risk areas may typically include areas where an environmental site investigation has demonstrated that hazardous materials have been released, stored, or disposed of but are present in quantities that probably require only limited to no response or remedial action to protect human health and the environment. Such designations mean that the levels of hazardous substances detected in a given area *do not* exceed media-specific action levels (e.g., risk-based concentrations); *do not* result in significant risks; *nor* otherwise exceed requisite regulatory standards.
- *High-risk areas*, represented by areas with extensive contamination; this may consist of areas/locations where the site records indicate or confirms that hazardous materials have been released and/or disposed. Typically, a significant level of corrective action response will be required for such sites. In general, the high-risk sites or problems will prompt the most concern—requiring immediate and urgent attention or corrective measures (that may include "time-critical" removal actions). A site or problem may be designated as "high risk" when site contamination or exposure represents a real or imminent threat to human health and/or the environment; in this case, an immediate action will generally be required to reduce the threat.

Indeed, to ensure the development of adequate and effectual risk management or corrective action strategies, potential contaminated site or chemical exposure problems may need to be categorized in a similar or other appropriate manner during the risk analysis.

Overall, site contamination problems may pose different levels of risk, depending on the type and extent of contamination present at any given location. The degree of hazard posed by the contaminants involved will generally be dependent on several factors such as physical form and composition of contaminants; quantities of contaminants; reactivity (i.e., fire, explosion); biological and ecological toxicity effects; mobility (i.e., transport in various environmental media, including leachability); persistence or attenuation in environment (viz., fate in environment, detoxification potential); indirect health effects (from pathogens, vectors, and so on); and local site conditions and environmental setting (e.g., temperature, soil type, groundwater flow conditions, humidity, and light). Consequently, it is very important to adequately characterize site contamination problems—that is, if cost-effective solutions are to be found for dealing with the prevailing hazards. Indeed, depending on the problem category, different corrective action response and risk management strategies may be utilized to address the particular contaminated site problem.

3 Legislative-Regulatory Considerations in Contaminated Site Decision-Making

In the past, some polluting behaviors were seen as normal and inevitable consequences of industrialization and national development. As a result, environmental protection in many countries was based on reliance on a nonpunitive, conciliatory style of securing compliance with environmental laws and regulations (UNICRI, 1993). Increasing damage to the environment has, however, changed perceptions of acts of pollution, especially in the industrialized countries. Indeed, among other things, there has been a growing tendency for the developed economies to steadily move toward criminal sanctions as a means of punishment and deterrence in the arena of environmental offences (Asante-Duah and Nagy, 1997). Furthermore, it is noteworthy that, to help abate potential problems to public health and the environment, several items of legislation have been formulated and implemented in most industrialized countries to deal with waste management issues and also the regulation of toxic substances present in our modern societies (Forester and Skinner, 1987). Some of the well-established environmental laws found in the different regions of the world are introduced in this chapter—with further examination of related legislation and regulations alluded to in the appropriate sections of subsequent chapters.

In a nutshell, this chapter presents a summary review of selected environmental legislation potentially relevant to the management of contaminated site problems. Depending on the type of program under evaluation, one or more of the existing regulations may dominate the decision-making process in a contaminated site management program. Meanwhile, it is worth mentioning here that several of the regulations are subject to changes/amendments. Also, depending on the specific situation or application, a combination of several regulatory specifications may be used. For further and up-to-date discussions on region-specific environmental regulations, the reader is encouraged to consult with their local regulatory agencies or applicable literature pertaining to the case region(s).

3.1 ENVIRONMENTAL CONTROL AND CLEANUP POLICIES IN DIFFERENT REGIONS OF THE WORLD

Generally, most industrialized countries incorporate a system for "cradle-to-grave" control in their hazardous materials management programs. The "cradle-to-grave" type of system monitors and regulates the movement of hazardous materials from manufacture through usage to the ultimate disposal of any associated hazardous wastes. The use of some kind of manifest system helps minimize abuses and violations of established national or regional control systems associated with hazardous materials movements and management. The manifest system serves as an identification form that accompanies each shipment of hazardous materials; the manifest is signed at each stage of transfer of responsibility, with each responsible personnel in the chain of custody keeping a record that is open to scrutiny and inspection by regulatory officials or other appropriate authorities and also most interested parties or stakeholders. It is expected that the use of such effective control systems will help minimize the creation of extensive land contamination problems. With that in mind, it is notable that several of the newly industrializing and other developing economies are also moving towards the establishment of workable environmental regulations—to help protect their often fragile environments.

Invariably, there tends to be improved contaminated site management practices in countries or regions where regulatory programs are well established—in comparison to those without appropriate regulatory and enforcement programs. The process used in the design of effectual contaminated site management programs must therefore incorporate several significant elements of all relevant environmental regulations. In fact, the establishment of national and/or regional regulatory control programs within the appropriate legislative regulatory and enforcement frameworks is a major step in developing effective waste and contaminated site management programs. In any case, it is noteworthy that variations in national legislation and controls do affect options available for the management of contaminated site problems in different regions of the world. Such variances may also affect the cleanup standards, restoration goals, and costs necessary to achieve the appropriate contaminated site management program objectives. The need to consult with local environmental regulations when faced with a contaminated site problem cannot, therefore, be overemphasized.

3.1.1 ENVIRONMENTAL LAWS IN NORTH AMERICA

North America, particularly the United States, has over the past several years come up with mountains of environmental legislation. Starting with a National Environmental Protection Act in 1969 that empowered the U.S. Environmental Protection Agency (EPA) and a National Environmental Quality Control Council (NEQCC) to lay the basis for regulations, a stream of environmental legislation followed in the next four decades and into the present time. Some of the relevant fundamental statutes and regulations which, directly or indirectly, may affect contaminated site management decisions, policies, and programs in the United States are briefly enumerated below—with further discussion offered elsewhere in the literature (e.g., Forester and Skinner, 1987; Holmes et al., 1993; USEPA, 1974, 1985a,c,e,f, 1987a–f,h,i, 1988a,b,c,e,f,h,j,k, 1989a–g,k–o).

- *Clean Air Act (CAA).* The CAA was originally conceived and adopted in 1963; however, a sweeping amendment that was effected in 1970 is widely recognized as the more powerful and important piece of this environmental legislation. The objective of the CAA of 1970 has been to protect and enhance air quality resources, in order to promote and maintain public health and welfare and the productive capacity of the population. This covers all pollutants that may cause significant risks. The CAA of 1970 has indeed undergone several amendments since its inception. In particular, the CAA Amendments of 1990 introduced further sweeping changes—including, for the first time, specific provisions addressing global air pollution problems.
- *Safe Drinking Water Act (SDWA).* The SDWA was enacted in 1974 in order to assure that all people served by public water systems would be provided with a supply of high quality water. The SDWA amendments of 1986 established new procedures and deadlines for setting national primary drinking water standards and established a national monitoring program for unregulated contaminants, among others. The statute covers public water systems, drinking water regulations, and the protection of underground sources of drinking water.
- *Clean Water Act (CWA).* The CWA was enacted in 1977, and an amendment to this introduced the Water Quality Act of 1987. The objective of the CWA has been to restore and maintain the chemical, physical, and biological integrity of the nation's waters. This objective is achieved through the control of discharges of pollutants into navigable waters—implemented through the application of federal, state and local discharge standards. The statute covers the limits on waste discharge to navigable waters, standards for discharge of toxic pollutants, and the prohibition on discharge of oil or hazardous substances into

navigable waters. Water quality criteria developed by the EPA under CWA authority is not by themselves enforceable standards but can be used in the development of enforceable standards. Another closely associated legislation, the Federal Water Pollution Control Act, deals solely with the regulation of effluent and water quality standards.

- *Resource Conservation and Recovery Act (RCRA).* The RCRA was enacted in 1976 (as an amendment to the Solid Waste Disposal Act of 1965, later amended in 1970 by the Resource Recovery Act) to regulate the management of hazardous waste, to ensure the safe disposal of wastes, and to provide for resource recovery from the environment by controlling hazardous wastes "from cradle to grave." This is a federal law that establishes a regulatory system to track hazardous substances from the time of generation to disposal. The law requires safe and secure procedures to be used in treating, transporting, sorting, and disposing of hazardous substances. Basically, RCRA regulates hazardous waste generation, storage, transportation, treatment, and disposal. Of special interest, the 1984 Hazardous and Solid Waste Amendments to RCRA, Subtitle C, covers a management system that regulates hazardous wastes from the time it is generated until the ultimate disposal—the so-called "cradle-to-grave" system. Thus, under RCRA, a hazardous waste management program is based on a "cradle-to-grave" concept that allows all hazardous wastes to be traced and equitably accounted for.
- *Comprehensive Environmental Response, Compensation, and Liability Act (CERCLA).* The CERCLA of 1980 ("Superfund") establishes a broad authority to deal with releases or threats of releases of hazardous substances, pollutants, or contaminants from vessels, containments, or facilities. This legislation deals with the remediation of hazardous waste sites, by providing for the cleanup of inactive and abandoned hazardous waste sites. The objective has been to provide a mechanism for the federal government to respond to uncontrolled releases of hazardous substances to the environment. The statute covers reporting requirements for past and present owners/operators of hazardous waste facilities—as well as the liability issues for owners/operators in relation to the cost of removal or remedial action and damages, in the event of a release or threat of release of hazardous wastes. Overall, CERCLA (or "Superfund") establishes a national program for responding to releases of hazardous substances into the environment, with the overarching mandate to protect human health and the environment from current and potential threats posed by uncontrolled hazardous substance releases. The Superfund Amendments and Reauthorization Act of 1986 (SARA) strengthens and expands the cleanup program under CERCLA, focusses on the need for emergency preparedness and community right-to-know, and changes the tax structure for financing the Hazardous Substance Response Trust Fund established under CERCLA to pay for the cleanup of abandoned and uncontrolled hazardous waste sites.
- *Toxic Substances Control Act (TSCA).* The TSCA of 1976 provides for a wide range of risk management actions to accommodate the variety of risk-benefit situations confronting the US EPA. The risk management decisions under TSCA would consider not only the risk factors (such as probability and severity of effects) but also non-risk factors (such as potential and actual benefits derived from use of the material and the availability of alternative substances). Broadly, TSCA regulates the manufacture, use, and disposal of chemical substances. It authorizes the US EPA to establish regulations pertaining to the testing of chemical substances and mixtures, premanufacture notification for new chemicals or significant new uses of existing substances, control of chemical substances or mixtures that pose an imminent hazard, and record-keeping and reporting requirements.
- *Federal Insecticide, Fungicide and Rodenticide Act (FIFRA).* The FIFRA of 1978 deals with published procedures for the disposal and storage of excess pesticides and pesticide containers. The US EPA has also promulgated tolerance levels for pesticides and pesticide residues in or on raw agricultural commodities under authority of the Federal Food, Drug,

and Cosmetic Act (40 CFR Part 180). FIFRA does indeed provide the US EPA with broad authorities to regulate all pesticides.

- *Endangered Species Act (ESA).* The ESA of 1973 (reauthorized in 1988) provides a means for conserving various species of fish, wildlife, and plants that are threatened with extinction. The ESA considers an endangered species as that which is in danger of extinction in all or significant portion of its range; a threatened species is that which is likely to become an endangered species in the near future. Also, the ESA provides for the designation of critical habitats (i.e., specific areas within the geographical area occupied by the endangered or threatened species) on which are found those physical or biological features essential to the conservation of the species in question.
- *Pollution Prevention Act.* The Pollution Prevention Act of 1990 calls pollution prevention as a "national objective" and establishes a hierarchy of environmental protection priorities as national policy. Under this act, it is the U.S. national policy that pollution should be prevented or reduced at the source whenever feasible; where pollution cannot be prevented, it should be recycled in an environmentally safe manner. In the absence of feasible prevention and recycling options or opportunities, pollution should be treated, and disposal should be used only as a last resort.

It is notable that comparable laws also exist in Canada. However, in Canada, the implementation and enforcement of environmental regulations have been principally the responsibilities of the provinces—with a few exceptions, of course. Indeed, in many ways, the Canadian system seems to be generally less formal than in the United States—albeit not any less stringent.

Indeed, additional laws and regulations exist in the United States and Canada that also work towards preventing or limiting the potential impacts of site contamination problems. For example, in the United States, the Hazardous Materials Transport Act of 1975 provides a legislation that deals with the regulation of transport of hazardous materials; the Fish and Wildlife Conservation Act of 1980 requires States to identify significant habitats and develop conservation plans for these areas; the Marine Mammal Protection Act of 1972 protects all marine mammals—some of which are endangered species; the Migratory Bird Treaty Act of 1972 implements many treaties involving migratory birds—in order to protect most species of native birds in the United States; the Wild and Scenic Rivers Act of 1972 preserves select rivers declared as possessing outstanding/remarkable scenic, recreational, geologic, fish and wildlife, historic, cultural, or other similar values; etc. Also, laws similar or comparable to those annotated above, together with similar legal provisions and regulations can be found in the legislative requirements for several other advanced economies—as reflected in the brief discussions provided further below.

3.1.1.1   Risk Assessment in the U.S. Regulatory Systems

The US EPA and several state regulatory agencies recognize the use of risk assessment as a facilitator of remedial action decisions and also in the enforcement of regulatory standards. Indeed, risk assessment techniques have been used, and continue to be used, in various regulatory programs used by federal, state, and local agencies. For instance, both the feasibility study (FS) process under CERCLA (or "Superfund") and the alternate concentration limit (ACL) demonstrations under RCRA involve the use of risk assessment to establish cleanup standards for contaminated sites.

The primary application of quantitative risk assessment in the Superfund program is to evaluate potential risks posed at each National Priorities List (NPL) facility, in order to assist in the identification of appropriate remedial action alternatives (Paustenbach, 1988); NPL is the list of uncontrolled or abandoned hazardous waste sites identified for possible long-term remedial actions under Superfund. The federal EPA also uses a risk-based evaluation method, the so-called Hazard Ranking System (HRS), to identify uncontrolled and abandoned hazardous waste sites falling under Superfund programs. The HRS allows the selection or rejection of a site for placement on the NPL; it is used for prioritizing sites so that those apparently posing the greatest risks receive quicker response.

In other risk assessment applications, ACLs (which can be considered as surrogate cleanup criteria) can be established, when hazardous constituents are identified in groundwater at RCRA facilities, by applying risk assessment procedures in the analytical processes involved. In fact, nearly every process for developing cleanup criteria incorporates some concept that can be classified as a risk assessment. Thus, all decisions on setting cleanup standards for potentially contaminated sites include, implicitly or explicitly, some risk assessment concepts.

Finally, it is noteworthy that, the basic steps involved in performing risk assessment at both Superfund and non-Superfund sites are fundamentally the same—except perhaps for the degree of detail typically utilized in the various procedural steps for the process.

Legislative–Regulatory Perspectives. The U.S. National Contingency Plan (NCP) (40 CFR 300) requires that an "applicable or relevant and appropriate requirements" (ARARs), standards or criteria be considered when developing remedial actions as part of a remedial investigation (RI)/FS process for contaminated site problems. *Applicable requirements* are those cleanup standards, standards of control, and other substantive environmental protection requirements, criteria, or limitations promulgated under federal or state laws that specifically address a hazardous substance, pollutant, contaminant, remedial action, location, or other circumstances at a CERCLA site (USEPA, 1988a). *Relevant and appropriate requirements* are those cleanup standards, standards of control, and other substantive environmental protection requirements, criteria, or limitations promulgated under federal or state laws that, while not "applicable" to a hazardous substance, pollutant, contaminant, remedial action, location, or other circumstances at a CERCLA site, address problems or situations sufficiently similar to those encountered at the CERCLA site, that their use is well suited to the particular site (USEPA, 1988a). The requirement can be either one of these categories but not both. Potential ARARs are identified through a review of current local, state, and federal guidelines available in several guidance documents and databases. In general, the identification of ARARs must be done on a site-specific basis.

Also, a so-called "to-be-considered materials" (TBCs) may be used, where appropriate in the RI/FS process. TBCs are non-promulgated advisories or guidance issued by federal or state government that are not legally binding and do not have the status of potential ARARs. However, in many circumstances, TBCs will be considered along with ARARs as part of the risk assessment and may be used in determining the necessary level of cleanup required for the protection of public health and the environment.

In fact, the cleanup requirements necessary to meet cleanup goals will generally have to be based not only on ARARs but also on risk-based criteria, TBCs, health advisories, and other guidance for the state or region in which a site is located. ARARs (and TBCs to some extent) may be chemical- (i.e., based on the nature and toxicity of the chemicals of potential concern), location- (e.g., based on sensitive habitats or ecosystems, wetlands, and floodplains), or action-specific (i.e., based on remedial actions). In any event, a comprehensive tabulation of the requirements, administering agencies, and summaries of potential federal and state ARARs, together with other federal, state, and local criteria; advisories; and TBC guidance will generally provide important input to contaminated site risk assessments. Indeed, requirements under several federal, state, and local laws and regulations (such as CAA, RCRA, SDWA, CWA, and TSCA) serve as typical sources of such information for contaminated site management programs. The details of procedures and criteria for selecting ARARs, TBCs, and other guidelines are discussed elsewhere in the literature (e.g., USEPA, 1988b,e, 1989b,n,o).

3.1.2 ENVIRONMENTAL LAWS WITHIN THE EUROPEAN COMMUNITY BLOC, INCLUDING THE UNITED KINGDOM

The European Community (EC) was created in 1957 by the Treaty of Rome. The European Union (EU) was more recently created by the Treaty of Maastricht and came into existence as an umbrella organization in November 1993. Fundamentally, the EC is a structural member of the EU.

The two most common forms of EU legislation are regulations and directives. Regulations become law throughout the EU as of their effective date, generally enforceable in each Member State, whereas directives are generally not necessarily enforceable in Member States but are meant to set out goals for the Member States to achieve through national legislation. In fact, of the five/major legislative acts enumerated in Article 189 of the EC Economic Treaty (i.e., Treaty Establishing the European Economic Community, Rome, 25 March, 1957)—namely, regulations, directives, decisions, recommendations, and opinions—the directive seems to play a unique role that affords the European Parliament the ability to respect indigenous national legislative traditions and philosophies while guiding the EU as whole towards a single common goal. The directive sets forth an objective and provides that each national government achieve this objective by the national means considered best able to accomplish the goal embodied in the directive.

The EU directive was designed to respect indigenous legislative traditions while providing a means for unification of goals. Specifically, the EU directives play an important role in environmental regulations within the member nations. For instance, the 1975 framework (Council Directive 75/442/EEC on Waste) established the general obligations of Member States for waste management, to ensure that wastes are disposed of without endangering human health or harming the environment. Toxic and dangerous wastes were covered under a 1978 directive (Council Directive 78/319/ EEC on Toxic and Dangerous Waste) that required these materials to be properly stored, treated, and disposed of in authorized facilities governed by proper regulatory authorities; this directive has more recently been replaced with the Directive on Hazardous Wastes (Council Directive 91/689/ EEC), adopted under the provisions of the framework Directive on Waste, and this new directive took effect at the start of 1994. Also, enacted in 1985, the environmental assessment directive (Council Directive 85/337/EEC of 27 June 1985 on the Assessment of the Effects of Certain Public and Private Projects on the Environment) provides the basis for environmental impact statements to be produced for those development projects that are likely to have significant effects on the environment by virtue of their nature, size or location. The environmental directives are designed to establish uniform result with respect to environmental integrity throughout the Member states. While striving towards a common goal of environmental consistency, the environmental directive respects national policy for the management, control, and regulation of environmental pollutants.

A key element of the relevant EU environmental Directives is the fairly recent concept of using the 'Best Available Techniques Not Entailing Excessive Costs' (BATNEEC) to prevent or reduce pollution and its effects. BATNEEC is interpreted as the technology for which operating experience had adequately demonstrated it to be the best commercially available to minimize releases, providing it had been proven to be economically viable.

It is also interesting to note the EU issuance of its Council Regulation No. 1836/93 on the voluntary participation of commercial enterprises in a common 'Eco-Management and Audit Scheme' (EMAS). The regulation entered into force on July 13, 1993, with its ultimate application having been dependent on the issuing of national regulations in Member states. In these regulations, the prevention, reduction and elimination of environmental burdens—possibly at their source— results in a sound management of resources on the basis of the 'Polluter Pays Principle', and the promotion of market-oriented contaminated site management are required. These are meant to stimulate and foster responsibility among commercial and industrial enterprises. The core elements of the Regulation are the introduction of an environmental management scheme at specific industrial sites and the execution of an environmental audit by independent auditors. In contrast to the environmental impact assessment that is carried out before a facility is installed, environmental audits take place during plant operations. After a successful participation in EMAS, the company obtains a participation certificate. It is noteworthy that, the EMAS requires a public statement—i.e., transparency in environmental terms.

For the most part, the individual European countries have consistently approached the issues of environmental management legislation from very different angles. Notwithstanding the interstate differences in environmental management legislation in Europe, however, there is a general tendency to develop hazardous waste minimization policies in most of the countries in the region. In fact, several European countries have already adopted, or are developing regulatory instruments that will directly affect the amount of hazardous waste produced—and therefore the potential concomitant site contamination problems.

Overall, the countries within the EU have aggressively worked on producing consistent environmental regulatory programs for their member nations, but that also recognizes the sovereign rights of the individual countries. As part of this process, the European Environment Agency was created in 1990 by a Community Regulation (Council Regulation 1210/90/EEC on the Establishment of the European Environment Agency and the European Environment Information and Observation Network), with the main purposes of collecting and disseminating information—rather than to have any enforcement authority. In fact, it is apparent that a vast and comprehensive legislative arsenal has been created within the EU, and therefore in each of the Member states—albeit the implementation of such legislation is yet to achieve the complete anticipated results.

When all is said and done, the EU directly affects member states through various legal instruments and several guiding principles—with the key principles that shape environmental policies of the EU's member states comprising: the 'polluter pays' principle; the need to avoid environmental damage at the source; and the need to take the environmental effects of energy and other EU policies into consideration. Some of the apparently unique legislative measures and relevant statutes and regulations which, directly or indirectly, may affect contaminated site management decisions, policies and programs within specific/selected EU nations are briefly annotated below—recognizing that similar or comparable frameworks exist in the other member nations. Further discussions on several aspects of environmental regulatory programs in the region can be found elsewhere in the legislative control literature (e.g., Garbutt, 1995a,b; Lister, 1996).

3.1.2.1 United Kingdom (UK)

Waste management has always been an issue visited by governments and policy-makers, past and present alike. Prior to the 19th century, however, pollution problems arising from the disposal of waste or refuse was frequently resolved through private litigation rather than by public prosecution under the Sanitary Acts. During the current era, the Public Health Act of 1936 enacted that any accumulation or deposition of refuse prejudicial to health or a nuisance is a Statutory Nuisance. The regulatory authorities have the power to serve Abatement Notices and to seek public prosecution; this power will remain extant, even after the full implementation of the Deposit of Poisonous Waste Act (DPWA) of 1972 and the Control of Pollution Act (CoPA) in 1974.

Early in 1972, the Deposit of Poisonous Waste Act (DPWA) had been issued, penalizing the deposit on land of poisonous, noxious or polluting wastes that could give rise to an environmental hazard, and to make offenders criminally and civilly liable for any resultant damage. The UK Control of Pollution Act of 1974 allows the government to restrict the production, importation, sale, or use of chemical substances in the UK; the Act controls the disposal of wastes on land by means of site licensing. Subsequently, the 1980 Control of Pollution (Special Waste) Regulations, which also repealed the DPWA, was established to ensure pre-notification for a limited range of the most hazardous types of wastes; to keep a 'cradle-to-grave' record of each disposal of special wastes; and to keep long-term records of locations of special waste disposal landfill sites.

The Environmental Protection Act (EPA 90) 1990 marked an important milestone in the development of the legislative philosophy and framework in the UK. The Act is seen as being of major importance, since it largely established Britain's strategy for pollution control and waste management for the foreseeable future (D. Slater, *in* Hester and Harrison, 1995). The Environmental Protection Act set up a system of integrated pollution control (IPC); prior to this, the regulation of pollution to land, air and water was dealt with under separate systems (Farmer, 1997).

More recently, the Environment Act 1995 sought to develop a new breed of environmental regulations in the UK. The significance of this Act includes—among other issues/things—a re-organization and concentration/integration of environmental regulatory functions, and whose adoption is complemented by pivotal reliance on cost-benefit and risk assessment techniques as mechanisms for its implementation (Jewell and Steele, 1996). The Act brings together in a single body, the existing environmental protection and pollution control functions of a number of regulators.

3.1.2.2 The Netherlands

The Netherlands has adopted a general ban on land disposal of hazardous waste. When exemptions from this ban are granted, the government may require the recipient of the exemption to conduct research on alternative technologies or management methods aimed at preventing potential needs for future land disposal. The ban itself acts, to some extent, as an incentive for waste minimization, although the relatively easy access to waste export to other countries for treatment or disposal has so far limited the impact of such a ban.

3.1.2.3 France

In France, the main goals of environmental policy in relation to the management of special industrial wastes are to develop an effective national treatment and disposal infrastructure, reduce the amount of wastes generated at source, and clean up sites contaminated by hazardous wastes. The French Ministry of the Environment's plan to attain these goals has been via providing clear definitions of the different classes of wastes, and imposing regulatory and financial constraints on waste generators.

Measures provided for in recent legislation include restrictions on the types of waste that can be landfilled, and a tax on the off-site treatment and disposal of special industrial waste—that will be used to fund the cleanup of abandoned contaminated sites. The special industrial wastes will include medical wastes and hazardous domestic wastes. The relevant Decree lists 51 groups of compounds, as well as 14 chemical and biological properties that are likely to render a waste hazardous.

The French government encourages waste avoidance and reduction by including waste production and management practices among the factors that are evaluated when the government has to make a decision on whether or not to authorize the operation of new or modified industrial facilities. An impact study is required prior to any authorization. However, the potential impact of these requirements on the reduction of hazardous waste is probably not being realized for the time being, because of an insufficient number of enforcement personnel and inadequate technical expertise among regulators and industry concerning alternative technologies and management methods.

From the year 2002, disposal by landfill in France is only permitted for wastes or treatment residues for which there is no commercially available recycling or treatment technology, and yet that requires disposal by landfill. The present legislation requires that, from 1998, all special industrial wastes be stabilized prior to disposal.

3.1.2.4 Federal Republic of Germany

In Germany, industrial waste is managed under the Federal Waste Disposal Act, which defines a category of hazardous waste as 'special'—requiring specific handling procedures. If a facility does not generate or handle 'special' wastes by definition, the agency may authorize the facility to handle the waste by itself, with only minimal regulatory requirements.

A regulatory instrument adopted by the German Parliament in 1986 provides an accelerated waste reduction policy, and authorizes the government to regulate the use, collection and disposal of products likely to cause environmental problems when discarded as waste. The regulation is coordinated with other efforts such as public financial support for developing and demonstrating new waste avoidance technologies, increasing the technical competence of regulators, and discouraging reliance on treatment and disposal as methods of choice for managing hazardous waste.

Overall, the federal system for the management of industrial waste consists of a basic legislative framework that delegates a significant authority to the state for the management and implementation of industrial waste regulations.

3.1.3 Environmental Policies and Regulations in Different Parts of the World

The legislative measures and relevant statutes and regulations which, directly or indirectly, affect contaminated site management decisions, policies and programs generally tend to be similar in their degree of sophistication according to the level of technological advancement. This means that, developed economies the world over usually will have regulatory systems comparable to that found in western Europe or North America (as exemplified by the brief reference that follows for Japan)—although this may not always be true, as for instance in the case with the former Soviet bloc countries of Central and Eastern Europe.

To illustrate the case for Japan, the Waste Management Law of 1970 requires anyone undertaking the collection, transport or disposal of industrial wastes to have obtained a permit. Waste disposal sites are also required to be designed and operated in a manner that prevents trespassing, and the disposal sites must be isolated from both surface and ground water resources, and measures taken to prevent leakage. In addition, the 1973 Chemical Substances Control Act requires a manufacturer or importer of a new chemical (i.e., one that is not enumerated in a 1973 pre-listed compilation) to provide the government with all available information concerning such a chemical; the substance can then be evaluated and classified as being dangerous or safe. As part of the nation's overall waste management program, Japan has put in place laws and regulations, as well as enforcement mechanisms that are comparable to that found in most other industrialized countries.

On the other hand, many developing economies tend to lack laws and/or enforcement tools that regulate environmental problems in an effective manner—albeit several nations in the developing regions of Latin America, the Caribbean, and South Africa often have reasonably sophisticated environmental legislation in place, even if they are with inadequate regulatory systems. In fact, information and environmental management regulations have been almost nonexistent and/or non-enforceable in most developing economy or newly industrializing countries. This is because most of these countries with economies in transition lack the laws and governmental institutions to deal with environmental regulatory issues. Many governments, however, recognize the need to protect their respective nation's health and environment, and are therefore seriously developing legislation that will deal with the insurmountable environmental management problems. The result is the development of systematic environmental control programs, most of which seem to follow the styles employed by the industrialized countries in Europe, North America, and elsewhere in the world.

3.2 A PARADIGM FOR INTERNATIONAL ENVIRONMENTAL LAWS

A disproportionately large part of environmental contamination problems in general may be attributed to waste management (or rather, waste mismanagement) cases. In one classic example of a waste mismanagement situation, when a committee of the British Parliament reported on waste disposal in 1981, reference was made to the fact that, 7 years earlier, an operator at a landfill site near London had died as a consequence of breathing toxic fumes resulting from the accidental disposal of two incompatible wastes at the same location. Another classic example a world away involved waste disposal at Love Canal in New York, USA; Love Canal was a disposal site for chemical wastes for about three decades. Subsequent use of the site apparently resulted in the residents in a township in the area suffering severe health impairments; it is believed that several children in the neighborhood apparently were born with serious birth defects. As alluded to earlier on (in Chapter 1), similar and/or comparable problems have occurred in several other locations globally. In fact, these types of cases/situations vis-à-vis the growing global awareness to the potential harms from exposures to the numerous and cocktails of chemicals within modern human environments in part prompted the 'World Summit on Sustainable Development' (held in Johannesburg in 2002) into making a global political commitment to effect sound chemicals management by 2020—albeit this seemingly noble effort may elude some economically-struggling countries or regions of the world. Further international efforts aimed at realizing this goal of 'sound chemicals management' subsequently resulted in the adoption of the 'Strategic Approach to International Chemicals Management' platform by the United Nations Environment Programme's Governing Council in February 2006, at Dubai.

In fact, increasing public concern about the several problems and potentially dangerous situations associated with land contamination and related problems, together with the legal provisions of various legislative instruments and regulatory programs, have all compelled both industry and governmental authorities to carefully formulate responsible waste management programs (see, e.g., Section 3.3). These programs include techniques and strategies needed to provide good waste management methods and technologies, and the development of cost-effective corrective action and risk management programs that will ensure public safety, as well as protect human health, the environment, and public and private properties.

In the formulation of environmental legislative instruments and the design of regulatory systems, it is important to recognize and appreciate the familiar saying that 'pollution knows and respects no geographical boundaries'. In fact, cross-boundary pollution problems have been around for a long time and this scenario is not about to change. Hence, to be truly effective, environmental protection usually must have an international dimension to it. Solutions to cross-boundary environmental issues will therefore tend to require international cooperation. This means that, despite the fact that individual countries will be exercising their sovereign rights in the formulation of environmental legislation for their nations, this should not necessarily happen in isolation; insofar as possible, this should be tied into a global environmental agenda for the international community. What is more, since solutions to global environmental problems generally require increased international cooperation—both regionally and globally—it should be expected that international environmental law would subjugate national sovereignty to this end. In this vein, it is apparent that the legislation and implementation of most international environmental laws have the tendency to neglect the social factors and the differing economic development status of States—whereas indeed the objective reality is that any global issue is an inseparable and organic whole (L. P. Cheng, *in* Weiss, 1992). On this basis, it becomes necessary for the international community, in dealing with global environmental issues, to consider the objective conditions of those States that are less developed economically and technologically as an almost separate 'entity' from the industrialized nations. This strategy will, hopefully, help alleviate part of the grim situation associated with the legislation and implementation of international environmental protection plans or laws by several nations/States. Indeed, the differing economic and political situations of States may have a dramatic impact on the implementation of international environmental laws in different countries; this fact should be recognized, if successful stories are to be told about existing and new international environmental laws.

Ultimately, an established environmental policy that is publicly available and understood at all appropriate levels is fundamental to attaining the requisite contaminated site management standards. The policy has to indicate commitments to: reduce waste, pollution and resource consumption; minimize environmental risks and hazard effects; design products with regard to their environmental effects throughout the life cycle; control the effects of raw material processing; and minimize the effects of new developments. Furthermore, a system of contaminated site management records should be established in order to demonstrate compliance with the requirements of the contaminated site management system, and to record the extent to which the planned environmental objectives and targets are being met.

3.3 MINIMIZING THE PRIMARY CAUSES OF ENVIRONMENTAL CONTAMINATION PROBLEMS AS THE ULTIMATE PARADIGM SHIFT

Industry has become an essential part of modern society, and waste production seems like an inevitable characteristic of most industrial activities. In this regard, waste management facilities apparently are responsible for most environmental contamination and contaminated site problems often encountered. In the past, hazardous waste management practices were indeed tantamount and synonymous to the simplistic rule of 'out of sight, out of mind'. This resulted in the creation of several contaminated sites that need to be addressed today. With the improved knowledge linking environmental contaminants to several human health problems and also to some ecological disasters, the 'new society' has come to realize the urgent need to clean up for the 'past sins'. Proactive actions are being taken to minimize the continued creation of more contaminated site problems. In fact, in recent years, and in several countries, social awareness of environmental problems and the need to reduce sources of contamination, as well as the urge to clean up contaminated sites, has been increasing. Nonetheless, some industries continue to generate and release large quantities of contaminants into the environment. All in all, a responsible system for dealing with the wastes generated by industry is essential to sustain the modem way of life. In particular, the effective management of hazardous wastes—and the associated TSDFs—is of major concern to both the industry generating wastes and to governments (representing the interests of individual citizens). This is because waste materials can cause adverse impacts on the environment and to public health. It is evident, however, that the proper management of hazardous wastes poses several challenges.

As a matter of fact, large quantities of wastes have always been generated by many industries, to the point that there is a near-crisis situation with managing such wastes. Consequently, waste/materials recycling and re-use, waste exchange, and waste minimization are becoming more prominent in the general waste management practices of several countries. For instance, the recovery of waste oils, solvents, and waste heat from incinerators has become a common practice in a number of locales. Also, operations exist in a number of countries to recycle heavy metals from various sources (such as silver from photofinishing operations, lead from lead-acid batteries, mercury from batteries and broken thermometers, and heavy metals from metal finishing wastes). Indeed, waste/materials recycling have become an integral part of many modem industrial processes. Furthermore, waste exchange schemes exist in some countries to promote the use of one company's byproduct or waste as another's raw material. Figure 3.1 illustrates the basic components of what seems to have become a typical waste management program for an industrial facility that chooses to adopt this *modus operandi*; in any case, the general trend of choice is for on-site waste

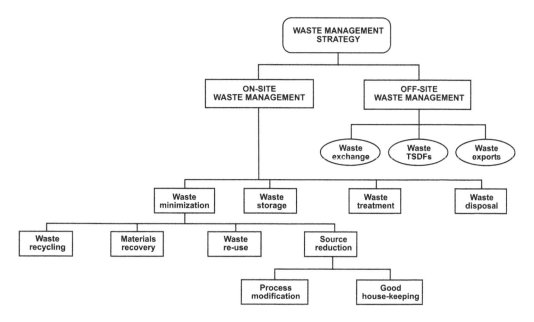

FIGURE 3.1 A framework for developing effectual/responsible waste management strategy/plan.

management insofar as possible—but with more emphasis for the future directed at waste minimization. Typical on-site waste minimization elements that can be applied to most waste management programs will consist of:

- Waste recycling (i.e., the recovery of materials used or produced by a process for separate use or direct re-use in-house)
- Material recovery (i.e., the processing of waste streams to recover materials which can be used as feedstock for conversion to another product)
- Waste re-use (i.e., the direct re-use of the wastes as is, without further processing)
- Reduction at source (that may comprise of process modifications and/or the introduction of good housekeeping methods)

All these aspects of the waste management program will likely dominate the scene in industrial waste management planning of the future. In particular, waste reduction at source will gain more prominence in the long-term. Reduction at source will be accomplished through process modifications, and through efficient or good housekeeping. The process modification aspect of the waste reduction at source will generally consist of the following typical elements:

- Modification of the process pathway
- Improved control measures
- Equipment modification
- Changes in operational setting/environment
- Increased automation
- Product substitution

Good and efficient housekeeping can be attained by implementing the following typical tasks:

- Material handling improvements
- Spill and leak monitoring, collection, and prevention
- Preventative maintenance
- Inventory control
- Waste stream segregation

Anyhow, even when an extensive effort is made at minimizing wastes, there still will be some residual wastes remaining when all is said and done. Such wastes can be permitted for long-term storage on-site; or for the permanent disposal at on-site landfills, surface impoundments, etc.; or for an end-of-pipe treatment on-site—in order to reduce waste quantity or contaminant concentrations prior to off-site disposal.

In closing, it is apparent that all waste management operations and hazardous materials handling or processing activities do constitute a potential source of land contamination—and therefore may be viewed as possible source of present or future risks. Effective regulatory and control systems should therefore become an integral part of all waste management programs in particular, and environmental management programs as a whole; the adoption of such a strategy will help minimize potential effects from the likely consequences of contaminated site problems. Indeed, the possibility of creating contaminated site problems will remain an inevitable concomitant of industrial activities in modern societies. However, their effects and extent can be minimized by embracing several of the proactive measures identified here.

Part II

Contaminated Site Problem Diagnosis

This part of the book is comprised of four sections—consisting of the following specific chapters:

- Chapter 4, *The Contaminated Site Characterization Process*, elaborates on the types of activities usually undertaken in the conduct of environmental site assessments at potentially contaminated sites. This includes the key elements and processes involved in streamlining both the contaminated site characterization and site restoration/management activities.
- Chapter 5, *Planning for Contaminated Site Investigation and Management Activities*, discusses the pertinent planning requirements that would assure a cost-effective implementation of a contaminated site investigation activity. This includes a general discussion of the steps typically taken to develop comprehensive workplans in the characterization and management of potentially contaminated site problems.
- Chapter 6, *Field Investigation Techniques for Potentially Contaminated Sites*, enumerates some of the key field investigation techniques that are commonly used to help achieve the ultimate goals of risk determination, risk management, and corrective action planning. This is in recognition of the fact that, typically, different levels of effort in the field investigation will generally be required for different contaminated site problems—and that a variety of field investigation methodologies and techniques may indeed be used during such investigation activities.
- Chapter 7, *Contaminant Fate and Behavior Assessment*, identifies and discusses the relevant phenomena influencing the fate and behavior of environmental contaminants, together with the important factors affecting the processes involved. This is in recognition of the fact that a good understanding of the contaminant fate and transport is necessary—in order to be able to properly characterize the potential risks associated with an environmental contamination and to further develop appropriate risk management and/or remedial action plans for a contaminated site problem.

4 The Contaminated Site Characterization Process

A complexity of processes may affect contaminant migration at contaminated sites and in the wider environment—often resulting in possible threats to human and ecological receptors even outside the source area. Indeed, upon a source release, contaminants may reach potential receptors via several environmental media (e.g., air, soils, groundwater, and surface water). Thus, it is imperative to adequately characterize a suspect site along with the source area and vicinity through a well-designed environmental assessment program (during which all contaminant sources and impacted media are thoroughly investigated)—in order to arrive at appropriate and cost-effective risk management and/or corrective action response decisions. In general, the characterization of a potentially contaminated site problem will classically consist of a process that helps to establish the presence or absence of contamination or hazards at a site, to delineate the nature and extent of the site contamination or hazards, and to determine possible threats posed by the contamination or hazard situation to human health and/or the environment. Subsequently, the affected site could be scheduled for restoration or other corrective action activities, as necessary. This chapter expounds the key elements and processes involved in streamlining both the contaminated site characterization and site restoration or corrective action activities.

4.1 GOALS AND REQUIREMENTS OF A CONTAMINATED SITE CHARACTERIZATION

The primary objective of a site characterization is to delineate the vertical and horizontal extent of a contamination, as well as evaluate the site conditions—especially recognizing the fact that a complete site characterization is critical for the proper design and installation of a remedial system that may be required to address a contaminated site problem. In fact, to begin developing an effectual site-specific remedial system, the design team needs to collect sufficient qualitative and quantitative information on site conditions and subsurface characteristics deemed pertinent to the contaminated site management and cleanup objectives. Ultimately, the packaging of the pertinent contaminant and site data (as necessary for the proper evaluation that would satisfy the stated purposes) can be achieved through a combination of contour maps (horizontal), boring logs (vertical), transect maps (vertical), compositional analyses, and concentration data (horizontal and vertical) for all constituents of concern; this information would typically be used in conjunction with the soil conditions and hydrogeological data to produce worthwhile design and implementation outcomes.

In general, the process of characterizing a contaminated site involves first determining the nature of contamination present, as well as the distribution of contaminants at the project site. Once it is clearly established as to what contaminants are present and their distribution across the site, then appropriate remedial and/or risk management actions can be evaluated for implementation. In the end, the site characterization efforts should provide answers to several pertinent questions—particularly the following:

- What kind of contamination exists at the site?
- Where and how is the contamination migrating?
- What hazards exist to public health and/or the environment?
- Will the selected or anticipated remediation technique likely work?

In the efforts involved here, an initial site characterization would typically provide data on the contaminants of concern, the media that is impacted, site conditions, local climate, remediation objectives, and cleanup standards. In addition, background information on the site could be gathered during this phase of the site assessment. Indeed, a combination of effectual technical approaches and forensic tools can help determine the historic causes, timings, and impacts of a site contamination problem (Ram et al., 1999)—and, oftentimes, this is a crucial starting point in developing effective corrective measures and risk management plans for contaminated site problems.

4.1.1 KEY INFORMATION REQUIREMENTS

The site characterization data associated with contaminated site assessments should generally provide, at a minimum, information on the site description, contamination assessment, soil conditions, hydrogeological conditions, aerial conditions, and risk assessments—as annotated in the following:

- *Site Description.* A general site description is a key requirement for managing contaminated site problems. The site description would typically include detailed maps of the location—comprising property boundaries, surrounding features and landmarks, residential or public areas, water bodies, roadways or other access ways, and descriptive or historical names for all relevant features. Furthermore, the site description should classically include maps illustrating the scale of the infrastructure, surface features, structures, buried services, and other obstacles that will need to be removed and/or accounted for in a site characterization or remedial action activity. Indeed, this may include buildings, structures, foundations, concrete pads, paved surfaces, tanks, pipes, drains, underground utility lines, monitoring/compliance wells, overhead power lines, and natural barriers. Other descriptive information that may prove useful include historical uses of the site, surrounding industrial or commercial sites, previous site investigations conducted, any previous remediation efforts, and the overseeing regulatory agencies.

- *Contamination Assessment.* The contaminant assessment carried out during site characterization efforts should include all relevant and potentially impacted media—which typically includes soils, sediments, surface waters, groundwater, and air emissions from the contaminated site. Indeed, a clear determination of the nature and types of contaminants released into the environment is a crucial requirement for managing contaminated site problems. In the efforts involved, the methods and frequency of sampling used to identify the potential contaminants of concern should be clearly established and agreed upon by all stakeholders. Ultimately, the information about the contaminant distribution in the media of concern (viz., soil, groundwater, surface water, etc.) is needed to complete the site characterization efforts; this is also utilized to facilitate the proper design of a prospective remedial system. For example, being able to delimit the "hot-spots" of contaminated soil may facilitate plans to have such areas removed prior to application of other types of final remedies; also, knowledge of the distribution of a plume that is contaminating groundwater will dictate the location of some system components used in a hydraulic barrier applications; etc.

  In general, sampling should ideally be initiated at the primary source or suspected source(s)—and this should then be extended away from the source(s) to downgradient (and analogous) locations relative to the primary source(s), until the plume is fully characterized. If the source is unknown, sampling will typically follow after conducting field screening using techniques such as soil gas surveys. Field screening may also be used to locate targets for exploratory drilling, or in determining the orientation of the water table, etc. Overall, field screening instruments and methods often produce fast, inexpensive information—and thus can prove to be virtually indispensable. For instance, geophysical field screening utilizes methods such as seismic refraction and other electromagnetic instruments to obtain data on the geologic and hydrogeological conditions at a site; in addition, geophysical methods are used to locate buried objects and delineate residual and floating products. Congruently, on-site analytical screening methods utilize on-site analytical instruments such as detector tubes, immunoassay, portable gas chromatograph (GC), photoionization detectors (PIDs), and X-ray fluorescence and ultraviolet meters to analyze the contamination at a site. Ultimately, therefore, common field sampling methods such as direct push technologies can be used with on-site geophysical and analytical screening methods to delineate contaminant plumes. (See Chapter 6 for further discussion on field investigation techniques.)

Finally, it is worth the mention here that, in choosing a site-specific approach for collecting environmental samples, one should keep in mind the need to have sufficient spatial distribution—so as to, at least, be able to interpolate contaminant concentrations between sampling locations (such as boreholes), when necessary. At any rate, the location and extent of the contaminant concentration must be accurately determined—and importantly, the site characterization should provide detailed data about the contaminants of concern.

- *Soil Conditions.* Knowledge of the soil conditions at a potentially contaminated site could provide a strong basis for contaminant fate and behavior analyses—among other things. For instance, an evaluation of soil conditions (viz., geological, geochemical, and micro-biological) will help determine whether a site is amenable to particular remedial systems and also provide insights into the amount of work required to prepare the site for such systems. The type of analysis may comprise a simple compilation of some basic geological characteristics—including the soil classification (e.g., sand, silt, and clay), salinity, electrical conductivity, cation exchange capacity, organic matter content, water holding capacity, and inorganic nutrient levels. Geochemical information may also be evaluated for its potential to affect the function and performance of potential remedial systems; for this, key parameters would include oxygen, carbon dioxide, and methane gas concentrations—plus redox potential, pH, and soil moisture. Such types of information are pertinent when evaluating the degradability of the contaminant under natural attenuation, active or passive biodegradation, or indeed variant remedial applications. Furthermore, additional information on the interaction between the native microbial populations, contaminants, and the soil system may have to be determined; much of this information is usually not available through routine site assessments/investigations—but would typically be assessed during "treatability" studies. Meanwhile, it is noteworthy that, native microbial consortia are often responsible for natural remediation processes (natural attenuation, bioremediation, and phytotechnologies)—and can indeed represent the primary mechanism for contaminant concentration reduction.
- *Hydrology and Hydrogeological Conditions.* A general evaluation of the hydrogeological conditions of a project site is an important requirement for managing contaminated site problems. Indeed, detailed information on the surface and subsurface hydrology must generally be available as part of the site characterization efforts; this is also necessary prior to the design and installation of likely remedial systems. The pertinent information typically would include groundwater levels, temperatures, flow velocity, porosity, hydraulic conductivity, soils heterogeneity, depth of aquitard, and the continuity and thickness of the aquitard. In addition, the groundwater chemistry (such as the pH and salinity) should be evaluated for its potential to affect the function and performance of anticipated remedial systems.

Additional site-specific hydrogeological data such as the physical setting, stratigraphy, aquifer and aquitard heterogeneity, structure, and sedimentology should be obtained during site characterization; among other things, aquifer tests should be performed to obtain hydrological data such as hydraulic conductivity and intrinsic permeability. This information will help characterize subsurface conditions, as well as provide indications of possible preferential flow paths, perched zones, and water-transmissive and resistive zones. Furthermore, all major controlling influences on groundwater recharge and flow should be defined (e.g., bedrock, production wells, tidal and seasonal influences, surface features, and infiltration). In addition, it is important to understand seasonal changes in the flow direction and flux due to vertical and lateral recharge.

• *Aerial Conditions.* Numerous aerial factors can affect contaminant fate and behaviors in the environment; as such, these conditions should be carefully examined. Among others, information relating to the seasonal changes in climate, including temperature, humidity, precipitation (viz., rain and snow), wind speed and prevailing direction, and the probabilities of floods or droughts (e.g., 25-, 50-, and 100-year events) should be available from local weather stations (often located at nearby cities, airports, major operating facilities, etc.). Indeed, these aerial characteristics can affect the design, operation, and maintenance of a number of remedial systems. Furthermore, these factors can be paramount to the successful design of some systems that have the potential to affect local hydrology (e.g., vegetative covers for infiltration reduction and hydraulic barriers for groundwater control).

• *Risk Assessment.* The risk assessment process is used to complete the site characterization process, as well as to provide an indication of whether or not site remediation may be required for a contaminated site problem after all. The contaminated site risk assessment would typically include identifying the ecological and human health receptors at the site, determining potential pathways for exposure, evaluating the possibility of contaminant migration towards those receptors, and determining the potential toxicity of the contaminants at the site. (See Part III for further discussion on the risk assessment concepts, principles, techniques, and methods of approach relevant to contaminated site characterization and restoration.)

Invariably, the first issue in any attempt to conduct a risk assessment relates to answering the seemingly straightforward question: "does a hazard exist?" In the exercises involved here, the hazard identification component of a risk assessment for contaminated site problems involves establishing the presence of an environmental contaminant or stressor that could potentially cause an adverse effect in potential receptors. The process involves a consideration of the major sources of hazards that may contribute to a variety of contaminated site problems and possible risk situations. Once the presence of potential environmental contaminants or hazards has been established, environmental samples will usually be gathered and analyzed for the likely hazards or chemicals of concern in the appropriate media of interest. The relevant data should then help identify the hazards or chemicals present at a given locale that are to be the focus of the risk assessment process. Finally, it is noteworthy that effective analytical protocols in relation to the sampling and laboratory procedures are typically required—in order to help minimize uncertainties associated with the data collection and evaluation aspects of the risk assessment.

It is apparent that an adequate site characterization is indeed essential to defining the nature and extent of contamination—in order that decisions regarding site cleanup can be formulated in a scientifically and legally defensible, yet cost-effective manner. Meanwhile, it ought to be emphasized here that, the site characterization data should be very carefully evaluated—especially, insofar as carrying out an analysis of the data for validity, sufficiency, and sensitivity. Furthermore, the site characterization data should be critically reviewed for its discernible applicability to the case-specific remedial project on hand.

4.2 ELEMENTS OF A SITE CHARACTERIZATION ACTIVITY

Credible site characterization programs tend to involve a complexity of activities that generally require careful planning. The initial step classically involves a data collection activity that is used to compile an accurate site description, history, and chronology of significant events. Subsequently, field sampling and laboratory analyses may be conducted to help define the nature and extent of contamination present at, or migrating from, a contaminated site. Overall, the process involved in the characterization of contaminated sites typically consists of the collection and analysis of a variety of environmental data necessary for the design of an effectual corrective action program. Consequently, it is important to use proven methods of approach for the sampling and analysis programs designed to tackle potentially contaminated site problems.

A wide variety of investigation techniques may indeed be used in the characterization of potentially contaminated sites—albeit the appropriate methods and applicable techniques will generally be dependent on the type of contaminants, the site geologic and hydrogeologic characteristics, site accessibility, availability of several technical resources, and budget constraints. Several methods of choice that may be used when aiming to conduct effective site characterization programs are described elsewhere in the literature of site assessments (e.g., Boulding and Ginn, 2003; BSI, 1988; Byrnes, 2000, 2008; CCME, 1993, 1994; CDHS, 1990; Driscoll, 1986; ITRC, 2003a–c, 2004, 2007a–d, 2012a, 2013a; Lesage and Jackson, 1992; OBG, 1988; Rong, 2018; Sara, 2003; USEPA, 1985a,d, 1987b,c,d,f, 1988e,f,g, 1989c,h,k,n,o,; WPCF, 1988). In fact, wide-ranging publications are available to provide technical standards and guidance for remedial investigation project scoping; data validation; surveying and mapping of contaminated site sampling locations; hydrogeologic characterization at contaminated sites; surface geophysical techniques; drilling, coring, sampling, and logging at contaminated sites; borehole geophysical techniques; design and construction of monitoring wells and piezometers; groundwater sampling; design and construction of extraction wells at contaminated sites; etc. These types of guidance will facilitate the implementation of a worthwhile site characterization plan.

Finally, it is worth the mention here that, to facilitate the overall site characterization process, workplans are usually needed to specify the administrative and logistic requirements of the site investigation activities. A typical workplan used to guide the investigation of contaminated site problems will commonly include the following key components: a sampling and analysis plan, a health and safety plan, and a quality assurance/quality control plan (QA/QC) (see Chapter 5). At any rate, when all is said and done, the information obtained from a site characterization activity should be adequate to predict the fate and behavior of the contaminants in the environment, as well as to determine the potential impacts associated with the site. In fact, the characterization of contaminant sources, migration patterns, and populations potentially at risk probably form the most important basis for determining the need for site remediation. Thus, the completion of an adequate site characterization is considered a very important module of any site restoration program that is designed to effectively remedy a contaminated site problem. Box 4.1 enumerates the several important elements of site characterization activities that are oftentimes used to support risk management and corrective action decisions at contaminated sites; the implementation of these typical action items will usually help realize the overall goal of a corrective action assessment.

**BOX 4.1   KEY ELEMENTS OF A SITE
CHARACTERIZATION ACTIVITY PROGRAM**

- Identify and describe the physical setting and/or geographical location of problem situation.
- Investigate site physical characteristics (to include surface features, geology, soils and vadose zone properties, drainage network and surface water hydrology [with particular emphasis on an assessment of the erosion potential of the site, and the potential impacts on any surface waters in the catchment], hydrogeology, meteorological conditions, land uses, and human and ecological populations).

(Continued)

BOX 4.1 (*Continued*)   KEY ELEMENTS OF A SITE CHARACTERIZATION ACTIVITY PROGRAM

- Determine sources of contamination, differentiating between naturally occurring and human-made sources.
- Design a sampling plan—that will allow for a characterization of both "hot-spots" and "representative areas"—through extensive sampling, field screening, visual observations, or a combination of these.
- Conduct field data collection and analysis.
- Determine the nature and extent of contamination (to include contaminant types and characteristics, contaminant concentrations, spatial and temporal distribution of contaminants, and impacted media [such as groundwater, soils, surface water, sediments, and air]).
- Assess the contaminant fate and transport properties/characteristics in the various environmental matrices.
- Identify potential migration pathways.
- Prepare a data summary that contains pertinent sampling information for each area of concern at the site—including the media sampled, chemicals analyzed, sampling statistics (such as frequency of detection of the contaminants of concern, range of detected concentrations, and average concentrations), background threshold concentrations in the local area, range of laboratory/sample quantitation limits, and sampling periods.
- Identify and compile regulatory limits for the target contaminants being investigated.
- Develop detailed iso-concentration plots to show the distribution of the chemicals of potential concern present at the site.
- Develop a site conceptual model for the case problem that can be refined as additional information about the site that is acquired—to include patterns of potential exposures.
- Conduct a risk assessment for the case problem—consisting of a baseline risk screening for the site (i.e., a preliminary risk assessment for a "no-action" scenario).
- Identify risk management and corrective action needs/requirements for the problem situation.

4.2.1 Designing a Cost-Effective Site Characterization Program

Several types of decisions are usually made very early in the design of a site characterization program, at which time only limited information may be available about the case site; indeed, reasonable decisions can still be made despite the uncertainties that may surround the process. In any event, this is best accomplished by the use of a phased or tiered type of approach to the overall investigation process. By adopting a hierarchical approach, each phase of a site investigation should, insofar as practicable, be based upon previously generated information about the site. A multilayered, iterative, and flexible survey that emphasizes *in situ* measurements and focuses on mapping contaminant boundaries tend to be critical to success of this process (Jolley and Wang, 1993). Such an approach will help define the specific information needed to understand the site conditions, as well as help optimize the cost of acquiring that information. In the end, the scope and detail of a site characterization activity should by and large be adequate to determine the following:

- Primary and secondary sources of contamination or hazards
- Nature, amount, and extent of contamination—or degree of hazard
- Fate and transport or behavior characteristics of the site contaminants, or hazards
- Pathways of contaminant migration, or hazard propagation

- Types of exposure scenarios associated with the hazard situation or site contamination problem
- Risk to both human and ecological receptors/populations
- Feasible solutions to mitigate potential receptor exposures to the hazards or site contaminants

In the efforts involved here, it is important that a preliminary identification of the types of contaminants, the chemical release potentials, and also the potential exposure pathways are made very early in the site characterization—especially because these are vital to decisions on the number, type, and location of environmental samples to be collected.

On the whole, it should be recognized that the investigation of huge complex sites may present increased logistical problems—making the design of site characterization programs for such situations more complicated. To allow for reasonably manageable situations in the investigation of extensive or large areas, it usually is prudent to divide the case problem or site up into what might be called "risk management units" (RMUs) [or "environmental management units" (EMUs) or "contaminated land zones" (CLZs) or "operable units" (OUs), etc.]. The RMUs (or EMUs, CLZs, OUs, etc.) would classically define the areas of concern—which are then used to guide the identification of the general sampling locations at or near the project site. The areas of concern, defined by the individual RMUs (or EMUs, CLZs, OUs, etc.), will typically include sections or portions of a facility/property or site that (i) are believed to have different chemical constituents, or (ii) are judged to have different anticipated concentrations or "hot spots," or (iii) are a suspected major contaminant release source, or (iv) differ from each other in terms of the anticipated spatial or temporal variability of contamination, or (v) have to be sampled using different field procedures and/or equipment and tools (USEPA, 1989d). All of the areas of concern (designated as RMUs or EMUs or CLZs or OUs) together should account for, or be representative of, the entire case-facility/property or site.

Finally, it is important to recognize the fact that, the key to a successful and cost-effective site characterization program would be to incorporate a credible optimization process in the overall design and implementation efforts—as, for example, in optimizing the number and length of field reconnaissance trips, the number of soil borings and wells, and the frequency of sampling and laboratory analyses. In fact, opportunities to effect cost savings begin during the initial field investigation conducted as part of a preliminary assessment. Thus, the site characterization activity should be very carefully planned prior to the start of any field activities. Ultimately, the characterization of a contaminated site problem should help determine the specific type(s) of site contaminants, their abundance or concentrations, the lateral and vertical extents of contamination, the volume of contaminated materials involved, and the background contaminant levels for native soils and water resources in the vicinity of the site.

4.2.2 THE NATURE OF CONTAMINATED SITE INVESTIGATION ACTIVITIES

Essentially, site investigations are conducted in order to characterize conditions at a project site; environmental samples collected from the site will generally be submitted to a certified analytical laboratory for analysis and interpretation. In all situations, however, before a subsurface investigation is initiated, it is crucial that the background information relating to the regional geology, hydrogeology, and site development history would have been fully researched. In any event, minimization of well installations and sample analyses during the initial phase of a site characterization can greatly reduce overall costs. Indeed, well installation can even add to the spread of contamination—caused by the possibility of penetrating impermeable layers containing contaminants and then allowing such contamination to migrate into previously uncontaminated subsurface (e.g., groundwater) systems (Jolley and Wang, 1993). Consequently, nonintrusive and screening measurements are generally preferred modes of operations whenever possible since that tends to minimize both sampling costs and the potential to spread contamination. Ultimately, however, sufficient information should be obtained that will reliably show the identity, areal and vertical extent, and the magnitude of contamination associated with the case facility/property or contaminated site problem.

Broadly speaking, and as an important starting point in the site investigation process, a realistic and conceptual representation of the contaminated site problem should be constructed—as discussed in the following (Section 4.3); additionally, the quality of data required from the specific study should be clearly defined. Once the level of confidence required for site data is established, strategies for sampling and analysis can then be developed (USEPA, 1988a). The identification of sampling requirements involves specifying the sampling design; the sampling method; sampling numbers, types, and locations; and the level of sampling QC. In fact, sampling program designs must seriously consider the quality of data needed—also recognizing that if the samples are not collected, preserved, and/or stored correctly before they are analyzed, the analytical data may be compromised. Additionally, if sufficient sample amounts are not collected, the method sensitivity requirements may not be achieved. Ultimately, the data derived from the site investigation may be used to perform a risk assessment (as outlined in Part III)—which then becomes a very important element in the site characterization, as well as the site restoration and/or risk management decisions.

Meanwhile, it is noteworthy that site investigation activities are quite often designed and implemented in accordance with several regulatory and legal requirements of the geographical region or area in which a potentially contaminated site or property is located. In a typical investigation, the influence of the responsible regulatory agencies may affect the several operational elements (such as site control measures; health and safety plans; soil borings, excavations, and monitoring well permitting and specifications; excavated materials control and disposal and the management of investigation-derived wastes in general; sample collection and analytical procedures; decontamination procedures; and traffic disruption/control) necessary to complete the site investigation program. Irrespective of the nature of regulatory authority involved, however, the basic site investigation strategy commonly adopted for contaminated site problems will typically comprise of the specific tasks and general elements summarized in Table 4.1 (Cairney, 1993; USEPA, 1988b).

TABLE 4.1
Major Tasks and Important Elements of a Site Investigation Program

Task	Elements
Problem definition	• Define project objectives (including the level of detail and topics of interest) • Determine DQOs
Preliminary evaluation	• Collect and analyze existing information (i.e., review available background information, previous reports, etc.) • Conduct visual inspection of the locale or site (i.e., field reconnaissance surveys) • Construct preliminary conceptual model of the site or problem situation
Sampling design	• Identify information required to refine conceptual model of the site or problem situation • Identify constraints and limitations (e.g., accessibility, regulatory controls, utilities, and financial limitations) • Define sampling, analysis, and interpretation strategies • Determine possible exploratory techniques and testing programs
Implementation of sampling and analysis plans	• Conduct necessary and appropriate exploratory work and perform appropriate testing at the problem location (e.g., site exploratory borings, test pits, and geophysical surveys) • Perform *in situ* testing, as appropriate • Carry out sampling activities • Compile record of investigation logs, photographs, and sample details • Perform laboratory analyses
Data evaluation	• Compile a database of relevant site information at the locale • Carry out logical analysis of site or environmental data • Refine conceptual model for the site or problem situation
Interpretation of results	• Enumerate implications of site investigation results • Prepare a report on findings

4.3 DEVELOPMENT OF CONCEPTUAL SITE MODELS FOR CONTAMINATED SITE PROBLEMS

Conceptual models are usually developed for contaminated sites in order to clearly and systematically identify and document likely contaminant source areas, migration and exposure pathways, potential receptors, and how these individual elements are interconnected. Specifically, a conceptual site model (CSM) provides a structured framework for characterizing possible threats posed by potentially contaminated sites. Indeed, the CSM aids in the organization and analysis of basic information relevant to developing likely corrective action decisions about a given site. Thus, the development of a comprehensive CSM is a generally recommended and vital part of the corrective action assessment or decision process for contaminated site problems.

Conceptualization principles are indeed an important starting point in formulating strategies to address most contaminated site problems, regardless of the source of origination for the chemicals of interest and anticipated impacts. Conceptual models generally establish a hypothesis about possible contaminant sources, contaminant fate and transport, and possible pathways of exposure to the populations potentially at risk (Figure 4.1). For an archetypical contaminated site or environmental contamination problem, evaluation of the CSM usually involves the following types of analyses:

- A *contaminant release analysis*—to determine contaminant release rates into specific environmental media over time. (This may include determining the spatial distribution of contaminants, appraising ambient conditions, analyzing site geology and hydrogeology, determining the extent to which contaminant sources can be adequately identified and characterized, determining the likelihood of releases if the contaminants remain on-site, determining the extent to which natural and artificial barriers currently contain contaminants and the adequacy of such barriers, identifying potential migration pathways, determining the extent to which site contaminants have migrated or are expected to migrate from their source(s), and estimating the contaminant release rates into specific environmental media over time.)
- A *contaminant transport and fate analysis*—to provide guidance for evaluating the transport, transformation, and fate of contaminants in the environment following their release; to identify off-site areas affected by contaminant migration; and to determine contaminant concentrations in these areas.
- An *exposed population analysis*—to determine the likelihood of human and ecological receptors coming into contact with the contaminants of concern.

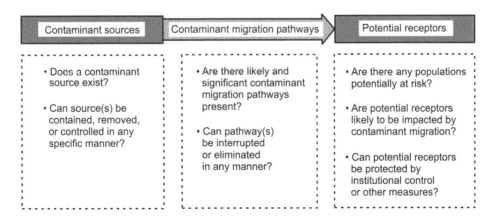

FIGURE 4.1 Conceptual elements for a contaminated site problem: hypothesis to test during an evaluation of the various elements of a CSM.

- An *integrated exposure analysis*—to provide guidance for calculating and integrating exposures to all populations affected by the various exposure scenarios associated with the contaminated site problem.

Indeed, the conceptual model helps to identify and document all known and suspected sources of contamination, types of contaminants and affected media, known and potential migration pathways, potential exposure pathways and routes, target receiving media, and known or potential human and ecological receptors. Such information can be used to develop a conceptual understanding of the contaminated site problem, so that potential risks to human and ecological health, as well as the environment at large can be evaluated more completely. Eventually, the CSM will usually help identify data gaps and further assist in developing strategies for data collection in support of con-taminated site risk management programs; this effort, in addition to assisting in the identification of appropriate sampling locations, will also assist in the screening of potential remedial options or technologies.

All in all, a conceptual evaluation model is used to facilitate a more holistic assessment of the nature and extent of a chemical release and/or contamination problem. It also identifies all known and suspected [or potential] contamination sources, the types of contaminants and affected media, the existing and potential exposure pathways, and the known or potential human and ecological receptors that might be threatened. This information is frequently summarized in pictorial or graphical form—and generally backed up by site/problem-specific data. The development of an adequate conceptual model is indeed a very important aspect of the technical evaluation scheme necessary for the successful completion of most environmental/site characterization programs. The framework integrates several types of information on the physical and environmental setting of the project site or specific issue on hand—which then forms a basis for human health and ecological risk assessments. The conceptual model is also relevant to the development and evaluation of corrective action or remedy programs identified for a variety of chemical release/exposure and contaminated site problems.

4.3.1 DESIGN REQUIREMENTS AND ELEMENTS OF A CSM

CSMs offer an integrated approach for assessing human and ecological population exposures to contaminants released from potentially contaminated sites. Typically, the CSM is developed using available field sampling data, historical records, aerial photographs, and hydrogeologic information about a project site. Once synthesized, this information may be presented in several different forms, such as map views (showing sources, pathways, receptors, and the distribution of contamination), cross-sectional views (illustrating sectional components hydrogeology), and/or tabular forms (sum-marizing and comparing contaminant concentrations against standards such as background thresh-olds, and/or regulatory or risk-based criteria). A typical CSM will usually incorporate the following basic elements:

- Identification of site contaminants and determination of their physical/chemical properties
- Characterization of the source(s) of contamination and site conditions
- Delineation of potential migration pathways
- Identification and characterization of all populations and resources that are potentially at risk
- Determination of the nature of interconnections between contaminant sources, contaminant migration pathways, and potential receptors.

Relationships among these elements provide a basis for testing a range of exposure hypotheses for a given contaminated site problem. For instance, in a representative scenario in which there is a source release, contaminants may be transported from the contaminated media by several processes

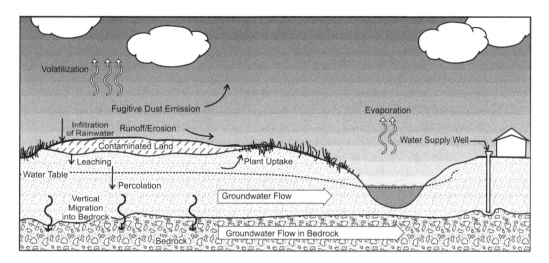

FIGURE 4.2   A diagrammatic conceptual representation of a contaminated site problem.

into other environmental compartments (as depicted by the diagrammatic representation shown in Figure 4.2). In this type of situation, precipitation may infiltrate into soils at a contaminated site and leach contaminants from the soil as it migrates through the contaminated material and the unsaturated soil zone. Infiltrating water may continue its downward migration until it encounters the water table at the top of the saturated zone; the mobilized contaminants may be diluted by the available groundwater flow. Once a contaminant enters the groundwater system, it is possible for it to be transported to a discharge point or to a water supply well location. There also is the possibility of continued downward migration of contaminants into a bedrock aquifer system. Contaminants may additionally be carried by surface runoff into surface water bodies. Air releases of particulate matter and vapors present additional migration pathways for the contaminants. Consequently, all these types of scenarios will typically be evaluated as part of a site characterization process for a given contaminated site problem.

Typically, the CSM is prepared early on in the project and used to guide site investigations and pertinent decision-making; however, this CSM should be updated periodically whenever new information that improves a further understanding of the particular site problem becomes available. Site history and preliminary assessment or site inspection data generally are very useful sources of information for developing preliminary CSMs. Subsequently, the CSM should be appropriately modified if the acquisition of additional data and new information necessitates a redesign. Several considerations and evaluations are indeed important to the design of a realistic and truly representative CSM that will meet the overall goals of a risk assessment and environmental management program (Box 4.2).

In general, the complexity and degree of sophistication of a CSM usually is consistent with the complexity of the particular problem or site and also the amount of data available. For instance, in attempts to characterize the source(s) of contamination, if there are multiple sites in proximity to one another such that it is almost impossible to differentiate between the individual source(s), the affected sites may be clustered together into a single "zone"—and then a corresponding CSM generated accordingly. Migration pathways and receptors can then be determined for the zone, rather than for individual sites. At any rate, all assumptions and limitations in this type of situation should be clearly identifiable and justifiable. Meanwhile, it is worth the mention here again that, as site characterization activities move forward, the CSM may have to be revised as necessary—and then this can be used to direct the next iteration of possible sampling activities needed to complete the site characterization efforts. The "finalized" CSM is then used to develop realistic exposure scenarios for the project site.

> ## BOX 4.2   MAJOR CONSIDERATIONS IN THE DESIGN OF A CSM
>
> - Determine the spatial distribution of contaminants
> - Analyze site geology and hydrogeology
> - Determine area climatic regime and hydrology
> - Determine the extent to which contaminant sources can be adequately identified and characterized
> - Determine the likelihood of releases if the contaminants remain on-site
> - Determine the extent to which natural and artificial barriers currently contain contaminants and the adequacy of such barriers
> - Identify potential migration pathways
> - Determine the extent to which site contaminants have migrated or are expected to migrate from their source(s)
> - Estimate the contaminant release rates into specific environmental media over time
> - Provide guidance for evaluating the transport, transformation, and fate of contaminants in the environment following their release
> - Identify off-site areas affected by contaminant migration
> - Determine contaminant amount, concentration, hazardous nature, and environmental fate properties of the constituents present in affected areas
> - Determine the human, ecological, and welfare resources potentially at risk
> - Identify the routes of contaminant exposure to populations potentially at risk
> - Determine the likelihood of human and ecological receptors coming into contact with the contaminants of concern
> - Assess the likelihood of contaminant migration posing a threat to public health, welfare, or the environment
> - Provide guidance for calculating and integrating exposures to all populations affected by the various exposure scenarios associated with the contaminated site
> - Determine the extent to which contamination levels could exceed relevant regulatory standards, in relation to public health or environmental standards and criteria

4.3.1.1   The Overarching Role of CSMs in Corrective Action Programs

It is apparent that the CSM facilitates an assessment of the nature and extent of contamination and also helps determine the potential impacts from such contamination. Consequently, in as early a stage as possible, during a site investigation, all available site information should be compiled and analyzed to develop a conceptual model for the site. This representation should incorporate contaminant sources and "sinks," the nature and behavior of the site contaminants, migration pathways, the affected environmental matrices, and potential receptors. On this basis, a realistic set of exposure scenarios can be developed for a site characterization program. Ultimately, the exposure scenarios developed for a given contaminated site problem can be used to support an evaluation of the risks posed by the site, as well as facilitate the implementation of appropriate decisions on the need for, and extent of, site restoration.

4.3.2   Development of Exposure Scenarios: Integrating Contaminant Sources with Exposure Pathways and Receptors

An exposure scenario, which is a description of the activity that brings a population into contact with a chemical source or a contaminated environmental medium, usually is the next logical development or outcome that follows after the design of a CSM. Exposure scenarios are developed based on the movement of chemicals or contaminants of interest in various environmental compartments into the potential receptor zones (Asante-Duah, 2017). In fact, a wide variety of *potential* exposure patterns can be anticipated from contaminated sites—pointing to a multiplicity of interconnected pathways through which populations may be exposed to site contamination. Some of the typical and commonly encountered exposure scenarios in relation to contaminated site problems are further identified in Chapter 10; those types of exposure scenarios will usually be evaluated as part of the site characterization process for a contaminated site problem.

In general, exposure scenarios are derived and modeled on the basis of the movements of chemicals in various environmental compartments. The exposure scenario associated with a given contaminated site problem may be well defined if the exposure is known to have already occurred. In most cases associated with the investigation of potentially contaminated sites, however, decisions typically have to be made about potential exposures that may not yet have occurred; consequently, hypothetical exposure scenarios are commonly developed for such applications. Anyhow, several tasks are usually undertaken to facilitate the development of complete and realistic exposure scenarios—including the following critical ones:

- Determine the sources of site contamination
- Identify the specific constituents of concern
- Identify the affected environmental media
- Delineate contaminant migration pathways
- Identify potential receptors
- Determine potential exposure routes
- Construct a representative conceptual model for the site
- Delineate likely and significant migration and exposure pathways.

Once the complete set of potential exposure scenarios have been fully determined, the range of critical exposure pathways can be identified. This information can then be used to design cost-effective sampling and investigation programs. The goal in this case will be to ensure a focused investigation—in order to be able to determine the specific potential exposure pathways of critical interest or that associated with the planned use of the site in a most cost-efficient manner. Subsequently, a range of engineering and/or institutional measures can be evaluated to help remove, immobilize, or otherwise control the site contaminants—or to mitigate the risks posed to potential human and ecological receptors associated with the site. Overall, the exposure scenarios developed for a given chemical release and/or exposure problem can be used to support an evaluation of the risks posed by the subject site, as well as facilitate the implementation of appropriate decisions regarding the need for, and extent of, possible corrective actions to undertake. Indeed, it is quite important to develop as realistic an exposure scenario as possible at all times; this can then be used to support an evaluation of the risks posed by the potentially contaminated site—as well as allow appropriate decisions to be made regarding the need for, and extent of, remediation.

Finally, it is noteworthy here that if numerous potential exposure scenarios exist, or if a complex exposure scenario has to be evaluated, it usually is helpful to use an "event-tree" model (or similar framework/structure) to clarify potential outcomes and/or consequences. The event tree concept, as exemplified by Figure 4.3, indeed offers an efficient way to develop exposure scenarios. By using such an approach, the various exposure contingencies can be identified and organized in a systematic manner. Once developed, priorities can be established to help focus the available resources on the aspects of greatest need. Table 4.2 illustrates an equivalent analytical protocol for developing the set of exposure scenarios; from the basic similarities, it is notable that this representation is analogous to the event tree structure shown in Figure 4.3. In any case, a wide variety of *potential* exposure patterns may generally be anticipated from a given chemical exposure situation—culminating in a multiplicity of interconnected pathways through which populations might become exposed to contamination. Invariably, the archetypal and commonly encountered

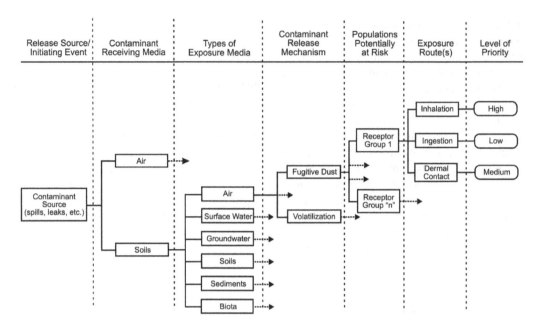

FIGURE 4.3   Diagrammatic representation of exposure scenarios using an event tree.

exposure scenarios will usually be evaluated as part of the exposure characterization process for a given chemical exposure problem.

4.4   DESIGN OF DATA COLLECTION AND EVALUATION PROGRAMS

The design and implementation of a substantive data collection and evaluation program is vital to the effective management of potentially contaminated site problems. To help meet such needs, data are typically collected at several stages of the site investigation—with initial data collection efforts usually limited to developing a general understanding of the site. At any rate, the common types of site data and information classically required in the investigation of potentially contaminated sites relate to the following (USEPA, 1989a):

- Contaminant identities
- Contaminant concentrations in the key sources and media of interest
- Characteristics of sources and contaminant release potential
- Characteristics of the physical and environmental setting that can affect the fate, transport, and persistence of the contaminants.

It is noteworthy that, because of the inherent variability in the materials and the diversity of processes used in industrial activities, it is not unexpected to find a wide variety of contaminants associated with any contaminated site problem. As a consequence, there is a corresponding variability in the range and type of hazards and risks that may be anticipated from different contaminated site problems.

In general, detailed record search of background information on the critical contaminants of potential concern should be compiled as part of the contaminated site investigation program. The investigation must then provide information on all contaminants known, suspected, or believed to be present at the site. Indeed, the investigation should cover all compounds for which the history of site activities, current visible contamination, or public concerns may suggest that such substances could be present. In addition to establishing the concentration of contaminants at a case

TABLE 4.2
Illustrative Tabular Analysis Chart Regarding the Nature of Exposure Scenarios That Might Be Associated with a Contaminated Site Problem

Contaminated Exposure Medium	Contaminant Release Source(s)	Contaminant Release Mechanism(s)	Potential Receptor Location	Receptor Groups Potentially at Risk	Potential Exposure Routes	Pathway Potentially Complete and Significant?
Air	Contaminated surface soils	Fugitive dust generation	On-site	On-site facility worker	Inhalation	Yes
					Incidental ingestion	No
					Dermal absorption	No
				Construction worker	Inhalation	Yes
					Incidental ingestion	No
					Dermal absorption	No
			Off-site	Downwind worker	Inhalation	No
					Incidental ingestion	No
					Dermal absorption	No
				Downwind resident	Inhalation	No
					Incidental ingestion	No
					Dermal absorption	No
		Volatilization	On-site	On-site facility worker	Inhalation	Yes
					Dermal absorption	No
				Construction workers	Inhalation	Yes
					Dermal absorption	No
			Off-site	Nearest downwind worker	Inhalation	No
					Dermal absorption	No
				Nearest downwind resident	Inhalation	No
					Dermal absorption	No

(Continued)

TABLE 4.2 (Continued)
Illustrative Tabular Analysis Chart Regarding the Nature of Exposure Scenarios That Might Be Associated with a Contaminated Site Problem

Contaminated Exposure Medium	Contaminant Release Source(s)	Contaminant Release Mechanism(s)	Potential Receptor Location	Receptor Groups Potentially at Risk	Potential Exposure Routes	Pathway Potentially Complete and Significant?
Soils	Contaminated soils and/or buried wastes	Direct contacting	On-site	On-site facility worker	Incidental ingestion	Yes
					Dermal absorption	Yes
				Construction worker	Incidental ingestion	Yes
					Dermal absorption	Yes
			Off-site	Downwind worker	Incidental ingestion	No
					Dermal absorption	No
				Downwind resident	Incidental ingestion	No
					Dermal absorption	No
Surface water	Contaminated surface soils	Surface runoff into surface impoundments	On-site	On-site facility worker	Inhalation	No
					Incidental ingestion	No
					Dermal absorption	No
				Construction worker	Inhalation	No
					Incidental ingestion	No
					Dermal absorption	No
		Erosional runoff	Off-site	Downslope resident	Inhalation	No
					Incidental ingestion	No
					Dermal absorption	No
				Recreational population	Inhalation	No
					Incidental ingestion	No
					Dermal absorption	No
	Contaminated groundwater	Groundwater discharge	Off-site	Downslope resident	Inhalation	No
					Incidental ingestion	No
					Dermal absorption	No
				Recreational population	Inhalation	No
					Incidental ingestion	No
					Dermal absorption	No

(Continued)

TABLE 4.2 (Continued)
Illustrative Tabular Analysis Chart Regarding the Nature of Exposure Scenarios That Might Be Associated with a Contaminated Site Problem

Contaminated Exposure Medium	Contaminant Release Source(s)	Contaminant Release Mechanism(s)	Potential Receptor Location	Receptor Groups Potentially at Risk	Potential Exposure Routes	Pathway Potentially Complete and Significant?
Groundwater	Contaminated soils	Infiltration/leaching	On-site	On-site facility worker	Inhalation	No
					Incidental ingestion	No
					Dermal absorption	No
				Construction worker	Inhalation	No
					Incidental ingestion	No
					Dermal absorption	No
			Off-site	Downgradient resident	Inhalation	Yes
					Incidental ingestion	Yes
					Dermal absorption	Yes
Drainage sediments	Contaminated surface soils	Surface runoff/episodic overland flow	On-site	On-site facility worker	Inhalation	No
					Incidental ingestion	No
					Dermal absorption	No
				Construction worker	Inhalation	No
					Incidental ingestion	No
					Dermal absorption	No
			Off-site	Nearest downgradient resident	Inhalation	No
					Incidental ingestion	No
					Dermal absorption	No

site, the site investigation should be designed to provide an indication of the naturally occurring or anthropogenic background levels of the target contaminants in the local environment, where appropriate. Ultimately, several chemical-specific factors (such as toxicity or potency, concentration, mobility, persistence, bioaccumulative or bioconcentration potential, synergistic or antagonistic effects, potentiation or neutralizing effects, frequency of detection, and naturally occurring or anthropogenic background thresholds) are used to further screen and select the specific target contaminants that will become the focus of a detailed site evaluation and risk management process. Typically, the selected target chemicals are those site contaminants that tend to be the most mobile and persistent; consequently, they reflect the likelihood of contamination at the problem location or case site and vicinity.

Further elaboration on how the overall data evaluation process affects the selection of significant chemicals of concern associated with a contaminated site problem is provided in Chapter 10 (Part III) of this volume; this also includes additional discussion of the use of proxy values for so-called "censored data."

4.4.1 Development of Data Quality Objectives

Data quality objectives (DQOs) are statements that specify the data needed to support environmental management and corrective action decisions. They are described to establish the desired degree of data reliability, the specific data requirements and considerations, and an assessment of the data applications as determined by the overall study objective(s). The DQO represents the full set of qualitative and quantitative constraints needed to specify the level of uncertainty that an analyst can accept when making a decision based on a particular set of data.

The DQO process consists of a planning tool that enables an investigator to specify the quality of the data required to support the objectives of a given study. In the course of the investigation of potentially contaminated site problems, DQOs are typically used as qualitative and quantitative statements that specify the quality of data required to support the risk assessment, corrective action or site restoration, and environmental management decisions. These are determined based on the end uses of the data to be collected.

DQOs are indeed an important aspect of the QA requirements in the entire environmental management process. Despite the fact that the DQO process is considered flexible and iterative, it generally follows a well-defined sequence of stages that will allow effective and efficient data management. In any event, it is apparent that, as the quantity and quality of data increases, the risk of making a wrong decision generally decreases (Figure 4.4) (USEPA, 1987a,b). Consequently, the DQO process should be carefully developed to allow for a responsible program that will produce adequate quantity and quality of information required for environmental management decisions. In fact, there are several benefits to establishing adequate DQOs for environmental management and corrective action programs, including the following (CCME, 1993):

- The DQO process ensures that the data generated are of known quality.
- All projects have some inherent degree of uncertainty, and DQOs help data users plan for uncertainty. By establishing DQOs, data users evaluate the consequences of uncertainty and specify constraints on the amount of uncertainty that can be tolerated in the expected study results. Thus, the likelihood of an incorrect decision is gauged *a priori*.
- The DQO process facilitates early communication among data users, data collectors, managers, and other technical staff—before time and money are spent collecting data; this will result in increased cost-efficiency in the data collection program.
- The DQO process provides a logistical structure for study planning that is iterative and that encourages the data users to narrow many vague objectives to one or a few critical questions.

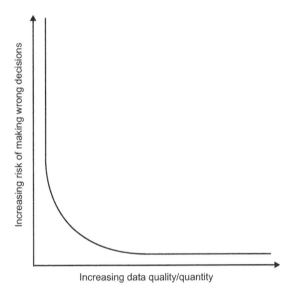

FIGURE 4.4   Relationship between decision risk and data quality/quantity.

- The structure of the DQO process provides a convenient way to document activities and decisions that can prove useful in litigation or administrative procedures.
- The DQO process establishes quantitative criteria for use as cutoff line, as to when to stop collecting environmental samples.

Overall, the DQO process tends to lead on to a well thought out sampling and analysis plan. Consequently, DQOs should preferably be established prior to data collection activities, to ensure that the data collected are sufficient and of adequate quality for their intended uses. The DQO should indeed be integrated with development of the sampling and analysis plan and should be revised as needed—based on the results of each data collection activity.

4.4.1.1   Data Quality Assessment

Data Quality Assessment (DQA) is the scientific and statistical evaluation of data to determine if the data obtained from environmental data operations are of the right type, quality, and quantity to support their intended use. It is noteworthy that, DQA is built on one fundamental premise—namely that data *quality*, as a concept, is meaningful only when it relates to the *intended use* of the data (USEPA, 2000c–e). The strength of the DQA is that, it is designed to promote an understanding of how well the data satisfy their intended use by progressing in a logical and efficient manner. In general, several statistical tools may be utilized in the quantitative aspect of the DQA; a number of such methodologies are discussed elsewhere in the literature (e.g., Gilbert, 1987; USEPA, 1989c,h,k,n,o,p, 1994e,h, 1996b,c,j,k, 2000c–e). Ultimately, the most appropriate procedure is used in summarizing and analyzing the data—based on the review of the DQOs, the sampling design, and the preliminary data review; as part of the process, it is quite important to identify the key underlying assumptions that must hold for the statistical procedures to be valid.

In general, careful planning improves the representativeness and overall quality of a sampling design, the effectiveness and efficiency with which the sampling and analysis plan is implemented, and the usefulness of subsequent DQA efforts. It is notable that the DQO process—which is a logical, systematic planning procedure based on scientific methods—emphasizes the planning and development of a sampling design to collect the right type, quality, and quantity of data needed to support the decision. In the efforts involved, using both the DQO process and the DQA will help to ensure that the decisions are supported by data of adequate quality—with the DQO process doing so *prospectively* and the DQA doing so *retrospectively* (USEPA, 1994e,h, 2000c–e). In the end, when good QA/QC procedures have been used in the overall process, the information derived from the investigation of a potentially contaminated site will be both reliable and of known quality. On the other hand, failure to follow good QA/QC procedures may seriously jeopardize the integrity of the data needed to make critical site restoration or risk management decisions—and which could adversely impact costs of possible remediation and/or risk management requirements for the contaminated site.

4.4.2 FIELD DATA COLLECTION AND ANALYSIS STRATAGEM

Data are generally collected during several stages of a site investigation—with initial data collection efforts usually limited to developing a general understanding of the site. For instance, a preliminary gas survey using subsurface probes and portable equipment will give an early indication of likely problem areas. Soil gas surveys are generally carried out as a precursor to exploratory excavations in order to identify areas that warrant closer scrutiny; they can also be used to assist in the delineation of previously identified plumes of contamination. Broadly speaking, this may be viewed as a rather important step to complete prior to the start of a full-scale site investigation. In fact, gases produced at contaminated sites will tend to migrate through the paths of least resistance—and thus, the presence of volatile contaminants or gas-producing materials can be determined by sampling the soil atmosphere within the ground. Installation of a gas-monitoring well network, in conjunction with sampling in buildings in the area, can be used to determine the need for a more comprehensive corrective action assessment and/or corrective measures. On-site vapor screening of soil samples during drilling can also provide indicators for presence of organic contamination; for example, organic vapor analyzer (OVA)/GC or GC/PID screening provides a relative measure of contamination by volatile organic chemicals. Additionally, predictive models can be used to estimate the extent of gas migration from a suspected subsurface source. This information can subsequently be used to identify apparent "hot-spots" and to select soil samples for detailed chemical analyses. The vapor analyses on-site can also be helpful in selecting screened intervals for monitoring wells. All these types of information can indeed be used to determine the possibility for human exposures and to determine appropriate locations for monitoring wells and gas collection systems.

In general, a phased approach to environmental sampling encourages the identification of key data requirements and needs as early in the site investigation process as possible (see Section 4.5). This ensures that the data collection effort is always directed at providing adequate information that meets the data quantity and quality requirements of the study. As a basic understanding of the particular problem situation is achieved, subsequent data collection efforts focus on identifying and filling in any remaining data gaps from a previous phase of the site investigation. Any additionally acquired data should be such as to further improve the understanding of the problem situation and also consolidate information necessary to manage the contaminated site problem in an effective manner. In this way, the overall site investigation effort can be continually re-scoped to minimize the collection of unnecessary data and to maximize the quality of data acquired. Meanwhile, in areas where the contamination source is known, the sampling strategy would characteristically be targeted around that source. Normally, sampling points would be located at regular distances along the lines radiating from the contaminant source. Provisions should also be made in the investigation to collect additional samples of small, isolated pockets of material that might appear visually suspect.

At the end of the day, the data gathering process should provide a logical, objective, and quantitative balance between the time and resources available for collecting the data and the quality of data, based on the intended use of such data. Meanwhile, it is notable that the analysis of previously acquired and newly generated data serves to provide an initial basis to understanding the nature and extent of contamination, which in turn aids in the design of appropriate risk management and corrective action programs.

4.4.2.1 Data Collection and Analysis Strategies

A variety of data collection and analysis protocols exist in the literature (e.g., Boulding, 1994; Boulding and Ginn, 2003; Byrnes, 1994, 2000, 2008; CCME, 1993, 1994; Csuros, 1994; Garrett, 1988; Hadley and Sedman, 1990; ITRC, 2003a–c, 2004, 2007a–d, 2012a, 2013a; Keith, 1992; Millette and Hays, 1994; O'Shay and Hoddinott, 1994; Rong, 2018; Sara, 2003; Schulin et al., 1993; Thompson, 1992; USEPA, 1982, 1985a–d, 1992a,e,f,i,j; Wilson et al., 1995) that may be adapted for the investigation of human exposure to chemical constituents found in the human and ecological environments. Regardless of the processes involved, however, it is important to recognize the fact that most chemical sampling and analysis procedures offer numerous opportunities for sample contamination from a variety of sources (Keith, 1988). To be able to address and account for possible errors arising from "foreign" sources, QC samples are typically included in the sampling and analytical schemes. The QC or "control" samples are analytical QC samples that are analyzed in the same manner as the "field" samples—and these are subsequently used in the assessment of any cross contamination that may have been introduced into a sample along its life cycle from the field (i.e., point of collection) to the laboratory (i.e., place of analysis).

Invariably, QC samples become an essential component of all carefully executed sampling and analysis program. This is because firm conclusions cannot be drawn from the investigation unless adequate controls have been included as part of the sampling and analytical protocols (Keith, 1988). To prevent or minimize the inclusion of "foreign" constituents in the characterization of chemical exposures and/or in a risk assessment, therefore, the concentrations of the chemicals detected in "control" samples must be compared with concentrations of the same chemicals detected in the "field" samples. In such an evaluation, the QC samples can indeed become a very important reference datum for the overall evaluation of the chemical sampling data.

In general, carefully designed sampling and analytical plans, as well as good sampling protocols, are necessary to facilitate credible data collection and analysis programs. Sampling protocols are written descriptions of the detailed procedures to be followed in collecting, packaging, labeling, preserving, transporting, storing, and tracking samples. The selection of appropriate analytical methods is also an integral part of the processes involved in the development of sampling plans—since this can strongly affect the acceptability of a sampling protocol. For example, the sensitivity of an analytical method could directly influence the amount of a sample needed in order to be able to measure analytes at prespecified minimum detection (or quantitation) limits. The analytical method may also affect the selection of storage containers and preservation techniques (Keith, 1988; Holmes et al., 1993). In any case, the devices that are used to collect, store, preserve, and transport samples must *not* alter the sample in any manner. In this regard, it is noteworthy that special procedures may be needed to preserve samples during the period between collection and analysis.

Finally, the development and implementation of an overall good QA/QC project plan for a sampling and analysis activity is critical to obtaining reliable analytical results. The soundness of the QA/QC program has a particularly direct bearing on the integrity of the sampling as well as the laboratory work. Thus, the general process for developing an adequate QA/QC program, as discussed elsewhere in the literature (e.g., CCME, 1994; USEPA, 1987a–g, 1992a,e,f,i,j), should be followed religiously. Also, it must be recognized that, the more specific a sampling protocol is, the less chance there will be for errors or erroneous assumptions.

4.4.2.2 Evaluation of QC Samples

To prevent the inclusion of "foreign" or non-site-related constituents in the characterization of a site contamination and/or in a risk assessment, the concentrations of the chemicals detected in "control" samples must be compared with concentrations of the same chemicals detected in the environmental samples from the impacted media present at the contaminated site. In general, QC samples containing common laboratory contaminants are evaluated differently from QC samples which contain chemicals that are not common laboratory contaminants. Thus, if the QC samples contain detectable levels of known common laboratory contaminants (e.g., acetone, 2-butanone [methyl ethyl ketone], methylene chloride, toluene, and the phthalate esters), then the environmental sample results may be considered as positive only if the concentrations in the sample exceed approximately ten times the maximum amount detected in any QC sample (DTSC, 1994; USEPA, 1989p,r, 1990g). For QC samples containing detectable levels of one or more organic or inorganic chemicals that are not considered to be common laboratory contaminants, environmental sample results may be considered as positive only if the concentration of the chemical in the environmental sample exceeds approximately five times the maximum amount detected in any QC sample (DTSC, 1994; USEPA, 1989p,r, 1990g). Invariably, QC samples become an essential component of any site investigation program. This is because firm conclusions cannot be drawn from the site investigation unless adequate controls have been included as part of the sampling and analytical protocols (Keith, 1988).

Environmental QC samples can indeed be a very important reference datum for the evaluation of environmental sampling data. The analysis of environmental QC samples provides a way to determine if contamination has been introduced into a sample set either in the field while the samples were being collected and transported to the laboratory or in the laboratory during sample preparation and analysis.

4.4.2.3 The Reporting and Handling of "Censored" Laboratory Data

Broadly speaking, data generated from chemical analysis may fall below the detection limit (DL) of the analytical procedure; these measurement data are generally described as not being detected or "non-detects" (NDs) (rather than as absolute zero or not present)—and the applicable or corresponding limit of detection is usually reported alongside. In cases where measurement data are described as ND, the concentration of the chemical is unknown—albeit it might well exist somewhere between zero and the DL. Data that includes both detected and non-detected results are called "censored data" in the statistical literature. This nomenclature is equally applicable to site characterization activities.

Indeed, it so happens that for a given set of laboratory samples acquired during site characterization, certain chemicals will often be reliably quantified in some (but not all) of the samples that were collected for analysis. Environmental data sets may therefore contain observations that are below the instrument or method DL or indeed its corresponding quantitation limit; such data are often referred to as "censored data" (or NDs). In general, the NDs do not necessarily mean that a chemical is not present at any level (i.e., completely absent)—but simply that any amount of such chemical potentially present was probably below the level that could be detected or reliably quantified using a particular analytical method. In other words, this situation may reflect the fact that either the chemical is truly absent at this location or sampled matrix at the time the sample was collected—or that the chemical is indeed present but only at a concentration below the quantitation limits of the analytical method that was used in the sample analysis.

Invariably, every laboratory analytical technique has detection and quantitation limits below which only "less than" values may be reported; the reporting of such values provides a degree of quantification for the censored data. In such situations, a decision has to be made as to how to treat such NDs and the associated "proxy" concentrations. Anyhow, there are a variety of ways to evaluate data that include values below the DL. However, there are no general procedures that are applicable in all cases. Indeed, all of the recommended procedures for analyzing data with NDs tend to depend on the amount of data below the DL (USEPA, 1989c,h,k,n,o,p, 2000c–e). For relatively small amounts below DL values, replacing the NDs with a small number and proceeding with the usual analysis may be satisfactory. For moderate amounts of data below the DL, a more detailed adjustment may be more appropriate. In situations where relatively large amounts of data below the DL exist, one may need only to consider whether the chemical was detected as above some level or not. Meanwhile, it is inarguable that the interpretation of small, moderate, and large amounts of data below the DL is rather subjective. Further yet, in addition to the percentage of samples below the DL, sample size influences which procedures should fittingly be used to evaluate the data; for example, a case where one sample out of four is not detected should generally be treated differently from another case where 25 samples out of 100 are not detected (USEPA, 2000)—that is, despite the fact that both cases seem to point to akin detection frequencies. That said, it is noteworthy that the appropriate procedure, which depends on the general pattern of detection for the chemical in the overall sampling events and investigation activities, may consist of the following determinations (HRI, 1995):

- If a chemical is rarely detected in any medium, and there is little to no reason to expect the chemical to be associated with the contaminated site problem, then it may be appropriate to exclude it from further analysis. This is particularly true if the chemical is a common laboratory contaminant.
- If a chemical is rarely detected in a specific medium, and there is no reason to expect the chemical to be significantly associated with the contaminated site problem, then it may be appropriate to exclude it from further analysis of that particular medium.
- If the pattern of a chemical's concentration in samples suggests that it was confined to well-defined "hot-spots" at the time of sampling, then the potential for contaminant migration should be considered. Consequently, it will be important to include such chemical in the analysis of both the source and receiving media.

In any case, it is customary to assign nonzero values to all sampling data reported as NDs. This is important because, even at or near their DLs, certain chemical constituents may be of considerable importance in the characterization of a chemical exposure problem. However, uncertainty about the actual values below the detection or quantitation limit can bias or preclude subsequent statistical analyses. Indeed, censored data do create significant uncertainties in the data analysis required of the chemical exposure characterization process; such data should therefore be handled in an appropriate manner—for instance, as elaborated in the example methods of approach provided in Section 10.2.

4.4.3 BACKGROUND SAMPLING CONSIDERATIONS

During corrective action assessments, it often becomes necessary to establish media-specific background thresholds for selected chemical constituents present at contaminated sites. The background threshold is meant to give an indication of the level of contamination in the environment that is not necessarily attributable to the project site under investigation. This serves to provide a "reference point-of-departure" that can be used to determine the magnitude of contamination in other environmental samples obtained from the "culprit" site or locale. In fact, background sampling is generally conducted to distinguish site-related contamination from naturally occurring or other non-site-related levels of select constituents or chemicals of concern. In most situations, *anthropogenic levels* (which are concentrations of chemicals that are present in the environment due to human-made, non-site sources, such as industry and automobiles), rather than naturally occurring levels, are preferably used as a basis for evaluating background sampling data in relation to contaminated site appraisals. Meanwhile, it is noteworthy that, even at background concentration levels, a given locale or site may still be posing significant risks to human health and/or the environment. And, whereas remedial action may not be required under such circumstances, it is still quite important to evaluate and document this situation—so that at least some risk management or institutional control measures can be implemented to protect populations potentially at risk in this area.

4.4.3.1 Using "Control" or "Reference" Sites to Establish Background Thresholds for Environmental Contaminants

Control (or reference) sites are important to understanding the significance of environmental sampling and monitoring data. Background samples collected at control sites are typically used to demonstrate whether or not a site is truly contaminated, by determining if the site samples are truly different from the background in the geographical area. In fact, some sort of background sampling is usually required as part of most corrective action assessments; this will then allow a technically valid scientific comparison to be made between environmental (suspected of containing site contaminants) and control site samples (possibly containing only naturally low or anthropogenic levels of the same chemicals or contaminants).

Broadly speaking, there are two types of background or control locations/sites—viz., "local" and "area"—whose differentiation is based primarily on the closeness of the control site to the environmental sampling or project site. Local control sites are usually adjacent or very near the potentially impacted project sites, whereas an area control site is simply in the same general area or region as the project site *but not adjacent to it*. In selecting and working with either type of control sites, the following general principles and factors should be observed insofar as possible (Keith, 1991; CCME, 1993):

- Control sites/locations generally should be upwind, upstream, and/or upgradient from the environmental sampling site.
- When possible, control site/location samples should be taken first to avoid possible cross contamination from the environmental sampling site.
- Travel between control sites/locations and environmental sampling areas should be minimized because of potential cross contamination caused by humans, equipment, and/ or vehicles.

As a general rule, local control sites are preferable to area control sites because they are physically closer. However, when a suitable local control site cannot be found, an area control site will reasonably provide for the requisite background sampling information. At any rate, locations selected as control sites should have similar characteristics (i.e., identical in their physical and environmental settings) as the potentially contaminated area under investigation—except for the pollution source.

Ideally, background samples (or control site samples) are preferably collected near the time and place congruent with environmental samples of interest. Ultimately, sampling information from the control sites are used to establish background thresholds.

4.4.3.2 Factors Affecting Background Sampling Design

Background samples are typically collected and evaluated to determine the possibility of a potentially contaminated site contributing to off-site contamination levels in the vicinity of the case site. The background samples would not have been significantly influenced by contamination from the project site. However, these samples are obtained from an environmental matrix that has similar basic characteristics as the matrix at the project site, in order to provide a justifiable basis for comparison. To satisfy acceptable criteria required of background samples, the following archetypical requirements should be carefully incorporated into the design of background sampling programs:

- *Significance of matrix effects on environmental contaminant levels.* Unless background samples are collected and analyzed under the same conditions as the environmental test samples, the presence and/or levels of the analytes of interest and the effects of the matrix on their analysis cannot be known or estimated with any acceptable degree of certainty. Therefore, background samples of each significantly different matrix must always be collected when different types of matrices are involved—such as various types of water, sediments, and soils in or near a sampling site area. For example, it has been observed in a number of investigations that the analysis of data from different soil types in the same background area reveals concentration levels of select inorganic constituents that are over twice as high in silt/clay as in sand (e.g., LaGoy and Schulz, 1993). Thus, it is important to give adequate consideration to the effects of natural variations in soil composition when one is designing a field sampling program.

* *Number of background samples.* A minimum of three background samples per medium will usually be collected, although more may be desired especially in complex environmental settings. In general, if the natural variability of a particular constituent present at a site or in the environment is relatively large, the sampling plan should reflect this site- or case-specific characteristic.
* *Background sampling locations.* In typical sampling programs, background air samples would consist of upwind air samples and, perhaps, different height samples; background soil samples would be collected near a site in areas upwind and upslope of the site—with the upwind locations typically being determined from a wind rose (i.e., a diagram or pictorial representation that summarizes pertinent statistical information about wind speed and direction at a specified location) for the geographical region in which the site is located; background groundwater samples generally come from upgradient well locations, in relation to groundwater flow direction(s) at the impacted area; and background surface water and sediment samples may be collected under both high and low flow conditions at upstream locations, and insofar as practicable, sample collection from nearby lakes and wetlands should comprise of shallow and deep samples (when sufficient water depth allows), to account for such differences potentially resulting from stratification or incomplete mixing.

More detailed background sampling considerations and strategies for the various environmental media of general interest can be found in the literature elsewhere (e.g., Keith, 1991; Lesage and Jackson, 1992; USEPA, 1988e, 1989k,n,o).

4.4.3.3 Statistical Evaluation of Background Sampling Data

The statistical evaluation of background sampling results will generally comprise of a determination of whether or not there is a difference between contaminant average concentrations in the background areas and the chemical concentrations at the problem location—and especially if indeed the concentrations are higher at the problem site location. Typically, the statistical evaluation of background sampling outcomes or data consists of the following one-tailed test of significance: the null hypothesis that there is no difference between contaminant concentrations in the background areas and the on-site chemical concentrations versus the alternative hypothesis that concentrations are higher on-site. Broadly speaking, however, a statistically significant difference between background samples and site-related local contamination should not, by itself, be a cause for alarm—nor should it necessarily trigger off a call for remedial or cleanup action; further evaluations (such as conducting a risk assessment) will ascertain the significance of the contamination. In cases where corrective action or remediation is required, the established background thresholds may be used as part of the process or basis for setting realistic cleanup or remedial action goals. Thus, the ability to estimate background or threshold concentrations can indeed become a critical factor in the formulation of reasonable cleanup or remedial action objectives at potentially contaminated sites.

It is noteworthy that auxiliary parameters—for example, geochemical and geotechnical parameters—might at times be needed to strengthen the background analysis; this may typically include evaluations such as determining if chemical concentration ranges vary as a function of soil type, sample type, or sample location, etc. Indeed, the application of geochemical association and enrichment analyses can become a very important aspect of background analyses; general discussions and description of how geochemical procedures may be used to help determine background ranges and related evaluations can be found elsewhere in the literature on geochemistry (e.g., Krauskopf and Bird, 1995; Li, 2000).

4.5   THE OVERALL SITE ASSESSMENT PROCESS: DESIGNING A HOLISTIC SITE ASSESSMENT PROGRAM

Site investigations consist of the planned and managed sequence of activities carried out to determine the nature and distribution of contaminants associated with potentially contaminated site problems. The activities involved usually are comprised of the following tasks (BSI, 1988):

- Identification of the principle hazards
- Design of sampling and analysis programs
- Collection and analysis of environmental samples
- Recording or reporting of laboratory results for further evaluation.

In order to get the most out of a site investigation, it must be conducted in a systematic manner. Systematic methods help focus the purpose, the required level of detail, and the several topics of interest—such as physical characteristics of the site, likely contaminants, extent and severity of possible contamination, effects of contaminants on populations potentially at risk, probability of harm to human health and the environment, and possible hazards during risk management and corrective action activities (Cairney, 1993). Indeed, in addition to establishing the concentration of contaminants at a case site, the site investigation should be designed to provide an indication of the general background level of the target contaminants in the local environment. The systematic process required for the site investigation essentially involves the early design of a representative conceptual model of the site, as detailed earlier on in Section 4.3. This conceptualization is used to assess the physical conditions at the site, as well as to identify the mechanisms and processes that could produce significant risks at the site and its vicinity.

In general, a phased approach to environmental sampling encourages the identification of key data requirements in the site investigation process. This ensures that the data collection effort is always directed at providing adequate information that meets the data quantity and quality require- ments of the study—and furthermore, that this is likely accomplished in a most cost-effective manner.

4.5.1   A PHASED APPROACH TO CONTAMINATED SITE ASSESSMENTS

A tiered or phased approach may be adopted in the investigation of potentially contaminated sites, in the hope of utilizing a more cost-effective strategy to determine the true extent of contamination at such a site. The use of a phased approach will generally result in an optimal data gathering and evaluation process that meets requisite data quantity and quality objectives. Consequently, pro- grams designed to investigate potentially contaminated site problems typically consist of a number of phases; these phases, reflecting the different degrees of detail, may be classified into the following archetypical successive tiers:

- *Tier 1—Preliminary Site Appraisal (PSA)*, consisting of records search together with a reconnaissance site appraisal and reporting.
- *Tier 2—Primary Site Investigation (PSI)*, comprising of a site investigation that involves contamination and environmental damage assessment.
- *Tier 3—Expanded Site Investigation (ESI)*, incorporating a comprehensive site assessment that also includes a preliminary feasibility study of site restoration measures.

Classically, successive tiers—which could perhaps be considered flexible, if not subjective—call for increasingly sophisticated levels of data collection and analysis. Table 4.3 summarizes the requirements of the different levels of effort associated with the typical site assessment process—as exemplified by the tiers identified previously.

TABLE 4.3

Summary Requirements of the Tiered Site Assessment Process

Level of Investigation	Purpose of Investigation	Typical "Add-On" Tasks Performed
Phase I: Preliminary site assessment	• To provide a qualitative indication of potential contamination at a site.	• Records search, including geologic and hydrogeologic literature search, aerial photo reviews, review of archival and regulatory agency records (to define the historical uses of site), and anecdotal reports (on site history and practices as made available by former employees, local residents, and local historians). • Site inspection/reconnaissance survey (to define the present condition of the site). • Personal interviews (to supplement historical use data). • Written report of findings (to document results and recommendations).
Phase IIA: Initial site investigation	• To identify the known or suspected source(s) of contamination. • To define the nature and extent of the contamination.	• Sampling of potentially impacted media at the surface. • Subsurface borings, well installations, and groundwater sample collection. • Sample analysis to identify and quantify contaminants. • Evaluation of sampling results and detailed report of findings, with conclusions.
Phase IIB: ESI (or remedial investigation/ feasibility study)	• To define the hydrogeology of the region and direction of contaminant plume migration. • To determine and evaluate those remedial options potentially applicable to the site restoration.	• Development of remediation goals and cleanup criteria. • Identification of alternative methods/ technologies for the site remediation. • Screening of alternatives to select those that are most feasible.
Phase III: Remedial measures investigation and corrective action implementation	• To recommend the site restoration method that seems most feasible in terms of technical performance and cost-effectiveness.	• Evaluation of alternatives in terms cost, probable effectiveness, etc. • Ranking of feasible alternatives and recommendations of the overall best method/technology for implementation. • Design and implementation of remedial options and monitoring programs.

In general, the objective of the Tier 1 or "initial phase" (which is comprised of basic background information gathering) should be to determine the history of the site with respect to contamination sources and any relevant characteristics of the site that are readily obtainable from historical records, reports, and interviews. The initial phase of investigation commonly may also use field-screening methods to delineate the approximate area and general magnitude of the problem. The Tier 2 or "intermediate phase" (involving a site characterization) has the primary objectives of defining the vertical and lateral extents of contamination, as well as understanding how the contaminants are affected by hydrogeologic conditions at the site. The investigation of hydrogeological conditions at the project site typically involves an evaluation of groundwater quality, groundwater flow directions, thickness and relative locations of aquifers and confining units, and potential discharge points for groundwater such as local abstraction wells or surface water bodies (Pratt, 1993). Such information is essential to the determination of potential migration pathways and exposure routes for contaminants in groundwater, especially in areas where groundwater is a significant source of portable water supply. Finally, the Tier 3 or "final phase" of the site investigation provides the data necessary to design appropriate and applicable site restoration measures. The key requirements of the different levels of effort associated with the site assessment process are elaborated in the following subsections.

4.5.1.1 Tier 1 Investigations—The PSA

A PSA is usually conducted to determine if a site is potentially contaminated as a result of past or current site activities, unauthorized dumping or disposal, or the migration of contaminants from adjacent or nearby properties.

The Purpose. The purpose of the PSA is to provide a qualitative indication of potential contamination at a site. The PSA typically is designed to document known or potential areas of concern due to the presence of contaminants, to establish the characteristics of the contaminants, to identify potential migration pathways, and to determine the potential for migration of the contaminant constituents.

The Tasks. The PSA will, at a minimum, involve record searching and a superficial physical survey that typically consist of the following activities:

- Review of historical records such as site history, past operation, and disposal practices.
- Review of readily available files and databases maintained by various regulatory and other governmental agencies in order to obtain information on site hydrogeology, characteristics of adjacent properties, known environmental problems in the general area, etc.
- Field reconnaissance of the subject site and adjacent properties in order to ascertain local soil and groundwater conditions, the proximity of the site to drinking water supplies and surface water discharges, existence of obvious contaminant sources or areas, etc.

Overall, data collection for the PSA is typically accomplished by sequentially performing a records search/review, a site reconnaissance, and where practicable and/or appropriate, a soil vapor contaminant assessment.

Representative specific tasks performed to meet the overall objective of a PSA may consist of a geologic and hydrogeologic literature search, aerial photo reviews, review of archival and regulatory or governmental agency records (to identify the historical uses of site), and review of anecdotal reports (on site history and practices that are made available by former employees, local residents, and local historians); visual site inspection/reconnaissance survey (to define the present condition of the site); personal interviews (to supplement historical use data); and a written report of findings (to document results, conclusions, and recommendations). After this initial site appraisal, the information so obtained is summarized, and then a preliminary sampling and analysis plan may be developed, as appropriate.

The Results. A Tier 1 investigation will generally conclude that either

1. No current or historic evidence of contamination considered likely to have affected the site is present, and therefore, no further investigation is required, *or*
2. Sources of contamination that may have affected the site have been identified, and therefore, further investigation is required at the Tier 2 level.

Thus, depending on the results of the PSA, a "further-response-action" or a "no-further-response-action" may be recommended for the project site.

Whereas the PSA provides some important site information, a more detailed characterization of the extent of soil and groundwater contamination may be necessary to properly assess long-term risks posed by a contaminated site. Consequently, where warranted, the next level of detail in the site investigation or assessment process is carried out to ascertain any initial indication of possible contamination.

4.5.1.2 Tier 2 Investigations—The PSI

The PSI should be designed to verify findings from the PSA and to furthermore determine the presence (and extent, as applicable) or absence of contamination at the site.

> *The Purpose.* The purpose of the PSI is to identify the known or suspected source(s) of contamination and also to define the nature and extent of the contamination. Typically, the investigation identifies specific contaminants, their concentrations, the areal extent of contamination, the fate and transport properties of the contaminants, and the potential migration pathways of concern.
>
> *The Tasks.* A Tier 2 investigation will normally be used to confirm whether or not a release has occurred. This is accomplished by implementing a limited program to collect and analyze appropriate site samples for some target contaminants. Typical tasks performed during this phase of the site investigation consist of those identified under the PSA, plus the following likely activities:
> - Sampling of potentially impacted media at the surface
> - Subsurface borings, well installations, and groundwater sample collection
> - Sample analysis to identify and quantify contaminants
>
> Subsequently, an evaluation of the sampling results and the preparation of a detailed report of findings, with conclusions, will help document the results of the Tier 2 investigation.
>
> *The Results.* A Tier 2 investigation will generally conclude that either:
> 1. No evidence of contamination was discovered, and therefore, no further investigation is required, *or*
> 2. Contamination that may require remediation has been found, and therefore, a Tier 3 level of investigation is recommended.

Consequently, the results of this initial comprehensive site assessment are used to determine the need for a "further-response-action" or a "no-further-response-action." In fact, if the site is determined to pose significant public health or environmental risks, then more extensive studies will typically be required to quantify the magnitude of contaminants present, to delineate the limits of contamination, to characterize in detail the specific chemical constituents present at the site, and to assess the fate and transport properties of the specific substances at the site.

4.5.1.3 Tier 3 Investigations—The ESI

The ESI strives to improve the prior site characterization efforts and also to identify—at least on a preliminary basis—the most cost-effective methods of remediation that will protect public health and the environment, as well as safeguard public and private property, under applicable and appropriate cleanup goals or regulations.

> *The Purpose.* An ESI is conducted with the main objective to characterize the contamination confirmed from the PSI. The overall purpose of the ESI is to define the regional hydrogeology and direction of contaminant plume migration and also to identify those remedial options potentially applicable to a site restoration program.
>
> *The Tasks.* The ESI process typically involves specifying the type of contamination present, evaluating the three-dimensional character of occurrence of the contamination, assessing the contaminant fate and transport, identifying possible human and ecological receptors potentially at risk, estimating the risks posed to the populations at risk, establishing a database to facilitate documentation of changes in the occurrences of the contamination, and conducting a preliminary screening of site restoration measures.

At a minimum, the ESI will include the collection and analysis of as many soil and water samples as necessary, in order to determine the full extent of site contamination. If contamination is determined as being confined to the unsaturated (vadose) zone, then no groundwater investigation may be necessary; otherwise, groundwater investigations are also carried out. As appropriate, other environmental media (e.g., surface water, sediments, and biota) may also have to be sampled. The level of detail for the data collection activities will be site-specific and dependent on the degree of soil and groundwater contamination found at the site. Typical tasks performed during the Tier 3 investigation comprise of those identified under the PSI, plus the following:

- Development of remediation goals and cleanup criteria
- Identification of alternative methods/technologies for a site restoration
- Screening of remediation alternatives to aid in the future selection of those that are most feasible

In general, comprehensive site assessments usually will involve sampling and testing to identify the types of contaminants, analyzing preselected or "priority" pollutants, and determining the horizontal and vertical extents of the contamination. This typically includes subsurface investigations, soil and water sampling, laboratory analyses, storage tank testing, and other relevant engineering investigations to quantify potential risks previously identified in the PSA and/or PSI.

The Results. A Tier 3 investigation will generally conclude that either:

1. No evidence of extensive contamination was discovered that requires remediation, and therefore, no further investigation is required, *or*
2. Contamination that may require remediation has been found, and therefore, a site restoration program should be developed and implemented.

Typically, the Tier 3 assessment will generate a report made up of a site characterization, a risk assessment, an evaluation of mitigation and remediation options, and an indication of the preferred remedial action plan.

4.6   THE ROADMAP FOR IMPLEMENTING AN OPTIMIZED AND EFFECTUAL SITE CHARACTERIZATION PROGRAM

The underlying goal in conducting site assessments is to determine an appropriate level of effort in any corrective action required for a site at which contamination is suspected or known to have occurred. The type of corrective action selected for the contaminated site problem will depend on the nature of contamination, the amount of contamination that could safely remain at the site following site restoration, and indeed several other site-specific factors. In any case, prior to the development of a corrective action plan for a contaminated site problem, a site assessment must be conducted to determine the true extent of contamination at the site. Invariably, the use of a phased approach will generally result in an optimal data gathering and evaluation process that meets uncompromising data quantity and quality objectives (Figure 4.5); such a strategy would indeed help address potentially contaminated site problems in a more cost-effective manner.

Overall, there are several important elements and requirements that should be duly considered during the implementation stage of a site characterization activity program for any potentially contaminated site problem. The initial step in the implementation of site characterization programs usually involves a data collection activity to compile an accurate property description, history, and chronology of significant events. This may consist of background information such as the following: location of facility or property, property boundaries, regulatory situation and related issues, physical characteristics of site or facility, environmental setting, zoning profile, current and historical land uses, facility operations, hazardous substances used or generated (together with waste management practices) at facility, and land uses in vicinity and/or proximity of facility that might influence

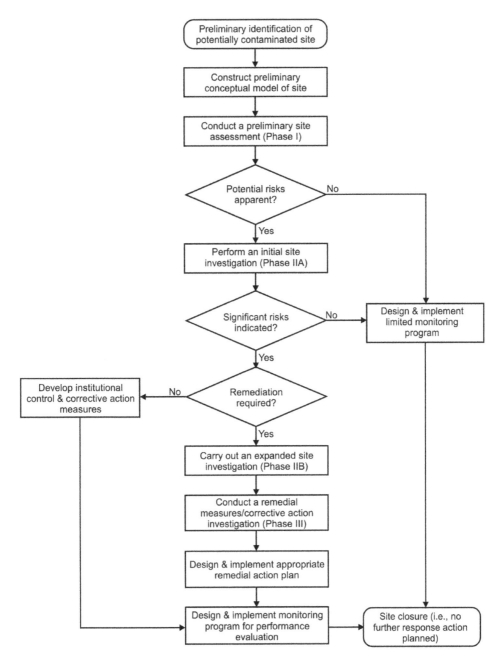

FIGURE 4.5 Overview of a decision process recommended for the systematic assessment of contaminated site problems.

conditions at the locale. Subsequently, the most important functions would characteristically consist of sampling and laboratory analyses. Field data are collected to help define the nature and extent of contamination associated with a locality. Consequently, it is important to use proven methods of approach for the sampling and analysis programs designed for the locale.

As a key implementation strategy, all sampling should be conducted in a manner that maintains sample integrity and encompasses adequate QA and QC. Specific sample locations should be chosen such that representative samples can be collected. Also, samples should almost certainly be collected from locations with visual observations of contamination, so that possible worst-case conditions may likely be identified. The use of field blanks and standards, as well as spiked samples can account for changes in samples that occur after sample collection. Sampling equipment should be constructed of inert materials; solid materials/samples collected are typically placed in resealable plastic bags, and fluid samples are placed in airtight glass or plastic containers. It is noteworthy that when samples are to be analyzed for organic constituents, glass containers are typically required. The samples are then labeled with an indelible marker. Each sample bag or container is labeled with a sample identification number, sample depth (where applicable), sample location, date and time of sample collection, preservation, and possibly a project number and the sampler's initials. A chain-of-custody form listing the sample number, date and time of sample collection, analyses requested, a project number, and persons responsible for handling the samples is then completed. Samples are generally kept on ice prior to and during transport/shipment to a certified laboratory for analysis; completed chain-of-custody records should accompany the samples to the laboratory. Indeed, the collection of representative samples generally involves different procedures for different situations. Several methods of choice that can be used to effectively complete site characterization programs in various types of situations are described elsewhere in the literature (e.g., Boulding, 1994; Boulding and Ginn, 2003; BSI, 1988; Byrnes, 1994, 2000, 2008; CCME, 1993, 1994; CDHS, 1990; Csuros, 1994; Driscoll, 1986; Holmes et al., 1993; ITRC, 2003a–c, 2004, 2007a–d, 2012a, 2013a; Jolley and Wang, 1993; Keith, 1988, 1991, 1992; Millette and Hays, 1994; Nielsen, 1991; OBG, 1988; O'Shay and Hoddinott, 1994; Perket, 1986; Rong, 2018; Sara, 2003; Singhroy et al., 1996; USEPA, 1985a,b, 1987a,b, 1988a,b, 1989a–c, 1991a–c; USEPA-NWWA, 1989; WPCF, 1988). At any rate, in every situation, all sampling equipment would be cleaned using a nonphosphate detergent, a tap water rinse, and a final rinse with distilled water prior to a sampling activity. Decontamination of equipment is necessary so that results from the samples collected do not show false positives; decontamination water generated from the site investigation activities is transferred into containers and sampled for analysis. Further details of the appropriate technical standards for sampling and sample-handling procedures can be found in the literature elsewhere (e.g., CCME, 1994; CDHS, 1990; Holmes et al., 1993).

Ultimately, decisions about the significance of observed site contamination should generally be formulated in the context of the site-specific scenarios associated with a project site. The mobility of the site contaminants, the location of populations potentially at risk, the size and shape of contaminant plumes, the presence or absence of free products or nonaqueous phase liquids, background concentrations, and the intended use of the site are all important factors to consider in assessing the significance of the contamination (Pratt, 1993). These factors can be integrated by developing a conceptual model for the site that shows the interrelationship between the contaminant source locations, transport media, and potential exposure point locations or compliance boundaries. Exposure scenarios can subsequently be developed based on the site-specific conditions and the anticipated land uses; the risk posed to the critical receptors is then determined and the cleanup goals or risk management plans defined in terms of an "acceptable" risk level. It is noteworthy that the development of an adequate CSM is an important aspect of the technical evaluation scheme necessary for the successful completion of a site characterization; this integrates geologic and hydrologic information—and provides a sound basis for human health and ecological risk assessments. The CSM is also relevant to the broader development and evaluation of corrective action programs for potentially contaminated sites.

In general, once a diagnostic assessment of a potentially contaminated site problem has been completed, plans can be made towards the implementation of effectual risk management and/or corrective actions, where warranted. Meanwhile, it is worth the mention here that a project manager needs to be cautious in accepting the results of any site investigation activity as being an absolute indicator of the true situation at a suspect site. In practice, rather than simply walk away from a site purported to be "clean," it often is a good idea to implement some form of monitoring program (usually for only a limited duration) for the candidate site even when no evidence of contamination has been found during the site assessment. This is important, even if for no other reason than to account for uncertainty—especially considering the possibility that even a carefully executed site investigation program could still miss some isolated pockets of contaminants that may become long-term release sources and, therefore, a potential liability issue.

All in all, the characterization of contaminant or hazard sources, contaminant migration or hazard propagation pathways, and potential receptors (that are most likely to be at risk) probably form the most important basis for determining the nature of corrective action response or risk management action required for any contaminated site problem situation. The completion of an adequate site characterization is therefore considered a very important component of any environmental management program that is designed to effectively remedy a contaminated site problem. In the end, flexibility and very careful planning aids in the ability to craft and successfully implement a cost-effective, streamlined site characterization and risk management or remedial response program for a given contaminated site problem.

5 Planning for Contaminated Site Investigation and Management Activities

There are numerous planning activities that generally should be carefully undertaken prior to carrying out contaminated site investigation and/or restoration activities; these are often documented in project workplans. Workplans are generally required to specify the administrative and logistic requirements of site investigation activities. Indeed, as part of any site characterization and/or the corrective action assessment program designed to address potentially contaminated site problems, a carefully executed investigative strategy or workplan should be developed to guide all relevant decisions. A typical workplan developed to facilitate the investigation of contaminated site problems will generally consist of the following key components:

- A sampling and analysis plan (SAP)
- A health and safety plan (HSP)
- A waste management plan (for investigation-generated wastes [IGWs])
- A site activity plan
- A quality assurance/quality control (QA/QC) plan

All the workplan elements, as represented by the summary listing in Box 5.1, should be adequately evaluated and appropriately documented.

BOX 5.1   GENERAL ELEMENTS OF A TYPICAL SITE INVESTIGATION/ CHARACTERIZATION AND MANAGEMENT WORKPLAN

- How site mapping will be performed (including survey limits, scale of site plan to be produced, horizontal and vertical control, and significant site features)
- Number of individuals to be involved in each field sampling task and estimated duration of work
- Identification of soil boring and test pit locations on a map to be provided in a detailed workplan
- Number of samples to be obtained in the field (including blanks and duplicates) and the sampling location (illustrated on maps to be included in a detailed workplan)
- An elaboration of how IGWs will be handled
- List of field and laboratory analyses to be performed
- A general discussion of data quality objectives (DQOs)
- Identification of pilot or bench-scale studies that will be performed, where necessary, in relationship to recommendations or expectations for remedial technologies screening, risk management strategies, and/or site stabilization processes
- A discussion of HSPs required for the site investigation or site restoration/corrective action activities, as well as that necessary to protect populations in the general vicinity of the site

---

This chapter expounds the pertinent planning requirements that would assure a cost-effective implementation of a contaminated site investigation activity; this includes a general discussion of the key steps typically undertaken to develop comprehensive workplans in the characterization and management of potentially contaminated site problems. The major components and tasks required for most site characterization and corrective action evaluation workplans are elaborated further in the proceeding sections—with greater details offered elsewhere in the literature (e.g., Boulding, 1994; CCME, 1993; CDHS, 1990; Keith, 1988, 1991; USEPA, 1985a,d, 1987b,c,e,f, 1988e,f,g, 1989c,h,p,r).

5.1   THE SAMPLING AND ANALYSIS PLAN

The SAP is an essential requirement of any environmental site investigation; SAPs generally are required to specify sample types, numbers, locations, and relevant procedures or strategies. Overall, SAPs provide a mechanism for planning and approving field activities (USEPA, 1988a,b, 1989b). The data necessary to meet the project objectives should be specified—including the selection of sampling methods and analytical protocols for the particular situation or site; this will also include an evaluation of any multiple-option approaches that will ensure timely and cost-effective data collection and evaluation. It is noteworthy that the required level of detail and the scope of the planned investigation generally will determine the DQOs. In any event, it is important that the sampling and analysis strategy is planned in such a manner so as to minimize the costs associated with any efforts directed at achieving the DQOs. In fact, the SAP typically will set the stage for developing cost-effective and effectual corrective action plans for potentially contaminated site problems—with an additional goal that is so aligned to ensure that sampling and data collection activities will be comparable to and compatible with previous (and possible future) data collection activities. Box 5.2 enumerates a checklist of the specific items that need to be ascertained in the development of a credible SAP (CCME, 1993; Holmes et al., 1993; Keith, 1988, 1991).

**BOX 5.2   CHECKLIST FOR DEVELOPING
SAMPLING AND ANALYSIS PROTOCOLS**

- What observations at sampling locations are to be recorded?
- Has information concerning DQOs, analytical methods, analytical detection limits, etc., been included?
- Have instructions for modifying protocols in case of unanticipated problems been specified?
- Has a list of all likely sampling equipment and materials been prepared?
- Are instructions for cleaning equipment before and after sampling available?
- Have instructions for each type of sample collection been prepared?
- Have instructions for completing sample labels been included?
- Have instructions for preserving each type of sample (such as preservatives to use and also maximum holding times of samples) been included?
- Have instructions for packaging, transporting, and storing samples been included?
- Have instructions for chain-of-custody procedures been included?
- Have HSPs been developed?
- Is there a waste management plan to deal with wastes generated during the site investigation activities?

An archetypical SAP will usually be comprised of two major components—namely (USEPA, 1988a,b, 1989b):

1. A *QA project plan* (QAPP)—that describes the policy, organization, functional activities, and QA and QC protocols necessary to achieve the DQOs dictated by the intended use of the data.
2. A *field sampling plan* (FSP)—that provides guidance for all fieldwork, by defining in detail the sampling and data-gathering methods to be used in a project. The FSP should be written so that a field sampling team unfamiliar with the site is still able to gather the samples and any field information required for the project.

On the whole, the methods by which data of adequate quality and quantity are to be obtained to meet the overall project goals should be specified and fully documented in the SAP developed as part of a detailed site/environmental investigation workplan. Among other things, an initial evaluation of a chemical release and consequential potential exposure problem associated with a contaminated site problem should provide some insight into the types of contaminants, the populations potentially at risk, and possibly an approximation of the magnitude of the risk. These factors can then be combined to design a sampling plan and to specify the size of sampling units to be addressed by each sample or set of samples. Also, it is notable that, in a number of situations, the laboratory designated to perform the sample analyses provides sample bottles, preservation materials, and explicit sample collection instructions; this is in part due to the complexity typically associated with the gathering many different samples from various matrices and that may also have to be analyzed using a wide range of analytical protocols.

Important issues to consider when one is making a decision on how to obtain reliable samples relate to the sampling objective and approach, sample collection methods, chain-of-custody documentation, sample preservation techniques, sample shipment methods, and sample holding times. In the end, the methods by which data of adequate quality and quantity are to be obtained to meet the overall project goals should be specified and fully documented in the SAP that is developed as part of a detailed environmental/site characterization workplan. Meanwhile, it should also be recognized that the selection of analytical methods is an integral part of the processes involved in the development of sampling plans, since this can substantially affect the acceptability of a sampling protocol. Furthermore, the use of appropriate sample collection methods can be as important as the use of appropriate analytical methods for sample analyses—and vice versa. Indeed, effective protocols may be viewed as a prerequisite in the sampling and laboratory procedures—in order to help minimize uncertainties in the environmental investigation process. A detailed discussion of pertinent sampling considerations and strategies for various environmental matrices can be found in the literature elsewhere (e.g., CDHS, 1990; Holmes et al., 1993; Keith, 1988, 1991; Lave and Upton, 1987; USEPA, 1988e,f,g, 1989c,h,p,r).

5.1.1 Purpose of the Sampling and Analysis Program

The principal objective of a sampling and analysis program is to obtain a small and informative portion of the statistical population being investigated, so that contaminant levels can be established as part of a site characterization and/or corrective action assessment program. Contaminated site sampling and analysis programs are typically designed and conducted for the following purposes (USEPA, 1989c,h,p,r):

- Determine the extent to which soils act as either contamination sources (i.e., situations when significant quantities of selected contaminants are found to be associated with soils initially and then released slowly over relatively long periods of time into other media) or as contamination sinks (i.e., situations where significant quantities of contaminants become permanently attached to soil and remain biologically unavailable).

- Determine the presence and concentration of specified contaminants in comparison to natural and/or anthropogenic background levels.
- Determine the concentration of contaminants and their spatial and temporal distribution.
- Obtain measurements for validation or use of contaminant transport and fate models.
- Identify pollutant sources, transport mechanisms or routes, and potential receptors.
- Determine the potential risks to human health and/or the environment (i.e., to flora and fauna) due to site contamination.
- Identify areas of contamination where action is needed and measure the efficacy of control or removal actions.
- Contribute to research technology transfer or environmental model development studies.
- Meet pertinent provisions and intent of environmental laws (such as Resource Conservation and Recovery Act (RCRA), Comprehensive Environmental Response, Compensation, and Liability Act (CERCLA), Federal Insecticide, Fungicide, and Rodenticide Act (FIFRA), and Toxic Substances Control Act (TSCA) under the U.S. environmental regulations; etc. – as discussed in Chapter 3, Section 3.1.1).

In fact, sampling and analysis of environmental pollutants is a very important part of the decision-making process involved in the management of potentially contaminated site problems. Yet, sampling and analysis could become one of the most expensive and time-consuming aspects of an environmental management or site characterization project. Even of greater concern is the fact that errors in sample collection, sample handling, or laboratory analysis can invalidate site characterization projects and/or add to the overall project costs. As such, all environmental samples that are intended for use in site characterization programs must be collected, handled, and analyzed properly—at least in accordance with all applicable/relevant methods and protocols. Box 5.3 provides a convenient checklist of the issues that should be verified when planning a sampling activity for a contaminated site problem (CCME, 1993), in order that the project goals are likely attained.

BOX 5.3   SAMPLING PLAN CHECKLIST

- What are the DQOs—and what corrective measures are planned if DQOs are not met (e.g., resampling or revision of DQOs)?
- Do program objectives need exploratory, monitoring, or both sampling types?
- Have arrangements been made for access to the case property?
- Is specialized sampling equipment needed and/or available?
- Are field crews who are experienced in the required types of sampling available?
- Have all analytes and analytical methods been listed?
- Have required good laboratory practice and/or method QA/QC protocols been listed?
- What type of sampling approach will be used (i.e., random, systematic, judgmental, or combinations thereof)?
- What type of data analysis methods will be used (e.g., geostatistical, control charts, and hypothesis testing)?
- Is the sampling approach compatible with data analysis methods?
- How many samples are needed?
- What types of QC samples are needed, and how many of each type of QC samples are needed (e.g., trip blanks, field blanks, and equipment blanks)?

5.1.2 Elements of a SAP: Sampling Requirements and Considerations

Environmental sampling at contaminated sites is generally carried out in order help characterize the site for enforcement and/or corrective actions. In such undertakings, several site-specific requirements are important to achieving the site characterization goals. Specific factors to consider when preparing the sampling plan include the following (WPCF, 1988):

- History of activity at the site
- Physical/chemical properties and hazardous characteristics of materials involved
- Topographic, geologic, pedologic, and hydrologic characteristics of the site
- Meteorologic conditions
- Flora and fauna of the site and vicinity (to include the identification of sensitive ecological species and systems, and the potential for bioaccumulation and biotransformation)
- Geographic and demographic information, including proximity of populations potentially at risk

A history of the site, including the sources of contamination and a preliminary conceptual model describing the apparent migration and exposure pathways should preferably be developed before a sampling plan is finalized. In certain situations, it may be necessary to conduct an exploratory study in order to give credence to the preliminary conceptual model—and also so that this preliminary conceptual model can be confirmed or modified, as appropriate. In general, as additional data help identify specific areas where an initial site conceptual model is somehow invalidated, the model should be modified to reflect the new information. In working towards the development of a representative conceptual model for a site, it is important to be aware of the fact that some significant model features may not necessarily be very apparent—but rather will tend to remain latent during much of the process. For example, consider the case of contaminated non-soil debris present in the soil mass at a potentially contaminated site; this debris must be appropriately taken into account when evaluating the hazard posed by the site. In fact, in some cases, the debris (e.g., wood chips) rather than the soil itself may be the major source of contaminant releases—recognizing that it is common to, for instance, use wood chips or shredded wood as absorbent for liquid wastes under certain reasonably appropriate circumstances. Consequently, screening the non-soil debris out of the soil material and excluding it from a hazard and/or risk analysis may bias the results and, therefore, the final conclusions reached about the site.

Finally, it is noteworthy that the key to a successful and cost-effective site characterization will be to optimize the number and length of field reconnaissance trips, the number of soil borings and wells, and the frequency of sampling and laboratory analyses. Indeed, opportunities to effect cost savings begin during the initial field investigation conducted as part of a preliminary assessment. For such reasons, it is imperative that, among other things, background information relating to the regional geology, hydrogeology, and site development history should have been fully researched before a subsurface investigation is initiated in any given situation. Regardless, minimization of well installations and sample analyses during the initial phase of site characterization activities can greatly reduce overall costs. Indeed, well installation can also potentially add to the spread of contamination by penetrating impermeable layers that contain contaminants—and then allowing such contamination to migrate into previously uncontaminated deeper zones (Jolley and Wang, 1993). Consequently, nonintrusive and screening measurements are generally preferred modes of operations, whenever possible, since that tends to minimize both sampling costs and the potential to spread contamination. Ultimately, however, sufficient information will have to be collected in order to reliably show the identity, areal and vertical extents, and the magnitude of contamination associated with the project site.

5.1.2.1 Sampling and Analysis Design Considerations

A preliminary identification of the types of contaminants, the chemical release potentials, and also the potential exposure pathways should be made very early in the site characterization and potential chemical exposure characterization efforts; this is because these are crucial to decisions on the number, type, and location of samples to be collected. Indeed, knowledge of the type of contaminants will generally help focus more attention on the specific media most likely to have been impacted or that remains vulnerable; thus, the type of chemicals at potentially contaminated sites may dictate the sections or areas and environmental media to be sampled. For instance, in the design of a sampling program, if it is believed that the contaminants of concern are relatively immobile (e.g., polychlorinated biphenyls and most metals), then sampling will initially focus on soils in the vicinity of the suspected source of contaminant releases. On the other hand, the sampling design for a site believed to be contaminated by more mobile compounds (e.g., organic solvents) will take account of the fact that contamination may already have migrated into groundwater systems and/or moved significant distances away from the original source area(s). Then also, some chemicals that bioaccumulate in aquatic life will likely be present in sediments—requiring that both sediments and biota be sampled and analyzed. Consequently, knowledge on the type of contaminants will generally help focus more attention on the specific media most likely to have been impacted or that remains vulnerable.

Similar decisions as mentioned previously will typically have to be made regarding analytical protocols as well. For instance, due to the differences in the relative toxicity of the different species of some chemicals (as, e.g., chromium may exist as trivalent chromium [Cr^{+3}] or as the more toxic hexavalent chromium [Cr^{+6}]), chemical speciation to differentiate between the various forms of the chemicals of potential concern at a contaminated site or in relation to a chemical release and potential exposure situation may sometimes be required in the design of analytical protocols.

Anyhow, regardless of the medium sampled, data variability problems may arise from temporal and spatial variations in field data. That is, sample composition may vary depending on the time of the year and weather conditions when the sample is collected. Ideally, samples from various media should be collected in a manner that accounts for temporal factors and weather conditions. If seasonal/temporal fluctuations cannot be characterized in the investigation, details of meteorological, seasonal, and climatic conditions during the sampling events must be carefully documented. For the most part, choosing an appropriate sampling interval that spans a sufficient length of time to allow one to obtain, for example, an independent groundwater sample will generally help reduce the effects of autocorrelation. Also, as appropriate, sampling both "background" and "compliance" locations at the same point-in-time should reduce attributions of likely temporal effects. Consequently, the ideal sampling scheme will typically incorporate a full annual sampling cycle. If this strategy cannot be accommodated in an investigation, then at least two sampling events should be considered—and these should probably take place during opposite seasonal extremes (such as high/low water and high/low recharge).

In general, an initial site evaluation usually provides insight into the types of site contaminants, the populations potentially at risk, and possibly the magnitude of the risk. These factors can be combined into the conceptual model, which is subsequently used to guide the design of a sampling plan, as well as to specify the size of sampling unit or support addressed by each sample or set of samples. Ultimately, all sampling plans should contain several fundamental elements—as noted in Box 5.4. Indeed, it is indisputable that several important issues usually will come into play when one is making a decision on how to obtain reliable samples—especially as relates to considerations of the sampling objective and approach, sample collection methods, chain-of-custody documentation, sample preservation techniques, sample shipment methods, sample holding times, and analytical protocols. A detailed discussion of key sampling considerations and strategies for various environmental matrices can be found elsewhere in the literature (e.g., CDHS, 1990; Holmes et al., 1993; Keith, 1988, 1991; USEPA, 1988e,f,g, 1989c,h,p,r).

BOX 5.4 KEY ELEMENTS OF A SAMPLING PLAN

- Background information about site or locale (that includes a description of the site or problem location and surrounding areas and a discussion of known and suspected contamination sources, probable migration pathways, and other general information about the physical and environmental setting)
- Sampling objectives (describing the intended uses of the data)
- Sampling location and frequency (that also identifies each sample matrix to be collected and the constituents to be analyzed)
- Sample designation (that establishes a sample numbering system for the specific project and should include the sample number, the sampling round, the sample matrix, and the name of the site or case property)
- Sampling equipment and procedures (including equipment to be used and material composition of equipment, along with decontamination procedures)
- Sample handling and analysis (including identification of sample preservation methods, types of sampling jars, shipping requirements, and holding times)

5.1.3 SAMPLING PROTOCOLS

Sampling protocols are written descriptions of the detailed procedures to be followed in collecting, packaging, labeling, preserving, transporting, storing, and documenting samples. In general, every sampling protocol must identify sampling locations—and this should include all of the equipment and information needed for sampling. Box 5.5 lists what might be considered the minimum documentation needed for most environmental sampling activities (CCME, 1993; Keith, 1988, 1991). In fact, the overall sampling protocol must identify sampling locations, as well as include all of the equipment and information needed for sampling, such as the types, number, and sizes of containers; labels; field logs; types of sampling devices; numbers and types of blanks, sample splits, and spikes; the sample volume; any composite samples; specific preservation instructions for each sample type; chain of custody procedures; transportation plans; field preparations (such as filter or pH adjustments); field measurements (such as pH and dissolved oxygen); and the reporting requirements (Keith, 1988). The sampling protocol should also identify those physical, meteorological, and related variables to be recorded or measured at the time of sampling. In addition, information concerning the analytical methods to be used, minimum sample volumes, desired minimum levels of quantitation, and analytical bias and precision limits may help sampling personnel make better decisions when unforeseen circumstances require changes to the sampling protocol.

At the end of the day, the devices used to collect, store, preserve, and transport samples must *not* alter the sample in any manner. In this regard, it is noteworthy that special procedures may be needed to preserve samples during the period between collection and analysis. In any case, the more specific a sampling protocol is, the less chance there will be for errors or erroneous assumptions. Several techniques and equipment that can be used in the sampling of contaminated soils, sediments, and waters are enumerated in the literature elsewhere (e.g., CCME, 1993; Keith, 1988, 1991).

**BOX 5.5 MINIMUM REQUIREMENTS FOR
DOCUMENTING ENVIRONMENTAL SAMPLING**

- Sampling date
- Sampling time
- Sample identification number

(*Continued*)

> **BOX 5.5 (*Continued*)   MINIMUM REQUIREMENTS FOR**
> **DOCUMENTING ENVIRONMENTAL SAMPLING**
>
> - Sampler's name
> - Sampling location
> - Sampling conditions or sample type
> - Sampling equipment
> - Preservation used
> - Time of preservation
> - Auxiliary data (i.e., relevant observations at sample location)

5.1.3.1   Sampling Strategies and Sample Handling Procedures

Broadly speaking, there are three basic sampling approaches—namely: random, systematic, and judgmental. There are also three primary combinations of each of these—that is, stratified-(judgmental)-random, systematic-random, and systematic-judgmental (CCME, 1993; Keith, 1991). Additionally, there are further variations that can be found among the three primary approaches and the three combinations thereof. For example, the systematic grid may be square or triangular; furthermore, samples may be taken at the nodes of the grid, at the center of the spaces defined by a grid, or randomly within the spaces defined by a grid. Anyhow, a combination of judgmental, systematic, or random sampling is often the most feasible approach to use in the investigation of potential environmental contamination and chemical release problems—albeit the sampling scheme should be flexible enough to allow relevant adjustments and/or modifications to take place during field activities.

In general, several different methods are available for acquiring data to support site characterization and chemical exposure characterization programs. The methodology used for sampling can indeed affect the accuracy of subsequent evaluations. It is therefore imperative to select the most appropriate methodology possible in order to obtain the most reliable results attainable; Holmes et al. (1993), among others, enumerate several important factors that should be considered when selecting a sampling method. Standard sampling practices often used in the investigation of commonly impacted environmental media include the following (Holmes et al., 1993):

- Three main types of water sampling schemes—viz., grab samples, composite samples, and continuous flowing samples. For the different sampling approaches, samples are collected with different types of water sampling equipment or tools. It is noteworthy that low-flow sampling—such as by the use of "micropurging"—reduces turbidity and is also more representative of dissolved concentrations.
- Two main types of soil sampling schemes—viz., surface sampling (0–15 cm) and soil samples from depth (>15 cm) (comprising of shallow subsurface sampling and deep soil samples). Samples can be collected with some form of core sampling or auger device or they may be collected by the use of excavations or trenches.
- Two main types of air sampling schemes—viz., instantaneous or grab samples (usually taken in a period of less than 5 min) and integrated air sampling. The choice of procedure for the air sampling is dependent on the contaminant to be measured.

Additional discussion on sampling strategies and requirements, together with several techniques and equipment that can be used in the sampling of contaminated air, soils, sediments, and waters are enumerated in Chapter 6 of this book—with further elaboration provided elsewhere in the literature (e.g., CCME, 1993; Holmes, et al., 1993).

5.1.4 LABORATORY ANALYTICAL PROTOCOLS

The selection of analytical methods is a key integral part of the processes involved in the development of sampling plans, since this can strongly affect the acceptability of a sampling protocol. For example, the sensitivity of an analytical method could directly influence the amount of a sample needed in order to be able to measure analytes at prespecified minimum detection (or quantitation) limits. The analytical method may also affect the selection of storage containers and preservation techniques (Keith, 1988; Holmes et al., 1993). Thus, the applicable analytical procedures, the details of which are outside the scope of this book, should be strictly adhered to.

Box 5.6 lists the minimum requirements for documenting laboratory work performed to support site characterization activities (CCME, 1993; USEPA, 1989p,r). Ultimately, in general, effective analytical programs and laboratory procedures are necessary to help minimize uncertainties in site investigation activities that are required to support contaminated site characterization programs as well as site restoration decisions. Guidelines for the selection of appropriate analytical methods are offered elsewhere in the literature (e.g., CCME, 1993; Keith, 1991; USEPA, 1989p,r). Invariably, analytical protocol and constituent parameter selection are usually carried out in a way that balances costs of analysis with adequacy of coverage.

**BOX 5.6   MINIMUM REQUIREMENTS FOR
DOCUMENTING LABORATORY WORK**

- Method of analysis
- Date of analysis
- Laboratory and/or facility carrying out analysis
- Analyst's name
- Calibration charts and other measurement charts (e.g., spectral)
- Method detection limits
- Confidence limits
- Records of calculations
- Actual analytical results

5.1.4.1 Selecting Laboratory Analysis Methods and Analytical Protocols: Laboratory and Analytical Program Requirements

The task of determining the essential analytical requirements involves specifying the most cost-effective analytical method that, together with the sampling methods, will meet the overall data quantity and quality objectives of a site investigation activity. Oftentimes, the initial analyses of environmental samples may be performed with a variety of field methods used for screening purposes. The rationale for using initial field screening methods is to help decide if the level of pollution at a site is high enough to warrant more expensive (and perhaps more specific and accurate) laboratory analyses. Indeed, methods that screen for a wide range of compounds, even if determined as groups or homologues, are useful because they allow more samples to be measured faster and far less expensively than with conventional la   boratory analyses. In the more detailed phase of a site assessment, the sampling analysis is generally performed by laboratory programs that comprise routine and nonroutine standardized analytical procedures and associated QC requirements managed under a broad QA program; these services are provided through routine analytical services, as well as by special analytical services.

In practice, analytical protocol and constituent parameter selection are usually carried out in a way that balances costs of analysis with adequacy of coverage. In general, if specific chemical constituents are known to be associated with previous site activities, they should most definitely be targeted for analysis; otherwise, a widely used and more broad-spectrum parameter listing to choose from might be the U.S. Environmental Protection Agency (USEPA) "Priority Pollutant List"—or indeed similar ones approved for use in other regions. For instance, if a metal processing and finishing operation facility is known to have existed at a derelict site, then some "priority pollutant" metals and perhaps some volatile organic compounds (likely associated with cleaning and degreasing) may become the target constituents or parameters of interest. In addition, a select but limited number of samples from the more decidedly suspect areas and media may be subjected to the full "priority pollutant" analyses.

Meanwhile, it is worth mentioning here that methods such as the extraction procedure (EP) toxicity and the toxicity characteristic leaching procedure (TCLP) testing should not be used as a primary indicator of contamination. Instead, total constituent analysis should be used to indicate the magnitude and extent of contamination. If significant contamination is confirmed, then it may become necessary to conduct supplemental testing (e.g., by using EP toxicity or TCLP testing) to determine if characteristic hazardous waste definitions (based on toxicity criteria) are indeed applicable to the particular situation. Another noteworthy point to make relates to the analyses of groundwater samples; in the procedures involved, it is always important to distinguish between total (i.e., without sample filtration) versus dissolved (i.e., with sample filtration) metal concentrations— especially because, among other things, the former (i.e., the total metal analyses, performed without filtration) might falsely suggest extensive groundwater contamination (possibly due to the presence of naturally occurring metals associated with suspended solids).

In general, effective analytical programs and laboratory procedures are necessary to help minimize uncertainties in site investigation activities. Usually there are several methods available for most environmental analytes of interest. Indeed, some analytes may have up to a dozen (if not more) methods to select from; on the other hand, some analytes may have no currently proven methods available per se. In the latter case, it usually means that some of the specific isomers that were selected as representative compounds for environmental pollution have not been verified to perform acceptably with any of the commonly used methods. General guidelines for the selection of analytical methods and strategies are offered elsewhere in the literature (e.g., CCME, 1993).

5.2 THE HEALTH AND SAFETY PLAN

Contaminated sites, by their nature and definition, contain concentrations of chemicals that may be harmful to various human population groups. One significant group potentially at risk from site contamination is the field crew who enters the contaminated site to collect samples and/or to monitor the extent of contamination. To minimize risks to site investigation personnel (and possible nearby community populations) as a result of the potential for exposure to site contamination, health and safety issues must always be addressed as part of any field investigation activity plan. Protecting the health and safety of the field investigation team, as well as the general public in the vicinity of the site, is indeed a major concern during the investigation of potentially contaminated sites. Proper planning and execution of safety protocols will help protect the site investigation team from accidents and needless exposure to hazardous or potentially hazardous chemicals. In the processes involved, health and safety data are generally required to help establish the level of protection needed for a project investigation crew. Such data are also used to determine if there should be immediate concern for any population living in proximity of the problem location. Specific items of special interest, as well as required health and safety issues and equipment are discussed in the following—with further details appearing elsewhere in the literature (e.g., Cheremisinoff and Graffia, 1995; Martin et al., 1992; OBG, 1988).

5.2.1 PURPOSE AND SCOPE OF A HSP

The purpose of a HSP is to identify, evaluate, and control health and safety hazards, as well as to provide for emergency response during site characterization and related fieldwork activities at a potentially contaminated site. The HSP specifies safety precautions needed to protect the populations potentially at risk during on-site or other site characterization activities. Consequently, a site-specific HSP should be prepared and implemented prior to the commencement of any site characterization or fieldwork activity at potentially contaminated sites. Among other things, all personnel entering the project site will generally have to comply with the applicable HSP. Also, the scope and coverage of the HSP may be modified or revised to encompass any changes that may occur at the site, or in the working conditions, following the development of the initial HSP.

Overall, the HSP should be so developed as to be in conformity with all requirements pertinent to occupational safety and health, as well as meet applicable national, state/provincial/regional and local laws, rules, regulations, statutes, and orders—as necessary to protect all populations potentially at risk. Furthermore, all personnel involved with on-site or other site characterization activities would have received adequate training, and there also should be a contingency plan in place that meets all safety requirements. For instance, in the United States, the HSP developed and implemented in the investigation of a potentially contaminated site should be in full compliance with all the requirements put out by the Occupational Safety and Health Administration (OSHA) (i.e., OSHA: 29 CFR 1910.120); the requirements of US EPA (i.e., EPA: Orders 1420.2 and 1440.3); and indeed any other relevant state or local laws, rules, regulations, statutes, and orders necessary to protect the populations potentially at risk. Moreover, all personnel involved with on-site activities would have received a 40-h OSHA Hazardous Waste Operations and Emergency Response Activities (HAZWOPER) training, along with a classically stipulated 8-h refresher course, where necessary.

5.2.2 Elements of an HSP

Health and safety data are generally required to help establish the level of protection needed for a site investigation crew. Such data are also used to determine if there should be immediate concern for any population living in proximity of the problem locale. Box 5.7 contains the relevant elements of a classic HSP that will satisfy the general requirements of a safe work activity (CDHS, 1990; USEPA, 1987b–f,h)—and Box 5.8 provides a generic example of a typical HSP format/outline that could be tailored to the needs of a specific site. As part of the documentation efforts, a description of pertinent site background information should also be fully elaborated in the HSP. Ultimately, it is important that these elements are fully evaluated to ensure compliance with local health and safety regulations and/or to avert potential health and safety problems during the site characterization activities. Details of specific items of required health and safety issues and equipment are discussed elsewhere in the literature (e.g., Cheremisinoff and Graffia, 1995; Martin et al., 1992; OBG, 1988).

Traditionally, a designated health and safety officer (HSO) establishes the level of protection required during a site characterization program—and then determines whether the level should be advanced or reduced at any future point in time during the investigation. With that established, it is noteworthy that every site characterization project would typically start with a health and safety review (during which all personnel sign a "review form"), a tailgate safety meeting (to be attended by all site investigation personnel), and a safety compliance agreement (that should be signed by all persons entering the case property or site—i.e., both site personnel and site visitors).

As a final important note here, emergency phone numbers should be compiled and included in the HSP. Also, the directions to the nearest hospital or medical facility, including a map clearly showing the shortest route from the site to the hospital or medical facility should be kept with the HSP at the site.

BOX 5.7   KEY ELEMENTS OF A SITE-SPECIFIC HSP

- Description of known hazards and risks associated with the planned site characterization activities (i.e., a health and safety risk analysis for existing conditions and for each task and operation)
- Listing of key personnel and alternates responsible for safety, emergency response operations, and public protection in the project location

(Continued)

> **BOX 5.7 (*Continued*)   KEY ELEMENTS OF A SITE-SPECIFIC HSP**
>
> - Description of the levels of protection to be worn by investigative personnel and visitors to the site or locality
> - Delineation of work and rest areas
> - Establishment of procedures to control access to the case property or site
> - Description of decontamination procedures for personnel and equipment
> - Establishment of emergency procedures, including emergency medical care for injuries and toxicological problems
> - Development of medical monitoring program for personnel (i.e., medical surveillance requirements)
> - Establishment of procedures for protecting workers from weather-related problems
> - Specification of any routine and special training required for personnel responding to environmental or health and safety emergencies
> - Definition of entry procedures for "confined" or enclosed spaces
> - Description of requirements for environmental surveillance program
> - Description of the frequency and types of air monitoring, personnel monitoring, and environmental sampling techniques and instrumentation to be used

> **BOX 5.8   TYPICAL OUTLINE OF A HSP**
>
> - Introduction (consisting of site information and the responsible personnel, such as the HSO and an onsite safety manager)
> - General safety requirements (including safety requirements to be met and the type of emergency equipment required)
> - Employee protection program
> - Decontamination procedures
> - Health and safety training
> - Emergency response (including emergency telephone numbers, personal injury actions, acute exposure to toxic materials responses, and directions to nearest hospital or medical facility)

5.2.3   The Health and Safety Personnel and Managers

The responsible personnel assigned to the implementation of the site-specific HSP usually will include a HSO and an "Onsite Health and Safety Manager" (OHSM)—albeit the responsibilities of the HSO and the OHSM may indeed be assumed by the same individual. Contractors, subcontractors, and other field investigation groups or teams, under the auspices of an HSO, are responsible for implementing the HSP during fieldwork site characterization and remedial activities.

5.2.3.1   Job Functions of the HSO

The HSO has the primary responsibility for ensuring that the policies and procedures of the HSP are fully implemented (by the OHSM); specifically, the HSO ensures that all personnel designated to work at the site meet the following benchmarks:

- Have been declared fit by a physician or other qualified health professional, for the specific tasks on hand
- Are able to wear air-purifying respirators (should it become necessary)

- Would have received adequate hazardous waste site operations and emergency response training

Prior to the commencement of any site activities, the HSO is responsible for providing copies of the HSP to the site crew (including subcontractors), as well as for advising the site crew on all health and safety matters. The HSO is also responsible for providing the appropriate safety monitoring equipment, along with any other resources necessary to implement the HSP. Finally, significant deviations/changes to the original HSP must be approved by the HSO.

5.2.3.2 Job Functions of the OHSM

The OHSM supervises all site activities and is responsible for implementing the HSP. The OHSM also has the more direct responsibility of providing copies of the HSP to the site crew (including subcontractors) and for advising the site crew on all health and safety matters. Specific tasks generally performed by the OHSM are shown in Box 5.9. Overall, the designated OHSM should take steps to protect personnel engaged in site activities from such physical hazards as falling objects (e.g., tools or equipment); tripping over hoses, pipes, tools, or equipment; slipping on wet or uneven surfaces; insufficient or faulty protective equipment; insufficient or faulty tools and equipment; overhead or belowground electrical hazards; heat stress and strain; insect and reptile bites; inhalation of dust; etc. The designated OHSM should also monitor and check the work habits of the site crew to ensure that they are safety conscious.

BOX 5.9   SPECIFIC TASKS REQUIRED OF THE OHSM

- Inspect site prior to start of work, especially for areas of concern that may have been omitted in the HSP or other areas that require special attention
- Conduct daily safety briefings and site-specific training for on-site personnel prior to commencing work
- Modify or develop health and safety procedures, after consultation with the HSO, when site or working conditions change
- Maintain adequate safety supplies and equipment on-site
- Maintain and supervise site control, decontamination, and contamination-reduction procedures
- Investigate all accidents and incidents that occur during site activities
- Discipline or dismiss personnel whose conduct do not meet the requirements of the HSP or whose conduct may jeopardize the health and safety of the site crew
- Immediate notification of the emergency response authorities (e.g., the fire and police departments) and the implementation of evacuation procedures during an emergency situation
- Coordination of emergency response equipment and also coordination of the transport of affected personnel to the nearest emergency medical care (including subsequent personnel decontamination) in the event of an emergency

---

As a rule, the OHSM will have the authority to resolve outstanding health and safety issues that come up during the site investigation activities or operations. Among other things, the OHSM has stop-work authority if a dangerous or potentially dangerous and unsafe situation exists at the site. Accordingly, the OHSM must be notified of any changes in site conditions that occur in the course of fieldwork activities—especially for those that may have significant impacts on the safety of personnel, the environment, or property.

5.2.4 Field Personnel Protection Program

In all cases of contaminated site investigation programs, the use of appropriate types of protection equipment will generally ensure the safety of the field crew; additional precautions deemed necessary to prevent exposures during drilling, sampling, and decontamination procedures should also be implemented—especially as site and weather conditions change. Decontamination and emergency procedures are indeed a very important part of the overall worker protection program. For instance, monitoring well installation activities, proposed soil sampling procedures, and other excavation operations will typically disturb potentially impacted soils—and thus allow attendant dusts to become airborne. Such dusts may migrate into the workers' breathing zone and then pose health threats to the affected individuals (by becoming deposited onto their lung tissues and mucous membranes). Under such circumstances, relevant criteria that meet stipulated threshold limit values should be instituted during the field operations—otherwise adequate corrective measures would have to be implemented, in order to circumvent potential risks to workers and other potential receptors in the vicinity of the site. Also, drilling operations will normally generate constant high decibel noise levels; under such circumstances, workers should generally be protected from high frequency hearing deficiency (otherwise known as tinnitus) that is caused by chronic exposures to high decibel noise and high level impact noise. The use of earplugs and other hearing protection apparatus should help minimize worker exposures to excessive and damaging noise from field activities. Of course, this is by no means the complete list of potential problems and their corresponding preventative or protective measures—but each and every concern should be carefully reviewed and addressed.

5.2.4.1 Levels of Protection

Site-specific health and safety data are often required to help establish the level of protection needed for the site investigation crew entering potentially contaminated sites. Such data are also used to determine if there should be immediate concern for any population living in proximity of the site. Typically, personnel protection for contaminated site problems is categorized into different groupings—concurrent with the level of suspected or anticipated hazards, etc.; for instance, within the United States, personal protective equipment (PPE) is traditionally categorized into the following four general "levels" for environmental site assessment operations:

1. *Level A*—representing the highest level of respiratory, skin, eye, and mucous membrane protection. This level is used if a chemical substance has been identified at concentrations that warrant using fully encapsulating equipment or the chemical substance presents a high degree of contact hazard that warrants using fully encapsulating equipment. Work performed in a confined or poorly ventilated area also requires "Level A" protection up until conditions change, and/or a lower level of protection is deemed safe and appropriate.
2. *Level B*—representing the highest level of respiratory protection but lesser level of skin and eye protection. This is the minimum level recommended during initial visits to a site up until when the nature of the site hazards is determined to demand less protection.
3. *Level C*—appropriate for situations where the criterion for using air-purifying respirators is met, but skin and eye exposure is unlikely. Generally, hazardous airborne substances are known and their concentrations have been measured.
4. *Level D*—for which no special safety equipment required other than those typically used at any construction site.

In general, the HSO establishes the level of protection required and then determines whether the level should be advanced or reduced in the course of site assessment operations. Specifically, the level of PPE to be worn by field personnel is identified and enforced by the designated HSO and OHSM—with a possibility that the levels of protection may change as additional information is acquired in the course of time.

It is notable that, for most of the typical environmental site activities conducted in the United States, direct worker contact with hazardous materials in soil can usually be mitigated by using the lowest level of personal protective equipment, that is, a "Level-D" personal protective equipment—consisting of coveralls, safety boots, glasses, and a hard hat. But to protect workers from unacceptable levels of airborne materials, at least a level of protection that includes a full-facepiece air-purifying respirator may be required—that is, a minimum of "Level-C" equipment (that includes a full-facepiece air-purifying respirator). In certain other cases, worker exposure to toxic materials may be at such levels as to warrant far more sophisticated equipment, in order to provide yet greater levels of protection against exposure—that is, a "Level-B," or even "Level-A" equipment in rare cases. Meanwhile, it should be emphasized here that all personnel utilizing most high-level PPEs must have successfully passed a thorough physical examination and should indeed have received the consent of a qualified physician prior to engaging in any on-site activities that utilize these types of equipment.

On the whole, if a higher level of protection above that originally specified in the HSP is needed, then approval by the HSO will be required. For instance, at a given facility or site, it may initially have been determined that work crews and field sampling teams will require Level C PPE—consisting of a full- or half-face respirators fitted with approved high efficiency particulate air (HEPA) filters, chemical-resistant coveralls (such as Tyvek or Saranex), hard hats, gloves with chemical-resistant outer shells and chemical-resistant linings, chemical-resistant boots, earplugs or other hearing protection apparatus, and safety glasses or goggles—during drilling and sampling activities. However, changes in site conditions may create a new situation in which a lower (e.g., Level D) or a higher level (e.g., Level B or Level A) PPE may be required; such changes should be closely monitored by the OHSM—and further communicated to the HSO.

5.2.4.2   Decontamination Procedures

All site personnel as well as equipment used on-site should be subject to a thorough decontamination process; separate decontamination stations should preferably be established for personnel, equipment, and other machinery. The decontamination area should be clearly delineated, highly visible, and accessible to all personnel engaged in site activities.

Personnel decontamination typically consists of portable showers and clean clothes that are provided to the employees. The OHSM should establish the equipment decontamination area—and then provide the decontamination station with a basin of soapy water, a rinse basin with plain water, thick plastic base sheeting, and a waste container with a disposable plastic bag. Small equipment such as hand tools may be decontaminated at this station. A decontamination area should also be established for large equipment; this area will consist of a thick plastic floor covering (of minimum thickness $\approx$10 mm) bermed to collect decontamination runoff (that will be disposed of appropriately) and wash and rinse solutions.

In general, a decontamination crew should be organized with the purpose of decontaminating personnel involved in site activities as well as the equipment used in such operations. The decontamination of the field activity personnel should be the primary focus of the decontamination crew. After decontamination of the site personnel, all equipment should then be decontaminated. Following the decontamination of the equipment, the decontamination crew would then decontaminate themselves.

5.2.4.3   Health and Safety Training

All personnel working at a project site should have participated in adequate health and safety training. This should be ascertained and certified by the designated HSO before an individual is allowed to enter into and work at a potentially contaminated site. For instance, in the United States, all personnel working at the site are traditionally required to have participated in a 40-h health and safety training course that is in accordance with a U.S. "Code of Federal Regulations"—viz., 29 CFR Part 1910.120(e)(2). Proof of current certification in this training must be presented to the HSO/OHSM before an individual is allowed to enter or work on the site. The HSO/OHSM should also conduct on-site health and safety training covering items required by the HSP. Furthermore, additional safety briefings should be provided by the HSO/OHSM if the scope of work changes in a manner that will potentially affect personal health and safety of the workers. All personnel engaged in site activities are indeed required to participate in the training by the HSO/OHSM.

Finally, the OHSM should make a copy of the HSP available to all personnel before any field investigations or remedial activities commence. All persons involved in these activities should sign a health and safety review sheet, certifying that they have read, understand, and agree to comply with the stipulations of the HSP.

5.2.4.4 Emergency Response

Several provisions for an emergency response plan should be carried out whenever there is a fire, explosion, or release of hazardous material that could threaten human health or the environment. The decision as to whether or not a fire, explosion, or hazardous material(s) release poses a real or potential hazard to human health or the environment is often to be made by the OHSM. If a situation requires outside assistance, the appropriate response parties should be contacted using mobile/cellular phones to be carried to the site and/or phones located in nearby facilities that should have been identified at start of the project.

To be able to effectively manage an emergency situation, emergency phone numbers should be compiled and included in the HSP. Also, the directions to the nearest hospital or medical facility, including a map clearly showing the shortest route from the site to the hospital or medical facility, should be kept with the HSP at the site. Finally, in all of the emergency response situations, the emergency transport and medical personnel should immediately be notified as to the type and degree of injury, as well as the extent and nature of contamination to the injured or affected individual(s).

5.2.4.5 Personal Injury

If injury occurring at a project site is not immediately life-threatening, the individual should be decontaminated and then emergency first aid provided. If the injury is serious, the appropriate emergency response agencies should be notified, who will then arrange transportation for the individual to the nearest medical facility; if required, life-saving procedures should be initiated.

5.2.4.6 Acute Exposure to Toxic Materials

The potential human exposure routes to toxic materials affecting site workers typically include inhalation, dermal contact, ingestion, and eye contact. It should be recognized that different procedures will usually be required to deal with worker exposures via the various routes—viz.:

- For *inhalation exposures*, the following steps should be followed: move the victim to fresh air, call for help, decontaminate as best as possible, and notify the appropriate emergency medical services to transport the individual to the nearest medical facility.
- For *dermal contact*, if life-saving procedures are required, then first notify the appropriate emergency medical services for transportation of the individual to the nearest medical facility; then decontaminate as best as possible and initiate life-saving procedures if qualified. In a dermal contact situation where life-saving procedures are not required, decontaminate the individual by using copious amounts of soap and water, wash/rinse affected area for at least 15 min, complete decontamination, and then provide appropriate medical attention.
- For *ingestion exposures*, the following steps should be followed: call for help, decontaminate as best as possible, and arrange for the appropriate emergency medical services to transport the individual to the nearest medical facility.
- In the case of *eye contact*, decontaminate as best as possible and arrange for the appropriate emergency medical services to transport the individual to the nearest medical facility. If life-saving procedures are not required, then rinse eyes using copious amounts of water for at least 15 min, complete decontamination, and then provide appropriate medical attention.

In general, the OHSM should provide information on any acute exposures and materials safety data sheets to appropriate medical personnel. Insofar as possible, the safety data sheets for the contaminants anticipated to be encountered at the site would already have been compiled and appended to the HSP.

5.2.5 IMPLEMENTATION OF THE HSP: GENERAL HEALTH AND SAFETY REQUIREMENTS FOR THE INVESTIGATION OF CONTAMINATED SITES

Several health and safety requirements are generally adopted during the investigation of potentially contaminated site problems. Congruently, a number of safety and emergency equipment are typically required on-site at all times during such investigation activities—including the following specific suggested items:

- First aid kits containing bandaging materials (e.g., band aids, adhesive tape, gauze pads and rolls, butterfly bandages, and splints); antibacterial ointments, oxygen, pain killers (e.g., aspirin and acetaminophen); etc.
- Plastic sheeting materials
- Shovels and tools
- Warning tapes
- Personal protection equipment, including respirators, personnel protective suits (e.g., Tyvek suits), hard hats, goggles, boots, gloves, and earplugs (or other hearing protection apparatus required to minimize worker exposures to excessive and damaging noise from field activities such as during drilling operations)
- Potable water with electrolytic solution

In general, safety plans should include requirements for hard hats, safety boots, safety glasses, respirators, self-contained breathing apparatus, gloves, and hazardous materials suits, if needed. The use of appropriate types of protection equipment will ensure the safety of the field crew. In addition, personal exposure monitoring and/or monitoring ambient air concentrations of some chemicals may be necessary to meet safety regulations. In all cases, the HSO establishes the level of protection required and determines whether the level should be elevated or reduced at any time throughout the course of a given site activity. Also, at all times, sufficient drinking water and vital safety equipment should be made available, as necessary. Box 5.10 identifies the minimum safety requirements to adopt during the investigation of potentially contaminated site problems. Further details of specific items of required safety equipment are discussed elsewhere in the literature (e.g., OBG, 1988).

BOX 5.10   MINIMUM SAFETY REQUIREMENTS FOR CONTAMINATED SITE INVESTIGATION ACTIVITIES

- At least one copy of the HSP should be available at the site at all times.
- In general, fieldwork will preferably be conducted during day-light hours only. The OHSM will have to grant special permission for any field activities beyond day-light hours.
- There should be at least two persons in the field/site at all times of site activities.
- Eating, drinking, and smoking should be restricted to a designated area, and all personnel should be required to wash their hands and face before eating, drinking, or smoking.

(Continued)

> **BOX 5.10** (*Continued*)   **MINIMUM SAFETY REQUIREMENTS FOR CONTAMINATED SITE INVESTIGATION ACTIVITIES**
>
> - Shaking off and blowing dust or other materials from potentially contaminated clothing or equipment should be prohibited.
> - The OHSM should take steps to protect personnel engaged in site activities from potential contacts with splash water generated from decontamination of samplers, augers, etc.

5.3   THE INVESTIGATION-GENERATED/DERIVED WASTE MANAGEMENT PLAN

Investigation-derived wastes (IDWs) [also, IGWs] are those wastes generated during environmental site characterization and/or remedial project activities—and are particularly important in most environmental contamination studies. Indeed, there are several ways by which IDWs may be produced—including from drill cuttings or core samples associated with soil boring or monitoring well installations; drilling muds; purge water removed from sampling wells before groundwater samples are collected; water, solvents, or other fluids used to decontaminate field equipment; groundwater and surface water samples that must be disposed of after analysis; and waste produced by on-site pilot-scale facilities constructed to test technologies best suited for remediation of a contaminated site. Other IDWs may arise from disposable sampling equipment, as well as disposable personal protective equipment (PPE) that are worn or held by an individual for protection against one or more health and safety hazards.

The overarching objective of an IDW management plan is to specify procedures needed to address the handling of both hazardous and nonhazardous IDWs. At any rate, the project/site-specific procedures should generally prevent contamination of clean areas and should comply with existing regional and/or local regulations. Specifically, the IDW management plan should include the characterization of IDW, delineation of any areas of contamination, and the identification of waste disposal methods.

At the end of the day, the project manager should generally select investigation methods that minimize the generation of IDWs. After all, minimizing the amount of wastes generated during a chemical release or site characterization activity tends to reduce the number of IDW/IGW handling problems and costs for disposal. Regardless, provisions should be made for the proper handling and disposal of IDWs/IGWs on-site or locally insofar as possible. In fact, most regulatory agencies do not recommend removal of IDWs from the place or region of origination, especially in situations where the wastes do not pose any immediate threat to human health or the environment; this is because removing wastes from such sites or areas usually would not benefit human health and the environment and could result in an inefficient spending of a significant portion of the total funds available for the site characterization and corrective action programs.

5.3.1   Elements of the IGW/IDW Management Plan

An IGW/IDW management plan, describing the storage, treatment, transportation, and disposal of any materials (both hazardous and nonhazardous) generated during a site characterization or related activity, should generally be included in the project workplan. In any case, to handle IDWs/IGWs properly, the site manager typically must, among other things, determine the waste types (e.g., soil cuttings, groundwater, decon fluids, or PPE); the waste characteristics; and the quantities of anticipated wastes. At the end of the day, the most important elements of the IDW management approach should be as reflected in the summary provided in Box 5.11 (USEPA, 1991d). It must indeed be accentuated here that, to the extent practicable, the handling, storage, treatment, and/or disposal of any IGWs/IDWs produced during site characterization and remedial activities must satisfy all regulatory requirements and stipulations that are applicable or relevant and appropriate to the particular site location or specific project (e.g., federal and state applicable or relevant and appropriate requirements (ARARs) under USEPA programs – as discussed in Chapter 3, Section 3.1.1.1). The procedures must also satisfy any limit requirements on the amount and concentration of the hazardous substances, pollutants, or contaminants involved.

BOX 5.11   REQUIREMENTS FOR THE IDW/IGW MANAGEMENT PROCESS

- Characterize IDW through the use of existing information (manifests, material safety data sheets, previous test results, knowledge of the waste generation process, and other relevant records) and best professional judgment.
- Leave a site in no worse condition than existed prior to the investigation.
- Remove those wastes that pose an immediate threat to human health or the environment.
- Delineate an "area of contamination" unit for leaving on-site, wastes that do not require off-site disposal or extended aboveground containerization.
- Comply with all regulatory requirements, to the extent practicable.
- Carefully plan and coordinate the IGW management program (e.g., containerize and dispose of hazardous groundwater, decontamination fluids, and PPE at permitted facilities *but* leave nonhazardous soil cuttings, groundwater, and decontamination fluids—preferably without containerization and testing—at the site of origination).
- Minimize the quantity of wastes generated.

5.3.1.1   Illustrative Decision-Making Elements of the Process for Screening IDWs

For illustrative purposes, consider an inactive site at which site characterization activities are underway. Initial sampling and analysis of some drummed IDWs indicate several contaminants in both soil and groundwater samples. The question has been raised as to whether these IDWs should be put back in place at the subject site or if they should be trucked out for off-site disposal.

The goal here is to present a decision process that will help manage the IDWs in a cost-effective manner. The process involved should ensure that the resulting decision or action does not increase site risks and that it contemporaneously minimizes the quantity of IDWs that require off-site handling. The rationale and justification for the proposed decision process include the following:

- Several important elements of an effective IDW management approach can be fulfilled by adopting a systematic process that allows cost-effective decisions to be made without compromising the technical effectiveness of disposal options selected to address the IDWs from the site.
- Usually, leaving IDWs on-site tends to result in a more cost-effective way to manage such wastes without increasing risks. Based on this premise, appropriate soil IDWs can be backfilled into shallow pits or spread around the locations where the waste materials had come from; appropriate liquid IDWs could possibly be discharged and allowed to infiltrate into soils at the site.
- Should a particular site be considered a candidate for "no further action" (NFA) or closure, the proposed methodology will not reduce the likelihood of an NFA or closure decision. Conversely, returning IDWs to a site for which remediation is a likely future activity should neither affect the feasibility study program nor remedial action selected.

Based on the above-indicated rationale, an appropriate implementation strategy can be developed to guide the screening and categorization of the IDWs, so that technically justifiable and cost-effective disposal practices can be selected. The methodologies involved will help determine if liquid IDWs could possibly be discharged at the site to grade or into unlined impoundments or if indeed such wastes require special treatment and/or handling. The process should also help determine whether solid IDWs (i.e., from soil cuttings) can be put back and spread out at the site of origination or if such wastes should be disposed of elsewhere (such as in a so-called Class III, Class II, or Class I landfill). The following discussion exemplifies the type of screening procedure that could be used to facilitate decisions on whether or not leaving soil cuttings and purged groundwater on-site will present significantly increased site risks.

Typically, a decision process is developed for use as a guide in the screening of IDWs containing several chemical constituents; Figure 5.1 offers a representative flowchart for the decision process proposed for use as a guide in such efforts. Foremost, it should be determined whether or not the IDWs constitute hazardous wastes under prevailing local regulations—in this case, the *California Code of Regulations* (CCR). As appropriate, IDWs determined to be hazardous wastes will usually be transferred to a waste-handling facility or other treatment, storage, and disposal facility (TSDF) for further management actions. For IDWs that are not considered hazardous following the initial screening, further screening is usually conducted to determine if the wastes can be left at the site, based on appropriate regulatory limits or action levels (such as preliminary remediation goals [PRGs], background threshold concentrations, and "total designated levels" [derived from attenuation and/or leachability factors for the waste constituents]). In general, where no standards are available,

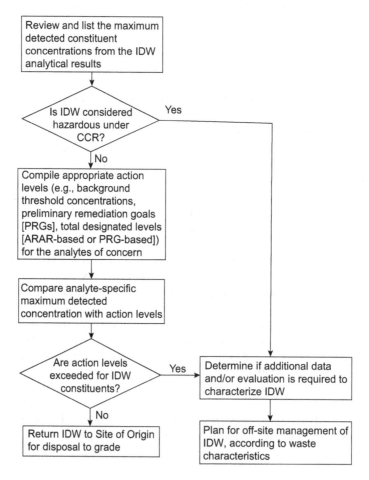

FIGURE 5.1 Example decision process for the screening of IDWs associated with contaminated site problems.

contaminant action levels can be established using the best scientifically justifiable information and professional judgment made on a site- and matrix-specific basis. If the contaminant concentrations in the IDWs fall within the range provided by the action levels (viz., background thresholds, PRGs, and total designated levels), then an "acceptable" contaminant concentration is indicated—and which should, in general, possibly allow the IDWs to be put back on to the site of origination.

The use of such an approach as presented here will indeed enable logical and consistent waste management decisions to be made—and, ultimately, will help determine the specific categories of IDWs that can be disposed of at the site of origination. The overall decision process will also allow a systematic determination to be made as to whether an IDW contains hazardous substances and whether the hazardous substances are present in such amounts as to be of significant concern or to be perceived as wastes requiring special handling. It is expected that this type of site- and matrix-specific decision process used to screen IDWs for on-site disposal will result in a more cost-effective way to manage such wastes than would otherwise be the case.

5.4 THE SITE ACTIVITY PLAN

The objective of the site activity plan is to review and critically evaluate previous site investigation activities, establish goals for proposed new activities, describe the proposed site activities (to include their relevance, possible impact, execution criteria, and associated logistical requirements), establish procedures/methods to be followed during the execution of site activities, and to confirm the format for reporting the results of site activities. Box 5.12 offers several specific elements required as part of a detailed site activity workplan (USEPA, 1985a,d, 1989c,h,p).

BOX 5.12   ELEMENTS OF A SITE ACTIVITY PLAN

- Site background information summary
- Workplan objectives
- Personnel required (including training, organization, and equipment)
- Nonstandard equipment description and contract services
- Hazards expected (physical/chemical) and project impact
- Site location/accessibility (including effect on sampling and remediation plans)
- Project schedule and budget
- Equipment and personnel mobilization/demobilization

5.5 THE QA AND QC PLAN

QA refers to a system for ensuring that all information, data, and resulting decisions compiled from an investigation (e.g., monitoring and sampling tasks) are technically sound, statistically valid, and properly documented. The QA program consists of a system of documented checks used to validate the reliability of a data set.

QC is the mechanism through which QA achieves its goals. QC programs define the frequency and methods of checks, audits, and reviews necessary to identify problems and corrective actions—thus verifying product quality. All QC measures should be performed for at least the most sensitive chemical constituents from each sampling event/date.

A detailed QA/QC plan, describing specific requirements for QA and QC of both laboratory analysis and field sampling/analysis, should be part of the site characterization or related project workplan. The plan requirements will typically relate to but not limited to the following: the use of blanks, spikes, and duplicates; sample scheduling and sampling procedures; cleaning of sampling equipment; storage; transportation; DQOs; chain-of-custody; reporting and documentation;

audits; and methods of analysis. The practices to be followed by the site investigation team and the oversight review—which will ensure that DQOs are met—must be clearly described in the QA/QC plan.

Several aspects of the site characterization or related programs can and should indeed be subjected to a quality assessment survey. In part, this is accomplished by submitting sample blanks (alongside the environmental samples) for analysis on a regular basis. The various blanks and checks that are recommended as part of the QA plan include the following particularly important ones—as elaborated in the following (CCME, 1994):

- *Trip Blank*—required to identify contamination of bottles and samples during travel and storage. To prepare the trip blank, the laboratory will fill containers with contaminant-free water and then deliver to the sampling crew; the field sampling crew subsequently ship and store these containers with the actual samples obtained from the site investigation activities. It is recommended to include one trip blank per shipment, especially where volatile contaminants are involved.
- *Field Blank*—required to identify contamination of samples during collection. This is prepared in the same manner as the trip blank (i.e., the laboratory fills containers with contaminant-free water and deliver to the sampling crew); subsequently, however, the field sampling crew expose this water to air in the locale (just like the actual samples obtained from the site investigation activities). It is recommended to include one field blank per locale or sampling event/day.
- *Equipment Blank*—required in identifying possible contamination from sampling equipment. To obtain an equipment blank, sampling devices are flushed with contaminant-free water, which is then analyzed. Typically, equipment blanks become important only if a problem is suspected (such as using a bailer to sample from multiple wells).
- *Blind Replicates*—required to identify laboratory variability. To prepare the blind replicate, a field sample is split into three containers and labeled as different samples before shipment to the laboratory for analyses. It is recommended to include one blind replicate in each day's activities—or an average of one per 10–25 samples, where large numbers of samples are involved.
- *Spiked Samples*—as required to help identify likely errors due to sample storage and analysis. To obtain the spiked sample, known concentration(s) are added to the sample bottle and then analyzed. It is recommended to include one spiked sample per locale—or an average of one per 25 samples, where a large number of samples are involved.

The development and implementation of a good QA/QC program during a sampling and analysis activity is indeed critical to obtaining reliable analytical results for the site characterization or related program. The soundness of the QA/QC program has a particularly direct bearing on the integrity of the environmental sampling and also the laboratory work. Thus, the general design process for assuring an adequate QA/QC program, as discussed elsewhere in the literature (e.g., CCME, 1994; USEPA, b,c,e, 1988e,f,g), should be adhered to in the strictest manner practicable. Finally, it must be accentuated here that, since data generated during a site characterization will provide a basis for risk management and site restoration decisions, such data should generally convey a valid representation of the true site conditions.

6 Field Investigation Techniques for Potentially Contaminated Sites

When it is suspected that some degree of hazard exists at a given locale or site, it becomes necessary to further investigate the situation and to fully account for the prevailing or anticipated hazards; this may be accomplished by the use of a well-designed field investigation program. For instance, in order to be able to characterize the nature and extent of suspected contamination, an environmental site investigation is generally required in the study of potentially contaminated site problems. Now, because of the inherent variability in the materials as well as the assortment of processes used in industrial activities, it is not unexpected to find a wide variety of environmental contaminants at any particular contaminated site. As a consequence, there is a corresponding variability in the range and type of hazards and risks that may be anticipated from different contaminated site problems. Contaminated sites should therefore be investigated very carefully and thoroughly—to ensure that risks to potentially exposed populations can be determined with a reasonably high degree of accuracy. To realize such goal, different levels of effort in the field investigation will typically be required for different contaminated site problems—and, indeed, a variety of field investigation methodologies and techniques may be used during such investigation activities. This chapter enumerates some of the major field investigation techniques that are commonly used to help achieve the ultimate site characterization goals—particularly in relation to the archetypical risk determination, risk management, and corrective action programs.

6.1 THE GENERAL TYPES OF FIELD DIAGNOSTIC TOOLS AND INVESTIGATION METHODOLOGIES

In formulating site investigation programs, it should be recognized that the most important primary sources of contaminant release to the various environmental media are usually associated with constituents in soils. The contaminated soils can subsequently impact various other environmental matrices by a variety of processes, as discussed in Chapter 7. The impacted media, having once served as "sinks," may eventually become secondary sources of contaminant releases into other environmental compartments as well. In any case, all relevant sources and "sinks" or impacted media at a potentially contaminated site should be thoroughly evaluated as part of any site investigation efforts; to attain this goal, a variety of field investigation tools and methodologies may be used during the investigation of potentially contaminated site problems—as selectively detailed below.

6.1.1 GEOPHYSICAL SURVEY METHODS

Geophysical techniques have been used in many "non-environmental" applications within other industries (principally the petroleum and mining industries) for several decades now. Increasingly, traditional geophysical technologies have found new and innovative uses in the investigation of contaminated site problems. In its environmental applications, geophysical survey techniques may, for instance, be used to determine the lateral extent of past landfilling activities, define areas that may have extensive disturbed soil formations, gage the possible presence of unknown buried metallic objects or voids, and to demarcate and clear utilities so that drilling activities may take place.

These techniques may also assist in the placement of groundwater monitoring wells, along with estimating the existence of preferential groundwater flow directions. In general, geophysical survey methods can be particularly useful when evaluating the following:

- Lithologic characteristics (e.g., stratigraphy, soil and rock properties, and fault location/ orientation)
- Aquifer properties (e.g., location of water table, moisture content, permeability, and water movement)
- Borehole casing characterization (e.g., construction details, borehole deviation, corrosion, and screen plugs).
- Borehole fluid characteristics (e.g., water chemistry and salinity).
- Characterizing contaminants (e.g., contaminant chemistry, hydrocarbon detection, and buried object detection).

It is noteworthy that, every geophysical method depends upon the ability to identify contrasts in the behavior of subsurface materials when subject to some form of induced energy. For instance, every soil and rock type has a unique dielectric constant, ability to transmit acoustic energy, and ability to maintain or affect a magnetic signal. Thus, the resolution of any geophysical technique directly depends on the capability to determine contrasts in these properties, as well as the capability to determine precise depth of detected contrasts.

In fact, the use of geophysical methods at large and surface geophysical techniques in particular can significantly reduce the amount of time and the cost involved in prototypical site characterization activities. It is therefore considered good practice to use surface geophysical surveys wherever feasible in the investigation of potentially contaminated site problems. In general, surface geophysical surveys are useful for providing information on subsurface geologic features in areas where limited stratigraphic data exists. When performed properly and utilized early in the site characterization process, surface geophysics can provide valuable information to use in the planning for monitoring well and piezometer placement. In addition, surface geophysics can be used to correlate the stratigraphy and hydrostratigraphy between wells, uncover buried structures, and in some instances, can directly detect underground contaminant plumes.

The more common geophysical survey methods used in the investigation of contaminated site problems include ground-penetrating radar (GPR), pipe and cable locator, electromagnetometry or electromagnetic (EM) surveys (using terrain conductivity meters), magnetometer/magnetometry, and geophysical gamma logging in boreholes (CDHS, 1990; Osiensky, 1995; Telford et al., 1976; USEPA, 1985a, 2000f). Indeed, borehole geophysical techniques can be of significant interest in their ability to provide efficient and cost-effective means of collecting lithologic and hydrologic information from wells and borings. These methods provide continuous measurements of physical properties along the entire length of a borehole—thus generally supplementing the discrete information typically gathered by coring. It is also noteworthy that geophysical technologies can be used in concert with each other to produce complementary results that increase both perspective and confidence.

6.1.1.1 Ground Penetrating Radar

The ground penetrating radar (GPR) method is a reflection technique that uses high-frequency radio waves, which are reflected off subsurface features, to determine the presence of subsurface objects and structures. In the surface operation mode, a small antenna is moved slowly across the ground surface, energy is radiated into the subsurface, then a receiving antenna continuously records the reflected energy; direct-push subsurface configurations are also available. Responses are caused by radar wave reflections from interfaces of materials having different electrical properties. As such, buried pipes and other discrete objects can be detected. Overall, the picture-like presentation associated with the radar method is highly useful for evaluating subsurface conditions.

GPR can be used to detect natural hydrogeologic conditions and the presence of both natural and anthropogenic artifacts, as well as for "clearing" locations prior to installing wells or pushing sensors into the subsurface. Depth of radar penetration is highly site-specific and limited by higher electrical conductivity of subsurface materials. In general, radar penetration ranges from 1 to 10 m from the source. Better overall penetration is achieved in dry, sandy, or rocky areas. The time the EM pulse takes to travel from the antenna to the buried object and back to the antenna (called the "two-way travel time") is proportional to the depth of the interface or object and the dielectric properties of the media. Dielectric properties are a complex function of composition and moisture content of the subsurface soil and rock. In most cases, moisture content exerts the greatest influence, since water has a very high relative dielectric value compared to common soils and rock, leading to lower travel velocity for moist soils.

It is noteworthy that GPR is a technique that belongs in the larger set of electromagnetometry— but then is often treated as a separate technology due to its widespread use.

6.1.1.2   Electromagnetometry/Electromagnetics

The EM method allows for measurement of subsurface electrical conductivities—including subsurface lithologic materials and porous fluids. Electrical conductivity is a function of the type of material, the porosity, the permeability, and the fluids in the pores; in general, conductivity of the pore fluids will dominate the measurement. Measurements can be made as stationary measurements or within a continuous profiling application.

It is noteworthy that, the term EMs has been used in contemporary literature as a descriptive term for various other geophysical methods (such as GPR and metal detection) which are based on EM principles. However, this discussion here uses EM to specifically imply the measurement of subsurface conductivities by low-frequency EM induction.

Variations in subsurface conductivity can be caused by changes in soil moisture content, groundwater-specific conductance, depth of soil cover over bedrock, and thickness of soil and rock layers. Changes in basic soil or rock types, structural features such as fractures or voids, and changes in salinity can also produce changes in conductivity. Contrasts between the conductivity of background soils and pore fluids can help identify such things as zones of saltwater intrusion, areas of groundwater seepage, and sometimes locations of organic fluids (which would have a relatively lower conductivity than water filled pores). Sequenced readings over multiple events can be used to estimate pore fluid flow parameters such as velocity and direction.

Overall, the EM method provides the means for delineating trenches, buried waste, utility lines, and in some cases, contaminant plumes. Surface applications are often referred to as "terrain conductivity"—and several hand-held devices are available. These can be very useful for initially assessing the location of burial pits, drums, utilities, identifying potential spill locations, and determining soil fracture patterns in depths ranging from 1 to 50 m. Typically, the operator sets up a grid and walks in a set directional pattern, often electronically marking reference areas which are later identified in the data file. This approach is often referred to as "profiling." Anomalies can be further investigated using more intrusive sampling and analytical techniques. Subsurface applications can consist of existing borehole or well deployment. However, direct-push devices are more commonly used for delineating soil type based on induced conductivity. For these applications, higher clay and water content typically display higher conductivity values.

6.1.1.3   Resistivity Methods

Geophysical resistivity methods measure the electrical resistivity of subsurface soil, rock, and groundwater—as, conversely, conductivity would measure the electrical conductivity of materials. The method is somewhat analogous to the EM method, in that resistivity is the reciprocal of conductivity.

A resistivity survey requires that an electrical current be transmitted into the ground via a pair of surface electrodes. The resulting potential field is measured at the surface between a second pair of electrodes. The subsurface resistivity is calculated from knowledge of the electrode separation and geometry of the electrode positions, applied current, and measured voltage. In general, most soil and rock minerals are electrical insulators (highly resistive), and the flow of current is primarily through the moisture-filled pore spaces. As a consequence, the resistivity of soils and rocks is controlled by the porosity and permeability of the system, the amount of pore water, and the concentration of dissolved solids in the pore water. Indeed, soils and rocks generally become less resistive as the moisture content increases; the material porosity and permeability increases; the dissolved solid and colloid content increases; and/or as the temperature increases (for some special situations or locales).

Measurements are typically made by stationary surface measurements but can also be deployed with direct-push platforms. In a surface-deployed configuration, the distance between the electrodes impacts the depth of measurement; in general, the farther apart they are, the greater the resistivity survey measurement domain (depth and volume) becomes—albeit this will depend on site-specific hydrogeology and the current transmitted into the subsurface. When deployed in a vertical configuration (often called "profiling"), lateral changes in subsurface properties can be mapped.

In general, the resistivity method can be used to assess lateral changes and vertical cross sections of the natural hydrogeologic settings—including depth to groundwater, depth to bedrock, and thickness of soils. In addition, it can be used to evaluate contaminant plumes and to locate buried wastes and trenches. Indeed, this technique is very well suited to delineation of contaminant plumes and the detection and location of changes in natural hydrogeologic conditions; and depth and thickness of subsurface layers can be assessed using these methods. Conductivity (reciprocal of resistivity) can also be used to assess the quality of ground and surface waters and delineation of differences between zones can yield useful information regarding the flow characteristics of the site under investigation.

6.1.1.4 Seismic Methods

Seismic methods involve transmission of acoustic waves into the subsurface and the measurement of travel times for the reflected and refracted waves. This data can yield useful information about thickness and depths of geological layers, depth of groundwater, location of burial pits and trenches, and structural information such as faults and bedding plane orientation (that may serve as potential surrogates for possible contaminant migration pathways) and depth. Typically, a strike plate is hit with a sledgehammer, or other equipment or explosive located at the ground surface is used to transmit a sound wave through the ground. An array of receivers is typically set up on the ground surface to monitor the reflected and refracted waves, which are recorded on a seismograph. Several subsurface configurations also exist, including direct-push devices and downhole transmitters and receiver arrays.

Seismic waves travel at different velocities depending on the soil and rock types present and their specific hardness, elasticity and density. In general, solid, denser, and water-saturated materials tend to display higher velocities. Refraction (or signal "bending") often occurs at the interface between layers with differing wave propagation properties. For instance, if a denser layer with a higher velocity (e.g., bedrock) exists below surface soils, a portion of the seismic waves will be refracted as they enter the denser layer, similar to the refraction of light in accordance with Snell's law. Some of these refracted waves cross the interface at a critical angle and then move parallel to the top of the dense layer at the higher velocity of the denser layer. The seismic wave traveling along this interface will continually release energy back into the upper layer by refraction. These waves may then be detected by geophones emplaced at the surface at various distances from the wave source along the path of propagation.

Since return signals for seismic techniques are typically measured in units of time, knowledge of the subsurface material seismic wave propagation characteristics is critical to determining the depth of the anomaly observed. For instance, when carrying out a surface seismic survey, one must determine the seismic signal propagation profile (commonly referred to as the vertical seismic profile

[VSP]), in order to estimate the depth of an observed anomaly. The signals can then be reconstructed or normalized to account for the differences in seismic wave propagation through the various materials, resulting in an estimation of distance (depth) to the observed anomaly. Generation of the VSP will typically require an additional step that includes drilling a borehole or well and lowering a seismic receiver device to successively deeper depths while transmitting a seismic signal from the surface. Indeed, although the general survey is mostly comprised of nonintrusive field efforts, the data tends to be useless without the VSP and subsurface confirmation efforts; thus, costs for these types of essential steps should be considered prior to committing to any seismic method.

Finally, it is noteworthy that seismic reflection, although popular in oil exploration activities, is not typically widely used in the investigation of contaminated site problems; this is because resolution in shallow (specifically less than 20 m depth) is not quite reliable per se.

6.1.1.5   Metal Detectors

The metal detector can provide information leading to location of buried drums, tanks, pipes, utility cables, and trench boundaries. They can also be used to help avoid buried utilities when planning drilling, pushing, and trenching activities. Metal detectors respond to electrical conductivity of metal targets, which is high compared to most soils; indeed, the metal detector responds to both ferrous and nonferrous metal objects.

In its application, a transmitter creates an alternating magnetic field around the transmitter coil. A balance condition is achieved to cancel the effect of this primary field at a receiver coil; typically, the balance (or "null") is accomplished by orienting the planes of the two coils perpendicular to one another. The primary field will induce eddy currents in a metal target within the instrument range. These eddy currents produce a secondary field that interacts with the primary field to upset the existing balance condition. This results in an audible signal. The most important factors influencing the detector response include the properties of the target, properties of the soil, target size (response is proportional to the area cubed), and depth. In general, the larger the target surface area, the greater the induced eddy currents, and therefore, the greater depth at which the target may be detected. Metal detector depth range decreases at a rate of the reciprocal of the target depth to the sixth power (i.e., signal intensity $\propto 1/[depth]^6$). Therefore, most metal detectors are limited to depths near the surface. In any case, metal detectors with coils of small diameter (less than 0.3 m), like those used by treasure hunters, are better than conventional "pipe locators" for locating small targets but have limited depth capabilities.

Data collection is typically made as stationary measurements or within a continuous profiling (horizontal) application. It is notable that metal detectors have a relatively short detection range—and this is indeed related to the target's surface area. For instance, small metal objects such as spray cans or quart-size oil containers can be detected to a distance of approximately 1 m; typical 55-gal drums can be detected to depths of approximately 1–3 m; and massive piles of metallic debris may be detected to depths of approximately 3–6 m.

6.1.1.6   Magnetometry

Magnetometry relates to the location of buried ferrous metals such as drums. By detecting anomalies in the earth's magnetic field caused by ferrous objects, a magnetometer provides a means of locating potential sources of waste release or obstacles for planned intrusive efforts. They can also be used for locating buried drums and trenches and for selecting drilling locations clear from potential obstacles. It must be emphasized here that, the magnetometer responds only to ferrous metal, such as iron or steel; unlike a metal detector, it does not respond to nonferrous materials such as copper, lead, or brass.

A magnetometer measures the intensity of the earth's magnetic field, which behaves as a large bar magnet. The presence of ferrous metals generates perturbations in the local strength of that field, which are detectable. The response is proportional to the mass of the ferrous target; for instance, a single drum can be detected to depths of up to 6 m, whereas massive piles of drums can be detected to depths of over 20 m. It is noteworthy that, the intensity of the magnetic field can change on a daily basis in response to sunspots and ionosphere conditions.

Two basic types of magnetometers are commonly used in the investigation of contaminated site problems—viz., fluxgate and proton devices. In a *fluxgate magnetometer*, the sensor consists of an iron core that undergoes changes in magnetic saturation in response to variations in the earth's magnetic field. Differences in saturation are proportional to field-strength variations, which are detected as electronic signals, amplified, and recorded. Two fluxgate elements can be mounted together to form a gradiometer, which measures the gradient of a directional component of the earth's magnetic field. In a *proton magnetometer*, an excitation voltage is applied to a coil around a container filled with a proton-rich fluid such as kerosene. The field produced reorients the protons in the fluid, allowing for measurement of a nuclear precession frequency, which is proportional to the strength of the field. Proton magnetometers measure the earth's total field intensity and are not sensitive to orientation.

Two types of magnetic measurements are typically made: a *total field measurement* is made from observing the value of the magnetic field at a selected point; and a *gradient measurement* is the difference between measurements taken at two points. Gradient measurements are less susceptible to noise. Data is typically made as stationary measurements or within a continuous profiling (horizontal) application. Factors influencing the response include mass of target, depth, permanent magnetism of target objects, the target's shape, orientation, and state of deterioration. Overall, if sensors of identical sensitivity are used, the total field system provides a greater working range and depth of detection.

For a successful deployment, the following factors must be carefully deliberated on project objectives, site conditions, site access, level of detail desired (reconnaissance or high resolution), estimated mass or quantity of targets, maximum depth of search, and safety requirements. For high-resolution surveys over large areas, continuous sensors and vehicle-mounted systems may be desired. Finally, as with most geophysical surveys, competent operators are essential for collection and interpretation of field data.

6.1.2 TRIAL PITTING

Trial pitting using a backhoe excavator is probably the most common technique for investigating the condition of shallow subsurface soil (Pratt, 1993). Typically, visual evidence of contamination (such as apparent staining or the presence of landfilled materials) as well as odors can be used to guide the selection of sampling locations, where samples are to be collected for laboratory analyses.

6.1.3 SOIL GAS SURVEYS

In some cases where organic chemicals represent the major source of contamination, a preliminary gas survey using subsurface probes and portable equipment can give an early indication of likely problem areas. Soil gas surveys are generally carried out as a precursor to exploratory excavations, in order to identify areas that warrant closer scrutiny. They can also be used to assist in the delineation of previously identified plumes of contamination. This can be a very important step that needs to be completed prior to the start of a full-scale site investigation that utilizes intrusive exploratory methods. Soil gas surveys can also be used to evaluate and identify large areas with organic contamination in a relatively short period of time (Pratt, 1993).

For a contaminant to be identifiable in a soil gas survey, it must be volatile—with a Henry's aw constant of at least 5×10^{-4} atm-m^3/mol and a vapor pressure of at least 1 mm Hg (at 20°C) (CCME, 1994). Notwithstanding, it should be recognized that, fine-grained materials of low gas permeability may serve as barriers to soil gas movement, resulting in such type of survey not detecting existing contamination because soil vapor is not able to reach the sampling point. High-water content materials (especially in perched water table zones) and sorptive organic-rich soils may also affect soil gas surveys in a similar manner—resulting in a possible reporting of "false negatives." Soil gas surveys may, therefore, not be used as an absolute indicator of site contamination—that is, not even for cases of suspected volatile organic contamination.

On the whole, soil gas surveys are well-established methods for assessing the subsurface distribution of volatile organic contamination. They are particularly useful in the vicinity of storage tanks that are suspected to be leaking or in areas where some form of release or spill of fuels or organic solvents is suspected. Classically, soil gas surveys are used to delineate the apparent extent of soil contamination and to identify locations for the collection of samples for more rigorous analysis conducted in an analytical laboratory. They can indeed be a powerful screening technique, if used in a professionally credible and responsible manner. This is because soil gas surveys can delineate source areas and track some contaminant plumes, allowing an investigator to more accurately place subsequent soil boring and monitoring well point locations.

Finally, it is noteworthy that soil gas sampling and analysis results tend to be more reliable at locations and depths where high constituent concentrations exist and where the soils are relatively permeable. Reliability of the results tends to be generally lesser in lower permeability settings, as well as when sampling shallow soil gas—since in both cases, leakage of atmospheric air into the samples can become a significant concern. In general, reliability is typically improved by using fixed probes and by ensuring that leakage of atmospheric air into the samples is avoided during purging or sampling.

6.1.4 GEOPROBES

The geoprobe represents the next step up from the soil–gas survey (Section 6.1.3)—that is, in terms of degree of sophistication and efforts; it is a hydraulically operated sampling tool that punches down into the soil. The geoprobe is indeed a rapid site assessment tool that is very effective for collecting soil–gas, soil, and groundwater samples. It generally consists of a rugged system that is capable of collecting samples at discrete depths to about 12 m (or 40 ft)—whereas, in contrast, most soil gas surveys can collect data from only about 120 to 300 cm (or 4–10 ft).

It is notable that the geoprobe system can work in most of the environmental media commonly associated with contaminated site problems—viz., pore space, soil, or groundwater. However, it does not perform well in competent clays or cobble—but may be useful in weathered bedrock if the overburden is well fractured.

6.1.5 FLUX CHAMBER MEASUREMENTS

There seems to be a general consensus that typical modeling results are usually inconclusive and may even be misleading. For instance, the Johnson and Ettinger and related models that had traditionally been used to evaluate contaminant intrusion into indoor air are considered to be "discovery" in nature and conservative by design; thus, they have commonly been used as the basis of conservative screening risk assessments in cases where the conceptual site model is consistent with the assumptions of the model. Risk assessments that conclude that there is no significant health risks based on this type of conservative predictive modeling approach may be used as a basis for proceeding with property redevelopment or site closure. However, risk assessments that indicate a potential for risk, based on the results of such conservative predictive models, should more reasonably call for further site characterization and/or a higher tier risk assessment—and this is where flux chamber measurements could become very important. Flux measurements determine mass emission rates using the concentration and flow rate of gases through a chamber and the surface area isolated by the chamber. The emission flux chamber testing approach can indeed be used to determine diffusion-based flux at land surfaces on open soil, and the resulting flux chamber data could find very useful applications in exposure and risk assessments. Surface emission flux chamber testing can be conducted in accordance with the USEPA's *Measurement of Gaseous Emission Rates from Land Surfaces Using an Emission Isolation Flux Chamber-User's Guide (1986)*. Indeed, this technology is believed to be the USEPA's preferred testing technique for the direct measurement of volatile compound vapor emissions; the method quantitatively measures vapor fluxes at the land surface emanating from subsurface volatile compound contamination.

In effect, the USEPA offers what may be considered as an appropriate "next step" for conducting site assessment for those sites requiring a more detailed or "in-depth" site assessment and higher tier exposure/risk assessment for the air pathway—that is, after possible completion of a conservative screening process. In support of such efforts, the USEPA describes some of its procedures in the technical guidance series, *Procedures for Conducting Air Pathway Analyses for Superfund Sites (1990)* and related updates; this technical guidance series describes technologies for collecting data for refined exposure assessment. One such technology used for this purpose is the USEPA surface emission isolation flux chamber. This technology can provide direct measurement data that can be used in the air pathway exposure evaluation, in order to reduce uncertainties inherent in predictive modeling approaches. The technology involves conducting emission measurements on open soil (e.g., a future building footprint), assuming non-advective flow conditions, and estimating a credible, reasonable maximum infiltration factor for soil gas.

In general, in order to determine if significant contaminant migration is occurring, surface emission flux chambers can be used to measure volatile contaminants emanating from the subsurface; these measured values can then be used to more accurately evaluate risk. It is noteworthy that, the original development of the principal USEPA-recommended flux chamber method came about because of the need to assess emissions of air-toxic contaminants at hazardous waste sites as part of remedial investigation efforts. The basic approach uses an enclosure or chamber of some design to isolate a surface emitting gas species; clean sweep air is added to the chamber at a controlled fixed rate, and the contents are sampled and analyzed for species of interest.

It is noteworthy that several investigators have characterized sites using data from both predictive modeling and direct measurements. Subsequently, comparative analysis methods have been used to demonstrate the conservative nature of most of the screening approach that uses predictive modeling. Indeed, in several cases of using the flux chamber measurements, several investigators determined that the residual volatile organic compounds (VOCs) in soil and groundwater at a site did not pose any significant risk to warrant remedial action prior to planned residential or similar developments—that is, contrary to results from screening predictive modeling outcomes. Overall, it appears that the flux chamber measurements approach often leads to more realistic and more "favorable" risk outcomes—and thus might be a worthwhile consideration for a number of projects.

6.1.6 Augering, Borehole Drilling, and Related Methods

Oftentimes, intrusive tools for environmental site investigation are required to penetrate the soil so that soil and groundwater samples can be collected at specified depths below the surface. These tools include traditional and innovative drilling methods, driven well points, and sampling equipment. Drilling techniques may include traditional geotechnical drilling methods using hollow stem augers (HSAs), as well as water well drilling methods using rotary rigs; more contemporaneous methods include sonic and directional drilling. It is noteworthy that, "drive-down" methods, including penetrometers, generally would not work in rock formations—but then can be used rather quickly and efficiently for rapid soil and groundwater sampling, whenever appropriate. The sampling equipment of interest includes fixed samplers and probes that are often placed downhole and can be used to transmit data or to bring water or vapor samples to the surface; coring is often used to collect subsurface materials for inspection and analysis.

Classically, soil samples for laboratory analyses are collected using hand-held equipment (including hand augers, split spoon samplers, and ring samplers) or by advancing soil borings by the use of a drilling rig—and the soil samples are usually collected from the apparent source areas, as well as from suspected margins of the source areas in an effort to estimate the extents of contamination. Meanwhile, it is notable that evidence of soil contamination in deeper borings that approach the water table suggest that groundwater is likely to be contaminated beneath and downgradient of the contaminant source area (Pratt, 1993). The volume of the release and the hydrogeological conditions are key factors in determining whether groundwater beneath the release site is truly contaminated. As necessary, the locations of groundwater monitoring or sampling wells, in relation to the potential on-site contaminant source areas and potential off-site receptors, are critical in the data collection program. As a general rule-of-thumb, a minimum of three wells is usually required to help determine the direction of groundwater flow at a project site; typically, one well is installed at the upgradient boundary of the site and at least two wells are positioned at locations downgradient from suspected source areas (Pratt, 1993). In practice, additional wells located at the downgradient site boundary may be necessary to determine whether contaminated groundwater is flowing off-site. In fact, if it is determined that there possibly are seasonal changes in the groundwater flow directions, as occurs at some hydrogeologically complex sites, then even more wells may be necessary to account for the changes in contaminant migration directions.

Classically, drilling activities should be accompanied by geophysical investigations (as discussed in Section 6.1.1) so that the location and depth of the resulting borehole can be optimized in relation to the drilling operation. In any event, continuous soil sampling is recommended to verify stratigraphic information. Meanwhile, it is notable that segments of the borehole can be "cased off" to protect uncontaminated zones—often achieved by driving larger diameter surface casing to the necessary depth and, in which case, the auger itself acts as a protective casing for the sampler installed just slightly ahead of the auger bit. A variety of efficient sampling systems (e.g., wireline) and sampling tools are available for use with HSA. Wells and the associated filter packs can be installed quickly and efficiently through the HSA, provided that the inside diameter of the auger has been properly chosen *a priori*.

6.1.6.1 Direct Push Technology

The use of direct push technology (DPT) for soil sampling is recommended because it is a lower cost alternative to drilling and can substantially increase the quality of the data obtained from the soil sampling program. DPT is also used for groundwater investigation—for example, as in the case of using the "Hydropunch" sampling technique to collect a depth-discrete groundwater sample inside a boring without installing a well. Indeed, these relatively recent tools in this category provide faster and cheaper ways to explore the subsurface characteristics than have been available in the past. Also, these methods are typically less intrusive, generate lesser investigation-derived wastes than "conventional" techniques, and generally allows for sample collection in areas with limited clearance.

In general, DPT is most applicable at sites with unconsolidated sediments and at depths less than 30 m (or approximately 100 ft)—albeit there are some that are not depth-constrained. In any case, if a DPT platform cannot be used, then drilling with a HSA method is recommended. Finally, it is noteworthy that a wide range of direct-push and limited access drilling techniques is indeed available for collecting soil, groundwater, and soil vapor grab samples—as well as for identifying stratigraphy or nonaqueous phase liquids (NAPLs) at a site.

6.1.6.2 Site Characterization and Analysis Penetrometer System

Traditional site characterization methods usually involve drilling, retrieving, and analyzing soil samples at some set vertical intervals—typically about 150 cm (or 5 ft). This data is then extrapolated to develop a picture of the subsurface site conditions. The analysis involved is usually performed at an off-site laboratory—and thus the analytical results can lag the fieldwork by several weeks. Also, because the initial sampling is not necessarily based on any prior knowledge of the subsurface conditions, this characterization process tends to involve some "guess-work"—and usually requires multiple iterations to adequately characterize the site. Indeed, the coarse vertical sampling intervals could very easily overestimate, underestimate, or even miss a plume completely. Furthermore, drill cuttings produced require special handling or management, and oftentimes, involves costly disposal. The site characterization and analysis penetrometer system (SCAPS) technology provides an alternative means of achieving similar objectives for certain categories of environmental contamination problems.

SCAPS generally installs direct push, micro wells for continued site monitoring. Because SCAPS is based on DPT, no drill cuttings are produced during the field operations. Also, the SCAPS technology classically incorporates a remotely operated decontamination system to clean all hardware before personnel contact. Furthermore, SCAPS destroys its investigation holes with pressure-injected, bottom-up style grouting. In addition, a miniature video microscope incorporated into the system will even allow a magnified, "gopher's eye view" of subsurface soils and features.

In a nutshell, the SCAPS is a "rolling laboratory" used to assist with subsurface screening of contamination at potentially contaminated sites. A classic SCAPS integrates a 20-ton cone penetrometer testing (CPT) truck with laser-induced fluorescence (LIF) technology to detect hydrocarbon-contaminated soil in real-time and *in situ*; soil, liquid, or vapor-phase samples can be retrieved from depth and analyzed onboard SCAPS or captured for further analysis. Laser energy is emitted into the soil, causing hydrocarbon compounds to fluoresce in its presence; the laser energy is delivered via a modified, 1-1/2 in. CPT probe, which is hydraulically pushed into the ground. As an example, most petroleum hydrocarbon products can be detected by either the nitrogen or the xenon-chloride LIF system. Semiquantitative contaminant data is reported every 2-1/2 in., and standard CPT soil classification data is reported every inch. Overall, SCAPS could be used as a "yardstick" for long-term monitoring of subsurface hydrocarbon contamination.

6.1.7 DIFFUSION SAMPLERS

The traditional procedure to sample for groundwater is to purge the well and extract the sample with a bailer or pump system. The purged water—typically about three well volumes—must be tested, handled, and properly disposed as investigative-derived waste (IDW). Also, the sampling equipment must be decontaminated and appropriately maintained. Additional logistics such as a pump power source and local IDW requirements increase the effort of a given sampling event. Furthermore, excessive pumping rates could be undesirable, since this increases the radius of influence around the well—and concomitantly result in sampling of well water that may be an integration of different water types. Low-flow sampling methods had been introduced to target a specific depth, reduce the water column disturbance, and to minimize the amount of purge water and the radius of influence; still, these methods do not quite eliminate typical pumping logistical issues or problems inherent in pumping water samples. The diffusion sampler technology seems to provide an alternative means of achieving similar objectives for some types of environmental contamination problems—and to eliminate problems associated with the traditional sampling process or low-flow sampling methods. In fact, the U.S. Geological Survey (USGS) has carried out extensive studies on various types of inexpensive, simple diffusion samplers as an alternative groundwater sampling method—in order to eliminate problems associated with the traditional sampling process.

Diffusion samplers have traditionally been made of polyethylene bags containing deionized water. These samplers are submerged in a well and can "target" specific depths within the screened or open interval. Indeed, sampler(s) can be suspended at one or many depths to better study the condition of different hydrogeologic features and to vertically profile the contaminant distribution. After allowing the samplers and the well water to equilibrate following deployment, usually for 14 days or so, the samplers are removed from the well, and then the water within the sampler can be analyzed by routine laboratory methods.

Water-filled passive diffusion bag (PDB) samplers have traditionally been suitable for obtaining concentrations of a variety of VOCs in groundwater at monitoring wells. The suggested application of the method has been for long-term monitoring of VOCs in groundwater wells at properly characterized sites. A typical PDB sampler consists of a low-density polyethylene (LDPE) lay-flat tube closed at both ends and containing deionized water. The sampler is positioned at the target horizon of the well by attachment to a weighted line or fixed pipe. The amount of time that the sampler should be left in the well prior to recovery depends on the time required by the PDB sampler to equilibrate with ambient water and the time required for the environmental disturbance caused by sampler deployment to return to ambient conditions. In any event, the samplers should be left in place long enough for the well water, contaminant distribution, and flow dynamics to restabilize following sampler deployment. Laboratory and field data suggest that about 2 weeks of equilibration probably is adequate for many applications; therefore, a minimum equilibration time of 2 weeks is suggested—and in less permeable formations, longer equilibration times may be required.

6.1.7.1 Diffusion Sampling

Diffusion sampling is a relatively new technology designed to use passive sampling techniques that eliminate the need for well purging. In fact, groundwater sample collection using diffusion samplers represents a relatively new technology that have traditionally utilized passive sampling methods for monitoring VOCs in groundwater. In this technique, a diffusive-membrane capsule is filled with deionized distilled water, sealed, mounted in a suspension device, and lowered to a specified depth in a monitoring well. Over time (generally no less than 72 h), VOCs in the groundwater diffuse across the capsule membrane and contaminant concentrations in the water inside the sampler attain equilibrium with the ambient groundwater. The sampler is subsequently removed from the well and the water within the diffusion sampler is transferred to a sample container and submitted for laboratory analysis. Once a diffusion sampler is placed in a well, it remains undisturbed until equilibrium is achieved between the water in the well casing and the water in the diffusion sampler. Depending on the hydrogeologic characteristics of the aquifer, the diffusion samplers can reach equilibrium within 3–4 days (Vroblesky and Campbell, 2001)—albeit up to 2 weeks or so may be a reasonable recommended *modus operandi*. Groundwater samples collected using the diffusion samplers are thought to be representative of water present within the well during the previous 24–72 h. Studies have indeed shown the VOC concentration in undisturbed water within the screened well interval can be representative of concentrations in the adjacent aquifer. Thus, a passive sampling method, such as diffusion samplers, has the potential to provide representative concentrations of aqueous contaminants as they exist in the undisturbed subsurface.

In general, the use of PDB samplers for collecting groundwater samples from wells offers a cost-effective approach to long-term monitoring of VOCs at well-characterized sites; further details on its applicability are provided elsewhere in the literature (e.g., Gefell et al., 1999; ITRC, 2004, 2007a; Jones et al., 1999; USEPA, 1997g; Vroblesky and Hyde, 1997; Vroblesky, 2000; Vroblesky and Campbell, 2001). The potential benefits and cost savings of using the diffusion sampler as an instrument for long-term monitoring are indeed significant—especially since no purge waters are generated, and labor requirements for sampler installation and retrieval are minimal. Overall, the inexpensive and simple diffusion sampler is an excellent groundwater sampling method for VOCs.

Finally, it is quite encouraging that the USGS has subsequently developed diffusion samplers constructed from regenerated cellulose dialysis membrane that have more extensive applications. Recall that, the traditional water-filled polyethylene PDB samplers have not been considered appropriate for all compounds; indeed, the samplers have not been considered suitable for inorganic ions and have had a limited applicability for non-VOCs and even for some VOCs. But now, the cellulose membrane allows for the collection of groundwater samples for inorganic compounds and nutrients—thus permitting an evaluation of the effectiveness of natural attenuation processes in reducing chemical contamination in groundwater systems. Indeed, it has been demonstrated at some project sites that this relatively new technology for using regenerated cellulose diffusion samplers can reduce the cost of long-term groundwater monitoring programs by about 50%. The regenerated cellulose dialysis membrane diffusion samplers offer all the benefits of conventional polyethylene diffusion samplers and the additional benefit of the ability to sample for inorganic compounds and nutrients.

6.2   FIELD CHEMICAL SCREENING METHODS

Field chemical screening methods offer a useful approach for rapid and/or real-time data collection during contaminated site assessment and remediation activities (see, e.g., ITRC, 2013a; Van Emon and Gerlach, 1995). Field chemical screening generally comprises a variety of chemical analytical methods that use portable versions of similar equipment to that found in the fixed-based laboratory. However, the screening methods are advantageous during site assessments because they are often less expensive than laboratory methods—as well as also allow the field crew to utilize real-time data to add to, or subtract from, sampling locations while already in the field. During remediation activities, screening methods can be used to segregate waste or soil piles—and to generally make the site operations more efficient. Field chemical screening is also used to protect workers via the monitoring of the ambient air—in order to enable the detection of such situations as relates to possible explosive environments or toxic vapors.

Field screening methods/tools often used for VOCs, semi-volatile organic compounds (SVOCs), and pesticides typically may be classified as *in situ* analysis methods (e.g., solid/porous fiber-optic chemical sensors and LIF) or *ex situ* analysis methods (e.g., photoionization detector [PID]; flame ionization detector [FID]; explosimeter; gas chromatography [GC]; mass spectrometry [MS]; GC/MS)—and these examples are briefly annotated as follows:

- *Solid/Porous Fiber Optic:* Fiber-optic chemical sensors are often used in cone penetrometers to detect hydrocarbons in dissolved or vapor phases. This method is low maintenance and cost but requires multiple sensors to discriminate between specific compounds.
- *LIF:* This spectroscopic technique identifies aromatic hydrocarbons with downhole detectors. Mixtures are difficult to identify due to spectral overlap.
- *PID:* These are used to detect aromatic hydrocarbons and hydrocarbons with carbon ranges between 6 and 10. This method is often combined with an FID and is sensitive to water vapor and methane.
- *FID:* These detect petroleum and polynuclear aromatic hydrocarbons (PAHs). The FID ionizes the analyte with a hydrogen flame, generating a voltage across a detector that is proportional to the concentration of the analyte. FIDs are not effective at differentiating between specific compounds (must be combined with another method).
- *Explosimeter:* These are used to determine if a flammable concentration of gas exists in the sample location. The result is reported as a percentage of the lower explosive limit.
- *GC:* A GC column is used along with selected detectors to analyze VOCs. The analyte separates into individual compounds within the long thin column. The compounds can be identified by their elution (exiting) time.
- *MS:* The MS ionizes the analyte and then identifies it by measuring the mass of its charged particles. This method works on extractions of VOCs, PAHs, SVOCs, pesticides, and herbicides.
- *GC/MS:* This method combines GC and MS techniques to identify and quantify VOCs, SVOC, PAHs, pesticides, and herbicides. This is the most expensive and most accurate of the commonly used analytical methods.

Field screening methods/tools commonly used for metals include the following examples:

- *X-Ray Fluorescence:* X-ray fluorescence (XRF) spectrometry is a nondestructive, analytical method used primarily to detect heavy metals in soil/solids samples. XRF spectrometry uses primarily X-rays to irradiate a sample, which causes elements in the sample to emit secondary radiation of a characteristic wavelength. Two basic types of detectors are used to detect and analyze the secondary radiation—namely, *wavelength-dispersive XRF spectrometry* and *energy-dispersive XRF spectrometry. Wavelength-dispersive XRF spectrometry* uses a crystal to diffract the X-rays, as the ranges of angular positions are scanned using a proportional or scintillation detector (an extremely sensitive instrument that can be used to detect alpha, beta, gamma, and X-radiation). *Energy-dispersive XRF spectrometry* uses a solid-state Si(Li) detector from which peaks representing pulse-height distributions of the X-ray spectra can be analyzed.

- *Chemical Colorimetric Kits:* These are self-contained portable kits for analyzing soil or water samples for the presence of a variety of inorganic and organic compounds. These tests require no instrumentation and can be performed in the field with little or no training. The two common forms of chemical colorimetric kits include *titrimetry kits* and *immunoassay colorimetric kits*—and these are generally only to be used as an indication or screening device. *Titrimetry kits*—are used for analyzing samples contaminated with heavy metals. Titrimetry is a wet chemistry procedure, by which a solution of known concentration is added to a water sample or soil-solute extract with an unknown concentration of the analyte of interest until the chemical reaction between the two solutions is complete (the equivalence point of titration). *Immunoassay colorimetric kits*—that utilize the immunoassay technology, relies on an antibody that is developed to have a high degree of sensitivity to the target compound. This antibody's high specificity is coupled within a sensitive colorimetric reaction that provides a visual result. The intensity of the color formed is inversely proportional to the concentration of the target analyte in the sample. The absence/determination is made by comparing the color developed by a sample of unknown concentration to the color formed with the standard containing the analyte at a known concentration. (See Section 6.2.1 for further discussion of this screening tool.)

Field screening methods/tools commonly used for radionuclides include the following examples:

- *Radiation Detectors:* These measure alpha, beta, or gamma radiation emitted by radionuclides. The instruments range from hand-held instruments to portable units with interchangeable probes. These methods produce real time or short turnaround data but cannot identify specific radionuclides.
- *Gamma Ray Spectrometry:* This method analyzes the energy spectrum of gamma quanta emissions to identify and quantify specific radionuclides. This method requires known standards and an operator with extensive experience.
- *Piezocone Magnetic Meter:* This meter is used with a cone penetrometer for downhole measurement of gross radiation. This method is inexpensive but is less accurate than radiation detectors and will not identify specific radionuclides.

Field screening methods/tools commonly used for explosives include the following examples:

- *GC:* A GC column is used along with selected detectors to analyze *low concentrations* of explosive organic compounds. The analyte separates into individual compounds within the long thin column. The compounds can be identified by their elution (exiting) time.
- *MS:* The MS ionizes *low concentrations* of the analyte and identifies it by measuring the mass of its charged particles. This method requires an extraction of the analyte.
- *GC/MS:* This method combines GC and MS techniques to identify and quantify *low concentrations* of the explosive material. This is the most expensive and most accurate of the commonly used analytical methods.
- *Ion Mobility Spectrometer (IMS):* The IMS technique is a laboratory-based or portable method for detecting organic vapors in air. Specific ionized compounds can be identified from their ionic mobility signature. This method is relatively expensive and is susceptible to interference.
- *Chemical Colorimetric Kits:* These kits provide portable analysis of soil or water samples for organic or inorganic compounds. Operators need minimal training and can read direct

results from color comparison charts after mixing the analyte with reagents. The technique loses accuracy when used with mixtures of contaminants.

- *Immunoassay Colorimetric Kits:* Immunoassays can be used to identify specific compounds for field screening. The method uses an antibody to generate a colorimetric reaction that can be compared to a color chart. Immunoassays do not work well with long-chain aliphatic hydrocarbons. (See Section 6.2.1 for further discussion of this screening tool.)

It must be acknowledged here that the above-mentioned listings are by no means complete and exhaustive. Furthermore, new tools and improvements to existing ones continue to be developed.

6.2.1 Immunoassay Screening Tools

Immunoassay is a technology for identifying and quantifying organic and inorganic compounds; it uses antibodies that have been developed to bind with a target compound or class of compounds (see, e.g., ITRC, 2013a; Van Emon and Gerlach, 1995). The immunoassay technique takes advantage of the ability of antibodies to bind selectively to a target analyte that is present in a sample matrix (such as soil or water) because it has a specific or distinctive physical structure. Working much like a key and lock, the binding sites on an antibody attach precisely and non-covalently to their corresponding target analyte (also known as the antigen). Because binding is based on the antigen's physical shape rather than its chemical properties, antibodies will not respond to substances that have dissimilar structures. Indeed, immunoassay kits have been developed that are compound-specific, detecting only the target analyte and its metabolites—albeit class-specific kits have been demonstrated to be the most useful in the environmental field. The technology has been used widely for field analysis in the environmental field because the antibodies can be highly specific to the target compound or group of compounds and also because immunoassay kits are relatively quick and simple to use. In fact, there are numerous advantages to using immunoassay in the field (i.e., rather than use formal analysis in a fixed laboratory)—including speed, portability, ease of use, relatively low cost per sample, and the range of contaminants that can be analyzed. It is notable that a beginner can learn how to use an immunoassay test kit in a day or less—and most people become proficient at using the test kit after analyzing just two or three batches of samples. Indeed, the test kits are designed specifically for easy operation, although a background in environmental science and chemistry is helpful to the operator.

Immunoassay testing can be used in the field in efforts to detect target chemicals in soil and other matrices or samples. Concentrations of analytes are identified through the use of a sensitive colorimetric reaction. The determination of the target analyte's presence is made by comparing the color developed by a sample of unknown concentration with the color formed by the standard containing the analyte at a known concentration. The concentration of the analyte is determined by the intensity of color in the sample. The concentration can be estimated roughly by the naked eye or can be determined more accurately with a photometer or spectrophotometer.

6.2.1.1 Types of Immunoassays

There are four principal types of immunoassay—namely: (i) enzyme immunoassay (EIA), (ii) radio-immunoassay (RIA), (iii) fluorescent immunoassay (FIA), and (iv) enzyme-linked immunosorbent assay (ELISA). Most immunoassay-based test kits for analyzing environmental contaminants are of the competitive ELISA type. Indeed, ELISA is most often used for environmental field analysis because the technology can be optimized for speed, sensitivity, and selectivity—and also because it contains no radioactive materials. Additionally, ELISA has a longer shelf life and is simpler to use than most other immunoassay methods. Indeed, neither RIA nor FIA seems to be in common use for the analysis of environmental samples.

In an ELISA immunoassay, antibodies are developed specifically to bind with a selected environmental contaminant or contaminants, and that selective response is used to confirm the presence of the contaminant in samples. First, the antibodies' binding sites must be available to bind with the antigen. The walls of a test tube can be coated with the antibodies, or they can be introduced into the test tube on coated magnetic or latex particles. With either method of introduction, the quantity of antibodies is known, and a limited number of antibody binding sites are available for the sample. Second, the test kit developer couples some of the contaminant, or antigen, with an enzyme that will react with a colorimetric agent to produce a color change but will not interfere with the antigen's ability to bind with the antibodies. The enzyme is referred to as the label because it allows detection of the antigen's presence—creating a *labeled antigen*. A solution that contains the labeled antigen for analysis, which is called the enzyme conjugate, is prepared. The third component is the colorimetric agent, or chromogen, that will react with the enzyme on the labeled antigen to cause a color to form.

During the analytical procedure utilized in an ELISA, a known amount of sample and a known amount of enzyme conjugate are introduced into a test tube that contains the antibodies, and the target analyte present in the sample competes with the labeled antigen in the enzyme conjugate for a limited number of antibody binding sites. A chromogen is then added to the test tube to react with the enzymes on the labeled antigen to cause the formation of a color. According to the law of mass action, the more the analyte present in the sample, the more enzyme conjugate it will displace from the binding sites—and the amount of bound conjugate is inversely proportional to the amount of analyte in the sample. The original concentration of the analyte can be determined by measuring the amount of enzyme conjugate bound to the antibody. Because the amount of bound enzyme conjugate determines the intensity of the color, the intensity of the color is inversely proportional to the amount of analyte present in the sample. Ultimately, the concentration of the target analyte can be determined by observing the color change visually or by using a photometer or spectrophotometer to measure the precise change in the color of the reaction.

6.2.1.2   Using the Immunoassay Test Kit

In practice, if the antibodies are coated on the inside surface of the test tube, the sample and enzyme conjugate are combined directly in the test tube; if the antibodies are coated on magnetic or latex particles, a carefully measured amount of the solution that contains the coated particles is added to the test tube. Measured amounts of both the enzyme conjugate and the actual sample containing the target analyte are added to the test tube. The action is a timed incubation step; during the incubation, the analyte in the sample competes with the known amount of labeled antigen in the enzyme conjugate for the limited number of antibody binding sites. After incubation, the excess unbound enzyme conjugate is washed (removed) from the test tube. The amount of the enzyme conjugate that remains in the test tube is measured through the use of a colorimetric reaction. An enzyme substrate and a chromogen are added to the test tube to cause the formation of the color; this action is also a timed step, after which a solution is added to stop the formation of color. Because the amount of bound enzyme conjugate determines the amount of color, the amount of color is inversely proportional to the amount of analyte present in the sample. The color of the sample can be compared visually with a zero solution or blank for a "yes or no," or qualitative, result. A semiquantitative result can be obtained by using a differential photometer to compare the degree of light absorbance of a sample with that of a standard or standards. Finally, a quantitative result can be obtained by generating a calibration curve of absorbance compared with concentration by using a spectrophotometer, computer software, calibration standards, and a zero solution. The light absorbance of the sample can be read from the spectrophotometer and converted into a concentration.

For the most part, typical test kits have been designed to analyze batches of 10–20 samples at a time. Indeed, it is not efficient to analyze two or three samples simultaneously because of the several timed steps involved. Furthermore, calibration standards, and sometimes control samples, must be analyzed with each batch. Once the process has begun, all samples must be carried through the timed steps in equal fashion. That requirement also limits the number of samples that should be analyzed simultaneously—namely, that no more than 30 should be analyzed simultaneously. Consequently, it is difficult to maintain the time schedule if large number of samples are being analyzed. In any event, consistency is crucial for all the timed steps and the user's pipetting technique; indeed, novices may require some practice to perfect their pipetting techniques. To achieve the greatest possible precision, the pipetting of reagents must be consistent for each sample. The analyst must also be careful to avoid cross contamination; it is noteworthy that the procedure can be monitored for consistency and cross contamination by duplicating standards, analyzing control samples, and analyzing method blanks.

6.2.1.3   Scope of Application of Immunoassays

Immunoassay kits are available for a wide variety of petroleum compounds and other classes of organic contaminants (see, e.g., ITRC, 2013a; Van Emon and Gerlach, 1995)—including total petroleum hydrocarbons (TPH) as gasoline; TPH as diesel fuel; jet fuels; benzene, toluene, ethylbenzene, and xylenes (BTEX); chlorinated compounds (such as trichloroethylene, tetrachloroethylene/perchloroethylene (PCE), carbon tetrachloride, and chloroform); PAH; various individual pesticides and classes of pesticides; explosives and propellants; dioxins; and individual Aroclors (polychlorinated biphenyls [PCBs]) and mixtures of PCBs in soil and water. There also are limited immunoassay kits available for inorganic contaminants—including lead and mercury, as well as other metals. Some kits are designed for classes of compounds (e.g., PAHs)—and thus will provide a concentration of total PAH but will not indicate the concentrations of individual compounds; it is worth the mention here that test kits specifically for carcinogenic PAH also is available, among others. In general, kits for various analytes have in the past been relatively slow to come to market because it has been technically challenging and time consuming to develop compound-specific antibodies. At any rate, because immunoassay is versatile, sensitive, specific, and rather inexpensive, its applications continue to expand.

Traditionally, immunoassay test kits have primarily measured lighter aromatic petroleum fractions because straight-chain hydrocarbons did not lend themselves to eliciting immune system responses. Thus, the test kits for petroleum hydrocarbons did not perform as well in analyzing for heavy petroleum products with few aromatic components, such as motor oil or grease, or for highly degraded petroleum fuels, since the lighter aromatic constituents have been driven off—albeit some new test systems on the market seem to overcome these apparent limitations. Also, immunoassay test kits are available for numerous pesticides and herbicides, such as alachlor, aldicarb, triazine herbicides, 2,4-dichlorophenoxyacetic acid (2,4-D), organophosphates, carbofuran, chlordane, cyclodienes, carbamates, dichlorodiphenyl trichloroethane(DDT), endosulfan, lindane, parathion, toxaphene, and many more. Some test kits for pesticides respond to only one compound, while others respond to an entire class of compounds. Additionally, immunoassay test kits are available that can detect PCBs in soil, water, and wipe samples; quantitative test kits have been developed for specific Aroclors, while several semiquantitative kits will measure the overall concentration of a mixture of Aroclors. Other kits can detect pentachlorophenol (PCP), a contaminant of soil and water commonly found at wood treating sites; immunoassay test kits that analyze for PCP will also respond in various degrees to other chlorophenols. In general, immunoassays can be designed to test single or many samples simultaneously. Results from the test can be measured with an analytical instrument, a portable field instrument, or interpreted visually.

It is noteworthy that detection limits for immunoassay often are comparable to or even lower than those used for conventional analytical methods—albeit the detection limits for immunoassay analytical techniques will vary according to the test kit used, the target analytes, the sample matrix, and any interference. Indeed, kits are available that can achieve parts per million (ppm), parts per billion (ppb), and even parts per trillion (ppt) detection limits in water samples, depending on the manufacturer and the target analyte. Detection limits tend to be higher for soils because extraction is necessary. In some cases, the range of detection for a particular target analyte actually may be too low to be useful at a site, and it will be necessary to perform one or more dilutions; for example, if the action level for a contaminant is 50 ppm, then it may be necessary to perform a 1:10 dilution of samples to be analyzed by a kit that has a detection limit of 50 ppb and an upper range of 5 ppm.

6.2.1.4 Limitations of the Immunoassay Technology

While there are many advantages to immunoassay, it is important that the user understands its limitations, if the technology is to be used properly to generate data that meets the needs of the project. First of all, to select the correct immunoassay test kits and use them effectively, prior knowledge of analytes and potential interferences is necessary; that is, the contaminants present or suspected to be present must be known, so that the correct test kit can be selected. Obtaining that information may require more than one trip to a site and possibly the collection of confirmation samples for off-site analysis. Under most circumstances, it may become quite time- and labor-intensive, before definitive quantitative results can be acquired.

Now, regarding the nature of target contaminants, it must be mentioned here that the petroleum hydrocarbon test kits have traditionally not performed well for heavy petroleum products (such as motor oil or grease or for highly degraded petroleum fuels). Also, methanol is not the best extraction solvent for heavy hydrocarbons, and the immunoassay test kits primarily measure lighter aromatic constituents. Under a number of circumstances for the analytes identified previously, there is a potential for false negative outcomes.

Finally, it is noteworthy here that some of the reagents used in these tests require refrigeration; it may, therefore, be necessary to have a cooler or refrigerator on site—and this need may well be one of the few burdensome logistical requirements for the use of immunoassay test kits. Also, some of the reagents are sensitive to sunlight—and, thus, it is sometimes impractical to analyze such samples outdoors. Furthermore, if a number of similar compounds are present at a site, it may be difficult to quantify certain compounds accurately because interferences may create false positive results. Additionally, whereas analysis with some kits can be accomplished quickly, it can be time-consuming to perform analyses with other kits.

6.3 FIELD DATA COLLECTION AND ANALYSIS STRATEGIES IN PRACTICE

Traditionally, the characterization of environmental (e.g., soil) contamination has been accomplished by acquiring several environmental samples (e.g., surface and subsurface soil samples), sealing them in sample containers, and shipping them for laboratory analyses. When the analytes of interest are VOCs, the sample may be extracted with solvent upon arrival at the laboratory—and the extract is subsequently analyzed in a GC or a GC/MS. Whereas these procedures tend to accommodate broad-spectrum analysis and low levels of detection, they are also known to suffer severe limitations with respect to precision and accuracy. The failings of this approach may be attributable to both the heterogeneity of the sampled (e.g., soil) medium as well as the ease with which VOCs vaporize and escape from the sample during the activities preceding the analysis. Consequently, environmental (e.g., soil) sampling and analysis can grossly misrepresent VOC concentrations in affected media (e.g., soil)—generally by biasing it to the low end. In some cases, the underestimation can be so significant as to seriously affect the results of a site investigation effort—and ultimately the development of a risk management and corrective action plan or site restoration program for the problem situation. It is noteworthy, however, that whereas individual environmental (e.g., soil) samples may underestimate VOC levels, a suite of samples from across an impacted area or site is likely to reliably provide an indication if a contamination problem exists and indeed will likely characterize the nature of the plume adequately.

A variety of data collection and analysis protocols exist in the literature (e.g., Boulding, 1994; Boulding and Ginn, 2003; Byrnes, 1994, 2000, 2008; CCME, 1993, 1994; Csuros, 1994; Garrett, 1988; Hadley and Sedman, 1990; Holmes et al., 1993; ITRC, 2003a,e,f, 2004, 2007a,b, 2012a, 2013a; Keith, 1992; Millette and Hays, 1994; O'Shay and Hoddinott, 1994; Rong, 2018; Sara, 2003; Schulin et al., 1993; Thompson, 1992; USEPA, 1982, 1985a,d; Wilson et al., 1995) that may be used or adopted in the site investigation program. Regardless of the method of choice, the following general rules must invariably be satisfied for all sampling procedures (Holmes et al., 1993):

- The discrete/point samples must be representative of the conditions that exist within the area where the selected samples were taken.
- The samples must be of sufficient volume or quantity and must be collected frequently enough to allow for reproducibility of testing necessary to meet the desired investigation objective and as dictated by the methods of analyses utilized.
- The samples must be appropriately collected, packaged, stored, shipped, and controlled prior to analysis—that is, in a manner that safeguards against change in the particular constituents or properties to be examined.

In about every situation, data is generally collected at several stages of the site investigation, with initial data collection efforts usually limited to developing a general understanding of the problem situation. In areas where the contamination source is known, the sampling strategy should be targeted around that source. Typically, sampling points should be located at regular distances along lines radiating from the contaminant source; provisions should also be made in the investigation to collect additional samples of small isolated pockets of material that seem visually suspect.

Further to collecting and analyzing samples from the impacted area, background (or control site) samples are typically collected and evaluated to determine the possibility of a potential contaminated site problem contributing to off-site contamination levels in the vicinity of the problem site. Control locations—generally considered not to have been impacted by a given contaminated site problem—are indeed important to understanding the significance of environmental sampling and monitoring data. Such locations should generally have similar characteristics (i.e., identical in their physical and environmental settings) as the potentially contaminated area or locale under investigation. In other words, the background samples would not have been significantly influenced by contamination from the project site; however, these samples are obtained from an environmental matrix that has similar basic characteristics as the matrix at the project site, in order to provide a justifiable basis for comparison. Indeed, the background sampling results allow a technically valid scientific comparison to be made between environmental samples (suspected of containing chemical contaminants associated with the problem location) and control location or site samples (possibly containing only naturally low or anthropogenic levels of the same chemicals). Ultimately, background samples collected at control locations or sites are typically used to demonstrate whether or not a project location or site is truly contaminated.

6.3.1 Soil Vapor Analysis

Soil vapor analysis offers a rapid and inexpensive alternative to soil analysis, by providing more representative data on concentrations. Using field GC equipment and hydraulically driven probes, soil vapors can be extracted and analyzed at various depths and several locations in a reasonably short period of time. Since results are obtained on a real-time basis, probe locations can be altered in the field to optimize positioning and complete characterization in a single mobilization. This reduces site characterization time and costs while assuring better coverage. In fact, an evaluation of data from a variety of sites contaminated with VOCs by several investigators has determined that the more up-to-date soil vapor techniques perform very well when properly utilized.

Soil vapor data does indeed provide a far more representative outline of chemical contamination patterns because the probe draws a volume of vapor from the soil that lies radially distributed out from the sample location. In doing so, the probe blends vapors from a volume of soil and yields an average concentration for that sector of soil. In contrast, a single soil sample may be high or low, depending on the probability that it incorporates a representative blend of particle sizes and mineral or organic content for the same sector of soil. Since VOCs generally concentrate on organic soil matters and clay, the soil sample needs to contain average levels of organics and clay to be representative.

6.3.1.1   Soil Vapor Analysis as a General Diagnostic Tool for Organic Contamination

Soil vapor analyses are often used in the investigation of contaminated sites—especially to help determine the possibility for human exposures, the need for corrective measures, and the appropriate locations for monitoring wells and gas collection systems. For instance, on-site vapor screening of soil samples by surface probes, and also during drilling, can provide indicators of organic contamination.

In general, it is expected that gases produced at contaminated sites will tend to migrate through the paths of least resistance—and the presence of volatile contaminants or gas-producing materials can be determined by sampling the soil atmosphere within the ground. For example, organic vapor analyzer (OVA)/GC or GC/PID screening provides a relative measure of contamination by VOCs. Also, predictive models can be used to estimate the extent of gas migration from a suspected sub-surface source. This information can subsequently be used to identify apparent "hot spots" and to select soil samples for detailed chemical analyses. The vapor analyses on-site can also be helpful in selecting screened intervals for monitoring wells, along with the installation of a gas-monitoring well network (and in conjunction with sampling in buildings in the area).

6.3.2   The Investigation of Sites with NAPLs

At most sites that are contaminated from spills and releases of organic solvents and/or fuels, the aqueous contaminant plume is usually detected first, whereas the so-called "free product" or NAPL—which actually serves as the source for long-term contaminant releases—tends to be elusive and rarely adequately defined. During any site investigation activity, it is important to use some practical indicators of NAPL presence, in order to either confirm or rule out the actual existence of such free product. Such indicators may include (but not necessarily limited to) information on known historical NAPL releases and/or usage, visible solvent product in subsurface samples (e.g., based on the appearance of saturated samples), high chemical concentrations in soil samples, very high OVA/PID readings, NAPL being detected in monitoring wells (as free product), and chemical concentrations detected in groundwater in excess of the specific compound's solubility limit (indicating possible presence of free product in well).

It is worth the mention here that, until the NAPL zone or area is identified and controlled or remediated, complete site characterization and subsequent cleanup of a project site is virtually impossible. This is because solvent droplets or pools in the NAPL zone would continue to serve as long-term contaminant sources that sustain any dissolved and/or vapor plumes. Indeed, concern about NAPLs exists because of their persistence in the subsurface, as well as their ability to contaminate large volumes of waters and soils due to the generally poor soil attenuating capability (Yong et al., 1992). Invariably, a greater level of detail in the assessment of the transport and dissolution of NAPLs will aid in the development of cost-effective techniques for the control and cleanup of this type of contamination.

6.3.2.1   Soil Gas Analysis as a Noninvasive Method for NAPL Site Characterization

Many NAPLs have high vapor pressures and will volatilize in the soil vadose zone to form a vapor plume around an NAPL source. VOCs dissolved in groundwater can also volatilize at the capillary fringe into soil gas. Soil gas surveys may therefore be used to help identify contaminated zones attributable to NAPL releases, which can then be subjected to further investigations as necessary to support site restoration decisions.

Overall, it is believed that soil gas contamination will usually be dominated by volatilization and vapor transport from contaminant sources in the vadose zone rather than from groundwater and that the upward transport of VOCs to the vadose zone from groundwater is probably limited to dissolved contaminants that are very near to the water table or to free product floating atop the water table.

6.3.3 THE USE OF ENVIRONMENTAL IMMUNOCHEMICAL TECHNOLOGIES

The advent of field-portable immunoassay methods has revolutionized many field and laboratory analyses that are used in the investigation of contaminated site problems. As part of this journey, environmental immunoassays have been developed and evaluated for a number of analytes (in a variety matrices)—including, but not limited to, major classes of pesticides, PCBs, PAHs, pentachlorophenol (PCP), the BTEX compounds (i.e., benzene, toluene, ethylbenzene, and xylenes), and some inorganic chemicals (ITRC, 2013a; Van Emon and Gerlach, 1995). In any case, all immunoassays rely on the interaction between an antibody (i.e., a protein that selectively recognizes and binds to a target analyte or group of related analytes) and a target analyte. Immunoassay test kits essentially package antibodies, reagents, standards, and substrates in field-transportable units that are ready for use. It is noteworthy that the various immunoassay kits and methods are tailored to specific classes of environmental contaminants.

The use of immunoassays was the direct result of some recognized shortcomings of the GC in its application to certain chemical compounds; it is apparent that the technique holds great promise for the future—and especially when used in conjunction with existing methods such as MS. Immunoassays can indeed be used in the laboratory as well as in the field. Its use in the pre-analysis of environmental samples prior to a GC analysis, for instance, can identify the need for dilution—thereby saving an expensive electron capture detector from contamination and down-time (Van Emon and Gerlach, 1995). It is expected that field immunoassay analysis will gain increasing popularity and recognition, especially because of the highly portable equipment and minimal setup requirements. Also, sampling for immunoassay analysis will generally yield near-real-time data for potentially contaminated sites. Furthermore, immunoassays can quickly and reliably map contamination present at contaminated sites—thus allowing subsequent sampling design to be more credible; this will then allow better use of in-laboratory analytical instrumentation and results.

Although designed for field use, most immunoassay kits usually are used in a sample trailer, mobile laboratory, or other fixed locations because of the amount of equipment required and the precision necessary to produce useful outcomes. In any case, to facilitate the processes involved, the manufacturer generally provides step-by-step instructions for the analytical method to be used. Indeed, most immunoassay test kits follow a "cook book" procedure—and, consequently, are designed to allow even a novice to use them proficiently. Still, some training may be necessary in the use of some test kits—particularly those intended for quantitative analysis.

6.4 PRACTICAL IMPLEMENTATION OF A FIELD INVESTIGATION PROGRAM

Site investigations are conducted in order to characterize site conditions—and environmental samples are typically collected from the site during such activities; samples collected will generally be submitted to a certified analytical laboratory for analysis. The sample characterization efforts typically consider detailed information in relation to the following:

- Field operations (including the role of individuals, chain-of-custody procedures, maintaining field logbook, and site monitoring)
- Sampling locations and rationale
- Field sampling and mapping
- Field quality control samples
- Sample decontamination procedures
- Analytical requirements and sample handling
- Sample delivery
- Data compilation and analyses
- Summary site evaluations

Essentially, the characterization of environmental samples helps determine the specific type(s) of contaminants, their abundance/concentrations, the lateral and vertical extent of contamination, the volume of materials involved, and the background contaminant levels for native soils and water resources in the vicinity of the project site.

6.4.1   An Initial Site Inspection

Typically, the goals of the initial site inspection include the completion of several tasks—as indicated in Box 6.1. All available information should be carefully reviewed and evaluated to provide the foundation for executing additional on-site activities. The preliminary data receive confirmation from observations made during site visits. The types of information generated serve as a useful database for project scoping. Ultimately, the review and initial site visit are used in a preliminary interpretation of site conditions.

It is noteworthy that geophysical surveys, limited field screening, or limited field analyses may be performed during an initial site inspection. These types of preliminary screening activities may help determine the variability of the environmental media, provide general interest background information, or determine if the site conditions have changed in comparison to what may have been reported in previous investigations.

BOX 6.1   TASKS REQUIRED OF AN INITIAL SITE INSPECTION

- Utilizing field analytical procedures, compile data on volatile chemical contaminants, radioactivity, and explosivity hazards in order to determine appropriate health and safety level requirements.
- Determine if any site condition could pose an imminent danger to public health and safety.
- Confirm, insofar as possible, all relevant information contained in previous documents and record any apparent discrepancies and/or any observable data that may be missing in previous documents.
- Update site conditions if undocumented changes have occurred.
- Investigate and inventory all possible off-site sources that may be contributing to contamination at the project site.
- Obtain information on location of access routes, sampling points, and site organizational requirements for the field investigation.

6.4.2   Sampling and Sample Handling Procedures

The collection of representative samples generally involves different procedures for unique situations; several sampling methods of approach that can be used in various types of situations are discussed elsewhere in the literature (e.g., CCME, 1994; CDHS, 1990; Keith, 1988, 1991; USEPA, 1985a,d, 1987b,c,e,f). Anyhow, in every situation, all sampling equipment (which should be constructed of inert materials) would be cleaned using a nonphosphate detergent, a tap water rinse, and a final rinse with distilled water prior to a sampling activity; decontamination of equipment is indeed necessary so that sample results do not show false positives. In the end, decontamination water generated from the site activities (e.g., during decontamination of hand auger and soil sampling equipment) usually is transferred into containers and treated as an IDW (and that are sampled for analysis).

Now, it cannot be emphasized enough that all sampling should be conducted in a manner that maintains sample integrity and encompasses adequate quality assurance and control. Among other things, specific sample locations should be chosen such that representative samples can be collected.

Also, samples should be collected from locations with visual observations of surface contamination so that possible worst-case conditions may be identified. Meanwhile, the use of field blanks and standards, as well as spiked samples can account for changes in samples that occur after sample collection. Broadly speaking, the following general statements apply to most sampling efforts:

- Refrigeration and protection of samples should minimize the chemical alteration of samples prior to analysis.
- Field analyses of samples will effectively avoid biases in determinations of parameters/constituents that do not store well (e.g., gases, alkalinity, and pH).
- Field blanks and standards will permit the correction of analytical results for changes that may occur after sample collection (i.e., during preservation, storage, and transport). Field blanks and standards enable quantitative correction for biases due to systematic errors arising from handling, storage, transport and laboratory procedures.
- Spiked samples and blind controls provide the means to correct combined sampling and analytical accuracy or recoveries for the actual conditions to which the samples have been exposed.

In any case, all sampling should be conducted in a manner that maintains sample integrity and should meet the requisite QA/QC criteria. The appropriate technical standards for sampling and sample-handling procedures can be found in the literature elsewhere (e.g., CCME, 1994; CDHS, 1990; Keith, 1988, 1991; USEPA, 1985a,d).

In general, sampling equipment should be constructed of inert materials. Soil samples collected are placed in resealable plastic bags; fluid samples are placed in air-tight glass or plastic containers. When samples are to be analyzed for organic constituents, glass containers are required. The samples are then labeled with an indelible marker. Each sample bag or container is labeled with a sample identification number, sample depth (where applicable), sample location, date and time of sample collection, preservation used (if any), and possibly a project number and the sampler's initials. A chain-of-custody form listing the sample number, date and time of sample collection, analyses requested, a project number, and persons responsible for handling the samples is then completed. Samples are generally kept on ice prior to and during transport/shipment to a certified or credible laboratory for analysis; completed chain-of-custody records should accompany all samples going to the laboratory.

6.4.2.1   General Nature of Sample Collection Tools/Equipment

A wide variety of unique sampling tools and equipment would typically be utilized in field investigation activities. For instance, simple core barrels are often used to collect soil for *ex situ* inspection and analysis—representing a "non-standard" practice. Of even greater interest insofar as sampling equipment is concerned are pumps, water extractors, and soil gas collectors that are particularly vital equipment for environmental site assessment operations. Pumps are used to carry water, free product, or vapors to the subsurface for collection, analysis, or treatment. It is notable that there are a variety of pumps available for case-specific situations—especially since conditions can vary considerable between sites and wells. Vadose zone water can be drawn from between soil pores above the groundwater table by vacuum extractors or by sponges that are left in the subsurface. Soil gas can be collected by passive charcoal vapor adsorbents that are left in the soil or by active soil gas vacuum pumping methods. Indeed, the use of a multitude of modified methods and tools tends to be the norm during the investigation or characterization of the usually wide-ranging landscape of contaminated site problems often encountered in practice.

6.4.3   Implementing the Field Data Collection Activities

A preliminary identification of the nature of site contaminants, the chemical release potentials, and also the likely migration and exposure pathways should be made very early in the site characterization—because these are crucial to decisions on the number, type, and location of samples to be collected. In fact, the nature of chemicals believed or known to be present at a contaminated site may dictate the areas and extent of environmental media to be sampled. For instance, in the design of a sampling program, if it is believed that the contaminants of concern are relatively immobile (e.g., most metals), then sampling may initially focus on soils in the vicinity of the suspected source of contaminant releases. On the other hand, the sampling design for a site believed to be contaminated by more mobile compounds (e.g., organic solvents) should necessarily take account of the fact that contamination may already have migrated into groundwater systems and/or moved significant distances away from the original source area(s). Consequently, knowledge on the type of contaminants will generally help focus more attention on the specific media most likely to have been impacted. Similar decisions as mentioned earlier will typically have to be made regarding the choice of analytical protocols. For instance, due to the differences in the relative toxicities of the different species of some chemicals (e.g., chromium may exist as trivalent chromium, Cr^{+3}, or as the more toxic hexavalent chromium, Cr^{+6}), chemical speciation to differentiate between the various forms of the same chemical of potential concern present at a contaminated site may be required in the design of some analytical protocols.

Finally, samples from various media should ideally be collected in a manner that accounts for temporal factors and weather conditions. If seasonal fluctuations cannot be characterized in the investigation, details of meteorological, seasonal, and climatic conditions during sampling must be well documented. In fact, the ideal sampling strategy will incorporate a full annual sampling cycle. If this strategy cannot be accommodated in an investigation, at least two sampling events should be considered which take place during opposite seasonal extremes (such as high/low flow and high/low recharge). It is noteworthy that, regardless of the medium sampled, data variability problems may arise from temporal and spatial variations in field data—that is, sample composition may vary depending on the time of the year and weather conditions when the sample is collected.

6.5   OPTIMIZATION STRATEGY FOR THE OVERALL FIELD INVESTIGATION PROCESS IN PRACTICE

Field studies are necessarily iterative processes—usually beginning with activities to determine both surface and subsurface conditions at a project site. Information collected from a review of the historical records of a potentially contaminated site, and also the current operating conditions, can be used to develop a systematic approach to the data collection and analysis process. In general, a phased approach to environmental sampling usually encourages the early identification of key data requirements in the site investigation process. This ensures that the data collection effort is always directed at providing adequate information that meets the data quantity and quality requirements of the study. As a basic understanding of the site characteristics is achieved, subsequent data collection efforts focus on identifying and filling in any remaining data gaps from a previous phase of the site investigation. Any additionally acquired data should be such as to further improve the understanding of site characteristics and also to consolidate information necessary to effectively manage the contaminated site problem. In this manner, the overall site investigation effort can be continually re-scoped to minimize the collection of unnecessary data and to maximize the quality of data acquired.

Finally, in order to successfully and cost-efficiently complete environmental site investigation and remediation projects, the best tools available should be used to instill efficiency and reliability in the efforts involved. For instance, the ability of field analytical systems to reduce costs and yet raise the level of quality of the available decision-making data should become a recognized metric. On the whole, the field data gathering process should provide a logical, objective, and quantitative balance between the time and resources available for collecting the data and the quality of data, based on the intended use of such data.

7 Contaminant Fate and Behavior Assessment

Contaminants released into the environment are controlled by a complex set of processes—consisting of transport, transformation, degradation and decay, cross-media transfers, and/or biological uptake and bioaccumulation. Environmental fate and transport analyses offer a way to assess the movement of chemicals between environmental compartments—further to the prediction of the long-term fate of such chemicals in the environment vis-à-vis potential receptor exposures. Indeed, once a chemical is suspected or determined to present a potential health or environmental hazard, then the first concern is one of the likelihood for and degree of exposure. The fate of chemical substances released into the environment forms a very important basis for evaluating the exposure of potential receptors to toxic chemicals. Consequently, the processes that affect the fate and transport of contaminants are identified as an important part of contaminated site assessment/characterization and restoration programs. In fact, a good understanding of the chemical fate and behavior is quite important—so as to be able to properly characterize the potential risks associated with an environmental contamination and to further develop appropriate risk management and/or remedial action plans for a contaminated site problem. The processes and phenomena that affect the fate and behavior of chemicals encountered in the environment should therefore be recognized as an important part of any contaminated site characterization, restoration, and/or risk management program. This chapter identifies and discusses several key phenomena that tend to significantly influence the fate and behavior of environmental contaminants, together with the important factors affecting the processes involved.

7.1 IMPORTANT PROPERTIES AND PARAMETERS AFFECTING CONTAMINANT FATE AND TRANSPORT IN THE ENVIRONMENT

As pollutants are released into various environmental media, several factors contribute to their uptake, transformation, and migration/transport from one environmental matrix into another or their phase change from one physical state into another. For example, in a typical groundwater system associated with a contaminated site problem, the solutes in the porous media will move with the mean velocity of the solvent by advective mechanism. In addition, other mechanisms governing the spread of contaminants include hydraulic dispersion and molecular diffusion (which is caused by the random Brownian motion of molecules in solution that occurs whether the solution in the porous media is stationary or has an average motion). Furthermore, the transport and concentration of the solute(s) are affected by reversible ion exchange with soil grains, the chemical degeneration with other constituents, fluid compression and expansion, and in the case of radioactive materials, by the radioactive decay. In general, examination of a contaminant's physical and chemical properties will often allow an estimation of its degree of environmental partitioning, migration, and/or attenuation. Qualitative analysis of the fate of a chemical can also be made by analogy with other chemicals whose fate are well documented; that is, if the chemical under investigation is structurally similar to a previously well-studied one, some parallel can be drawn to the environmental fate of the analogue. In addition, several site characteristics—including the amount of ambient moisture, humidity levels, temperatures, and wind speed; geologic, hydrologic, pedologic, and watershed characteristics; topographic features of the site and vicinity; vegetative cover of site and surrounding area; and land-use characteristics—may influence the environmental fate of chemicals. Other factors such as soil temperature, soil moisture content, initial contaminant concentration in the impacted media, and media pH may additionally affect the release of a chemical constituent from the environmental matrix in which it is found.

A number of important physical and chemical properties, processes, and parameters affecting the environmental fate and/or cross-media transfers of environmental contaminants are briefly annotated in the following sections—with further detailed discussions offered in the literature elsewhere (e.g., CDHS, 1986; Devinny et al., 1990; Evans, 1989; Haque, 2017; Hemond and Fechner, 1994; Lindsay, 1979; Lyman et al., 1990; Mahmood and Sims, 1986; Mansour, 1993; Neely, 1980; Sáez and Baygents, 2014; Samiullah, 1990; Selim, 2017a,b; Swann and Eschenroeder, 1983; Thibodeaux, 1979, 1996; USEPA, 1985a,c, 1989k; Yong et al., 1992).

7.1.1 PHYSICAL STATE

Contaminants present in the environment may exist in any or all of three major physical states—viz. solid (e.g., solids adsorbed onto soils at a contaminated site), liquid (e.g., free product or dissolved contaminants at a contaminated site), and vapor (e.g., vapor phase in the soil vadose zones of a contaminated site) states. Contaminants in the solid phase are generally less susceptible to release and migration than the fluids. However, certain processes (such as leaching, erosion and/or runoff, and physical transport of chemical constituents) can act as significant release mechanisms, irrespective of the physical state of a contaminant.

7.1.2 WATER SOLUBILITY

The solubility of a chemical in water is the maximum amount of the chemical that will dissolve in pure water at a specified temperature. Solubility is an important factor affecting a chemical constituent's release and subsequent migration and fate in the surface water and groundwater environments. In fact, among the various parameters affecting the fate and transport of organic chemicals in the environment, water solubility is one of the most important, especially with regards to hydrophilic compounds.

Typically, solubility affects mobility, leachability, availability for biodegradation, and the ultimate fate of a given constituent. In general, highly soluble chemicals are easily and quickly distributed by the hydrologic system. Such chemicals tend to have relatively low adsorption coefficients for soils and sediments and also relatively low bioconcentration factors (BCFs) in aquatic biota. Furthermore, they tend to be more readily biodegradable. Substances which are more soluble are more likely to desorb from soils and less likely to volatilize from water.

In combination with vapor pressure, water solubility yields a chemical's *Henry's Law Constant*—which determines whether or not the chemical will volatilize from water into air (see Section 7.1.5). In combination with the chemical's solubility in fats (obtained from the octanol–water partition coefficient), water solubility predicts whether or not a chemical will tend to concentrate in living organisms and also whether a soil contaminant will remain bound to soil or will leach from soil to contaminate groundwater or surface water bodies (see Section 7.1.6).

7.1.3 DIFFUSION

Diffusivity describes the movement of a molecule in a liquid or gas medium as a result of differences in concentrations. Diffusive processes create mass spreading due to molecular diffusion, in response to concentration gradients. Thus, diffusion coefficients are used to describe the movement of a molecule in a liquid or gas medium as a result of differences in concentration; it can also be used to calculate the dispersive component of chemical transport. In general, the higher the diffusivity, the more likely a chemical is to move in response to concentration gradients.

7.1.4 Dispersion

Dispersive processes create mass mixing due to system heterogeneities (e.g., velocity variations). Consequently, for example, as a pulse of contaminant plume migrates through a soil matrix, the peaks in concentration become decreased through spreading. Dispersion is indeed an important attenuation mechanism that results in the dilution of a contaminant; the degree of spreading or dilution is proportional to the size of the dispersion coefficients.

7.1.5 Volatilization

Volatilization is the process by which a chemical compound evaporates from one environmental compartment into the vapor phase. The volatilization of chemicals is indeed a very important mass-transfer process. The transfer process from the source (e.g., water body, sediments, or soils) into the atmosphere is dependent on the physical and chemical properties of the compound in question, the presence of other pollutants, the physical properties of the source media, and the atmospheric conditions.

Knowledge of volatilization rates is important in the determination of the amount of chemicals entering the atmosphere and the change of pollutant concentrations in the source media. Volatility is therefore considered a very important parameter for chemical hazard assessments. Some fundamental measures of a chemical's volatility or volatilization rate are noted in the following sections.

7.1.5.1 Boiling Point

Boiling point (BP) is the temperature at which the vapor pressure of a liquid is equal to the atmospheric pressure on the liquid. At this temperature, a substance transforms from the liquid into a vapor phase. Indeed, besides being an indicator of the physical state of a chemical, the BP also provides an indication of its volatility. Other physical properties, such as critical temperature and latent heat (or enthalpy) of vaporization may be predicted by use of a chemical's normal BP as an input.

7.1.5.2 Henry's Law Constant

Henry's Law Constant (H) provides a measure of the extent of chemical partitioning between air and water at equilibrium. It indicates the relative tendency of a constituent to volatilize from aqueous solution into the atmosphere, based on the competition between its vapor pressure and water solubility.

This parameter is important to determining the potential for cross-media transport into air. In general, contaminants with low Henry's law constant values will tend to favor the aqueous phase and will therefore volatilize into the atmosphere more slowly than would constituents with high values. As a general guideline, H values in the range of 10^{-7} to 10^{-5} (atm-m^3/mol) represent low volatilization; H between 10^{-5} and 10^{-3} (atm-m^3/mol) means volatilization is not rapid but possibly significant; and $H > 10^{-3}$ (atm-m^3/mol) implies volatilization is rapid. The variation in H between chemicals is indeed quite extensive.

7.1.5.3 Vapor Pressure

Vapor pressure is the pressure exerted by a chemical vapor in equilibrium with its solid or liquid form at any given temperature. It is a relative measure of the volatility of a chemical in its pure state and is an important determinant of the rate of volatilization. The vapor pressure of a chemical can be used to calculate the rate of volatilization of a pure substance from a surface or to estimate a Henry's law constant for chemicals with low water solubility.

In general, the higher the vapor pressure, the more volatile a chemical compound and therefore the more likely the chemical is to exist in significant quantities in a gaseous state. Thus, constituents with high vapor pressure are more likely to migrate from soil and groundwater to be transported into air.

7.1.6 PARTITIONING AND THE PARTITION COEFFICIENTS

The partitioning of a chemical between several phases within a variety of environmental matrices is considered a very important fate and behavior property for contaminant migration in the environment. Indeed, the *partition coefficient* (also called the *distribution coefficient*) is one of the most important parameters used in estimating the migration potential of contaminants present in aqueous solutions in contact with surface, subsurface, and suspended solids (USEPA, 1999j). The *partition coefficient* is a measure of the distribution of a given compound in two phases—and is expressed as a concentration ratio. Some fundamental measures of the partitioning phenomena are discussed in the subsections that follow.

7.1.6.1 Water/Air Partition Coefficient

The water/air partition coefficient (K_w) relates the distribution of a chemical between water and air. It consists of an expression that is equivalent to the reciprocal of Henry's law constant (H), that is,

$$K_w = \frac{C_{water}}{C_{air}} = \frac{1}{H} \tag{7.1}$$

where C_{air} is the concentration of the chemical in air (expressed in units of µg/L) and C_{water} is the concentration of the chemical in water (in µg/L).

7.1.6.2 Octanol/Water Partition Coefficient

The octanol/water partition coefficient (K_{ow}) is defined as the ratio of a chemical's concentration in the octanol phase (organic) to its concentration in the aqueous phase of a two-phase octanol/water system, represented by

$$K_{ow} = \frac{\text{concentration in octanol phase}}{\text{concentration in aqueous phase}} \tag{7.2}$$

This dimensionless parameter provides a measure of the extent of chemical partitioning between water and octanol at equilibrium. It has indeed become a particularly important parameter in studies of the environmental fate of organic chemicals.

In general, K_{ow} can be used to predict the magnitude of an organic constituent's tendency to partition between the aqueous and organic phases of a two-phase system, such as surface water and aquatic organisms. For instance, the higher the value of a K_{ow}, the greater would be the tendency of an organic constituent to adsorb to soil or waste matrices that contain appreciable organic carbon or to accumulate in biota. Indeed, this parameter has been found to relate to water solubility, soil/sediment adsorption coefficients, and bioaccumulation factors for aquatic life.

Broadly speaking, chemicals with low K_{ow} (<10) values may be considered relatively hydrophilic, whereas those with high K_{ow} (>10,000) values are very hydrophobic. Thus, the greater the K_{ow}, the more likely a chemical is to partition to octanol than to remain in water. In fact, high K_{ow} values are generally indicative of a chemical's ability to accumulate in fatty tissues and therefore bioaccumulate in the food chain. It is also a key variable in the estimation of skin permeability for chemical constituents. All in all, the hydrophilic chemicals tend to have high water solubilities, small soil or sediment adsorption coefficients, and small bioaccumulation factors for aquatic life.

7.1.6.3 Organic Carbon Adsorption Coefficient

The sorption characteristics of a chemical may be normalized to obtain a sorption constant based on organic carbon that is essentially independent of any soil material. The organic carbon adsorption coefficient (K_{oc}) provides a measure of the extent of partitioning of a chemical constituent between soil or sediment organic carbon and water at equilibrium.

Also called the organic carbon partition coefficient, K_{oc} is a measure of the tendency for organics to be adsorbed by soil and sediment and is expressed by the following relationship:

$$K_{oc}\left[\text{mL/g}\right] = \frac{\text{mg chemical adsorbed per g weight of soil or sediment organic carbon}}{\text{mg chemical dissolved per mL of water}} \qquad (7.3)$$

As an example, the extent to which an organic constituent partitions between the solid and solution phases of a saturated or unsaturated soil, or between runoff water and sediment, is determined by the physical and chemical properties of both the constituent and the soil (or sediment). It is notable that the K_{oc} is chemical-specific and largely independent of the soil or sediment properties. The tendency of a constituent to be adsorbed to soil is, however, dependent on its properties and also on the organic carbon content of the soil or sediment.

Values of K_{oc} typically range from 1 to 10^7—and the higher the K_{oc}, the more likely a chemical is to bind to soil or sediment than to remain in water. In other words, constituents with a high K_{oc} have a tendency to partition to the soil or sediment. In fact, this value is also a measure of the hydrophobicity of a chemical; in general, the more highly sorbed, the more hydrophobic (or the less hydrophilic) a substance.

7.1.6.4 Soil–Water Partition Coefficient

The mobility of contaminants in soil depends not only on properties related to the physical structure of the soil but also on the extent to which the soil material will retain, or adsorb, the pollutant constituents. The extent to which a constituent is adsorbed depends on the physicochemical properties of the chemical constituent and of the soil. The sorptive capacity must therefore be determined with reference to a particular constituent and soil pair.

The soil–water partition coefficient (K_d), also called the soil/water distribution coefficient, is generally used to quantify soil sorption. K_d is the ratio of the adsorbed contaminant concentration to the dissolved concentration at equilibrium, and for most environmental concentrations, it can be approximated by the following relationship:

$$K_d\left[\text{mL/g}\right] = \frac{\text{concentration of adsorbed chemical in soil (mg chemical per g soil)}}{\text{concentration of chemical in solution in water (mg chemical per mL water)}} \qquad (7.4)$$

Invariably, the distribution of a chemical between water and an adjoining soil or sediment may be described by this equilibrium expression that relates the amount of chemical sorbed to soil or sediment to the amount in water at equilibrium. As an example, it is notable that the K_d parameter is very important in estimating the potential for the adsorption of dissolved contaminants in contact with soils or similar geologic materials. K_d provides a soil- or sediment-specific measure of the extent of chemical partitioning between soil or sediment and water, unadjusted for dependence on organic carbon. On this basis, K_d describes the sorptive capacity of the soil and allows estimation of the concentration in one medium, given the concentration in the adjoining medium. For hydrophobic contaminants:

$$K_d = f_{oc}K_{oc} \qquad (7.5)$$

where: f_{oc} is the fraction of organic carbon in the soil.

In general, the higher the value of K_d, the less mobile is a contaminant; this is because, for large values of K_d, most of the chemical remains stationary and attached to soil particles due to the high degree of sorption. Thus, the higher the K_d, the more likely a chemical is to bind to soil or sediment than to remain in water. (By the way, to minimize the degree of uncertainties in contaminant behavior assessment and risk computations, site-specific K_d values should generally be utilized whenever possible.)

7.1.6.5 Bioconcentration Factor

The BCF is the ratio of the concentration of a chemical constituent in an organism or whole body (e.g., a fish) or specific tissue (e.g., fat) to the concentration in its surrounding medium (e.g., water) at equilibrium, expressed as follows:

$$
\begin{aligned}
\mathrm{BCF} &= \frac{[\text{concentration in biota}]}{[\text{concentration in surrounding medium}]} \\[6pt]
&= \frac{[\text{mg} - \text{chemical per g} - \text{biota (e.g., fish)}]}{[\text{mg} - \text{chemical per mL} - \text{medium (e.g., water)}]}
\end{aligned}
\tag{7.6}
$$

As a general example, the BCF indicates the degree to which a chemical residue may accumulate in aquatic organisms, coincident with ambient concentrations of the chemical in water; it is a measure of the tendency of a chemical in water to accumulate in the tissue of an organism. In this regard, the concentration of the chemical in the edible portion of the organism's tissue can be estimated by multiplying the concentration of the chemical in surface water by the fish BCF for that chemical. Thus, the average concentration in fish or biota is given by:

$$
C_{\text{fish-biota}}\,(\mu g/kg) = C_{\text{water}}\,(\mu g/L) \times \mathrm{BCF}
\tag{7.7}
$$

where C_{water} is the concentration in water. This parameter is indeed a very important determinant for human exposure to chemicals via ingestion of aquatic foods. The partitioning of a chemical between water and biota (e.g., fish) also gives a measure of the hydrophobicity of the chemical.

Values of BCF typically range from 1 to over 10^6. In general, constituents that exhibit a BCF greater than unity are potentially bioaccumulative, but those exhibiting a BCF greater than 100 trigger the greatest concern (USEPA, 1987a). Ranges of BCFs for various constituents and organisms can be used to predict the potential for bioaccumulation and therefore to determine whether, as an example, the sampling of biota is really a necessary part of an environmental characterization program. By and large, the accumulation of chemicals in aquatic organisms is indeed a growing concern as a significant source of environmental and health hazard.

7.1.7 Sorption and the Retardation Factors

Sorption, which collectively accounts for both adsorption and absorption, is the partitioning of a chemical constituent between the solution and solid phases. In this partitioning process, molecules of the dissolved constituents leave the liquid phase and attach to the solid phase; this partitioning continues until a state of equilibrium is reached. As an example of its real world application, the practical result of this type of partitioning process gives rise to a phenomenon called *retardation*— in which the effective velocity of the chemical constituents in a groundwater system is less than that of a "pure" groundwater flow.

Retardation is the chemical-specific, dynamic process of adsorption to and desorption from aquifer materials. It is typically characterized by a parameter called the *retardation factor* or *retardation coefficient*. In the assessment of the environmental fate and transport properties of chemical contaminants, reversible equilibrium and controlled sorption may be simulated by the use of the retardation factor or coefficient.

7.1.7.1 Retardation Factor

A contaminant retardation factor is the parameter commonly used in environmental transport models to describe the chemical interaction between the contaminant and typically the "surrounding"/"embedding" geological materials (such as soils, sediments, and similar geological formations). It includes processes such as surface adsorption, absorption into the soil structure or geological materials matrix, precipitation, and the physical filtration of colloids (USEPA, 1999). For the most part, it is used to describe the rate of contaminant transport relative to that of groundwater.

Mathematically, the retardation factor, R_f, is defined as the ratio of $[C_{mobile} + C_{sorbed}]$ to C_{mobile}, where C_{mobile} and C_{sorbed} are the mobile and sorbed chemical concentrations, respectively. Thus,

$$R_f = 1 + \frac{C_{sorbed}}{C_{mobile}} \tag{7.8}$$

It is noteworthy that, the chemical retardation term does not equal unity (1) when the solute interacts with the soil. Indeed, the retardation term is almost always greater than unity (1) due to solute sorption to soils—albeit there are extremely rare cases when the retardation factor is actually less than 1, and such circumstances are thought to be caused by "anion exclusion" (USEPA, 1999).

R_f can be calculated for a contaminant as a function of the chemical's soil–water partition coefficient (K_d) and also the bulk density (β) and porosity (n) of the medium through which the contaminant is moving. Typically, the retardation factors are calculated for linear sorption, in accordance with the following relationship:

$$R_f = 1 + \frac{\beta K_d}{n} = 1 + \frac{\beta K_{oc} f_{oc}}{n} \tag{7.9}$$

where $K_d = K_{oc} \times f_{oc}$, and f_{oc} is the organic carbon fraction.

It is notable that the velocity of a contaminant is one of the most important variables in any groundwater quality modeling study. Sorption affects the solute seepage velocity through retardation, which is a function of R_f. Estimating R_f is, therefore, very important if solute transport is to be adequately elucidated. In the aquifer system, the retardation factor gives a measure of how fast a compound moves in relation to groundwater (Hemond and Fechner, 1994; Nyer, 1993; USEPA, 1999). Defined in terms of groundwater and solute concentrations, therefore,

$$R_f = \frac{\text{groundwater velocity } [v]}{\text{solute velocity } [v^*]} \tag{7.10}$$

As an illustrative example, a retardation factor of two (2) indicates that the specific compound is traveling at one-half the groundwater flow rate, and a retardation factor of five (5) means that a plume of the dissolved compound will advance only one-fifth as fast as the groundwater parcel. In consequence, this will usually become a very important parameter in the design of groundwater remediation systems; in particular, sorption can have major effects on "pump-and-treat" cleanup times and volumes of water to be removed from a contaminated aquifer system. That is, the retardation factor may be used to determine how much the cleanup time might increase. Thus, if it would have taken 1 year to clean up a site under a "no-sorption" scenario, then this is going to take 2 years (for R_f of 2) or 5 years (for R_f of 5) due to the sorption effects. Similarly, if it would have taken 10 years to clean up the site under a "no-sorption" scenario, then this is now going to take 20 years (for R_f of 2) or 50 years (for R_f of 5) due to the sorption effects.

7.1.7.2 Sorption

Under equilibrium conditions, a sorbing solute will partition between the liquid and solid phases according to the value of R_f. The fraction of the total contaminant mass contained in an aquifer that is dissolved in the solution phase, $F_{dissolved}$, and the sorbed fraction, F_{sorbed}, can be calculated as follows:

$$F_{dissolved} = \frac{1}{R_f} \tag{7.11}$$

$$F_{\text{sorbed}} = 1 - \left[\frac{1}{R_{\text{f}}}\right] \qquad (7.12)$$

In general, if a compound is strongly adsorbed, then it also means this particular compound will be highly retarded.

7.1.8 Degradation

Degradation, whether biological, physical, or chemical, is often reported in the literature as a half-life—and this is typically measured in days. It is generally expressed as the time it takes for one-half of a given quantity of a compound to be degraded. A number of important measures of the degradation phenomena are described in the following subsections.

7.1.8.1 Chemical Half-Lives

Half-lives are used as measures of persistence, since they indicate how long a chemical will remain in various environmental media; long half-lives (e.g., greater than a month or a year) are characteristic of persistent constituents. In general, media-specific half-lives provide a relative measure of the persistence of a chemical in a given medium, although actual values can vary greatly depending on case-specific conditions. For example, the absence of certain microorganisms at a site, or the number of microorganisms, can influence the rate of biodegradation and, therefore, the half-life for specific compounds. As such, half-life values should be used only as a general indication of a chemical's persistence in the environment. On the whole, however, the higher the half-life value, the more persistent a chemical is likely to be.

7.1.8.2 Biodegradation

Biodegradation is one of the most important environmental processes affecting the breakdown of organic compounds. It results from the enzyme-catalyzed transformation of organic constituents, primarily by microorganisms. As a result of biodegradation, the ultimate fate of a constituent introduced into several environmental systems (e.g., soil and water) may be any compound other than the parent compound that was originally released into the environment. Biodegradation potential should therefore be carefully evaluated in the design of environmental monitoring programs—in particular for contaminated site assessment programs. It is noteworthy that, biological degradation may also initiate other chemical reactions, such as oxygen depletion in microbial degradation processes, creating anaerobic conditions, and the initiation of redox-potential-related reactions.

7.1.8.3 Chemical Degradation

Similar to photodegradation (see Section 7.1.8.4 below) and biodegradation (see preceding section), chemical degradation—primarily through hydrolysis and oxidation/reduction (redox) reactions—can also act to change chemical constituent species from what the parent compound used to be when it was first introduced into the environment. For instance, oxidation may occur as a result of chemical oxidants being formed during photochemical processes in natural waters. Similarly, reduction of constituents may take place in some surface water environments (primarily those with low oxygen levels). Hydrolysis of organics usually results in the introduction of a hydroxyl group (–OH) into a constituent structure; hydrated metal ions (particularly those with a valence ≥3) tend to form ions in aqueous solution, thereby enhancing species solubility.

7.1.8.4 Photolysis

Photolysis (or photodegradation) can be an important dissipative mechanism for specific chemical constituents in the environment. Similar to biodegradation (noted previously), photolysis may cause the ultimate fate of a constituent introduced into an environmental system (e.g., surface water, and soil) to be different from the constituent originally released. Hence, photodegradation potential should be carefully evaluated in designing sampling and analysis plans, as well as environmental monitoring programs.

7.1.9 Miscellaneous Equilibrium Constants from Speciation

Several equilibrium constants will usually be important predictors of a compound's chemical state in solution—and such parameters may indeed play key roles in appraising the fate and behavior attributes of chemicals often encountered in human and other biotic environments. For example, a constituent that is dissociated (ionized) in solution will be more soluble and, therefore, more likely to be released into the environment—and thence more likely to migrate in a surface water body, etc.; it is also noteworthy that ionic metallic species are likely to have a tendency to bind to particulate matter, if present in a surface water body—and furthermore, to settle out to the sediment over time and distance.

In general, many inorganic constituents, such as heavy metals and mineral acids, can occur as different ionized species depending on the ambient pH—and organic acids, such as the phenolic compounds, do indeed exhibit similar behavior. In the end, heavy metals are removed by ion exchange reactions during a natural attenuation, whereas trace organics are removed primarily by adsorption. That said, it is notable that metallic species also generally exhibit bioaccumulative properties; consequently, when metallic species are present in the environment, a study design that incorporates both sediment and biota sampling would typically be appropriate—perhaps even indispensable.

7.2 FACTORS AFFECTING CONTAMINANT MIGRATION

The degree of chemical migration from a contaminated site or an environmental compartment generally depends on both the physical and chemical characteristics of the individual constituents, as well as on the physical, chemical, and biological characteristics of the affected media. For example, physical characteristics of the contaminants such as solubility and volatility influence the rate at which chemicals may leach into groundwater or escape into the atmosphere. The characteristics of the environmental setting (such as the geologic or hydrogeologic features) also affect the rate of contaminant migration. In addition, under various environmental conditions, some chemicals will readily degrade to substances of relatively low toxicity whereas other chemicals may undergo complex reactions to become more toxic than the parent chemical constituent. Indeed, a number of natural attenuation processes such as biodegradation, dispersion, dilution, sorption, volatilization, ion exchange, precipitation, filtration, and gaseous exchange—that work to lessen or attenuate contaminant concentrations in the environment—can often be dominant factors in the fate and transport of contaminants. Thus, consideration and quantification of biodegradation processes in particular is essential to more thoroughly understand the contaminant fate and transport. Biodegradation is the most important "destructive" attenuation mechanism; "nondestructive" attenuation mechanisms include sorption, dispersion, dilution from recharge, and volatilization.

All other factors being equal, the extent and rate of contaminant movement are a function of the physical containment of the chemical constituents or the contaminated zone. A classical illustration of this concept pertains to the fact that a low permeability cap over a contaminated site will minimize water percolation from the surface—and therefore minimize leaching of chemicals into an underlying aquifer.

7.2.1 Contaminant Characteristics

The physical and chemical characteristics of constituents present in the environment and/or at contaminated sites determine the fate and transport properties of the contaminants, and thus, their degree of uptake, transformation, and/or migration through the environment. Some of the particularly important constituent properties affecting the fate and transport of contaminants in the environment include the following (Grisham, 1986):

- Solubility in water (which relates to leaching, partitioning, and mobility in the environment)
- Partitioning coefficients (relating to cross-media transfers, bioaccumulation potential, and sorption by organic matter)
- Hydrolysis (which relates to persistence in the environment or biota)
- Vapor pressure and Henry's law constant (relating to atmospheric mobility and the rate of vaporization or volatilization)
- Photolysis (which relates to persistence as a function of exposure to light)
- Degradation/half-life (relating to the degradation of contaminants and the resulting transformation products)
- Retardation factor (which relates to the sorptivity and mobility of the constituent within the solid-fluid media).

These parameters were discussed earlier in Section 7.1. Further details and additional parameters of possible interest are discussed elsewhere in the literature (e.g., CDHS, 1986; Devinny et al., 1990; Evans, 1989; Haque, 2017; Hemond and Fechner, 1994; Lindsay, 1979; Lyman et al., 1990; Mahmood and Sims, 1986; Mansour, 1993; Neely, 1980; Sáez and Baygents, 2014; Samiullah, 1990; Selim, 2017a,b; Swann and Eschenroeder, 1983; Thibodeaux, 1979, 1996; USEPA, 1985a,c 1989k; Yong et al., 1992).

7.2.2 Site Characteristics

Several characteristics of the physical environment (such as site characteristics) may generally influence the environmental fate of chemicals associated with contaminated site problems; such factors include the following key ones in particular:

- Amount of ambient moisture, humidity levels, temperatures, and wind speed
- Geologic, hydrologic, pedologic, and watershed characteristics
- Topographic features of the impacted location or site and its vicinity
- Vegetative cover of problem location or site and surrounding area
- Land-use characteristics

Indeed, a number of other factors such as soil temperature, soil moisture content, initial contaminant concentration in the impacted media, and media pH may additionally affect the release of a chemical constituent from the source environmental matrix in which it is found.

7.2.3 Nonaqueous Phase Liquid Fate and Transport

The process of determining the transport and fate of nonaqueous phase liquids (NAPLs) can present very tough challenges, not only because of the many types of NAPLs and different kinds of soil materials that provide the medium through which the NAPLs interact but also because of the mechanisms governing the fate of the NAPLs. The most important fluid properties that affect NAPL transport typically include volatility, relative polarity, affinity for soil organic matter or organic contaminants, density, viscosity, and interfacial tension (NRC, 1994a; Yong et al., 1992); at the interface between groundwater and a NAPL pool, spreading of the dissolved components occurs primarily as a result of molecular diffusion. Indeed, contact with the groundwater or infiltrating recharge water causes some of the chemical constituents of the NAPL to dissolve, resulting in aquifer contamination—and further to any trail of contamination left in the soil matrix. In any case, light nonaqueous phase liquids (LNAPLs) are usually found near the top of the saturated zone, with the rate of transport depending on the local groundwater gradients and viscosity of the specific substance. Dense nonaqueous phase liquids (DNAPLs), on the other hand, move downward through both the unsaturated and saturated zones, until they come to "rest" at an impermeable boundary; by definition, DNAPLs are liquids that are heavier than water and have a low aqueous solubility.

Classically, DNAPLs form when chlorinated solvents percolate through a soil column and migrate downward through an aquifer until they encounter an impermeable clay or mudstone aquitard. Indeed, when DNAPLs are released into the subsurface, they tend to rapidly migrate downward toward the bottom of an aquifer. The DNAPL initially conforms to the surface of the aquitard to form a network of pools and interlacing strings. As the solvent continues to move independently of the groundwater, it begins to form dendritic branching patterns in microporous zones throughout the aquitard surface and inside the soil column. Over a period of time, the DNAPL can diffuse into impermeable layers, resulting in distributed source terms that are extremely difficult to remove or access. These source terms in the aquitard slowly migrate and dissolve or disperse into the aquifer. In addition to directly contaminating large quantities of groundwater, the DNAPL constituents also degrade slowly to form other hazardous substances, such as vinyl chloride, which can threaten human health and inhibit land reuse. On the whole, DNAPLs characteristically have great penetration capability in the subsurface because of their relatively low solubility, high density, and low viscosity (Yong et al., 1992). In fact, the relatively high density of these liquids provides a driving force that can carry contamination deep into aquifer systems that may exist at depth—and as such DNAPL tends to be of very great concern, where they exist.

It is noteworthy that, NAPLs may be present at several contaminated sites globally. However, due to the numerous variables influencing NAPL transport and fate in the subsurface, they are likely to go undetected—and yet, they are likely to be a significant limiting factor in site restoration decisions (Charbeneau et al., 1992); this is especially true for DNAPLs, such as chlorinated solvents.

7.2.3.1 The Mechanics of NAPL Migration

The most important mechanisms of NAPL transport generally consist of aqueous transport within contaminant plumes, vapor-phase transport, and transport as a "free product" (NRC, 1994). The relative importance of each of these for the transport of a particular contaminant depends on the properties of the substance involved as well as the site characteristics. For example, as soil moisture increases, the importance of vapor-phase transport diminishes; this is because an increase in the moisture content reduces the pore space available for migration, decreasing the effective diffusivity and gas-phase permeability (NRC, 1994).

In general, as NAPL enters the subsurface environment, the liquid may dissolve into the water in the pore spaces, volatilize into the air in the pores, or remain behind in the pore spaces as an entrapped residual liquid. In the saturated zone, some of the chemicals from the NAPL will dissolve in the flowing groundwater, forming a contaminant plume to be transported further downgradient of the contamination source; the remainder of the NAPL will either float atop the water table (i.e., for LNAPL situations) or continue its downward migration (i.e., for DNAPL cases) until a relatively impermeable barrier or strata is encountered where the NAPL will pool on top of the obstruction—to become yet another potential contaminant release source.

As a practical example, when NAPLs (that are immiscible with water) are released at a site and then enter the subsurface environment, they tend to exist as distinct fluids that flow separately from the water phase. LNAPLs (i.e., fluids less dense than water) migrate downward through the vadose (unsaturated) zone but tend to form lenses that "float" on top of an aquifer upon reaching the water table. Typical examples of LNAPLs include hydrocarbon fuels such as gasoline, heating oil, kerosene, jet fuel, and aviation gas; such LNAPLs will pool and spread as a floating free product layer atop the water table if they are released to the subsurface in sufficient quantities. DNAPLs (i.e., denser-than-water NAPLs) will tend to "sink" into the aquifer. Typical examples of compounds commonly associated with DNAPL releases to the environment include chlorinated solvents/hydrocarbons (e.g., trichloroethylene [TCE], tetrachloroethene/tetrachloroethylene, TCE, and dichloroethylene, chlorophenols, chlorobenzenes, and polychlorinated biphenyls), coal tar wastes, creosote-based wood-treating oils, some pesticides, etc.; such DNAPLs can pass across the water table and may be found at greater depths within the saturated zone of a groundwater aquifer.

7.3 PHASE DISTRIBUTION AND CROSS-MEDIA TRANSFER OF CONTAMINANTS BETWEEN ENVIRONMENTAL COMPARTMENTS

Chemicals released into the environment and contaminants present in a given environmental matrix tend to be affected by several complex processes and phenomena—facilitating the transfer of contamination into other media. The potential for intermedia transfer of pollutants from the soil medium to other media is particularly significant; in fact, contaminated soil can be a major source of the contamination of groundwater, atmospheric air, subsurface soil gas, sediments, and surface water. Furthermore, the movement of contaminants through soils is generally very complex—with some pollutants moving rapidly while others move rather slowly from one environmental matrix into another.

Indeed, contaminated soil is often the main source of chemical repository for environmental pollutants. Consequently, soils often become the principal focus of attention in the investigation of contaminated site problems. This is reasonable because soils at such locales not only serve as a medium of exposure to potential receptors but also serve as a long-term reservoir for contaminants to be released into other media. It must also be recognized that the soil media is by no means an inert repository, since there is an active interchange of chemicals between soils and water, air, and biota (Figure 7.1)—with the main driving forces in contaminant transport in soils being advective and diffusive in nature (Yong et al., 1992).

In general, the affinity that contaminants have for soils can markedly affect their mobility by retarding transport. For instance, hydrophobic or cationic contaminants that are migrating in solution are subject to retardation effects. In fact, the hydrophobicity of a contaminant can greatly affect its fate—and which explains some of the different rates of contaminant migration that occur in the subsurface environments. Also, the phenomenon of adsorption is a major reason why the sediment zones of surface water bodies/systems may become highly contaminated with specific organic and inorganic chemicals. On the other hand, chemical constituents that have a moderate-to-high degree of mobility can leach from soils into groundwater; volatile constituents may contribute to subsurface gas in the

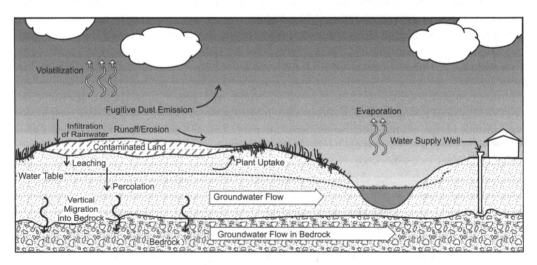

FIGURE 7.1   Illustrative sketch of the cross-media transfers of environmental contaminants.

vadose zone and also possible releases into the atmosphere. Conversely, it is possible for cross-media transport of constituents from other media into soils to take place; for example, chemical constituents may be transported into the soil matrix via atmospheric deposition (of suspended particulate matter from the atmosphere) and also through gas releases from prior-impacted subsurface air. The processes that affect the fate and transport of contaminants should therefore be recognized as an important part of any diagnostic and characterization program designed to address contaminated site problems.

7.3.1 Phase Distribution of Environmental Contaminants

Contamination occurring at contaminated sites may exist in different physical states and may generally be present in a variety of environmental matrices—especially in soils and groundwater. Overall, the contamination associated with contaminated sites may typically show up in the following phases:

- Adsorbed contamination (onto solid phase matter or soils)
- Vapor-phase contamination (present in the vadose zone due to volatilization into soil gas)
- Dissolved contamination (in water, present in both the unsaturated and saturated soil zones)
- Free product (as residual and mobile immiscible fluids, i.e., as LNAPL floating on the surface of the water table, or as DNAPL that sinks deep into the groundwater zone, or as NAPL persisting in the soil pore spaces)

This contamination system that exists in the soils, water, vapor phase, or the NAPLs tends towards a state of equilibrium, such that the chemical potential or fugacity is equal in all the phases that coexist in the environment. Changes in equilibrium between the phases can be occurring on a continuing basis as a result of several extraneous factors that affect the concentration gradients.

On the whole, the distribution of organic chemicals among environmental compartments can be defined in terms of such simple equilibrium expressions as illustrated in Figure 7.2 (Swann and Eschenroeder, 1983)—where the K_w, K_{oc}, and BCF symbols refer to partitioning coefficients discussed earlier on (in Section 7.1). The general assumption is that all environmental compartments are well mixed in order to achieve equilibrium between them. Meanwhile, it is noteworthy that the partitioning of inorganic chemicals is somewhat different from organic constituents. Typically, metals generally exhibit relatively low mobilities in soils (Evans, 1989). Also, relatively insignificant partitioning would be expected among environmental compartments for metals. Rather, the inorganic chemicals will tend to adsorb onto soils (which may become airborne or be transported by surface erosion) and sediments (that may be transported in water).

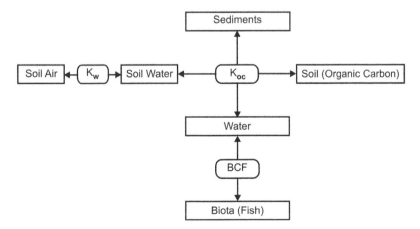

FIGURE 7.2 Major components of the partitioning of contaminants between environmental compartments.

7.4 MODELING CONTAMINANT FATE AND TRANSPORT

Environmental contamination can be transported far away from its primary source(s) of origination via a variety of natural processes (such as erosion and leaching)—culminating in the possible birth of new or secondary contaminated site problems. Conversely, some natural processes work to lessen or reduce contaminant concentrations in the environment through mechanisms of natural attenuation (such as dispersion/dilution, sorption and retardation, photolysis, and biodegradation). In any case, chemical contaminants entering the environment tend to be partitioned or distributed across various environmental compartments. Consequently, a good prediction of contaminant concentrations in the various environmental media is essential to adequately characterize environmental contamination or contaminated site and related chemical exposure problems—the results of which can also be used to support risk assessment and/or risk management decisions. Classically, environmental fate and transport analysis and modeling are used to assess the movement of chemicals between environmental compartments. For instance, simple mathematical models can be used to guide the decisions involved in estimating and managing the potential spread of contaminant plumes; on the basis of the modeling results, and as appropriate or necessary, wells or monitoring equipment or systems can then be located in areas expected to have elevated contaminant concentrations and/or in areas considered upgradient (or upwind), cross-gradient, and downgradient (or downwind) of a contaminant plume.

Mathematical algorithms are typically used to predict the potential for contaminants to migrate from one environmental media into another—or more importantly, from an environmental compartment into potential receptor locations or environmental compliance boundaries. For example, relevant exposure point concentrations associated with a contaminated site problem can be determined once the potentially affected populations are identified and the exposure scenarios are defined. Indeed, if the transport of compounds associated with this situation is under steady-state conditions, then monitoring data are generally adequate to determine potential exposure concentrations. On the other hand, if there are no data available, or if conditions are transient (such as pertains to a migrating plume in groundwater), then models are better used to predict exposure concentrations. Meanwhile, many factors—including the fate and transport properties of the chemicals of concern—must be considered in the model selection process. [By the way, it is noteworthy that, for this type of problem (and in lieu of an established trend in historical data that indicates the contrary), a potentially contaminated site problem may be considered to be in steady state with its surroundings.]

On the whole, mathematical models often serve as valuable tools for evaluating the behavior and fate of chemical constituents in various environmental media. The transport and fate of contaminants can indeed be predicted through the use of various methods—ranging from simple mass-balance and analytical procedures to multidimensional numerical solution of coupled differential equations. A select number of models of possible interest in environmental and contaminated site management programs are annotated in Appendix B; a broader selection of environmental models, together with model selection criteria and limitations, are discussed elsewhere in the literature of environmental and exposure modeling (e.g., CCME, 1994; CDHS, 1986; Clark, 1996; Feenstra et al., 1991; Ghadiri and Rose, 1992; Gordon, 1985; Haith, 1980; Haque, 2017; Honeycutt and Schabacker, 1994; Johnson and Ettinger, 1991; Jury et al., 1984; Mulkey, 1984; NRC, 1989b; Schnoor, 1996; Selim, 2017a,b; USEPA, 1985a,c, 1987f,g, 1988e,f,i; Williams et al., 1996). Meanwhile, it must be emphasized here that, the effective use of models in contaminant fate and behavior assessment depends greatly on the selection of models most suitable for this purpose.

7.4.1 APPLICATION OF MATHEMATICAL MODELS

Models can be used to address a wide range of questions that may need to be answered in environmental contamination and chemical exposure problems, as well as their associated environmental management programs—such as in helping answer the following types of questions:

- What are the prevailing and future chemical release or contamination levels?
- Are modeling predictions of pollution or chemical release from a process or situation met in reality?
- How do pollutants or chemical releases behave in the environment?
- What is the response of the environment to receiving pollution or chemical releases?

One of the major benefits associated with the use of mathematical models in environmental and contaminated site management programs relate to the fact that, environmental concentrations useful for exposure assessment and risk characterization can be estimated for several locations and time periods of interest. Indeed, since field data frequently are limited and insufficient for accurately and completely characterizing contaminated site problems, models can be particularly useful for studying spatial and temporal variabilities, together with potential uncertainties. In addition, sensitivity analyses can be performed—by varying specific parameters and then using models to explore the ramifications (as reflected by changes in the model outputs).

Models can indeed be used for several purposes in the study of contaminated site problems. More generally, mathematical models are usually used to simulate the response of a simplified version of a complex system. As such, their results are imperfect. Nonetheless, when used in a technically responsible manner, they can provide a very useful basis for making technically sound decisions about a contaminated site problem. In fact, they are particularly useful where several alternative scenarios are to be compared. In such cases, all the alternatives are compared on a similar basis; thus, whereas the numerical results of any single alternative may not be exact, the comparative results of showing that one alternative is superior to others will usually be valid. Ultimately, the effective use of models in contaminated site characterization and risk management programs depends greatly on the selection of models most suitable for its specified purpose.

By and large, the fate of chemical compounds released into the environment forms an important basis for evaluating the exposure of biological receptors to hazardous chemicals—because once contaminants are released into the environment, the pollutants may be transported into various media and environmental matrices occupied by the receptors. For instance, releases from potential contamination sources can cause human exposures to contaminants in a variety of ways, such as the following:

- Direct inhalation of airborne vapors and also respirable particulates
- Deposition of airborne contaminants onto soils, leading to human exposure via dermal absorption or ingestion
- Ingestion of food products that have been contaminated as a result of deposition onto crops or pasture lands and introduction into the human food chain
- Ingestion of contaminated dairy and meat products from animals consuming contaminated crops or waters
- Deposition of airborne contaminants on waterways, uptake through aquatic organisms, and eventual human consumption
- Leaching and runoff into water resources and consequential human exposures to contaminated waters.

Mathematical models tend to play prominent roles in the evaluation of the above-mentioned types of exposures. Multimedia transport models are generally used in the prediction of the long-term fate of the chemicals in the environment. In fact, a variety of mathematical algorithms and models are commonly used to support the determination of contaminant fate and transport in the environment—and whose results are then used in estimating the consequential exposures and risks to potential receptors. Some simple example models and equations that may be used in the estimation of the cross-media contaminant concentrations and the requisite exposure point concentrations are presented further later on in this chapter.

7.4.2 MODEL SELECTION

Numerous model classification systems with different complexities exist in practice—broadly categorized as analytical or numerical models, depending on the degree of mathematical sophistication involved in their formulation. Analytical models are models with simplifying underlying assumptions, often sufficient and appropriate for well-defined systems for which extensive data are available, and/or for which the limiting assumptions are valid. Whereas analytical models may suffice for some evaluation scenarios, numerical models (with more stringent underlying assumptions) may be required for more complex configurations and complicated systems. In any event, the choice of a model type that could be best used for specific applications is subject to numerous, sometimes convoluted, factors/constraints. Thus, simply choosing a more complicated model over a simple one will not necessarily ensure a better solution in all situations. In fact, since a model is a mathematical representation of a complex system, some degree of mathematical simplification usually must be made about the system being modeled. In these efforts, data limitations must be weighted appropriately, since it usually is not possible to obtain all of the input parameters due to the complexity (e.g., anisotropy and non-homogeneity) of natural systems.

Now, as a general word of caution, it is notable that the appropriateness of a particular model necessary to address environmental issues depends on the characteristics of the particular problem on hand; thus, the screening of models should be carefully tied to the project goals. Indeed, the wrong choice of models could result in the generation of false information—with consequential negative impacts on any decisions made thereof. On the other hand, the choice of appropriate fate and transport models that will give reasonable indications of the contaminant behavior will help produce a realistic conceptual representation of the problem—and this is important to the adequate characterization of any contaminated site or environmental contamination and related chemical exposure problem, which in turn is a prerequisite to developing reliable risk management policies and case-specific remedy strategies.

On the whole, the decisions about model selection can be a tricky one—often necessitating cautious warnings; this concern is best illustrated and summarized by the following interesting observation and note of comparison with "social models" made by Kaplan (Kaplan, 1964—as cited in Aris, 1994) that "Models are undeniably beautiful, and a man may justly be proud to be seen in their company. But they may have their hidden vices. The question is, after all, not only whether they are good to look at, but whether we can live happily with them." This illustrative and somehow analogous view held by much of society, for better or for worse, about "social models" does indeed compare very well with the underlying principles in the selection and use of environmental models—thus calling for the careful choice of such models to support chemical release and environmental management programs. Most importantly, it should be recognized that a given mathematical model that performs extremely well under one set of circumstances might not necessarily be appropriate for other similar or comparable situations for a variety of reasons.

In the end, the type of model selected to address any particular concern will be dependent on the overall goal of the assessment, the complexity of the problem, the type of contaminants of concern, the nature of the impacted and threatened media that are being considered in the investigation, and the type of corrective actions being considered. At any rate, it is noteworthy that in several environmental assessment situations, a "ballpark" or "order-of-magnitude" (i.e., a rough approximation) estimate of effectiveness for the contaminant behavior and fate is usually all that is required for most analyses—and in which case simple analytical models usually will suffice. General guidance for the effective selection of models in contaminated site characterization and risk management decisions is provided in the literature elsewhere (e.g., CCME, 1994; CDHS, 1990; Clark, 1996; Cowherd et al., 1985; DOE, 1987; Haque, 2017; NRC, 1989; Selim, 2017a,b; Schnoor, 1996; USEPA, 1985a,c, 1987f,g, 1988e,f,i,; Walton, 1984; Yong et al., 1992; Zirschy and Harris, 1986).

7.4.3   ILLUSTRATIVE EXAMPLE APPLICATION SCENARIOS FOR SIMPLE MATHEMATICAL MODELS: THE CASE OF AIR DISPERSION MODELING FOR ENVIRONMENTAL CHEMICALS

Atmospheric dispersion modeling has become an integral part of the planning and decision-making process in the assessment of public health and environmental impacts from various chemical release and contaminated site problems. It is an approach that can indeed be used to provide contaminant concentrations at potential receptor locations of interest based on emission rate and meteorological data. Naturally, the accuracy of the model predictions depends on the accuracy and representativeness of relevant input data. Broadly speaking, key model input data will include emissions and release parameters, meteorological data, and receptor locations. Typically, existing air monitoring data (if any) for the locale/area of interest can be utilized to facilitate the design of a receptor grid, as well as to select "indicator chemicals" to be modeled. This can also provide insight into likely background concentrations. Indeed, in all situations, case-specific data should be used whenever possible—in order to increase the accuracy of the emission rate estimates.

Overall, a number of general assumptions are normally made in the assessment of contaminant releases into the atmosphere, including the following key ones:

- Air dispersion and particulate deposition modeling of emissions adequately represent the fate and transport of chemical emission to ground level.
- The composition of emission products found at ground level is identical to the composition found at source, but concentrations are different.
- The potential receptors are exposed to the maximum annual average ground-level concentrations (GLCs) from the emission sources for 24 h/day, throughout a 70-year lifetime—a rather conservative assumption.
- There are no losses of chemicals through transformation and other processes (such as biodegradation or photodegradation)—a rather conservative assumption.

In the end, the combined approach of environmental fate analysis and field monitoring should provide an efficient and cost-effective strategy for investigating the impacts of air pathways on potential receptors, given a variety of meteorological conditions.

7.4.3.1   The Case of Human Exposures to Airborne Chemical Toxicants

Contaminants released into the atmosphere are subjected to a variety of physical and chemical influences—including transformation, deposition, and depletion processes—which are but secondary to transport and diffusion processes, albeit important in several situations. Although deposition depletes concentrations of the contaminants in air, it increases the concentrations on vegetation and in soils and surface water bodies. Furthermore, certain types of deposited contaminants are subject to some degree of resuspension, especially through wind erosion for wind speeds exceeding 15–25 km/h (≈10–15 miles/h).

Indeed, air emissions from contaminated sites often are a significant source of human exposure to toxic or hazardous substances. Contaminated sites can therefore pose significant risks to public health as a result of possible airborne release of soil particulate matter laden with toxic chemicals and/or volatile emissions. In fact, even very low-level air emissions could pose significant threats to exposed individuals, especially if toxic or carcinogenic contaminants are involved. Furthermore, remedial actions—especially ones involving excavation—may create much higher emissions than baseline or undisturbed conditions. Consequently, there is increased attention on the assessment of risks associated with air-toxic releases.

7.4.3.2   Classification of Air Emission Sources

Air emissions from contaminated sites may be classified as either point or area sources. Point sources include vents (e.g., landfill gas vents) and stacks (e.g., air stripper releases); area sources are generally associated with ground-level emissions (e.g., from landfills, lagoons, and contaminated surface areas). Area sources are released at ground level and disperse there, with less influence of winds and turbulence; point sources, generally, come from a stack and are emitted with an upward velocity, often at a height significantly above ground level. Thus, point sources tend to be more readily diluted by mixing and diffusion—that is, further to being at greater heights, so that GLCs are reduced. This means that, area source emissions may be up to 100 times as hazardous on a mass per time basis compared to point sources and stacks. This scenario demonstrates the importance of adequately describing the source type in the assessment of potential air impacts associated with contaminated site problems.

7.4.3.3 Categories of Air Emissions

Air contaminant emissions from contaminated sites fall into two major categories—namely, gas-phase emissions, and particulate matter emissions. The emission mechanisms associated with gas phase and particulate matter releases are quite different. Gas-phase emissions primarily involve organic compounds but may also include certain metals (such as mercury); these emissions may be released through several mechanisms such as volatilization, biodegradation, photodecomposition, hydrolysis, and combustion (USEPA, 1989e). Particulate matter emissions at contaminated sites can be released through wind erosion, mechanical disturbances, and combustion. For airborne particulates, the particle size distribution plays an important role in inhalation exposures. Large particles tend to settle out of the air more rapidly than small particles but may be important in terms of non-inhalation exposures. Very small particles (2.5–10 μm diameter) are considered to be respirable and, thus, pose a greater health hazard than the larger particles.

7.4.3.4 Determination of Air Contaminant Emissions

Once released to the ambient air, a contaminant is subject to simultaneous transport and diffusion processes in the atmosphere; these conditions are significantly affected by meteorological, topographical, and source factors. Additional fundamental atmospheric processes (other than atmospheric transport and diffusion) that affect airborne contaminants include transformation, deposition, and depletion. The extent to which all these atmospheric processes act on the contaminant of concern determines the magnitude, composition, and duration of the release; the route of human exposure; and the impact of the release on the environment.

Methods for estimating air emissions at contaminated sites may consist of emissions measurement (direct and indirect), air monitoring (supplemented by modeling, as appropriate), and predictive emissions modeling. Protocols for estimating emission levels for contaminants from several sources are available in the literature (e.g., CAPCOA, 1990; CDHS, 1986; Mackay and Leinonen, 1975; Mackay and Yeun, 1983; Thibodeaux and Hwang, 1982; USEPA, 1989a,d,e,f,g,l,m–g, 1990b-f,). In all cases, site-specific data should be used whenever possible, in order to increase the accuracy of the emission rate estimates. In fact, the combined approach of environmental fate analysis and field monitoring should provide an efficient and cost-effective strategy for investigating the impacts of air pathways on potential receptors, given a variety of meteorological conditions. Once the emission rates are determined, and the exposure scenarios are defined, decisions can be made regarding potential air impacts for specified activities.

7.4.3.5 Air Pathway Exposure Assessment

An air pathway exposure assessment (APEA) is a systematic approach involving the application of modeling and monitoring methods to estimate emission rates and concentrations of air contaminants. It can be used to assess actual or potential receptor exposures to atmospheric contamination. An APEA is basically an exposure assessment for the air pathway, consisting of the following key components:

- Characterization of air emission sources
- Determination of the effects of atmospheric processes on contaminant fate and transport

- Evaluation of populations potentially at risk (for various exposure periods)
- Estimation of receptor intakes and doses

The purpose of the exposure assessment is to estimate the extent of potential receptor exposures to the identified chemicals of concern (COCs). This involves emission quantification, modeling of environmental transport, evaluation of environmental fate, identification of potential exposure routes and populations potentially at risk from exposures, and the estimation of both short- and long-term exposure levels.

The following information, at a minimum, need to be collected and reviewed to support the design of an APEA program:

- Source data (to include contaminant toxicity factors and off-site sources)
- Receptor data (including identification of sensitive receptors, local land use, etc.)
- Environmental data (such as dispersion data, meteorological information, topographic maps, and soil and vegetation data)
- Previous APEA data (to include summary of existing environmental standards, air monitoring, emission rate monitoring and modeling, and dispersion modeling)

Overall, a first step in assessing air impacts associated with a contaminated site is to evaluate site-specific characteristics and the chemical contaminants present at the site and to determine if transport of hazardous chemicals to the air is of potential concern. Atmospheric dispersion and emission source modeling can be used with appropriate air sampling data as input to an atmospheric exposure assessment. Finally, exposure calculations are carried out using the appropriate relationships discussed in Section 9.3 of this title.

Some simple example model formulations that may be used in the estimation of the cross-media contaminant concentrations, as well as the requisite exposure point concentrations (that can be further used to facilitate responsible risk determinations), are presented further below for the air migration pathway. Specifically, some selected screening-level air emission modeling procedures are discussed further below for illustrative purposes only; these include the archetypical computational procedures for both volatile and nonvolatile emissions—with the nonvolatile compounds generally considered to be bound onto particulates by adsorption. [By the way, for the purposes of a screening evaluation, a volatile substance may be defined as any chemical with a vapor pressure greater than $[1 \times 10^{-3}]$ mm Hg or a Henry's law constant greater than $[1 \times 10-5]$ atm-m^3/mole (DTSC, 1994). Thus, chemicals with Henry's law constants less than or equal to these indicated values are generally considered as nonvolatile compounds.] General and specific protocols for estimating releases or emission levels for contaminants from several sources are available elsewhere in the literature (e.g., CAPCOA, 1990; CDHS, 1986; Mackay and Leinonen, 1975; Mackay and Yeun, 1983; Thibodeaux and Hwang, 1982; USEPA, 1989a,d,e,f,g,l,m, 1990b-f).

7.4.3.6 Screening-Level Estimation of Airborne Dust/Particulate Concentrations

Particulate emissions from potentially contaminated sites can cause human exposures to contaminants in a variety of ways, including the following:

- Direct inhalation of respirable particulates
- Deposition on soils, leading to human exposure via dermal absorption or ingestion
- Ingestion of food products that have been contaminated as a result of deposition on crops or pasture lands and introduction into the human food chain
- Ingestion of contaminated dairy and meat products from animals eating contaminated crops
- Deposition on waterways, uptake through aquatic organisms, and eventual human consumption

In the estimation of potential risks from particulate matter or fugitive dust inhalation, an estimate of respirable (oftentimes assumed to be <10 μm aerodynamic diameter, denoted by the symbol PM-10 or PM_{10}) fraction and concentrations are required. The amount of non-respirable (>10 μm aerodynamic diameter) concentrations may also be needed to estimate deposition of wind-blown emissions which will eventually reach potential receptors via other routes such as ingestion and dermal exposures.

In general, air models for fugitive dust emission and dispersion can be used to estimate the applicable exposure point concentrations of respirable particulates from chemical release sources, such as contaminated sites. In such models, fugitive dust dispersion concentrations evaluated are typically represented by a three-dimensional Gaussian distribution of particulate emissions from the source (e.g., CAPCOA, 1989; CDHS, 1986; DOE, 1987; USEPA, 1989a,d,e,f,g,l,m; USEPA, 1993a,b). For the screening analysis of particulate inhalation, the chemical concentration in air may be estimated in accordance the following equation:

$$C_a = C_s \times PM_{10} \times CF \tag{7.13}$$

where:

C_a = concentration in air (μg/m^3)
C_s = contaminant concentration in soil (mg/kg)
PM_{10} = airborne concentration of respirable dust (less than 10 μm in diameter), usually assumed to be 50 μg/m^3
CF = conversion factor, equal to 10^{-6} kg/mg

Typically, a screening-level assumption is made that, for non-volatile organic compounds (VOCs), particulate contamination levels are directly proportional to the maximum soil concentrations. The particulate concentration used in this type of characterization may be set at 50 μg/m^3; this value is the U.S. National Ambient Air Quality Standard for annual average respirable portion (PM_{10}) of suspended particulate matter (USEPA, 1993a,b). By using the above-mentioned assumptions, the non-VOC concentration in air is given as:

$$C_a = C_s \times \left[5 \times 10^{-5}\right] \tag{7.14}$$

It is noteworthy that this estimation procedure is not applicable to a site that is particularly dusty— as would be expected in situations where the air quality standard for suspended particulate matter is routinely exceeded (i.e., $PM_{10} > 50$ μg/m^3).

7.4.3.7 Screening-Level Estimation of Airborne Vapor Concentrations

The most important chemical parameters to consider in the evaluation of volatile air emissions are the vapor pressure and the Henry's law constant. Vapor pressure is a useful screening indicator of the potential for a chemical to volatilize from the media in which it currently exists. As a special example in relation to the utility of the Henry's law constant, it is notable that this is particularly important in estimating the tendency of a chemical to volatilize from a surface impoundment or water; it also indicates the tendency of a chemical to, for instance, partition between the soil and gas phase from soil water in the vadose zone or groundwater at a contaminated site.

As an example, in regards to the evaluation of a contaminated site problem, a vaporization model may be used to calculate flux from volatile compounds present in soils into the overlying air zone (DTSC, 1994; USEPA, 1990a–f, 1992a,g). Typically, the following general equation will be used to estimate the average emission over a residential lot (USEPA, 1990a–f, 1992a,g):

$$E_i = \frac{2AD_e P_a K_{as} C_i \times CF}{\left(3.14\,aT\right)} \tag{7.15}$$

where

E_i = average emission rate of contaminant i over a residential lot during the exposure interval (mg/s)

A = area of contamination (cm²), with a typical default value of 4.84×10^6 cm²

D_e = effective diffusivity of compound (cm²/s)

$\quad = D_i \left(P_a^{3.33} / P_t^2 \right)$

D_i = diffusivity in air for compound i (cm²/s)

P_t = total soil porosity (dimensionless)

$\quad = 1 - \left(\beta / \rho \right)$

β = soil bulk density (g/cm²), with typical default value of 1.5 g/cm³

ρ = particle density (g/cm³), with typical default value of 2.65 g/cm³

P_a = air-filled soil porosity (dimensionless)

$\quad = P_t - \theta_m \beta$

θ_m = soil moisture content (cm³/g), with typical default value of 0.1 cm³/g

K_{as} = soil/air partition coefficient (g/cm³)

$\quad = \left(H_c / K_d \right) \times \text{CF2} = \left(H_c / K_d \right) \times 41$

H_c = Henry's law constant (atm-m³/mol)

K_d = soil–water partition coefficient (cm³ water/g soil, or L/kg)

CF2 = 41, a conversion factor that change H_c into a dimensionless form

C_i = bulk soil concentration of contaminant i (i.e., chemical concentration in soil, mg/kg $\times 10^{-6}$ kg/mg) (g/g soil)

CF = 10^3 mg/g, a conversion factor

α = conversion factor composed of several quantities defined previously

$$\alpha = \frac{D_e \times P_a}{\left\{ P_a + \left[r(1 - P_a) / K_{as} \right] \right\}}$$

T = exposure interval (s), with a typical default value of 30 years = 9.5×10^8 s.

The equation for estimating emission rates of VOCs can be reduced to the following form:

$$E_i = \frac{1.6 \times 10^5 \times D_i \times \dfrac{H_c}{K_d} \times C_i}{\left(D_i \times \dfrac{0.023}{\left\{ 0.284 + \left[0.046 \times \dfrac{K_d}{H_c} \right] \right\}} \right)^{1/2}} \tag{7.16}$$

It is noteworthy that this equation is valid only in situations when no "free product" is expected to exist in the soil vadose zone. Analytical procedures for checking the saturation concentrations, and intended to help determine if free products exist, are available elsewhere in the literature (e.g., DTSC, 1994; USEPA, 1990a–f, 1992a,g). Under the circumstances when the soil contaminant concentration is greater than the calculated saturation concentration for the contaminant—implying the likely presence of free product—a more sophisticated evaluation scheme should be adopted.

Finally, the potential air contaminant concentration in the receptor's breathing zone that results from volatilization of chemicals through the soil surface is calculated over each discrete area of concern. A simple box model (Hwang and Falco, 1986; USEPA, 1990a–f; 1992a,g) can be used to provide an estimate of ambient air concentrations using the prior-calculated total emission rate calculated, as mentioned previously; in this case, the length dimensions of the hypothetical box within which mixing will occur is usually based on the minimum dimensions of a residential lot in the applicable locality/region (Hadley and Sedman, 1990). Consequently, a screening-level estimate of the ambient air concentration is given by the following:

$$C_a = \frac{E}{(LS \times V \times MH)}$$ (7.17)

where:

C_a = ambient air concentration (mg/m^3)

E = total emission rate (mg/s)

LS = length dimension perpendicular to wind (m), with a typical default value of 22 m based on the length dimension of a square residential lot with area 484 m^2 (Hadley and Sedman, 1990; USEPA, 1992a,g)

V = average wind speed within mixing zone (m/s), with typical default value of 2.25 m/s (Hadley and Sedman, 1990; USEPA, 1992a,g)

MH = mixing height (m), with typical default value of 2 m, or the height of the average breathing zone for an adult (Hadley and Sedman, 1990; USEPA, 1992a,g).

By using the above-indicated default values, the ambient air concentration can be estimated by:

$$C_a = \frac{E_i}{99}$$ (7.18)

where E_i would already have been estimated as shown previously.

7.4.3.8 Human Health Risk Effects and Implications of Air Contaminant Releases

Airborne chemical toxicants can impact human population via numerous trajectories. For instance, among several other possibilities and issues, volatile chemicals may be released into the gaseous phase from such sources as landfills, surface impoundments, contaminated surface waters, and open/ruptured chemical tanks or containers. Also, there is the potential for subsurface gas movements into underground structures such as pipes and basements and eventually into indoor air. Additionally, toxic chemicals adsorbed to soils may be transported to the ambient air as particulate matter or fugitive dust. Overall, chemical release sources can pose significant risks to public health as a result of possible airborne release of particulate matter laden with toxic chemicals and/or volatile emissions. In fact, even very low-level air emissions could pose significant threats to exposed individuals, especially if toxic or carcinogenic contaminants are involved. Consequently, there is increased concern and attention to the proper assessment of public health risks associated with chemical releases into air.

Finally, it is noteworthy that, to enable credible risk estimation in relation to human exposure to airborne chemical toxicants, there usually should be a reliable appraisal of the airborne concentrations of the target chemicals. The chemical concentration in air—oftentimes represented by the GLC—is a function of the source emission rate and the dilution factor at the points of interest (usually the potential receptor location and/or "breathing zone").

7.4.4   ILLUSTRATIVE EXAMPLE APPLICATION SCENARIOS FOR SIMPLE MATHEMATICAL MODELS: MODELING THE EROSION OF CONTAMINATED SOILS INTO SURFACE WATERS AS AN EXAMPLE

Contaminated runoff and overland flow of toxic contaminants constitutes one source of concern for surface water contamination at uncontrolled hazardous waste sites; surface runoff release estimation procedures (see, e.g., USEPA, 1988e,f,i–k) can be applied to such uncontrolled sites. A hypothetical problem is discussed here—to demonstrate the potential impacts that contaminated sites may have on surface water resources and also to illustrate the applicability of simple contaminant transport models in supporting a corrective action investigation at such sites.

7.4.4.1 Background to the Hypothetical Problem

This example problem involves an abandoned industrial process facility used for handling heavy/waste oils at the XYZ site located in eastern United States. Petroleum products have been released from this facility, leading to surface soil contamination. An environmental site assessment conducted for the XYZ site indicated the presence of the following organic constituents in the soils: benzene, benzo(a)anthracene, benzo(a)pyrene, ethylbenzene, naphthalene, phenanthrene, phenol, and toluene.

It is apparent from the physical setting of the site that contaminated soils may be carried off-site from the XYZ property almost exclusively by erosional/overland transport flow processes. A creek that adjoins the site is potentially threatened by possible contaminant loading. It is required by environmental laws/regulations to protect the surface water quality (from the migration of contaminated soil that may be washed into the nearby creek). However, current levels of chemicals in soil at the site would likely result in the exceedance of the water quality criteria to be met for this creek, that is, following contaminant loading from erosion runoff. The responsible parties are therefore required to clean up the site to such levels that will not significantly impact the creek after receiving runoff carrying contaminated soils from the XYZ site.

7.4.4.2 Modeling the Migration of Contaminated Soil to Surface Runoff: Soil/Sediment Loss and Contaminant Flux Calculations

Assuming that all contamination is attributable to adsorbed waste oil constituents present at the XYZ site, surface runoff release of chemicals can be estimated by using the modified universal soil loss equation (MUSLE) and sorption partition coefficients derived from the compounds octanol–water partition coefficients. The MUSLE allows the estimation of the amount of surface soil eroded in a storm event of given intensity, while sorption coefficients allow the projection of the amount of contaminant carried along with the soil, as well as the amount carried in dissolved form. The procedures used here are fully described in the literature elsewhere (e.g., DOE, 1987; USEPA, 1988e,f,i,j; Williams, 1975).

The soil loss caused by a storm event is given by the MUSLE (DOE, 1987; Williams, 1975):

$$Y(S)_E = a\left(V_r q_p\right)^{0.56} KLSCP \tag{7.19}$$

where
   $Y(S)_E$ = sediment yield (tons/event)
   a = conversion constant
   V_r = volume of runoff (acre-ft)
   q_p = peak flow rate (ft^3/s)
   K = soil erodibility factor (commonly expressed in tons/acre/dimensionless rainfall erodibility unit)
   L = the slope-length factor (dimensionless ratio)
   S = the slope-steepness factor (dimensionless ratio)
   C = the cover factor (dimensionless ratio)
   P = the erosion control practice factor (dimensionless ratio)

Computation of V_r. The storm runoff volume generated at the case site is calculated by the following equation (Mills et al., 1982):

$$V_r = 0.083 A Q_r$$

where

A = contaminated area ≈6 acres (estimated during remedial investigation)

Q_r = depth of runoff from site (in.)

The depth of runoff, Q_r, is determined by (Mockus, 1972):

$$Q_r = \frac{(R_t - 0.2S_w)^2}{(R_t + 0.8S_w)^2}$$

where

Q_r = the depth of runoff from the site (in.)

R_t = the total storm rainfall (in.)

S_w = water retention factor (in.)

The "Soil Conservation District" (or similar authority) may be consulted on a value for the total storm rainfall (R_t) to use. In this case, the 2-year 24-h event represents a typical annual storm; use of this suggested storm event yields a R_t value of approximately 3.5 in.. Also, an average annual storm of between 44 and 48 in. can be expected for this area. Thus,

$$R_t = 3.5 \text{ in.} \quad \text{and} \quad \text{average annual rainfall} \approx 46 \text{ in.}$$

This means that, the average number of average rainfall events per year $\approx \dfrac{46}{3.5} \approx 13$ events, yielding 920 storm events over a 70-year period.

The value of the water retention factor (S_w) is obtained as follows (Mockus, 1972):

$$S_w = \left\{ \frac{1,000}{CN} - 10 \right\}$$

where

S_w = water retention factor (in.)

CN = the Soil Conservation Services (SCS) runoff curve number (dimensionless)

The CN factor for the site is determined by the soil type at the XYZ site, its condition, and other parameters that establish a value indicative of the tendency of the soil to absorb and hold precipitation or to otherwise allow precipitation to run off from the site. CN values for uncontrolled hazardous waste sites can be estimated from tables in the literature (e.g., USEPA, 1988d; Schwab et al., 1966); charts have also been developed for determining CN values (e.g., USBR, 1977), and these may be used for estimating the CN. Based on the existing site conditions of the fill material that is very heterogeneous, higher infiltration and relatively low to moderate runoff is anticipated at the site. An estimate of CN = 74 for the soil type is taken as conservative enough for our purposes here (USEPA, 1988). This represents a moderately low runoff potential and an above-average infiltration rate of 4–8 mm/h seems reasonable for the site. In this case, S_w, Q_r, and V_r are estimated to be as follows:

$$S_w = \left\{ \frac{1,000}{74} - 10 \right\} = 3.51 \text{ in.}$$

$$Q_r = \frac{[3.5 - (0.2)(S_w)]^2}{[3.5 + (0.8)(S_w)]} = 1.24 \text{ in.}$$

$$V_r = (0.083)(5.9)(Q_r) = 0.61 \text{ acre-ft}$$

Computation of q_p. The peak runoff rate, q_p, is determined by (Haith, 1980) the following:

$$q_p = \frac{1.01 \, A \, R_t Q_r}{T_r \left(R_t - 0.2 S_w \right)}$$

where

q_p = the peak runoff rate (ft³/s)
A = contaminated area (acres)
R_t = the total storm rainfall (in.)
Q_r = the depth of runoff from the watershed area (in.)
T_r = storm duration (h)
S_w = water retention factor (in.)

For the typical storm represented by the 2-year 24-h rainfall event suggested for this scenario and given the following parameters

$A = 5.9$ acres
$R_t = 3.5$ in.
$Q_r = 1.24$ in.
$R_t = 24$ h
$S_w = 3.51$ in.

$$q_p = 0.38 \text{ cfs}$$

Estimation of K. The soil erodibility factors are indicators of the erosion potential of given soil types and are therefore site specific. The value of the soil erodibility factor, K, as obtained from the SCS is 0.32. This compares reasonably well with estimates given by charts and nomographs found in the literature (e.g., DOE, 1987; Erickson, 1977; Goldman et al., 1986). That is,

$$K = 0.32 \text{ tons/acre/rainfall erodibility unit}$$

Estimation of LS. The product of the slope-length and slope-steepness factors, LS, is determined from charts/nomographs given in the literature (e.g., USEPA, 1988; DOE, 1987; Mitchell and Bubenzer, 1980). This is based on a slope length of 570 ft and a slope of 0.5% obtained from information generated during the remedial investigations, yielding:

$$LS = 0.14$$

C and P Factors. A value of C equal to 1.0 is used here, assuming no vegetative cover exists at this site. This will help simulate a worst case scenario. Similarly, a worst-case (conservative) P value of 1 for uncontrolled sites is used.

Soil Loss Computation. The following is a summary of estimated parameters used as input to the model calculations:

$a = 95$
$V_r = 0.61$ acre-ft
$q_p = 0.38$ cfs
$K = 0.32$ tons/acre/rainfall erodibility unit
$LS = 0.14$
$C = 1.0$
$P = 1.0$

Thus, for a typical rainstorm event of 2-year 24-h magnitude, substitution of these estimated parameters into the MUSLE yields

$$Y(S)_E = 1.89 \text{ tons/event}$$

7.4.4.3 Dissolved/Sorbed Contaminant Loading

After computing the soil loss during a storm event, the amounts of adsorbed and dissolved substance loading on the creek can be calculated. The amounts of adsorbed and dissolved substances are determined by using the following equations (Haith, 1980; DOE, 1987):

$$S_s = \left[\frac{1}{\left(1 + q_c/K_d\beta\right)} \right] C_i A \tag{7.20}$$

$$D_s = \left[\frac{1}{\left(1 + K_d\beta/q_c\right)} \right] C_i A \tag{7.21}$$

where

S_s = sorbed substance quantity (kg)
q_c = available water capacity of the top cm of soil (dimensionless)
K_d = sorption partition coefficient (cm³/g)
  = $f_{oc} \times K_{oc}$, where f_{oc} is the organic carbon content/fraction of the soil an K_{oc} is the soil/water distribution coefficient, normalized for organic content
β = soil bulk density (g/cm³)
C_i = total substance concentration (kg/ha-cm)
  = $C_s \times \beta \times \phi$, where C_s is the chemical concentration in soil (mg/kg), β is the soil bulk density (g/cm³), and ϕ is a conversion factor
A = contaminated area (ha-cm)
D_s = dissolved substance quantity (kg)

The model assumes that only the contaminant in the top 1 cm of soil is available for release via runoff.

Estimating K_d. The soil sorption partition coefficient for a given chemical can be determined from known values of certain other physical/chemical parameters, primarily the chemical's octanol–water partition coefficient, solubility in water, or BCF. The sorption partition coefficient, K_d, is given by the following relationship:

$$K_d = f_{oc} K_{oc}$$

where

f_{oc} = the organic carbon content/fraction of the soil
K_{oc} = the soil/water distribution coefficient, normalized for organic content

An organic carbon fraction/content of 0.1% is assumed for this site.

Estimating β, θ_c. The contaminated site area, A, is 6 acres. The soil bulk density, β, is estimated to be about 1.4 g/cm³ (Walton, 1984). The available water capacity of the top 1 cm of soil is estimated at about $\theta_c \approx 120$ mm/m, that is, $\theta_c \approx 0.12$ (Walton, 1984).

Estimation of C_i. The total substance concentration, C_i [kg/ha-cm], is obtainable by multiplying the chemical concentration in soil, C_s [mg/kg], by the soil bulk density, β [g/cm³], and an appropriate conversion factor, ϕ:

$$C_i\left[\text{kg/ha-cm}\right] = C_s\,\beta\,\phi$$

where ϕ is a conversion factor, equal to 0.1. For the estimated value of $\beta = 1.40$ g/cm^3, this results in:

$$C_i\left[\text{kg/ha-cm}\right] = 0.1\,\beta\,C_s \quad \text{or} \quad C_i = 0.14\,C_s$$

Maximum concentrations found at the XYZ site are used in this computation to help simulate the possible worst-case scenario.

Computation of Dissolved/Sorbed Contaminant Loading. The following parameters are used to estimate the dissolved and sorbed quantities of contaminants reaching the creek:

$A = 5.9$ acres
$\theta = 0.12$
$\beta = 1.4$ g/cm^3
$f_{oc} = 0.1\%$

Consequently, the equation for sorbed substance quantity, S_s [kg], is simplified into the following:

$$S_s = \left\{\cfrac{1}{\left[\left(1+\cfrac{0.12}{\left[1.4\,K_d\right]}\right)\right]}\right\}(0.14\,C_s)(2.388) = \left\{\cfrac{1.4\,K_d}{1.4\,K_d + 0.12}\right\}(0.334\,C_s)$$

$$= \frac{0.468\,K_d\,C}{\left(0.12 + 0.028\,K_{oc}\right)} = \frac{0.009\,K_{oc}\,C_s}{\left(0.12 + 1.4\,K_d\right)}$$

Similarly, the equation for dissolved substance quantity, D_s [kg], is simplified into the following:

$$D_s = \left\{\cfrac{1}{1+\left[K_d\beta/\theta_c\right]}\right\}(0.334\,C_s) = \left\{\cfrac{1}{1+\left[1.4\,K_d/0.12\right]}\right\}(0.344\,C_s)$$

$$= \left(\cfrac{0.12}{1+\left\{1.4\,K_d/\left[0.12 + 1.4\,K_d\right]\right\}}\right)(0.344\,C_s) = \frac{0.04\,C_s}{\left[0.12 + 1.4\,K_d\right]}$$

$$= \frac{0.04\,C_s}{\left[0.12 + 0.028\,K_{oc}\right]}$$

7.4.4.4 Computation of Total Contaminant Loading on Creek

After calculating the amount of sorbed and dissolved contaminant, the total loading to the receiving water body is calculated as follows (Haith, 1980; DOE, 1987):

$$PX_i = \left[\frac{Y(S)_E}{100\,\beta}\right]S_s \tag{7.22}$$

$$PQ_i = \left[\frac{Q_r}{R_t}\right]D_s \tag{7.23}$$

where

PX_i = sorbed substance loss per event (kg)
$Y(S)_E$ = sediment yield (tons/event, metric tons)
β = soil bulk density (g/cm^3)
S_s = sorbed substance quantity (kg)
PQ_i = dissolved substance loss per event (kg)
Q_r = total storm runoff depth (cm)
R_t = total storm rainfall (cm)
D_s = dissolved substance quantity (kg)

PX_i and PQ_i can be converted to mass per volume terms for use in estimating contaminant concentration in the receiving water body, if divided by the site storm runoff volume, V_r (where $V_r = a\,A\,Q_r$). The contaminant concentrations in the surface runoff, C_{sr}, are then given by the following:

$$C_{sr} = \frac{(PS_i + PQ_i)}{V_r} \qquad (7.24)$$

Next, the contaminant concentrations in the creek are computed by a mass balance analysis according to the following relationship:

$$C_{cr} = \frac{(C_{sr} \times q_p)}{[q_p + Q_{cr}]} \qquad (7.25)$$

where

C_{cr} = concentration of contaminant in creek (mg/L)
C_{sr} = concentration of contaminant in surface runoff (mg/L)
q_p = peak runoff rate (cfs)
Q_{cr} = volumetric flow rate of creek (cfs)

7.4.4.5 Recommended Soil Cleanup Levels for Surface Water Protection

By performing back calculations, based on contaminant concentrations in creek as a result of the current constituents loading from the XYZ site, a conservative estimate is made as to what the maximum acceptable contaminant concentration should be on this site in order *not* to adversely impact the creek. The back modeling is carried out as follows:

$$C_{max} = \left[\frac{C_{std}}{C_{cr}}\right] \times C_s \qquad (7.26)$$

where

C_{max} = maximum acceptable soil concentration on site (mg/kg)
C_{std} = applicable surface water quality criteria (mg/L)
C_{cr} = constituent concentration in creek under current loading conditions (mg/L)
C_s = soil chemical concentration prior to cleanup (mg/kg)

Based on this modeling exercise, the site may be cleaned up to the appropriate maximum acceptable soil concentration value, C_{max}, so as *not* to impact the surface water quality under current contaminant loading conditions. The overall computational process for the analysis of surface water contamination is presented in Table 7.1—with the last column showing the restoration goals established for the cleanup of soils at the XYZ site. Thus, no significant/adverse impacts to the creek would be anticipated if site remediation is carried out to residual contaminant levels corresponding to the maximum acceptable soil concentrations estimated for this site.

TABLE 7.1

Computational Process for Modeling the Erosion of Contaminated Soils into Surface Water

K = 3.15E−01
L = 5.75E+02
S = 5.00E−03
LS = 1.40E−01
C = 1.00E+00
P = 1.00E+00
A = 5.95E+00

R_t = 3.50E+00
T_r = 2.40E+01
Average annual storm = 4.60E+01
River flow rate, QV_r = 8.45E−01
CN = 7.40E+01
Bulk density = 1.44E+00
Available water capacity = 1.20E−01
Organic fraction, F_{oc} = 1.00E−03

Strom events/year = 1.31E+01
Strom events in 70 years = 9.20E+02
S_w = 3.51E+00
Q_r = 1.24E+00
V_r = 6.12E−01
Peak flow, Q_p = 3.88E−01
$Y(S)_E$ = 1.87E+00

Chemical Constituent	K_{oc} (m³/kg)	K_{oc} (cm³/g)	K_d^a (cm³/g)	Soil Chemical Concentration C_a (mg/kg)	Sorbed Quantity S_a (kg)	Dissolved Quantity D_a (kg)	Sorbed Loss PX_t (kg)	Dissolved Loss PQ_t (kg)	Total Loss (kg)	Concentration in Runoff (mg/L)	Concentration in Surface Water (mg/L)	Chronic Water Quality Standard (mg/L)	Attenuation Factor	Maximum Acceptable Soil Concentration (mg/kg)
Benzene	8.30E−02	8.30E+01	8.30E−02	5.39E+01	9.33E+00	4.99E−03	1.21E−01	1.77E−03	1.23E−01	1.63E−01	5.13E−02	5.10E+00	9.52E−04	5,354
Benz(a) anthracene	1.38E+03	1.38E+06	1.38E+03	8.42E+01	2.92E+01	1.00E−02	3.50E+02	3.55E−03	3.50E+02	4.63E+02	1.46E+02	3.00E−01	1.73E+00	0.17
Benzo(a) pyrene	5.50E+03	5.50E+06	5.50E+03	1.08E+02	3.74E+01	1.00E−02	4.48E+02	3.55E−03	4.49E+02	5.94E+02	1.87E+02	3.00E−01	1.73E+00	0.17
Ethylbenzene	1.10E+00	1.10E+03	1.10E+00	6.52E+02	2.10E+02	9.31E−03	2.74E+00	3.30E−03	2.74E+00	3.63E+00	1.14E+00	4.30E−01	1.75E−03	246
Naphthalene	1.45E+00	1.45E+03	1.45E+00	4.80E+02	1.57E+02	9.47E−03	2.05E+00	3.36E−03	2.05E+00	2.72E+00	1.25E+00	3.00E−01	2.60E−03	115
Phenanthrene	1.40E+01	1.40E+04	1.40E+01	2.68E+02	9.24E+01	9.96E−03	1.20E+00	3.53E−03	1.21E+00	1.60E+00	7.34E−01	3.00E−01	2.74E−03	110
Phenol	1.42E−02	1.42E+01	1.42E−02	7.90E+02	3.99E+01	1.46E−03	5.19E−01	5.17E−04	5.20E−01	6.88E−01	2.17E−01	5.80E+00	2.74E−04	21,154
Toluene	3.00E−01	3.00E+02	3.00E−01	3.38E+03	9.17E+02	7.84E−03	1.19E+01	2.78E−03	1.19E+01	1.58E+01	4.98E+00	5.00E+00	1.47E−03	3,395

a A soil organic carbon fraction of 0.1% is assumed; $K_d = f_{oc} \times K_{oc}$.

7.5 SCOPE OF APPLICATION FOR THE USE OF CHEMICAL FATE AND BEHAVIOR MODELING IN EXPOSURE ANALYSES: A CASE FOR COUPLING CONTAMINANT FATE AND TRANSPORT MODELING WITH EXPOSURE MODELING

Regardless of how much environmental and/or exposure monitoring data is available, it is almost always desirable to generate one or more of the following attributes to support site characterization and/or corrective action decisions (Schnoor, 1996):

  i. An estimate of chemical concentrations under different sets of conditions
 ii. Results for a future chemical loading scenario
iii. A predicted "hindcast" or reconstructed history of chemical releases
 iv. Estimates at alternate [receptor or compliance] locations where field data do not exist

Under such circumstances, environmental models usually come in quite handy. Characteristically, multimedia mathematical models are often used to predict the potential for contaminant migration from a chemical release source to potential receptors, using pathways analyses concepts.

The general types of modeling practices used in exposure assessments for archetypical environmental chemical release scenarios commonly consist of a use of atmospheric, surface water, groundwater, multimedia, and food-chain models. In their practical applications, several modeling scenarios will typically be simulated and evaluated using the appropriate models for a given environmental contamination or chemical release problem. For example, the study of a contaminated site problem may require the modeling of infiltration of rain water, erosion/surface runoff release of chemicals, emission of particulate matter and vapors, chemical fate and transport through the unsaturated zone, chemical transport through the aquifer system, and/or mixing of groundwater with surface water—perhaps among several other things (Figure 7.3). In any case, due to the heterogeneity in environmental compartments and natural systems, models used for exposure assessments should be adequately tested, and insofar as possible, sensitivity runs should be carried out to help determine the most sensitive and/or critical parameters considered in the evaluation.

All in all, environmental models are typically designed to serve a variety of purposes—most importantly the following (Schnoor, 1996):

* To gain better understanding of the fate and transport of chemicals existing in, or to be introduced into, the environment
* To determine the temporal and spatial distributions of chemical exposure concentrations at potential receptor locations
* To predict future consequences of exposure under various chemical loading or release conditions, exposure scenarios, and/or management action alternatives
* To perform sensitivity analyses, by varying specific parameters, and then using models to explore the ramifications of such actions (as reflected by changes in the model outputs)

Ultimately, populations potentially at risk are designated, and then concentrations of the chemicals of concern are delineated or determined in each medium to which potential receptors may be exposed. Then, using the appropriate case-specific exposure parameter values, the intakes of the chemicals of concern can be estimated (see Part III). Indeed, such evaluations could be about past or current exposures or exposures anticipated in the future; this therefore makes mathematical modeling even more valuable—especially in the simulation of events and conditions that may not yet have occurred.

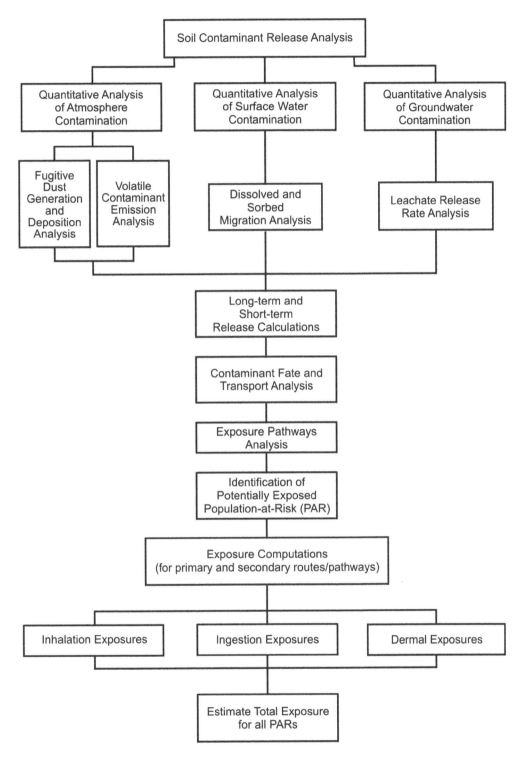

FIGURE 7.3 Scope of modeling applications: an example conceptual representation of the relationship between multimedia contaminant transfers and multipathway exposure analyses.

Part III

Contaminated Site Risk Assessment
Concepts, Principles, Techniques, and Methods of Approach

This part of the book is comprised of three sections—consisting of the following specific chapters:

- Chapter 8, *Principles, Concepts, and Procedural Elements in Contaminated Site Risk Assessment*, discusses key fundamental principles, concepts, and attributes that will facilitate the effectual application and interpretation of risk assessment information—and thus make it more suitable or useful in contaminated site risk management decisions; additionally, it presents a classical framework for carrying out the relevant evaluations in practice. This chapter further discusses the principal elements and activities necessary for obtaining and integrating the pertinent information that will eventually allow effective risk management and corrective action decisions to be made about a contaminated site and/or chemical exposure problem.
- Chapter 9, *Technical Approach to Human Health Endangerment Assessment*, presents a general discussion on methods of approach used for completing human health risk assessments as part of the investigation and management of contaminated site and chemical exposure problems. Specifically, it provides a procedural framework and an outline of the key elements of the health risk assessment process.
- Chapter 10, *Technical Approach to Ecological Endangerment Assessment*, elaborates the major components of the ecological risk assessment methodology, as may be applied to the evaluation of a contaminated site problem. Specifically, it provides a procedural framework and an outline of the key elements of the ecological risk assessment process.

8 Principles, Concepts, and Procedural Elements in Contaminated Site Risk Assessment

In planning for environmental and public health protection from the likely adverse effects from contaminated site problems, the first concern usually relates to whether or not the substance in question possesses potentially hazardous and/or toxic properties. As a corollary, once a chemical has been determined to present a potential health hazard, then the main concern becomes one of the likelihood for, and the degree of, exposure. In effect, risk to an exposed population is understood by examining the exposure that the population experiences relative to the hazard and the chemical potency information. Indeed, the assessment of health and environmental risks play an important role in site characterization activities, in corrective action planning, and also in risk mitigation and risk management strategies for contaminated site problems.

A major objective of any site-specific risk assessment is to provide an estimate of the baseline risks posed by the existing conditions at a contaminated site, and to further assist in the evaluation of site restoration options. Thus, appropriately applied, risk assessment techniques can be used to estimate the risks posed by site contaminants under various exposure scenarios and to further estimate the degree of risk reduction achievable by implementing various engineering remedies. Risk assessment methods commonly encountered in the literature of environmental and public health management and/or relevant to the management of contaminated site and chemical exposure problems characteristically require a clear understanding of several fundamental issues/tenets and related attributes. This chapter discusses key/fundamental principles, concepts, and attributes that will be expected to facilitate the effectual application and interpretation of risk assessment information—and thus make it more suitable or useful in contaminated site risk management decisions; additionally, it presents a classical framework for carrying out the relevant evaluations in practice. This chapter further discusses the principal elements and activities necessary for obtaining and integrating the pertinent information that will eventually allow effective risk management and corrective action decisions to be made about a contaminated site and/or chemical exposure problem.

8.1 FUNDAMENTAL PRINCIPLES OF CHEMICAL HAZARD, EXPOSURE, AND RISK ASSESSMENTS

Hazard is that object with the potential for creating undesirable adverse consequences; *exposure* is the situation of vulnerability to hazards; and *risk* is considered to be the probability or likelihood of an adverse effect due to some hazardous situation. Indeed, the distinction between hazard and risk is quite an important consideration in the overall appraisal of risk possibilities and/or scenarios; broadly speaking, it is the likelihood to harm as a result of exposure to a hazard that distinguishes risk from hazard. Accordingly, a substance is considered a hazard if it is capable of causing an adverse effect under any particular set of circumstance(s)—whereas risk generally reflects the probability that an adverse effect will occur under actual or realistic circumstances, also taking into account the potency of the specific substance and the level of exposure to that substance.

For example, a toxic chemical that is hazardous to human health does not constitute a risk unless human receptors/populations are exposed to such a substance. Thus, from the point of view of an organism or population exposure to chemicals, risk can be defined as the probability that the organism or population health could be affected to various degrees (including an individual or group suffering injury, disease, or even death) under specific set of circumstances.

The integrated and holistic assessment of hazards, exposures, and risks are indeed a very important contributor to any decision that is aimed at adequately managing any given hazardous situation. On the whole, a complete assessment of potential hazards posed by a substance or an object typically involves, among several other things, a critical evaluation of available scientific and technical information on the substance or object of concern, as well as the possible modes of exposure. In particular, it becomes increasingly apparent that potential receptors will have to be exposed to the hazards of concern before any risk could be said to exist. At the end of the day, the availability of an adequate and complete information set is an important prerequisite for producing sound hazard, exposure, and risk assessments.

Procedures for analyzing hazards and risks may typically be comprised of several steps and corresponding general elements (Asante-Duah, 2017). Some or all of the pertinent elements may have to be analyzed in a comprehensive manner, depending on the nature and level of detail of the hazard and/or risk analysis that is being performed. Anyhow, the analyses typically fall into two broad categories—namely: endangerment assessment (which may be considered as contaminant-based, such as human health and ecological risk assessment associated with chemical exposures) and safety assessment (which is system failure-based, such as probabilistic risk assessment of hazardous facilities or installations). At the end of the day, the final step will invariably be comprised of developing risk management and/or risk-prevention strategies for the problem situation.

8.1.1 What Is Risk Assessment?

Several somewhat differing definitions of risk assessment have been published in the literature by various authors to describe a variety of risk assessment methods and/or protocols (see, e.g., Asante-Duah, 1998, 2017; Cohrssen and Covello, 1989; Conway, 1982; Cothern, 1993; Covello et al., 1986; Covello and Mumpower, 1985; Crandall and Lave, 1981; Davies, 1996; Glickman and Gough, 1990; Gratt, 1996; Hallenbeck and Cunningham, 1988; Kates, 1978; Kolluru et al., 1996; LaGoy, 1994; Lave, 1982; Neely, 1994; Norrman, 2001; NRC, 1982a, 1983b, 1994b,d; Richardson, 1990, 1992; Rowe, 1977; Scheer et al., 2014; Turnberg, 1996; USEPA, 1984a,c; Whyte and Burton, 1980). In a generic sense, risk assessment may be considered to be a systematic process for arriving at estimates of all the significant risk factors or parameters associated with an entire range of "failure modes" and/or exposure scenarios in connection with some hazard situation(s). It entails the evaluation of all pertinent scientific information to enable a description of the likelihood, nature, and extent of harm to human and ecological health as a result of exposure to chemicals (and really other potential stressors) present in the human and ecological environments.

Risk assessment is indeed a scientific process that can be used to identify and characterize chemical exposure-related human and ecological health problems. In its application to the investigation and management of potentially contaminated site problems, the process encompasses an evaluation of all the significant risk factors associated with all feasible and identifiable exposure scenarios that are the result of contaminant releases into the environment. It may, for instance, involve the characterization of potential adverse consequences or impacts to human and ecological receptors that are potentially at risk from exposure to chemicals found at contaminated sites. All in all, risk assessment may be viewed as a powerful tool for developing insights into the relative importance of the various types of exposure scenarios associated with potentially hazardous situations. But as Moeller (1997) points out, it has to be recognized that a given risk assessment provides only a snapshot in time (and indeed in space as well) of the estimated risk of a given toxic agent at a particular phase of our understanding of the issues and problems. To be truly instructive and constructive, therefore, risk assessment should preferably be conducted on an iterative basis—being continually updated as new knowledge and information become available. Indeed, unless care is exercised and all interacting factors are considered, risk assessments directed at single issues, followed by ill-conceived risk management strategies, can create problems worse than those the management strategies were designed to correct. The single-issue approach can also create public myopia by excluding the totality of alternatives and consequences needed for an informed public choice (Moeller, 1997). Ultimately, it is quite important to examine the total system to which a given risk assessment is being applied.

It is expected that there will be growing applications of the risk assessment paradigm to several specific contaminated site problems, and this could affect the type of decisions made in relation to an environmental and/or public health risk management program. Such applications may cover a wide range of diverse problem situations, including several completely different and unique problems that may be resolved by the use of one form of risk assessment principle and methodology or another—as exemplified by the illustrative application scenarios annotated in Chapter 14 and elsewhere throughout the book.

As a final point here, it is noteworthy that, in general, some risk assessments may be classified as *retrospective*—that is, focusing on injury after the fact (e.g., nature and level of risks at a given contaminated site) or it may be considered as *predictive*—such as in evaluating possible future harm to human health or the environment (e.g., risks anticipated if a newly developed food additive is approved for use in consumer food products). Anyhow, in relation to the investigation of contaminated sites and chemical exposure problems, it is apparent that the focus of most risk assessments tends to be on a determination of potential or anticipated risks to the populations potentially at risk. Overall, the risk assessment process seeks to estimate the likelihood of occurrence of adverse effects resulting from exposures of human and/or ecological receptors to chemical, physical, and/or biological agents present in the receptor environments. [By the way, it is noteworthy that, human exposures to radiological contaminants may be evaluated in a manner similar to the chemical exposure problems—albeit certain unique issues may have to be taken into consideration for the radiological exposures (that occur via sources such as naturally occurring radioactive materials in soils, groundwater and ambient air.)] The process consists of a mechanism that utilizes the best available scientific knowledge to establish case-specific responses that will ensure justifiable and defensible decisions—as necessary for the management of hazardous situations in a cost-efficient manner. The process is also concerned with the assessment of the importance of all identified risk factors to the various stakeholders whose interests are embedded in a candidate problem situation (Petak and Atkisson, 1982).

8.1.2 Baseline Risk Assessments

Baseline risk assessments involve an analysis of the potential adverse effects (current or future) caused by receptor exposures to hazardous substances in the absence of any actions to control or mitigate these exposures—that is, under an assumption of "no-action." Thus, the baseline risk assessment provides an estimate of the potential risks to the populations at risk that follows from the receptor exposure to the hazards of concern, when no mitigative actions have been considered. Because this type of assessment identifies the primary threats associated with the situation, it also provides valuable input to the development and evaluation of alternative risk management and mitigative options. In fact, baseline risk assessments are usually conducted to evaluate the need for, and the extent of, corrective action in relation to a hazardous situation; that is, they provide the basis and rationale as to whether or not remedial action is necessary. Ultimately, the results of the baseline risk assessment are generally used for reach the following types of goals:

- Document the magnitude of risk at a given locale, as well as the primary causes of the risk.
- Help determine whether any response action is necessary for the problem situation.

- Prioritize the need for remedial action, where several problem situations are involved.
- Provide a basis for quantifying remedial action objectives.
- Develop and modify remedial action goals.
- Support and justify "no further action" decisions, as appropriate—by documenting the likely inconsequentiality of the threats posed by the hazard source(s).

On the whole, baseline risk assessments are designed to be case-specific—and therefore may vary in both detail and the extent to which qualitative and quantitative analyses are used. Also, it is noteworthy that the level of effort required to conduct a baseline risk assessment depends largely on the complexity and particular circumstances associated with the hazard situation under consideration.

8.1.3 RISK ASSESSMENT VERSUS RISK MANAGEMENT

Risk assessment has been defined as the "characterization of the potential adverse health effects of human exposures to environmental hazards" (NRC, 1983b). In a typical risk assessment, the extent to which a population has been or may be exposed to a certain chemical is determined; the extent of exposure is then considered in relation to the kind and degree of hazard posed by the chemical—thereby allowing an estimate to be made of the present or potential risk to the target population. Depending on the problem situation, different degrees of detail may be required for the process; in any event, the continuum of acute to chronic hazards and exposures would typically be fully investigated in a comprehensive assessment, so that the complete spectrum of risks can be defined for subsequent risk management decisions.

The risk management process—that utilizes prior-generated risk assessment information—involves making a decision on how to protect human and/or ecological health. Examples of risk management actions include deciding on how much of a given chemical of concern/interest an operating industry or company may discharge into a river; deciding on which substances may be stored at a hazardous waste disposal facility; deciding on the extent to which a hazardous waste site must be cleaned up; setting permit levels for chemical discharge, storage, or transport; establishing levels for air pollutant emissions; and determining the allowable levels of contamination in drinking water or food products. In a way, this generically portrays how risk management is distinct from risk assessment—but nevertheless maintains a fundamental relationship.

At the end of the day, risk assessment is generally conducted to facilitate risk management decisions; whereas risk assessment focuses on evaluating the likelihood of adverse effects, risk management involves the selection of a course of action in response to an identified risk—with the latter often based on many other factors (e.g., social, legal, political, or economic) over and above the risk assessment results. Essentially, risk assessment provides *information* on the likely risk, and risk management is the *action* taken based on that information (in combination with other "external" but potentially influential factors) to protect the affected populations.

8.2 FUNDAMENTAL CONCEPTS IN RISK ASSESSMENT PRACTICE

The general types of risk assessment often encountered in practice may range from an evaluation of the potential effects of toxic chemical releases known to be occurring, up through to evaluations of the potential effects of releases due to events whose probability of occurrence is uncertain (Moeller, 1997). Regardless, in order to adequately evaluate the risks associated with a given hazard situation, several concepts are usually used in the processes involved. Some of the fundamental concepts and definitions that will generally facilitate a better understanding of the risk assessment process and related application principles, and that may also affect risk management decisions, are introduced as follows (Asante-Duah, 2017):

- *Qualitative versus Quantitative Risk Assessment.* In contaminated site risk assessments, quantitative tools are often used to better define exposures, effects, and risks in the broad context of risk analysis. Such tools will usually use the plausible ranges associated with default exposure scenarios, toxicological parameters, and indeed other assumptions and policy positions. Although the utility of numerical risk estimates in risk analysis has to be appreciated, these estimates should be considered in the context of the variables and assumptions involved in their derivation—and indeed in the broader context of likely scientific or biomedical opinion, host factors, and actual exposure conditions. Consequently, directly or indirectly, qualitative descriptors also become part of a quantitative risk assessment process. For instance, in evaluating the assumptions and variables relating to both toxicity and exposure conditions for a chemical exposure problem, the risk outcome may be provided in qualitative terms—albeit the risk levels are expressed in quantitative terms.

  Indeed, measures used in risk analysis take various forms, depending on the type of problem, degree of resolution appropriate for the situation on hand, and the analysts' preferences. Thus, the risk parameter may be expressed in quantitative terms—in which case it could take on values from zero (associated with certainty for no-adverse effects) to unity (associated with certainty for adverse effects to occur). In several other cases, risk is only described qualitatively—such as by use of descriptors such as "high," "moderate," and "low; or indeed, the risk may be described in semiquantitative/semi-qualitative terms. In any case, the risk qualification or quantification process will normally rely on the use of several measures, parameters, and/or tools as reference yardsticks (Asante-Duah, 2017)— with "individual lifetime risk" (represented by the probability that the individual will be subjected to an adverse effect from exposure to identified hazards) being about the most commonly used measure of risk. Meanwhile, it is also worth mentioning here that the type or nature of "consuming/target audience" must be given careful consideration in choosing the type of risk measure or index to adopt for a given program or situation.

  In general, the attributable risk for any given problem situation can be expressed in qualitative, semiquantitative, or quantitative terms. For instance, in conveying qualitative conclusions regarding chemical hazards, narrative statements incorporating "weight-of-evidence" or "strength-of-evidence" conclusions may be used—that is, in lieu of alphanumeric designations alone being used. In other situations, pure numeric parameters are used—and yet in other circumstances, a combination of both numeric parameters and qualitative descriptors are used in the risk presentations/discussions.

- *Risk Categorization.* Oftentimes in risk studies, it becomes necessary to put the degree of hazards or risks into different categories for risk management purposes. A typical risk categorization scheme for potentially contaminated site problems might involve a grouping of the "candidate" problem sites on the basis of the potential risks attributable to various plausible conditions—such as high-, intermediate-, and low-risk sites, as conceptually depicted earlier by Figure 2.2 (Chapter 2). Under such classification scheme, a case-specific problem may be designated as "high-risk" when exposure represents real or imminent threat to human health or the environment; in general, the high-risk sites will prompt the most concern—requiring immediate and urgent attention or corrective measures to reduce the threat. Indeed, to ensure the development of adequate and effectual risk management or corrective action strategies, potential chemical exposure problems may need to be prudently categorized in a similar or other appropriate manner during the risk analysis. In the end, such a classification would likely facilitate the development and implementation of a more efficient risk management or corrective action program.

- *Conservatisms in Risk Assessments.* Many of the parameters and assumptions used in hazard, exposure, and risk evaluation studies tend to have high degrees of uncertainties associated with them—thereby potentially clouding the degree of confidence assigned to any estimated measures of safety. Conversely, "erring on the side of safety" tends to be the universal "mantra" of most safety designers and analysts. To facilitate a prospective safe design and analysis, it is a common practice to model risks such that risk levels determined for management decisions are preferably overestimated. Such "conservative" estimates (also, often cited as "worst-case," or "plausible upper bound" estimates) used in risk assessment are based on the supposition that pessimism in risk assessment (with resultant high estimates of risks) is more protective of public health and/or the environment.

Indeed, in performing risk assessments, scenarios have often been developed that will reflect the worst possible exposure pattern; this notion of *"worst-case scenario"* in the risk assessment generally refers to the event or series of events resulting in the greatest exposure or potential exposure. On the other hand, gross exaggeration of actual risks could lead to poor decisions being made with respect to the oftentimes very limited resources available for general risk mitigation purposes. Thus, after establishing a worst-case scenario, it is often desirable to also develop and analyze more realistic or "nominal" scenarios, so that the level of risk posed by a hazardous situation can be better bounded—via the selection of a "best" or "most likely" sets of assumptions for the risk assessment. But in deciding on what realistic assumptions are to be used in a risk assessment, it is imperative that the analyst chooses parameters that will, at the very worst, result in erring on the side of safety. Anyhow, it is notable that a number of investigators (see, e.g., Anderson and Yuhas, 1996; Burmaster and von Stackelberg, 1991; Cullen, 1994; Maxim, *in* Paustenbach, 1988) have been offering a variety of techniques that could help make risk assessments more realistic—that is, rather than the dependence on wholesale compounded conservative assumptions.

- *Individual versus Group Risks.* In the application of risk assessment to environmental and public health risk management programs, it often becomes important to distinguish between "individual" and "societal" risks—in order that the most appropriate metric/measure can be used in the analysis of case-specific problems. *Individual risks* are considered to be the frequency at which a given individual could potentially sustain a given level of adverse consequence from the realization or occurrence of specified hazards. *Societal risk*, on the other hand, relates to the frequency and the number of individuals sustaining some given level of adverse consequence in a given population due to the occurrence of specified hazards; the population risk provides an estimate of the extent of harm to the population or population segment under review.

  Individual risk estimates represent the risk borne by individual persons within a population—and are more appropriate in cases where individuals face relatively high risks. However, when individual risks are not inequitably high, then it becomes important during resources allocation, to deliberate on possible society-wide risks that might be relatively higher. Indeed, risk assessments almost always deal with more than a single individual. However, individual risks are also frequently calculated for some or all of the individual organisms in the population being studied, and these are then put into the context of where they fall in the distribution of risks for the entire population.

- *Deterministic versus Probabilistic Risk Assessments. Deterministic* risk assessment methods generally involve exclusive use of key data sets that lead to specific "singular" and/ or "monotonic" outcomes—often considered the "traditionalist" approach. *Probabilistic* [or, *Stochastic*] methods of approach typically entail the application of statistical tools that incorporate elements of random behavior in key data sets—often viewed as the more "contemporaneous" approach.

  In the application of risk assessment to environmental and human health risk management programs, it has become commonpractice to utilize either or both of deterministic and probabilistic methods of approach—in efforts to facilitate the most effectual decision-making processes and that would adequately support human health risk management needs. In practice, the deterministic approach to risk assessments can be said to be the classical or traditional tool preceding the development of stochastic or probabilistic methodologies. On the other hand, because deterministic models generally do not explicitly consider uncertainty in key variables and/or model parameters, such models provide a rather limited picture to support effectual risk management programs. Even so, deterministic models can be relied upon to a great extent for certain preliminary studies— that is, usually prior to a more detailed stochastic optimization or simulation study. Indeed, stochastic methods typically come into use when the deterministic approach is found to be somehow deficient. Regardless, stochastic processes may be conveniently evaluated in such a manner, and conclusions associated with them drawn and treated, as if the process was somehow deterministic.

- *Risk Acceptability and Risk Tolerance Criteria: de Minimis versus de Manifestis Risk Criteria.* An important concept in risk management is that there are levels of risk that are so great that they must not be allowed to occur at all cost, and yet there are other risk levels that are so low that they are not worth bothering with even at insignificant costs—known, respectively, as *de manifestis* and *de minimis* levels (Kocher and Hoffman, 1991; Suter, 1993; Travis et al., 1987; Whipple, 1987). Risk levels between these bounds are typically balanced against costs; technical feasibility of mitigation actions; and other socioeconomic, political, and legal considerations—in order to determine their acceptability or tolerability.

  On the whole, the concept of *de manifestis* risk is usually not seen as being controversial—because, after all, some hazard effects are clearly unacceptable. However, the *de minimis* risk concept tends to be controversial—in view of the implicit idea that some exposures to, and effects of, pollutants or hazards are acceptable (Suter, 1993). With that noted, it is still desirable to use these types of criteria to eliminate obviously trivial risks from further risk management actions—considering the fact that society cannot completely eliminate or prevent all human and environmental health effects associated with chemical exposure problems. Indeed, virtually all social systems have target risk levels—whether explicitly indicated or not—that represent tolerable limits to danger that the society is (or must be) prepared to accept in consequence of potential benefits that could accrue from a given activity. This tolerable limit is often designated as the *de minimis* or *acceptable* risk level. In simple terms, the *de minimis* principle assumes that extremely low risks are trivial and need not be controlled. The concept of *de minimis* or *acceptable* risk is essentially a threshold concept, in that it postulates a threshold of concern below which there would be indifference to changes in the level of risk. A *de minimis* risk level would therefore represent a cutoff, below which a regulatory agency could simply ignore related alleged problems or hazards. Thus, in the general process of establishing "acceptable" risk levels, it is possible to use *de minimis* levels below which one need not be concerned (Rowe, 1983); it is notable that current regulatory requirements are particularly important considerations in establishing such acceptable risk levels.

  At the end of the day, it is apparent that the concept of "acceptable risk level" relates to a very important issue in risk assessment—albeit the desirable or tolerable level of risk is not always attainable. Anyhow, it is noteworthy that risk acceptability (i.e., the level of risk that society can allow for a specified hazard situation) usually will have a spatial and temporal variability to it.

Further elaboration of the above-mentioned (and indeed related) concepts can be found elsewhere in the literature of risk assessment (e.g., Asante-Duah, 2017). Meanwhile, it is noteworthy that careful application of risk assessment and risk management principles and tools should generally help remove some of the fuzziness in defining the cutoff line between what may be considered a "safe level" and what apparently is a "dangerous level" for most chemicals—thus generating effectual management decisions for a given contaminated site problem.

8.3 GENERAL ATTRIBUTES OF CONTAMINATED SITE RISK ASSESSMENT

The conventional paradigm for risk assessment tends to lean towards its *predictive* nature—which generally deals with localized outcomes of a particular action that could result in adverse effects. However, there also has been increasing emphasis on assessments of the effects of environmental and public health hazards associated with "in-place" or existing chemical exposure problems; this assessment of past pollutions and exposures, with possible ongoing consequences, generally falls under the umbrella of what has been referred to as *retrospective* risk assessment (Suter, 1993). The impetus for a retrospective risk assessment may be a source, observed effects, or evidence of exposure. Source-driven retrospective assessments typically arise from observed pollution or exposures that requires elucidation of possible effects (e.g., hazardous waste sites, spills/accidental releases, and consumer product usage); effects-driven retrospective assessments usually ensue from the observation of perceptible effects in the field that requires explanation (e.g., localized public health indicators, fish or bird kills, and declining populations of a species); and exposure-driven retrospective assessments are normally prompted by evidence of exposure without prior evidence of a source or effects (e.g., the case of a scare over mercury found in the edible portions of dietary fish). In all cases, however, the principal objective of the risk assessment is to provide a basis for actions that will minimize the impairment of the environment and/ or of public health, welfare, and safety.

In general, risk assessment—which seems to be one of the fastest evolving tools for developing appropriate strategies in relation to environmental and public health management decisions—seeks to answer three basic questions:

- What could potentially go wrong?
- What are the chances for this to happen?
- What are the anticipated consequences, if this should indeed happen?

A complete analysis of risks associated with a given situation or activity will likely generate answers to these questions. Indeed, tasks performed during a risk assessment will generally help answer the infamous questions of "how safe is safe enough?" and/or "how clean is clean enough?" Subsequently, risk management becomes part of the overall evaluation process—in order to help address the archetypical follow-up question of "what can be done about the prevailing situation?" At this point in time, a decision would typically have to be made as to whether any existing risk is sufficiently high to represent an environmental and/or public health concern—and if so, to determine the nature of risk management actions. Appropriate mitigative activities can then be initiated by implementing the necessary corrective action and risk management decisions.

8.3.1 THE PURPOSE

The overall goal in a risk assessment is to identify potential "system failure modes" and exposure scenarios—and this is achieved via the fulfillment of several general objectives (Asante-Duah, 2017). This process is intended to facilitate the design of methods that will help reduce the probability of "failure," as well as minimize the attending public health, socioeconomic, and environmental consequences of any "failure" and/or exposure events. Its overarching purpose is to provide, insofar as possible, complete information sets to risk managers—so that the best possible decision can be made concerning a potentially hazardous situation. Indeed, as Whyte and Burton (1980) succinctly indicate, a major objective of risk assessment is to help develop risk management decisions that are more systematic, more comprehensive, more accountable, and more self-aware of appropriate programs than has often been the case in the past. The risk assessment process provides a framework for developing the risk information necessary to assist risk management decisions; information developed in the risk assessment will typically facilitate decisions about the allocation of resources for safety improvements and hazard/risk reduction. Also, the analysis will generally provide decision makers with a more justifiable basis for determining risk acceptability, as well as aid in choosing between possible corrective measures developed for risk mitigation programs.

When all is said and done, the information generated in a risk assessment is often used to determine the need for and the degree of mitigation required for chemical exposure problems. For instance, risk assessment techniques and principles are frequently utilized to facilitate the development of effectual site characterization and corrective action programs for contaminated sites scheduled for decommissioning and subsequent redevelopment for mixed uses (e.g., residential housing and commercial properties). In addition to providing information about the nature and magnitude of potential health and environmental risks associated with the contaminated site problem, the risk assessment also provides a basis for judging the need for any type of remedial action (Asante-Duah, 1998). Furthermore, risk assessment can be used to compare the risk reductions afforded by different remedial or risk control strategies. Indeed, the use of risk assessment techniques in contaminated site cleanup plans in particular, and corrective action programs in general, is becoming increasingly important and popular in so many localities. This is because the risk assessment serves as a useful tool for evaluating the effectiveness of remedies at contaminated sites and also for establishing cleanup objectives (including the determination of cleanup levels) that will produce efficient, feasible, and cost-effective remedial solutions. Typically, the general purpose for this type of problem situation is to gather sufficient information that will allow for an adequate and accurate characterization of the potential risks associated with the project site. In this case, the risk assessment process is used to determine whether the level of risk at a contaminated site warrants remediation and then to further project the amount of risk reduction necessary to protect public health and the environment. On this basis, an appropriate corrective action plan can then be developed and implemented for the case site and/or the impacted area.

8.3.2   THE ATTRIBUTES

The risk assessment process typically utilizes the best available scientific knowledge and data to establish case-specific responses in relation to hazard–receptor interactions. Depending on the scope of the analysis, the methods used in estimating risks may be either qualitative or quantitative— or indeed combinations thereof. Thus, the process may be one of data analysis or modeling or a combination of the two. In fact, the type and degree of detail of any risk assessment depends on its intended use; its purpose will generally shape the data needs, the protocol, the rigor, and the related efforts. In the end, the process of quantifying risks does, by its very nature, give a better understanding of the strengths and weaknesses of the potential hazards being examined. It also shows where a given effort can do the most good in modifying a system—in order to improve their safety and efficiency. Meanwhile, it is worth the mention here that the processes involved in any risk assessment usually require a multidisciplinary approach—often covering several areas of expertise in most situations.

The major attributes of risk assessment that are particularly relevant to environmental and public health risk management programs include the following:

- Identification and ranking of all existing and anticipated potential hazards
- Explicit consideration of all current and possible future exposure scenarios
- Qualification and/or quantification of risks associated with the full range of hazard situations, system responses, and exposure scenarios
- Identification of all significant contributors to the critical pathways, exposure scenarios, and/or total risks

- Determination of cost-effective risk-reduction policies, via the evaluation of risk-based remedial action alternatives and/or the adoption of efficient risk management and risk prevention programs
- Identification and analysis of all significant sources of uncertainties

Meanwhile, it is noteworthy that, there are inherent uncertainties associated with all risk assessments; this is due in part to the fact that the analyst's knowledge of the causative events and controlling factors usually is limited and also because the results obtained oftentimes tend to be dependent on the methodology and assumptions used. Furthermore, risk assessment can impose potential delays in the implementation of appropriate corrective measures—albeit the overall gain in program efficiency, as well as other potential advantages, is likely to more than compensate for any delays. But as Moeller (1997) points out, unless care is exercised and all interacting factors considered, the outcome could be a risk assessment directed at single issues, followed by ill-conceived management strategies—and this can create problems worse than those the management strategies were designed to correct in the first place. In fact, the single-issue approach can also create "public myopia" by excluding the totality of [feasible] alternatives and consequences essential for a more informed public/stakeholder preferred choice; consequently, it is imperative to ensure a more comprehensive evaluation that contemplates multiple feasible alternative management strategies.

8.3.3 RISK ASSESSMENT AS AN HOLISTIC GLOBAL TOOL FOR CONTAMINATED SITES RISK MANAGEMENT

Human populations and ecological receptors are continuously in contact with varying amounts of environmental contaminants present in air, water, soil, and food. Thus, methods for linking contaminant sources in the multiple environmental media to human and ecological receptor exposures are often necessary to facilitate the development of sound site characterization and credible corrective action or risk management programs. In particular, characterization of the risks associated with environmental contaminants usually requires an integrated model to effectively assess the environmental fate and behavior of the constituents of concern, as well as to determine potential human and ecosystem exposures.

Risk assessment is a process used to determine the magnitude and probability of actual or potential harm that a hazardous situation poses to human health and the environment. As an holistic approach to environmental and public health risk management, risk assessment integrates all relevant environmental and health issues and concerns surrounding a specific problem situation, in order to arrive at risk management decisions that are considered acceptable to all stakeholders. Among other things, the overall process should generally incorporate information that helps to answer the following pertinent questions (Asante-Duah, 2017):

- Why is the project/study being undertaken?
- How will results and conclusions from the project/study be used?
- What specific processes and methodologies will be utilized?
- What are the uncertainties and limitations surrounding the study?
- What contingency plans exist for resolving newly identified issues?

Also, effective risk communication should be recognized as a very important element of the holistic approach to managing chemical exposure and related environmental hazard problems. Thus, a system for the conveying of risk information derived from a risk assessment should be considered as a very essential integral part of the overall technique (Asante-Duah, 1998; 2017).

In its application to contaminated site and chemical exposure problems, the risk assessment process is used to compile and organize the scientific information that is necessary to support environmental and public health-risk management decisions. The approach is used to help identify potential problems, establish priorities, and provide a basis for regulatory actions. Indeed, it is apparent that the advancement of risk analysis in regulatory decision-making—among several others—has helped promote rational policy deliberations over the past several decades. Yet, as real-world practice indicates, risk analyses have often been as much the source of controversy in regulatory considerations as the facilitator of consensus (ACS and RFF, 1998). Anyhow, risk assessment can appropriately be regarded as a valuable tool for public health and environmental decision-making—albeit there tends to be disagreement among experts and policy makers about the extent to which its findings should influence decisions about risk. To help produce reasonable/pragmatic and balanced policies in its application, it is essential to explicitly recognize the character, strengths, and limitations of the analytical methods that are involved in the use of risk analyses techniques in the decision-making process.

Finally, to effectively utilize it as a public health and environmental management tool, risk assessment should be recognized as a multidisciplinary process that draws on data, information, principles, and expertise from many scientific disciplines—including biology, chemistry, earth sciences, engineering, epidemiology, medicine and health sciences, physics, toxicology, and statistics, among others. Indeed, risk assessment may be viewed as bringing a wide range of subjects and disciplines—from "archaeology to zoology"—together, to facilitate a more informed decision-making.

8.3.3.1 Diagnostic and Investigative Attributes of Risk Assessment

Risk assessment is generally considered an integral part of the diagnostic assessment of potentially contaminated site problems. In its application to the investigation of potentially contaminated site problems, the risk assessment process encompasses an evaluation of all the significant risk factors associated with all feasible and identifiable exposure scenarios that are the result of contaminant releases into the environment. It involves the characterization of potential adverse consequences or impacts to human and ecological receptors that are potentially at risk from exposure to site contaminants. Invariably, risk management and corrective action decisions about environmental contamination and contaminated site problems—including decisions about remediating contaminated sites—are made primarily on the basis of potential human health and ecological risks.

Procedures typically used in the risk assessment process will characteristically be comprised of the following key tasks:

- Identification of the sources of environmental contamination and chemical exposures
- Determination of the contaminant migration pathways and chemical exposure routes
- Identification of populations potentially at risk
- Determination of the specific chemicals of potential concern (CoPCs)
- Determination of frequency of potential receptor exposures to site contaminants
- Evaluation of contaminant exposure levels
- Determination of receptor response to chemical exposures
- Estimation of likely impacts or damage resulting from receptor exposures to the CoPCs

Ultimately, potential risks are estimated by considering the probability or likelihood of occurrence of harm; the intrinsic harmful features or properties of specified hazards; the populations potentially at risk; the exposure scenarios; and the extent of expected harm and potential effects. The results of the risk assessment are typically used to:

- Document the magnitude of risk or threats posed by a site, including the primary causes of the risk
- Facilitate a determination as to whether further response action is necessary at a site, or to support and justify a "no-further-action" decision, based on both existing and anticipated exposure scenarios associated with the site
- Prioritize the need for site restoration and provide a basis for quantifying remedial action objectives

Anyhow, in most typical applications, risk assessment is used to provide a baseline estimate of existing risks that are attributable to a specific agent or hazard; the baseline risk assessment consists of an evaluation of the potential threats to human health and the environment in the absence of any remedial or response action. Among several other things, the risk assessment process can also be used to determine the potential reduction in exposure and risk under various corrective action scenarios, as well as to support remedy selection in risk mitigation/abatement or control programs.

8.4 CONTAMINATED SITES AS A SOURCE OF WIDE-RANGING HUMAN EXPOSURES TO ENVIRONMENTAL CHEMICALS

Contaminated sites may arise in a number of ways—many of which are the result of manufacturing and other industrial activities or operations. In addition to the classical contaminated site situations involving direct releases at a given locale, several different physical and chemical processes can also affect contaminant migration from contaminated soils; thus, contaminated soils can potentially impact several other environmental matrices. For instance, atmospheric contamination may result from emissions of contaminated fugitive dusts and volatilization of chemicals present in soils; surface water contamination may result from contaminated runoff and overland flow of chemicals (from leaks, spills, and so on) and chemicals adsorbed to mobile sediments; groundwater contamination may result from the leaching of toxic chemicals from contaminated soils or the downward migration of chemicals from lagoons and ponds and so on. Indeed, several different physical and chemical processes affect chemical migration from contaminated sites—often resulting in a multitude of human exposures to various environmental chemicals. Accordingly, human exposures to chemicals at contaminated sites may occur in a variety/multiplicity of ways—including via the following more common example pathways (Asante-Duah, 1998, 2017):

- Direct inhalation of airborne vapors and also respirable particulates
- Deposition of airborne contaminants onto soils, leading to human exposure via dermal absorption or ingestion
- Ingestion of food products that have been contaminated as a result of deposition onto crops or pasture lands and subsequent introduction into the human food chain
- Ingestion of contaminated dairy and meat products from animals consuming contaminated crops or waters.
- Deposition of airborne contaminants onto waterways, uptake through aquatic organisms, and eventual human consumption of impacted aquatic foods.
- Leaching and runoff of soil contamination into water resources and consequential human exposures to contaminated waters in a water supply system

Contaminated sites, therefore, will usually represent a potentially long-term source for human exposure to a variety of chemical toxicants—making contaminated site issues a complex problem with worldwide implications; risk to public health arising from soils at contaminated sites is a matter of particularly greater concern.

8.4.1   A Classic Case Scenario Involving Chemical Vapor Intrusion
from Contaminated Sites into Overlying Buildings

There are a number of different kinds of indoor environments—with the most prominent consisting of offices or commercial buildings, homes, and schools—each with unique characteristics and associated problems. Among several others, indoor exposures can occur when substances are transported from outdoor sources into a building (as for example, when contaminated soil is tracked into buildings or gases volatilize from underlying contaminated soil or groundwater—usually referred to as "vapor intrusion" [VI]).

VI of chemicals generally refers to the migration of volatile chemicals from the subsurface into an overlying building—often constituting a potentially significant source of indoor volatile (organic) chemicals; more specifically, it is defined as the vapor-phase migration of (usually toxic) VOCs from a subsurface environment (e.g., contaminated soil and/or groundwater) into overlying or nearby structures/buildings (e.g., through floor slabs and foundation joints or cracks and gaps around utility lines), subsequently accumulating (to potentially "unacceptable" or "unsafe" levels) and potentially persisting in the indoor air—ultimately with consequential impacts on the indoor air quality and, thus, potentially posing risks to building occupants. Generally speaking, VOCs are characterized by relatively high vapor pressures that permit these compounds to vaporize and enter the atmosphere under normal conditions; because of these characteristics, the VI phenomenon is particularly unique or prevalent to this class of organic chemicals. Still, it is also notable that although VOCs typically present the most common concerns in regards to VI issues, there are a number of other contaminant families that may similarly engender VI problems—including other "vapor-forming" chemicals such as some semi-volatile organic compounds (semi-VOCs or SVOCs), elemental mercury, and radionuclides.

By and large, volatile chemicals in buried wastes or other subterranean contaminated soils/ groundwater can emit vapors that may in turn migrate through subsurface soils (and/or via sub-slabs, crawlspaces, and so on) into the indoor air spaces of overlying buildings. When this happens, the chemical concentrations in the released soil gas typically would decrease (or attenuate) as the vapors migrate through materials from the contamination sources into the overlying structures. This attenuation is usually the result of processes that control vapor transport in the soil materials (e.g., diffusion, advection, sorption, and potentially biotransformation), as well as processes that control the transport and dilution of vapors as they enter the building and mix with indoor air (e.g., pressure differential and building ventilation rates). Indeed, several other physicochemical and ambient environmental factors may generally affect the ultimate fate and behaviors of the chemicals of interest in any given VI problem situations.

As an archetypical illustrative example of a VI problem scenario, consider a situation whereby chlorinated solvents or petroleum products are accidentally released at an industrial or commercial facility—which then migrates downward and reaches groundwater where it can slowly dissolve and form contaminant plumes. Subsequently, the volatile compounds can volatilize and travel upward as soil vapors to reach the ground surface; in situations where buildings or other occupied structures sit atop such ground surface, contaminant vapors can seep through foundation cracks/joints and contaminate indoor air—presenting potentially serious public health concerns. Indeed, in view of the fact that many of the typical volatile compounds [such as benzene, tetrachloroethylene/perchloroethylene, and trichloroethylene (TCE)] are considered carcinogenic, there is always the concern that even relatively low levels of such chemicals inhaled by building occupants can pose unacceptable long-term health risks. On the other hand, evaluation of the VI pathway tends to be complicated by "background" volatile compound contributions (e.g., due to potential confounding effects of household VOC sources from consumer products), as well as considerable spatial and temporal variability in soil vapor and indoor air concentrations. Undeniably, vapor migration from subsurface environments into indoor air is often affected by many variables—not the least of which include building characteristics, anthropogenic conditions, and meteorological influences or seasonal changes; subsequent attenuation due to diffusion, advection, sorption, and potential degradation processes may also occur during movements from the contaminant source into the receptor exposure zones. Consequently, it makes more sense to use "multiple lines of evidence" to support and adequately/holistically evaluate the VI pathway and associated potential risks to public health.

Finally, it is worth mentioning that VI is considered an "emerging" and growing public health problem/concern that requires deliberate planning efforts—and even more importantly, careful assessment and management strategies to avert potential "hidden" but serious public health hazard situations. This might mean implementing aggressive VI pathway assessment at potentially contaminated sites or impacted structures—and then ensuring the implementation of appropriate vapor mitigation measures, as necessary.

8.5 THE RISK ASSESSMENT PROCESS: A GENERAL FRAMEWORK FOR CONDUCTING RISK ASSESSMENTS

Risk assessment is a scientific process that can be used to identify and characterize chemical exposure-related human and ecological health problems. Specific forms of risk assessment generally differ considerably in their levels of detail. Most risk assessments, however, share the same general logic—consisting of four basic elements, namely, hazard assessment, dose–response assessment, exposure assessment, and risk characterization (Figure 8.1).

Hazard assessment describes, qualitatively, the likelihood that a chemical agent can produce adverse health effects under certain environmental exposure conditions. *Dose–response assessment* quantitatively estimates the relationship between the magnitude of exposure and the degree and/or probability for the occurrence of a particular health effect. *Exposure assessment* determines the extent of receptor exposure. *Risk characterization* integrates the findings of the first three components to describe the nature and magnitude of health risk associated with environmental exposure to a chemical substance or a mixture of substances. A discussion of these

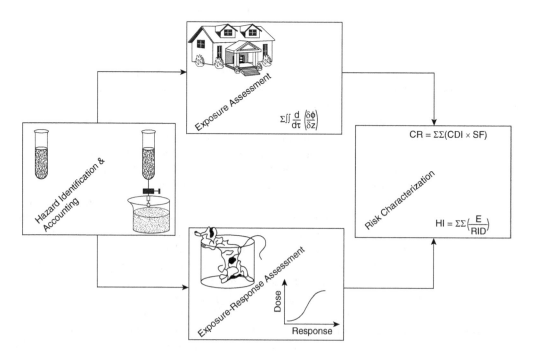

FIGURE 8.1 Illustrative elements of a risk assessment process.

fundamental elements follows—with more detailed elaboration provided in Chapters 9 and 10 of this title and also elsewhere in the risk analysis literature (e.g., Asante-Duah, 1998, 2017; Cohrssen and Covello, 1989; Conway, 1982; Cothern, 1993; Gheorghe and Nicolet-Monnier, 1995; Hallenbeck and Cunningham, 1988; Huckle, 1991; Kates, 1978; Kolluru et al., 1996; LaGoy, 1994; Lave, 1982; McColl, 1987; McTernan and Kaplan, 1990; Neely, 1994; NRC, 1982a, 1983b, 1994b,d; Paustenbach, 1988; Richardson, 1990; Rowe, 1977; Suter, 1993; USEPA, 1984a,c, 1989c,n,o; Whyte and Burton, 1980).

In the investigation of contaminated site problems, the focus of most risk assessments is on a determination of potential risks to human and ecological receptors. Although these represent different types of populations, the mechanics of the evaluation process are similar. It is noteworthy that, in general, much of the effort in the development of risk assessment methodologies has been directed at human health risk assessments (as reflected by the differences in the depth of coverage for human health vs. ecological risk assessment that can be found in the literature). However, the fundamental components of the risk assessment process for other biological organisms parallel those for human receptors and can indeed be described in similar terms.

8.5.1 HAZARD IDENTIFICATION AND ACCOUNTING

Hazard identification and accounting involves a qualitative assessment of the presence of and the degree of hazard that an agent or contaminant could have on potential receptors. The hazard identification consists of gathering and evaluating data on the types of health effects or diseases that may be produced by a chemical, and the exposure conditions under which human and ecological health damage, injury, or disease will be produced. It may also involve characterization of the behavior of a chemical within the body and the interactions it undergoes with organs, cells, or even parts of cells. Data of the latter types may be of value in answering the ultimate question of whether the forms of toxic effects determined to be produced by a substance in one population group or in experimental settings are also likely to be produced in the general population.

Hazard identification is not a risk assessment per se. This process involves simply determining whether it is scientifically correct to infer that toxic effects observed in one setting will occur in other settings—for example, whether substances found to be carcinogenic or teratogenic in experimental animals are likely to have the same results in humans. In the context of contaminated site risk management for potential chemical exposure problems, this may consist of the following:

- Identification of contaminant and chemical exposure sources
- Compilation of the lists of all chemical stressors present at the locale and impacting target receptors
- Identification and selection of the specific CoPCs (that should become the focus of the risk assessment) based on their specific hazardous properties (such as persistence, bioaccumulative properties, toxicity, and general fate and behavior properties)
- Compilation of summary statistics for the key constituents selected for further investigation and evaluation

Indeed, a major purpose of the hazard identification step of a human or ecological risk assessment is to identify a subset of CoPCs from all constituents detected during an investigation. The CoPCs are a subset of the complete set of constituents detected during an investigation that are exclusively carried through the quantitative risk assessment process. On the whole, the selection of CoPCs identifies those chemicals observed that have the most potential to be a significant contributor to human or ecological health risks—recognizing that most risk assessments tend to be dominated by a few compounds of significant concern (and indeed a few routes of exposure as well); as a matter of fact, the inclusion of all detected compounds in the risk assessment often has minimal influence on the total risk—and thus generally considered an unnecessary burden. In any case, several factors are typically considered in identifying CoPCs for risk assessments—including toxicity and magnitude of detected concentrations, frequency of detection, and essential nutrient status. The so-identified CoPCs are then carried forward for quantitative evaluation in the subsequent (baseline) risk assessment. Overall, the CoPC screening process is intended to identify the following:

i. Constituents that pose negligible risks—and therefore can be eliminated from further evaluation
ii. Constituents that merit further evaluation—either quantitatively or qualitatively—based on their potential to adversely affect target receptors, depending on specific types of exposures

Finally, it is noteworthy that, in identifying the CoPCs, an attempt is generally made to select all chemicals that could possibly represent the major part (usually, ≥95%) of the risks associated with the relevant exposures.

8.5.2 Exposure-Response Evaluation

The *exposure-response evaluation* (or the *effects assessment*) consists of a process that establishes the relationship between dose or level of exposure to a substance and the incidence-cum-severity of an effect. It considers the types of adverse effects associated with chemical exposures, the relationship between magnitude of exposure and adverse effects, and related uncertainties (such as the weight-of-evidence of a particular chemical's carcinogenicity in humans). In the context of contaminated site and chemical exposure problems, this evaluation will generally include a "dose-response evaluation" and/or a "toxicity assessment." Dose–response relationships are typically used to quantitatively evaluate the toxicity information and to characterize the relationship between dose of the contaminant administered or received and the incidence of adverse effects on an exposed population. From the quantitative dose–response relationship, appropriate toxicity values can be derived—and this is subsequently used to estimate the incidence of adverse effects occurring in populations at risk for different exposure levels. The toxicity assessment usually consists of compiling toxicological profiles for the CoPCs.

Dose–response assessment specifically involves describing the quantitative relationship between the amount of exposure to a substance and the extent of toxic injury or disease. Data are characteristically derived from animal studies or, less frequently, from studies in exposed human populations. There may be many different dose–response relationships for a substance if it produces different toxic effects under different conditions of exposure. Meanwhile, it is noteworthy that, even if the substance is known to be toxic, the risks of a substance cannot be ascertained with any degree of confidence unless dose–response relations are quantified.

8.5.2.1 Toxicity Assessment

A toxicity assessment is carried out as part of the contaminated site risk assessment—in order to both qualitatively and quantitatively determine the potential adverse health effects that could result from receptor exposure to environmental contaminants. This involves an evaluation of the types of adverse health effects associated with chemical exposures, the relationship between the magnitude of exposure and adverse effects, and the related uncertainties such as the weight-of-evidence of a particular chemical's carcinogenicity in humans.

A comprehensive toxicity assessment for chemicals found at contaminated sites is generally accomplished in two steps: hazard assessment and dose–response assessment. Hazard assessment is the process used to determine whether exposure to an agent can cause an increase in the incidence of an adverse health effect (e.g., cancer and birth defects); it involves a characterization of the nature and strength of the evidence of causation. Dose–response assessment is the process of quantitatively evaluating the toxicity information and characterizing the relationship between the dose of the contaminant administered or received (i.e., exposure to an agent) and the incidence of adverse health effects in the exposed populations; it is the process by which the potency of the compounds is estimated by use of dose–response relationships. These steps are discussed in more detail elsewhere in the literature (e.g., Klaassen et al., 1986; USEPA, 1989a).

8.5.3 Exposure Assessment and Analysis

An *exposure assessment* is conducted in order to estimate the magnitude of actual and/or potential receptor exposures to chemicals present in the human and ecological environments. The process considers the frequency and duration of the exposures, the nature and size of the populations potentially at risk (i.e., the risk group), and the pathways and routes by which the risk group may be exposed. Indeed, several physical and chemical characteristics of the chemicals of concern will provide an indication of the critical exposure features. These characteristics can also provide information necessary for determining the chemical's distribution, intake, metabolism, residence time, excretion, magnification, and half-life or breakdown to new chemical compounds.

The exposure assessment process is used to estimate the rates at which chemicals are absorbed by organisms. It generally involves several characterization and evaluation efforts (Box 8.1)—including contaminant distributions leading from release sources to the locations of likely exposure, the identification of significant migration and exposure pathways, the identification of potential receptors or the populations potentially at risk; the development of conceptual model(s) and exposure scenarios (including a determination of current and future exposure patterns and the analysis of environmental fate and persistence), the estimation/modeling of exposure point concentrations for the critical pathways and environmental media, and the estimation of chemical intakes for all potential receptors and significant pathways of concern (Asante-Duah, 2017).

BOX 8.1   GENERAL PROCEDURAL ELEMENTS OF AN INTEGRATED EXPOSURE ASSESSMENT PROCESS

- *Multimedia contaminant release analysis* (to include characterization of physical setting and monitoring/direct measurement data and modeling estimates; the results provide the basis for evaluating the potential for contaminant transport, transformation, and environmental fate).
- *Contaminant transport and fate analysis* (to include identification of migration and exposure pathways, environmental distribution, and concentrations; this analysis describes the extent and magnitude of environmental contamination).
- *Exposed population analysis* (to include evaluation of populations contacting chemicals released into the environment; this involves the identification, enumeration, and characterization of those population segments likely to be exposed).
- *Integrated exposure analysis* (to include development of exposure estimates for the selected exposure scenarios).
- *Uncertainty analysis* (to consist of identification of any uncertainties involved and an evaluation of their separate and cumulative impact on the assessment results).

In general, exposure assessments involve describing the nature and size of the population exposed to a substance and the magnitude and duration of their exposure. The evaluation could concern past or current exposures or exposures anticipated in the future. To complete a typical exposure analysis for a contaminated site or chemical exposure problem, populations potentially at risk are identified and concentrations of the chemicals of concern are determined in each medium to which potential receptors may be exposed. Finally, using the appropriate case-specific exposure parameter values, the intakes of the chemicals of concern are estimated. The exposure estimates can then be used to determine if any threats exist—based on the prevailing exposure conditions for the particular problem situation.

8.5.4 Risk Characterization and Consequence Determination

Risk characterization is the process of estimating the probable incidence of adverse impacts to potential receptors under a set of exposure conditions. Typically, the risk characterization summarizes and then integrates outputs of the exposure and toxicity assessments—in order to be able to qualitatively and/or quantitatively define risk levels. The process will usually include an elaboration of uncertainties associated with the risk estimates. Exposures resulting in the greatest risk can be identified in this process—and then mitigative measures can subsequently be selected to address the situation in order of priority and according to the levels of imminent risks.

In general, risk characterizations involve the integration of the data and information derived/ analyzed from the first three components of the risk assessment process (viz., hazard identification, dose–response assessment, and exposure assessment)—in order to ascertain the likelihood that humans might experience any of the various forms of toxicity associated with a substance. (By the way, in cases where exposure data are not available, hypothetical risks can be characterized by the integration of hazard identification and dose–response evaluation data alone.) In the final analysis, a framework to define the significance of the risk is developed, and all of the assumptions, uncertainties, and scientific judgments from the three preceding steps are also presented. Meanwhile, to the extent feasible, the risk characterization should include the distribution of risk among the target populations. When all is said and done, an adequate characterization of risks from hazards associated with chemical exposure problems allows risk management and corrective action decisions to be better focused.

8.6 RISK ASSESSMENT IMPLEMENTATION STRATEGY

Chemical risk assessment for contaminated site and chemical exposure problems may be defined as the characterization of the potential adverse health effects associated with human and/or ecological exposures to chemical hazards. In a typical contaminated site risk assessment process, the extent to which potential receptors have been or could be exposed to chemical hazards is determined. The extent of exposure is then considered in relation to the type and degree of hazard posed by the chemical(s) of concern—thereby permitting an estimate to be made of the present or future impacts to the populations at risk.

Several of the basic components and steps typically involved in a comprehensive chemical risk assessment that is designed for use in contaminated site risk management programs, as well as key aspects of the risk assessment methodology are presented in the proceeding chapters of this volume—with additional details provided elsewhere in the literature (e.g., Asante-Duah, 1998, 2017; Hoddinott, 1992; Huckle, 1991; NRC, 1983b; Patton, 1993; Paustenbach, 1988; Ricci, 1985; Ricci and Rowe, 1985; USEPA, 1984a,c, 1985e, 1986a–d, 1987d,b, 1989c,n,o, 1991c,e–h, 1992a,d,f,j; van Leeuwen and Hermens, 1995).

A number of techniques are indeed available for conducting risk assessments. Invariably, the methods of approach consist of the several basic procedural elements/components that are further outlined in Chapters 9 and 10 of this book. In any event, the key issues requiring significant attention in the processes involved will typically consist of finding answers to the following questions (Asante-Duah, 2017):

- What chemicals pose the greatest risk?
- What are the concentrations of the chemicals of concern in the exposure media?
- Which exposure pathways/routes are the most important?

- Which population groups, if any, face significant risk as a result of the possible exposures?
- What are the potential adverse effects of concern, given the exposure scenario(s) of interest?
- What is the range of risks to the affected populations?
- What are the environmental and public health implications for any identifiable corrective action and/or risk management alternatives?

For contaminated site problems, this will usually comprise of the selection of site contaminants of significant concern; an exposure assessment (consisting of a pathway analysis and the estimation of chemical intakes); a toxicity assessment (for human and ecological receptors); and a risk characterization (for both human health and ecological effects). The more commonly used risk assessment approaches that are relevant to the management of contaminated site problems are elaborated later in Chapters 9 and 10. It is noteworthy that, in general, much of the efforts in the development of risk assessment methodologies have been directed at human health risk assessments (as reflected by the differences in the depth of presentations on human health vs. ecological risk assessment commonly found in the literature). However, the fundamental components of the risk assessment process for most other biological organisms parallel those for human receptors and can be described in similar terms.

As a general guiding principle, risk assessments should be carried out in an iterative fashion and in a manner that can be appropriately adjusted to incorporate new scientific information and regulatory changes—but with the ultimate goal being to minimize public health and socioeconomic consequences associated with a potentially hazardous situation. Typically, an iterative approach would start with relatively inexpensive screening techniques—and then for hazards suspected of exceeding the *de minimis* risk, further evaluation is conducted by moving on to more complex and resource-intensive levels of data-gathering, model construction, and model application (NRC, 1994b,d).

In effect, risk assessments will normally be conducted in an iterative manner that grows in depth with increasing problem complexity (Asante-Duah, 2017). Consider, as an example, a site-specific risk assessment that is used to evaluate/address potential health impacts associated with chemical releases from industrial facilities or hazardous waste sites. A tiered approach is generally recommended in the conduct of such site-specific risk assessments. Usually, this will involve two broad levels of detail—that is, a "screening" (or "Tier 1") and a "comprehensive" (or "Tier 2") evaluation. In the screening evaluation, relatively simple models, conservative assumptions, and default generic parameters are typically used to determine an upper-bound risk estimate associated with a chemical release from the case facility. No detailed/comprehensive evaluation is warranted if the initial estimate is below a preestablished reference or target level (i.e., the *de minimis* risk). On the other hand, if the screening risk estimate is above the "acceptable" or *de minimis* risk level, then the more comprehensive/detailed evaluation (that utilizes more sophisticated and realistic data evaluation techniques than were used in the "Tier 1" screening) should be carried out. This more comprehensive next step will confirm the existence (or otherwise) of significant risks—which then forms the basis for developing any risk management action plans. The rationale for such a tiered approach is to optimize the use of resources—in that it makes efficient use of time and resources, by applying more advanced and time-consuming techniques to CoPCs and scenarios only where necessary. In other words, the comprehensive/detailed risk assessment is performed only when truly warranted. Irrespective of the level of detail, however, a well-defined protocol should always be used to assess the potential risks. Ultimately, a decision on the level of detail (e.g., qualitative, quantitative, or combinations thereof) at which an analysis is carried out will usually be based on the complexity of the situation, as well as the uncertainties associated with the anticipated or predicted risk.

Finally, it should be recognized that, there are several direct and indirect legislative issues that affect contaminated site risk assessment programs in different regions of the world. Differences in legislation among different nations (or even within a nation) tend to result in varying types of contaminated site risk management strategies being adopted or implemented. Indeed, legislation remains the basis for the administrative and management processes in the implementation of most contaminated site risk management policy agendas. Despite the good intents of most regulatory controls, however, it should be acknowledged that, in some cases, the risk assessment seems to be carried out simply to comply with the prevailing legislation—and may not necessarily result in any significant hazard or risk reduction.

8.6.1 General Considerations in Contaminated Site Risk Assessments

Invariably, the management of all contaminated site and chemical exposure problems starts with hazard identification and/or a data collection and data evaluation activity. The data evaluation aspect of a contaminated site risk assessment consists of an identification and analysis of the chemicals associated with a chemical exposure problem that should become the focus of the contaminated site risk management program. In this process, an attempt is generally made to select all chemicals that could represent the major part of the risks associated with case-related exposures; typically, this will consist of all constituents contributing ≥95% of the overall risks. Chemicals are screened based on such parameters as toxicity, ecotoxicity, carcinogenicity, concentrations of the detected constituents, and the frequency of detection in the sampled matrix.

The exposure assessment phase of the contaminated site risk assessment is used to estimate the rates at which chemicals are absorbed by potential receptors. Since most potential receptors tend to be exposed to chemicals from a variety of sources and/or in different environmental media, an evaluation of the relative contributions of each medium and/or source to total chemical intake could be critical in a multi-pathway exposure analysis. In fact, the accuracy with which such exposures are characterized could be a major determinant of the ultimate validity of the risk assessment.

The quantitative evaluation of toxicological effects consists of a compilation of toxicological profiles (including the intrinsic toxicological properties of the chemicals of concern, which may include their acute, subchronic, chronic, carcinogenic, and/or reproductive effects) and the determination of appropriate toxicity or ecotoxicity indices (see Chapter 9 and Appendix C).

Finally, the risk characterization consists of estimating the probable incidence of adverse impacts to potential receptors under various exposure conditions. It involves an integration of the toxicity or ecotoxicity and exposure assessments, resulting in a quantitative estimation of the actual and potential risks and/or hazards due to exposure to each key chemical constituent and also the possible additive effects of exposure to mixtures of the CoPCs.

8.6.1.1 The General Dynamics of Characterizing Chemical Exposure Problems and Conducting Contaminated Site Risk Assessments

The characterization of chemical exposure problems is a process used to establish the presence or absence of chemical hazards, to delineate the nature and degree of the hazards, and to determine possible threats posed by the exposure or hazard situation to human and/or ecological health. The exposure routes (which may consist of inhalation, ingestion, and/or dermal contacts) and duration of exposure (that may be short term [acute] or long term [chronic]) will significantly influence the degree of impacts on the affected receptors. The nature and behavior of chemical substances also form a very important basis for evaluating the potential for human or ecological exposures to the possible toxic or hazardous constituents of the substance.

Several chemical-specific, receptor-specific, and even environmental factors need to be recognized and/or evaluated as an important part of any environmental and public health risk management program that is designed to address problems that arise from exposure of potential receptors to various chemical substances. The general types of data and information necessary for the investigation of potential chemical exposure problems relate to the following (Asante-Duah, 2017):

- Identities of the chemicals of concern
- Concentrations contacted by potential receptors of interest
- Receptor characteristics
- Characteristics of the physical and environmental setting that can affect behavior and degree of exposure to the chemicals
- Receptor response upon contact with the target chemicals

In addition, it is necessary to generate information on the chemical intake rates for the specific receptor(s), together with numerous other exposure parameters. Indeed, all parameters that could potentially impact the human and ecological health outcomes should be carefully evaluated—including the following especially important categories (Asante-Duah, 2017):

- Exposure duration and frequency
- Exposure media and routes
- Target receptor attributes
- Potential receptor exposures history

Meanwhile, it is noteworthy that the above-mentioned listing is by no means complete for the universe of potential exposure possibilities—but certainly represents the critical ones that must certainly be examined rather closely.

Finally, it is notable that traditionally, contaminated site endangerment assessments have focused almost exclusively on risks to human health, often ignoring potential ecological effects because of the common but mistaken belief that protection of human health automatically protects nonhuman organisms (Suter, 1993). In fact, it is true that human health risks in most situations are more substantial than ecological risks, and mitigative actions taken to alleviate risks to human health are often sufficient to mitigate potential ecological risks at the same time. However, in some other situations, nonhuman organisms, populations, or ecosystems may be more sensitive to site contaminants than are human receptors. Consequently, ecological risk assessment programs should be considered as an equally important component of the management of contaminated site problems—in order to be able to arrive at site restoration decisions that offer an adequate level of protection for both human and ecological populations that are potentially at risk.

8.7 CONTAMINATED SITE RISK ASSESSMENT IN PRACTICE: THE SCOPE OF RISK ASSESSMENT UTILIZATION AND APPLICATIONS TO CONTAMINATED SITE PROBLEMS

Risk assessment is a process used to evaluate the collective demographic, geographic, physical, chemical, biological, and related factors associated with environmental contamination problems; this helps to determine and characterize possible risks to public health and the environment. The overall objective of such an assessment is to determine the magnitude and probability of actual or potential harm that the environmental contamination problem poses to human health and the environment. Almost invariably, every process for developing effectual environmental risk management programs should incorporate some concepts or principles of risk assessment. In particular, all decisions on restoration plans for potential environmental contamination problems will include, implicitly or explicitly, some elements of risk assessment. Indeed, the risk assessment process is intended to give the risk management team the best possible evaluation of all available scientific data, in order to arrive at justifiable and defensible decisions on a wide range of issues. For example, to ensure public safety in all situations, contaminant migration beyond a compliance boundary into the public exposure domain must be below some stipulated risk-based maximum exposure level—as typically established through a risk assessment process. Ultimately, based on the results of a risk assessment, decisions can be made in relation to the types of risk management actions necessary to address a given environmental contamination problem or hazardous situation. In fact, risk-based decision-making will generally result in the design of better environmental risk management programs. This is because risk assessment can produce more efficient and consistent risk reduction policies. It can also be used as a screening device for setting priorities.

Quantitative risk assessment often becomes an integral part of most contaminated site risk management and corrective action programs. In general, a risk assessment process is utilized to determine whether the level of risk at a contaminated site warrants remediation and to further project the amount of risk reduction necessary to protect public health and the environment. In particular, baseline risk assessments are usually conducted to evaluate the need for and the extent of remediation required at potentially contaminated sites. That is, they provide the basis and rationale as to whether or not remedial action is necessary. Overall, the baseline risk assessment contributes to the adequate characterization of contaminated site problems. It further facilitates the development, evaluation, and selection of appropriate corrective action response alternatives. In the processes involved, four key elements are important in arriving at appropriate risk management solutions—namely, the chemical hazard identification, the chemical toxicity assessment or exposure-response evaluation, the exposure assessment, and the risk characterization. Each of these elements typically will, among other things, help answer the following fundamental questions:

- *Chemical hazard identification step*—"what chemicals are present in the human or ecological environments of interest?" and "is the chemical agent likely to have an adverse effect on the potential human or ecological receptor?"
- *Chemical toxicity assessment or exposure-response evaluation step*—"what is the relationship between human or ecological exposure/dose to the chemical of potential concern and the response, incidence, injury, or disease as a result of the receptor exposure?" In other words, "what harmful effects can be caused by the target chemicals, and at what concentration or dose?"
- *Exposure assessment step*—"what individuals, sub-populations, or population groups may be exposed to the chemical of potential concern?" and "how much exposure is likely to result from various activities of the potential receptor—that is, what types and levels of exposure are anticipated or observed under various scenarios?"
- *Risk characterization step*—"what is the estimated incidence of adverse effect to the exposed individuals or population groups—that is, what risks are presented by the chemical hazard source?" and "what is the degree of confidence associated with the estimated risks?"

All in all, the fundamental tasks involved in most contaminated site risk assessments will consist of the key components shown in Box 8.2—revealing a methodical framework; a careful implementation of this framework should generally provide answers to the previous questions. The risks and/or hazards associated with residual contamination to be left at a contaminated site following the implementation of a remedial action can also be evaluated as part of the site restoration program. This is accomplished via the simulation of risk characterization scenarios for future land use and exposure conditions at the restored site. [By the way, it is noteworthy that the development of cleanup criteria (elaborated in Chapter 11) is not necessarily an integral component of the risk assessment process, since this belongs more so to the realm of risk management. Oftentimes, however, this is included in corrective action assessment programs.]

Risk assessment has several specific applications that could affect the type of decisions to be made in relation to environmental risk management programs. For instance, the application of the risk assessment process to contaminated site problems will generally serve to document the fact that risks to human health and the environment have been evaluated and incorporated into a set of appropriate response actions. Properly applied, risk assessment techniques can indeed be used to estimate the risks posed by environmental hazards under various exposure scenarios and to further estimate the degree of risk reduction achievable by implementing various scientific remedies. In the end, risk assessment classically serves as a management tool to facilitate effective decision-making on the control of environmental pollution problems. In most applications, it is used to provide a baseline estimate of existing risks that are attributable to a specific agent or hazard; the baseline risk assessment consists of an evaluation of the potential threats to human health and the environment in the absence of any remedial action. Risk assessment can also be used to determine the potential reduction in exposure and risk under various corrective action scenarios. In particular, risk assessment can be effectively used to support remedy selection in site restoration programs. Illustrative examples of the practical application of the processes involved are provided in Chapter 14. Meanwhile, it cannot be stated enough that there are many uncertainties associated with risk assessments. These uncertainties are due in part to the complexity of the exposure-dose-effect relationship, and also the lack of, or incomplete knowledge/information about the physical, chemical, and biological processes within and between human or ecological exposure to chemical substances and health effects.

BOX 8.2   ILLUSTRATIVE BASIC OUTLINE FOR A CONTAMINATED SITE RISK ASSESSMENT REPORT

Section Topic	Basic Subject Matter
General overview	
	• Background information on the case problem or locale
	• The risk assessment process
	• Purpose and scope of the risk assessment
	• The risk assessment technique and method of approach
	• Legal and regulatory issues in the risk assessment
	• Limits of application for the risk assessment
Data collection	
	• Chemical exposure sources of potential concern
	• General case-specific data collection considerations
	• Assessment of the data quality objectives
	• Identification of data gathering uncertainties
Data evaluation	
	• General case-specific data evaluation considerations
	• Identification, quantification, and categorization of target chemicals
	• Statistical analyses of relevant chemical data
	• Screening and selection of the CoPCs
	• Identification of uncertainties associated with data evaluation
Exposure assessment	
	• Characterization of the exposure setting (to include the physical setting and populations potentially at risk)
	• Identification of the chemical-containing sources/media, exposure pathways, and potentially affected receptors
	• Determination of the important fate and behavior processes for the CoPCs
	• Determination of the likely and significant exposure routes
	• Development of representative conceptual model(s) for the problem situation
	• Development of realistic exposure scenarios (to include both current and potential future possibilities)
	• Estimation/modeling of exposure point concentrations for the CoPCs

(Continued)

**BOX 8.2 (*Continued*)   ILLUSTRATIVE BASIC OUTLINE FOR A
CONTAMINATED SITE RISK ASSESSMENT REPORT**

Section Topic	Basic Subject Matter
	• Quantification of exposures (i.e., computation of potential receptor intakes/doses for the applicable exposure scenarios)
	• Identification of uncertainties associated with exposure parameters
Toxicity assessment	
	• Compilation of the relevant toxicological profiles of the CoPCs
	• Determination of the appropriate and relevant toxicity index parameters
	• Identification of uncertainties relating to the toxicity information
Risk characterization	
	• Estimation of the human carcinogenic risks from carcinogens
	• Estimation of the noncarcinogenic effects for systemic toxicants
	• Estimation of ecological risk quotients of ecological receptors
	• Sensitivity analyses of relevant parameters
	• Identification and evaluation of uncertainties associated with the risk estimates
Risk summary discussion	
	• Summarization of risk information
	• Discussion of all identifiable sources of uncertainties

9 Technical Approach to Human Health Endangerment Assessment

Human health risk assessment is defined as the characterization of the potential adverse health effects associated with human exposures to environmental hazards (NRC, 1983b). In a typical human health endangerment assessment process, the extent to which potential human receptors have been, or could be, exposed to chemical(s) associated with an environmental contamination problem is determined. The extent of exposure is considered in relation to the type and degree of hazard posed by the chemical(s) of concern, thereby permitting an estimate to be made of the present or future health risks to the populations-at-risk. Indeed, risks to human health as a result of exposure to toxic materials present or introduced into our living and work environments are a matter of grave concern to modern societies—calling for the use of credible techniques that will facilitate effectual decisions on this front. In general, systematic human health risk assessment methods can be employed to better facilitate responsible risk management programs.

This chapter presents a general discussion in relation to the classical methods of approach used for completing human health risk assessments as part of the investigation and management of contaminated site and chemical exposure problems. Specifically, it provides a procedural framework and an outline of the key elements of the health risk assessment process; this also includes some concise discussions of the specific elements identified in the framework. It is noteworthy that the scope of applications and common uses for the health risk assessment methodology may indeed vary greatly; some commonly encountered types of example applications and uses are provided later in Chapter 14 (Part IV) of this volume.

9.1 DETERMINING RISKS FROM CONTAMINATED SITES: A GENERAL FRAMEWORK FOR HUMAN HEALTH RISK ASSESSMENTS

Quantitative human health risk assessment is often an integral part of most risk management and corrective action response programs that are designed to address contaminated site problems. Figure 9.1 shows the basic components and steps typically involved in a comprehensive human health risk assessment designed for use in contaminated site management programs; the elements involved in the typical human health risk assessment usually will comprise the following tasks:

- Data Evaluation
  - Assess the quality of available data
  - Identify, quantify, and categorize environmental contaminants
  - Screen and select chemicals of potential concern
  - Carry out statistical analysis of relevant environmental or site data
- Exposure Assessment
  - Compile information on the physical setting of the site or problem location
  - Identify source areas, significant migration pathways, and potentially impacted or receiving media
  - Determine the important environmental fate and transport processes for the chemicals of potential concern, including cross-media transfers

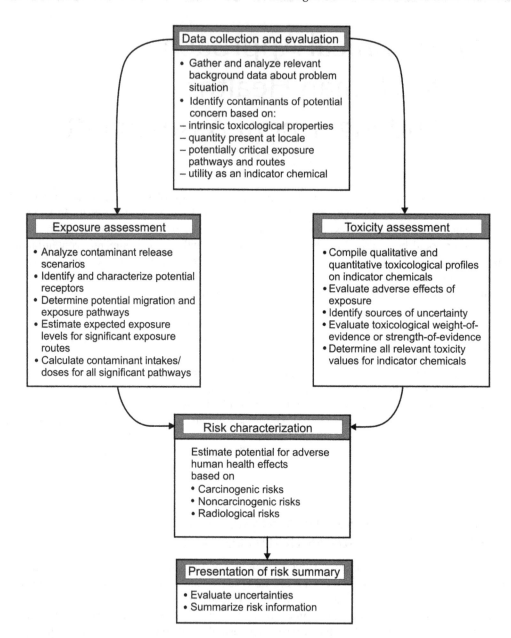

FIGURE 9.1 A general framework for the human health risk assessment process: fundamental procedural components and principal elements of a contaminated site risk assessment.

- Identify populations potentially at risk
- Determine likely and significant receptor exposure pathways
- Develop representative conceptual model(s) for the site or problem situation
- Develop exposure scenarios (to include both the current and potential future land-uses for the site or locale)
- Estimate/model exposure point concentrations (EPCs) for the chemicals of potential concern found in the significant environmental media
- Compute potential receptor intakes and resultant doses for the chemicals of potential concern (for all potential receptors and significant pathways of concern)

- Toxicity Assessment
  - Compile toxicological profiles (to include the intrinsic toxicological properties of the chemicals of potential concern, such as their acute, subchronic, chronic, carcinogenic, and reproductive effects)
  - Determine appropriate toxicity indices (such as the acceptable daily intakes (ADIs) or reference doses (RfDs), cancer slope or potency factors)
- Risk Characterization
  - Estimate carcinogenic risks from carcinogens
  - Estimate noncarcinogenic hazard quotients HQs and indices for systemic toxicants
  - Perform sensitivity analyses, evaluate uncertainties associated with the risk estimates, and summarize the risk information

The principal elements of the human endangerment assessment process are elaborated in the subsequent sections—with additional details to be found elsewhere in the literature (e.g., Asante-Duah, 1998, 2017; Hoddinott, 1992; Huckle, 1991; NRC, 1983b; Patton, 1993; Paustenbach, 1988; Ricci, 1985; Ricci and Rowe, 1985; USEPA, 1984a,c 1985e, 1986a–d, 1987d, 1989c,n,o, 1991c,e–h, 1992a,d,f,j; van Leeuwen and Hermens, 1995).

Invariably, all contaminated site management programs start with hazard identification and/or a data collection-cum-data evaluation phase. The data evaluation aspect of a human health risk assessment consists of an identification and analysis of the chemicals associated with a chemical exposure problem that should become the focus of the site characterization and risk management program. In this process, an attempt is generally made to select all chemicals that could represent the major part of the risks associated with case-related exposures; typically, this will consist of all constituents contributing ≥95% of the overall risks. Chemicals are screened based on such parameters as toxicity, carcinogenicity, concentrations of the detected constituents, and the frequency of detection in the sampled matrix.

The exposure assessment phase of the human health risk assessment is used to estimate the rates at which chemicals are absorbed by potential receptors. Since most potential receptors tend to be exposed to chemicals from a variety of sources and/or in different environmental media, an evaluation of the relative contributions of each medium and/or source to total chemical intake could be critical in a multi-pathway exposure analysis. In fact, the accuracy with which exposures are characterized could be a major determinant of the ultimate validity of the risk assessment.

A toxicity assessment is conducted as part of the human health risk assessment, in order to be able to both qualitatively and quantitatively determine the potential adverse health effects that could result from a human exposure to the chemicals of potential concern. This involves an evaluation of the types of adverse health effects associated with chemical exposures, the relationship between the magnitude of exposure and adverse effects, and related uncertainties such as the weight-of-evidence of a particular chemical's carcinogenicity in humans. The quantitative evaluation of toxicological effects typically consists of a compilation of toxicological profiles (including the intrinsic toxicological properties of the chemicals of concern, which may include their acute, subchronic, chronic, carcinogenic, and/or reproductive effects) and the determination of appropriate toxicity indices (see Appendix C).

Finally, the risk characterization consists of estimating the probable incidence of adverse impacts to potential receptors under various exposure conditions. It involves integrating the toxicity and exposure assessments, resulting in a quantitative estimation of the actual and potential risks and/or hazards due to exposure to each key chemical constituent—and also due to the possible additive effects of exposure to mixtures of the chemicals of potential concern. Typically, the risks to potentially exposed populations resulting from exposure to the site contaminants are characterized through a calculation of noncarcinogenic HQs and indices and/or carcinogenic risks (CAPCOA, 1990; CDHS, 1986; USEPA, 1989a). These parameters can then be compared with benchmark standards in order to arrive at risk decisions about a contaminated site and/or chemical exposure problem.

9.2 CHEMICAL HAZARD DETERMINATION VIA DATA COLLECTION AND EVALUATION: SCREENING AND SELECTING THE CHEMICALS OF POTENTIAL CONCERN

In order to make an accurate determination of the level of hazard potentially posed by a chemical, it is very important that the appropriate set of exposure data is collected during the hazard identification and accounting processes. It is also very important to use appropriate data evaluation tools in the processes involved. In fact, statistical procedures used for the evaluation of environmental data can significantly affect the conclusions of a given environmental characterization and risk assessment program. Consider, for instance, the use of a normal distribution (whose central tendency is measured by the arithmetic mean) to describe environmental contaminant distribution, rather than lognormal statistics (whose central tendency is defined by the geometric mean); the former will often result in significant overestimation of contamination levels. Appropriate statistical methods should therefore be utilized in the evaluation of environmental sampling data (e.g., in relation to the choice of proper averaging techniques).

This section provides a summary discussion of the principal activities involved in the acquisition and manipulation of the pertinent chemical hazard information (Asante-Duah, 2017)—that will then allow effective environmental and public health risk management decisions to be made about contaminated site and chemical exposure problems. It should be acknowledged and emphasized, however, that the discussion here focuses on a premise that human health risks tend to drive most key decisions made about potentially contaminated sites—albeit there are some situations when ecological receptors would become the reason for any hazard/risk determination. In any event, the concepts presented here are fundamentally and generally applicable to ecological receptors as well—with elements more or less exclusive to ecological receptors having been identified in Chapter 10 of this title.

9.2.1 STATISTICAL EVALUATION OF CHEMICAL SAMPLING/CONCENTRATION DATA

Several statistical methods of approach can serve to facilitate the evaluation of data associated with contaminated site characterization and related programs. For instance, statistical procedures can aid in the determination of sampling size requirements that are necessary to produce statistically reliable data set required for corrective action evaluations; also, statistical procedures used for data evaluation can significantly affect a corrective action response decision. Indeed, statistical procedures used for the evaluation of the chemical exposure data can significantly affect the conclusions of a given exposure characterization and risk assessment program. Consequently, appropriate statistical methods (e.g., in relation to the choice of proper averaging techniques) should be utilized in the evaluation of chemical sampling data. Of special interest, it is noteworthy that, over the years, extensive technical literature has been developed regarding the "best" probability distribution to utilize in different scientific applications—and such sources should be consulted for appropriate guidance on the statistical tools of choice.

In general, statistical analysis procedures used in the evaluation of environmental data should reflect the character of the underlying distribution of the data set. The appropriateness of any distribution assumed or used for a given data set should preferably be checked prior to its application; this can be accomplished by using some goodness-of-fit methods, such as the Chi-Square Test for Goodness-of-Fit (see, e.g., Cressie, 1994; Freund and Walpole, 1987; Gilbert, 1987; Miller and Freund, 1985; Sharp, 1979; Wonnacott and Wonnacott, 1972). The choice of statistical parameters for contaminated site characterization and related decisions is also critical to the effectiveness of the overall risk management plan. Because of the uncertainty associated with any estimate of exposure concentration (EC), the upper confidence interval (i.e., the 95% upper confidence limit) on the average is frequently used in corrective action assessment programs and indeed most environmental impact evaluation programs. Furthermore, contamination levels and exposures may have temporal variations and the dynamic nature of such parameters should, insofar as possible, be incorporated in the evaluation of the environmental data.

Several of the available statistical methods and procedures finding widespread use in corrective action assessment and environmental management programs can be found in the literature on statistics (e.g., Berthouex and Brown, 1994; Cressie, 1994; Freund and Walpole, 1987; Gibbons, 1994; Gilbert, 1987; Hipel, 1988; Miller and Freund, 1985; Ott, 1995; Sachs, 1984; Sharp, 1979; Wonnacott and Wonnacott, 1972; Zirschy and Harris, 1986). It should be recognized and acknowledged, however, that the logic of classical statistics and environmental sampling tend to be quite different. This is because environmental studies/investigations usually utilize nonrandom sampling—and thus it becomes difficult to attempt to fit a classical statistical distribution (e.g., a normal distribution) to most environmental sampling data (since they tend to be "biased" because of their "nonrandom" nature). Even so, classical statistics offer reasonable and pragmatic analytical procedures for use in environmental data evaluations. Some commonly used examples of methods of approach that find general application in the evaluation of environmental data are briefly discussed below.

9.2.1.1 Choice of Statistical Distribution

Of the many statistical distributions available, the Gaussian (or normal) distribution has been widely utilized to describe environmental data; however, there is considerable support for the use of the lognormal distribution in describing such data. Consequently, chemical concentration data for environmental samples have been described by the lognormal distribution, rather than by a normal distribution (Gilbert, 1987; Leidel and Busch, 1985; Rappaport and Selvin, 1987; Saltzman, 1997). Basically, the use of lognormal statistics for the data set $X_1, X_2, X_3, \ldots, X_n$ requires that the logarithmic transform of these data (i.e., $\ln[X_1], \ln[X_2], \ln[X_3], \ldots, \ln[X_n]$) can be expected to be normally distributed.

In general, the statistical parameters used to describe the different distributions can differ significantly; for instance, the central tendency for the normal distributions is measured by the arithmetic mean, whereas the central tendency for the lognormal distribution is defined by the geometric mean. Ultimately, the use of a normal distribution to describe environmental chemical concentration data, rather than lognormal statistics, will often result in significant overestimation, and may be overly conservative—albeit some investigators may argue otherwise (e.g., Parkhurst, 1998). In fact, Parkhurst (1998) argues that geometric means are biased low and do not represent components of mass balances properly, *whereas* arithmetic means are unbiased, easier to calculate and understand, scientifically more meaningful for concentration data, and more protective of public health. Even so, this same investigator still concedes to the non-universality of this school of thought—and these types of arguments and counter-arguments only go to reinforce the fact that no one particular parameter or distribution may be appropriate for every situation. Consequently, care must be exercised in the choice of statistical methods for the data manipulation exercises carried out during the hazard accounting process—and indeed in regards to other aspects of a risk assessment.

9.2.1.2 Statistical Evaluation of "Non-Detect" Values: Derivation and Use of "Proxy" Concentrations for "Censored" Data

"Proxy" concentrations are usually employed when a chemical is not detected in a specific sampled medium per se. A variety of approaches are offered in the literature for deriving and using proxy values in environmental data analyses, including the following relatively simpler ones (Asante-Duah, 1998, 2017; HRI, 1995; USEPA, 1989a, 1992a,j):

- Set the sample concentration to zero—albeit it is inappropriate to convert non-detects into zeros without specific justification (e.g., the analyte was not detected above the detection limit in any sample at the locale or site being investigated)
- Drop the sample with the non-detect for the particular chemical from further analysis
- Set the proxy sample concentration to the sample quantitation limit (SQL)
- Set the proxy sample concentration to one-half the SQL

Other methods of approach to the derivation of proxy concentrations may involve the use of "distributional" or "bounding" methods; unlike the simple substitution methods shown above, distributional methods make use of the data above the reporting limit in order to extrapolate below it (Ferson, 2002; Gleit, 1985; Hass and Scheff, 1990; Helsel, 1990; Kushner, 1976; Millard, 1997; Rowe, 1983; Singh and Nocerino, 2001; Smith, 1995; USEPA, 1992a,j, 2000d, 2002a). In general, selecting the appropriate method requires consideration of the degree of censoring, the goals of the assessment, and the degree of accuracy required.

In any event, during the analysis of environmental sampling data that contains some non-detects (NDs), a fraction of the SQL is usually assumed (as a proxy or estimated concentration) for non-detectable levels—instead of assuming a value of zero, or neglecting such values. This procedure is typically used, provided there is at least one detected value from the analytical results, and/or if there is reason to believe that the chemical is possibly present in the sample at a concentration below the SQL. The approach conservatively assumes that some level of the chemical could be present (even though a ND has been recorded) and arbitrarily sets that level at the appropriate percentage of the SQL. In general, the favored approach in the calculation of the applicable statistical values during the evaluation of data containing NDs involves the use of a value of one-half of the SQL (i.e., $\frac{SQL}{2}$ if data set is assumed to be normally distributed). This approach assumes that the samples are equally likely to have any value between the detection limit and zero, and can be described by a normal distribution. However, when the sample values above the ND level are lognormally distributed, it generally may be assumed that the ND values are also lognormally distributed. The best estimate of the ND values for a lognormally distributed data set is the reported SQL divided by the square root of two (i.e., $\frac{SQL}{\sqrt{2}} = \frac{SQL}{1.414}$) (CDHS, 1990; USEPA, 1989a). Also, in some situations, the SQL value itself may be used if there is strong enough reason to believe that the chemical concentration is closer to this value, rather than to a fraction of the SQL.

Several other methods for the statistical analysis of environmental sampling data containing various proportions of NDs, such as using the Cohen method, are discussed elsewhere in the literature (e.g., Gilbert, 1987; USEPA, 1996b,c, 1998b,f,g). Indeed, no general procedures exist for the statistical analyses of censored data sets that can be used in all applications of statistical analysis of environmental sampling data evaluation—thus the need for caution in the implementation of the procedure of choice (USEPA, 1996b,c). Where it is apparent that serious biases could result from the use of any of the preceding methods of approach, more sophisticated analytical and evaluation methods may be warranted.

Finally, it is noteworthy that, notwithstanding the options available from the above procedures of deriving and/or using "proxy" concentrations, re-sampling and further laboratory analysis should always be viewed as the preferred approach to resolving uncertainties that surround ND results obtained from sampled media. Thence, if the initially reported data represent a problem in sample collection or analytical methods rather than a true failure to detect a chemical of potential concern, then the problem could be rectified (e.g., by the use of more sensitive analytical protocols) before critical decisions are made based on the earlier results.

9.2.1.3 The Ultimate Selection of Statistical Averaging Techniques

Reasonable discretion should generally be exercised in the selection of an averaging technique during the analysis of environmental sampling data. This is because the selection of specific methods of approach to average a set of environmental sampling data can have profound effects on the resulting concentration—especially for data sets from sampling results that are not normally distributed. For example, when dealing with log-normally distributed data, geometric means are often used as a measure of central tendency—in order to ensure that a few very high values do not exert excessive influence on the characterization of the distribution. However, if high concentrations do indeed represent "hotspots" in a spatial or temporal distribution of the data set, then using the geometric mean could inappropriately discount the contribution of these high chemical concentrations present in the environmental samples. This is particularly significant if, for instance, the spatial pattern indicates that areas of high concentration for a chemical release are in close proximity to compliance boundaries or near exposure locations for sensitive populations (such as children and the elderly).

The geometric mean has indeed been extensively and consistently used as an averaging parameter in the past. Its principal advantage is in minimizing the effects of "outlier" values (i.e., a few values that are much higher or lower than the general range of sample values). Its corresponding disadvantage is that discounting these values may be inappropriate when they represent true variations in concentrations from one part of an impacted area or group to another (such as a "hot-spot" vs. a "cold-spot" region). As a measure of central tendency, the geometric mean is most appropriate if sample data are lognormally distributed, and without an obvious spatial pattern.

The arithmetic mean—commonly used when referring to an "average"—is more sensitive to a small number of extreme values or a single "outlier" compared to the geometric mean. Its corresponding advantage is that true high concentrations will not be inappropriately discounted. When faced with limited sampling data, however, this may not provide a conservative enough estimate of environmental chemical impacts.

In fact, none of the above measures, in themselves, may be appropriate in the face of limited and variable sampling data. Current applications tend to favor the use of an upper confidence level (UCL) on the average concentration. Even so, if the UCL exceeds the maximum detected value among a data pool, then the latter is used as the source term or EPC. It is noteworthy that, in situations where there is a discernible spatial pattern to chemical concentration data, standard approaches to data aggregation and analysis may usually be inadequate, or even inappropriate.

9.2.1.4 Illustrative Example Computations Demonstrating the Potential Effects of Variant Statistical Averaging Techniques

To demonstrate the possible effects of the choice of statistical distributions and/or averaging techniques on the analysis of environmental data, consider a case involving the estimation of the mean, standard deviation, and confidence limits from monthly laboratory analysis data for groundwater concentrations obtained from a potential drinking water well. The goal here is to compare the selected statistical parameters based on the assumption that this data is normally distributed versus an alternative assumption that the data is lognormally distributed. To accomplish this task, the several statistical manipulations enumerated below are carried out on the "raw" and log-transformed data for the concentrations of benzene in the groundwater samples shown in Table 9.1.

1. *Statistical Manipulation of the "Raw" Data.* Calculate the following statistical parameters for the "raw" data: mean, standard deviation, and 95% confidence limits (95% CL) (see standard statistics textbooks for details of applicable procedures involved). The arithmetic mean, standard deviation, and 95% CL for a set of n values are defined, respectively, as follows:

$$X_m = \frac{\sum_{i=1}^{n} X_i}{n} \tag{9.1}$$

$$SD_x = \sqrt{\frac{\sum_{i=1}^{n} (X_i - X_m)^2}{n-1}} \tag{9.2}$$

$$CL_x = X_m \pm \frac{ts}{\sqrt{n}} \tag{9.3}$$

TABLE 9.1

Environmental Sampling Data Used to Illustrate the Effects of Statistical Averaging Techniques on Exposure Point Concentration Predictions

	Concentration of Benzene in Drinking Water (µg/L)	
Sampling Event	Original "Raw" Data, X	Log-Transformed Data, $Y = \ln(X)$
January	0.049	−3.016
February	0.056	−2.882
March	0.085	−2.465
April	1.200	0.182
May	0.810	−0.211
June	0.056	−2.882
July	0.049	−3.016
August	0.048	−3.037
September	0.062	−2.781
October	0.039	−3.244
November	0.045	−3.101
December	0.056	−2.882

where X_m = arithmetic mean of "raw" data; SD_x = standard deviation for "raw" data; CL_x = 95% confidence interval (95% CI) of "raw" data; t is the value of the Student t-distribution (as expounded in standard statistical books) for the desired confidence level (e.g., 95% CL, which is equivalent to a level of significance of $\alpha = 5\%$) and degrees of freedom, $(n - 1)$; and s is an estimate of the standard deviation from the mean (X_m). Thus,

$$X_m = 0.213\,\mu g/L$$

$$SD_x = 0.379\,\mu g/L$$

$$CL_x = 0.213 \pm 0.241 \left(i.e., -0.028 \leq CI_x \leq 0.454\right) \text{ and } UCL_x = 0.454\,\mu g/L$$

where UCL_x = 95% UCL of "raw" data.

Note that the computation of the 95% CL for the untransformed data produces a confidence interval of $0.213 \pm 0.109t = 0.213 \pm 0.241$ (where $t = 2.20$, obtained from the Student t-distribution for $(n - 1) = 12 - 1 = 11$ degrees of freedom)—and which therefore indicates a nonzero probability for a negative concentration value; indeed, such value may very well be considered meaningless in practical terms—consequently revealing some of the shortcomings of this type of computational method of approach.

2. *Statistical Manipulation of the Log-Transformed Data.* Calculate the following statistical parameters for the log-transformed data: mean, standard deviation, and 95% CL (see standard statistics textbooks for details of applicable procedures involved). The geometric mean, standard deviation, and 95% CL for a set of n values are defined, respectively, as follows:

$$X_{gm} = \text{antilog}\left\{\frac{\sum_{i=1}^{n} \ln X_i}{n}\right\} \tag{9.4}$$

$$SD_x = \sqrt{\dfrac{\sum\limits_{i=1}^{n}\left(X_i - X_{gm}\right)^2}{n-1}} \qquad (9.5)$$

$$CL_x = X_{gm} \pm \dfrac{ts}{\sqrt{n}} \qquad (9.6)$$

where X_{gm} = geometric mean for the "raw" data; SD_x = standard deviation of "raw" data (assuming lognormal distribution); CL_x = 95% CI for the "raw" data (assuming lognormal distribution); t is the value of the Student t-distribution (as expounded in standard statistical books) for the desired confidence level and degrees of freedom, $(n-1)$; and s is an estimate of the standard deviation of the mean (X_{gm}). Thus,

$$Y_{a\text{-mean}} = -2.445$$

$$SD_y = 1.154$$

$$CL_y = -2.445 \pm 0.733 \left(\text{i.e., a confidence interval from} -3.178\,\text{to}-1.712\right)$$

where $Y_{a\text{-mean}}$ = arithmetic mean of log-transformed data; SD_y = standard deviation of log-transformed data; and CL_y = 95% CI) of log-transformed data. In this case, computation of the 95% CL for the log-transformed data yields a confidence interval of $-2.445 \pm 0.333t = -2.445 \pm 0.733$ (where $t = 2.20$, obtained from the student t-distribution for $(n-1) = 12 - 1 = 11$ degrees of freedom).

Now, transforming the average of the logarithmic Y values back into arithmetic values yields a geometric mean value of $X_{gm} = e^{-2.445} = 0.087$. Furthermore, transforming the confidence limits of the log-transformed values back into the arithmetic realm yields a 95% confidence interval of 0.042–0.180 µg/L; recognize that these consist of positive concentration values only. Hence,

$$X_{gm} = 0.087\,\mu g/L$$

$$SD_x = 3.171\,\mu g/L$$

$$0.042 \le CI_x \le 0.180\,\mu g/L$$

$$UCL_x = 0.180\,\mu g/L$$

where UCL_x = 95% UCL for the "raw" data (assuming lognormal distribution).

In consideration of the above, it is obvious that the arithmetic mean, $X_m = 0.213$ µg/L, is substantially larger than the geometric mean of $X_{gm} = 0.087$ µg/L. This may be attributed to the two relatively higher sample concentration values in the data set (namely, sampling events for the months of April and May shown in Table 9.1)—which consequently tend to strongly bias the arithmetic mean; on the other hand, the logarithmic transform acts to suppress the extreme values. A similar observation can be made for the 95% UCL of the normally and lognormally distributed data sets. In any event, irrespective of the type of underlying distribution, the 95% UCL is generally a preferred statistical parameter to use in the evaluation of environmental data, rather than the statistical mean values.

The results from the above example analysis illustrate the potential effects that could result from the choice of one distribution type over another, and also the implications of selecting specific statistical parameters in the evaluation of environmental sampling data. In general, the use of arithmetic or geometric mean values for the estimation of average concentrations would tend to bias the target receptor concentration or other related estimates; the 95% UCL characteristically offers a better value to use—albeit may not necessarily be a panacea in all situations.

More sophisticated methods, such as geostatistical techniques that account for spatial variations in concentrations may also be employed for estimating the average environmental concentrations (e.g., USEPA, 1988a; Zirschy and Harris, 1986). For example, a technique called block kriging is frequently used to estimate soil chemical concentrations in sections of contaminated sites in which only sparse sampling data exist. In this case, the site is divided into blocks (or grids), and concentrations are determined within blocks by using interpolation procedures that incorporate sampling data in the vicinity of the block. The sampling data are weighted in proportion to the distance of the sampling location from the block. Also, weighted moving-average estimation techniques based on geostatistics are applicable for estimating average contaminant levels present at a site.

9.2.2 Determining "EPCs": Estimation of the Concentration of a Chemical in the Environment

Once the decision is made to undertake a contaminated site risk assessment, the available chemical exposure data has to be carefully analyzed, in order to arrive at a list of chemicals of potential concern (CoPCs); the CoPCs represent the target chemicals of focus in the risk assessment process. In general, the target chemicals of significant interest to chemical exposure problems may be selected for further detailed evaluation on the basis of several specific important considerations (Asante-Duah, 2017); indeed, the use of appropriate selection criteria will allow an analyst to continue with the exposure and risk characterization process only if the chemicals represent potential threats to human or ecological health. For such chemicals, general summary statistics would commonly be compiled; meanwhile, it is worth the mention here that, where applicable, data for samples and their duplicates are typically averaged before summary statistics are calculated—such that a sample and its duplicate are ultimately treated as one sample for the purpose of calculating summary statistics (including maximum detection and frequency of detection). Where constituents are not detected in both a sample and its duplicate, the resulting values are the average of the sample-specific quantitation limits (SSQLs). Where both the sample and the duplicate contain detected constituents, the resulting values are the average of the detected results. Where a constituent in one of the pair is reported as not detected and the constituent is detected in the other, the detected concentration is conservatively used to represent the value of interest. On the whole, the following summary statistics are typically generated as part of the key statistical parameters of interest:

- *Frequency of detection*—reported as a ratio between the number of samples reported as detected for a specific constituent and the total number of samples analyzed.
- *Maximum detected concentration*—for each constituent/receptor/medium combination, after duplicates have been averaged.
- *Mean detected concentration*—typically the arithmetic mean concentration for each constituent/receptor/medium combination, after duplicates have been averaged, based on detected results only.
- *Minimum detected concentration*—for each constituent/area/medium combination, after duplicates have been averaged.

Next, the proper EPC for the target populations potentially at risk from the CoPCs is determined; an EPC is the concentration of the CoPC in the target material or product at the point of contact with the human or ecological receptor.

Finally, it is worth the mention here that, in the absence of adequate and/or appropriate field sampling data, a variety of mathematical algorithms and models are often employed to support the determination of chemical ECs in human and/or ecological exposure media (Asante-Duah, 2017). Such forms of chemical exposure models are typically designed to serve a variety of purposes; a general guidance for the effective selection of models used in chemical exposure characterization and risk management decisions is provided in the literature elsewhere (e.g., Asante-Duah, 1998, 2017; CCME, 1994; CDHS, 1990; Clark, 1996; Cowherd et al., 1985; DOE, 1987; NRC, 1989b; Schnoor, 1996; USEPA, 1987d,f,g, 1988d–f,i,j; Yong et al., 1992; Zirschy and Harris, 1986). The results from the modeling are generally used to estimate the consequential exposures and risks to potential receptors associated with a given chemical exposure problem. One of the major benefits associated with the use of mathematical models in human and ecological health risk management programs relate to the fact that, environmental concentrations useful for exposure assessment and risk characterization can be estimated for several locations and time-periods of interest. Indeed, since field data are often limited and/or insufficient to facilitate an accurate and complete characterization of chemical exposure problems, models can be particularly useful for studying spatial and temporal variability, together with potential uncertainties. In addition, sensitivity analyses can be conducted by varying specific exposure parameters—and then using models to explore any ramifications reflected by changes in the model outputs.

9.2.2.1   The EPC Determination Process

The EPC determination process typically will consist of an appropriate statistical evaluation of the exposure sampling data—especially when large data sets are involved; this concentration term is a conservative estimate of the average chemical concentration in an environmental medium. The EPC is determined for each individual exposure unit within a site; an exposure unit is the area throughout which a receptor moves and encounters an environmental medium for the duration of the exposure. Unless there is site-specific evidence to the contrary, an individual receptor is assumed to be equally exposed to media within all portions of the exposure unit over the time frame of the risk assessment (USEPA, 2002a). For normally distributed data, an UCL on the mean based on the Student's t-statistic is recommended—and for lognormal data, the Land method using the H-statistic is recommended (USEPA, 2002a); Singh et al. (1997) and Schulz and Griffin (1999) suggest several alternate methods for calculating a UCL for non-normal data distributions.

In the final analysis, the process/approach used to estimate a potential receptor's EPC to a chemical constituent associated with a contaminated site problem will comprise of the following elements (Asante-Duah, 2017):

- Determining the distribution of the chemical exposure/sampling data, and fitting the appropriate distribution to the data set (e.g., normal and lognormal);
- Developing the basic statistics for the exposure/sampling data—to include calculation of the relevant statistical parameters, such as the upper 95% confidence limit (UCL_{95}); and
- Calculating the EPC—usually defined as the minimum of either the UCL or the maximum exposure/sampling data value—conceptually represented as follows: EPC = min (UCL_{95} or Max-Value)

By and large, the so-derived EPC (that may indeed be significantly different from any field-measured chemical concentrations) represents the "true" or reasonable exposure level at the potential receptor location of interest—and this value is used in the calculation of the chemical intake/dose for the populations potentially at risk.

It is noteworthy that, because some of the averaging methods (such as the Land method) can produce very high estimates of the UCL, the maximum observed concentration may be used as the EPC rather than the calculated UCL in cases where the UCL exceeds the maximum concentration (USEPA, 1992a,j, 2002a). It is important to note, however, that defaulting to the maximum observed concentration may not be protective when sample sizes are very small because the observed maximum may be smaller than the population-mean. Thus, it is important to collect sufficient samples in accordance with the data quality objectives (DQOs) for a site. In general, the use of the maximum as the default EPC is reasonable only when the data samples have been collected at random from the exposure unit and the sample size is large (USEPA, 2002a).

9.3 EXPOSURE ASSESSMENT: ANALYSIS OF POPULATIONS INTAKE OF CHEMICALS

Once an environmental chemical has been determined to present a potential human or ecological health hazard, the main concern then shifts to the likelihood for, and degree of, human and/ or ecological exposure that might ensue. The exposure assessment phase of the risk assessment is used to estimate the rates at which chemicals are absorbed by potential (human or ecological) receptors. More specifically, it is generally used to determine the magnitude of actual and/or potential receptor exposures to chemical constituents, the frequency and duration of these exposures, and the pathways via which the target receptor is potentially exposed to the chemicals that they contact from a variety of sources. The exposure assessment also involves describing the nature and size of the population exposed to a substance (i.e., the risk group, which refers to the actual or hypothetical exposed population) and the magnitude and duration of their exposure. Since most potential humans (and indeed ecological receptors also) tend to be exposed to chemicals from a variety of sources and/or in different environmental media, an evaluation of the relative contributions of each medium and/or source to total chemical intake could be critical in a multi-pathway exposure scenario associated with most exposure analyses. In fact, the accuracy with which exposures are characterized could be a major determinant of the ultimate validity of a risk assessment. On this journey, several techniques may be used for the exposure assessment, including the modeling of anticipated future exposures; environmental monitoring of current exposures; and biological monitoring to determine past exposures. The physico-chemical properties of the contaminants of potential concern and the impacted media are important considerations in the exposure modeling. In general, a variety of exposure models and conservative but realistic assumptions regarding contaminant migration and equilibrium partitioning are used to facilitate the exposure quantification process. Site-related EPCs can be determined once the potentially affected populations are identified and the exposure scenarios are defined. If the transport of compounds associated with the site is under steady-state conditions, monitoring data are generally adequate to determine potential ECs. If there are no data available, or if conditions are transient (such as pertains to a migrating plume in groundwater), models are better used to predict ECs. Many factors—including the fate and transport properties of the chemicals of concern—must be considered in the model selection process. In any case, in lieu of an established trend in historical data indicating the contrary, a potentially contaminated site may be considered to be in steady state with its surroundings. In the end, the accuracy with which exposures are characterized can indeed become a major determinant of the ultimate validity of a risk assessment.

All things considered, there are three fundamental steps for most exposure assessments—namely:

i. Characterization of the exposure setting at the case-site—to include the physical environment and potentially exposed populations;
ii. Identification of the significant exposure pathways—to include sources or origins of release, exposure points, and exposure routes; and
iii. Quantification of exposure—to include efforts directed at determining ECs and intake variables.

On the whole, the interconnectivity of the exposure routes to the hazard sources are typically determined by integrating information from an initial environmental characterization with knowledge about potentially exposed populations and their likely behaviors. In the final analysis, the significance of the chemical hazard is evaluated on the basis of whether the target chemical could cause significant adverse exposures and impacts. Accordingly, the exposure assessment process would typically involve several characterization and evaluation efforts—including the following key tasks:

- Determination of chemical distributions and behaviors—traced from a "release" or "originating" source to the locations for likely human exposure;
- Identification of significant chemical release, migration, and exposure pathways;
- Identification of potential receptors—i.e., the populations potentially at risk;
- Development of conceptual site model(s) (CSM) or conceptual exposure model(s) and exposure scenarios—including a determination of current and future exposure patterns;
- Analysis of the environmental fate and persistence/transport of the CoPCs—including intermedia transfers;
- Estimation/modeling of EPCs for the critical exposure pathways and media; and
- Estimation of chemical intakes for all potential receptors, and for all significant exposure pathways associated with the CoPCs.

In the end, as part of a consequential and holistic exposure characterization effort, populations potentially at risk are defined, and concentrations of the chemicals of potential concern (CoPCs) are determined in each medium to which potential receptors may be exposed. Then, using the appropriate case-specific exposure parameters, the intakes of the CoPCs can be estimated. The evaluation could indeed concern past or current exposures, as well as exposures anticipated in the future.

This section discusses the principal exposure evaluation tasks that, upon careful implementation, should allow effective risk management decisions to be made about contaminated sites and related chemical exposure problems. It must be acknowledged and emphasized here that, although the discussions that follow are biased towards human exposure situations, fundamentally and in general, the concepts presented are equally applicable to ecological receptors as well—with elements very specific to ecological receptors having been identified in Chapter 10 of this title.

9.3.1 Factors Affecting Exposures to Chemical Hazards

The characterization of chemical exposure problems is a process used to establish the presence or absence of chemical hazards, to delineate the nature and degree of the hazards, and to determine possible (human and ecological) health threats posed by the exposure or hazard situation. The routes of chemical exposure (which may consist of inhalation, ingestion, and/or dermal contacts), as well as the duration of exposure (that may be short term [acute], intermediate term, or long term [chronic]) will significantly influence the level of impacts on the affected receptors. The nature and behavior of the chemical substances of interest also form a very important basis for evaluating the potential for human and ecological exposures to its possible toxic or hazardous constituents.

By and large, the assessment of human or ecological receptor exposure to environmental contaminants/chemicals requires translating concentrations found in the target environmental matrix or receptor environments into quantitative estimates of the amount of chemical that comes into contact with the individual potentially at risk—usually, an individual selected at random from the larger population-at-risk (PAR). The PAR refers to the receptors that do, or plausibly could, inhabit or traverse the location that is nearest to the source of contamination. Contact is expressed by the amount of material per unit body weight (mg/kg-day) that, typically, enters the lungs (for an inhalation exposure); enters the gastrointestinal tract (for an ingestion exposure); or crosses the stratum corneum of the skin (for a dermal contact exposure). This quantity is used as a basis for projecting the incidence of health or ecological detriment within the populations or receptors of interest.

To accomplish the task of exposure determination, several important exposure parameters and/or information will typically be acquired (Asante-Duah, 2017)—also recognizing that, in terms of chemical exposures, the amount of contacted material that is bioavailable for absorption is a very important consideration. At any rate, it is noteworthy that conservative estimates in the exposure evaluation oftentimes assume that a potential receptor is always in the same location, exposed to the same ambient concentration, and that there is 100% absorption upon exposure. These assumptions hardly represent any real-life situation. In fact, lower exposures will generally be expected under most circumstances—especially due to the fact that potential receptors will typically be exposed to lower or even near-zero levels of the CoPCs for the period of time spent outside the impacted areas or "chemical-laden" settings.

In the final analysis, the extent of a receptor's exposure is estimated by identifying realistic exposure scenarios that describe the potential pathways of exposure to CoPCs, as well as the specific activities and behaviors of individuals that might lead to contact with the CoPCs encountered in the environment. The evaluation could indeed concern past or current exposures, as well as exposures anticipated in the future. In any case, it is also noteworthy that, because of the differences in activity patterns and sensitivity to exposures, multiple (typically three) age groups are normally considered in most evaluations involving human receptors—e.g., young child age 1–6 years (i.e., from 1 up to the 7th birthday); older child age 7–18 years (i.e., from 7 up to the 19th birthday); and adult (>18 years of age) (USEPA, 2014a).

9.3.1.1   Chemical Intake versus Dose

Intake (also commonly called "exposure," or "applied dose") is defined as the amount of chemical coming into contact with a receptor's visible exterior body (e.g., skin and openings into the body such as mouth and nostrils), or with the "abstract/conceptual" exchange boundaries (such as the skin, lungs, or gastrointestinal tract); and *dose* (also commonly called "absorbed dose," or "internal dose") is the amount of chemical absorbed by the body into the bloodstream. In fact, the *internal dose* (i.e., absorbed dose) tends to differ significantly from the (externally) *applied dose* (i.e., exposure or intake)—recognizing that the *internal dose* of a chemical is the amount of a chemical that directly crosses the barrier at the absorption site into the systemic circulation.

The intake value quantifies the amount of a chemical contacted during each exposure event—where "event" may have different meanings depending on the nature of exposure scenario being considered (e.g., each day's inhalation of an air contaminant may constitute one inhalation exposure event). The quantity of a chemical absorbed into the bloodstream per event—represented by the dose—is calculated by further considering pertinent physiological parameters (such as gastrointestinal absorption rates). Overall, the internal dose of a chemical is considered rather important for predicting the potential toxic effects of the chemical; this is because, among other things, once in the systemic circulation, the chemical is able to reach all major target organ sites.

It is noteworthy that, in general, when the systemic absorption from an intake is unknown, or cannot be estimated by a defensible scientific argument, intake and dose are considered to be the same (i.e., a 100% absorption into the bloodstream from contact is assumed). Such an approach provides a conservative estimate of the actual exposures. In any case, intakes and doses are normally calculated during the same step of the exposure assessment; the former multiplied by an absorption factor yields the latter value.

9.3.1.2   Chronic versus Subchronic Exposures

Event-based intake values are converted to final intake values by multiplying the intake per event by the frequency of exposure events, over the timeframe being considered in an exposure assessment. *Chronic daily intake* (CDI), which measures long-term (chronic) exposures, are based on the number of events that are assumed to occur within an assumed lifetime for potential receptors; *subchronic daily intake* (SDI), which represents projected receptor exposures over a short-term period, consider only a portion of a lifetime (USEPA, 1989b). The respective intake values are calculated by multiplying the estimated exposure point chemical concentrations by the appropriate receptor exposure and body weight factors.

SDIs are generally used to evaluate subchronic noncarcinogenic effects, whereas CDIs are used to evaluate both carcinogenic risks and chronic noncarcinogenic effects. It is noteworthy that the short-term exposures can result when a particular activity is performed for a limited number of years or when, for instance, a chemical with a short half-life degrades to negligible concentrations within several months of its presence in a receptor's exposure setting.

9.3.2 The Role of CSMs and Exposure Scenarios

The CSM provides the classic framework for the risk assessment; it is generally used to identify appropriate exposure pathways and receptors in order to engender a more focused evaluation during the risk assessment process. Indeed, the conceptual model generally enables a better and more comprehensive assessment of the nature and extent of exposure, as well as helps determine the potential impacts from such exposures. Consequently, in as early a stage as possible during a chemical exposure investigation, all available information should be compiled and analyzed to help develop a representative CSM for the problem situation. With that said, it is also notable that the CSM is generally meant to be a "living paradigm" that can (and perhaps must) be updated and modified as appropriate when additional data or information become available—in order to properly exhibit its typically continuously evolving nature.

In general, the information contained in a conceptual model is developed during the various stages of an environmental characterization process, and also from controlled field and laboratory experiments that may be conducted in studies pertaining to the potential environmental contamination problem. As environmental investigations progress, the conceptual model may be revised as necessary and used to direct the next iteration of characterization activities. The updated or finalized conceptual model is then used to develop realistic exposure scenarios for the project, and which then forms an important basis for completing an effectual and credible risk assessment or environmental management program. Principally, the purpose of the CSM is to identify: (i) potential chemical sources; (ii) potential migration pathways of constituents from source areas to environmental media where exposure can occur; (iii) potential receptors; and (iv) potential exposure pathways by which constituent uptake into the body of an organism might occur. Ultimately, potentially complete exposure pathways are identified for possible further evaluation within the risk assessment framework. Each potentially complete exposure pathway for any CoPC is generally evaluated quantitatively in the risk assessment—also recognizing that some receptor populations may be potentially exposed to CoPCs by more than one pathway. Indeed, the conceptual model facilitates an assessment of the nature and extent of exposure, as well as helps determine the potential impacts from such exposures. Consequently, in as early a stage as possible during a chemical exposure or contaminated site investigation, all available information should be compiled and analyzed to help develop a representative CSM for the problem situation. Further elaboration on this topic is provided in Chapter 4 of this book.

Finally, it is noteworthy that the exposure scenario associated with a given hazardous situation may be better defined if the exposure is known to have already occurred. In most cases associated with the investigation of potential chemical exposure or contaminated site problems, however, important decisions may have to be made about exposures that may not yet have occurred—and in which case hypothetical exposure scenarios are generally developed to facilitate the problem solution. Ultimately, the type/nature of human exposure scenarios associated with a given exposure situation provides clear direction for the exposure assessment. Also, the exposure scenarios developed for a given chemical exposure or contaminated site problem can be used to support an evaluation of the risks posed by the situation, as well as facilitate the development of appropriate risk management and/or corrective action decisions.

9.3.2.1 The General Types of Human Exposure Scenarios
for Contaminated Site Problems

An examination of potential human exposures associated with contaminated site problems often involves a complexity of integrated evaluations and issues related to the variant contaminant migration and exposure pathways (see Figure 7.3). Anyhow, human exposures to chemicals may predominantly occur via the inhalation, dermal, and/or oral routes. Under such circumstances, a wide variety of *potential* exposure patterns can be anticipated from any form of human exposures to chemicals. As an illustrative example, a select list of typical or commonly encountered exposure scenarios in relation to environmental contamination problems might include the following (Asante-Duah, 1998, 2017; HRI, 1995):

- Inhalation Exposures
  - Indoor air—resulting from potential receptor exposure to contaminants (including both volatile constituents and fugitive dust) found in indoor ambient air
  - Indoor air—resulting from potential receptor exposure to volatile chemicals in domestic water that may volatilize inside a house (e.g., during hot water showering), and then contaminate indoor air
  - Outdoor air—resulting from potential receptor exposure to contaminants (including both volatile constituents and fugitive dust) found in outdoor ambient air
  - Outdoor air—resulting from potential receptor exposure to volatile chemicals in irrigation water, or other surface water bodies, that may volatilize and contaminate outdoor air
- Ingestion Exposures
  - Drinking water—resulting from potential receptor oral exposure to contaminants found in domestic water used for drinking or cooking purposes
  - Swimming—resulting from potential receptor exposure (via incidental ingestion) to contaminants in surface water bodies
  - Incidental soil ingestion—resulting from potential receptor exposure to contaminants found in dust and soils
  - Crop consumption—resulting from potential receptor exposures to contaminated foods (such as vegetables and fruits produced in household gardens that utilized contaminated soils, groundwater, or irrigation water during the cultivation process)
  - Dairy and meat consumption—resulting from potential receptor exposure to contaminated foods (such as locally grown livestock that may have become contaminated through the use of contaminated domestic water supplies, or from feeding on contaminated crops, and/or from contaminated air and soils)
  - Seafood consumption—resulting from potential receptor exposure to contaminated foods (such as fish and shellfish harvested from contaminated waters or that have been exposed to contaminated sediments, and that consequently have bioaccumulated toxic levels of chemicals in their edible portions)
- Dermal Exposures
  - Showering—resulting from potential receptor exposure (via skin absorption) to contaminants in domestic water supply
  - Swimming—resulting from potential receptor exposure (via skin absorption) to contaminants in surface water bodies
  - Direct soils contact—resulting from potential receptor exposure to contaminants present in outdoor soils

These types of exposure scenarios will typically be evaluated as part of an exposure assessment component of a public health risk management program. It should be emphasized, however, that this listing is by no means complete, since new exposure scenarios are always possible for case-specific situations; still, this demonstrates the multiplicity and inter-connectivity nature of the numerous pathways via which populations may become exposed to chemical constituents. Indeed, whereas the above-listed exposure scenarios may not all be relevant for every chemical exposure problem encountered in practice, a number of other exposure scenarios not listed or even alluded to here may have to be evaluated for the particular local conditions of interest—all the while recognizing that comprehensive human exposure assessments must include both direct and indirect exposure from ingredients found in various environmental compartments (such as in ambient air, water, soil, the food-chain, and consumer products) In any event, once the complete set of potential exposure scenarios has been fully determined for a given situation, the range of critical exposure pathways can then be identified to support subsequent evaluations.

In the end, careful consideration of the types and extent of potential human exposures, combined with hazard assessment and exposure-response information, is necessary to enable the completion of a credible human health risk assessment. Additionally, the exposure assessment (which is very critical to determining potential risks) requires realistic data to determine the extent of possible skin, inhalation, and ingestion exposures to products and components (Corn, 1993). Subsequent efforts are then directed at reaching the mandated goal of a given case-specific risk determination—recognizing that the goal of a human health risk assessment under any given set of circumstances would typically be to describe, with as little uncertainty as possible, the anticipated/projected risk (or indeed an otherwise lack of risk) to the populations potentially at risk (e.g., a given consumer or population group); this is done in relation to their exposure to potentially hazardous/toxic chemicals that may be found within their inhabited/occupied environments. Ultimately, the resulting information generated can then be used to support the design of cost-effective public health risk management programs and/or corrective action response programs.

9.3.3   Potential Exposure Quantification: The Exposure Estimation Model

In order to determine health risk arising from CoPCs for a given problem situation, it is invariably necessary to estimate the potential exposure dose for each CoPC. In fact, the exposure dose is estimated for each CoPC, and for each exposure pathway/route by which the likely receptor is assumed to be exposed. In the processes involved, exposure dose equations generally combine the estimates of CoPC concentrations in the target medium of interest with assumptions regarding the type and magnitude of each receptor's potential exposure—so as to arrive at a numerical estimate of the exposure dose (intake); the exposure dose is defined as the amount of CoPC taken up into the receptor—and this is generally expressed in units of milligrams of CoPC per kilogram of body weight per day (mg/kg-day) (USEPA, 1989a). Meanwhile, it is noteworthy that exposure doses are defined differently for potential carcinogenic versus noncarcinogenic effects. The "chronic daily intake" is generally used to estimate a receptor's potential average daily dose (ADD) from exposure to a CoPC with respect to noncarcinogenic effects—and generally calculated by averaging the exposure dose over the period of time for which the receptor is assumed to be exposed; thus, the averaging period is the same as the exposure duration for CoPCs with noncarcinogenic effects. For CoPCs with potential carcinogenic effects, however, the "CDI" is calculated by averaging the exposure dose over the receptor's assumed lifetime (e.g., usually 70 years for human receptors); therefore, the averaging period is the same as the receptor's assumed lifetime. Ultimately, these potential receptor exposures can be evaluated via the calculation of the so-called *average daily dose* (ADD) and/or the *lifetime average daily dose* (LADD). The standardized equations for estimating a receptor's intake on this basis are presented later on, below.

On the whole, the analysis of potential exposures to chemicals in the environment often involves several complex issues; invariably, potential receptors may become exposed to a variety of environmental chemicals via several different exposure routes—represented primarily by the inhalation, ingestion, and dermal exposure routes for human receptors (Asante-Duah, 2017). The carcinogenic effects (and sometimes the chronic noncarcinogenic effects) associated with a chemical exposure or contaminated site problem involve estimating the LADD; for noncarcinogenic effects, the ADD is usually used. The ADD differs from the LADD in that the former is not averaged over a lifetime; rather, it is the ADD pertaining to the actual duration of exposure. The *maximum daily dose* (MDD) will typically be used in estimating acute or subchronic exposures.

At the end of the day, receptor exposures to chemical materials may be conservatively quantified according to the generic equation shown in Box 9.1. The various exposure parameters used

**BOX 9.1   GENERAL EQUATION FOR ESTIMATING
POTENTIAL RECEPTOR EXPOSURES TO CHEMICALS**

$$EXP = \frac{C_{medium} \times CR \times CF \times FI \times ABS_f \times EF \times ED}{BW \times AT}$$

where

EXP = intake (i.e., the amount of chemical at the exchange boundary), adjusted for absorption (mg/kg-day)

C_{medium} = average or reasonably maximum EC of chemical contacted by potential receptor over the exposure period in the medium of concern (e.g., $\mu g/m^3$ [air]; or $\mu g/L$ [water]; or mg/kg [solid materials, such as food and soils])

CR = contact rate, i.e., the amount of "chemical-based" medium contacted per unit time or event (e.g., inhalation rate, IR, in m^3/day [air]; or ingestion rate in mg/day [food; soil], or L/day [water])

CF = conversion factor (10^{-6} kg/mg for solid media, or 1.00 for fluid media)

FI = fraction of intake from "chemical-based" source (dimensionless)

ABS_f = bioavailability or absorption factor (%)

EF = exposure frequency (days/years)

ED = exposure duration (years)

BW = body weight, i.e., the average body weight over the exposure period (kg)

AT = averaging time (period over which exposure is averaged—days)

  = ED × 365 days/year, for noncarcinogenic effects of human exposure

  = LT × 365 days/year = 70 years × 365 days/year, for carcinogenic effects of human exposure (assuming an average lifetime, LT, of 70 years)

in this model may be derived on a case-specific basis, or they may be compiled from regulatory guidance manuals and documents, and indeed other related scientific literature (e.g., Binder et al., 1986; Calabrese et al., 1989; CAPCOA, 1990; DTSC, 1994; Finley et al., 1994; Hrudey et al., 1996; Ikegami et al., 2014; LaGoy, 1987; Lepow et al., 1974, 1975; OSA, 1992; Sedman, 1989; Smith, 1987; Stanek and Calabrese, 1990; Travis and Arms, 1988; USEPA, 1987d, 1989f,g,n,o, , 1991a,e,f,g, 1992a,b,f,h,j, 1997e,f,h,i,k 1998e,f,g, 2000a–e, 2011a, 2014a; van Wijnen, 1990); these parameters are usually based on information relating to the maximum exposure level that results from specified categories of receptor activity and/or exposures.

The methods by which each specific type of chemical exposure might be estimated—including the relevant exposure estimation algorithms/equations for specific major routes of exposure (viz., inhalation, ingestion, and skin contacting)—are discussed in greater detail below. These algorithms and related ones are elaborated in an even greater detail elsewhere in the literature (e.g., Asante-Duah, 1998, 2017; CAPCOA, 1990; CDHS, 1986; DTSC, 1994; McKone, 1989; McKone and Daniels, 1991; NRC, 1991a,b; USEPA, 1986b–e, 1988d,e,i,j, 1989f,g,n,o, 1991c,e–h, 1992a,b,h,j, 1997c–f,h, 1998d–g, 2000a,b,g,h, 2011a, 2014a). Further illustration of the computational steps involved in the calculation of human receptor intakes and doses is also presented below.

9.3.3.1   Potential Receptor Inhalation Exposures

Two major types of inhalation exposures are generally considered in the investigation of potential chemical exposure or contaminated site problems (see Figure 7.3)—broadly categorized into the inhalation of airborne fugitive dust/particulates, in which all individuals within approximately 80 km ($\cong$50 miles) radius of a chemical release source are potentially impacted; and the inhalation of volatile compounds (i.e., airborne, vapor-phase chemicals). In general, potential inhalation intakes may be estimated based on the length of exposure, the IR of the exposed individual, the concentration of constituents in the inhaled air, and the amount retained in the lungs. [By the way, it is generally recommended that when estimating risk via inhalation, risk assessors should use the concentration of the chemical in air as the exposure metric (e.g., mg/m^3)—i.e., rather than a use of inhalation intake of a contaminant in air based on IR and BW (e.g., mg/kg-day). Under this set of circumstances, the general approach involves the estimation of ECs for each receptor exposed to contaminants via inhalation in the risk assessment—where the ECs are time-weighted average concentrations derived from measured or modeled contaminant concentrations in air at a locale or within an "exposure object" (and possibly further adjusted based on the characteristics of the exposure scenario being evaluated)].

9.3.3.2 Receptor Inhalation Exposure to Particulates from Constituents in Fugitive/Airborne Dust

Box 9.2 shows an algorithm that can be used to calculate potential receptor intakes resulting from the inhalation of constituents in wind-borne fugitive dust (CAPCOA, 1990; DTSC, 1994; USEPA, 1988d,e,i,j, 1989f,g,n,o, 1992a,b,h,j, 1997c–h, 1998d–g, 2000a,b,g,h, 2004a–g, 2011a, 2014a). The constituent concentration in air, Ca, is defined by the ground-level concentration (GLC)—usually represented by the respirable (PM-10) particles—expressed in $\mu g/m^3$. The PM-10 particles consist of particulate matter with physical/aerodynamic diameter of less than 10 microns (i.e., $<10\,\mu m$)—and it represents the respirable portion of the particulate emissions; this portion is capable of being deposited in thoracic (tracheobronchial and alveolar) portions of the lower respiratory tract. It is noteworthy that fine particulate matter has also been characterized by $PM_{2.5}$ (i.e., $\leq 2.5\,\mu m$ aerodynamic diameter). Finally, it should be recognized that the total PM exposure for an individual during a given period of time usually consists of exposures to many different particles from various sources whiles the receptor is in different microenvironments. As such, these different human microenvironments should be carefully identified so that the corresponding exposures can be properly appraised.

BOX 9.2   EQUATION FOR ESTIMATING INHALATION EXPOSURE TO CHEMICAL CONSTITUENTS IN FUGITIVE/AIRBORNE DUST

$$INH_a = \frac{C_a \times IR \times RR \times ABS_s \times ET \times EF \times ED}{BW \times AT}$$

where

INH_a = inhalation intake (mg/kg-day)

C_a = chemical concentration of airborne particulates (defined by the ground-level concentration [GLC], and represented by the respirable, PM-10 particles) (mg/m^3)

IR = inhalation rate (m^3/h)

RR = retention rate of inhaled air (%)

ABS_s = percent of chemical absorbed into the bloodstream (%)

ET = exposure time (h/day)

EF = exposure frequency (days/years)

ED = exposure duration (years)

BW = body weight, i.e., the average body weight over the exposure period (kg)

AT = averaging time (period over which exposure is averaged—days)

   = ED × 365 days/year, for noncarcinogenic effects of human exposure

   = LT × 365 days/year = 70 years × 365 days/year, for carcinogenic effects of human exposure (assuming an average lifetime, LT, of 70 years)

9.3.3.3 Receptor Inhalation Exposure to Volatile Compounds

Box 9.3 shows an algorithm that can be used to calculate potential receptor intakes resulting from the inhalation of airborne vapor-phase chemicals (CAPCOA, 1990; DTSC, 1994; USEPA, 1988d,e,i,j, 1989f,g,n,o, 1992a,b,h,j, 1997c–h, 1998d–g, 2000a,b,g,h, 2004a–g, 2011a, 2014a). The vapor-phase contaminant concentration in air is assumed to be in equilibrium with the concentration in the release source. Meanwhile, it is noteworthy that showering generally seems to represent a prominent activity that promotes the release of volatile organic chemicals (VOCs) from water—especially because of the high turbulence, high surface area, and small droplets of water involved. In fact, some contemporary studies have shown that risks from inhalation while showering can be comparable to—if not greater than—risks from drinking contaminated water (Jo et al., 1990a,b; Kuo et al., 1998; McKone, 1987; Richardson et al., 2002; Wilkes et al., 1996). Thus, this exposure scenario represents a particularly important one to carefully examine/evaluate in a human health risk assessment, whenever applicable.

9.3.3.4 Potential Receptor Ingestion Exposures

Exposure through ingestion is a function of the concentration of the constituents in the material ingested (e.g., soil, water, or food products such as crops, dairy and beef), the gastrointestinal absorption of the constituent in solid or fluid matrix, and the amount ingested. Indeed, the total dose received by the potential receptors from chemical ingestion will, in general, be dependent on the absorption of the chemical across the gastrointestinal (GI) lining; the scientific literature provides some estimates of such absorption factors for various chemical substances—but for chemicals without published absorption values and for which absorption factors are not implicitly accounted for in toxicological parameters, absorption may conservatively be assumed to be 100%.

Potential receptor ingestion exposures specific to the oral intake of chemical-impacted waters, the consumption of chemicals in food products, and the incidental ingestion of other contaminated solid matrices (such as soils/sediments) are annotated below.

BOX 9.3   EQUATION FOR ESTIMATING INHALATION EXPOSURE TO VAPOR-PHASE CHEMICAL CONSTITUENTS

$$\text{INH}_{av} = \frac{C_{av} \times \text{IR} \times \text{RR} \times \text{ABS}_s \times \text{ET} \times \text{EF} \times \text{ED}}{\text{BW} \times \text{AT}}$$

where

INH_{av} = inhalation intake (mg/kg-day)

C_{av} = chemical concentration in air (mg/m³) (The vapor-phase contaminant concentration in air is assumed to be in equilibrium with the concentration in the release source)

IR = inhalation rate (m³/h)

RR = retention rate of inhaled air (%)

ABS_s = percent of chemical absorbed into the bloodstream (%)

ET = exposure time (h/day)

EF = exposure frequency (days/years)

ED = exposure duration (years)

BW = body weight, i.e., the average body weight over the exposure period (kg)

AT = averaging time (period over which exposure is averaged—days)

  = ED × 365 days/year, for noncarcinogenic effects of human exposure

  = LT × 365 days/year = 70 years × 365 days/year, for carcinogenic effects of human exposure (assuming an average lifetime, LT, of 70 years)

9.3.3.5 Receptor Exposure through Ingestion of Constituents in Drinking Water

Exposure to contaminants via the ingestion of contaminated fluids may be estimated using the algorithm shown in Box 9.4 (CAPCOA, 1990; DTSC, 1994; USEPA, 1988d,e,i,j, 1989f,g,n,o, 1992a,b,h,j, 1997c–h, 1998d–g, 2000a,b,g,h, 2004a–g, 2011a, 2014a). This is comprised of the applicable relationship for estimating the chemical exposure intake that occurs through the ingestion of drinking water.

As a special case, receptor exposure through incidental ingestion of constituents in water *during swimming activities* (i.e., the result of the ingestion of contaminated surface water during recreational activities) may be estimated by using the algorithm shown in Box 9.4A.

BOX 9.4 EQUATION FOR ESTIMATING INGESTION EXPOSURE TO CONSTITUENTS IN WATER USED FOR CULINARY PURPOSES

$$\text{ING}_{dw} = \frac{C_w \times \text{WIR} \times \text{FI} \times \text{ABS}_s \times \text{EF} \times \text{ED}}{\text{BW} \times \text{AT}}$$

where

ING_{dw} = ingestion intake, adjusted for absorption (mg/kg-day)
C_w = chemical concentration in drinking water (mg/L)
WIR = average ingestion rate (L/day)
FI = fraction ingested from contaminated source (unitless)
ABS_s = bioavailability/gastrointestinal (GI) absorption factor (%)
EF = exposure frequency (days/year)
ED = exposure duration (years)
BW = body weight (kg)
AT = averaging time (period over which exposure is averaged—days)

BOX 9.4A EQUATION FOR ESTIMATING INCIDENTAL INGESTION EXPOSURE TO CONTAMINATED SURFACE WATER DURING RECREATIONAL ACTIVITIES

$$\text{ING}_r = \frac{C_w \times \text{CR} \times \text{ABS}_s \times \text{ET} \times \text{EF} \times \text{ED}}{\text{BW} \times \text{AT}}$$

where

ING_r = ingestion intake, adjusted for absorption (mg/kg-day)
C_w = chemical concentration in water (mg/L)
CR = contact rate (L/h)
ABS_s = bioavailability/gastrointestinal (GI) absorption factor (%)
ET = exposure time (h/event)
EF = exposure frequency (days/year)
ED = exposure duration (years)
BW = body weight (kg)
AT = averaging time (period over which exposure is averaged—days)

9.3.3.6 Receptor Exposure through Ingestion of Constituents in Food Products

Typically, exposure from the ingestion of food can occur via the ingestion of plant products, fish, animal products, and mother's milk. A general algorithm for estimating the exposure intake through the ingestion of foods is shown in Box 9.5—with corresponding relationships defined below for specific types of food products.

- *Ingestion of Plant Products*—Exposure through the ingestion of plant products, ING_p, is a function of the type of plant, gastrointestinal absorption factor, and the fraction of plants ingested that are affected by the chemical constituents of concern. The exposure estimation is performed for each plant type in accordance with the algorithm presented in Box 9.5A (CAPCOA, 1990; USEPA, 1989f,g,n,o, 1992a,b,h,j, 1997c–h, 1998d–g, 2000a,b,g,h, 2004a–g, 2011a, 2014a).
- *Bioaccumulation and Ingestion of Seafood*—Exposure from the ingestion of chemical constituents in fish (e.g., obtained from contaminated surface water bodies) may be estimated using the algorithm shown in Box 9.5B (USEPA, 1987a, 1988d,e,i,j, 1989f,g,n,o, 1992a,b,h,j, 1997c–h, 1998d–g, 2000a,b,g,h, 2004a–g, 2011a, 2014a).
- *Ingestion of Animal Products*—Exposure resulting from the ingestion of animal products, ING_a, is a function of the type of meat ingested (including animal milk products and eggs), gastrointestinal absorption factor, and the fraction of animal products ingested that are affected by the constituents of concern. The exposure estimation is carried out for each animal product type by using the form of relationship shown in Box 9.5C (CAPCOA, 1990; USEPA, 1989f,g,n,o, 1992a,b,h,j, 1997c–h, 1998d–g, 2000a,b,g,h, 2004a–g, 2011a, 2014a).
- *Ingestion of Mother's Milk*—Exposure through the ingestion of a mother's milk, ING_m, is a function of the average chemical concentration in the mother's milk, the amount of mother's milk ingested, and gastrointestinal absorption factor—estimated according to the relationship shown in Box 9.5D (CAPCOA, 1990; USEPA, 1989f,g,n,o, 1992a,b,h,j, 1997c–h, 1998d–g, 2000a,b,g,h, 2004a–g, 2011a, 2014a).

**BOX 9.5   EQUATION FOR ESTIMATING INGESTION
EXPOSURE TO CONSTITUENTS IN FOOD PRODUCTS**

$$ING_f = \frac{C_f \times FIR \times CF \times FI \times ABS_s \times EF \times ED}{BW \times AT}$$

where

ING_f = ingestion intake, adjusted for absorption (mg/kg-day)

C_f = chemical concentration in food (mg/kg or mg/L)

FIR = average food ingestion rate (mg or L/meal)

CF = conversion factor (10^{-6} kg/mg for solids and 1.00 for fluids)

FI = fraction ingested from contaminated source (unitless)

ABS_s = bioavailability/gastrointestinal (GI) absorption factor (%)

EF = exposure frequency (meals/year)

ED = exposure duration (years)

BW = body weight (kg)

AT = averaging time (period over which exposure is averaged—days)

BOX 9.5A EQUATION FOR ESTIMATING INGESTION EXPOSURE TO CONSTITUENTS IN PLANT PRODUCTS

$$ING_p = \frac{CP_Z \times PIR_Z \times FI_Z \times ABS_s \times EF \times ED}{BW \times AT}$$

where

ING_p = exposure intake from ingestion of plant products, adjusted for absorption (mg/kg-day)

CP_Z = chemical concentration in plant type Z (mg/kg)

PIR_Z = average consumption rate for plant type Z (kg/day)

FI_Z = fraction of plant type Z ingested from contaminated source (unitless)

ABS_s = bioavailability/gastrointestinal (GI) absorption factor (%)

EF = exposure frequency (days/years)

ED = exposure duration (years)

BW = body weight (kg)

AT = averaging time (period over which exposure is averaged—days)

BOX 9.5B EQUATION FOR ESTIMATING INGESTION EXPOSURE TO CONSTITUENTS IN CONTAMINATED SEAFOOD

$$ING_{sf} = \frac{C_w \times FIR \times CF \times BCF \times FI \times ABS_s \times EF \times ED}{BW \times AT}$$

where

ING_{sf} = total exposure, adjusted for absorption (mg/kg-day)

C_w = chemical concentration in surface water (mg/L)

FIR = average fish ingestion rate (g/day)

CF = conversion factor ($=10^{-3}$ kg/g)

BCF = chemical-specific bioconcentration factor (L/kg)

FI = fraction ingested from contaminated source (unitless)

ABS_s = bioavailability/gastrointestinal (GI) absorption factor (%)

EF = exposure frequency (days/years)

ED = exposure duration (years)

BW = body weight (kg)

AT = averaging time (period over which exposure is averaged—days)

9.3.3.7 Receptor Exposure through Pica and Incidental Ingestion of Soil/Sediment

Exposures that result from the incidental ingestion of contaminants sorbed onto soils are determined by multiplying the concentration of the constituent in the medium of concern by the amount of soil/material ingested per day and the degree of absorption. The applicable relationship for estimating the resulting exposures is shown in Box 9.6 (CAPCOA, 1990; USEPA, 1988d,e,i,j, 1989f,g,n,o, 1992a,b,h,j, 1997c–h, 1998d–g, 2000a,b,g,h, 2004a–g, 2011a, 2014a). In general, it is usually assumed that all ingested soil during receptor exposures comes from a contaminated source, so that the FI term becomes unity.

BOX 9.5C   EQUATION FOR ESTIMATING INGESTION EXPOSURE TO CONSTITUENTS IN ANIMAL PRODUCTS

$$ING_a = \frac{CAP_Z \times APIR_Z \times FI_Z \times ABS_s \times EF \times ED}{BW \times AT}$$

where

ING_a = exposure intake through ingestion of plant products, adjusted for absorption (mg/kg-day)

CAP_Z = chemical concentration in food type Z (mg/kg)

$APIR_Z$ = average consumption rate for food type Z (kg/day)

FI_Z = fraction of product type Z ingested from contaminated source (unitless)

ABS_s = bioavailability/gastrointestinal (GI) absorption factor (%)

EF = exposure frequency (days/years)

ED = exposure duration (years)

BW = body weight (kg)

AT = averaging time (period over which exposure is averaged—days)

BOX 9.5D   EQUATION FOR ESTIMATING INGESTION EXPOSURE TO CHEMICALS IN MOTHER'S MILK USED FOR BREAST-FEEDING

$$ING_m = \frac{CMM \times IBM \times ABS_s \times EF \times ED}{BW \times AT}$$

where

ING_m = exposure intake through ingestion of mother's milk, adjusted for absorption (mg/kg-day)

CMM = chemical concentration in mother's milk—which is a function of a mother's exposure through all routes and the contaminant body half-life (mg/kg)

IBM = daily average ingestion rate for breast milk (kg/day)

ABS_s = bioavailability/gastrointestinal (GI) absorption factor (%)

EF = exposure frequency (days/years)

ED = exposure duration (years)

BW = body weight (kg)

AT = averaging time (period over which exposure is averaged—days)

9.3.3.8   Potential Receptor Dermal Exposures

The major types of dermal exposures that could affect chemical exposure decisions consist of dermal contacts with contaminants adsorbed onto or within solid matrices (e.g., soils), and dermal absorption from contaminated waters and other fluids. In general, dermal intake is a function of the chemical concentration in the medium of concern, the body surface area in contact with the medium, the duration of the contact, flux of the medium across the skin surface, and the absorbed fraction.

Potential receptor dermal exposures via dermal contacts with solid matrices containing chemical constituents, and from the dermal absorption of chemicals present in contaminated water media, are annotated below.

BOX 9.6   EQUATION FOR ESTIMATING PICA AND INCIDENTAL INGESTION EXPOSURE TO CONTAMINATED SOILS/SEDIMENTS

$$ING_s = \frac{C_s \times SIR \times CF \times FI \times ABS_s \times EF \times ED}{BW \times AT}$$

where
   ING_s = ingestion intake, adjusted for absorption (mg/kg-day)
   C_s = chemical concentration in soil (mg/kg)
   SIR = average soil ingestion rate (mg soil/day)
   CF = conversion factor (10^{-6} kg/mg)
   FI = fraction ingested from contaminated source (unitless)
   ABS_s = bioavailability/gastrointestinal (GI) absorption factor (%)
   EF = exposure frequency (days/years)
   ED = exposure duration (years)
   BW = body weight (kg)
   AT = averaging time (period over which exposure is averaged—days)

9.3.3.9   Receptor Exposure through Contact/Dermal Absorption from Solid Matrices

The dermal exposures to chemicals in solid materials (e.g., soils and sediments) may be estimated by applying the equation shown in Box 9.7 (CAPCOA, 1990; DTSC, 1994; USEPA, 1988d,e,i,j, 1989f,g,n,o, 1992a,b,h,j, 1997c–h, 1998d–g, 2000a,b,g,h, 2004a–g, 2011a, 2014a).

BOX 9.7   EQUATION FOR ESTIMATING DERMAL EXPOSURES THROUGH CONTACTS WITH CONSTITUENTS IN SOLID MATRICES (E.G., CONTAMINATED SOILS)

$$DEX_s = \frac{C_s \times CF \times SA \times AF \times ABS_s \times SM \times EF \times ED}{BW \times AT}$$

where
   DEX_s = absorbed dose (mg/kg-day)
   C_s = chemical concentration in solid materials (e.g., contaminated soils) (mg/kg)
   CF = conversion factor (10^{-6} kg/mg)
   SA = skin surface area available for contact, i.e., surface area of exposed skin (cm^2/event)
   AF = solid material to skin adherence factor (e.g., soil loading on skin) (mg/cm^2)
   ABS_s = skin absorption factor for chemicals in solid matrices (e.g., contaminated soils) (%)
   SM = factor for solid materials matrix effects (%)
   EF = exposure frequency (events/year)
   ED = exposure duration (years)
   BW = body weight (kg)
   AT = averaging time (period over which exposure is averaged—days)

**BOX 9.8   EQUATION FOR ESTIMATING DERMAL EXPOSURES
THROUGH CONTACTS WITH CONTAMINATED WATERS**

$$\mathrm{DEX_w} = \frac{C_w \times \mathrm{CF} \times \mathrm{SA} \times \mathrm{PC} \times \mathrm{ABS_s} \times \mathrm{ET} \times \mathrm{EF} \times \mathrm{ED}}{\mathrm{BW} \times \mathrm{AT}}$$

where

$\mathrm{DEX_w}$ = absorbed dose from dermal contact with chemicals in water (mg/kg-day)
C_w = chemical concentration in water (mg/L)
CF = volumetric conversion factor for water (1 L/1,000 cm³)
SA = skin surface area available for contact, i.e., surface area of exposed skin (cm²)
PC = chemical-specific dermal permeability constant (cm/h)
$\mathrm{ABS_s}$ = skin absorption factor for chemicals in water (%)
ET = exposure time (h/day)
EF = exposure frequency (days/year)
ED = exposure duration (years)
BW = body weight (kg)
AT = averaging time (period over which exposure is averaged—days).

9.3.3.10   Receptor Exposure through Dermal Contact with Waters and Seeps

Dermal exposures to chemicals in water may occur during domestic use (such as bathing and washing), or through recreational activities (such as swimming or fishing). As a specific example, the dermal intakes of chemicals in ground or surface water and/or from seeps from a contaminated site may be estimated by using the type of equation shown in Box 9.8 (USEPA, 1988d,e,i,j, 1989f,g,n,o, 1992a,b,h,j, 1997c–h, 1998d–g, 2000a,b,g,h, 2004a–g, 2011a, 2014a).

9.3.4   ESTABLISHING "EXPOSURE INTAKE FACTORS" FOR USE IN THE COMPUTATION OF CHEMICAL INTAKES AND DOSES DURING EXPOSURE ASSESSMENTS

Several exposure parameters are normally required so as to be able to model the various exposure scenarios typically associated with chemical exposure or contaminated site problems. Oftentimes, default values are obtainable from the scientific literature for some of the requisite parameters used in the estimation of chemical intakes and doses. Table 9.2 shows typical parameters that exemplify a generic set of values commonly used in some applications; indeed, this is by no means complete—and more detailed information on such parameters can be obtained from various scientific sources (e.g., Calabrese et al., 1989; CAPCOA, 1990; Lepow et al., 1974, 1975; OSA, 1992; USEPA, 1987d, 1988d,e,i,j, 1989f,g,n,o, 1992a,b,h,j, 1997c–h, 1998d–g, 2000a,b,g,h, 2004a–g, 2011a, 2014a).

A spreadsheet to help in automatically calculating exposure "intake factors" for varying input parameters that reflect case-specific problem scenarios may be developed (based on the algorithms presented in the preceding sections) to facilitate the computational efforts involved in the exposure assessment (Table 9.3). Some example evaluations—in some cases offered only as an illustration of the computational mechanics involved in the exposure assessment process—for potential receptor groups purportedly exposed through inhalation, soil ingestion (viz., incidental or pica behavior), and dermal contact are elaborated elsewhere in the literature (e.g., Asante-Duah, 1998, 2017).

TABLE 9.2

An Example Listing of Case-Specific Exposure Parameters

Parameter	Child Aged up to 6 Years	Child Aged 6–12 Years	Adult
Physical Characteristics			
Average body weight (kg)	16	29	70
Average total skin surface area (cm²)	6,980	10,470	18,150
Average lifetime (years)	70	70	70
Average lifetime exposure period (years)	5	6	58
Activity Characteristics			
Inhalation rate (m³/h)	0.25	0.46	0.83
Retention rate of inhaled air (%)	100	100	100
Frequency of fugitive dust inhalation (days/year)			
• off-site residents, schools, and by-passers	365	365	365
• off-site workers	–	–	260
Duration of fugitive dust inhalation (outside) (h/day)			
• off-site residents, schools, and by-passers	12	12	12
• off-site workers	–	–	8
Amount of incidentally ingested soils (mg/day)	200	100	50
Frequency of soil contact (days/year)			
• off-site residents, schools, and by-passers	330	330	330
• off-site workers	–	–	260
Duration of soil contact (h/day)			
• off-site residents, schools, and by-passers	12	8	8
• off-site workers	–	–	8
Skin area contacted by soil (%)	20	20	10
Material Characteristics			
Soil-to-skin adherence factor (mg/cm²)	0.75	0.75	0.75
Soil matrix attenuation factor (%)	15	15	15

Note: The exposure factors represented here are considered to project potential maximum exposures (and therefore these are expected to produce conservative estimates). Indeed, these could be modified, as appropriate—to reflect the most reasonable exposure patterns anticipated for a project-specific situation; for instance, realistically, soil exposure is generally reduced from snow cover and rainy days—thus reducing potential exposures for children playing outdoors in a contaminated area. In any case, the sources and/or rationale for the choice of the exposure parameters should be very well supported and adequately documented.

9.3.5 REFINING THE RECEPTOR CHEMICAL EXPOSURE ESTIMATES

To be certain realistic exposure estimates are generated to support risk assessments, a variety of refinements may have to be undertaken during an exposure determination phase of a study; this may be particularly important when one is carrying out comprehensive exposure assessments. Some of the key attributes recommended for serious consideration in any efforts to refine chemical exposure estimates (and therefore the consequential risk estimates) include the following (Asante-Duah, 2017):

- *Incorporating Chemical Bioavailability Adjustments into Exposure Calculations. Bioavailability* is defined as the fraction of a chemical that is taken up by the body's circulatory system relative to the amount that an organism is exposed to during, for instance,

TABLE 9.3
Example Spreadsheet for Calculating Case-Specific "Intake Factors" for an Exposure Assessment

Fugitive Dust Inhalation Pathway

Receptor Group	IR	RR	ET	EF	ED	BW	AT	INH Factor
C(1–6)@NCancer	0.25	1	12	365	5	16	1,825	1.88E-01
C(1–6)@Cancer	0.25	1	12	365	5	16	25,550	1.34E-02
C(6–12)@NCancer	0.46	1	12	365	6	29	2,190	1.90E-01
C(6–12)@Cancer	0.46	1	12	365	6	29	25,550	1.63E-02
ResAdult@NCancer	0.83	1	12	365	58	70	21,170	1.42E-01
ResAdult@Cancer	0.83	1	12	365	58	70	25,550	1.18E-01
JobAdult@NCancer	0.83	1	8	260	58	70	21,170	6.76E-02
JobAdult@Cancer	0.83	1	8	260	58	70	25,550	5.60E-02

Soil Ingestion Pathway

Receptor Group	IR	CF	FI	EF	ED	BW	AT	ING Factor
C(1–6)@NCancer	200	1.00E-06	1	330	5	16	1,825	1.13E-05
C(1–6)@Cancer	200	1.00E-06	1	330	5	16	25,550	8.07E-07
C(6–12)@NCancer	100	1.00E-06	1	330	6	29	2,190	3.12E-06
C(6–12)@Cancer	100	1.00E-06	1	330	6	29	25,550	2.67E-07
ResAdult@NCancer	50	1.00E-06	1	330	58	70	21,170	6.46E-07
ResAdult@Cancer	50	1.00E-06	1	330	58	70	25,550	5.35E-07
JobAdult@NCancer	50	1.00E-06	1	260	58	70	21,170	5.09E-07
JobAdult@Cancer	50	1.00E-06	1	260	58	70	25,550	4.22E-07

Soil Dermal Contact Pathway

Receptor Group	SA	CF	AF	SM	EF	ED	BW	AT	DEX Factor
C(1–6)@NCancer	1,396	1E-06	0.75	0.15	330	5	16	1,825	8.87E-06
C(1–6)@Cancer	1,396	1E-06	0.75	0.15	330	5	16	25,550	6.34E-07
C(6–12)@NCancer	2,094	1E-06	0.75	0.15	330	6	29	2,190	7.34E-06
C(6–12)@Cancer	2,094	1E-06	0.75	0.15	330	6	29	25,550	6.30E-07
ResAdult@NCancer	1,815	1E-06	0.75	0.15	330	58	70	21,170	2.64E-06
ResAdult@Cancer	1,815	1E-06	0.75	0.15	330	58	70	25,550	2.19E-06
JobAdult@NCancer	1,815	1E-06	0.75	0.15	260	58	70	21,170	2.08E-06
JobAdult@Cancer	1,815	1E-06	0.75	0.15	260	58	70	25,550	1.72E-06

Notes:

Notations and units are same as defined in the text.

INH factor = Inhalation factor for calculation of doses and intakes.

ING factor = Soil ingestion factor for calculation of doses and intakes.

DEX factor = Dermal exposure (via skin absorption) factor for calculation of doses and intakes.

C(1–6)@NCancer; C(6–12)@NCancer; ResAdult@NCancer; JobAdult@NCancer = Noncarcinogenic effects for a child aged 1–6 years; child aged 6–12 years; resident adult; and adult worker, respectively.

C(1–6)@Cancer; C(6–12)@Cancer; ResAdult@Cancer; JobAdult@Cancer = Carcinogenic effects for a child aged 1–6 years; child aged 6–12 years; resident adult; and adult worker, respectively.

the ingestion of a chemical-laden material of interest. Incontrovertibly, bioavailability is a rather important concept in risk determination—especially because exposure and risk are more closely related to the bioavailable fraction of a chemical than to its total concentration in any given media/matrix; thus, this would tend to have significant implications in determining any "safe" levels of chemicals in an exposure medium. Invariably, the amount of contacted material that is bioavailable for absorption is very important in the evaluation of any receptor's exposure to chemicals.

Bioavailability can be influenced by external physical/chemical factors such as the form of a chemical in the exposure media, as well as by internal biological factors such as absorption mechanisms within a living organism. Broadly speaking, the bioavailability of a CoPC may be estimated by multiplying the fraction of the chemical that is bioaccessible and the fraction that is absorbed. Meanwhile, it is notable that bioavailablity can be media-specific; for instance, the bioavailability of metals ingested in a soil matrix is generally believed to be considerably lower than the bioavailability of the same metals ingested in water.

Overall, bioavailability has a direct and significant relationship to exposure dose and risk; among other things, a lower bioavailability means a decrease in exposure dose and risk—and, conversely, higher bioavailability implies an increased exposure dose and risk. Indeed, bioavailability generally refers to how much of a chemical is "available" to have an adverse effect on humans or other organisms. Consequently, knowledge of chemical bioavailability can play key roles in risk management decisions. For example, bioavailability adjustments in risk assessment can help establish reduced time and cost necessary for site remediation; in this case, bioavailability would be inversely related to risk-based cleanup levels—i.e., lower bioavailability results in increased risk-based cleanup levels. In fact, when risk assessments are adjusted to account for lower case-specific bioavailability, the resulting increase in cleanup levels can, in some cases, reduce remediation costs substantially. This is because, determining the site-specific bioavailability can allow for a revising of the exposure estimates—so as to more realistically and pragmatically reflect the conditions at a project site.

• *Incorporating Chemical Degradation into Exposure Calculations.* Many chemicals are transformed to structurally related degradation/daughter products in the environment before they are mineralized (e.g., dichlorodiphenyl dichloroethylene, DDE formed out of dichlorodiphenyl trichloroethane, DDT)—with each of the resultant transformation products tending to display their own toxicity and persistence characteristics. Indeed, when certain chemical compounds undergo degradation, potentially more toxic daughter products result (such as is the case when trichloroethylene (TCE) biodegrades to produce vinyl chloride). On the other hand, there are situations where the end products of degradation are less toxic than the parent compounds. Consequently, it is often imperative to include pertinent data on such transformation products into chemical exposure and risk assessments—albeit this often adds another layer of complexity to the overall exposure and risk assessment process, especially because, among other things, toxicity data for the daughter products may often be lacking.

In fact, since receptor exposures could be occurring over long time periods, a more valid approach in exposure modeling will be to take chemical degradation (or indeed other transformation processes) into consideration during an exposure assessment. Under such circumstances, if significant degradation is likely to occur, then exposure calculations become much more complicated. In that case, chemical concentrations at exposure or release sources are calculated at frequent and short time intervals, and then summed over the exposure period.

To illustrate the concept of incorporating chemical degradation into exposure assessment, let us assume first-order kinetics for a hypothetical chemical exposure problem. An approximation of the degradation effects for this type of scenario can be obtained by multiplying the chemical concentration data by a degradation factor, DGF, defined by:

$$\mathrm{DGF} = \frac{\left(1 - e^{-kt}\right)}{kt}$$

where k is a chemical-specific degradation rate constant [days^{-1}] and t is the time period over which exposure occurs [days]. For a first-order decaying substance, k is estimated from the following relationship:

$$T_{1/2}[\text{days}] = \frac{0.693}{k} \quad \text{or} \quad k[\text{days}^{-1}] = \frac{0.693}{T_{1/2}}$$

where $T_{1/2}$ is the chemical half-life, which is the time after which the mass of a given substance will be one-half its initial value.

It is noteworthy that the degradation factor is usually ignored in most exposure calculations; this is especially justifiable if the degradation product is of potentially equal toxicity, and is present in comparable amounts as the parent compound. In any case, although it cannot always be proven that the daughter products will result in receptor exposures that are at comparable levels to the parent compound, the DGF term is still ignored in most screening-level exposure assessments. Still, as necessary, various methods of approach may be utilized to incorporate transformation products into exposure and risk assessment of the parent compounds. For instance, Fenner et al. (2002) offer some elaborate procedures that integrate the chemical transformation kinetics into the overall assessment—by calculating the environmental exposure to parent compounds and daughter products as they are being formed in the degradation/transformation cascade, and then subsequently developing a corresponding risk quotient.

- *Receptor Age Adjustments to Population Exposure Factors.* Age adjustments are often necessary when receptor exposures to a chemical occur from childhood through the adult life. Such adjustments are meant to account for the transitioning of a potential receptor from childhood (requiring one set of intake assumptions and exposure parameters) into adulthood (that requires a different set of chemical intake assumptions and exposure parameters). Indeed, in the processes involved especially in human exposure assessments, it frequently becomes very apparent that contact rates can be significantly different for children versus adults. Consequently, carcinogenic risks (that are averaged over a receptor's lifetime) should preferably be calculated by applying the appropriate age-adjusted factors. Further details on the development of age-adjusted factors are provided elsewhere in the literature (e.g., DTSC, 1994; OSA, 1992; USEPA, 1989f,g,n,o, 1997c–h, 1998d–g, 2000a,b,g,h, 2004a–g, 2011a, 2014a).

  The use of age-adjusted factors are especially important in certain specific situations—such as those involving human soil ingestion exposures, which are typically higher during childhood and decrease with age. For instance, because the soil ingestion rate is generally different for children and adults, the carcinogenic risk due to direct ingestion of soil should preferably be calculated using an age-adjusted ingestion factor. This takes into account the differences in daily soil ingestion rates, body weights, exposure fraction, and exposure duration for the two exposure groups—albeit exposure frequency may be assumed the same for the two "quasi-divergent" groups. If calculated in this manner, then the estimated exposure/intake factor will result in a more realistic, yet health protective, risk evaluation—compared to, for instance, using an "adult-only" type of assumption. Indeed, in a refined and comprehensive evaluation, it is generally recommended to incorporate age-adjustment factors in the chemical exposure assessment, wherever appropriate. On the contrary, and for the sake of simplicity, such types of age adjustment will usually not be made part of most screening-level computational processes in an exposure/risk assessment.

- *Spatial and Temporal Averaging of Chemical Exposure Estimates.* Oftentimes, in major environmental policy decisions, it becomes necessary to evaluate chemical exposure situations for population groups—rather than for individuals only. In such type of more practical and realistic chemical exposure assessment, it usually is more appropriate (and indeed less conservative) to estimate chemical exposure to a specific population subgroup over an exposure duration of less than a lifetime (Asante-Duah, 2017). Further details on the evaluation processes involved in the spatial and temporal averaging techniques for chemical exposure problems can be found elsewhere in the literature (e.g., Asante-Duah, 1998, 2017; CDHS, 1990; OSA, 1992; USEPA, 1992a,b,h,j, 1998d–g, 2000a,b,g,h, 2004a–g, 2011a, 2014a).

Finally, it is also noteworthy that, in developing EPCs from environmental sampling data, appropriate concepts of areal and temporal averaging can often be incorporated in the evaluation process (Washburn and Edelmann, 1999). For instance, since exposures to soils over a period of time would generally be expected to occur over an area, rather than at point locations, it is usually more appropriate to expect EPCs to be associated with an exposure area, rather than at every discrete point within the area. Similarly, since exposures from groundwater is associated with a well that generally draws the water from an area, rather than a single point in the aquifer, it is usually is more appropriate to average groundwater concentration over the well's recharge area in order to predict the EPCs—rather than assume that the EPCs are necessarily unique at each point in the aquifer.

With all the above-stated in mind, it is notable that chemical contaminants entering the environment tend to be partitioned or distributed across various environmental media and biota—with the distribution of chemicals entering the environmental compartments being the result of a number of complex processes. In any event, the potential hazards and/or risks associated with the individual chemicals are very much dependent on the extent of multimedia exposures to potential receptors. Thus, a good prediction of chemical concentrations in the various compartmental media *together with* a carefully executed exposure assessment is essential for the completion of a credible risk assessment.

9.4 EVALUATION OF CHEMICAL TOXICITY

In planning for (public health and/or environmental) protection from the likely adverse effects of (human and/or ecological) exposure to chemicals, a primary concern generally relates to whether or not the substance in question possesses potentially hazardous and/or toxic properties. Indeed, an organism's response to chemical exposures is as much dependent on the toxicity of the contacted substance as it is on the degree of exposure—among other factors. Chemical toxicity may be characterized using variant nomenclatures—but generally done in relation to the duration and location of exposure to an organism, and/or in accordance with the timing between exposure to the toxicant and the first appearance of symptoms associated with toxicity (Asante-Duah, 2017). In any case, a good understanding of the time-dependent behavior of a toxicant as related to its absorption, distribution, storage, biotransformation, and elimination is necessary to explain how such toxicants are capable of producing "acute" or "chronic" toxicity, "local" or "systemic" toxicity, and "immediate" or "delayed" toxicity (Asante-Duah, 2017; Hughes, 1996). It is noteworthy that among other things, the terms "acute" and "chronic" as applied to toxicity may also be used to describe the duration of exposure—namely, "acute exposure" and "chronic exposure." Indeed, it has been established that acute and chronic exposure to many a toxicants will parallel acute and chronic toxicity—albeit, in some cases, acute exposure can lead to chronic toxicity (Hughes, 1996).

Toxicity represents the state of being poisonous—and therefore may be said to indicate the state of adverse effects or symptoms being produced by toxicants in an organism. In general, toxicity tends to vary according to both the duration and location of the receptor that is exposed to the toxicant, as well as the receptor-specific responses of the exposed organism (Hughes, 1996; Renwick et al., 2001; WHO, 2010). The prototypical processes that would usually be anticipated following the exposure of an organism to a toxic substance, up through the realization of a toxic response on such organism, may be exhibited by the use of varying degrees/levels of detail appropriate for the case-specific study or program (Asante-Duah, 2017). Indeed, the more detailed the discussion or display of the intermediary processes that occur between the "external dose" and the consequential or potentially "toxic response," the better the chance to foster clearer understanding among most audiences—further to facilitating a more comprehensive and comprehensible risk determination.

In practice, an evaluation of the toxicological effects typically consists of a compilation of toxicological profiles of the chemicals of concern (including the intrinsic toxicological properties of the chemicals—that may include their acute, subchronic, chronic, carcinogenic, and/or reproductive effects) and a determination of the relevant toxicity indices. This section calls out the major underlying concepts, principles, and procedures that are often employed in the evaluation of the hazard effects or toxicity of various chemical constituents found at contaminated sites and/or in the wider environment. It must be acknowledged and emphasized here that, although the discussions that follow are biased towards human exposure situations, the concepts presented are generally applicable to ecological receptors as well—with elements that are very specific to ecological settings and receptors having been identified in Chapter 10 of this title.

9.4.1 Categorization of Human Toxic Effects from Chemical Exposures: Carcinogenicity versus Noncarcinogenicity

The toxic characteristics of a substance are usually categorized according to the organs or systems they affect (e.g., kidney, liver, and nervous system), or the disease they cause (e.g., birth defects and cancer). In any case, chemical substances generally fall into one of the two broad categories of "carcinogens" versus "noncarcinogens'—customarily based, respectively, on their potential to induce cancer and their possession of systemic toxicity effects. Indeed, for the purpose of human health risk determination, chemical toxicants are usually distinctly categorized into carcinogenic and noncarcinogenic groups.

In general, chemicals that give rise to toxic endpoints other than cancer and gene mutations are often referred to as "systemic toxicants" because of their effects on the function of various organ systems; the toxic endpoints are referred to as "non-cancer" or "systemic" toxicity. Most chemicals that produce non-cancer toxicity do not cause a similar degree of toxicity in all organs, but usually demonstrate major toxicity to one or two organs; these are referred to as the target organs of toxicity for the chemicals (Klaassen et al., 1986; USEPA, 1989i,n). Also, it is apparent that chemicals that cause cancer and gene mutations would commonly evoke other toxic effects (viz., systemic toxicity) as well.

9.4.1.1 Identification of Carcinogens

Carcinogenesis is the process by which normal tissue becomes cancerous—i.e., the production of cancer, most likely via a series of steps—viz., initiation, promotion, and progression; the carcinogenic event modifies the genome and/or other molecular control mechanisms of the target cells, giving rise to a population of altered cells. An important issue in chemical carcinogenesis relates to the concepts of "initiators" and "promoters." An *initiator* is a chemical/substance or agent capable of starting but not necessarily completing the process of producing an abnormal uncontrolled growth of tissue, usually by altering a cell's genetic material; thus, initiated cells may or may not be transformed into tumors. A *promoter* is defined as an agent that results in an increase in cancer induction when it is administered at some time after a receptor has been exposed to an *initiator*; thus, this represents an agent that is not carcinogenic in itself, but when administered after an initiator of carcinogenesis, serves to dramatically potentiate the effect of a low dose of a carcinogen—by stimulating the clonal expansion of the initiated cell to produce a neoplasm. Further yet, a *co-carcinogen* is an agent that is not carcinogenic on its own, but enhances the activity of another agent that is carcinogenic when administered together with the carcinogen; it is noteworthy that a *co-carcinogen* differs from a *promoter* only in that the former is administered at the same time as the initiator. It is believed that initiators, co-carcinogens, and promoters do not usually induce tumors when administered separately. Indeed, it has become apparent that a series of developmental stages is required for carcinogenesis (OSTP, 1985). Many chemical carcinogens are believed to be *complete carcinogens*—i.e., chemicals that are capable of inducing tumors in animals or humans without supplemental exposure to other agents; thus, these chemicals function as both initiators and promoters. Generally speaking, the term "complete" refers to the three stages of carcinogenesis (namely: initiation, promotion, and progression) that need to be present in order to induce a cancer. It should be acknowledged, however, that promoters themselves are usually not necessarily carcinogens; these may include dietary fat, alcohols, saccharin, halogenated solvents, and estrogen. Even so, most regulatory agencies in many different jurisdictions do not usually distinguish between initiators and promoters, especially because it is often very difficult to confirm whether a given chemical acts by promotion alone, etc. (OSHA, 1980; OSTP, 1985; USEPA, 1984a,c).

Both human and animal studies are used in the evaluation of whether chemicals are possible human carcinogens. The strongest evidence for establishing a relationship between exposure to a given chemical and cancer in humans comes from epidemiological studies. These studies of human exposure and cancer must consider the latency period for cancer development, because apparently the exposure to the carcinogen often occurs many years (sometimes 20–30 years, or even more) before the first sign of cancer appears. On the other hand, the most common method for identifying substances as potential human carcinogens is by long-term animal bioassays. These bioassays provide accurate information about dose and duration of exposure, as well as interactions of the substance with other chemicals or modifiers. In these studies, the chemical, substance, or mixture is administered to one or, usually, two laboratory rodent species over a range of doses and durations of exposure with all experimental conditions carefully chosen to maximize the likelihood of identifying any carcinogenic effects (Huff, 1999).

In general, experimental carcinogenesis research is based on the scientific assumption that chemicals causing cancer in animals will have similar effects in humans. It must be acknowledged, however, that it is not possible to predict with complete certainty from animal studies alone which agents, substances, mixtures, and/or exposure circumstances will be carcinogenic in humans. Conversely, all known human carcinogens that have been tested adequately also produce cancers in laboratory animals. In many cases, an agent was found to cause cancer in animals and only subsequently confirmed to cause cancer in humans (Huff, 1993). In any event, it is noteworthy that laboratory animals' adverse responses to chemicals (of which cancer is only one) do not always strictly correspond to similar or equivalent human responses. Yet still, laboratory animals remain the best tool for detecting potential human health hazards of all kinds, including cancer (OTA, 1981; Tomatis et al., 1997).

9.4.2 Carcinogen Classification Systems

Two prominent carcinogenicity evaluation philosophies—one based on "weight-of-evidence" and the other on "strength-of-evidence"—seem to have found the most common acceptance and widespread usage. Systems that employ the weight-of-evidence evaluations consider and balance the negative indicators of carcinogenicity with those showing carcinogenic activity; and schemes using the strength-of-evidence evaluations consider combined strengths of all positive animal tests (vis-à-vis human epidemiology studies and genotoxicity) to rank a chemical without evaluating negative studies, nor considering potency or mechanism (Huckle, 1991). On the basis of the preceding, carcinogenic chemicals are generally classified into several categories, depending on the "weight-of-evidence" or "strength-of-evidence" available on a particular chemical's carcinogenicity (Hallenbeck and Cunningham, 1988; Huckle, 1991; IARC, 1982, IARC, 2006; USDHS, 1989, 1996; USEPA, 1986b–f, 2005a–c, 2012).

A chemical's potential for human carcinogenicity is inferred from the available information relevant to the potential carcinogenicity of the chemical, and from judgments regarding the quality of the available studies. On the whole, carcinogens may be categorized into the following broad identifiable groupings (IARC, 1982; Theiss, 1983; USDHS, 1989, 1996):

- *"Known human carcinogens"*—defined as those chemicals for which there exists sufficient evidence of carcinogenicity from studies in humans to indicate a causal relationship between exposure to the agent, substance, or mixture and human cancer.
- *"Reasonably anticipated to be human carcinogens"*—referring to those chemical substances for which there is limited evidence for carcinogenicity in humans and/or sufficient evidence of carcinogenicity in experimental animals. Sufficient evidence in animals is demonstrated by positive carcinogenicity findings in multiple strains and species of animals; in multiple experiments; or to an unusual degree, with regard to incidence, site or type of tumor, or age of onset; or there is less than sufficient evidence of carcinogenicity in humans or laboratory animals.
- *"Sufficient evidence of carcinogenicity"* and *"Limited evidence of carcinogenicity"*— used in the criteria for judging the adequacy of available data for identifying carcinogens; it refers only to the amount and adequacy of the available evidence, and not to the potency of carcinogenic effect on the mechanisms involved.

Other varying carcinogen classification schemes also exist globally within various regulatory and legislative groups. Even so, it is apparent that the numerous agencies around the world and in various jurisdictions have schemes that are conceptually similar, but may vary in the specific descriptors and criteria used—e.g., as observed with systems maintained by the International Agency for Research on Cancer (IARC), US EPA, and Health Canada; in actual fact, most of these nomenclatural systems are adapted or modified from the IARC classifications.

In the final analysis, carcinogenicity classifications are used for a wide range of purposes— including human health risk assessments, regulatory decision-making, risk management, and cancer prevention measures. These classifications are based on an evaluation of both human and animal studies as well as supporting mechanistic data (i.e., studies at the cellular or molecular level). Because human evidence is scarce for most substances, animal studies generally provide most of the evidence for classification—albeit positive animal results are not always evidence of human carcinogenicity. At the end of it all, each of the classification schemes identifies the potential for a substance to cause cancer, but not necessarily how likely it is to occur at typical human exposure levels.

9.4.2.1 Weight-of-Evidence Classification and Narratives

A weight-of-evidence approach has been widely used by the US EPA regulatory body to classify the likelihood that an agent in question is a human carcinogen—ultimately producing a five-level classification scheme and corresponding narrative (USEPA, 2012). This is a classification system for characterizing the extent to which available data indicate that an agent is a human carcinogen (or possesses some other toxic effects such as developmental toxicity). A three-stage procedure has typically been utilized in the process—namely:

- *Stage 1*—the evidence is characterized separately for human studies and for animal studies.
- *Stage 2*—the human and animal evidence are integrated into a presumptive overall classification.
- *Stage 3*—the provisional classification is modified (i.e., adjusted upwards or downwards), based on analysis of the supporting evidence.

The outcome of this process is that, chemicals are placed into one of five general categories— namely, Groups A–E (Box 9.9)—further discussed below. It is worth mentioning here that, the guidelines for classification of the weight-of-evidence for human carcinogenicity published by the US EPA (e.g., USEPA, 1984a,c, 1986b–f, 2005a–c)—which basically consists of the categorization of the weight-of-evidence into the five groups—are indeed general adaptations from those maintained by the IARC (IARC, 1984, 1987, 1988).

BOX 9.9 THE US EPA WEIGHT-OF-EVIDENCE CLASSIFICATION SYSTEM AND DESCRIPTORS FOR POTENTIAL CARCINOGENS

US EPA Group	Reference Category
A	Human carcinogen (i.e., Carcinogenic—or known human carcinogen)
B	Probable human carcinogen (i.e., Likely to be carcinogenic):
	• B1 indicates limited human evidence
	• B2 indicates sufficient evidence in animals and inadequate or no evidence in humans
C	Possible human carcinogen (viz., Suggestive evidence)
D	Not classifiable as to human carcinogenicity (viz., Inadequate information)
E	No Evidence of carcinogenicity in humans (or, Evidence of noncarcinogenicity for humans—i.e., Not likely to be carcinogenic)

Further to the above, the following descriptors have more recently been recommended along with the corresponding weight-of-evidence narratives (USEPA, 2005a–c):

- *Carcinogenic to Humans*—this descriptor indicates strong evidence of human carcinogenicity (including presence of convincing epidemiological evidence demonstrating causality between human exposure and cancer, or existence of compelling evidence of carcinogenicity in animals alongside mechanistic information that demonstrates a similar mode(s) of carcinogenic action in animals and in humans).
- *Likely to be Carcinogenic to Humans*—this descriptor is appropriate when the weight-of-evidence is adequate to demonstrate carcinogenic potential to humans.
- *Suggestive Evidence of Carcinogenic Potential*—this descriptor is appropriate when the weight-of-evidence is suggestive of carcinogenicity; a concern for potential carcinogenic effects in humans is raised, but the data are judged not sufficient for a stronger conclusion.
- *Inadequate Information to Assess Carcinogenic Potential*—this descriptor is appropriate when available data are judged inadequate for applying one of the other descriptors.
- *Not Likely to be Carcinogenic to Humans*—this descriptor is appropriate when the available data are considered robust for deciding that there is no basis for human hazard concern.

It is noteworthy that more than one descriptor can indeed be used when the effects of a constituent differ by dose or exposure route. While these narrative descriptions represent important advances in carcinogen risk assessment, the alphanumeric system still offers some very useful attributes.

9.4.2.1.1 Group A—Human Carcinogen, or "Carcinogenic to Humans"

For this group, there is sufficient evidence from epidemiologic studies to support a causal association between exposure to the agent and human cancer; in general, the following three criteria must be satisfied before a causal association can be inferred between exposure and cancer in humans (Hallenbeck and Cunningham, 1988; USEPA, 1986b–f, 2005a–c):

- No identified bias which could explain the association;
- Possibility of confounding factors (i.e., variables other than chemical exposure level which can affect the incidence or degree of the parameter being measured) has been considered and ruled out as explaining the association; and
- Association is unlikely to be due to chance.

Indeed, this group tends to be used only when there is sufficient evidence from epidemiologic studies to support a causal association between exposure to the agents and cancer—albeit, exceptionally, it may be used for lesser weight of epidemiological evidence, strengthened by other lines of evidence.

9.4.2.1.2 Group B—Probable Human Carcinogen, or "Likely to be Carcinogenic to Humans"

This group includes agents for which the weight-of-evidence of human carcinogenicity based on epidemiologic studies is "limited"—and also includes agents for which the weight-of-evidence of carcinogenicity based on animal studies is "sufficient." The category consists of agents for which the evidence of human carcinogenicity from epidemiologic studies ranges from almost sufficient to inadequate. Thus, there would be a demonstration of plausible (not definitively causal) association between human exposure and cancer.

Traditionally, this group has been divided into two subgroups—reflecting higher (Group B1) and lower (Group B2) degrees of evidence. Usually, category B1 is reserved for agents with which there is limited evidence of carcinogenicity to humans from epidemiologic studies; limited evidence of carcinogenicity indicates that a causal interpretation is credible—but then alternative explanations such as chance, bias, or confounding factors could not be excluded. Inadequate evidence indicates that one of the following two conditions prevailed: (i) there were few pertinent data; or (ii) the available studies, while showing evidence of association, did not exclude chance, bias, or confounding factors (Hallenbeck and Cunningham, 1988; USEPA, 1986b–f, 2005a–c). When there are inadequate data for humans, it is reasonable to consider agents for which there is sufficient evidence of carcinogenicity in animals as if they presented a carcinogenic risk to humans. Therefore, agents for which there is "sufficient" evidence from animal studies and for which there is "inadequate" evidence from human (epidemiological) studies or "no data" from epidemiologic studies would usually result in a classification as B2 (CDHS, 1986; Hallenbeck and Cunningham, 1988; USEPA, 1986b–f, 2005a–c).

9.4.2.1.3 Group C—Possible Human Carcinogen, or "Suggestive Evidence of Carcinogenic Potential"

This group has been used for agents with limited evidence of carcinogenicity in animals in the absence of human data. Limited evidence means that the data suggest a carcinogenic effect, but are generally limited for the following reasons (Hallenbeck and Cunningham, 1988; USEPA, 1986b–f, 2005a–c):

- The studies involve a single species, strain, or experiment; or
- The experiments are restricted by inadequate dosage levels, inadequate duration of exposure to the agent, inadequate period of follow-up, poor survival, too few animals, or inadequate reporting; or
- An increase in the incidence of benign tumors only.

On the whole, Group C classification essentially relies on a wide variety of evidence—including the following (Hallenbeck and Cunningham, 1988; USEPA, 1986b–f, 2005a–c): definitive malignant tumor response in a single well conducted experiment that does not meet conditions for "sufficient" evidence; tumor response of marginal statistical significance in studies having inadequate design or reporting; benign but not malignant tumors, with an agent showing no response in a variety of short-term tests for mutagenicity; and responses of marginal statistical significance in a tissue known to have a high and/or variable background tumor rate.

9.4.2.1.4 Group D—Not Classifiable as to Human Carcinogenicity, or "Inadequate Information to Assess Carcinogenic Potential"

This group has generally been used for agents with inadequate animal evidence of carcinogenicity, and also inadequate evidence from human (epidemiological) studies. Inadequate evidence means that, because of major qualitative or quantitative limitations, the studies cannot necessarily be interpreted as showing either the presence or absence of a carcinogenic effect.

9.4.2.1.5 Group E—No Evidence of Carcinogenicity in Humans, or "Not Likely to be Carcinogenic to Humans"

This group has been used to describe agents indicating evidence of noncarcinogenicity for humans, together with no evidence of carcinogenicity in at least two adequate animal tests in different species, or no evidence in both adequate animal and human (epidemiological) studies. The designation of an agent as being in this group is based on the available evidence, and should not be interpreted as a definitive conclusion that the agent will not be a carcinogen under any circumstances.

9.4.2.2 Strength-of-Evidence Classification

The IARC bases its classification on the so-called strength-of-evidence philosophy. Procedurally, the IARC assembles Working Groups for specific substances to make scientific judgments on the evidence for or against carcinogenicity (IARC, 2006); the evidence from human and animal studies is evaluated separately, and then the full body of evidence is considered as a whole to categorize a substance into one of five groups. The corresponding IARC classification system (somehow comparable or equivalent to the US EPA system description presented above) is shown in Box 9.10—and further discussed below.

9.4.2.2.1 Group 1—Known Human Carcinogen

This group is generally used for agents with sufficient evidence from human (epidemiological) studies as to human carcinogenicity. Thus, the Group 1 agent is essentially considered carcinogenic to humans.

9.4.2.2.2 Group 2—Probable or Possible Human Carcinogens

This category includes agents for which, at one extreme, the degree of evidence of carcinogenicity in humans is almost sufficient—as well as agents for which, at the other extreme, there are no human data but for which there is experimental evidence of carcinogenicity. Agents are assigned to either 2A (probably carcinogenic) or 2B (possibly carcinogenic) on the basis of epidemiological, experimental and other relevant data. These two subgroups are elaborated further in the proceeding sections below.

Group 2A—Probable Human Carcinogen. This group is generally used to represent agents for which there is sufficient animal evidence, evidence of human carcinogenicity, or at least limited evidence from human (epidemiological) studies. Indeed, these are probably carcinogenic to humans—and usually have at least limited human evidence.

BOX 9.10   THE IARC STRENGTH-OF-EVIDENCE CLASSIFICATION SYSTEM AND DESCRIPTORS FOR POTENTIAL CARCINOGENS

IARC Group	Category
1	Human carcinogen (i.e., Carcinogenic—Known human carcinogen)
2	Probable or Possible human carcinogen:
	• 2A indicates limited human evidence (i.e., Probably carcinogenic)
	• 2B indicates sufficient evidence in animals and inadequate or no evidence in humans (i.e., Possibly carcinogenic)
3	Not classifiable as to human carcinogenicity
4	No Evidence of carcinogenicity in humans (i.e., Probably not carcinogenic)

On the whole, this category is used when there is limited evidence of carcinogenicity in humans and sufficient evidence of carcinogenicity in experimental animals. Exceptionally, an agent may be classified into this category solely on the basis of limited evidence of carcinogenicity in humans or of sufficient evidence of carcinogenicity in experimental animals, strengthened by supporting evidence from other relevant data.

Group 2B—Possible Human Carcinogen. This group is generally used to represent agents for which there is sufficient animal evidence but inadequate evidence from human (epidemiological) studies, or where there is limited evidence from human (epidemiological) studies in the absence of sufficient animal evidence. These are viewed as possibly carcinogenic to humans—but usually have no human evidence.

On the whole, this category is generally used for agents that indicate limited evidence in humans, in the absence of sufficient evidence in experimental animals. It may also be used when there is inadequate evidence of carcinogenicity in humans, or when human data are nonexistent but there is sufficient evidence of carcinogenicity in experimental animals. In some instances, an agent for which there is inadequate evidence or no data in humans but limited evidence of carcinogenicity in experimental animals together with supporting evidence from other relevant data may be placed in this group.

9.4.2.2.3   Group 3—Not Classifiable

This group is generally used for agents for which there are inadequate animal evidence and inadequate evidence from human (epidemiological) studies—but where there is sufficient evidence of carcinogenicity in experimental animals. Overall, the Group 3 agent is not classifiable as to its carcinogenicity to humans—and agents are typically placed in this category when they do not fall into any other group.

9.4.2.2.4   Group 4—Noncarcinogenic to Humans

This group is generally used for an agent or substance for which there is evidence to support a lack of carcinogenicity. The Group 4 agent is probably not carcinogenic to humans—and this category is essentially used for agents for which there is evidence suggesting lack of carcinogenicity in humans, together with evidence suggesting lack of carcinogenicity in experimental animals. Meanwhile, it is notable that under some circumstances, agents for which there is inadequate evidence of (or no data on) carcinogenicity in humans but for which there is evidence suggesting lack of carcinogenicity in experimental animals, consistently and strongly supported by a broad range of other relevant data, may also be placed into this group.

9.4.3   Dose–Response Relationships

The dose–response relationship is about the most fundamental concept in toxicology. A dose–response relationship exists when there is a consistent mathematical relationship that describes the proportion of test organisms responding to a specific dose of a toxicant/substance for a given exposure period. A number of assumptions usually will need to be considered when attempting to establish a dose–response relationship—most importantly, that the following hold true (Hughes, 1996):

- The observed response is caused by the substance administered to the organism;
- The magnitude of the response is directly related to the magnitude of the dose; and
- It is possible to correctly observe and measure a response.

In general, the relationship between the degree of exposure to a chemical (viz., the dose) and the magnitude of chemical-induced effects (viz., the response) is typically described by a dose–response curve—often referred to by "stressor-response profile" in the characterization of ecological effects. The typical dose–response curve is sigmoidal—but can also be linear, concave, convex, or bimodal;

indeed, the shape of the curve can offer clues to the mechanism of action of the toxin, indicate multiple toxic effects, and identify the existence and extent of sensitive sub-populations (Derelanko and Hollinger, 1995). In any case, dose–response curves fall into the following two broad categories/groups (Figure 9.2):

1. Those in which no response is observed until some minimum (i.e., threshold) dose is reached; and
2. Those in which no threshold is manifest—meaning that some type of response is expected for any dose, no matter how small.

In essence, for some chemicals, a very small dose causes no observable effects whereas a higher dose will result in some toxicity, and still higher doses cause even greater toxicity—up to the point of fatality; such chemicals are called *threshold chemicals* ("Curve B" in Figure 9.2). For other chemicals, such as most carcinogens, the threshold concept may not be applicable—in which case no minimum level is required to induce adverse and overt toxicity effects ("Curve A" in Figure 9.2).

Meanwhile, it should be acknowledged here that the most important part of the dose–response curve for a threshold chemical is the dose at which significant effects first begin to show. The highest dose that does not produce an observable adverse effect is the "no-observed-adverse-effect-level" (NOAEL), and the lowest dose that produces an observable adverse effect is the "lowest-observed-adverse-effect-level" (LOAEL). For non-threshold chemicals, the dose–response curve behaves differently, in that there is no dose that is free of risk. Anyhow, at the end of the day, several important variables (e.g., as noted in Asante-Duah, 2017) may help determine the characteristics of dose–response relationships—and these parameters should be given careful consideration when performing toxicity tests, and also when interpreting toxicity data (USEPA, 1985a,b). [By the way, when one is dealing specifically with the characterization of ecological effects, the dose–response curves (also, stressor-response profiles) may be represented as functional relationships between the amounts of a chemical substance and its morbidity/lethality (Figure 9.3).]

9.4.3.1 "Threshold" versus "Non-Threshold" Concepts

Noncarcinogens commonly/traditionally are believed to operate by "threshold" mechanisms—i.e., the manifestation of systemic effects requires a threshold level of exposure or dose to be exceeded

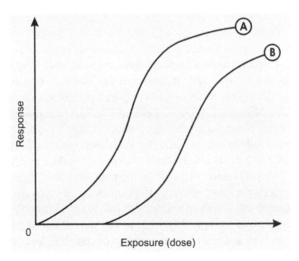

FIGURE 9.2 Schematic representation of exposure-response relationships: Illustration of dose–response relationship for (A) = non-threshold chemicals & (B) = threshold chemicals.

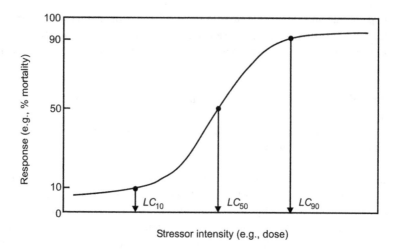

FIGURE 9.3   A schematic of a stressor-response curve, showing effect levels corresponding to: 10% (LC$_{10}$), 50% (LC$_{50}$) and 90% (LC$_{90}$) lethalities or mortalities.

during a continuous exposure episode. Thus, non-cancer or systemic toxicity is generally treated as if there is an identifiable exposure threshold below which there are no observable adverse effects—and this means that, in general, continuous exposure to levels below the threshold will produce no adverse or noticeable health effects. In fact, for many noncarcinogenic effects, protective mechanisms are believed to exist in the mammalian physiological system that must be overcome before the adverse effect of a chemical constituent is manifested. Consequently, a range of exposures exist from zero to some finite value—called the *threshold level*—that can be tolerated by the exposed organism with essentially no likelihood of adverse effects. This characteristic distinguishes systemic endpoints from carcinogenic and mutagenic endpoints, which are often treated as "non-threshold" processes; in other words, the threshold concept and principle is not quite applicable for carcinogens, since it is believed that no thresholds exist for this group. Indeed, carcinogenesis, unlike many noncarcinogenic health effects, is generally thought to be a phenomenon for which risk evaluation based on presumption of a threshold may be inappropriate (USEPA, 1989i,n).

On the whole, it is usually assumed in risk assessments that, any finite exposure to carcinogens could result in a clinical state of disease. This hypothesized mechanism for carcinogenesis is referred to as "non-threshold"—because it is believed that there is essentially no level of exposure to such a chemical that does not pose a finite probability, however small, of generating a carcinogenic response. Indeed, cancer effects have traditionally been considered to have no threshold—and thus, any exposure is associated with some risk. It is noteworthy, however, that among some professional groups, there is the belief that certain carcinogens require a threshold exposure level to be exceeded to provoke carcinogenic effects (e.g., Wilson, 1996, 1997). In fact, opinion among regulatory scientists seems to have been returning to the somewhat "ancient" presumption that at least some cancer-causing substances induce effects through a threshold process (Wilson, 1997). This implies that, for such substances, there exists a finite level of exposure or dose at which no finite response is necessarily indicated. This perspective is based on the understanding among a number of toxicologists that most substances do not cause adverse effects unless exposures are sufficient to overwhelm the body's normal processes and defenses; under this view, the human body is able to accommodate various chemical, physical, and biological stresses at the subcellular and biochemical levels with control processes that adapt to and minimize the impact of chemical and other stressors—and only when the capacity of these protections is exceeded by high and sustained doses is an adverse impact to be expected. For instance, among others, Health Canada has noted the potential for a "practical" threshold for genotoxic effects, for the most part attributable to the interplay between the genotoxicity and cellular DNA-repair mechanisms for the receptor/organism—albeit it is also assumed that all exposure levels have some concomitant risk.

Finally, it is noteworthy that cancer risks to humans are generally assessed differently for substances that act through a threshold mechanism compared to those acting via a non-threshold mechanism; yet still, most regulatory agencies in many different jurisdictions have only rarely considered the evidence to be strong enough to designate a threshold mechanism for carcinogens, and to support deviating from the conservative approach of assuming no-threshold for cancer risk assessments. As a case in point in relation to the "no-threshold" concept, carcinogens have historically been regulated in the U.S. as though there is no dose below which there is zero risk of developing cancer. In fact, this seems to have remained the contemporary general practice despite the evidence that many substances probably contribute to cancer development only at doses above a certain threshold, that is, at levels above low human exposures.

9.4.4 Dose–Response Assessment and Quantification

Dose–response assessment is the process of quantitatively evaluating toxicity information and characterizing the relationship between the dose of the chemical administered or received (i.e., exposure to an agent) and the incidence of adverse health effects in the exposed populations. The process consists of estimating the potency of the specific compounds by the use of dose–response relationships. In the case of carcinogens, for example, this involves estimating the probability that an individual exposed to a given amount of chemical will contract cancer due to that exposure; potency estimates may be given as "unit risk factor" (URF, expressed in $\mu g/m^3$) or as "potency slopes" (in units of $[mg/kg\text{-}day]^{-1}$). Data are generally derived from animal studies or, less frequently, from studies in exposed human populations.

The dose–response assessment first addresses the relationship of dose to the degree of response observed in an experiment or a human study. When chemical exposures are outside the range of observations, extrapolations are necessary—in order to be able to estimate or characterize the dose relationship. The extrapolations will typically be made from high to low doses, from animal to human responses, and/or from a specific route of exposure to a different one. The details of the extrapolation mechanics are beyond the scope of this discussion, but are elaborated elsewhere in the literature (e.g., Brown, 1978; CDHS, 1986; Crump, 1981; Crump and Howe, 1984; Gaylor and Kodell, 1980; Gaylor and Shapiro, 1979; Hogan, 1983; Krewski and Van Ryzin, 1981).

In general, even if a substance is known to be toxic, the risks associated with the substance cannot be ascertained with any degree of confidence unless dose–response relationships are quantified. The dose–response relationships are typically used to determine what dose of a particular chemical causes specific levels of toxic effects to potential receptors. In fact, there may be many different dose–response relationships for any given substance if it produces different toxic effects under different conditions of exposure. In any case, the response of a given toxicant depends on the mechanism of its action; for the simplest scenario, the response, R, is directly proportional to its concentration, (C), so that:

$$R = k \cdot [C] \tag{9.7}$$

where k is a rate constant. This would be the case for a chemical that metabolizes rapidly. Even so, the response and the value of the rate constant would tend to differ for different risk groups of individuals and for unique exposures. If, for instance, the toxicant accumulates in the body, the response is better defined as follows:

$$R = k \cdot [C] \cdot t^n \tag{9.8}$$

where t is the time and n is a constant. For cumulative exposures, the response would generally increase with time. Thus, the cumulative effect may show as linear until a threshold is reached, after which secondary effects begin to affect and enhance the responses. Also, the cumulative effect may be related to what is referred to as the "body burden" (BB). The BB is determined by the relative rates of absorption (ABS), storage (STR), elimination (ELM), and biotransformation (BTF)—in accordance with the following relationship (Meyer, 1983):

$$BB = ABS + STR - ELM - BTF \qquad (9.9)$$

Each of the factors involved in the quantification of the BB is dependent on a number of biological and physiochemical factors. In fact, the response of an individual to a given dose cannot be truly quantitatively predicted since it depends on many extraneous factors, such as general health and diet of individual receptors. Nonetheless, from the quantitative dose–response relationship, toxicity values can be derived and used to estimate the incidence of adverse effects occurring in potential receptors at different exposure levels.

In general, even if a substance is known to be toxic, the risks associated with the substance cannot be ascertained with any degree of confidence unless dose–response relationships are appropriately quantified. Meanwhile, it is noteworthy that the fundamental principles underlying the dose–response assessment for carcinogenic chemicals remain arguable—especially in relation to the tenet that there is some degree of carcinogenic risk associated with every potential carcinogen, no matter how small the dose. This is because, the speculation and/or belief that chemically-induced cancer is a non-threshold process/phenomenon may be false after all—albeit it represents a conservative default policy necessary to ensure adequate protection of human health; accordingly, this potential shortcoming should be kept in perspective in consequential policy decisions about the assessment.

Three major classes of mathematical extrapolation models are often used for relating dose and response in the sub-experimental dose range, namely:

- Tolerance Distribution models—including Probit, Logit, and Weibull;
- Mechanistic models—including One-hit, Multi-hit, and Multi-stage; and
- Time-to-Occurrence models—including Lognormal and Weibull.

Indeed, other independent models—such as linear, quadratic, and linear-cum-quadratic—may also be employed for this purpose. The details on all these wide-ranging types models are beyond the scope of this discussion, but are elaborated elsewhere in the literature (e.g., Brown, 1978; CDHS, 1986; Crump, 1981; Crump and Howe, 1984; Derelanko and Hollinger, 1995; Gaylor and Kodell, 1980; Gaylor and Shapiro, 1979; Hogan, 1983; Jolley and Wang, 1993; Krewski and Van Ryzin, 1981; Tan, 1991). In any event, the choice of model is generally determined by its consistency with the current understanding of the mechanisms of carcinogenesis.

In general, mathematical models fitted to high-dose experimental data are often used to help characterize the low-dose risks typically encountered in chemical risk quantifications; in such situations, different dose–response models are likely to yield substantially varying results for the same estimated effective dose. Thus, the concept of "model averaging" has been suggested for use in contemporary times, so as to provide some robustness under such circumstances; specifically, this approach will more likely provide outcomes that are generally better than attainable from the use of a single dose–response model during the processes involved in estimating an effective dose associated with a representative or pragmatic chemical exposure situation (i.e., the typical dose levels to which targeted human populations are potentially exposed)—especially in cases involving extremely small risks (Faes et al., 2007; Kim et al., 2014; Wheeler and Bailer, 2007, 2009). Exceptions may occur, however, for cases of chemicals that have not been sufficiently studied.

9.4.5 DETERMINATION OF TOXICOLOGICAL PARAMETERS FOR HUMAN HEALTH RISK ASSESSMENTS

In the processes involved in the assessment of human health risks arising from chemical exposures, it often becomes necessary to compare receptor chemical intakes with doses shown to cause adverse effects in humans or experimental animals. Correspondingly, the dose at which no effects are observed in human populations or experimental animals is referred to as the "no-observed-effect-level" (NOEL); where data identifying a NOEL are lacking, a "lowest-observed-effect-level" (LOEL) may be used as the basis for determining safe threshold doses.

For acute effects, short-term exposures/doses shown to produce no adverse effects characterize the parameters of interest needed to support the risk assessment process; this is called the "no-observed-adverse-effect-level" (NOAEL). A NOAEL is an experimentally determined dose at which there has been no statistically or biologically significant indication of the toxic effect of concern. In cases where a NOAEL has not been demonstrated experimentally, the "lowest-observed-adverse-effect level" (LOAEL) is used.

In general, for chemicals possessing carcinogenic potentials, the LADD is typically compared with the NOEL identified in long-term bioassay experimental tests; for chemicals with acute effects, the MDD is compared with the NOAEL observed in short-term animal studies. An elaboration on the derivation of the relevant toxicological parameters commonly used in human health risk assessments follows below, with further in-depth discussions to be found in the literature elsewhere (e.g., Dourson and Stara, 1983; Faes et al., 2007; Kim et al., 2014; Wheeler and Bailer, 2007, 2009; USEPA, 1985b, 1986a, 1989a,c,d).

All things considered, chemical hazard effects evaluation is characteristically conducted as part of a risk assessment in order to, qualitatively and/or quantitatively, determine the potential for adverse effects from receptor exposures to chemical stressor(s). For most chemical exposure problems, this usually will comprise of intricate toxicological evaluations, the ultimate goal of which is to derive reliable estimates of the amount of chemical exposure that may be considered "tolerable" (or "acceptable" or "reasonably safe") for humans. The relevant toxicity parameters that are generated during this process usually will be dependent on the mechanism and/or mode of action for the particular toxicant, on the receptor of interest.

By and large, public health risk assessments for chemical exposure problems would typically rely heavily on "archived" toxicity indices/information developed for specific chemicals. A summary listing of such toxicological parameters—predominantly represented by the cancer slope factor (SF) (for carcinogenic effects) and the RfD (for non-cancer effects)—is provided in Table C.1 (Appendix C) for some representative chemicals commonly found in the human living and work environments. A more complete and up-to-date listing may be obtained from a variety of toxicological databases—such as the Integrated Risk Information System (IRIS) database (developed and maintained by the US EPA); the International Register of Potentially Toxic Chemicals (IRPTC) database (from UNEP); and the International Toxicity Estimates for Risks (ITER) database (from TERA) (see Appendix B.2). Where toxicity information does not exist at all, a decision may be made to estimate toxicological data from that of similar compounds (i.e., with respect to molecular weight and structural similarities; etc.).

9.4.5.1 Toxicity Parameters for Noncarcinogenic Effects

Traditionally, risk decisions on systemic toxicity are made using the concept of "acceptable daily intake" (ADI), or by using the so-called reference dose (RfD). The ADI is the amount of a chemical (in mg/kg body weight/day) to which a receptor can be exposed to on a daily basis over an extended period of time—usually a lifetime—without suffering a deleterious effect. The RfD is defined as the maximum amount of a chemical (in mg/kg body weight/day) that the human body can absorb without experiencing chronic health effects. Thus, for exposure of humans to the noncarcinogenic effects of environmental chemicals, the ADI or RfD may be used as a measure of exposure that is considered to be without adverse effects. Meanwhile, it is worth the mention here that, although often used interchangeably, RfDs are based on a more rigorously defined methodology—and is therefore generally preferred over ADIs.

The reference concentration (RfC) of a chemical—like the RfD—represents an estimate of the exposure that can occur on a daily basis over a prolonged period, with a reasonable anticipation that no adverse effect will occur from that exposure. In contrast to RfDs, however, RfCs are expressed in units of concentration in an environmental medium (e.g., mg/m^3 or $\mu g/L$). RfCs customarily presuppose continuous exposure, with an average IR and body weight; it may therefore be inappropriate to use them in "non-standard" exposure scenarios.

In general, both the RfD and RfC represent estimates of the exposure that can occur on a daily basis over a prolonged period, with a reasonable expectation that no adverse effect will occur from that exposure. In assessing the chronic and subchronic effects of non-carcinogens and also noncarcinogenic effects associated with carcinogens, the experimental dose value (e.g., NOEL) is usually divided by a safety (or uncertainty) factor to yield the RfD—as elaborated and illustrated further below.

Finally, it should be mentioned here that, when no toxicological information exists for a chemical of interest or concern, concepts of structure-activity relationships may have to be employed to help derive acceptable intake levels by influence and analogy analysis in comparison with closely related or similar compounds. In such cases, some reasonable degree of conservatism is typically recommended in any judgment call to be made.

9.4.5.1.1 Reference Doses (RfDs)

A RfD is defined as an estimate (with uncertainty spanning perhaps an order of magnitude) of a daily (oral) exposure to the human population (including sensitive subgroups) that is likely to be without an appreciable risk of deleterious effects during a lifetime. In general, it provides an estimate of the continuous daily exposure of a noncarcinogenic substance for the general human population (including sensitive subgroups) which appears to be without an appreciable risk of deleterious effects. Indeed, RfDs are established as thresholds of exposure to toxic substances below which there should be no adverse health impact. Broadly speaking, these thresholds are established on a substance-specific basis for oral and inhalation exposures, taking into account evidence from both human epidemiologic and laboratory toxicologic studies. Correspondingly, subchronic RfD is typically used to refer to cases involving only a portion of the lifetime, whereas chronic RfD is associated with lifetime exposures.

In general, RfDs can be derived from a NOAEL, LOAEL, or benchmark dose (BMD)—or indeed by using categorical regression, with uncertainty factors commonly applied to reflect limitations of the data used. Various types of RfDs are available depending on the critical effect (developmental or other) and the length of exposure being evaluated (chronic or subchronic).

Chronic oral RfDs are specifically developed to be protective for long-term exposure to a compound. As a typical guideline for most risk assessments, chronic oral RfDs generally should be used to evaluate the potential noncarcinogenic effects associated with exposure periods greater than 7 years (approximately 10% of an average human lifetime). However, this is not a bright-line; for instance, it is noteworthy that the U.S. ATSDR (Agency for Toxic Substances and Disease Registry) defines chronic exposure as greater than 1 year for use of their typical decision values.

Subchronic oral RfDs are specifically developed to be protective for short-term exposure to a compound. As a typical guideline for most risk assessments, subchronic oral RfDs should generally be used to evaluate the potential noncarcinogenic effects of exposure periods between 2 weeks and 7 years. However, this is not a bright-line; for instance, it is noteworthy that the U.S. ATSDR defines subchronic exposure as less than 1 year for use of their typical decision values.

9.4.5.1.2 Reference Concentrations (RfCs)

A RfC is defined as an estimate (with uncertainty spanning perhaps an order of magnitude) of a continuous inhalation exposure to the human population (including sensitive subgroups) that is likely to be without an appreciable risk of deleterious effects during a lifetime. It can be derived from a NOAEL, LOAEL, or benchmark concentration—or indeed by using categorical regression with uncertainty factors generally applied to reflect limitations of the data used. Various types of RfCs are available depending on the critical effect (developmental or other) and the length of exposure being evaluated (chronic or subchronic).

The *chronic inhalation RfC* is generally used for continuous or near continuous inhalation exposures that occur for 7 years or more. However, this is not a bright-line—as, for example, the U.S. ATSDR chronic equivalent values are based on exposures longer than 1 year. It is also noteworthy that the USEPA chronic inhalation RfCs are typically expressed in units of (mg/m^3)—albeit other agencies present same values in $\mu g/m^3$.

The *subchronic inhalation RfC* is generally used for exposures that are between 2 weeks and 7 years. However, this is not a bright-line—as, for example, the U.S. ATSDR subchronic equivalent values are based on exposures less than 1 year. It is also notable that the USEPA subchronic inhalation RfCs are usually expressed in units of (mg/m^3)—albeit some agencies present these in $\mu g/m^3$.

9.4.5.1.3   Derivation of RfDs and RfCs

The RfD is a "benchmark" dose operationally derived from the NOAEL by consistent application of general "order-of-magnitude" "uncertainty factors" (UFs) (also called "safety factors") that reflect various types of data sets used to estimate RfDs. In addition, a so-called modifying factor (MF) that is commonly based on professional judgment of the entire database associated with the specific chemical under review is sometimes applied. Broadly stated, RfDs (and ADIs) are calculated by dividing a NOEL (i.e., the highest level at which a chemical causes no observable changes in the species under investigation), a NOAEL (i.e., the highest level at which a chemical causes no observable adverse effect in the species being tested), or a LOAEL (i.e., that dose rate of chemical at which there are statistically or biologically significant increases in frequency or severity of adverse effects between the exposed and appropriate control groups) derived from human or animal toxicity studies by one or more uncertainty and MFs. Corresponding statements can also be made in relation to the derivation of RfCs.

Typically, to derive a RfD or RfC for a non-cancer critical effect, the common practice is to apply standard UFs to the NOAEL, LOAEL, or indeed a benchmark dose/concentration (BMD/BMC). The UFs are used to account for the extrapolation uncertainties (e.g., inter-individual variation, interspecies differences, and exposure duration) and database adequacy. A MF is also used to account for the general level of confidence in the critical study(s) used in the derivation of the RfD or RfC. Although a use of default values tends to be the norm, replacements for default UFs are used when chemical-specific data are available to modify such standard default values; this is known as the "data-derived" approach. Also, the use of pharmacokinetic or dosimetry models can obviate the need for an UF to account for differences in toxicokinetics across species. Anyhow, it is noteworthy that a number of related factors can indeed result in significant uncertainties in the RfD or RfC values. Among these is the selection of different observed effects as a critical effect—which may vary within and across available studies. Also significant is the choice of different data sets for the identification of the NOAEL, LOAEL, or BMD analysis; the use of different values for the various UFs; and additional judgments that impact the MF.

On the whole, RfDs are usually calculated using a single exposure level together with uncertainty factors that account for specific deficiencies in the toxicological database. Both the exposure level and uncertainty factors are selected and evaluated in the context of the available chemical-specific literature. After all the toxicological, epidemiologic, and supporting data have been reviewed and evaluated, a key study is selected that reflects optimal data on the critical effect. Dose–response data points for all reported effects are typically examined as part of this review. Additional general and specific issues of particular significance in this endeavor—including the types of response levels (ranked in order of increasing severity of toxic effects as NOEL, NOAEL, LOAEL, and FEL [the Frank effect level, defined as overt or gross adverse effects]) that are considered in deriving

RfDs are discussed elsewhere in the literature (e.g., USEPA, 1989b). Ultimately, the RfD (or ADI) can be determined from the NOAEL (or LOAEL, or BMD) for the critical toxic effect by consistent application of UFs and a MF, in accordance with the following relationship:

$$\text{Human dose}\left(\text{e.g., ADI or RfD}\right) = \frac{\text{Experimental dose}\left(\text{e.g., NOAEL}\right)}{\text{UF} \times \text{MF}} \tag{9.10}$$

or, specifically:

$$\text{RfD} = \frac{\text{NOAEL}}{\left(\text{UF} \times \text{MF}\right)} \tag{9.11}$$

or, more generally:

$$\text{RfD} = \frac{\left[\text{NOAEL or LOAEL or BMD}\right]}{\left[\sum_{i=1}^{n} \text{UF}_i \times \text{MF}\right]} \tag{9.12}$$

The derivation of a RfC is a parallel process that is appropriately based on a "no-observed-adverse-effect-concentration" (NOAEC) or "lowest-observed-adverse-effect-concentration" (LOAEC)—or indeed the BMC. Alternatively, a RfC may be derived from a RfD, taking into account the exposure conditions of the study used to derive the RfD.

Determination of the Uncertainty and Modifying Factors. The uncertainty factors used in the derivation of an RfD generally reflect the scientific judgment regarding the various types of data used to estimate the RfD values; it is basically used to offset the uncertainties associated with extrapolation of data, etc. Generally speaking, the UF consists of multipliers of 10 (although values less than 10 could also be used)—each factor representing a specific area of uncertainty inherent in the extrapolations from the available data. For example, a factor of 10 may be introduced to account for the possible differences in responsiveness between humans and animals in prolonged exposure studies. Indeed, for interspecies extrapolation of toxic effects seen in experimental animals to what might occur in exposed humans, an UF of up to tenfold is generally recommended; this is usually viewed as consisting of two components, viz.: one that accounts for metabolic or pharmacokinetic differences between the species, and another that addresses pharmacodynamic differences (i.e., differences between the response of human and animal tissues to the chemical exposure). Next, a second factor of 10 may be used to account for variation in susceptibility among individuals in the human population; indeed, exposed humans are known to vary considerably in their response to toxic chemical and drug exposures due to age, disease states, and genetic makeup—particularly in genetic polymorphisms for enzymes (isozymes) for detoxifying chemicals. In such types of cases, the resultant UF of 100 has been judged to be appropriate for many chemicals. For other chemicals with databases that are less complete (for example, those for which only the results of subchronic studies are available), an additional factor of 10 (leading to a UF of 1,000) might be judged to be more appropriate. For certain other chemicals, such as those associated with well-characterized responses in sensitive humans (as, for example, regarding the effect of fluoride on human teeth), an UF as small as 1 might be selected (Dourson and Stara, 1983). Finally an additional tenfold UF may be used to account for possible carcinogenicity effects. Meanwhile, it is notable that, within the US EPA, the maximum cumulative UF for any given database tends to be 3,000; thus, databases weaker than this are judged too uncertain to estimate RfDs or RfCs.

Box 9.11 provides the general guidelines for the process of selecting uncertainty and MFs during the derivation of RfDs (Dourson and Stara, 1983; USEPA, 1986b, 1989a,b, 1993b); it is noteworthy that the uncertainty factors shown are as typically used by the US EPA—also recognizing that,

---

**BOX 9.11   GENERAL GUIDELINES FOR SELECTING UNCERTAINTY
AND MODIFYING FACTORS IN THE DERIVATION OF RFDS**

- Standard Uncertainty Factors (UFs):
  - Use a tenfold factor when extrapolating from valid experimental results in studies using prolonged exposure to average healthy humans. This factor is intended to account for the variation in sensitivity among the members of the human population, due to heterogeneity in human populations, and is referenced as "10H". Thus, if NOAEL is based on human data, a safety factor of 10 is usually applied to the NOAEL dose to account for variations in sensitivities between individual humans.
  - Use an additional tenfold factor when extrapolating from valid results of long-term studies on experimental animals when results of studies of human exposure are not available or are inadequate. This factor is intended to account for the uncertainty involved in extrapolating from animal data to humans and is referenced as "10A". Thus, if NOAEL is based on animal data, the NOAEL dose is divided by an additional safety factor of 10, to account for differences between animals and humans.
  - Use an additional tenfold factor when extrapolating from less than chronic results on experimental animals when there are no useful long-term human data. This factor is intended to account for the uncertainty involved in extrapolating from less than chronic (i.e., subchronic or acute) NOAELs to chronic NOAELs and is referenced as "10S".
  - Use an additional tenfold factor when driving an RfD from a LOAEL, instead of a NOAEL. This factor is intended to account for the uncertainty involved in extrapolating from LOAELs to NOAELs and is referenced as "10L".
  - Use an additional up to tenfold factor when extrapolating from valid results in experimental animals when the data are "incomplete." This factor is intended to account for the inability of any single animal study to adequately address all possible adverse outcomes in humans, and is referenced as "10D."
- Modifying Factor (MF):
  - Use professional judgment to determine the MF, which is an additional uncertainty factor that is greater than zero and less than or equal to 10. The magnitude of the MF depends upon the qualitative professional assessment of scientific uncertainties of the study and data base not explicitly treated above— e.g., the completeness of the overall data base and the number of species tested. The default value for the MF is 1.

---

although other health and environmental organizations or agencies may use similar guidelines, they do not necessarily subdivide these elements to the same extent. Additionally, it is worth mentioning here that the UFs can indeed include various "chemical-specific adjustment factors" (CSAFs); CSAFs represent part of a broader continuum of approaches which incorporate increasing amounts of data to reduce uncertainty—generally ranging from default ("presumed protective") to more "biologically-based predictive" methodologies (Meek, 2001; WHO, 2010).

In general, the choice of the UF and MF values reflects the uncertainty associated with the estimation of a RfD from different human or animal toxicity databases. For instance, if sufficient data from chronic duration exposure studies are available on the threshold region of a chemical's critical toxic effect in a known sensitive human population, then the UF used to estimate the RfD may be set at unity (1); this is because, under such circumstances, these data are judged to be sufficiently predictive of a population sub-threshold dose—so that additional UFs are not needed after all (USEPA, 1989b).

Illustrative Examples of the RfD Derivation Process. Some hypothetical example situations involving the determination of RfDs based on information on NOAEL, and then also on LOAEL, are provided below.

- *Determination of the RfD for a Hypothetical Example Using the NOAEL.* Consider the case involving a study carried out on 250 animals (e.g., rats) that is of subchronic duration—and subsequently yielding a NOAEL dosage of 5 mg/kg/day. Then, in this case

$$UF = 10H \times 10A \times 10S = 1,000$$

In addition, there is a subjective adjustment (represented by the MF), based on the high number of animals (250) per dose group, as follows:

$$MF = 0.75$$

These factors then give $UF \times MF = 750$, and consequently:

$$RfD = \frac{NOAEL}{UF \times MF} = \frac{5}{750} = 0.007 \text{ mg/kg/day}$$

- *Determination of the RfD for a Hypothetical Example Using the LOAEL.* If the NOAEL is not available, and if 25 mg/kg/day had been the lowest dose from the test that showed adverse effects, then

$$UF = 10H \times 10A \times 10S \times 10L = 10,000$$

Using again the subjective adjustment of MF = 0.75, one obtains:

$$RfD = \frac{LOAEL}{UF \times MF} = \frac{25}{7,500} = 0.003 \text{ mg/kg/day}$$

9.4.5.1.4   The Benchmark Dose (BMD) Approach

The RfD or RfC for humans is often derived from animal experiments—with the NOAEL (which represent the highest experimental dose for which no adverse health effects have been documented) often being a starting point for its calculation. However, using the NOAEL in the derivation of RfDs and RfCs has long been recognized as having significant limitations—especially because: it is limited to only one of the doses in the study, and is also dependent on the particular study design; it does not account for variability in the estimate of the dose–response; it does not account for the slope of the dose–response curve; and it cannot be applied when there is no NOAEL, except through the application of an uncertainty factor (Crump, 1984; Kimmel and Gaylor, 1988; USEPA, 1995e). As an alternative to the use of NOAEL and LOAEL in the determination of the RfD or RfC in the non-cancer risk evaluation, therefore, other methodologies have become increasingly popular—such as has happened with the so-called "benchmark dose" (BMD) approach. An important goal of the BMD approach is to define a starting point of departure for the computation of a reference value (viz., RfD or RfC), or a slope factor, that is more independent of study design.

The BMD is an estimate of the dose or concentration that produces a predetermined change in response rate of an adverse effect (called the "benchmark response" or BMR) compared to background—and this is typically defined for a given exposure route and duration. The BMR generally should be near the low end of the range of increased risk that can be detected by a bioassay; indeed, low BMRs can impart high model dependence. In general, if there is an accepted (minimum) level of change in the endpoint that is considered to be biologically significant, then that amount is the BMR—like, for instance, a 10% body weight decrease compared to the control; in the absence of any other idea of what level of response to consider as being adverse, a change in the mean that equals one standard deviation of the control from the mean of the control may be utilized in this case (Crump, 1995).

Overall, the use of BMD methods involve fitting mathematical models to dose–response data, and then using the different results to select an appropriate BMD or BMC (as represented by, say, a $BMCL_x$) that is associated with a predetermined BMR (such as a 10% increase in the incidence of a particular lesion, or a 10% decrease in body weight changes). In practice, the BMD represents a lower confidence limit on the effective dose associated with some defined level of effect—e.g., a 5% or 10% increase in response. In other words, it is the confidence limit on the dose that elicits adverse responses in a fraction (often 5% or 10%) of the experimental animals; the confidence limits characterize uncertainty in the dose that affects a specified fraction of the animals. Thus, a $BMCL_x$ is defined as the lower 95% confidence limit of the dose that will result in a level of x% response; for example, $BMCL_{10}$ is the lower 95% confidence limit of a dose for a 10% increase in a particular response (US EPA, 1995c).

Ultimately, a dose–response relationship is fitted to the bioassay data points, and a confidence limit for that relationship is determined. The BMD is the dose yielding the desired response rate (e.g., 5% or 10%) based on the curve representing the confidence limit. Unlike the NOAEL, the BMD makes use of all the bioassay data and is not constrained to dose values administered in the experiment. On the other hand, there is no scientific basis for selecting a particular response rate for the BMD (e.g., 5% vs. 10%), and neither does the BMD easily address "continuous" responses (e.g., as reflected by changes in body weight)—since it depends on identifying the dose at which a certain fraction of animals is "affected" by the toxicant.

Deriving an RfD or RfC Using a BMD. In utilizing the BMD approach, the equation for an RfD or RfC becomes:

$$RfD = \frac{BMDL}{UF} \tag{9.13}$$

$$RfC = \frac{BMCL}{UF} \tag{9.14}$$

where, in this case, the lower confidence bound on the BMD (viz., the BMDL or BMCL) may be considered as simply being a procedural replacement for the NOAEL that gives the same "comfort level" of minimal risk—albeit this may not necessarily be construed as the NOAEL equivalent *per se.* Also, in this case, there is no UF for LOAEL to NOAEL extrapolation, etc.

9.4.5.2 Inter-Conversions of Noncarcinogenic Toxicity Parameters

Usually, the RfD for inhalation exposure is reported both as a concentration in air (in units of, e.g., mg/m³) and as a corresponding inhaled dose (mg/kg-day). Anyhow, when determining the toxicity value for inhalation pathways, the inhalation RfC [mg/m³] should be used whenever available. Broadly speaking, the RfC can also be converted to equivalent RfD values (in units of dose [mg/kg-day]) via multiplying the RfC term by an average human IR of 20 m³/day (for adults), and dividing it by an average adult body weight of 70 kg—as follows:

$$RfD_i \left[mg/kg\text{-}day \right] = \frac{RfC \left[mg/m^3 \right] \times 20\ m^3/day}{70\ kg} = 0.286\ RfC \tag{9.15}$$

Correspondingly, RfD values associated with oral exposures (and reported in mg/kg-day) can also be converted to a corresponding concentration in drinking water (usually called the "drinking water equivalent level," DWEL), as follows:

$$\text{DWEL}\left[\text{mg/L in water}\right] = \frac{\text{oral RfD}\left(\text{mg/kg-day}\right) \times \text{body weight(kg)}}{\text{ingestion rate}\left(\text{L/day}\right)}$$

$$= \frac{\text{RfD}_o\left(\text{mg/kg-day}\right) \times 70(\text{kg})}{2\left(\text{L/day}\right)} = 35\,\text{RfD}_o$$

(9.16)

The above derivation assumes a 2 L/day of water consumption by a 70-kg adult.

9.4.5.3 Toxicity Parameters for Carcinogenic Effects

Under a no-threshold assumption for carcinogenic effects, exposure to any level of a carcinogen is considered to have a finite risk of inducing cancer. An estimate of the resulting excess cancer per unit dose (called the unit cancer risk, or the cancer slope factor/cancer potency factor) is typically used to develop risk decisions for chemical exposure problems. On the whole, two specific toxicity parameters for expressing carcinogenic hazards based on the dose–response function find common application in human health risk assessments, namely:

- Cancer slope factor ([C]SF)—that expresses the slope of the dose–response function in dose-related units (i.e., [mg/kg-day]$^{-1}$); and
- URF—that expresses the slope in concentration-based units (i.e., [μg/m^3]$^{-1}$).

Typically, the [C]SFs are used when evaluating risks from oral or dermal exposures, whereas the URFs are used to evaluate risks from inhalation exposures. *Oral slope factors* are toxicity values for evaluating the probability of an individual developing cancer from oral exposure to contaminant levels over a lifetime; oral slope factors are expressed in units of (mg/kg-day)$^{-1}$. Generally used for the inhalation exposure route, *inhalation unit risk* toxicity values are expressed in units of (μg/m^3)$^{-1}$.

It is noteworthy that cancer dose–response assessment generally involves many scientific judgments regarding the following: the selection of different data sets (e.g., benign and malignant tumors, or their precursor responses) for extrapolation; the choice of low dose extrapolation approach based on the interpretation and assessment of the mode of action for the selected tumorigenic response(s); the choice of extrapolation models and methods to account for differences in dose across species; and the selection of the point of departure for low dose extrapolation. Indeed, many judgments usually have to be made in the many steps of the assessment process in the face of data variability. Also, different science policy choices and default procedures or methods are used to bridge data and knowledge gaps. Consequently, it is generally recognized that significant uncertainty exists in the cancer risk estimates.

Finally, it should be mentioned here that when no toxicological information exist for a chemical of interest or concern, structural similarity factors, etc. can be used to estimate cancer potency units for the chemicals that are suspected carcinogens but lack such pertinent values. For instance, in the case of a missing URF, this concept may be used to derive a surrogate parameter for the chemical with unknown URF—by, for example, estimating the geometric mean of a number of similar compounds with known URFs, and then using this as the surrogate value. Also, in contemporary times, PBPK modeling has become a preferred approach to both dose estimation and interspecies scaling of inhalation exposures, wherever data are available to support such efforts.

9.4.5.3.1 Slope Factors (SFs)

The SF, also called cancer potency factor (CPF) or potency slope, is a measure of the carcinogenic toxicity or potency of a chemical. It is a plausible upper-bound estimate of the probability of a response per unit intake of a chemical over a lifetime—represented by the cancer risk (proportion affected) per unit of dose (i.e., risk per mg/kg/day). In general, the CPF is used in human health risk assessments to estimate an upper-bound lifetime probability of an individual developing cancer as a result of exposure to a given level of a potential carcinogen. This represents a slope factor derived from a mathematical function (e.g., the LMS model) that is used to extrapolate the probability of incidence of cancer from a bioassay in animals using high doses to that expected to be observed at the low doses, likely to be found in chronic human exposures.

In general, slope factors should always be accompanied by the weight-of-evidence classification to indicate the strength of the evidence that the agent is a human carcinogen. A slope factor and the accompanying weight-of-evidence determination are the toxicity data most commonly used to evaluate potential human carcinogenic risks.

9.4.5.3.2 Inhalation Unit Risk (IUR)

The IUR is defined as the upper-bound excess lifetime cancer risk estimated to result from continuous exposure to an agent at a concentration of 1 $\mu g/m^3$ in air. In evaluating risks from chemicals found in certain human environmental settings, dose–response measures may generally be expressed as risk per concentration unit—yielding the URF (also called unit cancer risk, UCR or unit risk, UR) values. These measures may include the URF for air (viz., an inhalation URF), and the unit risk for drinking water (viz., an oral URF). In essence, the continuous lifetime ECs units for air and drinking water are usually expressed in micrograms per cubic meter ($\mu g/m^3$) and micrograms per liter ($\mu g/L$), respectively.

9.4.5.3.3 Derivation of SFs and URFs

The determination of carcinogenic toxicity parameters often involves the use of a variety of mathematical extrapolation models. In fact, scientific investigators have developed numerous models to extrapolate and estimate low-dose carcinogenic risks to humans from the high-dose carcinogenic effects usually observed in experimental animal studies. Such models yield an estimate of the upper limit in lifetime risk per unit of dose (or the unit cancer risk). On the whole, the nature of extrapolation employed for a given chemical during the estimation of carcinogenic potency very much depends on the existence of data to support linearity or nonlinearity—or indeed a biologically-based, or a case-specific model (USEPA, 1996f). In any event, the more popular approach among major regulatory agencies (such as the US EPA) involves a use of the LMS model—particularly because of the conservative attributes of this model. The linearized multistage model uses animal tumor incidence data to compute maximum likelihood estimates and upper 95% confidence limits (UCL$_{95}$) of risk associated with a particular dose. In general, the true risk is very unlikely to be greater than the UCL, may be lower than the UCL, and could be even as low as zero. In fact, the linearized multistage model yields upper bound estimates of risks that are a linear function of dose at low doses, and are frequently used as a basis for a number of regulatory decisions.

Overall, the linearized multistage model is known to make several conservative assumptions that result in highly conservative risk estimates—and thus yields over-estimates of actual URFs for carcinogens; in fact, the actual risks may perhaps be substantially lower than that predicted by the upper bounds of this model (Paustenbach, 1988). Even so, such approach is generally preferred since it allows analysts to err on the side of safety—and therefore would likely offer better protection of public health.

Deriving CSF Using a BMD. The BMD is an estimate of the dose or concentration that produces a predetermined change in response rate of an adverse effect (called the "benchmark response" or BMR) compared to background—and this is typically defined for a given exposure route and duration. The BMR generally should be near the low end of the range of increased risk that can be detected by a bioassay; indeed, low BMRs can impart high model dependence. In general, if there is an accepted (minimum) level of change in the endpoint that is considered to be biologically significant, then that amount is the BMR—like, for instance, a 10% body weight decrease compared to the control; in the absence of any other idea of what level of response to consider as being adverse, a change in the mean that equals one standard deviation of the control from the mean of the control may be utilized in this case (Crump, 1995).

Overall, the use of BMD methods involve fitting mathematical models to dose–response data, and then using the different results to select an appropriate BMD or BMC (as represented by, say, a BMCL$_x$) that is associated with a predetermined benchmark response (such as a 10% increase in the incidence of a particular lesion, or a 10% decrease in body weight changes). In practice, the BMD represents a lower confidence limit on the effective dose associated with some defined level of effect—e.g., a 5% or 10% increase in response. In other words, it is the confidence limit on the dose that elicits adverse responses in a fraction (often 5% or 10%) of the experimental animals; the confidence limits characterize uncertainty in the dose that affects a specified fraction of the animals. Thus, a BMCL$_x$ is defined as the lower 95% confidence limit of the dose that will result in a level of x% response; for example, BMCL$_{10}$ is the lower 95% confidence limit of a dose for a 10% increase in a particular response (US EPA, 1995c,e).

In utilizing the BMD approach here, the equation for an CSF becomes:

$$CSF = \frac{BMR}{BMDL} \tag{9.17}$$

As an illustrative example, this may be represented by: CSF = 0.1/BMDL$_{10}$.

9.4.5.4 Inter-Conversion of Carcinogenic Toxicity Parameters

Generally speaking, the URF estimates the upper-bound probability of a "typical" or "average" person contracting cancer when continuously exposed to one microgram per cubic meter (1 μg/m^3) of the chemical over an average (70-year) lifetime. Potency estimates are also given in terms of the potency slope factor (SF), which is the probability of contracting cancer as a result of exposure to a given lifetime dose (in units of mg/kg-day). That said, the SF can be converted to URF (also, unit risk, UR or unit cancer risk, UCR), by adopting several assumptions. The most critical requirement here is that the endpoint of concern must be a systematic tumor—in order that the potential target organs will experience the same blood concentration of the active carcinogen, regardless of the method of administration. This implies an assumption of equivalent absorption by the various routes of administration. Lastly, the basis for such conversions is the assumption that, at low doses, the dose–response curve is linear—so that the following holds true:

$$P(d) = SF \times [DOSE] \tag{9.18}$$

where $P(d)$ is the response (probability) as a function of dose; SF is the cancer potency slope factor ([mg/kg-day]$^{-1}$); and [DOSE] is the amount of chemical intake (mg/kg-day). The inter-conversions between URF and SF are expounded below.

- *Inter-conversions of the Inhalation Potency Factor.* Risks associated with a unit chemical concentration in air may be estimated in accordance with following mathematical relationship:

$$\text{Air unit risk} = \text{risk per } \mu g/m^3 (\text{air})$$

$$= \left[\text{slope factor (risk per mg/kg/day)} \right] \times \left[\frac{1}{\text{body weight (kg)}} \right]$$

$$\times \left[\text{inhalation rate} \left(m^3/day \right) \right] \times \left[10^{-3} \left(mg/\mu g \right) \right]$$

Thus, the inhalation potency can be converted to an inhalation URF by applying the following conversion factor:

$$\left[(\text{kg-day})/\text{mg}\right]\times\left[1/70\text{ kg}\right]\times\left[20\text{ m}^3/\text{day}\right]\times\left[1\text{ mg}/1,000\text{ μg}\right]=2.86\times10^{-4}$$

Accordingly, the lifetime excess cancer risk from inhaling 1 μg/m³ concentration for a full lifetime is estimated as:

$$\text{URF}_i\left(\text{μg}/\text{m}^3\right)^{-1}=\left(2.86\times10^{-4}\right)\times\text{SF}_i \tag{9.19}$$

Conversely, the SF_i can be derived from the URF_i as follows:

$$\text{SF}_i=\left(3.5\times10^3\right)\times\text{URF}_i \tag{9.20}$$

The assumptions used in the above derivations involve a 70-kg body weight, and an average IR of 20 m³/day.

- *Inter-conversions of the Oral Potency Factor.* Risks associated with a unit chemical concentration in water may be estimated in accordance with following mathematical relationship:

  Water unit risk = risk per μg/L (water)

  $$=\left[\text{slope factor (risk per mg/kg/day)}\right]\times\left[\frac{1}{\text{body weight (kg)}}\right]$$

  $$\times\left[\text{ingestion rate}\left(\text{L}/\text{day}\right)\right]\times\left[10^{-3}\left(\text{mg}/\text{μg}\right)\right]$$

Thus, the ingestion potency can be converted to an oral URF value by applying the following conversion factor:

$$\left[(\text{kg-day})/\text{mg}\right]\times\left[1/70\text{ kg}\right]\times\left[2\text{ L}/\text{day}\right]\times\left[1\text{ mg}/1,000\text{ μg}\right]=2.86\times10^{-5}$$

Accordingly, the lifetime excess cancer risk from ingesting 1 μg/L concentration for a full lifetime is:

$$\text{URF}_o\left(\text{μg}/\text{L}\right)^{-1}=\left(2.86\times10^{-5}\right)\times\text{SF}_o \tag{9.21}$$

Conversely, the potency, SF_o, can be derived from the unit risk as follows:

$$\text{SF}_o=\left(3.5\times10^4\right)\times\text{URF}_o \tag{9.22}$$

The assumptions used in the above derivations involve a 70-kg body weight, and an average water ingestion rate of 2 L/day.

9.4.5.5 The Use of Surrogate Toxicity Parameters

Risk characterizations generally should consider every likely exposure route in the evaluation process. However, toxicity data may not always be available for each route of concern—and in which case the use of surrogate values might become necessary; the pertinent surrogate values may indeed include extrapolation from data for a different exposure route. Broadly speaking, extrapolations may be reasonable for some cases where there is reliable information on the degree of absorption of materials by both routes of exposure in question—and assuming the substance is not locally more active by one route.

In principle, it is generally not possible to extrapolate between exposure routes for some substances that produce localized effects dependent upon the route of exposure. For example, a toxicity value based on localized lung tumors that result only from inhalation exposure to a substance would not be appropriate for estimating risks associated with dermal exposure to the substance. Thus, it may be appropriate to extrapolate dermal toxicity values *only* from values derived for oral exposure. In fact, it is recommended that oral toxicity reference values *not* be extrapolated casually from inhalation toxicity values, although some such extrapolations may be performed on a case-by-case basis (USEPA, 1989b). At any rate, these types of extrapolation can become useful approximations to employ—at least for preliminary risk assessments.

9.4.5.6 Exposure Route-Specificity of Toxicological Parameters

Toxicity parameters used in risk assessments are dependent on the route of exposure. However, as an example, oral RfDs and SFs have been used in practice for both ingestion and dermal exposures to some chemicals that affect receptors through a systemic action—albeit this will be inappropriate if the chemical affects the receptor contacts through direct local action at the point of application. In fact, in several (but certainly not all) situations, it is quite appropriate to use oral SFs and RfDs as surrogate values to estimate systemic toxicity as a result of dermal absorption of a chemical (DTSC, 1994; USEPA, 1989b, 1992). It is noteworthy, however, that direct use of the oral SF or oral RfD does not account for differences in absorption and metabolism between the oral and dermal routes—leading to increasingly uncertain outcomes. Also, in the evaluation of the inhalation pathways, when an inhalation SF or RfD is not available for a compound, the oral SF or RfD may be used in its place—but mostly for screening-level types of analyses; similarly, inhalation SFs and RfDs may be used as surrogates for both ingestion and dermal exposures for those chemicals lacking oral toxicity values—also adding to the uncertainties in the evaluations.

In addition to the uncertainties caused by route differences, further uncertainty is introduced by the fact that the oral dose–response relationships are based on potential (i.e., administered) dose, whereas dermal dose estimates are absorbed doses. Ideally, these differences in route and dose type should be resolved via pharmacokinetic modeling. Alternatively, if estimates of the gastrointestinal absorption fraction are available for the compound of interest in the appropriate vehicle, then the oral dose–response factor, unadjusted for absorption, can be converted to an absorbed dose equivalent as follows:

$$RfD_{absorbed} = RfD_{administered} \times ABS_{GI} \tag{9.23}$$

$$SF_{absorbed} = \frac{SF_{administered}}{ABS_{GI}} \tag{9.24}$$

On the average, absorption fractions corresponding to approximately 10% and 1% are typically applied to organic and inorganic chemicals, respectively.

It is noteworthy that, for the most part, direct toxic effects on the skin have not been adequately evaluated for several chemicals encountered in the human work and living environments. This means that, it may be inappropriate to use the oral slope factor to evaluate the risks associated with a dermal exposure to carcinogens such as benz(*a*)pyrene, that are believed to cause skin cancer through a direct action at the point of application—i.e., unless proper adjustments are made accordingly. Indeed, depending on the chemical involved, the use of an oral SF or oral RfD for the dermal route is likely to result in either an over- or under-estimation of the risk or hazard. Consequently, the use of the oral toxicity value as a surrogate for a dermal value will usually tend to increase the uncertainty in the estimation of risks and hazards. However, this approach is not generally expected to significantly under-estimate the risk or hazard relative to the other routes of exposure that are evaluated in most risk assessments (DTSC, 1994; USEPA, 1992).

Certainly, other methods of approach that are quite different from the above may be used to generate surrogate toxicity values. For instance, in some situations, toxicity values to be used in characterizing risks are available only for certain chemicals within a chemical class—and this may require a different evaluation approach. In such cases, rather than simply eliminating those chemicals without toxicity values from a quantitative evaluation, it usually is prudent to group data for such class of chemicals into well-defined categories (e.g., according to structure-activity relationships, or indeed other similarities) for consideration in the risk assessment. Such grouping should not be based solely on toxicity class or carcinogenic classifications. Regardless, it still must be acknowledged that significant uncertainties will likely result by using this type of approach as well. Hence, if and when this type of grouping is carried out, the rationale should be explicitly stated and adequately documented in the risk assessment summary—emphasizing the fact that the action may have produced over- or under-estimates of the true risk.

As a final point here, the introduction of additional uncertainties in an approach that relies on surrogate toxicity parameters cannot be over-emphasized—and such uncertainties should be properly documented and adequately elucidated as part of the overall risk evaluation process.

9.4.5.7 Route-to-Route Extrapolation of Toxicological Parameters

For systemic effects away from the site of entry, an inhalation toxicity parameter, TP_{inh} [mg/m³], may be converted to an oral value, TP_{oral} [mg/kg-day)], or vice versa, by using the following type of relationship (van Leeuwen and Hermens, 1995):

$$TP_{inh} \times IR \times t \times BAF_{inh} = TP_{oral} \times BAF_{oral} \times BW \tag{9.25}$$

where IR is the inhalation rate [m³/h]; t is the time [h]; BAF_r is the bioavailability for route r, for which default values should be used if no data exists [e.g., use 1 for oral exposure, 0.75 for inhalation exposure, and 0 (in the case of very low or very high lipophilicity or high molecular weight) or 1 (in the case of intermediate lipophilicity and low molecular weight) for dermal exposure]; and BW is the body weight [kg]. A dermal toxicity parameter for systemic effects, TP_{derm} [mg/kg-day] can also be derived from the TP_{oral} [mg/kg-day] or the TP_{inh} [mg/m³] values as follows (van Leeuwen and Hermens, 1995):

$$TP_{derm} = TP_{oral} \times \frac{BAF_{oral}}{BAF_{derm}} \tag{9.26}$$

$$TP_{derm} = \frac{TP_{inh} \times IR \times t}{BW} \times \frac{BAF_{inh}}{BAF_{derm}} \tag{9.27}$$

It is noteworthy that route-to-route extrapolation introduces additional uncertainty into the overall risk assessment process; such uncertainty can be reduced by utilizing physiologically based pharmacokinetic (PBPK) models. Indeed, PBPK models are particularly useful for predicting disposition differences due to exposure route differences—i.e., if sufficient pharmacokinetic data is available (van Leeuwen and Hermens, 1995).

Illustrative examples of the route-to-route extrapolation process are well documented in the literature elsewhere for both the carcinogenic and noncarcinogenic effects of chemical constituents (e.g., Asante-Duah, 2017). Anyhow, in general, and insofar as possible, inhalation values should *not* be extrapolated from oral values—i.e., if at all avoidable. Yet, situations do sometimes arise when it becomes necessary to rely on such approximations to make effective environmental and public health risk management decisions.

9.4.5.8 Toxicity Equivalence Factors and Toxicity Equivalency Concentration

Some chemicals are members of the same family and exhibit similar toxicological properties; however, these chemicals would generally differ in their degrees of toxicity. To carry out a hazard effects assessment for such chemicals, a "toxicity equivalence factor" (TEF) may first be applied to adjust the measured concentrations to a toxicity equivalent concentration—i.e., prior to embarking on the ultimate risk determination goal. In fact, the TEF approach has been extensively used for the hazard assessment of different classes of toxic chemical mixtures.

Broadly speaking, a TEF procedure is one used to derive quantitative dose–response estimates for substances that are members of a certain category or class of agents. The assumptions implicit in the utilization of the TEF approach include the following significant ones (NATO/CCMS, 1988; Safe, 1998):

- The individual compounds all act through the same biologic or toxic pathway;
- The effects of individual chemicals in a mixture are essentially additive at sub-maximal levels of exposure;
- The dose–response curves for different congeners should be parallel; and
- The organotropic manifestations of all congeners must be identical over the relevant range of doses.

In essence, a basic premise of the TEF methodology is the presence of a common biologic end-point, or in the case of multiple end-points, a common mechanism of action. A second key assumption is the additivity of effects. In fact, these assumptions are inherent in all TEF-schemes—and thus, the accuracy of all TEF-schemes will be affected by situations where such assumptions are not applicable. It is also noteworthy that, for more complex mixtures containing compounds that act through multiple pathways to give both similar and different toxic responses, the TEF/TEQ approach may *not* be appropriate *per se* (Safe, 1998).

On the whole, TEFs are based on shared characteristics that can be used to hierarchically rank-order the class members by carcinogenic potency when cancer bioassay and related data are inadequate for this purpose. The rank-ordering is by reference to the characteristics and potency of a well-studied member or members of the class under review. Other class members are then indexed to the reference agent(s) by using one or more shared characteristics to generate their TEFs. Examples of shared characteristics that may be used include receptor-binding characteristics; results of biological activity assays related to carcinogenicity; or structure-activity relationships. The TEFs are usually indexed at increments of a factor of 10; very good data, however, may permit a smaller increment to be used. Further elaboration of the processes involved in this approach is well documented in the literature elsewhere (e.g., Asante-Duah, 2017).

9.5 CHEMICAL RISK CHARACTERIZATION AND UNCERTAINTY ANALYSES

Fundamentally, risk characterization consists of estimating the probable incidence of adverse impacts to potential receptors, under the various exposure conditions associated with a chemical hazard situation. It involves an integration of the hazard effects and exposure assessments—in order to arrive at an estimate of the health and/or environmental risk to the exposed population. In general, all information derived from each step of a chemical exposure-cum-hazard assessment are integrated and utilized during the risk characterization—so as to help project the degree and severity of adverse health effects in the populations potentially at risk.

Risk characterization is indeed the final step in the risk assessment process, and it becomes the first input into risk management programs. Thus, risk characterization serves as a bridge between risk assessment and risk management—and is therefore a key factor in the ultimate decision-making process that is developed to address contaminated site and chemical exposure problems. Classically, risk characterization commonly will entail a statement regarding the "response" or "risk of harm" that is expected in the population under an associated set of exposure conditions, together with a description of uncertainties (NRC, 1983). Through probabilistic modeling and analyses, uncertainties associated with the risk evaluation process can be assessed properly, and their effects on a given decision accounted for systematically. In this manner, the risks associated with given decisions may be delineated—and then appropriate corrective measures taken accordingly.

It is noteworthy that, depending on the nature of populations potentially at risk from a contaminated site and/or an environmental contamination problem, different types of risk measures or parameters may be employed in the risk characterization process. Typically, the cancer risk and HQ-cum-hazard index (HI) estimates are used to define potential risks to human health (see Sections 9.5.2 and 9.5.3); the ecological quotient (similar to the human health HQ) estimate is used to define risks to potential ecological receptors (see Chapter 10); and the risk costs associated with a "pathway probability" concept is utilized in probabilistic risk characterizations. The general details of various risk characterization models and measures finding application in commonly encountered environmental contamination problems is discussed further below.

This section elaborates the mechanics of the risk characterization process, together with example risk presentation modalities that would tend to, among several other things, facilitate effective risk management and/or risk communication efforts; it also includes an elaboration of the uncertainties that surround the overall process. It must be acknowledged and emphasized here that, although the discussions that follow are biased towards human exposure situations, the concepts presented are equally applicable to ecological receptors as well—with elements that are very specific to ecological settings and receptors having been identified in Chapter 10 of this title.

9.5.1 SOME FUNDAMENTAL ISSUES AND CONSIDERATIONS AFFECTING THE RISK CHARACTERIZATION PROCESS

The risk characterization process consists of an integration of the toxicity and exposure assessments—resulting in a quantitative estimation of the actual and potential risks and/or hazards associated with a chemical exposure problem. Broadly stated, risk from a receptor or organism exposure to chemicals is a function of dose or intake and potency, viz.:

$$\text{Risk from chemical exposure} = [\text{Dose of chemical}] \times [\text{Chemical potency}] \qquad (9.28)$$

Overall, chemical risk characterization is viewed as a process by which dose–response information is integrated with quantitative estimates of a receptor or organism exposure derived in an exposure assessment; the result is a quantitative estimate of the likelihood that the exposed organisms will experience some form of adverse health effects under a given set of exposure assumptions. During the risk characterization, chemical-specific toxicity information is traditionally compared against both field measured and estimated chemical exposure levels (and in some cases, those levels predicted through fate and behavior modeling) in order to determine whether concentrations associated with a chemical exposure problem are of significant concern. In principle, the process should also consider the possible additive or cumulative and related effects of exposure to mixtures of the chemicals of potential concern.

Two general types of health risk are typically characterized for each potential exposure pathway considered—viz.: potential carcinogenic risk, and potential noncarcinogenic hazard. Broadly speaking, characterization of the potential health effects of potential carcinogenic versus noncarcinogenic chemicals are approached very differently. A key difference in the approaches arises from the conservative assumption that substances with possible carcinogenic action typically behave via a no-threshold mechanism, whereas other toxic actions may have a threshold (i.e., a dose below which few individuals would be expected to show a response of concern)—albeit this viewpoint has been challenged, and remains in debate. Thus, under the no-threshold assumption, it becomes necessary to calculate a risk number—whereas for chemicals with a threshold, it is possible to simply characterize an exposure as above or below the designated threshold level (generally termed a reference dose or reference concentration). Also, potential carcinogenic risk is evaluated by averaging exposure over a "normal" human or organism lifetime, whereas potential noncarcinogenic hazard is evaluated by averaging exposure over the total exposure period considered in practice. Indeed, depending on the nature of populations potentially at risk from a chemical exposure problem, different types of risk metrics or parameters may be employed in the risk characterization process. At any rate, the cancer risk estimates and HQ-cum- HI estimates are the measures of choice typically used to define potential risks to human health in particular (see Sections 9.5.2 and 9.5.3). Undeniably, it is almost indispensable to have these measures available to support effectual risk management programs. Consequently, the health risks to potentially exposed populations resulting from chemical exposures are characterized through a calculation of noncarcinogenic HQs/indices and/or carcinogenic risks (CAPCOA, 1990; CDHS, 1986; USEPA, 1986a, 1989a).

At the end of the day, an effective risk characterization should be carried out in such a manner that it fully, openly, and clearly characterize risks as well as disclose the scientific analyses, uncertainties, assumptions, and science policies that underlie decisions utilized throughout the risk assessment and risk management processes. In fact, every risk assessment should clearly delineate the strengths and weaknesses of the data, the assumptions made, the uncertainties in the methodology, and the rationale used in reaching the conclusions (e.g., similar or different routes of exposure, and metabolic differences between humans and test animals). Furthermore, the hazard and risk assessment of human exposure to chemicals must take a miscellany of other critical issues into account—especially as relates to scenarios whereby chemical interactions may significantly influence toxic outcomes; chemical interactions are indeed very important determinants in evaluating the potential hazards and risks of exposure to chemical mixtures (Safe, 1998).

Lastly, it is noteworthy that a health risk assessment/characterization is only as good as its component parts—i.e., the hazard characterization, the dose–response analysis, and the exposure assessment. Confidence in the results of a risk assessment is thus a function of the confidence in the results of the analysis of these distinct key elements, and indeed their corresponding ingredients. In the end, several important issues usually will have very significant bearing on the processes involved in completing risk characterization tasks designed to support effective risk management programs; a number of the particularly important topics/issues are discussed below.

9.5.1.1 Adjustments for Chemical Absorption: Administered versus Absorbed Dose

Oftentimes, absorption adjustments may become necessary during the risk estimation process—in order to ensure that the exposure estimate and the toxicity value being compared during the risk characterization are both expressed as absorbed doses, or both expressed as administered doses (i.e., intakes). Adjustments may also be required for different vehicles of exposure (e.g., water, food, or soil)—albeit, in most cases, the unadjusted toxicity value will provide a reasonable or conservative estimate of risk. Furthermore, adjustments may be needed for different absorption efficiencies, depending on the medium of exposure; in general, correction for fractional absorption is particularly appropriate when interaction with environmental media or other chemicals may alter absorption from what would typically be expected for the pure compound. Correction may also be necessary when assessment of exposure is via a different route of contact than what was utilized in the experimental studies used to establish the toxicity parameters (i.e., the SFs, RfDs, etc. discussed in Section 9.4). For instance, only limited toxicity reference values generally exist for dermal exposure; consequently, oral values are frequently used to assess risks from dermal exposures (USEPA, 1989d). On the other hand, most RfDs and some carcinogenic SFs usually are expressed as the amount of substance administered per unit time and unit body weight, *whereas* exposure estimates for the dermal route of exposure are eventually expressed as absorbed doses. Thus, for dermal exposures, it may become particularly important to adjust an oral toxicity value from an administered to an absorbed dose.

Absorption efficiency adjustment procedures are elaborated further elsewhere in the literature (e.g., Asante-Duah, 2017; USEPA, 1989d, 1992). Meanwhile, it is noteworthy that for evaluations of the dermal exposure pathway, if the oral toxicity value is already expressed as an absorbed dose, then it is not necessary to adjust the toxicity value. Also, exposure estimates should not be adjusted for absorption efficiency if the toxicity values are based on administered dose. Furthermore, in the absence of reliable information, 100% absorption is usually used for most chemicals; for metals, an approximately 10% absorption may be considered a reasonable upper-bound for other than the inhalation exposure route.

In general, absorption factors should not be used to modify exposure estimates in those cases where absorption is inherently factored into the toxicity/risk parameters used for the risk characterization. Thus, "correction" for fractional absorption is appropriate only for those values derived from experimental studies based on absorbed dose. In other words, absorbed dose should be used in risk characterization only if the applicable toxicity parameter (e.g., SF or RfD) has been adjusted for absorption; otherwise, intake (unadjusted for absorption) are used for the calculation of risk levels.

9.5.1.2   Aggregate Effects of Chemical Mixtures and Multiple Exposures

Oftentimes in the study of human or other organism exposures to chemical hazards, it becomes necessary to carry out aggregate and cumulative exposure and risk assessments. In fact, in most situations, it is quite important to consider both aggregate and cumulative exposures—to facilitate the making of effectual risk assessment and risk management decisions, as well as help the process of setting chemical tolerance or safe exposure levels. In general, aggregate exposures may occur across different pathways and media that contribute to one or more routes of an individual receptor's exposure—which then becomes the basis for determining cumulative risks.

Cumulative risk refers to effects from chemicals that have a common mode of toxicological action—and thus have aggregate exposure considerations as part of the assessment process (Clayton et al., 2002). Indeed, whereas some chemical hazard situations involve significant exposure to only a single compound, most instances of chemical exposure problems can involve concurrent or sequential exposures to a mixture of compounds that may induce similar or dissimilar effects over exposure periods ranging from short-term to a lifetime (USEPA, 1984a, 1986b). Meanwhile, it is notable that evaluating mixtures of chemicals is one of the areas of risk assessment with obviously many uncertainties; this is especially so, because several types of interactions in chemical mixtures are possible—including the following key distinct attributes:

- Additive—wherein the effects of the mixture equals that of adding the effects of the individual constituents.
- Synergistic—wherein the effects of the mixture is greater than obtained by adding the effects of the individual constituents.
- Antagonistic—wherein the effects of the mixture is less than obtained by adding the effects of the individual constituents.

Of particular concern are those mixtures where the effects are synergistic. Unfortunately, the toxicology of complex mixtures is not very well understood—complicating the problem involved in the assessment of the potential for these compounds to cause various health effects. Nonetheless, there is the need to assess the cumulative health risks for the chemical mixtures, despite potential large uncertainties that may exist. The risk assessment process must, therefore, address the multiple endpoints or effects, and also the uncertainties in the dose–response functions for each effect.

Finally, in combining multi-chemical risk estimates for multiple chemical sources, it should be noted here that, if two sources do not affect the same individual or subpopulation, then the sources' individual risk estimates (and/or hazard indices) do not quite influence each other—and, therefore, these risks should not be combined. Thus, one should not automatically sum risks from all sources evaluated for a chemical exposure problem—i.e., unless if it has been determined/established that such aggregation is appropriate. On the other hand, potential receptors are typically exposed not to isolated chemical sources, but rather to a complex, dilute mixture of many origins. Considering how many chemicals are present in the wide array of environmental compartments, there are virtually infinite number of combinations that could constitute potential synergisms and antagonisms. In the absence of any concrete evidence of what the interactive effects might be, however, an additive method that simply sums individual chemical effects on a target organ is usually employed in the evaluation of chemical mixtures.

9.5.1.3 Carcinogenic Chemical Effects

The common method of approach in the assessment of chemical mixtures assumes additivity of effects for carcinogens when evaluating multiple carcinogens—albeit alternative procedures that are more realistic and/or less conservative have been proposed for certain situations by some investigators (e.g., Bogen, 1994; Chen et al., 1990; Gaylor and Chen, 1996; Kodell and Chen, 1994; Slob, 1994). In any case, prior to a summation for aggregate risks, estimated cancer risks should perhaps be (preferably) segregated by weight-of-evidence (or strength-of-evidence) category for the chemicals of concern—the goal being to provide a clear understanding of the risk contribution of each category of carcinogen.

9.5.1.4 Systemic (Non-Cancer) Chemical Effects

For multiple chemical exposures to non-carcinogens and the noncarcinogenic effects of carcinogens, constituents should be grouped by the same mode of toxicological action (i.e., those that induce the same physiologic endpoint—such as liver or kidney toxicity). Cumulative noncarcinogenic risk is evaluated through the use of a HI that is generated for each health or physiologic 'endpoint'. Physiologic/toxicological endpoints that will normally be considered with respect to chronic toxicity include: cardiovascular systems (CVS); central nervous system (CNS); gastrointestinal (GI) system; immune system; reproductive system (including teratogenic and developmental effects); kidney (i.e., renal); liver (i.e., hepatic); and the respiratory system.

In fact, in a strict sense, constituents should not be grouped together unless they induce/affect the same toxicological/physiologic endpoint. Thus, in a well-defined risk characterization exercise, it becomes necessary to segregate chemicals by organ-specific toxicity—since strict additivity without consideration for target-organ toxicities could over-estimate potential hazards (USEPA, 1986b, 1989d). Accordingly, the "true" HI is preferably calculated only after putting chemicals into groups with same physiologic endpoints. Listings of chemicals with their associated noncarcinogenic toxic effects on specific target organ/system can be found in such databases as IRIS (Integrated Risk Information System), as well as in the literature elsewhere (e.g., Cohrssen and Covello, 1989a; USEPA, 1996).

9.5.2 CARCINOGENIC RISK EFFECTS: ESTIMATION OF CARCINOGENIC RISKS TO HUMAN HEALTH

For potential carcinogens, risk is defined by the incremental probability of an individual developing cancer over a lifetime as a result of exposure to a carcinogen. This risk of developing cancer can be estimated by combining information about the carcinogenic potency of a chemical and exposure to the substance. Specifically, carcinogenic risks are estimated by multiplying the route-specific cancer slope factor (which is the upper 95% confidence limit of the probability of a carcinogenic response per unit intake over a lifetime of exposure) by the estimated intakes; this yields the excess or incremental individual lifetime cancer risk.

Broadly speaking, risks associated with the "inhalation" and "non-inhalation" pathways may be estimated in accordance with some adaptations of the following generic relationships:

Risk for 'inhalation pathways'

$$= \text{Ground-level concentration (GLC) or 'Exposure Concentration' (EC)} \left[\mu g/m^3 \right] \quad (9.29)$$

$$\times \text{Inhalation Unit Risk} \left[\left(\mu g/m^3 \right)^{-1} \right]$$

$$\text{Risk for 'non-inhalation pathways'} = \text{Dose} \left[mg/kg\text{-day} \right] \times \text{Potency Slope} \left[\left(mg/kg\text{-day} \right)^{-1} \right] \quad (9.30)$$

The resulting estimates can then be compared with benchmark criteria/standards in order to arrive at risk decisions about a given chemical exposure problem.

In practice, a customarily preferred first step in a cancer risk assessment (i.e., when appraising health risks for cancer endpoints) is to characterize the hazard using a "weight-of-evidence" (or perhaps a "strength-of-evidence") narrative—e.g., by using one of the following five standard hazard descriptors: "Carcinogenic to Humans"; "Likely to Be Carcinogenic to Humans"; "Suggestive Evidence of Carcinogenic Potential"; "Inadequate Information to Assess Carcinogenic Potential"; and "Not Likely to Be Carcinogenic to Humans." The narrative describes the available evidence, including its strengths and limitations, and "provides a conclusion with regard to human carcinogenic potential" (USEPA, 2005a–c). Depending on how much is known about the "mode-of-action" of the agent of interest, one of two methods is used for completing any pertinent extrapolations, viz.: linear or nonlinear extrapolation. A linear extrapolation is used in the "absence of sufficient information on modes-of-action" or when "the mode-of-action information indicates that the dose–response curve at low dose is or is expected to be linear"; for a linear extrapolation, the "slope factor" is considered "an upper-bound estimate of risk per increment of dose"—and this is used to estimate risks at different exposure levels (USEPA, 2005a–c). A nonlinear approach would be used "when there is sufficient data to ascertain the mode of action—with the conclusion that it is not linear at low doses, and the agent does not demonstrate mutagenic or other activity consistent with linearity at low doses"; details of the computational approaches are offered elsewhere (e.g., USEPA, 2005a–c).

On the whole, the carcinogenic effects of the constituents associated with potential chemical exposure problems are typically calculated using the linear low-dose and one-hit models, represented by the following relationships (USEPA, 1989d):

$$\text{Linear low-dose model, } CR = CDI \times SF \quad (9.31)$$

$$\text{One-hit model, } CR = 1 - \exp(-CDI \times SF) \quad (9.32)$$

where CR is the probability of an individual developing cancer (dimensionless); CDI is the chronic daily intake for long-term exposure (i.e., averaged over receptor lifetime) (mg/kg-day); and SF is the cancer slope factor ([mg/kg-day]$^{-1}$). The linear low-dose model is based on the so-called "linearized multistage" (LMS) model—which assumes that there are multiple stages for cancer; the "one-hit" model assumes that there is a single stage for cancer, and that one molecular or radiation interaction induces malignant change—making it very conservative. In reality, and for all practical purposes, the linear low-dose cancer risk model is valid only at low risk levels (i.e., estimated risks <0.01); for situations where chemical intakes may be high (i.e., potential risks >0.01), the one-hit model represents the more appropriate algorithm to use.

As a simple illustrative example calculation of human health carcinogenic risk, consider a situation where PCBs from abandoned electrical transformers have leaked into a groundwater reservoir that serves as a community water supply source. Environmental sampling and analysis conducted in a routine testing of the public water supply system showed an average PCB concentration of 2 µg/L.

Thence, the pertinent question here is: "what is the individual lifetime cancer risk associated with a drinking water exposure from this source?" Now, assuming that the only exposure route of concern here is from water ingestion, and using a cancer oral SF of 7E-02 (obtained from Table C.1 in Appendix C) and applicable/appropriate "intake factor" (see Section 9.3), then the cancer risk attributable to this exposure scenario is estimated as follows:

$$\text{Cancer risk} = \text{SF}_o \times \text{CDI}_o$$

$$= \text{SF}_o \times C_w \times 0.0149$$

$$= \left(7 \times 10^{-2}\right) \times \left(2\,\mu g/L \times 10^{-3}\,mg/\mu g\right) \times 0.0149 = 2.1 \times 10^{-6}$$

Similar evaluations can indeed be carried out for the various media and exposure routes of potential concern or possible interest.

Anyway, the method of approach for assessing the cumulative health risks from chemical mixtures generally assumes additivity of effects for carcinogens when evaluating chemical mixtures or multiple carcinogens. Thus, for multiple carcinogenic chemicals and multiple exposure routes/pathways, the aggregate cancer risk for all exposure routes and all chemicals of concern associated with a potential chemical exposure problem can be estimated using the algorithms shown in Boxes 9.12A and 9.12B. The combination of risks across exposure routes is based on the assumption that the same receptors would consistently experience the reasonable maximum exposure via the multiple routes. Hence, if specific routes do not affect the same individual or receptor group, risks should not be combined under those circumstances.

Finally, as a rule-of-thumb, incremental risks of between 10^{-4} and 10^{-7} are generally perceived as being reasonable and adequate for the protection of human health—with 10^{-6} often used as the "point-of-departure." In reality, however, populations may be exposed to the same constituents from sources unknown or unrelated to a specific study. Consequently, it is preferable that the

---

BOX 9.12A   THE LINEAR LOW-DOSE MODEL FOR THE ESTIMATION OF LOW-LEVEL CARCINOGENIC RISKS

$$\text{Total Cancer Risk, TCR}_{\text{lo-risk}} = \sum_{j=1}^{p} \sum_{i=1}^{n} \left(\text{CDI}_{ij} \times \text{SF}_{ij}\right)$$

and

$$\text{Aggregate/Cumulative Total Cancer Risk, ATCR}_{\text{lo-risk}} = \sum_{k=1}^{s} \left\{ \sum_{j=1}^{p} \sum_{i=1}^{n} \left(\text{CDI}_{ij} \times \text{SF}_{ij}\right) \right\}$$

where
   TCR = probability of an individual developing cancer (dimensionless)
   CDI_{ij} = chronic daily intake for the ith chemical and jth route (mg/kg-day)
   SF_{ij} = slope factor for the ith chemical and jth route ([mg/kg-day]$^{-1}$)
   n = total number of carcinogens
   p = total number of pathways or exposure routes
   s = total number for multiple sources of exposures to receptor (e.g., dietary, drinking water, occupational, and residential, recreational)

BOX 9.12B   THE ONE-HIT MODEL FOR THE ESTIMATION OF HIGH-LEVEL CARCINOGENIC RISKS

$$\text{Total Cancer Risk, TCR}_{\text{hi-risk}} = \sum_{j=1}^{p}\sum_{i=1}^{n}\left[1 - \exp\left(-\text{CDI}_{ij} \times \text{SF}_{ij}\right)\right]$$

and

Aggregate/Cumulative Total Cancer Risk, $\text{ATCR}_{\text{hi-risk}}$

$$= \sum_{k=1}^{s}\left\{\sum_{j=1}^{p}\sum_{i=1}^{n}\left[1 - \exp\left(-\text{CDI}_{ij} \times SF_{ij}\right)\right]\right\}$$

where
  TCR = probability of an individual developing cancer (dimensionless)
  CDI_{ij} = chronic daily intake for the ith chemical and jth route (mg/kg-day)
  SF_{ij} = slope factor for the ith chemical and jth route ($[\text{mg/kg-day}]^{-1}$)
  n = total number of carcinogens
  p = total number of pathways or exposure routes
  s = total number for multiple sources of exposures to receptor (e.g., dietary, drinking water, occupational, residential, and recreational)

---

estimated carcinogenic risk is well below the 10^{-6} benchmark level—in order to allow for a reasonable margin of protectiveness for populations potentially at risk. Surely, if a calculated cancer risk exceeds the 10^{-6} benchmark, then the health-based criterion for the chemical mixture has been exceeded, and the need for corrective measures and/or risk management actions must be given serious consideration.

It is noteworthy that the carcinogenic risk (i.e., the probability of developing cancer) depends on many things—including the intensity, route, and duration of exposure to a carcinogen people experience. Different people may respond differently to similar exposures, depending on their age, sex, nutritional status, overall health, genetics, and many other factors. Only in a few instances can risk be estimated with complete confidence, and these estimations require studies of long-term human exposures and cancer incidence in restricted environments, which are rarely available (USDHS, 2002).

9.5.2.1   Population Excess Cancer Burden

The two important parameters or measures often used for describing carcinogenic effects are the individual cancer risk and the estimated number of cancer cases (i.e., the cancer burden). The individual cancer risk from simultaneous exposure to several carcinogens is assumed to be the sum of the individual cancer risks from each individual chemical. The risk experienced by the individual receiving the greatest exposure is referred to as the "maximum individual risk."

Now, to assess the population cancer burden associated with a chemical exposure problem, the number of cancer cases due to an exposure source within a given community can be estimated by multiplying the individual risk experienced by a group of people by the number of people in that group. Thus, if 10 million people (as an example) experience an estimated cancer risk of 10^{-6} over their lifetimes, it would be estimated that 10 (i.e., 10 million $\times$ 10^{-6}) additional cancer cases could occur for this group. The number of cancer incidents in each receptor area can be added to estimate the number of cancer incidents over an entire region. Hence, the excess cancer burden, B_{gi}, is given by:

$$B_{gi} = \sum \left(R_{gi} \times P_g \right) \tag{9.33}$$

where B_{gi} is the population excess cancer burden for ith chemical for exposed group, G; R_{gi} is the excess lifetime cancer risk for ith chemical for the exposed population group, G; P_g is the number of persons in exposed population group, G. Assuming cancer burden from each carcinogen is additive, the total population group excess cancer burden is given by:

$$B_g = \sum_{i=1}^{N} B_{gi} = \sum_{i=1}^{N} \left(R_{gi} \times P_g \right) \tag{9.34}$$

and the total population burden, B, is represented by:

$$B = \sum_{g=1}^{G} B_g = \sum_{g=1}^{G} \left\{ \sum_{i=1}^{N} B_{gi} \right\} = \sum_{g=1}^{G} \left\{ \sum_{i=1}^{N} \left(R_{gi} \times P_g \right) \right\} \tag{9.35}$$

Insofar as possible, cancer risk estimates are expressed in terms of both individual and population risk. For the population risk, the individual upper-bound estimate of excess lifetime cancer risk for an average exposure scenario is simply multiplied by the size of the potentially exposed population.

9.5.3 Non-Cancer Risk Effects: Estimation of the Noncarcinogenic Hazards to Human Health

The potential non-cancer health effects resulting from a chemical exposure problem are usually expressed by the HQ and/or the HI. The HQ is defined by the ratio of the estimated chemical exposure level to the route-specific RfD, represented as follows (USEPA, 1989d):

$$\text{Hazard Quotient, HQ} = \frac{E}{\text{RfD}} \tag{9.36}$$

where E is the chemical exposure level or intake (mg/kg-day); and RfD is the reference dose (mg/kg-day) (note that the HQ associated with the inhalation pathway may preferably be represented as follows: $\text{HQ} = \text{EC/RfC}$, where EC is the exposure concentration in $\mu g/m^3$ and RfC is the inhalation toxicity value in $\mu g/m^3$).

As a simple illustrative example calculation of human health noncarcinogenic risk, consider a situation where an aluminum container is used for the storage of water meant for household consumption. Laboratory testing of the water revealed that some aluminum consistently gets leached and dissolved into this drinking water—with average concentrations of approximately 10 mg/L. The question here then is: "what is the individual non-cancer risk for a person who uses this source for drinking water?" Now, assuming the only exposure route of concern is associated with water ingestion (a reasonable assumption for this situation), and using a non-cancer toxicity index (i.e., an RfD) of 1.0 (obtained from Table C.1 in Appendix C), then the non-cancer risk attributable to this exposure scenario is calculated to be:

$$\text{Hazard Index} = \left(1/\text{RfD}_o \right) \times \text{CDI}_o$$

$$= \left(1/\text{RfD}_o \right) \times C_w \times 0.0639$$

$$= 1.0 \times 10 \text{ mg/L} \times 0.0639 = 0.6$$

Similar evaluations can indeed be carried out for the various media and exposure routes of potential concern or possible interest.

Anyway, for multiple chemical exposures to non-carcinogens and the noncarcinogenic effects of carcinogens, constituents are normally grouped by the same mode of toxicological action. Cumulative non-cancer risk is then evaluated using a HI that is generated for each health or toxicological "endpoint." Chemicals with the same endpoint are generally included in a HI calculation. Thus, for multiple noncarcinogenic effects of several chemical compounds and multiple exposure routes, the aggregate non-cancer risk for all exposure routes and all constituents associated with a potential chemical exposure problem can be estimated using the algorithm shown in Box 9.13. It is noteworthy that, the combination of HQs across exposure routes is based on the assumption that the same receptors would consistently experience the reasonable maximum exposure via the multiple routes. Thus, if specific sources do not affect the same individual or receptor group, HQs should not be combined under those circumstances. Furthermore, and in the strictest sense, constituents should not be grouped together unless the physiologic/toxicological endpoint is known to be the same—otherwise the efforts will likely over-estimate and/or overstate potential health effects.

Finally, in accordance with general guidelines on the interpretation of hazard indices, for any given chemical, there may be potential for adverse health effects if the HI exceeds unity (1)—albeit it is possible that no toxic effects will occur even if this benchmark level is exceeded, since the RfD incorporates a large margin of safety. At any rate, as a rule-of-thumb in the interpretation of the results from HI calculations, a reference value of less than or equal to unity (i.e., HI ≤1) should be taken as the acceptable benchmark. Also, it is noteworthy that, for HI values greater than unity (i.e., HI >1), the higher the value, the greater is the likelihood of adverse noncarcinogenic health impacts. In the final analysis, since populations may be exposed to the same constituents from sources unknown or unrelated to a case-problem, it is preferred that the estimated noncarcinogenic HI be well below the benchmark level of unity—in order to allow for additional margin of protectiveness for populations potentially at risk. Indeed, if any calculated HI exceeds unity, then the

BOX 9.13   GENERAL EQUATION FOR CALCULATING NONCARCINOGENIC RISKS TO HUMAN HEALTH

$$\text{Total Hazard Index} = \sum_{j=1}^{p}\sum_{i=1}^{n}\frac{E_{ij}}{\text{RfD}_{ij}} = \sum_{j=1}^{p}\sum_{i=1}^{n}[\text{HQ}]_{ij}$$

and

$$\text{Aggregate/Cumulative Total Hazard Index} = \sum_{k=1}^{s}\left\{\sum_{j=1}^{p}\sum_{i=1}^{n}\frac{E_{ij}}{\text{RfD}_{ij}}\right\} = \sum_{k=1}^{s}\left\{\sum_{j=1}^{p}\sum_{i=1}^{n}[\text{HQ}]_{ij}\right\}$$

where
   E_{ij} = exposure level (or intake) for the ith chemical and jth route (mg/kg-day)
   RfD_{ij} = acceptable intake level (or reference dose) for the ith chemical and jth exposure route (mg/kg-day)
   $[\text{HQ}]_{ij}$ = hazard quotient for the ith chemical and jth route
   n = total number of chemicals showing noncarcinogenic effects
   p = total number of pathways or exposure routes
   s = total number for multiple sources of exposures to receptor (e.g., dietary, drinking water, occupational, residential, and recreational)

health-based criterion for the chemical mixture or multiple routes has been exceeded, and the need for corrective measures must be given serious consideration.

9.5.3.1 Chronic versus Subchronic Noncarcinogenic Effects

Human receptor exposures to chemicals can occur over long-term periods (i.e., chronic exposures), or over short-term periods (i.e., subchronic exposures). Chronic exposures for humans usually range in duration from about 7 years to a lifetime; sub-chronic human exposures typically range in duration from about 2 weeks to 7 years (USEPA, 1989a)—albeit shorter-term exposures of less than 2 weeks could also be anticipated. Accordingly, appropriate chronic and subchronic toxicity parameters and intakes should generally be used in the estimation of noncarcinogenic effects associated with the different exposure duration—as reflected in the relationships shown below.

The chronic non-cancer HI is represented by the following modification to the general equation presented earlier on in Box 11.3:

$$\text{Total Chronic Hazard Index} = \sum_{j=1}^{p} \sum_{i=1}^{n} \frac{\text{CDI}_{ij}}{\text{RFD}_{ij}} \tag{9.37}$$

where CDI_{ij} is chronic daily intake for the ith constituent and jth exposure route, and RfD_{ij} is chronic reference dose for ith constituent and jth exposure route.

The subchronic non-cancer HI is represented by the following modification to the general equation presented earlier on in Box 11.3:

$$\text{Total Subchronic Hazard Index} = \sum_{j=1}^{p} \sum_{i=1}^{n} \frac{\text{SDI}_{ij}}{\text{RfD}_{s_{ij}}} \tag{9.38}$$

where SDI_{ij} is subchronic daily intake for the ith constituent and jth exposure route, and $\text{RfD}_{s_{ij}}$ is subchronic reference dose for ith constituent and jth exposure route.

9.5.3.2 Interpreting the Non-Cancer Risk Metric

The "HQ" (viz., the ratio of the environmental exposure to the RfD or RfC) and the "HI" (viz., the sum of HQs of chemicals to which a person is exposed—and that affect the same target organ, or operate by the same mechanism of action) are generally used as indicators of the likelihood of harm arising from the noncarcinogenic effects of chemicals encountered in human environments (USEPA, 2000a–e). In such usage, an HI less than unity (1) is commonly understood as being indicative of a lack of appreciable risk, whereas a value over unity (1) would indicate a likely increased risk; thus, the larger the HI, the greater the risk—albeit the index is not related to the likelihood of adverse effect except in qualitative terms. In fact, the HI cannot be translated into a probability realm that would necessarily suggest that adverse effects will occur—and also, is not likely to be proportional to risk *per se* (USEPA, 2006a; NRC, 2009). As such, this RfD-based risk characterization does not quite provide information on the fraction of a population adversely affected by a given dose, or on any other direct measure of risk for that matter (USEPA, 2000a; NRC, 2009).

Meanwhile, it is worth the mention here that, in more recent times, some investigators have been advocating for the development and use of a "hazard range" concept (rather that the "simplistic" point value) to facilitate better and more informed decision-making about exposures and likely effects to humans of the noncancer attributes of chemicals; this would somehow parallel the practices that already exist for the cancer effects from chemicals (viz., the 10^{-6} to 10^{-4} risk range concept for carcinogenicity). In fact, although the RfD and RfC have generally been defined in terms of metrics that carry with them uncertainties that perhaps span an order of magnitude, risk managers have generally not implemented their decisions by necessarily accounting for this implicit uncertainty; consequently, non-cancer hazards have frequently been evaluated and/or regulated in such a manner that the HQ or index of one (1) is more or less interpreted as a "bright line" for risk management decision-making.

9.5.4 Risk Characterization in Practice: The Typical Nature of Risk Presentations

A primary aim of risk assessment should be to inform decision-makers about the public health implications of various strategies for reducing receptor/populations exposures to the totality of environmental stressors. And yet, oftentimes, risk assessment applications seem centered on simply evaluating risks associated with individual chemicals in the context of regulatory requirements or isolated actions. Indeed, after going through all the requisite computational exercises, the risk values are often stated simply as numerals—such as is expressed in the following statements:

- Risk probability of occurrence of additional cases of cancer—e.g., a cancer risk of 1×10^{-6}, which reflects the estimated number of excess cancer cases in a population.
- Hazard index of non-cancer health effects such as neurotoxicity or birth defects—e.g., a HI of 1, reflecting the degree of harm from a given level of exposure.

One of the most important points to remember in all cases of risk presentation, however, is that the numbers by themselves may not tell the whole story (Asante-Duah, 2017). For instance, a human cancer risk of 10^{-6} for an "average exposed person" (e.g., someone exposed via food products only) may not necessarily be interpreted to be the same as a cancer risk of 10^{-6} for a "maximally exposed individual" (e.g., someone exposed from living in a highly contaminated area)—i.e., despite the fact that the numerical risk values may be identical. In fact, omission of the qualifier terms—e.g., "average" or "maximally/most exposed"—could mean an incomplete description of the true risk scenarios, and this could result in poor risk management strategies and/or a failure in risk communication tasks. Thus, it is very important to know, and to recognize such seemingly subtle differences in the risk summarization—or indeed throughout the risk characterization process.

To ultimately ensure an effective risk presentation, it must be recognized that the qualitative aspect of a risk characterization (which may also include an explicit recognition of all assumptions, uncertainties, etc.) may be as important as its quantitative component (i.e., the estimated risk numbers). The qualitative considerations are indeed essential to making judgments about the reliability of the calculated risk numbers, and therefore the confidence associated with the characterization of the potential risks.

Finally, it is worth mentioning here that, although it is generally preferable to have quantitative information as the primary health risk characterization/assessment outputs, it will often be useful enough to provide qualitative information about potential health effects when risks cannot be fully quantified. Furthermore, it should prove quite useful to incorporate appropriate terminologies that distinguish the full discussion of possible health effects from the myriad other effects that may be considered in a cumulative impact assessment (NRC, 2009); indeed, any such undertakings should be such that, at the end of the day, it would be seen as serving a reasonably important role with regards to the decision on hand.

9.5.5 Uncertainty and Variability Issues and Concerns in Risk Assessment

Uncertainty and variability are almost an omnipresent aspect of risk assessments—and tackling these in a reasonably comprehensive manner is crucial to the overall risk assessment process. Indeed, risk assessments tend to be highly uncertain, as well as highly variable; due to the oftentimes limited availability of data for most scientific endeavors, uncertainty in particular tends to be rather pervasive in so many studies. Broadly stated, *uncertainty* stems from lack of knowledge—and thus can be characterized and managed but not necessarily eliminated, whereas *variability* is an inherent characteristic of a population—inasmuch as people vary substantially in their exposures and their susceptibility to potentially harmful effects of exposures to the stressors of concern/interest (NRC, 2009). In general, uncertainty can be reduced by the use of more or better data; on the other hand, variability cannot be reduced, but it can be better characterized with improved information. In any event, when all is said and done, uncertainty (alongside variability) analyses become key factors in the ultimate decision-making process that is typically developed to address chemical exposure problems. By way of probabilistic modeling and analyses, uncertainties associated with the risk evaluation process can be assessed properly and their effects on a given decision accounted for systematically. In this manner the risks associated with given decisions may be aptly delineated, and then appropriate corrective measures taken accordingly. An elaborate discussion of the key issues and evaluation modalities regarding uncertainty and variability matters that surround the overall risk assessment process can be found elsewhere in the literature of risk assessment (see, e.g., Asante-Duah, 2017).

On the whole, uncertainties are difficult to quantify, or at best, the quantification of uncertainty is itself uncertain. Thus, the risk levels generated in a risk assessment are useful only as a yardstick, and as a decision-making tool for the prioritization of problem situations—rather than to be construed as actual expected rates of disease, or adversarial impacts in exposed populations. For such reasons, it is used only as an estimate of risks, mostly based on current level of knowledge coupled with several assumptions. Quantitative descriptions of uncertainty, which could take into account random and systematic sources of uncertainty in potency, exposure, intakes, etc. would usually help present the spectrum of possible true values of risk estimates, together with the probability (or likelihood) associated with each point in the spectrum.

Finally, it is noteworthy that uncertainty is invariably embedded in most risk evaluation processes. Indeed, many areas of science or scientific works involve uncertainty—and broadly speaking, uncertainty can become an obstacle to effective decision-making, that is, unless effectually addressed. Anyhow, by acknowledging (and hopefully characterizing or addressing) uncertainty issues associated with a given project or undertaking, there just might be the chance of making a decision that would likely yield the greatest benefits for public health. Broadly stated in rather simplistic terms, the characterization of uncertainty during risk assessments generally implies that "lower bounds," "central estimates," and "upper bounds" of risk can all be appropriately defined or identified and properly utilized in the risk-based decision-making processes—i.e., rather than a blind focus simply on so-called conservative or "health protective" estimates of risk on only one end of the "risk spectrum." After all, uncertainty has to be seen more so as the characterization of our "state of knowledge" of the problem on hand—and not as a barrier to effective decisions and actions. At any rate, for all practical purposes, uncertainties are generally propagated through the analysis under consideration. To the extent possible, a "sensitivity analysis" provides insight into the possible range of results. *Sensitivity analysis* entails the determination of how rapidly the output of an analysis changes with respect to variations in the input. Meanwhile, it is also notable that sensitivity studies do not usually incorporate the error range or uncertainty of the input parameters—thus serving as a distinguishing element from uncertainty analyses.

9.5.5.1   The Uncertainty Analysis in Practice

Within any of the major steps of the health risk assessment process, assumptions must be made due to a lack of absolute scientific knowledge. Some of the assumptions may be supported by reasonable amounts of scientific evidence, whiles others may not necessarily be supported to same level of confidence; regardless, every assumption likely introduces some degree of uncertainty into the risk assessment process. Traditionally, and especially in the regulatory realm of things, the risk assessment methodology tends to require that conservative assumptions be made throughout the risk assessment—at least to ensure that risks are not underestimated; on the other hand, when all of the conservative assumptions and approaches are combined, it is more likely that risk results/outcomes would be overestimated, rather than underestimated. Anyhow, insofar as possible, the assumptions that introduce the greatest amount of uncertainty in the risk assessment would tend be quantified and/or comprehensively discussed as part of the overall risk determination process; meanwhile, the assumptions for which there may not be enough information available to assign a numerical value to the uncertainty *per se* (and thus cannot be factored into the risk quantification/calculations) are typically discussed in qualitative terms. Ultimately, these uncertainties may also be properly incorporated into an overall risk management plan for pragmatic action. Practical example discussions relating to the concepts offered in this section and elsewhere in the book can be found in the broader scientific literature (e.g., Asante-Duah, 2017; Fagerlin et al., 2007; IOM, 2013; Lipkus, 2007; Morgan, 2009; Nelson et al., 2009; Spiegelhalter et al., 2011; Visschers et al., 2009).

9.5.5.2 The Role of Sensitivity Analyses in Presenting and Managing Uncertain Risks

Inevitably, some degree of uncertainties remains in quantitative risk estimates in virtually all fields of applied risk analysis. A carefully executed analysis of uncertainties therefore plays a very important role in all risk assessments. On the other hand, either or both of a comprehensive qualitative analysis and a rigorous quantitative analysis of uncertainties will be of little value if the results of such analysis are not clearly presented for effective use in the decision-making process. To facilitate the design of an effectual process, a number of methods of approach have been suggested by some investigators (e.g., Cox and Ricci—see, Paustenbach, 1988) for presenting risk analysis results to decision-makers, including the following:

- Risk assessment results should be presented in a sufficiently disaggregated form (to show risks for different subgroups) so that key uncertainties and heterogeneities are not lost in the aggregation.
- Confidence bands around the predictions of statistical models can be useful, but uncertainties about the assumptions of the model itself should also be presented.
- Both individual (e.g., the typical and most threatened individuals in a population) and population/group risks should be presented, so that the equity of the distribution of individual risks in the population can be appreciated and taken into account.
- Any uncertainties, heterogeneities, or correlations across individual risks should be identified.
- Population risks can be described at the "micro" level (namely, in terms of frequency distribution of individual risks), and/or at the "macro" level (namely, by using decision-analytic models, in terms of attributes such as equivalent number of life-years).

On the whole, uncertainty is typically expressed in terms of the probability or likelihood of an event, and can indeed be presented numerically, verbally, and/or graphically—with each approach having its unique advantages and disadvantages (IOM, 2013). It is noteworthy that, the uncertainty analysis can also be achieved via sensitivity analyses for key assumptions.

Sensitivity analysis is generally defined as the assessment of the impact of changes in input values on model outputs. Often a useful adjunct to the traditional uncertainty analysis, sensitivity analysis is comprised of a process that examines the relative change or response of output variables caused by variation of the input variables and parameters (Calabrese and Kostecki, 1991; Iman and Helton, 1988; USEPA, 1992a, 1997a). It is indeed a technique that tests the sensitivity of an output variable to the possible variation in the input variables of a given model. Accordingly, the process serves to identify the sensitivity of the calculated result vis-à-vis the various input assumptions—and thus identify key uncertainties, as well as help bracket potential risks so that policy-makers can make more informed decisions or choices.

Typically, the performance of sensitivity testing requires data on the range of values for each relevant model parameter. The intent of sensitivity analysis is then to identify the influential input variables, and to develop bounds on the model output. When computing the sensitivity with respect to a given input variable, all other input variables are generally held fixed at their "nominal" values. By identifying the influential or critical input variables, more resources can then be directed to reduce their uncertainties—and thence reduce the output uncertainty. Thus, as an example, the main purpose of sensitivity analyses in an exposure characterization would be to determine which variables in the applicable model equations, as well as the specific pathways or scenarios, would likely affect the consequential exposure estimates the most. These techniques can also be used to assess key sources of variability and uncertainty for the purpose of prioritizing additional data collection and/or research efforts.

In the end, notwithstanding the added value of sensitivity analyses, several factors may still contribute to the over- or under-estimation of risks. For example, in human health risk assessments, some factors will invariably underestimate health impacts associated with the chemicals evaluated in the assessment. These may include: lack of potency data for some carcinogenic chemicals; risk contributions from compounds produced as transformation byproducts, but that are not quantified; and the fact that all risks are assumed to be additive, although certain combinations of exposure may potentially have synergistic effects. Conversely, another set of factors would invariably cause the process to overestimate risks. These may include the fact that: many unit risk and potency factors are often considered plausible upper-bound estimates of carcinogenic potency, when indeed the true potency of the chemical could be considerably lower; exposure estimates are often very conservative; and possible antagonistic effects, for chemicals whose combined presence reduce toxic impacts, are not accounted for properly.

9.6 A CALL FOR ACTION ON THE OVERARCHING IMPACTS OF HUMAN EXPOSURE TO CHEMICALS ORIGINATING FROM CONTAMINATED SITES

Human exposure to chemicals may occur via different human contact sites and target organs, and also under a variety of exposure scenarios. The contact sites represent the physical areas of chemical contacting, and the target organ are the internal body organs that tend to transport, process, and/ or store the absorbed chemicals; an exposure scenario is a description of the activity that brings a human receptor into contact with a chemical material, product, or medium. Human populations may indeed become exposed to a variety of chemicals via several different exposure routes—represented primarily by the inhalation, ingestion, and dermal exposure routes. Congruently, human chemical uptake occurs mainly through the skin (from dermal contacts), via inhalation (from vapors/gases and particulate matter), and/or by ingestion (through oral consumptions). Under such circumstances, a wide variety of *potential* exposure patterns can be anticipated from any form of human exposures to chemicals (Asante-Duah, 2017).

To evaluate potential receptor impacts upon chemical contacting, chemical exposure investigations—typically consisting of the planned and managed sequence of activities carried out to determine the nature and distribution of hazards associated with potential chemical exposure problems—can be systematically designed and effectively used to address human exposure and response to the chemical toxicants so-encountered. Several characteristics of the chemicals of concern as well as the human contact sites will provide an indication of the critical features of exposure; these will also provide information necessary to determine the chemical's distribution, uptake, residence time, magnification, and breakdown to new chemical compounds. In particular, the physical and chemical characteristics of the chemicals as well as the target organs involved can significantly affect the intake, distribution, half-life, metabolism, and excretion of such chemicals by potential receptors. The major human contact sites, target organs, and exposure scenarios that can be expected to become key players in the assessment of human exposure to and response from chemical hazards are elaborated elsewhere in the scientific literature (e.g., Al-Saleh and Coate, 1995; Asante-Duah, 2017; Berlow et al., 1982; Berne and Levy, 1993; Brooks et al., 1995; Brum et al., 1994; Corn, 1993; Davey and Halliday, 1994; Dienhart, 1973; Frohse et al., 1961; Guyton, 1968, 1971, 1982, 1986; Homburger et al., 1983; Hughes, 1996; OECD, 1993; Roberts, 2014; Scanlon and Sanders, 1995; Willis, 1996; USEPA, 1992a).

Overall, chemical exposure problems may pose significant risks to the general public because of the potential health effects. Risks to human health as a result of exposure to toxic materials present or introduced into our living and work environments are, therefore, a matter of grave concern to modern societies; this calls for a use of systematic and well thought-out processes to address any rippling effects from contaminated sites and similar sources. The use of proper risk analysis and decision tools can, at the very least, help verify the risks to which a population or organism is exposed in an effectual manner, as well as enable the development of appropriate techniques to properly manage such risks—in order to ensure an acceptable quality to the lives of the populations potentially at risk.

10 Technical Approach to Ecological Endangerment Assessment

An ecological risk assessment (ERA) is comprised of a process that evaluates the likelihood that adverse ecological effects may occur, or are occurring as a result of exposure to one or more environmental stressors—with such stressors usually having been imposed by way of human activities (Bartell et al., 1992; Linthurst et al., 1995; NRC, 1983; Richardson, 1995; Solomon, 1996; USEPA, 1986a, 1988h, 1992d, 1994d, 1995b). The process evaluates the potential adverse effects of human activities on the plants and animals that make up ecosystems. Indeed, a good ERA helps determine if living organisms and/or their environment have been adversely affected, or may be affected in the future due to existing conditions.

A typical ERA involves the qualitative and/or quantitative appraisal of the actual or potential impacts of stressors (e.g., contaminants associated with contaminated site problems) on plants and animals at a given site (USEPA, 1989c, 1990i). It is primarily concerned with the adverse effects of risk agents on populations of particular animal, plant, or microbial species, as well as on the structure and function of ecosystems (Cohrssen and Covello, 1989). On the whole, ERA techniques can be employed to better develop responsible contaminated site risk management and corrective action response programs. This chapter elaborates the major components of the ERA methodology—as may be applied to the evaluation of a contaminated site problem. Specifically, it provides a procedural framework and an outline of the key elements of the ERA process.

10.1 ECOLOGICAL IMPACTS ASSESSMENT FOR ENVIRONMENTAL CONTAMINATION PROBLEMS

Knowledge of environmental effects is important in analyzing both potential nonhuman and human health risks associated with chemical releases into the environment. An ERA is a process for organizing and analyzing data, information, assumptions, and uncertainties in order to evaluate the likelihood of adverse ecological effects from a contaminated site problem. ERAs typically evaluate the ecological effects arising from human-related activities that have resulted in the release of chemicals into the environment, or in habitat destruction or modifications (including wetland destruction), or indeed several other similar outcomes. In general, a contaminant entering the environment may cause adverse ecological effects only if: (i) the contaminant exists in a form and concentration sufficient to cause harm; (ii) the contaminant comes in contact with organisms or environmental media with which it can interact; and (iii) the interaction that takes place is detrimental to life functions. Typically, an ecological assessment activity is used to determine the nature, magnitude, and transience or permanence of observed or expected effects of contaminants introduced into an ecosystem. This type of assessment is usually directed at investigating the loss of habitat, reduction in population size, changes in community structure, and changes in ecosystem structure and function.

10.1.1   THE GENERAL PURPOSE AND SCOPE OF AN *ERA*

The objectives of an ERA in relation to contaminated site problems classically would consist of identifying and estimating the potential ecological impacts associated with chemicals emanating from the subject site—with specific focus being to determine the following attributes in particular:

- Biological and ecological characteristics of the study area
- Types, forms, amounts, distribution, and concentration of the contaminants of concern
- Migration pathways to, and exposure of ecological receptors to pollutants
- Habitats potentially affected and populations potentially exposed to contaminants
- Actual and/or potential ecological effects/impacts, and overall nature of risks

In practice, ERAs have generally involved the application of the science of ecotoxicology to public policy (Suter, 1993). Ideally, the ERA would estimate the potential for occurrence of adverse effects that are manifested as changes in the diversity, health, and behavior of the constellation of organisms that share a given environment over time. Ecological areas included in an ERA should therefore not be limited by property boundaries of a study region or space, if affected environments or habitats are located beyond the property boundaries.

ERAs can be divided into two general types based on habitats—namely, terrestrial (land) and aquatic (freshwater and marine), with avifauna belonging into either or both groups. Typical ecological effects include changes in the aquatic and terrestrial natural resources brought about by exposure to environmental contaminants. Although the assessment focuses on the impacts of contaminants on the terrestrial and aquatic flora and fauna that inhabit an impacted locale and vicinity, ecological assessments may also identify new or unexpected exposure pathways that could potentially affect human populations through the food chain or through changes in the ecosystem.

Classically, ERAs evaluate the likelihood that adverse ecological effects are occurring or may occur as a result of exposure to physical (e.g., site cleanup activities) or chemical (e.g., release of hazardous substances) stressors at a site. These assessments often contain detailed information regarding the interaction of these "stressors" with the biological community at the site. Part of the assessment process includes creating exposure profiles which: (i) describe the sources and distribution of harmful entities; (ii) identify sensitive organisms or populations; (iii) characterize potential exposure pathways; and (iv) estimate the intensity and extent of exposures at a site. Overall, exposure profiles are developed to identify ecological receptors (tissues, organisms, populations, communities, and ecosystems), habitats, and pathways of exposure. The sources and distribution of stressors in the environment are also characterized. Other information contained in ERAs may include evaluations of individual species, populations of species, general trophic levels, communities, habitat types, ecosystems, or landscapes.

In summary, an ERA evaluates the potential adverse effects that human activities have on the living organisms that make up ecosystems. The risk assessment process provides a way to develop, organize, and present scientific information so that it is relevant to environmental decisions. When conducted for a particular place such as a watershed, the ERA process can be used to identify vulnerable and valued resources, prioritize data collection activity, and link human activities to their potential effects. Ultimately, the findings of an ERA provide a basis for comparing different management options—thus enabling decision-makers, stakeholders, and the public to make better informed decisions about the management of ecological resources.

10.1.2   GENERAL CONSIDERATIONS IN ECOLOGICAL RISK INVESTIGATIONS

ERA allows for the identification of habitats and organisms that may be affected by the chemicals of potential concern associated with a contaminated site problem. Because ecosystems are complex and involve both structural and functional components, assessment of ecological risk involves consideration of how stressors affect the organisms that inhabit a site and the chemical, physiological, and behavioral interactions among organisms and their surrounding environment. In particular, the following elements are typically given in-depth consideration during an ecological investigation associated with contaminated site problems:

- Definition and role of the ERA within the context of the contaminated site problem assessment and restoration
- Establishment of a concept of acceptable ecological risk
- Evaluation and selection of appropriate ecological endpoints at the population, community, and ecosystem levels
- Evaluation of exposure and biomarkers of exposure
- Validation of strategy adopted for the ERA (including the basis for its acceptability and appropriateness for the case-specific problem)
- Design of field sampling programs, data analysis and evaluation plans, and ecological monitoring programs
- Determination of acute and chronic risks and secondary hazards
- Application of ERA results to environmental restoration plans (e.g., for derivation of site-specific remediation objectives)

Meanwhile, it is noteworthy that the environmental transport media of greatest interest in ERAs usually relate to surface water and soil—especially because these are the media that are most frequently contacted by the organisms of interest. Whereas surface water is of primary interest to aquatic ecosystems, terrestrial ecosystems involve both soil and surface water; the reason that both media are of concern in terrestrial assessments is because many terrestrial receptors contact surface water bodies for such reasons as drinking, development through some of the life stages (e.g., tadpole stage of frogs and toads, and larval stage of dragonflies), and living in or near the water (e.g., beaver, muskrat, and some snakes). Consequently, areas of contaminated soil and territories near contaminated surface water bodies as well as near contaminated soils will normally require carefully crafted ERAs.

Finally, it is notable that ecological risk analysis can be performed at several hierarchical levels, and risks can be expressed as a qualitative or quantitative estimate, depending on the available data (Zehnder, 1995).

10.1.2.1 The Role of ERA in Corrective Action Decisions at Contaminated Sites

Oftentimes, and more so in the past, only limited attention has been given to the ecosystems associated with contaminated sites, and also to the protection of ecological resources during site remediation activities. Instead, much of the focus has been on the protection of human health and resources directly affecting public health and safety. In contemporary times, however, the ecological assessment of contaminated sites is gaining considerable attention. This is the result of prevailing knowledge or awareness of the intricate interactions between ecological receptors/systems and contaminated site cleanup processes.

Although human health is frequently the major concern in corrective action assessments, an ecological assessment may serve to expand the scope of the investigation for a potentially contaminated site problem—usually by enlarging the area under consideration, or re-defining remediation criteria, or both. In fact, a detailed assessment may be required to determine whether or not the potential ecological effects of the contaminants at a site warrant remedial action. Thus, ecological data gathered before and during remedial investigations are used as follows: (i) to help determine the appropriate level of detail for an ecological assessment; (ii) to decide if remedial action is necessary based on ecological considerations; (iii) to evaluate the potential ecological effects of relevant remedial options; (iv) to provide information necessary for mitigation of site threats; and (v) to design monitoring strategies used to assess the progress and effectiveness of remediation. Meanwhile, it has to be recognized that in a number of situations, site remediation can destroy or otherwise affect uncontaminated ecological resources. For instance, soil removal techniques, alteration of site hydrology, and site preparation are examples of remediation activities that can result in inadvertent damage to ecological resources. Thus, the ecological impacts of a site restoration activity must be understood by decision-makers before remediation plans are approved and implemented. In situations where adverse ecological effects are identified, corrective action alternatives with potentially less damaging impacts must be evaluated as preferred methods of choice—and therefore means that, the assessment of potential ecological impacts should be performed for all feasible remedial alternatives.

In general, there usually are several important ecological concerns associated with contaminated site cleanup programs that should be addressed early enough in site characterization programs. Furthermore, there often are a number of legislative requirements for incorporating ecological issues as part of site investigation and cleanup efforts for contaminated sites. To achieve adequate ecological protection and/or regulatory compliance, ecological assessments should address the overall site contamination issues; this should also be coordinated with all aspects of site cleanup— including human health concerns, engineering feasibility, and economic considerations (de Serres and Bloom, 1996; Maughan, 1993; Suter et al., 1995). Indeed, absent of a human health risk, it should be determined as early as possible in the ERA whether an active remedial action would likely be more harmful to the environment than a passive remedy. In any case, it is apparent that ecological resources must be appropriately evaluated in order to achieve the mandate of any comprehensive program designed to ensure the effective management of contaminated site problems.

10.1.2.2 Ecological Risk Management Principles

An ultimate goal of most contaminated site risk management programs is to select remedies that are protective of human health and the environment, both in the short term and long term. Since ecological receptors at sites exist within a larger ecosystem context, remedies selected for protection of these receptors should also assure protection of the ecosystem components upon which they depend or which they support. Except at a few very large sites, most ERAs typically do not address effects on entire ecosystems; instead, the efforts usually consists of gathering effects data on individuals in order to predict or postulate potential effects on local wildlife, fish, invertebrate, and plant populations and communities that occur or that could occur in specific habitats at sites (e.g., wetland, floodplain, stream, estuary, and grassland). ERAs incorporate a wide range of tests and studies to either directly estimate community effects (e.g., benthic species diversity) or indirectly predict local population-level effects (e.g., toxicity tests on individual species), both of which can contribute to estimating ecological risk. Indeed, contaminated site remedial actions generally should not be designed to protect organisms on an individual basis (the exception being designated protected status resources, such as listed or candidate threatened and endangered species or treaty-protected species that could be exposed to site releases), but to protect local populations and communities of biota. Levels that are expected to protect local populations and communities can be estimated by extrapolating from effects on individuals and groups of individuals using a "lines-of-evidence" approach. Indeed, the performance of multiyear field studies at contaminated sites in attempts to quantify or predict long-term changes in local populations is not necessarily required for proper risk management decisions to be made. On the other hand, data from discrete field and laboratory studies, if properly planned and appropriately interpreted, can be used to estimate local population or community-level effects.

The likelihood of the response alternatives to achieve success and the time frame for a biological community to fully recover should be considered in remedy selection. Although most receptors and habitats can recover from physical disturbances, risk managers should carefully weigh both the short- and long-term ecological effects of active remediation alternatives and passive alternatives when selecting a final response. This does not imply that there should necessarily be a preference for passive remediation; all reasonable alternatives should be considered. For example, the resilience and high productivity of many aquatic communities allows for aggressive remediation, whereas the removal of bottomland hardwood forest communities in an area in which they cannot be restored due to water management considerations may argue heavily against extensive action in all but the most highly contaminated areas.

10.2   THE ERA PROCESS AND METHODOLOGY

Figure 10.1 illustrates the fundamental elements of an ERA process—in similar terms to the human health risk assessment presented earlier on in Chapter 9; the key fundamental elements (Box 10.1) are discussed in subsequent sections below. Indeed, the procedural components of an ERA program will generally consist of several tasks (Box 10.2)—usually conducted in phases to ensure program cost-effectiveness. Where it is deemed necessary, a more detailed assessment that is comprised of biological diversity analysis and population studies may become part of the overall ERA process.

It is noteworthy that, in general, the uniqueness of ecological systems makes it difficult to comprehensively prescribe and standardize procedures for ERAs (Kolluru et al., 1996). This uniqueness does indeed argue in support of flexibility in implementing any standard sets of protocols for ecological assessment programs. Although several general considerations familiar to many risk assessment procedures may apply similarly to ERAs, it must be recognized that, in assessing ecosystem risks, the process typically becomes much more complex. For the most part, this added

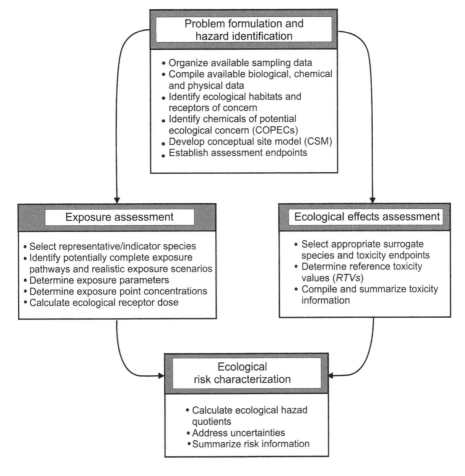

FIGURE 10.1   The ecological risk assessment process.

complexity may be attributed to the fact that toxicity effects assessment (which is a very important component of the overall risk assessment process) usually must be determined for several different families of organisms; in addition, the risk assessment must also consider the role of the various organisms in the ecosystem structure and function.

In general, ERAs will normally be conducted in phases. The phased approach facilitates the utilization of cost-effective programs—by eliminating unnecessary and expensive sampling and analysis efforts. For instance, a "Phase I" level of effort for an ERA will generally consist of the elements shown in Box 10.2. Where deemed necessary, a "Phase II" assessment that comprises of a biological diversity analysis, and then also a "Phase III" assessment consisting of population studies may become part of the overall ERA process.

BOX 10.1   KEY FUNDAMENTAL ELEMENTS OF AN ERA

- Evaluation of Site Characteristics—to comprise of:
  - Nature and extent of contaminated area
  - Sensitive environments
- Contaminant Evaluation—to comprise of:
  - Identification and characterization of contaminants of interest
  - Biological and environmental concentrations
  - Toxicity of contaminants
  - Potential benchmark/reference criteria
- Potential for Exposure—to comprise of:
  - Actual or potential sources of contaminant release
  - Media into which contaminants can be, or are being released
  - Organisms that can come into contact with the contaminants
  - Environmental conditions under which transport and/or exposure may be taking place
- Selection of Assessment and Measurement Endpoints—to comprise of:
  - Ecological endpoints
  - Evaluation of potentially affected habitats
  - Evaluation of potentially affected populations
- Ecological Risk Characterization—to comprise of:
  - Qualitative risk characterization
  - Quantitative risk characterization

BOX 10.2   GENERAL BASIC TASKS IN AN ERA PROGRAM

- Compilation of relevant site data for the study area
- Determination of background (ambient) concentrations of the site contaminants in the study area
- Identification of the contaminants of potential ecological concern
- Identification and location of habitats and environments at the study area and its vicinity
- Selection of indicator species or habitats
- Ecotoxicity assessment and/or bioassay of selected indicator species
- Development of an appropriate conceptual site model (CSM) and/or food chain diagram

(Continued)

BOX 10.2 (*Continued*) GENERAL BASIC TASKS IN AN ERA PROGRAM

- Establishment of appropriate assessment endpoints for all chemicals of potential ecological concern
- Characterization of exposure based on environmental fate-and-transport of the chemicals of potential ecological concern, as predicted using the CSM and ecological food chain considerations
- Development of ecological risk characterization parameters for the indicator species/habitats

---

Broadly speaking, the process used to evaluate environmental or ecological risk parallels that used in the evaluation of human health risks (as discussed in the preceding chapter). In both cases, potential risks are determined by the integration of information on chemical exposures with toxicological data for the contaminants of potential concern. Unlike endangerment assessment for human populations, however, ERAs often lack significant amount of critical and credible data necessary for a comprehensive quantitative evaluation. Nonetheless, the pertinent data requirements should be identified and categorized insofar as practicable, so that reliable risk estimates can be generated.

10.2.1 Ecological Hazard Evaluation

The design of an ERA program for a contaminated site problem typically involves a process to clearly define the common elements of populations, communities, and ecosystems—which then forms a basis for the development of a logical framework that can be used to characterize risks at the project site. Invariably, the development of an ERA requires the identification of one or several ecological assessment endpoints; these endpoints define the environmental resources which are to be protected, and which, if found to be impacted, determine the need for corrective actions. The selection of appropriate site-specific assessment endpoints is therefore crucial to the development of a cost-effective ecological characterization and/or contaminated site management program that is protective of potential ecological receptors.

As noted previously, a contaminant entering the environment will cause adverse effects if, and only if it exists in a form and concentration sufficient to cause harm; it comes in contact with organisms or environmental media with which it can interact; and the ensuing interaction is detrimental to life functions. The hazard evaluation process sets out to make such qualitative determinations; the general elements of the hazard assessment process needed for an ERA are discussed below.

10.2.1.1 Identification of the Nature of Ecosystems

The different types of ecosystems have unique combinations of physical, chemical, and biological characteristics—and thus may respond to contamination in their own unique ways. The physical and chemical structure of an ecosystem may determine how contaminants affect its resident species, and the biological interactions may determine where and how the contaminants move in the environment and which species are exposed to particular concentrations. It is therefore imperative to clearly identify the types of ecosystem(s) associated with the particular contaminated site problem.

Ecosystems may be classified into two broad categories—namely: terrestrial and aquatic ecosystems, with wetlands serving as a zone of transition between terrestrial and aquatic environments. The following specific ecosystems will normally be investigated in an ERA:

- Terrestrial ecosystems (to be categorized according to the vegetation types that dominate the plant community and terrestrial animals)
- Wetlands (which are areas in which topography and hydrology create a zone of transition between terrestrial and aquatic environments)

- Freshwater ecosystems (in which environment, the dynamics of water temperature, and movement of water can significantly affect the availability and toxicity of contaminants)
- Marine ecosystems (which are of primary importance because of their vast size and critical ecological functions)
- Estuaries (which support a multitude of diverse communities, are more productive than their marine or freshwater sources, and are important breeding grounds for numerous fish, shellfish, and bird species)

Indeed, the types of ecosystems vary with climatic, topographical, geological, chemical, and biotic factors. In any event, the ecosystem types pertaining to a case-specific study should be adequately defined and holistically integrated into the overall ERA.

10.2.1.2 Evaluation of Ecological Habitats and Community Structure

Different evaluation strategies are generally employed in ERAs, depending on the level of refinement required to define the conditions within an ecological community. For instance, an evaluation of the condition of aquatic communities may proceed from two directions—as discussed below.

- *Examination of lower trophic levels.* The first direction could consist of examining the structure of the lower trophic levels as an indication of the overall health of the aquatic ecosystem. This approach emphasizes the base of the aquatic food chain, and may involve studies of plankton (microscopic flora and fauna), periphyton (including bacteria, yeast, molds, algae, and protozoa), macrophyton (aquatic plants), and benthic macroinvertebrates (e.g., insects, annelid worms, mollusks, flatworms, roundworms, and crustaceans). Benthic macroinvertebrates are commonly used in studies of aquatic communities. These organisms usually occupy a position near the base of the food chain. Just as importantly, however, their range within the aquatic environment is restricted, so that their community structure may be referenced to a particular stream reach or portion of a lake substrate. By contrast, fish are generally mobile within the aquatic environment, and evidence of stress or contaminant load may not be amenable to interpretation with reference to specific releases. The presence or absence of particular benthic macroinvertebrate species, sometimes referred to as "indicator species," may provide evidence of a response to environmental stress. A "species diversity index" provides a quantitative measure of the degree of stress within the aquatic community; this is an example of the common basis for interpreting the results of studies pertaining to aquatic biological communities. Measures of species diversity are most useful for comparison of streams with similar hydrologic characteristics, or for the analysis of trends over time within a single stream (USEPA, 1989c).
- *"Focused" examination of select species.* The second approach to evaluating the condition of an aquatic community could focus on a particular group or species, possibly because of its commercial or recreational importance or because a substantial historic database already exists. This is done through selective sampling of specific organisms (most commonly fish), and an evaluation of standard "condition factors" (e.g., length, weight, and girth). In many cases, receiving water bodies are recreational fisheries, monitored by some government regulatory agencies or similar entity. In such cases, it is common to find some historical record of the condition of the fish population, and it may be possible to correlate contaminant release records with alterations in the status of the fish population.

In any case, a wide variety of other possible measures of community structure can be employed in ERA programs. Additional detail regarding the application of other measures of community structure can be found in the literature of ecological assessments and related subjects (e.g., Barnthouse and Suter, 1986; Carlsen, 1996; NRC, 1989; USEPA, 1973).

In general, the different levels of an ecological community are studied to determine if they exhibit any evidence of stress. If the community appears to have been disturbed, the goal will be to characterize the source(s) of the stress and, specifically, to focus on the degree to which the release of environmental contaminants has caused the disturbance or possibly exacerbated an existing problem.

10.2.1.3 Identification and Selection of Ecological Assessment Endpoints

Assessment endpoints are explicit expressions of actual environmental value that is to be protected (Suter, 1993; USEPA, 1992d). The principal criteria used in the selection of assessment endpoints include the following: (i) their ecological relevance; (ii) their susceptibility to the stressor; and (iii) whether they represent management and stakeholder goals (to include a representation of societal values). Indeed, ecologically relevant endpoints reflect important characteristics of the system, and are functionally related to other endpoints (USEPA, 1992d)—and these are endpoints that help sustain the natural structure and function of an ecosystem.

Although assessment endpoints must be defined in terms of measurable attributes, selection is not dependent on the ability to measure those attributes directly or on whether methods, models, and data are currently available. If the response of an assessment endpoint cannot be directly measured, it may be predicted from measures of responses by surrogate or similar entities (Suter, 1993; USEPA, 1992d). Measures that will be used to evaluate assessment endpoint response vis-à-vis the types of exposures that are being considered in the risk assessment are often identified during conceptual model development, and further specified in the analysis plan.

Overall, the development of an ERA requires the identification of one or several ecological assessment endpoints. These endpoints define the environmental resources which are to be protected, and which, if found to be impacted, determine the need for corrective actions. The selection of appropriate site-specific assessment endpoints is therefore crucial to the development of a cost-effective site characterization and/or site restoration program that will be protective of potential ecological receptors. Meanwhile, it is notable that ecological resources are considered susceptible when they are sensitive to a human-induced stressor to which they are exposed. Delayed effects and multiple stressor exposures add complexity to evaluations of susceptibility. Conceptual models need to reflect these factors; if a species is unlikely to be exposed to the stressor of concern, it is inappropriate as an assessment endpoint.

10.2.1.4 Selection of Ecological Indicator/Target Species

It generally is not feasible to evaluate every species that may be present at a locale that is affected by a contaminated site problem. Consequently, selected target or indicator species will normally be chosen in an ERA study. Then, by using reasonably conservative assumptions in the overall assessment, it is rationalized that adequate protection of selected indicator species will provide protection for all other significant environmental species as well. That said, it is noteworthy that, not every organism may be suitable for use as indicator species in the evaluation of contaminant impacts on ecological systems. Thus, several general considerations and specific criteria should be used to guide the selection of target species in an ERA (Box 10.3) (USEPA, 1989c, 1990i).

In general, it is important to carefully consider the effects of environmental contaminants on both endangered populations as well as on the habitats critical to their survival. Consequently, the presence of threatened or endangered species, and/or habitats critical to their survival should be well documented, and the location of such species appropriately determined. Similarly, sensitive sport or commercial species—and indeed habitats essential for their reproduction and survival— should be properly identified. Information on these types of metrics may be obtained from appropriate national, federal, provincial, state, regional, local, and/or private institutions and other organizations.

BOX 10.3   GUIDING CRITERIA FOR THE
SELECTION OF ERA TARGET SPECIES

- Species that are threatened, endangered, rare, or of special concern
- Species that are valuable for several purposes of interest to human populations (i.e., of economic and societal values)
- Species critical to the structure and function of the particular ecosystem in which they inhabit
- Species that serve as indicators of important changes in the ecosystem
- Relevance of species at the site and its vicinity

10.2.1.5   Screening for Chemicals of Potential Ecological Concern

A very important early step in the assessment of ecological risks associated with contaminated site problems is the screening of chemicals detected at the impacted locale, in order to identify the specific constituents that do indeed represent a potential risk. Part of this screening process usually will involve a comparison of measured contaminant concentrations to "benchmark" values (e.g., a national ambient water quality criteria [NAWQC] for protection of aquatic life and a sediment quality criteria [SQC]); the "benchmark" values generally represent constituent concentrations that are regarded to be "nontoxic" and/or "nonimpactful" to the receptors of interest. In fact, the most appropriate screening strategy is to use multiple benchmark values along with background threshold concentrations, knowledge of contaminant composition and its nature, and physicochemical properties to identify the chemicals of potential ecological concern (CoPECs) (Suter, 1996).

Several alternative approaches for calculating ecotoxicological screening benchmarks that will facilitate the CoPECs selection process are presented in the literature elsewhere (e.g., Ankley et al., 1996; Suter, 1996; USEPA, 1988h; van Leeuwen, 1990). Naturally, the relative utility of any given benchmark depends on the reliability of the source information on which such value is based or derived from. Indeed, the choice of method for calculating benchmarks can significantly influence their sensitivity and utility—and thus the need for careful evaluation of the alternative methods of choice in the derivation of such benchmark/reference values.

10.2.1.6   Identification of the Nature and Ecological Effects of Environmental Contaminants

The introduction of contaminants into an ecosystem can cause direct and/or indirect harm to organisms. Typical major consequences from ecosystem exposures to contaminants may include the specific effects identified below (Calmano and Forstner, 1996; NRC, 1981; Pickering and Henderson, 1966; USEPA, 1989c).

- *Reduction in population size.* Contaminants can cause reductions in populations of organisms through numerous mechanisms affecting species births, mortalities, and migratory tendencies.
- *Changes in Community Structure.* Contaminants introduced into ecosystems may create opportunities for unanticipated and unpredictable changes in community with respect to species composition and relative abundance. Because most environmental contaminants of concern exhibit toxic effects, they often reduce the number and kinds of species that can survive in the habitat. This may then result in a community dominated by large numbers of a few species that are tolerant of the contaminant, or a community in which no species predominate but most of the component populations contain fewer organisms. In fact, a contaminant need not be directly toxic to affect community structure.

- *Changes in Ecosystem Structure and Function.* As contaminants modify the species com-position and relative abundance of populations in a community, the oftentimes complex patterns of matter and energy flow within the ecosystem may also change. If certain key species are reduced or eliminated, this may interrupt the flow of energy and nutrients to other species not directly experiencing a toxic effect.

Indeed, knowledge of such environmental effects is important in analyzing potential risks from chemical releases, migration pathways, and potential receptor exposures. Based on the nature of exposures to both human and nonhuman receptors, responsible contaminated site management programs can be developed to address the culprit environmental contamination problem.

Meanwhile, it is worth the mention here that although a contaminant may cause illness and/or death to individual organisms, its effects on the structure and function of ecological assemblages or interlinkages may be measured in terms quite different from those used to describe individual effects. Consequently, an ecological hazard evaluation should include a wider spectrum of eco-logical effects on individual organisms as well as the ecological interlinkages. Furthermore, the biological, chemical, and environmental factors perceived to influence the ecological effects of con-taminants should be identified and succinctly described. On the whole, a variety of environmental variables (Box 10.4) can significantly influence the nature and extent of the effects of a contaminant on ecological receptors, and may therefore determine the degree of impacts that contaminants exert on ecological systems; environmental effects commonly include changes in aquatic and terrestrial natural resources brought about by exposure to chemical substances.

BOX 10.4   KEY FACTORS INFLUENCING THE ECOLOGICAL EFFECTS OF ENVIRONMENTAL CONTAMINANTS

- Nature of Contamination
  - Chemical category
  - Physical and chemical properties
  - Frequency of release
  - Toxicity
- Physical/Chemical Characteristics of the Environment
  - Temperature, pH, salinity, hardness, etc.
  - Soil composition, etc.
- Biological Factors
  - Susceptibility of species
  - Characteristics governing population abundance and distribution
  - Temporal variability in communities
  - Movement of chemicals in food chains

10.2.2   EXPOSURE ASSESSMENT: THE CHARACTERIZATION OF ECOLOGICAL EFFECTS

The objectives of the exposure assessment are to define contaminant behaviors; identify potential ecological receptors; determine exposure routes by which contaminants may reach ecological recep-tors; and estimate the degree of contact and/or intakes of the CoPECs by the potential receptors. Overall, the exposure characterization process describes the contact or co-occurrence of stressors with ecological receptors. The characterization is based on measures of exposure, and of ecosystem and receptor characteristics. These measures are used to analyze sources, the distribution of the stressor in the environment, and the extent and pattern of contact or co-occurrence.

In general, exposure is analyzed by describing the sources and releases, the distribution of the stressor in the environment, and the extent and pattern of contact or co-occurrence of the CoPECs with ecological receptors of interest. There are a large number of methods that can be employed in the exposure and effects assessments used to evaluate ecological risks, depending upon conditions that exist at a particular site. The selection of particular methods will depend upon the types of habitats and organisms, the size of the site, and the contaminants (or other stressors) present. A clear understanding of the relevant methods and procedures that can be employed in the conduct of ERAs would help to better evaluate work and sampling plans.

10.2.2.1   Development of Ecological CSMs

The CSM identifies the complete exposure pathways that will be evaluated in the ERA. The model identifies contaminant sources, release and transport mechanisms, exposure routes, and receptors— as well as includes the risk hypotheses to be addressed by the assessment. The CSM functions as a dynamic planning tool, helping in the identification of data gaps, site knowledge, and assessment methods.

In a typical contaminated site situation, a chemical may be released into the environment— which is then subject to physical dispersal into the air, water, soils, and/or sediments. The chemical may then be transported spatially, and into the biota—and perhaps be chemically or otherwise modified or transformed and degraded by abiotic processes (such as photolysis and hydrolysis), and/or by microorganisms present in the environment. The resulting transformation products may have different environmental behavior patterns and toxicological properties from the parent chemical. Nonetheless, it is the specific nature of the exposure scenarios that determines the potential for any adverse impacts—and CSMs are used to facilitate the development of credible exposure scenarios.

Overall, the exposure analysis process evaluates the interaction of identified environmental stressors with the ecological component. As part of this process, a food chain (also called foodweb)—that gives a simplified and generic conceptual representation of typical interlinkages resulting from the consumption, uptake, and absorption processes associated with an ecological community—is normally constructed for the target species, to facilitate the development of realistic exposure scenarios (Figure 10.2). Subsequently, the relevant exposure routes are selected

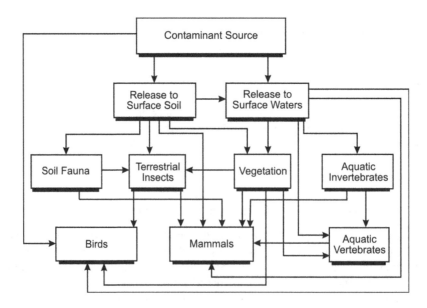

FIGURE 10.2   A simplified schematic of a food chain diagram showing select contaminant migration pathways within an ecosystem.

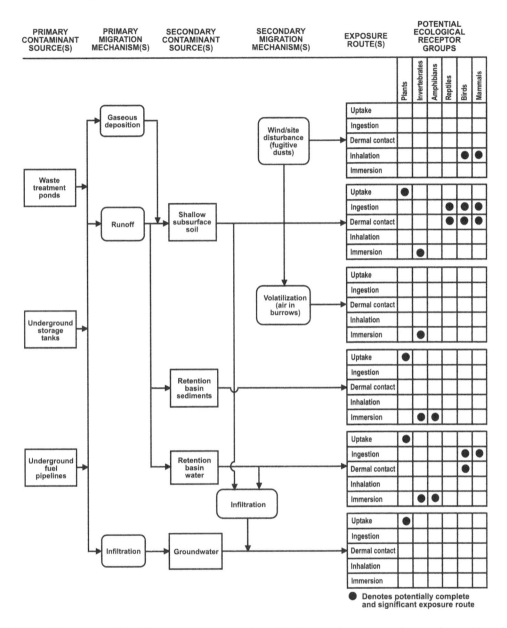

FIGURE 10.3 A simplified illustrative conceptual model diagram for a contaminated site problem, in relation to an ecological system.

based on the behavior patterns and/or ecological niches of the target species and communities (Figure 10.3). This means that the nature of the target organisms (e.g., birds and fish) must be identified together with the nature of exposure (such as acute, chronic, or intermittent), as part of the ERA process.

Finally, it is worth the mention here that CSMs for ERAs are developed based on information about stressors, potential exposures, and predicted effects on an ecological entity (the assessment endpoint). The complexity of the CSM depends on the complexity of the problem, number of stressors, number of assessment endpoints, nature of effects, and characteristics of the ecosystem. The general design process/protocol for developing CSMs that was discussed previously in Chapter 4 (Part II) will normally be employed to complete the appropriate CSM diagram.

10.2.2.2   Estimation of Chemical Intakes by Potential Ecological Receptors

The amount of a target species exposure to environmental contamination is based on the maximum plausible exposure concentrations of the chemicals in the affected environmental matrices. The total daily exposure (in mg/kg-day) of target species can be estimated by summing the amounts of constituents ingested and absorbed from all sources (e.g., soil, vegetation, surface water, fish tissue, and other target species)—and also that portion absorbed through inhalation and dermal contacts.

Analytical procedures used to estimate the receptor exposures to chemicals in various contaminated media (such as wildlife or a game species' daily chemical exposure and the resulting body burden) are similar to that discussed under human health risk assessment in Chapter 9. For example, wildlife or game daily chemical exposure may be estimated by applying the algorithm shown in Box 10.5. As an illustration of the practical application of this algorithm, it is recognized that a deer's daily exposure to a chemical (e.g., dioxin) in an ecological setup (together with the resulting body burden) will usually need to be estimated before a human consumer's (e.g., a hunter) oral exposure from eating deer meat can be estimated. In this case, the deer's average daily uptake of dioxins will be given by:

$$E = C \times F \times [G_d G_c G_u + I_d I_c I_b] \times \text{BAF} \times [1/\text{BW}] \tag{10.1}$$

where G_d is the grass component of diet; G_c is the grass consumption factor; G_u is the soil chemical uptake coefficient for grass; I_d is the rodent component of diet; I_c is the rodent consumption factor; and I_b is the rodent bioavailabilty factor. Further details on the relevant models and algorithms can be found elsewhere in the literature of endangerment assessment (e.g., CAPCOA, 1990; Paustenbach, 1988; Sutter, 1993; USEPA, 1993).

**BOX 10.5   ESTIMATION OF WILDLIFE OR GAME
EXPOSURE TO ENVIRONMENTAL CHEMICALS**

$$E = C \times F \times \left\{ \sum_{i=1}^{n} D_{d_i} D_{c_i} D_{u_i} \right\} \times \text{BAF} \times \frac{1}{\text{BW}}$$

where

E = exposure or average daily dose (μg/kg-day)
C = chemical concentration in media (e.g., soil) averaged over appropriate exposure period (70-year if to be subsequently consumed by humans, etc.)
F = food consumption rate (g/day)
D_{d_i} = component i of diet
D_{c_i} = consumption factor of i
D_{u_i} = bioaccumulation or uptake coefficient for i
BAF = bioavailability factor
BW = body weight (kg)

10.2.3   Ecotoxicity Assessment: The Characterization of Ecological Effects

The ecotoxicological assessment phase of an ERA comprises of a determination of the ecological response to potential environmental stressors. Characterization of ecological effects describes the effects that are elicited by a stressor, links these effects with the assessment endpoints, and then evaluates how the effects change with varying stressor levels. The conclusions of the ecological effects characterization are typically summarized in a stressor–response profile (similar to the dose–response curve in a human health risk assessment)—and also as noted in Chapter 9.

In general, evaluating the ecotoxicity of a particular substance requires careful specification of: the endpoints of concern (which entails describing the organism tested or observed); the nature of the effects; the concentration or dose needed to produce the effect; the duration of exposure needed to produce the effect; and the environmental conditions under which the effects were observed (Calow, 1993; MacCarthy and Mackay, 1993; USEPA, 1989c). During the ecological response analysis, the relationship between the stressor and the magnitude of ecological effects involved is quantified, and cause-and-effect relationships are evaluated. In addition, extrapolations will generally be made from measurement endpoints to assessment endpoints—resulting in the generation of a stressor–response profile which becomes an input to the risk characterization (Zehnder, 1995).

Similar to the human health endangerment assessment (discussed in Chapter 9), the scientific literature is reviewed to obtain ecotoxicity information for the chemicals of potential ecological concern that are associated with the contaminated site problem—albeit existing ecotoxicity data tends to be rather limited in so many ways. Thus, additional data would typically have to be developed via field sampling and analysis and/or bioassays. Subsequently, critical toxicity values for the contaminants of concern are derived for the target ecological receptor species and ecological communities of concern, to be used in characterizing risks associated with the locale.

10.2.3.1 Determination of Toxicological Parameters for Ecological Health Hazard Effects and ERAs

Several—but certainly not all—of the concepts and principles discussed in Section 9.4 for the determination of toxicological parameters for human health risk assessments are fundamentally applicable to the determination of ecological toxicity parameters. In any case, it is noteworthy here that extensive discussions on ecological toxicity parameters are purposively excluded in this presentation—especially because they tend to be more unique to the widely varying receptors addressed in ERAs. It is therefore suggested that the reader consults with other sources that deal directly, and in detail, with the particular ecological receptors of interest for the particular project site being evaluated. In general, information on ecological toxicity parameters—often represented by the toxicity reference values, TRVs—may be obtained from sources such as the ECOTOX database (see Appendix B.2.5). Where toxicity information does not exist at all, a decision may be made to estimate toxicological data from that of similar compounds (i.e., with respect to molecular weight and structural similarities, etc.).

10.2.4 Ecological Risk Characterization

Ecological risk characterization brings together the knowledge and data developed during the course of evaluating the measurement endpoints and addressing data gaps—and then integrating effects and exposure of the site contaminants with the ecological resources that are the focus of the assessment. The exposure and effects information is considered together with associated uncertainties to arrive at an estimate of the risks posed by site conditions, along with a projection of the likelihood that adverse impacts would be incurred. It is noteworthy that a variety of approaches are available for estimating and describing risks and identifying associated uncertainties, and for integrating this information into a risk characterization.

During the risk characterization process, the likelihood of adverse effects occurring as a result of exposure to an environmental stressor is evaluated; it is notable that, overall, the ecological risk characterization process consists of steps similar to that discussed under the human endangerment assessment (Chapter 9). That is, the doses determined for the ecological receptors and community during the exposure assessment are integrated with the appropriate toxicity values and information derived in the ecotoxicity assessment, in order to arrive at plausible ecological risk estimates. This entails both temporal and spatial components—requiring an evaluation of the probability or likelihood of an adverse effect occurring; the degree of permanence and/or reversibility of each effect; the magnitude of each effect; and receptor populations or habitats that will be affected.

Classically, a quantitative ecological risk characterization is accomplished by using the ecological risk quotient (ErQ) method—similar to the hazard quotient employed in a human health risk characterization (see Chapter 9). In the ErQ approach, the exposure point concentration or estimated daily dose is compared to a benchmark critical toxicity parameter (e.g., "NAWQC," or a threshold reference value) as follows:

$$\text{Ecological risk Quotient (ErQ)} = \frac{\left[\text{exposure point concentration or estimated daily dose}\right]}{\left[\text{benchmark critical ecotoxicity parameter; or a surrogate}\right]}$$

(10.2)

The denominator represents the concentration that produces an assessment endpoint response (e.g., toxic effects) in target species. Where data for specific species or endpoints are unavailable, other toxicity data (e.g., LC_{50} or LOEL) may be used to derive a surrogate parameter. When NOEL-type values are used as surrogate, however, the ErQ will have a significantly more conservative meaning as an indicator of risk.

In the final analysis, the ErQ provides an estimate of the risk of an environmental contaminant to indicator species, independent of the interactions among species or between different chemicals of potential ecological concern. In general, if ErQ <1, then no unacceptable risk is indicated; conversely, an ErQ >1 calls for action or further refined investigations, due to the possibility of unacceptable levels of risk to potential ecological receptors.

10.2.4.1 Consideration of Bioaccumulation Effects

The uptake of sediment-bound or soil-bound chemicals by organisms (i.e., bioaccumulation) may either be measured directly by collecting and analyzing the tissues of representative organisms, or it may be estimated. In the initial stages of a risk assessment, estimates are typically derived in accordance with the following equation:

$$C_t = C_s \times \text{BAF}$$

(10.3)

where
  C_t = concentration in tissue (mg/kg)
  C_s = concentration in sediment or soil (mg/kg)
  BAF = bioaccumulation factor ([mg/kg − tissue] / [mg/kg − sediment or soil])

The BAF values, defined as the ratios of the concentration of the chemical in the tissues of the organism to the concentration of the chemical in sediment or soil, would typically have been derived for various chemicals and species—and would then be available in the literature. In the event that BAF values for relevant chemicals or species of interest are not available in the literature, they may be derived using tissue and soil or sediment data available in the literature or determined experimentally at the project site. It is noteworthy, however, that this relationship may not be valid for those metals that are essential trace nutrients for plants and animals.

In the event that tissue-based TRVs are available, then C_t can be used to derive a hazard quotient (HQ)—as defined by the following equation:

$$HQ = C_t/TRV \tag{10.4}$$

Furthermore, C_t can be used to represent the exposure point concentration in the estimation of ingested doses for upper trophic level species—so that, for example:

$$\text{Dose Ingested} = [C_t \times IR]/BW \tag{10.5}$$

where
  IR = ingestion rate of receptor species (kg/day)
  BW = body weight of receptor species (kg)

For upper tropic level species, quantitative data can additionally be used to modify ingested doses for use in calculating risk estimates. These data may be so-incorporated as described for the noncarcinogenic human health risk assessment (see Chapter 9). For example, when evaluating exposures resulting from the ingestion of contaminated prey items, the following simplified equation may be used to determine the risk from food ingested by the ecological receptor:

$$\text{Risk} = [\text{Intake} \times ABS]/TRV \tag{10.6}$$

where
  Intake = dose ingested (mg/kg/day)
  ABS = absorption factor (unitless)
  TRV = toxicity reference value (mg/kg/day)

In any case, for screening-level evaluations, the ABS is typically assumed to be 1 (i.e., absorption is 100%). However, as the investigation progresses through the ERA process, it may be possible to refine this value to reflect actual conditions either through a review of the relevant literature or through bioassays, as described for human receptor exposures.

10.2.4.2 Uncertainty Analysis

Three common qualitatively distinct sources of uncertainty to evaluate in virtually all risk assessments are the inherent variability, parameter uncertainty, and model errors. Invariably, all ecological risk estimates rely on numerous assumptions and consideration of the many uncertainties that are inherent in the ERA process (Box 10.6). In particular, it is noteworthy that, virtually all risk assessments have data gaps that must be addressed, but it is not always possible to obtain more information—especially due to lack of time and/or monetary resources, or a practical means to acquire more data. Under such circumstances, extrapolations (e.g., between taxa, between responses, from laboratory to field data, between geographical areas, between spatial scales, and/or between temporal scales) may be the only way to bridge gaps in available data—in order to link measures of effect with assessment endpoints. Consequently, it is important to address the level of confidence or the degree of uncertainty associated with the estimated risk attributable to a contaminated site problem.

Generally speaking, uncertainties are typically associated with both toxicity information (such as ecotoxicity values and site-specific dose–response assessments) and exposure assessment information. In any event, factors that may significantly increase the uncertainty of the ERA should be identified and addressed, at least qualitatively, and where possible, in a quantitative manner.

> ## BOX 10.6   MAJOR SOURCES OF UNCERTAINTY AND VARIABILITY IN ERAs
>
> - Uncertainty due to natural complexity of ecosystems—i.e., uncertainty associated with response of ecosystems (or their components) to anthropogenic stress—which may involve numerous factors
> - Uncertainty due to stochasticity of ecosystem measures
> - Source term uncertainty—i.e., associated with estimates of the rate and the spatial and temporal pattern of release of a chemical from a source or set of sources
> - Uncertainty in extrapolation models—i.e., extrapolating effects data (including, extrapolation of toxicity data from one species to another or from low dose to high dose for same species, etc.)
> - Uncertainty concerning magnitude of effects
> - Uncertainty associated with different test endpoints
> - Environmental variability in time and space
> - Variations in sensitivity among individuals and life stages, and between species
> - Stochastic birth and death processes

10.3   A GENERAL FRAMEWORK FOR ERAs

The framework for ERA is conceptually similar to the approach used for human health risk assessments (presented in Chapter 9), but is distinctive in its emphasis in three areas. First, ERA can consider effects beyond the individual or species level—and may indeed examine a variety of assessment endpoints, an entire population, community, or ecosystem. Second, the ecological values to be protected are selected from a wide range of possibilities, based on both scientific and policy considerations. Finally, ERAs consider nonchemical stressors to the environment, such as loss of wildlife habitat.

A general framework for the conduct of an ERA will typically consist of the following three fundamental phases (USEPA, 1992d): *problem formulation*, *problem analysis*, and *risk characterization* (Figure 10.4). The ERA process is indeed iterative by nature; for instance, it may take more than one pass through problem formulation to complete planning for the risk assessment, or information gathered in the analysis phase may suggest further problem formulation activities such as modification of the previously selected endpoints. To maximize efficient use of the oftentimes limited resources, therefore, ERAs are frequently designed in sequential tiers that proceed from simple, relatively inexpensive evaluations to more costly and complex assessments. Ultimately, the value of the risk assessment depends on whether it is used to make quality environmental and/or risk management decisions.

10.3.1   The Problem Formulation Phase

Problem formulation is a formal process for generating and evaluating preliminary hypotheses about why ecological effects have occurred, or may occur, from ensuing human activities. It involves identifying and delimiting goals and assessment endpoints, preparing the CSM, and developing an analysis plan. Successful completion of the problem formulation phase depends on the quality of several investigatory elements—especially relating to assessment endpoints that adequately reflect management goals and the ecosystem they represent, as well as conceptual models that describe key relationships between a stressor and assessment endpoint or among several stressors and assessment endpoints (USEPA, 1992d). Essential to the development of these problem formulation elements are the effective integration and evaluation of available information on sources of stressors and stressor characteristics, exposure characteristics, ecosystem potentially at risk, and ecological effects.

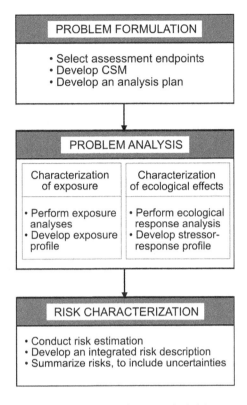

FIGURE 10.4 A generalized procedural framework for ecological risk assessments.

10.3.1.1 The Essential Role of Assessment Endpoints

An assessment endpoint is an explicit expression of the environmental value (such as relates to species, ecological resource, or habitat type) that is to be protected. Assessment endpoints relate to statutory mandates (with regards to protection of the environment), but must be specific enough to guide the development of the risk assessment study design at a particular site. Useful assessment endpoints define both the valued ecological resource and a characteristic of the resource to protect (e.g., reproductive success, production per unit area, and areal extent). Assessment endpoints are indeed critical to problem formulation because they link the risk assessment to management concerns—and they are central to conceptual model development.

10.3.1.2 Utility of the CSM

A conceptual model, as employed in problem formulation, is a written description and visual representation of predicted responses by ecological entities to stressors to which they are exposed; the model includes ecosystem processes that influence these responses.

On the whole, the conceptual model describes a series of working hypotheses of how exposures might affect the ecological components of an environment. The ecosystem or ecosystem components potentially at risk and the relationships between assessment and measures of effects and exposure scenarios also are described in the conceptual model; measures of effects are changes in attributes of assessment endpoints or their surrogates in response to the stressors to which they were exposed. Two additional types of measures are used since data other than those used to evaluated responses (i.e., measures of effects) is often required for an ERA—namely: measures of exposures (which include the stressor and source measurements) and measures of ecosystems and receptor characteristics (which include water quality conditions, soil parameters, and habitat measures).

10.3.1.3 Formulating an Analysis Plan

The analysis plan specifies the data required to evaluate the impacts to the assessment endpoints and the methods that will be used to analyze the data. The design of an analysis plan is the final stage of the problem formulation. It includes the most important pathways and relationships identified during problem formulation that will be pursued in the problem analysis phase of the ERA process. Several criteria—including the availability of information; the strength of information about relationships between stressors and effects; the assessment endpoints and their relationship to ecosystem function; relative importance or influence and mode of action of stressors; and completeness of known exposure pathways—become an important basis for the selection of critical relationships in the CSM that are to be pursued in the analysis. The analysis plan does indeed provide the basis for making selections of data sets that will be used for the ERA.

10.3.2 THE PROBLEM ANALYSIS PHASE

The problem analysis phase of the ERA is composed of two principal activities—the characterization of exposure, and the characterization of ecological effects; ultimately, these will result in the development of exposure and stressor–response profiles. This phase consists of hazard identification plus dose–response and exposure assessments—and involves evaluating exposure to stressors (consisting of the creation of profiles to evaluate the exposure of ecological receptors to stressors), as well as the relationship between stressor levels and ecological effects. Specifically, this comprises the technical evaluation of data to facilitate the development of conclusions about ecological exposure and relationships between the stressor and ecological effects. During this phase, measures of exposure (e.g., source attributes, stressor levels in the environment), measures of effects (e.g., results of laboratory or field studies), and measures of ecosystem and receptor attributes (e.g., life history characteristics) are used to evaluate issues that were identified in the problem formulation phase.

It is noteworthy that the CSM and analysis plan developed during the problem formulation phase provides the framework for the problem analysis phase. The end result of the problem analysis phase typically will consist of summary profiles that describe exposure and the stressor–response relationship. When combined, these profiles provide the basis for reaching conclusions about risk during the subsequent risk characterization phase.

10.3.3 THE RISK CHARACTERIZATION PHASE

The purpose of the risk characterization phase of the ERA is to evaluate the likelihood that adverse effects have occurred, or will occur, as a result of exposure to a stressor. Its key elements consist of estimating risk through integration of exposure and stressor–response profiles, describing risk by discussing lines of evidence, and determining ecological adversity.

On the whole, the risk estimation is used to determine the likelihood of adverse effects to assessment endpoints by integrating exposure and effects data, and then evaluating any associated uncertainties. The process uses exposure and stressor–response profiles derived *a priori* in the problem analysis phase of the ERA.

10.4 ERA IN PRACTICE

Invariably, a qualitative and/or quantitative ERA often becomes an integral part of most environmental risk management programs that are designed to address contaminated site problems. Similar to the human health endangerment assessment presented earlier in Chapter 9, the basic components and tasks involved in a comprehensive ERA may consist of the following tasks:

- Data Evaluation
  - Assess the quality of available data
  - Identify, quantify, and categorize environmental contaminants
  - Screen and select chemicals of potential ecological concern
  - Carry out statistical analysis of relevant environmental data
- Exposure Assessment
  - Compile information on the physical setting of the locale
  - Identify source areas, significant migration pathways, and potentially impacted or receiving media
  - Determine the important environmental fate-and-transport processes for the chemicals of potential concern, including cross-media transfers
  - Identify populations potentially at risk
  - Determine likely and significant receptor exposure pathways
  - Develop representative CSM(s)
  - Develop exposure scenarios (to include both the current and potential future land-uses for the site)
  - Estimate/model exposure point concentrations for the chemicals of potential concern found in the significant environmental media
  - Compute potential receptor intakes and resultant doses for the chemicals of potential ecological concern (for all potential receptors and significant pathways of concern)
- Ecotoxicity Assessment
  - Compile toxicological profiles (to include the intrinsic toxicological properties of the chemicals of potential ecological concern, such as their acute, subchronic, chronic, and reproductive effects)
  - Determine appropriate toxicity indices (such as NAWQC, SQC, and TRVs)
- Risk Characterization
  - Quantify risks or hazards, as appropriate, by estimating ErQs or similar risk parameters
  - Perform sensitivity analyses, evaluate uncertainties associated with the risk estimates, and summarize the risk information

Figure 10.5 shows the general nature of the complete decision-making framework—illuminating the relevant tasks typically undertaken for most ERAs; several particularly important criteria to adopt in various significant aspects of an ERA process are enumerated in Box 10.7.

Meanwhile, it is notable that unlike endangerment assessment for human populations, ERAs often lack readily available and/or significant amount of critical and credible data necessary for a comprehensive quantitative evaluation. Nonetheless, the pertinent data requirements should be identified and categorized insofar as practicable. Indeed, the collection and review of the existing information on terrestrial and aquatic ecosystems, wetlands and floodplains, threatened and endangered species, soils, and other topics relevant to the study should form a prime basis for identifying any data gaps. For instance, survey information on soil types, vegetation cover, and residential migratory wildlife may be required for terrestrial habitats, whereas comparative information needed for freshwater and marine habitats will most likely include survey data on abundance, distribution, and kinds of populations of plants and animals living in the water column and in or on the bottom. The biological and ecological information collected should include a general identification of the flora and fauna associated within and around the site, with particular emphasis placed on sensitive environments, especially endangered species and their habitats and those species consumed by humans or found in human food chains. Furthermore, the biological, chemical, and environmental factors perceived to influence the ecological effects of contaminants should be completely identified and adequately described.

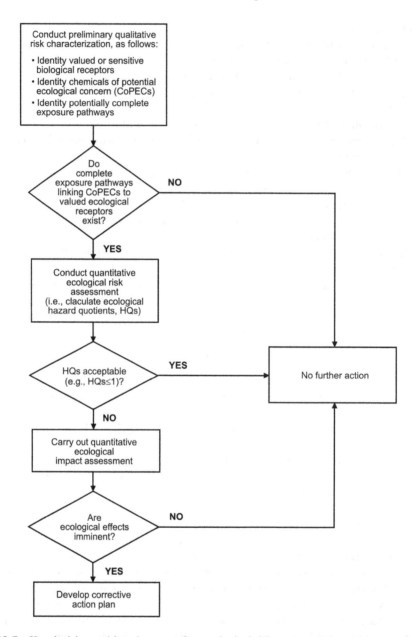

FIGURE 10.5   Key decision-making elements of an ecological risk assessment in practice.

BOX 10.7   SELECTION CRITERIA FOR VARIOUS
MAJOR ASPECTS OF AN ERA

ERA Issue/Item	Selection Criteria
• Valued Biological Receptors	• Regulatory importance • Ecological importance • Commercial importance • Recreational importance

(Continued)

BOX 10.7 (*Continued*) SELECTION CRITERIA FOR VARIOUS MAJOR ASPECTS OF AN ERA

- Chemicals of Potential Ecological Concern (CoPECs)—considered during hazard identification

  - Accessibility
  - Concentration
  - Bioaccumulation/bioconcentration potential
  - Potency/toxicity
  - Bioavailability
  - Persistence

- Potentially Complete Exposure Pathways/Exposure Scenario

  - Location of CoPECs
  - Migration potential of CoPECs
  - Behavior and ecology of valued biological receptors

- Establishing [Exposure] Assessment Endpoints—which forms the basis for collecting data, for determination of contaminant effects, and for evaluation of migration needs from releases or mitigation activities (i.e., the explicit expressions of values that are to be protected)

  - Societal relevance
  - Biological relevance
  - Measurability or predictability
  - Susceptibility to chemicals

- Establishing [Exposure Assessment] Measurement Endpoints—which link the existing conditions on-site to the goals expressed by the assessment endpoints (i.e., are more easily collected data that act as a surrogate for the assessment endpoints)

  - Relationship to assessment endpoint
  - Ability to be measured
  - Availability of existing data
  - Relationship to known contaminants and pathways
  - Degree of natural variability
  - Temporal and spatial relational scale

10.4.1 THE ERA IMPLEMENTATION STRATEGY

Most ERAs are divided into a screening-level (Tier 1) ERA and a baseline (Tier 2) ERA—with a possibility for remedy-directed (Tier 3) ERA (Figure 10.6). A screening-level ecological risk assessment (SERA) is a simplified risk assessment that can be conducted with limited data and uses conservative assumptions to minimize the chances of concluding that there is no risk when in fact a risk exists. In a baseline ecological risk assessment (BERA), the conservative assumptions are eliminated and replaced with best estimates to more accurately assess the site's risk. Regardless of whether a SERA or BERA is being conducted, the general process always consists of the planning, problem formulation, analysis, risk characterization, and risk management tasks previously identified in Section 10.3. Depending on the outcomes of the prior tiers, execution of the third tier may have to be invoked or brought into play.

Broadly speaking, ERAs should be conducted at the screening level using best available, cost-effective chemical and biological screening technologies. If analyses at the initial screening level show contaminant concentrations below levels of concern, then a finding of no further action will likely be an acceptable course of action; the reference screening levels should be regional or national levels/criteria that have been agreed to in advance with regulators. If screening determines areas of high contamination or significant toxicity, then a more thorough ERA may be necessary to define the spatial distribution of the contaminants, evaluate what populations are at risk, and determine contaminant mobility and bioavailability. In this latter case, it may be more appropriate to do an initial cost/benefit analysis of an active or passive remediation before deciding on an expensive site-specific ERA.

On the whole, ERAs are performed at contaminated sites to determine if contaminants are causing adverse ecological effects. They are conducted for both terrestrial and aquatic environments, including the marine environment; although the general approach for each type is similar, the complexity and cost involved in the latter (that primarily focus on the risk to ecological resources

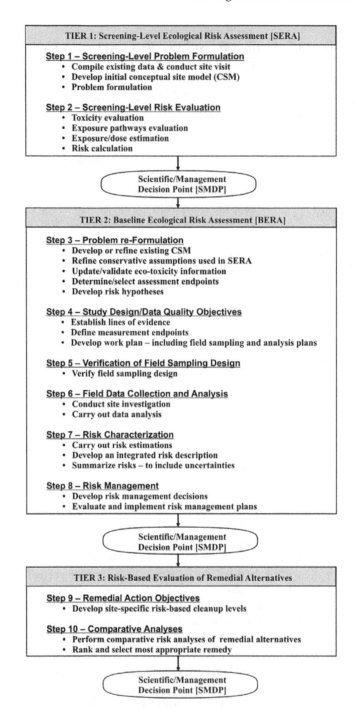

FIGURE 10.6   The tiered procedural framework for ecological risk assessment in practice.

posed by contaminated sediments) can be significant. Unlike terrestrial risk assessments, where ERAs often play a supporting role to human health risk assessments, marine assessments are driven mostly by ecological concerns (with the exception of recreation/subsistence fishing and shell-fishing concerns). In any case, ERAs are classically performed as part of the baseline risk assessment process involved in contaminated site investigation and characterization activities—and the risk estimate typically provided by an ERA is often more qualitative than the quantitative assessment provided by a human health risk assessment (see Chapter 9).

Broadly, the ERA process is composed of the eight steps shown in Figure 10.6—and further annotated below; the process starts with moving from a screening-level risk assessment to a full baseline risk assessment. The US EPA guidance, as an example, incorporates points where the risk managers are specifically included in the risk assessment process; these points, called scientific management decision points (SMDPs), assist in the coordination of the risk assessment with the risk management goals (USEPA, 1997).

10.4.1.1 Step 1: Screening-Level/Preliminary Problem Formulation and Ecological Effects Evaluation/Characterization

A problem formulation and ecological effects evaluation is part of this initial step. Even though site-specific data will be limited in most cases, the following information should be available upon the completion of this step:

- Description of the environmental setting, including habitat types, observed species and species likely to be present based on habitat types documented, and threatened, rare, and endangered species
- Description of contaminants known or suspected to exist at the site and the maximum concentrations present in each medium
- Contaminant fate-and-transport mechanisms that might exist
- Mechanisms of ecotoxicity associated with contaminants and categories of receptors that may be affected
- Complete exposure pathways that might exist from contaminant sources to receptors that could be affected
- Screening ecotoxicity values equivalent to chronic "no-observable-adverse-effects-levels" (NOAELs) based on conservative assumptions

The screening-level problem formulation and ecological effects evaluation is indeed part of the initial ecological risk screening assessment.

For this initial step (of the SERA), it is likely that site-specific information for determining the nature and extent of contamination and for characterizing ecological receptors at the site is limited. Anyhow, this step includes all the functions of problem formulation (more fully described in Steps 3 and 4) as well as ecological effects analyses—albeit on a screening level. Also, assessment endpoints will include any likely adverse ecological effects on receptors for which exposure pathways are complete, as determined from the information listed above; and measurement endpoints will be based on the available literature regarding mechanisms of toxicity—and this will be used to establish the screening ecotoxicity values (that will be used with estimated exposure levels to screen for ecological risks, as described in Step 2). Overall, the results of this step are used in conjunction with exposure estimates in the preliminary risk calculation in Step 2.

10.4.1.2 Step 2: Screening-Level/Preliminary Exposure Estimate and Risk Calculations

In this step, risk is estimated by comparing maximum documented exposure concentrations with the ecotoxicity screening values developed in Step 1. Based on the outcome, the risk manager will decide that: either the SERA is adequate to determine that ecological threats are negligible, or the process should continue to the more detailed ERA outlined in Steps 3–7 below. If the process continues, the screening-level assessment serves to identify exposure pathways and preliminary contaminants of concern for the baseline risk assessment—by eliminating those contaminants and exposure pathways that pose negligible risks.

At the conclusion of the exposure estimate and screening-level risk calculation step, the following information would have been compiled:

1. Exposure estimates based on conservative assumptions and maximum concentrations present
2. Hazard quotients (or hazard indices) indicating which, if any, contaminants and exposure pathways might pose ecological threats

Based on the results of the screening-level ecological risk calculation, the risk manager, and lead risk assessor will determine whether or not contaminants from the site pose an ecological threat. If there are sufficient data to determine that ecological threats are negligible, the ERA will be complete at this step—with a finding of negligible ecological risk. If the data indicate that there is (or might be) a risk of adverse ecological effects, the ERA process will continue.

It is noteworthy that conservative assumptions have been used for each step of the SERA; thus, simply requiring a cleanup action based solely on this information would not be considered technically defensible. In fact, to stop the assessment at this stage, the conclusion of negligible ecological risk must be adequately documented and technically defensible. Among other things, a lack of information on the toxicity of a contaminant or on complete exposure pathways will typically result in a decision to continue with the ERA process (Steps 3–7).

10.4.1.3 Step 3: Problem Formulation

This step refines the screening-level problem formulation and, with input from stakeholders and other involved parties, expands on the ecological issues that are of concern at the particular site. The results of the screening assessment and additional site-specific information are used to determine the scope and goals of the BERA, and then form the basis of the conceptual model, which is completed in Step 4; a conceptual model describes a series of working hypotheses of how an exposure might affect the ecological components of an environment.

In the screening-level assessment, conservative assumptions were used where site-specific information was lacking. In Step 3, however, the results of the screening assessment and additional site-specific information are used to determine the scope and goals of the BERA—also recognizing that Steps 3–7 are required only for sites for which the screening-level assessment indicated a need for further ecological risk evaluation.

In general, Step 3 of the eight-step process initiates the problem formulation phase of the BERA—and this problem formulation at Step 3 includes several activities, viz.:

- Refining preliminary contaminants of ecological concern
- Further characterizing ecological effects of contaminants
- Reviewing and refining information on contaminant fate-and-transport, complete exposure pathways, and ecosystems potentially at risk
- Selecting assessment endpoints
- Developing a conceptual model with working hypotheses or questions that the site investigation will address

At the conclusion of Step 3, there is a "SMDP," which consists of agreement on four items: the assessment endpoints, the exposure pathways, the risk questions, and conceptual model integrating these components. The products generated in Step 3 are used to select measurement endpoints, and to develop the ERA work plan (WP) and sampling and analysis plan (SAP) for the site in Step 4. Steps 3 and 4 are, effectively, the data quality objective (DQO) process for the BERA. In the end, by combining information on the following: (i) the potential contaminants present; (ii) the ecotoxicity of the contaminants; (iii) environmental fate-and-transport; (iv) the ecological setting; and (v) complete exposure pathways, an evaluation is made of what aspects of the ecosystem at the site could be at risk, and what the adverse ecological response could be.

In the above evaluation, decisions on what might constitute "critical exposure pathways" are based on the following: (i) exposure pathways to sensitive species' populations or communities; and (ii) exposure levels associated with predominant fate-and-transport mechanisms at a site. On the basis of these types of information, assessment endpoints and specific questions or testable hypotheses that, together with the rest of the conceptual model, form the basis for the site investigation may be established. At this stage, site-specific information on exposure pathways and/or the presence of specific species is likely to be incomplete. In any case, by using the conceptual model developed thus far, measurement endpoints can be selected, and a plan for closing the information gaps can be developed and written into the ecological WP and SAP as described in Step 4.

10.4.1.4   Step 4: Study Design and Data Quality Objective Process

Step 4 completes the conceptual model, which was initiated in Step 3; the site conceptual model, which includes assessment endpoints, exposure pathways, and risk questions or hypotheses, is completed in Step 4 with the development of measurement endpoints. The conceptual model is then used as the basis to develop the study design and data quality objectives (DQOs). The end products of Step 4 are the ERA WP and the SAP, which describe the details of the site investigation, as well as the data analysis methods and DQOs. The WP documents the decisions and evaluations made during problem formulation and identifies additional research tasks needed to fully evaluate the risks to ecological resources. The SAP provides a detailed description of sampling and data-gathering procedures, as well as a description of the steps required to achieve the study objectives. As part of the DQO process, the SAP specifies acceptable levels of decision errors that will be used as the basis for establishing the quantity and quality of data needed to support ecological risk management decisions.

The WP and/or the SAP should specify the methods by which the collected data will be analyzed—and the plan(s) should include all food-chain-exposure-model parameters, data reduction techniques, data interpretation methods, and statistical analyses that will be used. Meanwhile, it is notable that once this step is completed, most of the professional judgment needed for the ERA would have been incorporated into the design and details of the WP and SAP. This does not limit the need for qualified professionals in the implementation of the investigation, data acquisition, or data interpretation. However, there should be no fundamental changes in goals or approach to the ERA once the WP and SAP are finalized.

At the conclusion of Step 4, there will be an agreement on the contents of the WP and SAP—and these plans can be parts of a larger WP and SAP that are developed to meet other remedial investigation needs, or they can be separate documents. When possible, any field sampling efforts for the ERA should overlap with other site data collection efforts to reduce sampling costs, and to prevent redundant sampling.

10.4.1.5   Step 5: Verification of Field Sampling Design

In this step, the sampling plan, exposure pathways, and measures of effects, are evaluated to verify that the SAP is appropriate for the site. During field verification of the sampling design, the testable hypotheses, exposure pathway models, and measurement endpoints are evaluated for their appropriateness and implementability. The assessment endpoint(s), however, should not be under evaluation in this step; indeed, the appropriateness of the assessment endpoint should have been resolved in Step 3. If an assessment endpoint is changed at this step, the risk assessor must return to Step 3, because the entire process leading to the actual site investigation in Step 6 assumes the selection of appropriate assessment endpoints.

Overall, field verification of the sampling plan is very important to ensuring that the DQOs of the site investigation can be met. This step verifies that the selected assessment endpoints, testable hypotheses, exposure pathway model, measurement endpoints, and study design from Steps 3 and 4 are appropriate and implementable at the site. By verifying the field sampling plan prior to conducting the full site investigation, well-considered alterations can be made to the study design and/or implementation, if necessary; any such changes will ensure that the ERA meets the study objectives. Meanwhile, if changing conditions force changes to the sampling plan in the field (e.g., selection of a different reference site), the changes should be carefully documented and evaluated.

10.4.1.6   Step 6: Site Investigation and Data Analysis

This step involves the collection of information to characterize exposures and ecological effects at the site. Whereas much of the data for characterizing potential ecological effects would have been collected during the problem formulation stage, the site investigation provides evidence of existing ecological impacts and additional exposure effects and response information. In any case, both the site investigation and data analysis should be conducted according to the WP and SAP developed in Step 4.

In general, information collected during the site investigation is used to characterize exposures and ecological effects; the site investigation includes all of the field sampling and surveys that are conducted as part of the ERA. The site investigation step of the ERA should indeed be a straight-forward implementation of the study designed in Step 4, and as verified in Step 5. In instances where unexpected conditions arise in the field that indicate a need to change the study design, the project team should reevaluate the feasibility or adequacy of the sampling design. Any proposed changes to the WP or SAP must be fully documented and evaluated in the baseline risk assessment. Ultimately, the site investigation and analysis of exposure and effects should be straightforward, following the WP and SAP developed in Step 4 and tested in Step 5.

In the end, the analysis phase of the ERA consists of the technical evaluation of data on existing and potential exposures and ecological effects, and is based on the information collected during Steps 1–5 and the site investigation in Step 6. As part of the overall process, the exposure characterization relies heavily on data from the site investigation and can involve fate-and-transport modeling. Much of the information for characterizing potential ecological effects would have been gathered from the literature review during problem formulation, but the site investigation might provide evidence of existing ecological impacts and additional exposure-response information. On the whole, analyses of exposure and effects are performed interactively—and indeed follow the data interpretation and analysis methods specified in the WP and SAP. Site-specific data obtained during Step 6 replace many of the assumptions that were made for the screening-level analysis in Steps 1 and 2. Evidence of an exposure-response relationship between contamination and ecological responses at a site helps to establish causality. The results of Step 6 are used to characterize ecological risks in Step 7.

10.4.1.7   Step 7: Risk Characterization

The Risk Characterization step integrates the results of the exposure profile and exposure-effects information (or stressor–response analysis)—and is the final phase of the ERA process. Risk characterization includes two major components: risk estimation and risk description; together, these provide information to help judge the ecological significance of risk estimates in the absence of remedial activities. Risk estimation involves integrating exposure profiles with the exposure-effects information and summarizing the associated uncertainties; and risk descriptions provide information important for interpreting the risk results—and also identifies a threshold for adverse effects on the assessment endpoints (as a range between contamination levels identified as posing no ecological risk and the lowest contamination levels identified as likely to produce adverse ecological effects). To ensure that the risk characterization is transparent, clear, and reasonable, information regarding the strengths and limitations of the assessment must be clearly identified and described.

On the whole, during the risk characterization, data on exposure and effects are integrated into a statement about risk to the assessment endpoints established during problem formulation. A weight-of-evidence approach is used to interpret the implications of different studies or tests for the assessment endpoints. In a well-designed study, risk characterization should be straightforward, because the procedures were basically established in the WP and SAP. In any case, the risk characterization section of the BERA should include a qualitative and quantitative presentation of the risk results and associated uncertainties.

10.4.1.8   Step 8: Risk Management

Risk management is a distinctly different process from risk assessment (NRC, 1983, 1994; USEPA, 1984a, 1995b). The risk assessment establishes whether a risk is present and defines a range or magnitude of the risk. In risk management, the results of the risk assessment are integrated with other considerations to make and justify risk management decisions; additional risk management considerations can include the implications of existing background levels of contamination, available cleanup technologies, tradeoffs between human and ecological concerns, costs of alternative actions, and remedy selections. In the end, risk management decisions are the responsibility of the risk manager (or the site manager), not the risk assessor—albeit the risk manager should have been involved in planning the risk assessment; after all, likely knowing the options available for reducing risks, the risk manager can help to frame questions during the problem formulation phase of the risk assessment.

In Step 7, the risk assessment team would have identified a threshold for effects on the assessment endpoint as a range between contamination levels identified as posing no ecological risk and the lowest contamination levels identified as likely to produce adverse ecological effects. In Step 8, the risk manager evaluates several factors in deciding whether or not to clean up to within that range. In fact, by this time, the risk assessment should have established whether a risk is present, as well as defined a range or magnitude of that risk. With this information, a site risk manager must integrate the risk assessment results with other considerations to make and/or justify ensuing risk management decisions.

In the end, it is quite important that the risk manager understands the risk assessment (including its uncertainties, assumptions, and level of resolution) carried out at a given project site. With an understanding of potential adverse effects posed by residual levels of site contaminants and also posed by the remedial actions themselves, the risk manager can balance the ecological costs and benefits of the available remedial options—also recognizing that a good understanding of the uncertainties associated with the risk assessment is critical to evaluating the overall protectiveness of any remedy.

10.4.2   A TIERED APPROACH TO *ERA*s: MOVING FROM *SERA* TO *BERA*, AND BEYOND

Commonly, three tiers may be adopted in the ERAs designed to address contaminated site problems—namely: (i) SERA; (ii) BERA; and (iii) corrective measure evaluation ecological risk assessment (CoMERA). Indeed, for large or complex sites, the further use of sub-tiers may help focus the ERA process as well as improve the quality of the risk characterizations.

SERAs provide a general indication of the *potential* for ecological risk (or lack thereof); this may be conducted for several purposes—including: (i) to estimate the likelihood that the particular ecological risk exists; (ii) to identify the need for site-specific data collection efforts; or (iii) to focus the site-specific ERAs, where warranted. SERAs typically comprise the Steps 1 and 2 elaborated above in Section 10.4.1. If the SERA risk characterization indicates the need for further assessment, then this would generally trigger the BERA process.

BERAs typically comprise the Steps 3–8 elaborated above in Section 10.4.1. BERAs are generally used as follows: (i) to identify and characterize the current and potential threats to the environment and its inhabitant organisms as a result of the presence of hazardous substances and (ii) to develop risk management and site restoration plans that will be protective of those natural resources and populations at risk. Invariably, a BERA may be followed by a CoMERA—except in situations where alternate risk management and institutional control measures have been determined to be more appropriate.

CoMERAs are generally used as follows: (i) to evaluate the ecological impacts of alternative remediation strategies and (ii) to establish cleanup levels in the selected remedy that will be protective of those natural resources and populations at risk.

10.4.2.1 Tier 1—SERA

The "Tier 1" ecological risk assessment consists of a screening-level ecological risk assessment (SERA) that employs existing site data and conservative assumptions to support risk management decisions. This "Tier 1" process consists of two key steps: an exposure evaluation, and a risk characterization. In the exposure evaluation, site visits and existing data are appraised in order to ascertain if complete exposure pathways exist that would likely link the potential contaminants of concern with the ecological receptors of interest. The screening assessment continues only for those contaminants and receptors for which complete exposure pathways are indicated. In the risk characterization step, conservative assumptions regarding exposure (such as maximum contaminant concentrations, 100% exposure, and maximum uptake rates) are used to estimate potential exposure to the contaminants of potential concern. Estimated exposures are then compared to screening ecotoxicity values (which are representative of "safe" contaminant concentrations or doses), with the aim of calculating a hazard quotient; typically, this is then used to support one of the following three risk management decisions: (i) site conditions do not pose unacceptable risks, and no further action is warranted; (ii) site conditions pose a potentially unacceptable risk that requires additional evaluation with a "Tier 2" BERA; or (iii) site conditions pose a potentially obvious and significant unacceptable risk, and accelerated site remediation is warranted.

Broadly speaking, SERAs would tend to include the following essential components:

1. Screening-level problem formulation and ecological effects characterization (Step 1 from Section 10.4.1)
2. Screening-level exposure estimates and risk calculation (Step 2 from Section 10.4.1)
3. SMDP, signifying either negligible risk or a need for continuation on to a BERA

Meanwhile, it is noteworthy that SERAs are neither designed nor intended to provide definitive estimates of actual risks, generate cleanup goals, and are usually not based on site-specific assumptions. Rather, the purpose of SERAs is to assess the need, and if required, the level of effort necessary to conduct a detailed "baseline" ERA (BERA) for a particular site or facility. Thus, any necessary refinement of the contaminants of potential ecological concern occurs during the BERA, rather than during the SERA.

10.4.2.2 Tier 2—BERA

Unlike human health risk assessment (see Chapter 9), ERA typically would not necessarily have a well-defined "unified" or "singular" target population group that would be the focus of a given study per se. On the other hand, the complex process for identifying ecological risk target populations is crucial to the success and defensibility of any risk impact evaluation carried out during this process. In any case, at this point in an ERA, it is apparent that the conservative screening carried out during "Tier 1" for potential ecological impacts associated with site chemical constituents would have been exceeded; thus, there should now be an evaluation of the more realistic site-specific factors to better determine if risk is indeed likely. Among other things, a number of key issues may receive the closest attention during an ERA implementation program—including a refinement of any conservative exposure assumptions used in the process (as discussed below for an illustrative case/project).

Classically, the "Tier 1" SERA identified a number of contaminants that may be of potential concern (viz., the CoPECs) with regards to ecological resources. As a result, a risk management decision has been made to move into the "Tier 2" BERA process. However, because of the very conservative assumptions employed in the "Tier 1" assessment, some of the CoPECs identified for further evaluation in "Tier 2" may actually be an inaccurate representation—and thus further evaluation of these CoPECs in "Tier 2" may be unwarranted. A key role of the initial steps of the 'Tier 2' evaluation (see 'Step 3' in Figure 10.6) is to refine the exposure scenarios, replacing the conservative assumptions required by "Tier 1" with more realistic and defensible exposure parameters. Using less conservative assumptions, the risk assessor recalculates the "Tier 1" risk estimates, and uses these results to refine the CoPECs identified by the "Tier 1" screen. The risk manager then selects the appropriate exit criteria—namely: (i) the refinement supports an acceptable risk determination, and the site exits the BERA process; or (ii) the refinement does not support an acceptable risk determination, and the site continues into the BERA process.

In general, the above concerns regarding ecological risk may warrant a baseline risk consideration in order to better evaluate the magnitude and breadth of risk; however, potential for ecological impacts may be addressed by refinement to one or more site risk assumptions. If the revision of assumptions results in an "acceptable" level of eco-risk, a more streamlined evaluation to address the remaining elements in this step of the risk evaluation may be carried out to document the site conditions, and to reach no further action decisions—albeit refinement may still lead to the conduct of a BERA.

In the final analysis, ecological risk characterization brings together the knowledge and data developed during the course of evaluating the measurement endpoints and addressing data gaps, as well as integrating effects and exposure of the site contaminants with the ecological resources that are the focus of the assessment. The exposure and effects information is considered together with associated uncertainties—in order to provide an estimate of the risks posed by site conditions, and of the likelihood that adverse impacts would be incurred or otherwise. Subsequently, risk management takes the results of the risk assessment together with other key considerations to arrive at a decision on whether remediation will be needed and/or implemented.

10.4.2.3   Tier 3—CoMERA

The "Tier 3" involves an "alternatives evaluation"—and considers the type of remediation that could potentially be implemented. Factors other than risk that are considered in the alternative evaluation typically would include balancing environmental impacts of remediation with overall risk reduction; cost and implementability of remediation; remediation effectiveness and residual risks; and public acceptance. Effective communication of the remedy evaluation process and final remedy selection is also critical to the acceptance of the final management decision by the public, regulators, and stakeholders.

In practical terms, one of the most important determinants relates to the issue of what cleanup levels may be considered protective of the target populations. Indeed, when a decision is made that a response action should be taken at a site based on unacceptable ecological risk, the risk manager normally then selects chemical-specific cleanup levels that are acceptable—i.e., cleanup levels that provide adequate protection of the ecological receptors (as represented by the selected assessment endpoints) at risk. Congruently, the risk assessor can use the same toxicity tests, population or community-level studies, or bioaccumulation models that were used to determine if there was an unacceptable ecological risk to identify appropriate cleanup levels. As part of the overall process, sufficient testing and interpretation should be performed at various site locations to quantify the relationship between chemical concentrations and effects. The data can then be used to establish a concentration and response correlation, to help define the concentration that represents an acceptable (i.e., protective) level of risk. Meanwhile, it is worth the mention here that at some relatively small sites, it may indeed be more cost-effective to remove, treat, or contain all contamination rather than to generate a concentration-response correlational curve for such efforts—since the latter can be quite demanding and costly.

In closing, it must be recognized that there tends to be significant difficulty in determining the acceptable level of adverse effects for the receptors to be protected—as, for instance, deciding on what percent reduction in fish survival or in benthic species diversity may be deemed reasonable, etc.; indeed, there is no "magic" number that can be used—as this this is more so dependent on the assessment endpoints selected and the risk assessment measures used, including chemical and biological data gathered from the range of contaminated locations and compared to the reference locations. In fact, whereas it may be desirable to identify a standard numerical level of risk reduction that is protective, it is impracticable to do this for each possible species that could be exposed.

It is for this reason that surrogate measures or representative species may be used to evaluate the ecological risks to the assessment endpoints at a given project site. At sites in locations where large amounts of data exist relating abundance or population/community indices with chemical concentrations, biotic indices (i.e., instead of chemical concentrations) may also be used to generate acceptable/protective levels, and to delineate the area needing remediation. In any case, the acceptable level of adverse effects should be discussed by the risk assessor and risk manager as early as possible in the risk assessment process—and should also be coordinated with all stakeholders.

10.5 THE COMPARATIVE ROLES OF ECOLOGICAL AND HUMAN HEALTH ENDANGERMENT ASSESSMENTS

Traditionally, most endangerment assessments have focused almost exclusively on risks to human health, often ignoring potential ecological effects. This bias has resulted in part from anthropocentrism, and in part from the common but mistaken belief that protection of human health automatically protects nonhuman organisms at all times (Suter, 1993). Indeed, human health risks in most situations are more substantial than ecological risks; thus, considering that the mitigative actions taken to alleviate risks to human health are often sufficient to mitigate potential ecological risks at the same time, extensive ecological investigations are usually not required for most environmental management programs (including contaminated site problems). However, it should also be recognized that, in some situations nonhuman organisms, populations, or ecosystems might be more sensitive than human receptors. Consequently, ERA programs have become an important part of the management of contaminated site problems—in order to facilitate the attainment of an adequate level of protectiveness for both human and ecological populations potentially at risk. Its purpose is to contribute to the protection and management of the environment through scientifically justifiable and credible evaluation of the ecological effects of human activities. Ultimately, ecological risk analysis becomes a very important instrument that is used to integrate ecological issues during the formulation of the broader environmental management decisions often encountered in practice.

It is notable that the process used to evaluate environmental or ecological risks are analogous to, and indeed parallels that used in the evaluation of human health risks that was presented in Chapter 9 (see Table 10.1 for comparison of analogous tasks/activities). In both cases, potential risks are determined by the integration of information on chemical exposures with toxicological data for the contaminants of potential concern. In fact, numerous and diverse risk assessment methodologies

TABLE 10.1
Typical General Elements of the Parallel Activities Involved in Human Health and Ecological Risk Assessments

Human Health Risk Assessment	Ecological Risk Assessment
• Identify chemicals of potential concern (CoCs) to human health	• Identify CoPECs for ecological receptors
• Quantify release, migration, and fate of contaminants that could affect human receptors	• Quantify release, migration, and fate of contaminants that could affect ecological systems and/or receptors
• Determine human populations potentially-at-risk (PARs)	• Determine potentially affected habitats and potentially exposed ecological populations
• Identify exposure routes for PARs	• Identify exposure routes for potentially exposed ecological populations
• Conduct health effects studies	• Conduct ecological effects studies and tests
• Characterize human health risks, recognizing and addressing associated uncertainties	• Characterize ecological risks, recognizing and addressing associated uncertainties

have been developed for the protection of human health—and most of these applications can be useful to ERAs as well. Similar to the human health risk assessment process, the ERA process is also based on two major fundamental components—namely: characterization of exposure, and characterization of ecological effects. In adapting the relevant tools from human health risk assessment to ERA problems, however, it must be recognized that nonhuman organisms, populations, or ecosystems may be more sensitive than humans under similar types of circumstances; such difference may be attributable to a variety of reasons, including the following (Suter, 1993):

- Some types of exposure scenarios are peculiar only to nonhuman receptors (e.g., root uptake and drinking from waste sumps).
- Certain chemicals are likely to be more toxic to some nonhuman species than to humans—considering the heterogeneity in the former.
- There are mechanisms of action at the ecosystem level (such as eutrophication by nutrient chemicals) that have no human analogs.
- Nonhuman organisms may be exposed more intensely to chemicals, even when the routes of exposure are similar. After all, in general, humans are not immersed in a particular ambient environment.
- Most birds and mammals have higher metabolic rates than humans—thus resulting in the former group receiving a larger dose per unit body weight.
- Nonhuman organisms are highly coupled to their environments, so that even when they are resistant to a chemical, they may still experience secondary effects—such as loss of food or physical habitat.

In either case of human health and ecological endangerment assessment, the process paradigm offers a rigorous form of evaluation to qualify and/or quantify potential effects of environmental contaminants on a variety of populations potentially at risk.

Lastly, it is noteworthy that establishing remediation goals for ecological receptors tends to be considerably more difficult than establishing such goals for the protection of human health due to the paucity of broadly applicable and quantifiable toxicological data. What is more, protective exposure levels are best established on a site-specific basis—especially because of the often large variation in the kinds and numbers of receptor species typically present at most sites; the differences in their susceptibility to contaminants; their recuperative potential following exposure; and the tremendous variation in environmental bioavailability of many contaminants in different media.

Part IV

Development, Design, and Implementation of Risk-Based Solutions and Restoration Plans for Contaminated Site Problems

This part of the book is comprised of four sections—consisting of the following specific chapters:

- Chapter 11, *Risk-Based Restoration Goals and Cleanup Criteria for Contaminated Site Problems*, presents discussions of how risk assessment may be used to help answer the infamous/ever-daunting "how clean is clean enough" question and indeed to facilitate a determination of what constitutes a reasonably "safe" or "acceptable" concentration of chemical constituents found at contaminated sites and in other human environments. This also includes an elaboration of a number of analytical relationships that can be adapted or used to estimate such "safe" cleanup levels that are necessary for environmental contamination restoration and risk management decisions.
- Chapter 12, *General Types of Contaminated Site Restoration Methods and Technologies*, offers a broad overview of several site restoration techniques that may be used in a variety of contaminated site cleanup programs; this consists of a general discussion and synoptic presentation of contaminated site restoration methods and technologies. This is in recognition of the fact that in order to have adequately performed site investigation activities and evaluate the feasibility of site remediation alternatives, a complete understanding of available remedial technologies is necessary.
- Chapter 13, *Contaminated Site Risk Management Stratagems: The Design of Effectual Corrective Action Programs for Contaminated Site Problems*, presents strategies and relevant protocols that can be used to aid the overall contaminated site restoration and management decision-making process. This also includes an elaboration of important

concepts, analytical tools, and key steps in the effectual design of typical contaminated site risk assessment and risk management programs.

- Chapter 14, *Turning Brownfields into Greenfields: Illustrative Paradigms of Risk-Based Solutions to Contaminated Site Problems*, enumerates the general scope for the application of risk assessment and risk-based solutions, as pertains to the management of potentially contaminated site problems. It also discusses specific practical example situations for the utilization of the risk assessment paradigm.

11 Risk-Based Restoration Goals and Cleanup Criteria for Contaminated Site Problems

An important and yet controversial issue that comes up in attempts to establish restoration goals and cleanup limits at contaminated sites relates to the notion of an "acceptable chemical exposure level" (ACEL). The *ACEL* may be considered as the concentration of a chemical in a particular medium or matrix that, when exceeded, presents significant risk of adverse impact to potential receptors. In fact, in a number of situations, the ACEL concept tends to drive the site restoration and risk management decision made about several contaminated site problems. However, the ACELs may not always result in "safe" or "tolerable" risk levels per se—in part due to the nature of the critical exposure scenarios, receptor-specific factors, and other conditions that are specific to the particular hazard situation. Under such circumstances, it becomes necessary to develop more stringent and health-protective levels that will meet the "safe" or "tolerable" risk level criteria.

This chapter presents discussions of how risk assessment may be used to help answer the infamous/ever-daunting "how clean is clean enough" question, and indeed to facilitate a determination of what constitutes a reasonably "safe" or "acceptable" concentration of chemical constituents found at contaminated sites and in other human environments. This also includes an elaboration of a number of analytical relationships that can be adapted or used to estimate such "safe" cleanup levels that are necessary for contaminated site restoration and risk management decisions. It should be acknowledged and emphasized, however, that the discussion here in this chapter focuses on a premise that human health risks tend to drive most cleanup decisions at potentially contaminated sites—albeit there are some situations when ecological receptors would become the reason for any site restoration decisions. In any event, the concepts presented here are fundamentally and generally applicable to ecological receptors as well—with elements very specific to ecological receptors having been identified in Chapter 10 of this book.

11.1 GENERAL CONSIDERATION IN DETERMINING "SAFE" LEVELS FOR CHEMICALS OF INTEREST

Traditionally, as a default assumption, most analysts have worked on the premise that for most types of chemical effects, there is a dose level below which a response is unlikely—mainly because homeostatic, compensation, and adaptive mechanisms in the cell of the affected organism will be expected to protect against toxic effects. For such reasons then, all chemical effects excluding cancer/genotoxicity have conventionally been assumed to have a "threshold"—i.e., a dose below which there is no probability of harm. Accordingly, a so-called "safe dose" has often been derived for these types of threshold effects. Indeed, even for the so-called "non-threshold" chemicals, a "quasi safe dose" may be derived based on the specific level of risk that is considered acceptable to the target populace.

11.1.1 REQUIREMENTS AND CRITERIA FOR ESTABLISHING RISK-BASED CHEMICAL EXPOSURE LEVELS AND CLEANUP OBJECTIVES

Risk-based chemical exposure levels (RBCELs) and target cleanup levels may generally be derived for various chemical sources and environmental matrices by manipulating the exposure and risk models previously presented in Chapter 9, and further elaborated in the literature elsewhere (e.g., Asante-Duah, 2017; ASTM, 1994a, 1995a,b; USEPA, 1991a,e–h, 1996j,k). Basically, this involves a "back calculation" process that yields a media concentration predicated on health-protective exposure parameters; as an example involving potential human exposures, the RBCEL generally should result in a target cumulative non-cancer hazard index of ≤ 1 and/or a target cumulative carcinogenic risk $\leq 10^{-6}$ for human receptors. On the whole, since risk is a function of both the chemical exposure and that chemical's toxicity, a complete understanding of the exposure scenarios together with an accurate determination of the constituent toxicity are key to developing "permissible" exposure levels that will be protective of human and/or ecological health.

By and large, the target RBCELs are typically established for both the carcinogenic and noncarcinogenic effects of the constituents of concern—with the more stringent value or outcome usually being selected as the acceptable health criterion or restoration goal; invariably, the carcinogenic limit tends to be more stringent in most situations where both values exist—albeit this is not necessarily true in all situations. Indeed, until recently, cancer risk was typically the driver in risk management decisions for any chemical evaluated with respect to both cancer risk and non-cancer hazard, particularly when risk management decisions emphasized the lower end of the excess lifetime cancer risk spectrum. However, some more recent experiences indicate that risk managers should be cognizant of the fact that there can be situations where risk management decisions could be driven by non-cancer endpoints; for instance, the reference dose (RfD) and reference concentration (RfC) values of some chemicals (e.g., trichloroethylene, or TCE) had to be revised to lower levels (i.e., indicative of potentially greater toxicity) in recent times.

Further to the above, and within the general procedural framework, the following criteria and general guidelines may additionally be used to facilitate the process of establishing media-specific RBCELs and/or site restoration goals:

- Assuming dose additivity, $\displaystyle\sum_{j=1}^{p}\sum_{i=1}^{n}\frac{\mathrm{CMAX}_{ij}}{\mathrm{RBCEL}_{ij}} < 1$

  where CMAX_{ij} is the prevailing maximum concentration of constituent i in product or matrix j, and RBCEL_{ij} is the risk-based chemical exposure level for constituent i in product or matrix j.
- In developing site restoration goals or EQC, it usually is necessary to establish a target level of risk for the constituents of concern; such standards are generally established within the cancer risk range of 10^{-7} to 10^{-4} (with a lifetime excess cancer risk of 10^{-6} normally used as a point-of-departure) and a non-cancer hazard index of 1.
- It is recommended that the cumulative risk posed by multiple chemical constituents not exceed a 10^{-4} cancer risk and/or a hazard index of unity for human receptors.
- If sensitive populations (including vulnerable persons, such as children and the sick—and also sensitive ecosystems and habitats, or threatened or endangered species) are to be protected, then more stringent standards may be required.
- If nearby populations are exposed to hazardous constituents from other sources, lower target levels may generally be required than would ordinarily be necessary.
- If exposures to certain hazardous constituents occur through multiple pathways and/or routes, lower target levels should generally be prescribed.

- If sensitive or critical ecological receptors could become the driving force behind remediation decisions, then the ecological risk quotients (see Chapter 11) should be more carefully evaluated in the decisions about the site restoration goals.

Indeed, if/when the above conditions are satisfied, the corresponding RBCEL may be viewed as representing a maximum acceptable constituent level that will likely be sufficiently protective of public and environmental health. At the end of the day, exceeding the RBCEL will usually call for the development and implementation of a corrective action and/or effectual risk management plan.

11.1.2 THE MECHANICS OF DETERMINING RBCELS

After defining the critical exposure pathways/routes and exposure scenarios associated with a contaminated site problem and/or appropriate for a given a chemical exposure problem, it generally becomes possible to estimate a corresponding RBCEL that would not pose significant risks to an exposed population. To determine the RBCEL for a chemical compound, algebraic manipulations of the hazard index and/or carcinogenic risk equations together with the exposure estimation equations discussed in Chapter 9 can be used to arrive at the appropriate analytical relationships. The step-wise computational efforts involved in this exercise consist of a "back calculation" process that yields a media concentration predicated on health-protective exposure parameters; as an example, the RBCEL generally results in a cumulative non-cancer hazard index of ≤ 1 and/or a cumulative carcinogenic risk $\leq 10^{-6}$.

The processes involved in the determination of the RBCELs are summarized in the proceeding sections. In practice, for chemicals with carcinogenic effects, a target risk of (1×10^{-6}) is typically used in the "back calculation" exercise, and a target hazard index of 1.0 is typically used for non-carcinogenic effects; for substances that are both carcinogenic as well possess systemic toxicity properties, the lower of the resulting carcinogenic or noncarcinogenic criterion would typically be used for the relevant public health risk management action or decision.

11.1.2.1 Human RBCELs for Carcinogenic Constituents

As discussed in Chapter 9, the cancer risk (CR) associated with the principal human exposure routes (comprised of inhalation, ingestion, and dermal exposures) may be represented by the following equation:

$$\begin{aligned}
CR &= \left\{ \sum_{i=1}^{p} CDI_p \times SF_p \right\} \\
&= \left[CDI_i \times SF_i \right]_{\text{inhalation}} + \left[CDI_o \times SF_o \right]_{\text{ingestion}} + \left[CDI_d \times SF_o \right]_{\text{dermal contact}} \\
&\equiv C_m \left\{ \left[INHf \times SF_i \right] + \left[INGf \times SF_o \right] + \left[DEXf \times SF_o \right] \right\}
\end{aligned} \tag{11.1}$$

where the CDIs represent the chronic daily intakes, adjusted for absorption (mg/kg-day); INHf, INGf, and DEXf represent the inhalation, ingestion, and dermal contact "intake factors," respectively (see Chapter 9); C_m is the chemical concentration in environmental/exposure matrix of concern; and the SFs are the route-specific cancer slope factors; and the subscripts i, o, and d refer to the inhalation, oral ingestion, and dermal contact exposures, respectively.

The above model can be reformulated to calculate the carcinogenic RBCEL (viz., RBCELc) for the environmental/exposure media of interest. This involves "back calculating" from the chemical intake equations presented in Chapter 9 for inhalation, ingestion, and dermal contact exposures. Hence,

$$\text{RBCEL}_c = C_m = \frac{\text{CR}}{\left\{\left[\text{INHf} \times \text{SF}_i\right] + \left[\text{INGf} \times \text{SF}_o\right] + \left[\text{DEXf} \times \text{SF}_o\right]\right\}} \qquad (11.2)$$

For illustrative purposes, let us assume that there is only one chemical constituent present in soils at a hypothetical contaminated site; furthermore, assume that exposures via the dermal and ingestion routes are the only pathways contributing to, or at least dominating, the total target carcinogenic risk (of, say $\text{CR} = 10^{-6}$). Hence,

$$\text{CDI} = \frac{\text{CR}}{\text{SF}_o} = \text{RSD}$$

or,

$$\left(\text{CDI}_{ing} + \text{CDI}_{der}\right) = \frac{\text{CR}}{\text{SF}_o}$$

i.e.,

$$\frac{\left(\text{RBC}_c \times \text{SIR} \times \text{CF} \times \text{FI} \times \text{ABS}_{si} \times \text{EF} \times \text{ED}\right)}{\left(\text{BW} \times \text{AT} \times 365\right)}$$

$$+ \frac{\left(\text{RBC}_c \times \text{CF} \times \text{SA} \times \text{AF} \times \text{ABS}_{sd} \times \text{SM} \times \text{EF} \times \text{ED}\right)}{\left(\text{BW} \times \text{AT} \times 365\right)} = \frac{\text{CR}}{\text{SF}_o}$$

Consequently,

$$\text{RBCEL}_c = \frac{\left(\text{BW} \times \text{AT} \times 365\right) \times \left(\text{RSD}\right)}{\left(\text{CF} \times \text{EF} \times \text{ED}\right)\left\{\left(\text{SIR} \times \text{FI} \times \text{ABS}_{si}\right) + \left(\text{SA} \times \text{AF} \times \text{ABS}_{sd} \times \text{SM}\right)\right\}}$$

where RSD represents the risk-specific dose, defined by the ratio of the target risk to the slope factor. Indeed, the estimated RBCEL may serve as surrogate for a health-based ACEL—albeit some case-specific adjustments will usually be required, in order to arrive at a true ACEL used in public health risk management decisions.

11.1.2.2 Health-Based ACELs for Carcinogenic Chemicals

As health-based criteria, ACELs for carcinogens may be determined in a similar manner to the so-called "virtually safe dose" (VSD) of a carcinogenic chemical constituent. A *VSD* is the daily dose of a carcinogenic chemical that, over a lifetime, will result in an incidence of cancer at a specified risk level; usually, this is calculated based on the appropriate de minimis risk level.

The governing equation for calculating ACELs for carcinogenic constituents is shown in Box 11.1. This model—developed from algorithms and concepts presented earlier on in Chapter 9—assumes that there is only one chemical constituent involved in the problem situation. In other situations where several chemicals may be of concern, it is assumed (for simplification purposes) that each carcinogen has a different mode of biological action and target organs. Each of the carcinogens is, therefore, assigned 100% of the "acceptable" excess carcinogenic risk (typically equal to $[1 \times 10^{-6}]$) in calculating the health-based ACELs; in other words, the excess carcinogenic risk is not allocated among the carcinogens.

BOX 11.1   GENERAL EQUATION FOR CALCULATING ACCEPTABLE CHEMICAL EXPOSURE LEVELS FOR CARCINOGENIC CONSTITUENTS

$$\text{ACEL}_c = \frac{(R \times \text{BW} \times \text{LT} \times \text{CF})}{(\text{SF} \times I \times A \times \text{ED})}$$

where

ACEL_c = acceptable chemical exposure level (equivalent to the VSD) in medium of concern (e.g., mg/kg in food; mg/L in water)

R = specified benchmark risk level, usually set at 10^{-6} (dimensionless)

BW = body weight (kg)

LT = assumed lifetime (years)

CF = conversion factor (equals 10^6 for ingestion exposure from solid materials; 1.00 for ingestion of fluids)

SF = cancer slope factor ($[\text{mg/kg-day}]^{-1}$)

I = intake assumption (mg/day for solid material ingestion rate; L/day for fluid ingestion)

A = absorption factor (dimensionless)

ED = exposure duration (years)

11.1.2.3   Human RBCELs for Noncarcinogenic Effects of Chemical Constituents

As discussed in Chapter 9, the hazard index (HI) associated with the principal human exposure routes (comprised of inhalation, ingestion, and dermal exposures) may be represented by the following equation:

$$
\begin{aligned}
\text{HI} &= \left\{ \sum_{i=1}^{p} \frac{\text{CDI}_p}{\text{RfD}_p} \right\} \\
&= \left[\frac{\text{CDI}_i}{\text{RfD}_i} \right]_{\text{inhalation}} + \left[\frac{\text{CDI}_o}{\text{RfD}_o} \right]_{\text{ingestion}} + \left[\frac{\text{CDI}_d}{\text{RfD}_o} \right]_{\text{dermal contact}} \\
&\equiv C_m \left\{ \left[\frac{\text{INHf}}{\text{RfD}_i} \right] + \left[\frac{\text{INGf}}{\text{RfD}_o} \right] + \left[\frac{\text{DEXf}}{\text{RfD}_o} \right] \right\}
\end{aligned}
\tag{11.3}
$$

where the CDIs represent the chronic daily intakes, adjusted for absorption (mg/kg-day); INHf, INGf, and DEXf represent the inhalation, ingestion, and dermal contact "intake factors," respectively (see Chapter 9); C_m is the chemical concentration in environmental/exposure matrix of concern; and the RfDs are the route-specific reference doses; the subscripts i, o, and d refer to the inhalation, oral ingestion and dermal contact exposures, respectively.

The above model can be reformulated to calculate the noncarcinogenic RBCEL (viz., RBCEL_{nc}) for the environmental/exposure media of interest. This is derived by "back calculating" from the chemical intake equations presented in Chapter 9 for inhalation, ingestion, and dermal contact exposures. Hence,

$$
\text{RBCEL}_{nc} = C_m = \frac{1}{\left\{ \left[\dfrac{\text{INHf}}{\text{RfD}_i} \right] + \left[\dfrac{\text{INGf}}{\text{RfD}_o} \right] + \left[\dfrac{\text{DEXf}}{\text{RfD}_o} \right] \right\}}
\tag{11.4}
$$

For illustrative purposes, assume that there is only one chemical constituent present in soils at a hypothetical contaminated site, and that only exposures via the dermal and ingestion routes contribute to, or at least dominate the total target hazard index (of HI = 1). Then,

$$CDI = RfD$$

or

$$\left(CDI_{ing} + CDI_{der}\right) = RfD_o$$

i.e.,

$$\frac{\left(RBC_{nc} \times SIR \times CF \times FI \times ABS_{si} \times EF \times ED\right)}{\left(BW \times AT \times 365\right)}$$

$$+ \frac{\left(RBC_{nc} \times CF \times SA \times AF \times ABS_{sd} \times SM \times EF \times ED\right)}{\left(BW \times AT \times 365\right)} = RfD_o$$

Consequently,

$$RBCEL_{nc} = \frac{\left(BW \times AT \times 365\right) \times \left(RfD_o\right)}{\left(CF \times EF \times ED\right)\left\{\left(SIR \times FI \times ABS_{si}\right) + \left(SA \times AF \times ABS_{sd} \times SM\right)\right\}}$$

assuming a benchmark hazard index of unity. Indeed, the estimated RBCEL may serve as surrogate for a health-based ACEL—albeit some case-specific adjustments will usually be required, in order to arrive at a true ACEL used in public health risk management decisions.

11.1.2.4 Health-Based ACELs for Noncarcinogenic Chemicals

As health-based criteria, ACELs for noncarcinogens may be determined in a similar manner to the so-called "allowable daily intakes" (ADIs) of a noncarcinogenic chemical constituent. The ADI represents the threshold exposure limit below which no adverse effects are anticipated.

The governing equation for calculating ACELs for noncarcinogenic effects (i.e., the systemic toxicity) of chemical constituents is shown in Box 11.2. This model—derived from algorithms and concepts presented earlier on in Chapter 9—assumes that there is only one chemical constituent involved. In situations where several chemicals may be of concern, it is assumed (for simplification purposes) that each chemical has a different organ-specific noncarcinogenic effect. Otherwise, the right hand side may be multiplied by a "percentage factor" to account for contribution to hazard index by each noncarcinogenic chemical subgroup—or may indeed be appropriately manipulated by other methods.

**BOX 11.2   GENERAL EQUATION FOR CALCULATING
ACCEPTABLE CHEMICAL EXPOSURE LEVELS FOR
NONCARCINOGENIC EFFECTS OF SYSTEMIC TOXICANTS**

$$ACEL_{nc} = \frac{\left(RfD \times BW \times CF\right)}{\left(I \times A\right)}$$

where

$ACEL_{nc}$ = acceptable chemical exposure level in medium of concern (e.g., mg/kg in food; mg/L in water)

RfD = reference dose (mg/kg-day)

BW = body weight (kg)

CF = conversion factor (equals 10^6 for ingestion exposure from solid materials; 1.00 for fluid ingestion)

I = intake assumption (mg/day for solid material ingestion rate; L/day for fluid ingestion)

A = absorption factor (dimensionless)

11.1.3 MISCELLANEOUS METHODS FOR ESTABLISHING ENVIRONMENTAL QUALITY GOALS

Several possibilities exist to use various analytical tools in the development of alternative and/or media-specific chemical exposure concentration limits and environmental quality or restoration goals. Some select general procedures commonly employed in establishing environmental quality goals are briefly annotated below. Broadly speaking, these approaches represent reasonably conservative ways of setting environmental quality goals. Thus, the use of such methods will generally ensure that risks are not underestimated—which tantamount to situations that invariably result in a reasonably adequate protection of public health and the environment.

11.1.3.1 Determination of Risk-Specific Concentrations in Air

The estimation of health-protective concentrations of chemical constituents in air must generally take into account the toxicity of the chemicals of potential concern (CoPCs), as well as the potential exposure scenarios and parameters of individuals breathing the impacted air. By employing the risk assessment concepts and methodologies discussed in Chapter 9, risk-specific concentrations of chemicals in air may be estimated from the unit risk in air as follows:

$$
\text{Air concentration}\left[\mu g/m^3\right] = \frac{[\text{specified risk level}] \times [\text{body weight}]}{SF_i \times [\text{inhalation rate}] \times 10^{-3}}
$$

$$
= \frac{[\text{specified risk level}]}{URF_i} = \frac{1 \times 10^{-6}}{URF_i}
\tag{11.5}
$$

The assumptions generally used for such computations involve a stipulated risk level of 10^{-6}, a 70-kg body weight, and an average inhalation rate of $20\,m^3/day$.

11.1.3.2 Determination of Risk-Specific Concentrations in Water

The estimation of health-protective concentrations of chemical constituents in drinking water must generally take into account the toxicity of the chemicals of potential concern, as well as the potential exposure scenarios and parameters of individuals using the water. By employing the risk assessment concepts and methodologies discussed in Chapter 9, risk-specific concentrations of chemicals in drinking water can be estimated from the oral slope factor. The water concentration corrected for an upper-bound increased lifetime risk of R $(=10^{-6})$ is given by:

$$
\text{Water concentration}\left[\text{mg/L}\right] = \frac{[\text{specified risk level}] \times [\text{body weight}]}{SF_o \times [\text{ingestion rate}]}
$$

$$
= \frac{\text{specified risk level}}{URF_o}
\tag{11.6}
$$

The assumptions generally used for such computations involve a stipulated risk level of 10^{-6}, a 70-kg body weight, and an average water ingestion rate of 2 L/day so that:

$$
\text{Water concentration}\left[\text{mg/L}\right] = \frac{1 \times 10^{-6} \times 70\,\text{kg}}{SF_o \left(\text{mg/kg/day}\right)^{-1} \times 2\,\text{L/day}} = \frac{3.5 \times 10^{-5}}{SF_o}
$$

It is noteworthy that, in general, the estimation of health-protective concentrations of chemical constituents in drinking water that results in negligible risk outcomes must also account for the fact that tap water is typically used directly as drinking water, as well as for preparing foods and beverages, etc. Indeed, the water may also be typically used for bathing/showering, in washing clothes and dishes, flushing of toilets, and in a variety of other household uses—some of which could result in potential dermal and inhalation exposures as well. To allow for these additional exposures, therefore, the assumed daily volume of water consumed by an adult may typically be increased from the default value of 2 L/day indicated above, to say 3 L-equivalents/day (L_{eq}/day).

11.2 "HOW CLEAN IS CLEAN ENOUGH" FOR A CONTAMINATED SITE PROBLEM?

Numerous groups of peoples or individuals around the world may become exposed to a barrage of chemical compounds emanating from contaminated sites that exist in their communities. Typically, risk assessments (which allow the population exposures to be estimated by measurements and/ or models) assist in the determination and management of potential health and environmental problems that could be expected or anticipated from the presence of various chemicals in the potential receptor environments—including the ecosystems. Meanwhile, it is apparent that the exposure assessment component of the processes involved tends to be particularly complicated, although not insurmountable—especially because of the huge diversity in exposure patterns and chemical constituents. There is also the additional issue of intermittent exposures to variable amounts and types of environments containing varying concentrations of chemical compounds (van Veen, 1996; Vermeire et al., 1993). Notwithstanding, risk-based analyses can be carefully designed to help evaluate the "safe levels" of chemicals that appear in various human and ecological environments.

Generally speaking, contaminated site risk is a function of exposure and toxicity—determined primarily based on the exposure patterns/rates and the toxicity of the chemical components of concern. This can be represented by the following conceptual expression:

$$\text{Risk} = f\left(\text{Exposure}, \text{Toxicity}\right) \tag{11.7}$$

or,

$$\text{'Safe Level'} \propto \frac{1}{\text{Risk}} = \frac{1}{f(\text{Exposure}, \text{Toxicity})} \tag{11.8}$$

At the end of the day, for a particular contaminated site to be classified as being reasonably safe for reuse, the chemical-specific exposure dose should generally be less than the chemical's "acceptable" daily intake—defined as the daily intake level for a chemical that represents no anticipated significant risk to the exposed receptor. In practical terms for most contaminated site problems, soils and groundwater tend to represent the most significant media of concern; the computational procedures for developing risk-based restoration goals for these environmental media are therefore elaborated in Sections 11.4 and 11.5. The same principles can indeed be extended to the formulation and development of target levels for a variety of other environmental matrices, as and when necessary.

11.2.1 ESTABLISHING "TOLERABLE" CHEMICAL CONCENTRATIONS FOR THE PROTECTION OF HUMAN HEALTH

The 16th century Swiss philosopher, physician, and alchemist, Paracelsus, indicated once upon a time that only the dose of a substance determines its toxicity; this notion makes it even more difficult to ascertain the levels that constitute hazardous human exposure to chemicals. However, careful application of risk assessment and risk management principles and tools should generally help remove some of the fuzziness in defining the cutoff line between what may be considered a "safe level" and what apparently might be a "dangerous level" for most chemicals (Asante-Duah, 2017).

In general, chemicals traditionally found in human environments (including those occurring at contaminated sites) may be classified into two broad categories—viz.: carcinogenic and noncarcinogenic materials. The methods for deriving the "acceptable" daily intakes and/or "tolerable" concentrations for such chemicals are generally based on procedures/protocols presented earlier on in Chapter 9; the general concepts are briefly annotated below.

11.2.1.1 "Acceptable" Daily Intake and "Tolerable" Concentration for Carcinogens

The "acceptable" daily intake for carcinogenic materials appearing in the human environment may be estimated by using the following approximate relationships:

$$\text{ADI}_{\text{carcinogen}} = \frac{\left[\text{TR} \times \text{AT} \times 365 \text{ day/year} \right]}{[\text{ED} \times \text{EF} \times \text{SF}]} \tag{11.9}$$

Thence, the "tolerable" chemical concentration for carcinogens [$\text{TC}_{\text{carcinogen}}$] (mg/kg or mg/L) in the human environment will be defined by,

$$\text{TC}_{\text{carcinogen}} = \frac{\left[\text{ADI}_{\text{carcinogen}} \times \text{BW} \right]}{[\text{FR} \times \text{CR} \times \text{ABS}]} \times \text{CF} \tag{11.10}$$

where $\text{ADI}_{\text{carcinogen}}$ is the "acceptable" daily intake for the carcinogenic materials (mg/kg-day); TR is the generally acceptable risk level (usually set at 10^{-6}); AT is the averaging time (years); ED is the exposure duration (year); EF is the exposure frequency (day/year); SF is the cancer potency or slope factor ($[\text{mg/kg-day}]^{-1}$); BW is the average body weight (kg); FR is the fraction of consumed material that is assumed to be contaminated; CR is the consumption rate (kg/day or L/day); ABS is the % absorption rate; and CF is a conversion factor to help maintain the dimensional tractability of the algorithm.

11.2.1.2 "Acceptable" Daily Intake and "Tolerable" Concentration for Noncarcinogens

The "acceptable" daily intake for noncarcinogenic materials appearing in the human environment may be estimated by using the following approximate relationship:

$$\text{ADI}_{\text{noncarcinogen}} = \frac{[\text{HQ} \times \text{AT} \times 365 \text{ day/year} \times \text{RfD}]}{[\text{ED} \times \text{EF}]} \tag{11.11}$$

Thence, the "tolerable" chemical concentration for noncarcinogens [$\text{TC}_{\text{noncarcinogen}}$] (mg/kg or mg/L) in the human environment will be defined by:

$$\text{TC}_{\text{noncarcinogen}} = \frac{\left[\text{ADI}_{\text{noncarcinogen}} \times \text{BW} \right]}{\text{FR} \times \text{CR} \times \text{ABS}} \times \text{CF} \tag{11.12}$$

where $\text{ADI}_{\text{noncarcinogen}}$ is the "acceptable" daily intake for the noncarcinogenic materials (mg/kg-day); HQ is the generally acceptable hazard level (usually set at 1); AT is the averaging time (years); ED is the exposure duration (year); EF is the exposure frequency (day/year); RfD is the non-cancer reference dose or acceptable daily intake (mg/kg-day); BW is the average body weight (kg); FR is the fraction of consumed material that is assumed to be contaminated; CR is the consumption rate (kg/day or L/day); ABS is the % absorption rate; and CF is a conversion factor to help maintain the dimensional tractability of the algorithm.

11.2.2 Determination of "Allowable" Ecological Limits

Similar to the preceding discussions (i.e., Section 11.2.1), the methods for deriving the "acceptable" daily intakes and/or "tolerable" concentrations for chemicals of potential ecological concern (CoPECs) are generally based on procedures/protocols presented earlier on in Chapter 9 and also Chapter 10. Thus, the general concepts for the derivation of risk-based ecological limits are also fundamentally similar to those annotated in the proceeding sections for human receptors. Still, it must be reiterated here that several of certain elements that are very specific to ecological receptors (such as have been identified in Chapter 10 of this title) should be carefully integrated into the overall process. Furthermore, it must be acknowledged that to enable good modeling results, the derivation of the ecological limits usually will call for relatively more experimental and/or laboratory studies (including bioassays and related toxicity testing)—i.e., in comparison to the requirements for deriving risk-based levels predicated on human exposures only.

11.2.3 Miscellaneous Methods for Deriving Site Cleanup Criteria

Several possibilities exist to use various analytical tools in the development of alternative and media-specific contaminant concentration limits and environmental restoration goals. Some select general procedures commonly employed in establishing environmental quality goals are briefly annotated below. Broadly speaking, these approaches represent reasonably conservative ways of setting environmental restoration goals—and which use tantamount to being protective of public health and the environment, by ensuring that risks are generally not underestimated.

11.2.3.1 Use of Contaminant Equilibrium Partitioning Coefficients

EQC may be derived from correlational analyses of the partitioning of chemicals between environmental compartments. For instance, based on the assumption that the distribution of contaminants among various compartments in sediment is controlled by continuous equilibrium exchanges, chemical-specific partition coefficients can be used to predict contaminant concentrations in sediment, biota, and/or water. That is, sediment–water equilibrium partitioning concepts can be used to predict contaminant concentrations in sediment and/or water. Similarly, sediment–biota partition coefficient can be determined and used to predict distribution of the contaminant between sediment and benthic organism and/or interstitial water, and benthic organism (assuming bioaccumulation factors are constant and independent of organism or sediment). Other equilibrium relationships can indeed be employed for the estimation of various EQC.

Of special interest in regards to this concept is the propensity for contaminated soils to function as a reservoir of contaminated mass that can degrade groundwater quality over the long term (Calabrese et al., 1996). While generic soil cleanup standards are typically not available, generic groundwater standards usually exist; thus, the latter can be used in the framework of equilibrium partitioning concepts with the intention of deriving appropriate soil cleanup standards. The model most commonly used for this type of evaluation has been predicated on the assumption that the contaminated soils exist in equilibrium with the affected groundwater such that:

$$K_d = \frac{C_s}{C_{gw}}$$

where K_d is the soil–water partitioning coefficient—and $K_d = K_{oc} \times f_{oc}$; K_{oc} is the compound-specific organic carbon partitioning coefficient; f_{oc} is the soil organic carbon content; C_s is source soil concentration; C_{gw} is receiving groundwater media concentration. By carrying out an algebraic manipulation of the equilibrium partitioning equation, a soil cleanup level that should help meet applicable groundwater standards can be derived as follows:

$$C_{s\text{-limit}} = K_d \times C_{gw\text{-std}}$$

where $C_{\text{s-limit}}$ is required source soil cleanup limit; $C_{\text{gw-std}}$ is receiving groundwater media standard or guideline to be maintained. It must be acknowledged that this represents a rather simplistic conservative screening approach—since, among other things, the method is not quite sensitive to the site-specific conditions that govern contaminant transport through the unsaturated zone into groundwater.

11.2.3.2 Use of Attenuation–Dilution Factors

In the past, environmental restoration goals have been established that either account for overall attenuation (except for dilution effects) or simply dilution only (e.g., Brown, 1986; Dawson and Sanning, 1982; Santos and Sullivan, 1988; USEPA, 1987a,d,f,g). To take account of overall contaminant attenuation along with dilution effects, a relationship that integrates total attenuation and net dilution concurrently seems appropriate, if this is to become the basis for a realistic environmental restoration program. For example, in the application of attenuation-dilution factors as the basis for developing site cleanup limits for a potentially contaminated site, the following relationship that accounts for both total attenuation and net dilution concurrently may be adopted for such purposes:

$$\text{SCL} = \{\text{Std}\} \times \Pi_m \{\text{AF}\} \times \Pi_m \{\text{DF}\}$$

where SCL is source/soil cleanup level (mg/kg); Std is receiving media criteria or regulatory standard to be attained (e.g., drinking water standard); $\Pi_m\{\text{AF}\}$ is cumulative attenuation factor, that defines the loss of contaminant during transport (i.e., product of the intermedia distribution constants); and $\Pi_m\{\text{DF}\}$ is cumulative dilution factor during transport (i.e., product of the ratios of receiving media to source concentrations). For a single intermedia transfer, the soil cleanup level may be estimated by a more simplified relationship; if, for instance, leachate migrates from soil into groundwater that is a source of a drinking water supply, the SCL may be estimated as follows:

$$\text{SCL} = \text{DWS} \times \text{AF} \times \text{DF}$$

where DWS is drinking water standard; AF is attenuation of contaminant in soil (typical values in the range 1–1,000); DF is dilution of contaminant by groundwater (typical values in the range 1–100).

In fact, several parameters should be carefully evaluated when establishing a site-specific attenuation-dilution factor—including distance to the point of exposure or nearest downgradient well; groundwater flow rates; soil organic content; and contaminant degradation rates. Meanwhile, it is worth the mention here that alternate cleanup goals based simply on attenuation or dilution mechanisms may not quite be acceptable unless if accompanied by an in-depth evaluation and discussion of all pertinent processes involved.

11.2.3.3 Mass Balance Analyses

Simple mass balance analytical relationships can be applied between various environmental compartments in order to derive environmental restoration goals. For intermedia transfer of contaminants, current contaminant loadings may be coupled with allowable loadings (as represented by available media standards), and a back-modeling procedure can then be used to obtain the required EQC, according to the following simple relationship:

$$C_{\max} = \frac{C_{\text{std}}}{C_r} \times C_s$$

where $C_{\max}$ is the maximum acceptable source concentration; C_{std} is receiving media environmental quality criteria for target receptors (i.e., regulatory standard); C_r is receiving media concentration; and C_s is prevailing source concentration.

For example, consider a situation where water quality standards exist that should be met for a creek adjoining a contaminated site. By performing back calculations, based on contaminant concentrations in the creek as a result of the current constituents loading from the site, a conservative estimate can be made for C_{max}. The use of such concentration limit in a corrective action process ensures that the creek is not adversely impacted, based on the exposure scenario defined for the site. Consequently, if the site is cleaned up to such levels, then the surface water quality is not likely to be impacted.

11.2.4 What Is the "Desirable"/"Preferable" Health-Protective Residual Chemical Level for Corrective Action Decisions?

Oftentimes, the RBCEL that has been established based on an acceptable risk level or hazard index are for a single contaminant in one environmental matrix or exposure medium. Consequently, the risk and hazard associated with multiple contaminants in a multi-media setting are not fully accounted for during the "back-modeling" process used to establish the RBCELs. In contrast, the evaluation of risks associated with a given chemical exposure problem usually involves a set of equations designed to estimate hazard and risk for several chemicals, and for a multiplicity of exposure routes. Under this latter type of scenario, the computed "acceptable" risks could indeed exceed the health-protective limits; accordingly, it becomes necessary to establish a modified RBCEL for the requisite environmental or public health risk management decision. To obtain the "modified RBCEL," the "acceptable" chemical exposure level is estimated in the same manner as elaborated earlier on in the preceding (and related) sections—but with the cumulative effects of multiple chemicals being taken into account through a process of apportioning the target risks and hazards among all the CoPCs.

11.2.4.1 The "Modified RBCEL" for Carcinogenic Chemicals

A modified RBCEL for carcinogenic constituents may be derived by the application of a "risk disaggregation factor"—that allows for the apportionment of risk among all CoPCs. That is, the new RBCEL may be estimated by proportionately aggregating (or perhaps rather disaggregating) the target cancer risk among the CoPCs, and then using the corresponding target risk level in the equation presented earlier on in Box 11.1. The assumption used for apportioning the excess carcinogenic risk may be one that considers all carcinogens as having the same mode of biological actions and target organs; otherwise, excess carcinogenic risk is not apportioned among carcinogens, but rather each assumes the same value in the computational efforts. A more comprehensive approach to "apportioning" or "allotting" risks would involve more complicated mathematical manipulations—such as by the use of linear programming algorithms.

In general, the acceptable risk level may be apportioned between the chemical constituents contributing to the overall target risk by assuming that each constituent contributes equally or proportionately to the total acceptable risk. The "risk fraction" obtained for each constituent can then be used to derive the modified RBCEL—by working from the relationships established previously for the computation of RBCELs (Section 11.1.2) or comparable ones; by utilizing the approach for estimating media RBCELs, the modified RBCEL is derived in accordance the following approximate relationship:

$$RBCEL_{c\text{-mod}} = \frac{[\%] \times CR}{\left\{ \left[INHf \times SF_i \right] + \left[INGf \times SF_o \right] + \left[DEXf \times SF_o \right] \right\}}$$

(11.13)

All the terms are the same as defined previously in Section 11.1.2, and [%] represents the proportionate contribution from a specific chemical constituent to the overall target risk level. One may also choose to use "weighting factors" in apportioning the chemical contributions to the target risk levels; for instance, this could be based on carcinogenic classes—such that "Class A" carcinogens are given twice as much weight as "Class B," etc., or chemicals posing carcinogenic risk via all exposure routes are given more weight than those presenting similar risks via specific routes only. Overall, the use of the modified RBCEL approach will likely ensure that the sum of risks from all the chemicals involved over all exposure pathways is less than or equal to the set target de minimis risk (e.g., $\leq 10^{-6}$).

11.2.4.2 The "Modified RBCEL" for Noncarcinogenic Constituents

A modified RBCEL for noncarcinogenic constituents may be derived by application of a "hazard disaggregation factor"—that allows for the apportionment of target hazard index among all CoPCs. That is, the new RBCEL may be estimated by proportionately aggregating (or perhaps rather disaggregating) the non-cancer hazard index among the CoPCs, and then using the corresponding target hazard level in the equation presented earlier on in Box 11.2.

In general, the acceptable hazard level may be apportioned between the chemical constituents contributing to the overall hazard index by assuming that each constituent contributes equally or proportionately to the total acceptable hazard index—all the while accounting for commonality in endpoint effects as well. The "hazard fraction" obtained for each constituent can then be used to derive the modified RBCEL—by working from the relationships established previously for the computation of RBCELs (Section 11.1.2) or comparable ones. By using the approach to estimating media RBCELs, the modified RBCEL is derived in accordance the following approximate relationship for noncarcinogenic effects of chemicals having the same toxicological endpoints:

$$RBCEL_{nc\text{-}mod} = \frac{[\%] \times 1}{\left\{ \left[\frac{INHf}{RfD_i} \right] + \left[\frac{INGf}{RfD_o} \right] + \left[\frac{DEXf}{RfD_o} \right] \right\}} \qquad (11.14)$$

All the terms are the same as defined previously in Section 11.2.1, and [%] represents the proportionate contribution from a specific chemical constituent to the overall target hazard index for the noncarcinogenic effects of chemicals with same physiologic endpoint. Overall, the use of the modified RBCEL approach will ensure that the sum of hazard quotients over all exposure pathways for all chemicals (with the same physiologic endpoints) is less than or equal to the hazard index criterion of 1.0.

11.2.4.3 Incorporating Degradation Rates into the Estimation of Environmental Restoration/Quality Criteria

The effect of chemical degradation is not incorporated into estimated RBCELs often enough. However, since exposure scenarios used in calculating the RBCELs or similar criteria usually make the assumption that exposures could be occurring over long time periods (up to a lifetime of 70 years), it is prudent, at least in a detailed analysis, to consider the fact that degradation or other transformation of the CoPC could occur. Under such circumstances, the degradation properties of the CoPCs should be carefully evaluated. Subsequently, an adjusted RBCEL (or its equivalent) can be estimated—that is based on the original RBCEL (or equivalent), a degradation rate coefficient, and the specified exposure duration. The new adjusted RBCEL is then given by:

$$RBCEL_a = \frac{RBC}{\text{degradation factor (DGF)}} \qquad (11.15)$$

where $RBCEL_a$ is the adjusted RBCEL or its equivalent, and that incorporates a degradation rate coefficient. Assuming first-order kinetics, as an example, an approximation of the degradation effects can be obtained as follows:

$$\text{DGF} = \frac{\left(1 - e^{-kt}\right)}{kt} \qquad (11.16)$$

where k is a chemical-specific degradation rate constant [days^{-1}], and t is time period over which exposure occurs [days]. For a first-order decaying substance, k is estimated from the following relationship:

$$T_{1/2}[\text{days}] = \frac{0.693}{k} \quad \text{or} \quad k\left[\text{days}^{-1}\right] = \frac{0.693}{T_{1/2}} \qquad (11.17)$$

where $T_{1/2}$ is the half-life, which is the time after which the mass of a given substance will be one-half its initial value. Consequently,

$$\text{RBCEL}_a = \text{RBCEL} \times \frac{kt}{\left(1 - e^{-kt}\right)} \qquad (11.18)$$

This relationship assumes that a first-order degradation/decay is occurring during the complete exposure period; decay/degradation is initiated at time, $t = 0$ years; and the RBCEL is the average allowable concentration over the exposure period. In fact, if significant degradation is likely to occur, the RBCEL$_a$ calculations become much more complicated; in that case, predicated source chemical levels must be calculated at frequent intervals and summed over the exposure period.

11.3 ESTABLISHING RISK-BASED CLEANUP LIMITS FOR SOILS AT CONTAMINATED SITES

In addressing potentially contaminated site problems, soils can become the major focus of attention in the risk management decisions involved; this is because soils at such sites could serve as a major long-term reservoir for chemical contaminants—usually with the capacity to release contamination into several other environmental media. As such, the importance of soil cleanup for such contaminated sites cannot be overemphasized. In fact, the soil media typically requires particularly close attention in most risk-based evaluations carried out for contaminated sites—albeit groundwater contaminant plumes underlying such sites are proving to be equally, if not more, problematic in some situations.

In order to determine the risk-based cleanup level for a chemical compound present in soils at a contaminated site, algebraic manipulations of the hazard index and/or carcinogenic risk equations together with the exposure estimation equations discussed in Chapter 9 can be used to arrive at the appropriate analytical relationships. The step-wise computational efforts involved in this exercise consist of a "back calculation" process that yields an acceptable soil concentration (ASC) predicated on health-protective exposure parameters; as a classic example, the ASC generally results in a target cumulative non-cancer hazard index of ≤ 1 and/or a target cumulative carcinogenic risk $\leq 10^{-6}$. Indeed, for chemicals with carcinogenic effects, a target risk of $[1 \times 10^{-6}]$ is typically used in the "back calculation," and a target hazard index of 1.0 is typically used for noncarcinogenic effects. The processes involved in the determination of the ASCs are summarized in the sections that follow directly below. For substances that are both carcinogenic and possess systemic toxicity properties, the lower of the carcinogenic or noncarcinogenic criterion would characteristically be used for the relevant site restoration and/or risk management decisions.

11.3.1 SOIL CHEMICAL LIMITS FOR CARCINOGENIC CONTAMINANTS

Box 11.3 shows a general equation for calculating the risk-based site restoration criteria for a single carcinogenic chemical present in soils at a contaminated site. This has been derived by "back calculating" from the risk and chemical exposure equations associated with the inhalation of soil emissions, ingestion of soils, and dermal contact with soils. It is noteworthy that, where appropriate and necessary, this general equation may also be reformulated to incorporate the receptor age-adjustment exposure factors developed and presented earlier on in Chapter 9.

11.3.1.1 An Illustrative Example

In a simplified example of the application of the above equation (for calculating media-specific ASC for a carcinogenic chemical), consider a hypothetical site located within a residential setting where children might become exposed to site contamination during recreational activities. It has been found that soil at this playground for young children in the neighborhood is contaminated with methylene chloride. It is expected that children aged 1–6 years could be ingesting approximately 200 mg of the contaminated soils per day during outdoor activities at the impacted playground. The ASC associated with the *ingestion only exposure* of 200 mg of soil (contaminated with methylene chloride, with an oral SF of 2.0×10^{-3} [mg/kg-day]$^{-1}$) on a daily basis, by a 16-kg child, over a 5-year exposure period is conservatively estimated to be:

$$\mathrm{ASC_{mc}} = \frac{\left[10^{-6} \times 16 \times 70 \times 365\right]}{\left[0.002 \times 200 \times 1 \times 1 \times 365 \times 5 \times 10^{-6}\right]} \approx 560 \text{ mg/kg}$$

That is, the allowable exposure concentration (represented by the ASC) for methylene chloride in soils within this residential setting, assuming a benchmark excess lifetime cancer risk level of 10^{-6}, is estimated to be approximately 560 mg/kg. Thus, if environmental sampling and analysis indicates contamination levels in excess of 560 mg/kg at this residential playground, then immediate risk control action (such as restricting access to the playground as an interim measure) should probably be implemented.

It is noteworthy that other potentially significant exposure routes (e.g., dermal contact and inhalation) as well as other sources of exposure (e.g., via drinking water and food) have not been accounted for in this illustrative example. Meanwhile, all such other exposure routes and sources may require the need to further lower the calculated ASC for any site restoration decisions. Indeed, regulatory guidance would probably require reducing the contaminant concentration, $\mathrm{ASC_{mc}}$, to only a fraction (e.g., 20%) of the calculated value in view of the fact that there could be other sources of exposure (e.g., air and food). Anyhow, this kind of thinking should generally be factored into the overall risk management decisions about contaminated site management problems.

BOX 11.3   GENERAL EQUATION FOR CALCULATING RISK-BASED SOIL CLEANUP LEVEL FOR A CARCINOGENIC CHEMICAL CONSTITUENT

$$\mathrm{ASC_c} = \frac{\mathrm{TCR}}{\left(\left(\dfrac{\mathrm{EF} \times \mathrm{ED} \times \mathrm{CF}}{\mathrm{BW} \times \mathrm{AT} \times 365}\right) \times \left\{\begin{array}{l}\left[\mathrm{SF_i} \times \mathrm{IR} \times \mathrm{RR} \times \mathrm{ABS_a} \times \mathrm{AEF} \times \mathrm{CF_a}\right] \\ + \left[\left(\mathrm{SF_o} \times \mathrm{SIR} \times \mathrm{FI} \times \mathrm{ABS_{si}}\right) + \left(\mathrm{SF_o} \times \mathrm{SA} \times \mathrm{AF} \times \mathrm{ABS_{sd}} \times \mathrm{SM}\right)\right]\end{array}\right\}\right)}$$

$$= \frac{(\mathrm{TCR}) \times (\mathrm{BW} \times \mathrm{AT} \times 365)}{(\mathrm{EF} \times \mathrm{ED} \times \mathrm{CF}) \times \left\{\begin{array}{l}\left[\mathrm{SF_i} \times \mathrm{IR} \times \mathrm{RR} \times \mathrm{ABS_a} \times \mathrm{AEF} \times \mathrm{CF_a}\right] \\ + \mathrm{SF_o}\left[\left(\mathrm{SIR} \times \mathrm{FI} \times \mathrm{ABS_{si}}\right) + \left(\mathrm{SA} \times \mathrm{AF} \times \mathrm{ABS_{sd}} \times \mathrm{SM}\right)\right]\end{array}\right\}}$$

where
$\mathrm{ASC_c}$ = acceptable soil concentration (i.e., acceptable risk-based cleanup level) of carcinogenic contaminant in soil (mg/kg)

(Continued)

BOX 11.3 (*Continued*)   GENERAL EQUATION FOR CALCULATING RISK-BASED SOIL CLEANUP LEVEL FOR A CARCINOGENIC CHEMICAL CONSTITUENT

TCR = target cancer risk, usually set at 10^{-6} (dimensionless)

SF_i = inhalation slope factor ($[\text{mg/kg-day}]^{-1}$)

SF_o = oral slope factor ($[\text{mg/kg-day}]^{-1}$)

IR = inhalation rate (m^3/day)

RR = retention rate of inhaled air (%)

ABS_a = percent chemical absorbed into bloodstream (%)

AEF = air emissions factor, i.e., PM_{10} particulate emissions or volatilization (kg/m^3)

CF_a = conversion factor for air emission term (10^6)

SIR = soil ingestion rate (mg/day)

CF = conversion factor (10^{-6} kg/mg)

FI = fraction ingested from contaminated source (dimensionless)

ABS_{si} = bioavailability absorption factor for ingestion exposure (%)

ABS_{sd} = bioavailability absorption factor for dermal exposures (%)

SA = skin surface area available for contact, i.e., surface area of exposed skin (cm^2/event)

AF = soil to skin adherence factor, i.e., soil loading on skin (mg/cm^2)

SM = factor for soil matrix effects (%)

EF = exposure frequency (days/year)

ED = exposure duration (years)

BW = body weight (kg)

AT = averaging time (i.e., period over which exposure is averaged) (years)

11.3.2   Soil Chemical Limits for the Noncarcinogenic Effects of Site Contaminants

Box 11.4 shows a general equation for calculating the risk-based site restoration criteria for the noncarcinogenic effects of a single chemical constituent found in soils at a contaminated site. This has been derived by "back calculating" from the hazard and chemical exposure equations associated with the inhalation of soil emissions, ingestion of soils, and dermal contact with soils.

BOX 11.4   GENERAL EQUATION FOR CALCULATING RISK-BASED SOIL CLEANUP LEVEL FOR THE NONCARCINOGENIC EFFECTS OF A CHEMICAL CONSTITUENT

$$ASC_{nc} = \cfrac{\text{Target Hazard Quotient}}{\left(\cfrac{EF \times ED \times 10^{-6}}{BW \times AT \times 365}\right) \times \left\{ \left[\cfrac{IR \times RR \times ABS_a}{RfD_i} \times AEF \times CF_a\right] + \left[\left(\cfrac{SIR}{RfD_o} \times FI \times ABS_{si}\right)\right] + \left[\left(\cfrac{SA \times AF \times ABS_{sd} \times SM}{RfD_o}\right)\right] \right\}}$$

$$= \cfrac{(THQ) \times (BW \times AT \times 365)}{(EF \times ED \times CF) \times \left\{ \left[\cfrac{IR \times RR \times ABS_a}{RfD_i} \times AEF \times CF_a\right] + \cfrac{1}{RfD_o}\left[(SIR \times FI \times ABS_{si}) + (SA \times AF \times ABS_{sd} \times SM)\right] \right\}}$$

(*Continued*)

> **BOX 11.4 (*Continued*)   GENERAL EQUATION FOR CALCULATING**
> **RISK-BASED SOIL CLEANUP LEVEL FOR THE NONCARCINOGENIC**
> **EFFECTS OF A CHEMICAL CONSTITUENT**
>
> where
>
> ASC_{nc} = acceptable soil concentration (i.e., acceptable risk-based cleanup level) of noncarcinogenic contaminant in soil (mg/kg)
>
> THQ = target hazard quotient (usually equal to 1) (unitless)
>
> RfD_i = inhalation reference dose (mg/kg-day)
>
> RfD_o = oral reference dose (mg/kg-day)
>
> IR = inhalation rate (m³/day)
>
> RR = retention rate of inhaled air (%)
>
> ABS_a = percent chemical absorbed into bloodstream (%)
>
> AEF = air emission factor, i.e., PM_{10} particulate emissions or volatilization (kg/m³)
>
> CF_a = conversion factor for air emission term (10^6)
>
> SIR = soil ingestion rate (mg/day)
>
> CF = conversion factor (10^{-6} kg/mg)
>
> FI = fraction ingested from contaminated source (dimensionless)
>
> ABS_{si} = bioavailability absorption factor for ingestion exposure (%)
>
> ABS_{sd} = bioavailability absorption factor for dermal exposures (%)
>
> SA = skin surface area available for contact, i.e., surface area of exposed skin (cm²/event)
>
> AF = soil to skin adherence factor, i.e., soil loading on skin (mg/cm²)
>
> SM = factor for soil matrix effects (%)
>
> EF = exposure frequency (days/year)
>
> ED = exposure duration (years)
>
> BW = body weight (kg)
>
> AT = averaging time (i.e., period over which exposure is averaged, equals ED for noncarcinogens) (years)

11.3.2.1   An Illustrative Example

In a simplified example of the application of the ASC equation (for calculating media-specific ASC for the noncarcinogenic effects of a chemical constituent), consider a hypothetical site located within a residential setting where children might become exposed to site contamination during recreational activities. It has been found that soil at this playground for young children in the neighborhood is contaminated with ethylbenzene. It is expected that children aged 1–6 years could be ingesting approximately 200 mg of contaminated soils per day during outdoor activities at the impacted playground. The ASC associated with the *ingestion only exposure* of 200 mg of soil (contaminated with ethylbenzene, with an oral RfD of 0.1 mg/kg-day) on a daily basis, by a 16-kg child, over a 5-year exposure period is conservatively estimated to be:

$$ASC_{ebz} = \frac{0.1 \times [1 \times 16 \times 5 \times 365]}{[200 \times 1 \times 1 \times 365 \times 5 \times 10^{-6}]} \approx 8,000 \text{ mg/kg}$$

That is, the allowable exposure concentration (represented by the ASC) for ethylbenzene in soils within this residential setting is estimated to be approximately 8,000 mg/kg. Thus, if environmental sampling and analysis indicates contamination levels in excess of 8,000 mg/kg at this residential playground, then immediate risk control action (such as restricting access to the playground as an interim measure) should probably be implemented.

It is noteworthy that other potentially significant exposure routes (e.g., dermal contact and inhalation) as well as other sources of exposure (e.g., via drinking water and food) have not been accounted for in this illustrative example. Meanwhile, all such other exposure routes and sources may require the need to further lower the calculated ASC for any site restoration decisions. Indeed, regulatory guidance would probably require reducing the contaminant concentration, ASC_{ebz}, to only a fraction (e.g., 20%) of the calculated value in view of the fact that there could be other sources of exposure (e.g., air and food). Anyhow, this kind of thinking should generally be factored into the overall risk management decisions about contaminated site management problems.

11.4 ESTABLISHING RISK-BASED CLEANUP LIMITS FOR CONTAMINATED WATERS AS A PRACTICAL EXAMPLE

To determine the risk-based cleanup level for a chemical compound present in water, algebraic manipulations of the hazard index and/or carcinogenic risk equations together with the exposure estimation equations discussed in Chapter 9 can be used to arrive at the appropriate analytical relationships. The step-wise computational efforts involved in this exercise consist of a "back calculation" process that yields an acceptable water concentration (AWC) predicated on health-protective exposure parameters; as a classic example, the AWC generally results in a target cumulative noncancer hazard index of ≤1 and/or a target cumulative carcinogenic risk ≤10^{-6}. Indeed, for chemicals with carcinogenic effects, a target risk of $[1 \times 10^{-6}]$ is typically used in the "back calculation," and a target hazard index of 1.0 is typically used for noncarcinogenic effects. The processes involved in the determination of the AWCs are summarized in the sections that follow directly below. For substances that are both carcinogenic and possess systemic toxicity properties, the lower of the carcinogenic or noncarcinogenic criterion would characteristically be used for the relevant corrective action and/or risk management decisions.

11.4.1 WATER CHEMICAL LIMITS FOR CARCINOGENIC CONTAMINANTS

Box 11.5 shows a general equation for calculating the risk-based restoration criteria for a single carcinogenic constituent present in potable water at a contaminated site. This has been derived by "back calculating" from the risk and chemical exposure equations associated with the inhalation of contaminants in water (for volatile constituents only), ingestion of water, and dermal contact with water. It is noteworthy that, where appropriate and necessary, this general equation may also be reformulated to incorporate the receptor age-adjustment exposure factors developed and presented earlier on in Chapter 9.

BOX 11.5   GENERAL EQUATION FOR CALCULATING RISK-BASED WATER CLEANUP LEVEL FOR A CARCINOGENIC CHEMICAL CONSTITUENT

$$AWC_c = \frac{TCR}{\left(\dfrac{EF \times ED}{BW \times AT \times 365}\right) \times \left\{ \begin{array}{l} [SF_i \times IR_w \times RR \times ABS_a \times CF_a] + \\ \left[(SF_o \times WIR \times FI \times ABS_{si}) + (SF_o \times SA \times K_p \times ET \times ABS_{sd} \times CF) \right] \end{array} \right\}}$$

$$= \frac{TCR \times (BW \times AT \times 365)}{(EF \times ED) \times \left\{ \begin{array}{l} [SF_i \times IR_w \times RR \times ABS_a \times CF_a] \\ + SF_o \left[(WIR \times FI \times ABS_{Si}) + (SA \times K_p \times ET \times ABS_{sd} \times CF) \right] \end{array} \right\}}$$

(Continued)

BOX 11.5 (*Continued*)   GENERAL EQUATION FOR CALCULATING RISK-BASED WATER CLEANUP LEVEL FOR A CARCINOGENIC CHEMICAL CONSTITUENT

where

AWC_c = acceptable water concentration (i.e., acceptable risk-based cleanup level) of carcinogenic contaminant in water (mg/L)

TCR = target cancer risk, usually set at 10^{-6} (dimensionless)

SF_i = inhalation slope factor ($[mg/kg\text{-}day]^{-1}$)

SF_o = oral slope factor ($[mg/kg\text{-}day]^{-1}$)

IR_w = intake from the inhalation of volatile compounds (sometimes equivalent to the amount of ingested water) (m^3/day)

RR = retention rate of inhaled air (%)

ABS_a = percent chemical absorbed into bloodstream (%)

CF_a = conversion factor for volatiles inhalation term ($1,000\ L/1\ m^3 = 10^3\ L/m^3$)

WIR = water ingestion rate (L/day)

CF = conversion factor ($1\ L/1,000\ cm^3 = 10^{-3}\ L/cm^3$)

FI = Fraction ingested from contaminated source (unitless)

ABS_{si} = bioavailability absorption factor for ingestion exposure (%)

ABS_{sd} = bioavailability absorption factor for dermal exposures (%)

SA = skin surface area available for contact, i.e., surface area of exposed skin (cm^2/event)

K_p = chemical-specific dermal permeability coefficient from water (cm^2/h)

ET = exposure time during water contacts (e.g., during showering/bathing activity) (h/day)

EF = exposure frequency (days/years)

ED = exposure duration (years)

BW = body weight (kg)

AT = averaging time (i.e., period over which exposure is averaged) (years).

11.4.1.1   Illustrative Example Calculations

In a simplified example of the application of the above equation (for calculating media-specific AWC for a carcinogenic chemical), consider the hypothetical case of a contaminated site that is impacting an underlying water supply aquifer as a result of contaminant migration into groundwater. This groundwater resource is used for culinary water supply purposes. The AWC associated with the *ingestion only exposure* to 2 L of water (contaminated with methylene chloride, with an oral SF of 2.0×10^{-3} [mg/kg-day]$^{-1}$) on a daily basis, by a 70-kg adult, over a 70-year lifetime is given by the following approximation:

$$AWC_{mc} = \frac{[R \times BW \times LT \times CF]}{[SF \times I \times A \times ED]}$$

i.e.,

$$AWC_{mc} = \frac{\left[10^{-6} \times 70 \times 70 \times 365\right]}{\left[0.002 \times 2 \times 1 \times 365 \times 70\right]} \approx 0.0175\ \text{mg/L} = 17.5\ \mu\text{g/L}$$

That is, assuming a benchmark excess lifetime cancer risk level of 10^{-6}, the allowable exposure concentration for methylene chloride (represented by the AWC) is estimated at 17.5 μg/L.

Next, consider another situation of a contaminated site impacting a multipurpose surface water body due to overland flow; this surface water body is used both as a culinary water supply source and for recreational purposes. Now, if the same potential receptor identified above is exposed through both water and fish consumption (with the fish intake quantified by using its bioconcentration factor, BCF), then a lower AWC may be anticipated. In this case, assuming an average daily consumption of aquatic organisms, DIA, of 6.5 g/day, and a BCF of 0.91 L/kg (i.e., in addition to the drinking water intake), then human exposure levels for ingestion of both water and fish is determined from the following modified equation:

$$AWC_{mc} = \frac{[R \times BW \times LT \times CF]}{[SF \times (I + (DIA \times BCF)) \times A \times ED]}$$

$$= \frac{[10^{-6} \times 70 \times 70 \times 1]}{[0.002 \times (2 + (0.0065 \times 0.91)) \times 1 \times 70]} \approx 0.0175 \text{ mg/L} = 17.5 \text{ µg/L}$$

Thus, the allowable exposure concentration (represented by the AWC) for drinking water and eating aquatic organisms contaminated with methylene chloride is also approximately 17.5 µg/L in this particular case.

Obviously, the inclusion of other pertinent exposure routes (such as inhalation of vapors and dermal contacts during showering/bathing activities) would likely call for a lower AWC in any aquifer restoration decision. Indeed, regulatory guidance would probably require reducing the contaminant concentration, AWC_{mc}, to only a fraction (e.g., 20%) of the calculated value in view of the fact that there could be other sources of exposure (e.g., air and food). Anyhow, this kind of thinking should generally be factored into the overall risk management decisions about contaminated water management problems.

11.4.2 WATER CHEMICAL LIMITS FOR THE NONCARCINOGENIC EFFECTS OF SITE CONTAMINANTS

Box 11.6 shows a general equation for calculating the risk-based restoration criteria for a single noncarcinogenic constituent present in potable water at a contaminated site. This has been derived by "back calculating" from the risk and chemical exposure equations associated with the inhalation of contaminants in water (for volatile constituents only), ingestion of water, and dermal contact with water.

BOX 11.6 GENERAL EQUATION FOR CALCULATING RISK-BASED WATER CLEANUP LEVEL FOR NONCARCINOGENIC EFFECTS OF A CHEMICAL CONSTITUENT

$$AWC_{nc} = \frac{THQ}{\left(\frac{EF \times ED}{BW \times AT \times 365}\right) \times \left\{ \left[\frac{IR_W \times RR \times ABS_a \times CF_a}{RfD_i}\right] + \left[\left(\frac{WIR}{RfD_o} \times FI \times ABS_{si}\right)\right] + \left[\left(\frac{SA \times K_p \times ET \times ABS_{sd} \times CF}{RfD_o}\right)\right]\right\}}$$

$$= \frac{THQ \times (BW \times AT \times 365)}{(EF \times ED) \times \left\{ \left[\frac{IR_W \times RR \times ABS_a \times CF_a}{RfD_i}\right] + \frac{1}{RfD_o}\left[(WIR \times FI \times ABS_{si}) + (SA \times K_p \times ET \times ABS_{sd} \times CF)\right]\right\}}$$

(Continued)

BOX 11.6 (*Continued*) GENERAL EQUATION FOR CALCULATING RISK-BASED WATER CLEANUP LEVEL FOR NONCARCINOGENIC EFFECTS OF A CHEMICAL CONSTITUENT

where

AWC_{nc} = acceptable water concentration (i.e., acceptable risk-based cleanup level) of noncarcinogenic contaminant in water (mg/L)

THQ = target hazard quotient (usually equal to 1)

RfD_i = inhalation reference dose (mg/kg-day)

RfD_o = oral reference dose (mg/kg-day)

IR_w = inhalation intake rate (m^3/day)

RR = retention rate of inhaled air (%)

ABS_a = percent chemical absorbed into bloodstream (%)

CF_a = conversion factor for volatiles inhalation term (1,000 L/1 m^3 = 10^3 L/m^3)

WIR = water intake rate (L/day)

CF = conversion factor (1 L/1,000 cm^3 = 10^{-3} L/cm^3)

FI = fraction ingested from contaminated source (dimensionless)

ABS_{si} = bioavailability absorption factor for ingestion exposure (%)

ABS_{sd} = bioavailability absorption factor for dermal exposures (%)

SA = skin surface area available for contact, i.e., surface area of exposed skin (cm^2/event)

K_p = chemical-specific dermal permeability coefficient from water (cm^2/h)

ET = exposure time during water contacts (e.g., during showering/bathing activity) (h/day)

EF = exposure frequency (days/years)

ED = exposure duration (years)

BW = body weight (kg)

AT = averaging time (i.e., period over which exposure is averaged) (years).

11.4.2.1 Illustrative Example Calculations

In a simplified example of the application of the above equation (for calculating media-specific AWC for a noncarcinogenic chemical), consider the hypothetical case of a contaminated site that is impacting an underlying water supply aquifer as a result of contaminant migration into groundwater. This groundwater resource is used for culinary water supply purposes. The AWC associated with the *ingestion only exposure* to 2 L of water (contaminated with ethylbenzene, with an oral RfD of 0.1 mg/kg-day) on a daily basis, by a 70-kg adult is approximated by:

$$AWC_{ebz} = \frac{0.1 \times [1 \times 70 \times 70 \times 365]}{[2 \times 1 \times 1 \times 365 \times 70]} \approx 3,500 \ \mu g/L$$

That is, the allowable exposure concentration (represented by the AWC) for ethylbenzene is estimated to be 3,500 μg/L. Of course, additional exposures via inhalation and dermal contacts during showering/bathing and washing activities may also have to be incorporated to yield an even lower AWC, in order to arrive at a more responsible water restoration decision. Indeed, regulatory guidance would probably require reducing the contaminant concentration, AWC_{ebz}, to only a fraction (e.g., 20%) of the calculated value in view of the fact that there could be other sources of exposure (e.g., air and food). Anyhow, this kind of thinking should generally be factored into the overall risk management decisions about contaminated water management problems.

Next, consider another situation of a contaminated site impacting a multipurpose surface water body due to overland flow; this surface water body is used both as a culinary water supply source and for recreational purposes. Now, if this same potential receptor as identified above is exposed through both contaminated water and fish consumptions (whereby the fish intake quantified by using its BCF), then a lower AL might be anticipated. In this case, assuming an average adult daily consumption of aquatic organisms, DIA, of 6.5 g/day, and a BCF of 37.5 L/kg, human exposure levels for ingestion of both water and fish is determined from the following modified equation:

$$AWC_{ebz}\left[mg/L\right] = \frac{\left[RfD\ mg/kg\text{-}day \times BW\ kg\right]}{\left[2\,L/day + \left(0.0065\ kg \times BCF\ L/kg\right)\right] \times 1}$$

$$= \frac{\left[0.1\ mg/kg\text{-}day \times 70\ kg\right]}{\left[2\,L/day + \left(0.0065\ kg \times 37.5\ L/kg\right)\right]} = 3{,}120\ \mu g/L$$

Thus, the allowable exposure concentration (represented by the water AL) for drinking water and eating aquatic organisms contaminated with ethylbenzene is approximately 3,120 μg/L.

Of course, additional exposures via inhalation and dermal contacts during showering/bathing and washing activities may also have to be incorporated to yield an even lower AWC, in order to arrive at a more responsible water restoration decision. Indeed, regulatory guidance would probably require reducing the contaminant concentration, AWC_{ebz}, to only a fraction (e.g., 20%) of the calculated value in view of the fact that there could be other sources of exposure (e.g., air and food). This thinking should generally be factored into the overall risk management decisions about contaminated water management problems.

11.5 AN HOLISTIC PROTOCOL FOR ESTABLISHING SITE RESTORATION GOALS FOR CONTAMINATED SITES

By utilizing methodologies that establish cleanup criteria based on risk assessment principles, corrective action programs can generally be conducted in a cost-effective and efficient manner. Risk assessment has indeed become particularly useful in determining the level of cleanup most appropriate for potentially contaminated sites. As part of a holistic approach, once realistic risk reduction levels potentially achievable by various remedial alternatives are known, the decision-maker can then use other scientific criteria (such as implementability, reliability, operability, and cost) to select a final design alternative. Subsequently, an appropriate corrective action plan can be developed and implemented for the contaminated site in question. In fact, a major consideration in developing a remedial action plan for a contaminated site is the level of cleanup to be achieved—which could become the driving force behind remediation costs. The site cleanup limit concept generally facilitates decisions as to the effective use of limited funds to clean up a site to a level appropriate/safe for its intended use. It is therefore be prudent to allocate adequate resources to develop appropriate and defensible cleanup criteria.

In principle, the cleanup criteria selected for a potentially contaminated site may vary significantly from one site to another—due especially to the prevailing site-specific conditions. Similarly, mitigation measures may be case-specific for various hazardous situations and problems. Now, consider for illustrative purposes, a potentially contaminated site that is being envisioned for remediation so that it could possibly be re-developed for either residential or industrial purposes. Contaminant levels in residential soils in which children might play (which allows for pica behavior in toddlers and other infants) must necessarily be lower than the same contaminant levels in soils present at a site designated for large industrial complexes (which effectively prevent direct exposures to contaminated soils). Also, the release potential of several chemical constituents will usually be different from sandy soils vs. clayey soils—and this will invariably affect the possible exposure scenarios, and therefore the acceptable soil contaminant levels that are designated for the different types of soils. Consequently, it is generally preferable to establish and use site-specific cleanup criteria for contaminated site problems encountered in practice, especially where soil exposures is critical to the site restoration decisions.

In general, preliminary remediation goals (PRGs) are usually established as part of the cleanup objectives early in a site characterization process—and the development of PRGs typically requires site-specific data relating to the impacted media of interest, the CoPCs, and the probable future land uses. Indeed, if site-specific cleanup criteria are to be developed for a site, substantial wealth of information must also be collected on site soil and groundwater characteristics. Typical soil characteristics required to determine site-specific cleanup criteria include porosity, particle size, moisture content, organic carbon content, partition coefficients, soil pH, depth of contamination; general aquifer characteristics required to determine site-specific cleanup criteria include effective porosity, hydraulic conductivity, bulk density, longitudinal and transverse dispersivities, aquifer saturated thickness, hydraulic gradient, depth to water table, average groundwater velocity (Lesage and Jackson, 1992). These parameters are also useful, in a general sense, for interpreting the results of a site investigation—which are used to characterize the site, and subsequently to develop the corrective action objectives.

In the final analysis, it is apparent that an early determination of remediation goals tends to facilitate the development of a range of feasible corrective action decisions, which in turn helps focus remedy selection on the most effective remedial alternative(s). Meanwhile, it is noteworthy that an initial list of PRGs may have to be revised when new data becomes available during the site characterization process. In fact, PRGs may have to be refined into final remediation goals throughout the process leading up to the final remedy selection. Consequently, it is important to iteratively review and re-evaluate the media and CoPCs, future land uses, and exposure assumptions originally identified during project formulation.

11.5.1 Factors Affecting the Development of Risk-Based Remediation Goals

Several exposure- and technical-related factors will tend to affect the processes involved in the development of remediation objectives and cleanup goals (see, e.g., ASTM, 1994a, 1995a,b; Bowers et al., 1996; CCME, 1991; CRWQCB, 1989; Fitchko, 1989; Liptak and Lombardo, 1996; LUFT, 1989; Odermatt and Menatti, 1996; Pierzynski et al., 1994; USEPA, 1989c,f,g,k,n,o, 1991a,e–h, 1996j,k; WPCF, 1990)—including the key determinants identified in Box 11.7. In particular, the type of exposure scenarios envisioned for a contaminated site and its vicinity usually would significantly affect whatever is considered to be an acceptable cleanup level. Thus, entirely different cleanup levels may be needed for similar pieces of equally contaminated sites, based on the differences in the exposure scenarios. That is, the same amount of contamination at similar sites does not necessarily call for the same level of cleanup. In general, however, the cleanup must attain contaminant levels that are protective of all receptors for both current and future land uses.

BOX 11.7 IMPORTANT FACTORS AFFECTING THE PROCESS OF ESTABLISHING CONTAMINATED SITE CLEANUP CRITERIA

- Nature and level of risks involved
- Regulatory requirements and/or guidelines
- Migration and exposure pathways (from contaminant sources to receptors)
- Individual site characteristics affecting exposure
- Current and future beneficial uses of the affected land and subsurface resources
- Variability in exposure scenarios
- Probability of occurrence of exposure to the populations potentially at risk
- Possibilities of receptor exposures to elevated levels of other contamination not related to site activities
- Sensitivity and vulnerability of the populations potentially at risk
- Potential effects of site contamination on human and ecological receptors
- Reliability of scientific and technical information/data relating to exposure assessment, toxicity data, risk models, and potential remediation strategies

In general, due to the possibility for different cleanup levels to be imposed on similar sites potentially contaminated to the same degree, the final remediation costs for such sites may be found to be significantly different. It is therefore imperative that a systematic approach is adopted in developing site-specific cleanup criteria for contaminated sites. It is also important that the determination of the extent of cleanup required at a contaminated site is based on an assessment of the potential risks to both human health and the environment. In fact, the use of risk-based cleanup levels would likely result in timely, cost-effective, and adequate site restoration programs. As a rule-of-thumb, remedies whose cumulative effects fall within the risk range of approximately 10^{-4} to 10^{-7} for carcinogens, or meet an acceptable hazard level of unity for noncarcinogenic effects, are generally considered protective of human health. Where necessary, however, the potential ecological impacts should also be determined before a final site restoration decision is made. Indeed, media cleanup goals should generally be established at contaminant levels protective of both human health and the environment. Oftentimes, however, cleanup levels established for the protection of human health will also be protective of the environment at the same time. But there may be instances where adverse environmental effects may occur at or below contaminant levels that adequately protect human health. Consequently, sensitive ecosystems (e.g., wetlands) as well as threatened and endangered species or habitats that may be affected by releases of hazardous contaminants or constituents should, insofar as possible, be evaluated separately as part of the process used to establish media cleanup criteria needed for site restoration initiatives.

Ultimately, after defining the critical pathways and exposure scenarios associated with a project site, it often is possible to calculate the various media concentrations at or below which potential receptor exposures will pose no significant risks to the exposed populations. Figure 11.1 shows a general protocol that can be adapted in developing appropriate cleanup goals as part of a contaminated site restoration activity. Meanwhile, it is noteworthy that, in developing risk-based cleanup levels, appropriate concepts of areal and temporal averaging can often be incorporated (Bowers et al., 1996; Washburn and Edelmann, 1999). For instance, since exposures to soils over a period of time would generally be expected to occur over an area, rather than at point locations, it is usually more appropriate to expect cleanup levels to be met over an exposure area, rather than at every discrete point within the area. Similarly, since exposures from groundwater is associated with a well that generally draws the water from an area, rather than a single point in the aquifer, it is usually is more appropriate to average groundwater concentration over the well's recharge area in order to meet the cleanup levels—rather than to require that the cleanup levels necessarily be met at each point in the aquifer. Indeed, remediation is generally not required for every individual location that has a concentration above the target cleanup level—recognizing that if every location with a concentration above the target cleanup level were to be remediated, then the average remaining would likely be well below the stipulated cleanup target. Thus, exceedances of the cleanup target at individual locations within a given exposure area should be viewed as being acceptable, as long as the average concentration over the entire exposure area meets the cleanup target.

11.5.2 THE ROLE OF RISK-BASED CLEANUP LEVELS IN SITE RESTORATION DECISIONS

Typically, pre-established PRGs are often used to define acceptable chemical exposure limits for human exposure—i.e., if they are determined to represent "safe" or "tolerable" benchmark levels for the case-specific situation. However, such generic PRGs may not always be available, or may not even offer adequate public health and environmental protection under certain circumstances. For instance, the presence of multiple constituents, multiple exposure routes, or other extraneous factors could result in "unacceptable" aggregate risk being associated with a PRG for the particular situation. Under such circumstances, a new "acceptable" or "safe" level may be better represented by the RBCEL (or similar appropriate criterion)—that are derived for the various exposure routes, and

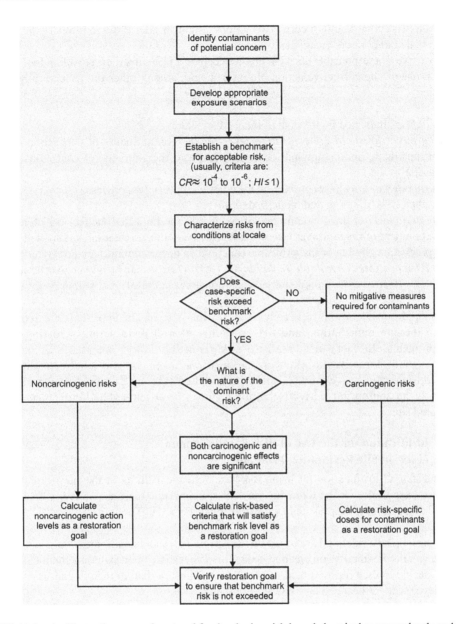

FIGURE 11.1 An illustrative general protocol for developing risk-based chemical exposure levels and public health or site restoration goals.

from elaborately defined exposure scenarios. Fundamentally, risk-based benchmarks are developed via "back-modeling" from a target risk level that produces an acceptable RBCEL—which can serve as a surrogate PRG; invariably, the type of exposure scenarios envisioned as well as the exposure assumptions used may predicate the new benchmark level. As the preferred risk-based benchmark, the RBCEL can then be used as a surrogate or replacement for the PRG of the CoPC. It is noteworthy that, as an example, when the calculated RBCEL based on non-cancer toxicity is less protective of public health than the cancer-based value, the surrogate PRG for the CoPC is set at the lower of the two—usually the one based on the cancer effects. Ultimately, the risk-based benchmarks predicated on RBCELs may be used to determine the degree of contamination and/or chemical exposures; evaluate the need for intervention and receptor monitoring or further investigation; provide guidance on the need for risk control and/or corrective actions; establish safer PRGs; and verify the adequacy of possible remedial/corrective actions.

Now, to arrive at a responsible decision on the acceptable cleanup criteria to adopt for a contaminated site problem, the corrective action program should, among other things, carefully evaluate the following:

- *Level of risk* indicated by the contaminants of concern
- *Background threshold concentration levels* of the contaminants of concern at upgradient, upstream, and/or upwind locations relative to the source(s) of contamination or release(s)
- *Natural attenuation effects* of the contaminants of concern (via processes such as evaporation, photolysis, dilution, and biodegradation)
- *Asymptotic level* of the contaminants of concern in the impacted media—which denotes the attainment of contaminant levels below which continued remedial actions will generally produce negligible or insignificant reductions in target contaminant levels
- *Best [Demonstrated] Available Technologies* (B[D]ATs) that can be proven to offer feasible and cost-effective remediation methods and processes in the site restoration program

Other principal considerations generally relate to the cost of cleanup, the time required for completing the site remediation, and the possibility of a cleanup activity creating potential liability problems. In fact, once realistic risk reduction levels potentially achievable by various remedial alternatives are known, the decision-maker can use other scientific criteria (such as implementability, reliability, operability, and cost) to select a final design alternative. Subsequently, an appropriate corrective action plan can be developed and implemented for the contaminated site.

11.5.2.1 Justification for the Use of Risk-Based Cleanup Levels in Site Restoration Programs

There is substantial justification for using risk-based chemical limits in the site restoration decision process—especially for the following particularly important reasons (ASTM, 1994a, 1995a,b; Lesage and Jackson, 1992; Liptak and Lombardo, 1996; Pratt, 1993):

- The cleanup levels defined are based on site-specific data and conditions, which will likely have significant impacts on cleanup costs; this is because, the risk-based chemical limits are specific to the proposed land-use or are set to ensure that a range of end-uses can be achieved without unacceptable risk to future land uses.
- The levels defined are sensitive to human health and environmental effects, without necessarily being sensitive to regulatory changes.
- The methodology allows for the derivation of an *asymptotic level* of the contaminants of concern in the impacted media—which marks the cleanup level corresponding to a point of diminishing returns (i.e., the point when monitoring indicates that only little additional progress can be made in reducing the contaminant levels); this represents the attainment of contaminant levels below which continued remediation produces negligible reductions in contamination.

Overall, it is notable that the cleanup criteria could become the driving force behind remediation costs—and it has indeed been demonstrated that the development and use of risk-based cleanup guidelines results in timely and cost-effective remediation (e.g., Liptak and Lombardo, 1996). It is therefore prudent to allocate adequate resources to develop site-specific cleanup criteria that will facilitate the selection and design of cost-effective remedial action alternatives.

11.6   TOWARDS AN INFORMED DECISION-MAKING ON ACCEPTABLE CLEANUP LEVELS FOR CONTAMINATED SITE PROBLEMS

This chapter has elaborated a number of analytical relationships that can be adopted or used to estimate EQC and/or benchmark restoration goals necessary for site restoration and management decisions. Indeed, oftentimes pre-established EQC are used to define environmental target goals if they are determined to represent "acceptable" benchmark levels for the case-specific situation. However, such EQC may not always be available or they may not be adequate if the presence of multiple contaminants, multiple pathways, or other extraneous factors could result in "unacceptable" aggregate risk for the particular situation. Under such circumstances, the "acceptable" level may be represented by a risk-based benchmark—classically derived for the various pathways from elaborately defined exposure scenarios. In fact, the use of risk assessment principles to establish case-specific benchmarks and cleanup objectives for contaminated site problems represent an even better and more sophisticated approach to designing cost-effective site restoration and management programs—i.e., in comparison with the use of generic benchmarks. The contaminant-specific target levels are usually established for all affected environmental media that require remediation. Ultimately, the use of such an approach aids in the development and/or selection of appropriate site restoration and risk management strategies capable of achieving a relevant set of performance goals—such that public and ecological health are not jeopardized. In fact, the use of site-specific cleanup criteria will normally result in significant cost-savings because this allows a site management team to employ cost-effective corrective action strategies to achieve significant risk reduction for the particular situation.

Typically, the risk-based benchmark is developed by "back-modeling" from a target risk level, in order to obtain an acceptable risk-based concentration or the maximum acceptable concentration. In general, once the degree of risk or hazard due to existing levels of contamination has been determined, then by working backwards using information on contaminant dilution-attenuation concepts, degradation properties, partitioning coefficients, and/or mass balance analyses, media cleanup criteria may be established for a potentially contaminated site. Invariably, the type of exposure scenarios envisioned and the exposure assumptions used may predicate the concomitant benchmark level, and will ensure that public health and/or the environment are not jeopardized by any residual contamination. Indeed, the type of exposure scenarios envisioned for a contaminated site and its vicinity tends to significantly affect the selection and acceptance of appropriate site cleanup criteria. Thus, entirely different cleanup levels may be needed for similar pieces of contaminated sites, based on the differences in the exposure scenarios. That is, the specifics of two contaminated sites may be very similar, and yet two radically different site restoration or cleanup philosophies may have to be implemented for these sites. This also means that the same amount of contamination at similar sites does not necessarily call for the same level of cleanup. In general, however, the cleanup must attain contaminant levels that are protective of both current and future land uses.

In the final analysis, the benchmarks may be used to determine the degree of contamination; evaluate the need for further investigation of problem situation; provide guidance on the need for corrective actions; establish environmental restoration goals and strategies; and verify the adequacy of possible remedial actions. In any case, the scale and urgency of response actions at contaminated sites depend on the degree to which contaminant levels exceed their respective benchmarks or risk-based criteria. But then, where site remediation is not feasible, the EQC or risk-based criteria can be used to guide land-use restrictions or other forms of risk management actions that are protective of human health and the environment.

12 General Types of Contaminated Site Restoration Methods and Technologies

Several cleanup techniques are universally available for contaminated site restoration programs. However, no one particular technology or process is usually appropriate for all contaminant types and/or under the variety of site-specific conditions that exist at different project sites. Indeed, the choice of cleanup strategy and technology is driven by site conditions, contaminant types, contamination sources, source control measures, existing and/or anticipated land uses, and potential impacts of possible remedial actions. In any case, an in-place or on-site technology will generally be given preference as a remedial alternative of choice in most site restoration programs.

This chapter offers a broad overview of several site restoration techniques that may be used in a variety of contaminated site cleanup programs. A more detailed description of the processes, equipment, and controls, as well as the detailed design elements of the various technologies, can be found elsewhere in the literature (e.g., ARB, 1991; Bansal and Goyal, 2005; Bharagava, 2017; Cairney, 1987, 1993; Chandra, 2015; Chandra et al., 2017; Charbeneau et al. 1992; Chen et al., 2016; Cho et al., 2002; Dragan, 2014; Gavaskar, 1998; Hauser, 2008; Hyman and Dupont, 2001; ITRC, 2003b,c,d,g, 2005a,c,b, 2009a-c, 2011a-f; Jolley and Wang, 1993; Karol, 2003; Kuo, 2014; Naidu and Birke, 2014; Niessen, 2010; NRC, 1994a; Nyer, 1992, 1993; OBG, 1988; Payne et al., 2008; Pratt, 1993; Rai, 2018; Russell, 2011; Selim, 2017; Siegrist et al., 2011; Sims, 1990; Sims et al., 1986; Suthersan, 2001, 2005; Suthersan and Payne, 2004; Suthersan et al., 2016; Tabbaa and Stegemann, 2005; Talley, 2005; Tang, 2003; USEPA, 1984b, 1985c, 1988f, 1989e,k,q, 1990a–d; Vandegrift et al., 1992; Winegardner and Testa, 2000; Yong and Mulligan, 2003). Meanwhile, it is notable that, in order to adequately perform site investigation activities and evaluate the feasibility of site remediation alternatives, a complete understanding of available remedial technologies is necessary.

12.1 CONTAMINANT TREATMENT PROCESSES AND TECHNOLOGIES

There are several proven cleanup technologies and processes that may be used in the management of contaminated site problems. The most widely used restoration methods for contaminated site remediation programs that are applicable to the frequently encountered environmental contaminants may be classified as variations or combinations of the following general and broad categories: physical, chemical, biological, and/or thermal treatment techniques. Within these, most of the specific contaminant treatment processes and site restoration technologies commonly encountered in contaminated site cleanup programs fall into the following wide-ranging categories and related examples:

- Biological treatment
  - Bioremediation (*in situ* or *ex situ*)
  - Composting
  - Landfarming
  - Phytoextraction
- Chemical treatment
  - Hydrogen peroxide catalysis
  - *In situ* chemical treatment (various)

- Physical treatment
  - Activated carbon adsorption
  - Soil flushing or washing
  - Vacuum extraction
  - Solvent extraction
  - Air or steam stripping
- High-temperature treatment
  - Incineration
  - Infrared thermal treatment
  - *In situ* vitrification (ISV)
  - Plasma-fired reactor
  - Electric reactor
- Low-temperature treatment
  - Low volatilization
  - *In situ* radio frequency heating (RFH)
  - Low-temperature thermal desorption (LTTD)
- Isolation and control by geotechnical methods
  - Hydraulic isolation (by pumpage and gradient control)
  - Physical isolation by installation of barriers (e.g., capping and slurry walls)
  - Physical isolation by hydraulic control measures (e.g., trenches and wells)
  - Excavation and containment (e.g., landfilling)
- Immobilization techniques
  - Encapsulation
  - Stabilization/solidification
  - Asphalt incorporation
- Material reuse without treatment
  - Recycling of contaminated materials

Overall, physical operations and chemical processes or their close variants make up several of the contemporary methods used in hazardous waste treatment, as well as for groundwater and soil remediation. Indeed, most of the treatment technologies consist of *in situ* and *ex situ* systems that use the physical and chemical properties of the media (soil, groundwater, gas, sediment) and those of the contaminants within the media to achieve successful remediation.

Meanwhile, it is worth mentioning here that containment and removal types of remediation technologies immobilize contaminants through stabilization, solidification, removal and placement in a secure landfill, or the application of passive/active containment control systems; these processes focus not on remediating or destroying the contaminants *per se*, but rather, by minimizing the rate of contaminant migration from a waste source, reducing a contaminant's mobility, controlling the potential migration pathways in a contaminated environment, and reducing a waste's risk and threat to human health and the environment.

12.1.1 Physical Processes

Physical processes consist of operations used to remediate contaminants through the application of physical forces to an impacted medium, in order to generally change the form or constitution of a contaminant, but not necessarily its fundamental chemical identity. For instance, some chemical constituents present at contaminated sites can be treated through separation and purification processes, which consist of techniques such as filtration, centrifugation, floatation, distillation, evaporation, solvent extraction, reverse osmosis, ion exchange, activated carbon adsorption, decantation, and constituent immobilization by solidification; these types of techniques generally do not alter the chemical composition of the contaminants *per se*.

In general, the primary objectives of most physical treatment methods are to separate hazardous materials from those that are considered to be less hazardous, to separate different types of hazardous materials into various streams that require different treatment methods, or to pretreat a contaminated material prior to final disposal. Indeed, insofar as reasonable, physical processes are rarely used as the final treatment option for any contaminated material.

12.1.2 CHEMICAL PROCESSES

Chemical treatment processes change the actual characteristics of contaminants and classically transform them into less harmful by-products. Indeed, certain types of chemicals or contaminated materials can be either separated or rendered less hazardous through chemical treatment processes. For instance, certain solvents can be used to remove chlorine atom(s) from chlorinated hazardous materials, with the result that toxic compounds are converted to less toxic, more water-soluble compounds; typically, the reaction products are more easily removed from soil, as well as more easily treated.

Chemical treatment technologies commonly available or often applied to contaminated materials include neutralization, coagulation–precipitation, and oxidation–reduction (redox).

12.1.3 BIOLOGICAL PROCESSES

Biological treatment has become a viable and cost-effective alternative technology for treating a wide variety of organic chemical contamination problems in contemporary times. Basically, by using microorganisms (natural or engineered) to degrade chemical constituents, biological processes may transform toxic materials into nontoxic elements such as water, carbon dioxide, and other innocuous products.

Biological treatment may be possible for just about any organic hazardous waste because most organic chemicals can be degraded if the proper microbial communities are established, maintained, and controlled. Some of the key factors used to classify biological treatment systems include the following: flow configuration, biomass state, and free oxygen availability. In general, biological remediation technologies degrade organic materials by the action of microorganisms. Degradation alters the molecular structure of organic compounds and either reduces the compounds into daughter products or completely breaks down the organic molecules into cellular mass, carbon dioxide, water, and inert inorganic residuals.

It is noteworthy here that biological treatment is highly sensitive to changes in the organic composition and concentrations of the material being treated. For instance, biodegradation has been found to be less efficient at low substrate concentrations; this is because, if the concentration is too low, the compound may not be metabolized by a microbial population that may favor another substrate that is available in higher concentrations. On the other hand, very high substrate concentrations may be toxic to the microbial community. Also, it has become apparent that biological treatment generally has no effect on dissolved inorganic substances; on the other hand, significant levels of some inorganic chemicals may inhibit biological activity or may even kill off the microbes. As a whole, careful monitoring and control of dissolved oxygen (DO), nutritional factors, and potentially toxic substances must be carried out during biological treatments to ensure the general viability and effectiveness of the microbial populations of interest.

12.1.4 THERMAL PROCESSES

Thermal treatment technologies use heat to destroy or change the contaminants of concern. Thermal destruction processes control temperature and oxygen availability and then convert hazardous materials into carbon dioxide, water, and other products of combustion. In general, thermal degradation is applicable to contaminated materials that contain significant concentrations of organic compounds; the applicable techniques can be implemented through different types of incineration (i.e., controlled, high-temperature burning in the presence of oxygen) or pyrolysis (i.e., chemical degradation due to elevated temperatures, not requiring the presence of oxygen).

It is noteworthy that depending on the type of contaminated material being treated, a wide variety of end products may result from thermal treatments. Thermal degradation of organic compounds primarily results in the formation and atmospheric releases of by-products such as water, nitrogen, oxygen, carbon dioxide, acid gases, and particulate matter. If metals are present in the incinerated materials or wastes, some portion of these metals may also be emitted into the atmosphere. Furthermore, products of incomplete combustion may be formed and released. In the end, if significant quantities of inorganic materials are present in the original waste materials, then residual ash and slag will additionally be produced from the thermal degradation processes.

12.2 CONTAMINATED SITE RESTORATION METHODS AND TECHNIQUES

A number of site restoration methods and techniques finding relatively widespread applications in the management of contaminated site problems are discussed later, with the depth of discussion shown here in part reflecting the degree of sophistication and/or interest in the particular site restoration method or technology. Meanwhile, it is worth mentioning here that when several types of contaminants are of concern, the possibility of combining two or more in-place technologies to achieve the overall remediation goals becomes virtually inevitable.

12.2.1 ACTIVATED CARBON ADSORPTION

The *activated carbon adsorption* technology is based on the principle that certain chemical constituents preferentially adsorb to organic carbon. The target contaminant groups for carbon adsorption are hydrocarbons, semivolatile organic compounds (SVOCs), and explosives. Limited effectiveness may be achieved on halogenated volatile organic compounds (VOCs) (e.g., vinyl chloride [VC] or dichloroethylenes [DCEs]) and pesticides. Meanwhile, it is noteworthy that the capacity of carbon to adsorb contaminants depends on the properties of the contaminants. For instance, large polar molecules tend to adsorb more strongly than small nonpolar molecules, and thus, some common chlorinated solvents, such as VC, are poorly adsorbed.

Adsorption by activated carbon has a long history of use in treating municipal, industrial, and hazardous wastewaters. The two most common reactor configurations for carbon adsorption systems are the fixed bed and the pulsed or moving bed, with the fixed-bed configuration being the most widely used for adsorption from liquids. It is notable that pretreatment for the removal of suspended solids from liquid streams to be treated is an important design consideration. Indeed, carbon adsorption is particularly effective for polishing water discharges from other remedial technologies in order to attain regulatory compliance.

Granular activated carbon (GAC) systems typically consist of one or more vessels filled with carbon connected in series and/or parallel and operating under atmospheric, negative, or positive pressure. In GAC adsorption, contaminants are adsorbed to the carbon, and spent carbon (i.e., carbon that has reached its maximum adsorption capacity) is typically regenerated by incineration; the carbon can then be regenerated in place, regenerated at an off-site regeneration facility, or disposed of, depending upon economic considerations. In most cases, the highest cost involved in using a carbon adsorption system is the disposal of used carbon and its replacement with new carbon (Nyer, 1993). In fact, a complete carbon adsorption treatment system design must carefully account for the final disposition of the spent carbon.

Overall, carbon adsorption systems can be deployed rapidly, and contaminant removal efficiencies tend to be high. However, logistic and economic disadvantages usually arise from the need to manage the spent carbon. Modifications to the GAC system, such as a use of silicone-impregnated carbon, could increase the removal efficiency and extend the length of operation; it may also be safer to regenerate.

12.2.1.1 Liquid-Phase GAC Adsorption

Liquid-phase GAC adsorption is typically used to treat pumped groundwater or wastewater residual from a treatment process; it is a full-scale treatment technology in which groundwater is pumped through one or more vessels containing activated carbon to which dissolved organic contaminants (as well as certain metal and inorganic molecules) adsorb. When the concentration of contaminants in the effluent from the bed reaches or exceeds a certain level, the carbon can be regenerated in place; removed and regenerated at an off-site facility; or removed and appropriately disposed. It is noteworthy that carbon used for explosive- or metal-contaminated groundwater probably cannot be regenerated and would typically be removed and properly disposed.

In general, liquid-phase GAC can be used to treat halogenated and nonhalogenated VOCs and SVOCs, polychlorinated biphenyls (PCBs), and also explosive compounds. It is worth mentioning here that liquid-phase carbon adsorption is effective for removing contaminants at low concentrations (typically less than 1 mg/L) from water at nearly any flow rate, and for removing contaminants at higher concentrations (typically about 40 L/min or 10 gpm) from water at low flow rates.

In the process involved in a liquid-phase carbon adsorption, GAC is packed in vertical columns, and contaminated water flows through it by gravity. GAC has a high surface area-to-volume ratio, and many compounds readily bond to the carbon surfaces; contaminants from water are therefore adsorbed to the carbon, and the effluent water has a lower contaminant concentration. In practice, water may be passed through several of these columns to ensure complete contaminant removal.

Activated carbon systems are indeed capable of efficiently removing very low concentrations of dissolved organic compounds from contaminated groundwater. In general, the duration of operation and maintenance of a GAC system are dependent on contaminant type, concentration, and volume; regulatory cleanup requirements; and metal concentrations.

12.2.1.2 Vapor-Phase GAC Adsorption

Vapor-phase GAC treatment is performed by passing an off-gas stream through one or more vessels containing activated carbon, which then removes contaminants from the gas stream by sorption until available active sites are occupied. Classically, vapor-phase adsorption is used to treat off-gas from a primary treatment process. Commercial grades of activated carbon are available for specific use in vapor-phase applications. The GAC is typically used in packed beds through which the contaminated air flows until the concentration of contaminants in the effluent from the carbon bed exceeds an acceptable level.

In general, vapor-phase carbon adsorption is primarily used to treat halogenated and nonhalogenated VOCs and SVOCs, including PCBs, in a gas stream. GAC treatment is most efficiently applied when the contaminant concentration is less than 200 ppmv or the off-gas flow rate is low.

12.2.2 Asphalt Batching

Asphalt batching (also referred to as *asphalt incorporation*) is a method for treating hydrocarbon-contaminated soils. It is a remedial technique that involves the incorporation of petroleum-laden soils into hot asphalt mixes as a partial substitute for stone aggregate. This mixture can then be used for pavements and similar constructions. The use as aggregate in asphalt mix is indeed a recycling tactic for managing granular soils and solids such as abrasive blasting media; this approach involves using the waste as the fine aggregate in asphalt, with the waste matrix replacing some or all of the sand used to prepare an asphalt paving mixture. The bitumen binder that forms the paving also immobilizes the contaminants. The waste material can be used in hot mix or cold mix asphalt. Hot mix asphalt is prepared by blending and drying fine aggregate and coarse aggregate; the dried blend is then mixed with bitumen at 150°C–180°C (300°F–350°F). Cold mix asphalt is prepared by blending sand, gravel, and a bitumen/water emulsion at ambient temperature. A typical finished paving mixture using the hot or cold mix method contains about 45% fine aggregate, 50% coarse aggregate, and 5% bitumen. The contaminated soil or sand usually replaces about half or less of the fine aggregate.

The process of asphalt batching consists of excavating the contaminated soils that then undergo an initial thermal treatment, followed by incorporation of the treated soil into aggregate for asphalt. During the incorporation process, the mixture including the impacted soils is heated, resulting in the volatilization of the more volatile hydrocarbon constituents at various temperatures. The remaining compounds become incorporated into the asphalt matrix during cooling, thereby limiting constituent migration.

A wide variety of wastes containing petroleum hydrocarbons or moderate to low concentrations of metal contaminants can be used to make asphalt. For the reuse option to be acceptable, the asphalt prepared using the contaminated matrix must provide adequate paving performance and immobilize the contaminants. Sandy soils containing petroleum contaminants and metal-contaminated sand are particularly suitable for reuse in asphalt. Petroleum contaminants are similar to the bitumen binder and are easily accommodated in the asphalt mixture. Foundry casting sand and spent sand blasting grit also contain metal contaminants that can be effectively immobilized in the bitumen and sand matrix by replacing some of the clean sand normally used in asphalt. It is noteworthy, however, that certain hazardous wastes and/or highly toxic wastes are unlikely to be acceptable for reuse in asphalt.

12.2.3 BIOREMEDIATION

Bioremediation (or *biorestoration*) is a process in which indigenous or inoculated microorganisms (viz., fungi, bacteria, and other microbes) metabolize/degrade organic contaminants found in an environmental medium such as soil and/or groundwater. The bioremediation process usually attempts to accelerate the natural biodegradation process by providing nutrients, electron acceptors, and competent degrading microorganisms that may otherwise be limiting the rapid conversion of organic compounds to innocuous end products.

Figure 12.1 is a diagrammatic representation of the general design elements of an archetypal bioremediation system. Fundamentally, the biorestoration process relies on microorganisms (especially bacteria and fungi) to transform hazardous compounds found in soil and groundwater systems (or even airstreams) into innocuous or less toxic metabolic products. In general, an optimized biotransformation condition may be attained by manipulating the physical environment and controlling nutrient supplements. In fact, the technique requires careful process control to establish the appropriate microbial population. By using microorganisms (natural or engineered) to degrade contaminants in soil, groundwater, or air, bioremediation transforms hazardous/toxic materials into nontoxic elements such as water, carbon dioxide, and other innocuous products.

In practice, the biodegradation of a compound under field conditions is affected by temperature, type of target compounds, soil DO levels, soil moisture content, soil permeability, oxidation–reduction potential, soil pH, availability and concentrations of a compound, availability of nutrients, and the natural microbial community. These factors generally act together to determine the biodegradability of the contaminants in a particular setting. For instance, the biodegradation rate of most petroleum hydrocarbons increases with increasing temperatures due to increased biological activity. Also, aerobic conditions are required for the degradation of hydrocarbons; thus, reduced soil oxygen levels lead to sharply reduced hydrocarbon utilization by microbes. In fact, because oxygen transfer is a key factor in *in situ* biodegradation processes, the soils must be fairly permeable to allow this transfer to occur. Additionally, because of its limited solubility in water, oxygen often becomes rapidly limiting in the presence of excess biodegradable organic carbon. However, the development of anoxic conditions does not necessarily mean that metabolism of organic contaminants ceases. Moreover, it is believed that microorganisms can also affect inorganic contaminant mobility by both direct (e.g., oxidation, reduction, methylation, and denitrification) and indirect (e.g., biosorption, pH change, and sulfide production) processes (Fredrickson et al., 1993). Consequently,

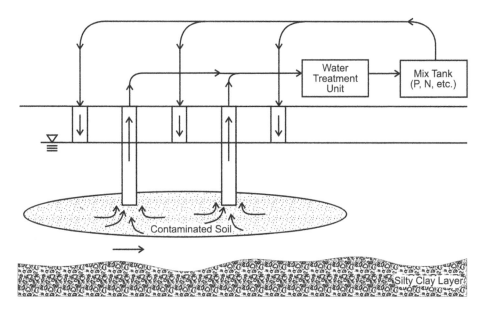

FIGURE 12.1   A simplified illustrative example of the bioremediation technology concept.

microbial reduction reactions can influence heavy metal solubility and hence may serve as rather useful remediation strategies to employ.

On the whole, bioremediation has become a viable and cost-effective site restoration technology for treating a wide variety of contaminants such as petroleum and aromatic hydrocarbons, chlorinated solvents, and pesticides. As an innovative technology, it offers the very important advantage of permanently removing contaminants by biodegradation and therefore reduces potential long-term liabilities since the contamination is destroyed, rather than removed, in typical biorestoration processes.

12.2.3.1   General Types of Biorestoration Programs

There are three major schools of thought in relation to the approach one takes to biorestoration programs: *natural in situ bioremediation*, *biostimulation*, and *bioaugmentation* (Barnhart, 1992; Fredrickson et al., 1993).

Natural in situ bioremediation involves the attenuation of contaminants by indigenous (native) microorganisms without any manipulation. In fact, in a number of situations, it is more economical to allow bioremediation to proceed naturally at the sacrifice of higher removal rates and greater effectiveness than are often achieved with the addition of electron acceptors and other nutrients.

Biostimulation is a process whereby the addition of selected amounts of nutrient materials stimulates or encourages the growth of the indigenous bacteria in the soil, resulting in the degradation of the target contaminant(s).

Bioaugmentation is a process in which specially selected bacteria cultures that are predisposed to metabolize the target compound(s) are added to the impacted media/soil, along with the nutrient materials, to encourage degradation of the contaminants of concern.

Invariably, biorestoration can be enhanced by using the native microorganisms and injecting nutrients, including oxygen, or indeed by injecting microorganisms to the subsurface environment. By and large, it should be recognized that special methods may be necessary to enhance the biorestoration process for certain compounds. In any case, as part of an effectual design process, the toxicity of degradation by-products should also be carefully evaluated and taken into consideration, recognizing that in some cases such as the degradation of trichloroethylene (TCE) to VC, the daughter products are more toxic than the parent compound.

12.2.3.2 Enhanced Bioremediation

Since most contaminated sites require prompt remedial actions, an acceleration of the natural biodegradation process (i.e., enhanced biodegradation) is generally desirable. Under such circumstances, the soil system may have to be modified to promote the activity of the naturally found organisms. Promotion methods may include the addition of nutrients and/or microbes, as well as the aeration of the soil. Indeed, if the intrinsic microorganism flora does not work on specific types of contaminants, selectively adapted inoculants can also be added to the soil to facilitate the process. On the whole, the process typically involves delivery of oxygen to the aquifer and may also require the addition of other nutrients and/or cometabolites. Common methods for oxygen delivery include air or oxygen sparging, injection of a dilute solution of hydrogen peroxide, or use of oxygen release compounds (ORCs) such as magnesium oxides. Some amendments such as the ORCs can be passively introduced in wells or trenches.

In practice, the enhanced biodegradation process may be achieved as a conventional waterborne biodegradation, in which water is used to uniformly transport nutrients (nitrogen and phosphorous) and oxygen through the unsaturated zone; or as a soil-venting enhanced biodegradation, in which a system is engineered to increase the microbial biodegradation in the vadose zone using pumped air as the oxygen source; or indeed as a variation thereof the previously stated. Oxygen enhancement, for instance, can be achieved by either sparging air or oxygen below the water table or circulating hydrogen peroxide (H_2O_2) throughout the contaminated groundwater zone. Under anaerobic conditions, nitrate is circulated throughout the groundwater contamination zone to enhance bioremediation. Additionally, solid-phase peroxide products (e.g., oxygen-releasing compounds) can also be used for oxygen enhancement and to increase the rate of biodegradation. Select variants of the processes involved are identified, which are follows:

Enhanced in situ aerobic bioremediation: This is primarily used to treat nonhalogenated SVOCs such as diesel fuel and heavy fuel oil. These contaminants can be readily and directly metabolized and often require only oxygen addition to stimulate biodegradation.

Cometabolic biotreatment: This is primarily used for chlorinated VOC (CVOC) and SVOC contamination; treatment of chlorinated solvents such as TCE usually requires the addition of a co-substrate such as methane, propane, or toluene. Cometabolic treatment of other SVOCs can require structural analogs for the stimulation of contaminant degradation. The technology is applicable at sites where the aquifer characteristics are such as to allow effective delivery and mixing of the amendments, and where regulatory constraints do not preclude the use of such variant of the biorestoration technology.

In fact, the cometabolic biotreatment procedure is quite unique in that a primary substrate must be added to support the growth of the microorganisms and promote the degradation of contaminants. The amendments are circulated through the contaminated zone to provide mixing and intimate contact between the oxygen, nutrients, contaminant, and microorganisms. This contact is required to enhance the rate of aerobic biodegradation of the organic contaminants by the microorganisms distributed throughout the contaminated volume in the aquifer.

Oxygen enhancement with air sparging: Air sparging below the water table increases groundwater oxygen concentration and enhances the rate of biological degradation of organic contaminants by naturally occurring microbes. Air sparging also causes an increase in mixing in the saturated zone, which then increases the contact between groundwater and soil. The ease and low cost of installing small-diameter air injection points allows considerable flexibility in the design and construction of such types of remediation system. It is noteworthy that oxygen enhancement with air sparging is typically used in conjunction with soil vapor extraction (SVE) or bioventing to enhance removal of the volatile component under consideration. (VOC stripping enhanced by air sparging is addressed in greater detail in Section 12.2.19.)

Oxygen enhancement with hydrogen peroxide (H_2O_2): During hydrogen peroxide enhancement, a dilute solution of hydrogen peroxide is circulated through the contaminated groundwater zone to increase the oxygen content of groundwater and thus to enhance the rate of aerobic biodegradation of organic contaminants by naturally occurring microbes.

Nitrate enhancement: This consists of enhancing the anaerobic biodegradation through the addition of nitrate. In the process involved, solubilized nitrate is circulated throughout groundwater contamination zones to provide an alternative electron acceptor for biological activity and to enhance the rate of degradation of organic contaminants. For instance, most fuels have been shown to degrade rapidly under aerobic conditions, but success is often limited by the inability to provide sufficient oxygen to the contaminated zones as a result of the low water solubility of oxygen, and because oxygen is rapidly consumed by aerobic microbes. Since nitrate can also serve as an electron acceptor and is more soluble in water than in oxygen, the addition of nitrate to an aquifer results in the anaerobic biodegradation of toluene, ethylbenzene, and xylenes. Meanwhile, it is notable that the benzene component of fuel has been found to generally biodegrade slowly under strictly anaerobic conditions; thus, a mixed oxygen/nitrate system would prove advantageous in that the addition of nitrate would supplement the demand for oxygen rather than replace it, allowing for benzene to be biodegraded under microaerophilic conditions.

12.2.4   CHEMICAL FIXATION/IMMOBILIZATION

Chemical fixation (or *immobilization*) is a technique to chemically fix or modify the chemical structure of contaminants, via the application of specific reagents. The fixation process consists of an immobilization of the contaminants of interest in-place, thereby preventing migration into unaffected environmental media. The process is generally designed to immobilize contaminants within the contaminated matrix, instead of removing them through chemical or physical treatment. In its application, generally, it is necessary to perform bench-scale treatability studies prior to a field/full-scale chemical fixation treatment; among other things, leachability testing is typically performed to measure the immobilization of contaminants.

In the chemical fixation technology, the contaminated soil or material is blended with precise amounts of reagent(s) to stabilize and/or encapsulate chemical constituents that are then stockpiled and allowed to cure. For example, soil is mixed with a binder such as portland cement to reduce contaminant mobility by a combination of physical entrapment (e.g., encapsulation or porosity reduction) and chemical reaction (e.g., hydroxide precipitation). Oftentimes, the treated material is rendered nonhazardous and may be backfilled and left on the site. Also, the process generally improves soil condition (e.g., increases compressive strength, and decreases permeability), and there is ease of permitting for proven methods.

Chemical fixation can indeed be applied to many organic and metal-bearing contaminants or waste streams—albeit *in situ* solidification/stabilization technology seems to have limited effectiveness against SVOCs and pesticides; in any case, binder formulations designed to be more effective in treating organics continue to be developed and tested.

12.2.5   CHEMICAL REDUCTION/OXIDATION

Reduction/oxidation (*redox*) reactions can be used to chemically convert hazardous contaminants into nonhazardous or less toxic compounds that are more stable, less mobile, and/or inert. Redox reactions involve the transfer of electrons from one compound to another. Specifically, one reactant is oxidized (loses electrons) and the other is reduced (gains electrons). The oxidizing agents most commonly used for the treatment of hazardous contaminants are ozone, hydrogen peroxide, hypochlorites, chlorine, and chlorine dioxide; the reducing agents most commonly used for the treatment of hazardous waste are ferrous sulfate, sodium bisulfite, and sodium hydrosulfite.

Chemical redox treatment is implemented by mixing treatment chemicals with the water stream to promote a redox reaction. In remediation applications, chemical redox of inorganic contaminants is usually a wastewater pretreatment step performed to prepare a specific contaminant for removal, or to reduce the toxicity of a contaminant by altering its chemical form. The most common applications of chemical redox are as follows: reducing chromium (VI) to chromium (III) in preparation for hydroxide precipitation; oxidizing arsenic (III) to arsenic (V) to reduce toxicity and improve removal by subsequent processes; and oxidizing cyanide to produce CO_2 and N_2. Chemical redox is indeed a full-scale, well-established technology used for disinfection of drinking water and wastewater, and it is also a common treatment for cyanide wastes. Enhanced redox systems are being used quite frequently to treat contaminants in soils in recent times.

The target contaminant group for chemical redox is inorganics. Although the technology can be used on nonhalogenated VOCs and SVOCs, fuel hydrocarbons, and pesticides, it tends to be less effective against these types of contaminants.

12.2.6 ELECTROKINETIC [SOIL] DECONTAMINATION

Electrokinetic [soil] decontamination (or *electrokinetic separation*) uses electrodes to pass a direct current through a contaminated media/material (e.g., soils, sludge, and sediments), thus causing ions and water to migrate through the subsurface toward the electrodes (Figure 12.2). Small pumps placed at the cathodes remove the metal-containing fluids for treatment.

The principle of electrokinetic remediation (ER) relies upon the application of a low-intensity direct current through the soil between ceramic electrode pairs that are divided into a cathode array and an anode array, and ceramic electrode pairs that have been implanted in the ground on each side of the contaminated soil mass. The electrical current mobilizes charged species, causing ions and water to move toward the electrodes; that is, it causes electroosmosis and ion migration. The aqueous phase or contaminants desorbed from the soil surface are transported toward respective electrodes, depending on their charge. In general, metal ions, ammonium ions, and positively

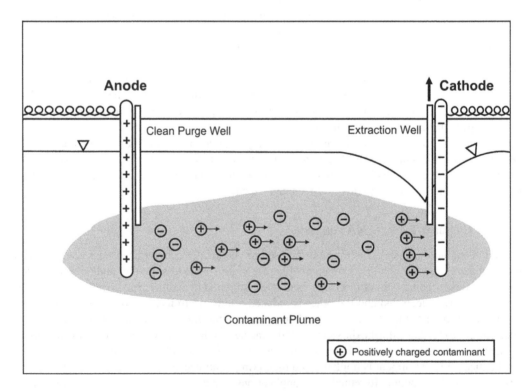

FIGURE 12.2 A general schematic of ER process. (Adapted from, USEPA, 1995d.)

charged organic compounds move toward the cathode, and anions such as chloride, cyanide, fluoride, nitrate, and other negatively charged organic compounds move toward the anode. The current creates an acid front at the anode and a base front at the cathode; this generation of acidic condition *in situ* may help to mobilize sorbed metal contaminants for transport to the collection system at the cathode. The contaminants may then be extracted to a recovery system or deposited at the electrode. Surfactants and complexing agents can be used to increase solubility and assist in the movement of the contaminant. Also, reagents may be introduced at the electrodes to enhance contaminant removal rates.

Overall, two primary mechanisms transport contaminants through the soil toward one or the other electrodes: electromigration and electroosmosis. In electromigration, charged particles are transported through the substrate; in contrast, electroosmosis is the movement of liquids containing ions relative to a stationary charged surface. Of the two, electromigration is the main mechanism for the ER process. The direction and rate of movement of an ionic species will depend on its charge, both in magnitude and polarity, and in the magnitude of the electroosmosis-induced flow velocity. Nonionic species, both inorganic and organic, will also be transported along with the electroosmosis-induced water flow.

Classically, two approaches are usually taken during ER: *enhanced removal* and *treatment without removal*. *Enhanced removal* is achieved by electrokinetic transport of contaminants toward the polarized electrodes to concentrate the contaminants for subsequent removal and *ex situ* treatment. Removal of contaminants at the electrode may be accomplished by several means, for example, electroplating at the electrode, precipitation or coprecipitation at the electrode, pumping of water near the electrode, or complexing with ion-exchange resins. Enhanced removal is widely used in the remediation of soils contaminated with metals. *Treatment without removal* is achieved by electroosmotic transport of contaminants through treatment zones placed between electrodes. The polarity of the electrodes is reversed periodically, which reverses the direction of the contaminants back and forth through treatment zones. The frequency with which electrode polarity is reversed is determined by the rate of transport of contaminants through the soil. This approach can be used for *in situ* remediation of soils contaminated with organic compounds.

The target contaminant groups for the ER technology are dissolved polar species such as heavy metals, anions, and polar or water-soluble organics in soil, mud, sludge, and marine dredging. Concentrations that can be treated range from rather low to very high contaminant levels. Also, electrokinetics is most applicable in low-permeability soils; such soils are typically saturated and partially saturated clays and silt–clay mixtures, and are not readily drained. Electrokinetics has indeed been used for decades in the oil recovery industry, and to remove water from soils, but *in situ* application of electrokinetics to remediate contaminated soil is relatively new; in particular, there has been significant focus on developing *in situ* electrokinetic techniques for the treatment of low-permeability soils, which tends to be more resistant to remediation with traditional technologies because of their low hydraulic conductivity.

12.2.7 Encapsulation

Encapsulation (using physical barrier systems) is a remedial approach comprising the physical isolation and containment of the contaminated material, as typical of well-engineered landfills. The technique often includes a physical barrier installation at the sides of the contaminated area because its purpose is to leave the contaminant safely in-place at the location of choice.

Overall, the encapsulation technique consists of isolating the impacted soils or matrix through the use of low-permeability caps, slurry walls, grout curtains, or cutoff walls. The contaminant source is covered with low-permeability layers of synthetic textiles or clay cap; the cap is designed to limit infiltration of precipitation and thus prevent leaching and migration of contaminants away from the site and into groundwater. In general, isolation and containment systems can work adequately, but there is no guarantee as to the destruction of the encapsulated contaminants.

12.2.8 INCINERATION

Incineration is a thermal treatment/degradation process by which contaminated materials are exposed to excessive heat in a variety of incinerator types; the different types of incinerators are generally designed and built for different purposes. Incineration is primarily used to treat halogenated and nonhalogenated SVOCs, PCBs, and ordnance compounds. It is typically used in the remediation of soils contaminated with explosives and hazardous wastes, particularly chlorinated hydrocarbons, PCBs, and dioxins. In general, the destruction and removal efficiency for properly operated incinerators tends to exceed the 99.99% requirement for hazardous wastes.

The incineration process typically involves the thermal destruction of contaminants by burning. Depending on the intensity of the heat, the contaminants of concern are volatilized and/or destroyed during the incineration process. Incinerator off-gases and combustion residuals generally require treatment, and the ash remaining from incineration of hazardous materials is usually disposed of in landfills. Classically, the incinerator off-gas requires treatment by an air pollution control system to remove particulates and neutralize and remove acid gases (HCl, NO_x, and SO_x); baghouses, venturi scrubbers, and wet electrostatic precipitators remove particulates, whereas packed-bed scrubbers and spray driers remove acid gases. Ash discharged from the incinerator is cooled and collected for disposal. In fact, ash remaining after incineration may require special treatment (e.g., solidification/ stabilization to reduce metal leachability) prior to disposal.

Broadly speaking, the incineration process is performed by supplying heat from fuel combustion or electrical input source, which then causes thermal decomposition of organic contaminants through cracking and oxidation reactions at high temperatures (usually between 760°C and 1,550°C or 1,400°F and 3,000°F). The organic contaminants are primarily converted into carbon dioxide and water vapor. Other products of incineration can include nitrite oxides, nitrates, and ammonia (for nitrogen-containing wastes), sulfur oxides and sulfate (for sulfur-containing wastes), and halogen acids (for halogenated wastes).

Incineration of high calorific value wastes may indeed be regarded as a form of recycling if the heat generated is used for other economic purposes. Even so, incineration is typically a very expensive treatment alternative, with the cost being very much dependent on the specific processes involved. Of particular interest is *catalytic conversion*, which is an incineration process that uses a catalyst to reduce the usually high incineration temperature requirements. Used in special treatment applications, catalytic units can be built for distinctive contaminants, but the cost of such units tends to be rather high. In practice, various types of incineration facilities exist, with the more common ones identified in the following sections. It is noteworthy that contaminated soils are typically treated in a "rotary kiln" or a "fluidized bed" incinerator (see below).

12.2.8.1 Circulating Bed Combustor

Circulating bed combustor (CBC) uses high-velocity air to entrain circulating solids and create a highly turbulent combustion zone that destroys toxic hydrocarbons. The CBC operates at lower temperatures than conventional incinerators (viz., 1,450°F–1,600°F or 780°C–870°C). The CBC's high turbulence produces a uniform temperature around the combustion chamber and hot cyclone. The CBC also completely mixes the waste material during combustion. Effective mixing and low combustion temperature reduce operating costs and potential emissions of gases such as nitrogen oxides (NO_x) and carbon monoxide (CO).

12.2.8.2 Fluidized Bed

Fluidized bed incinerators involve a single chamber that contains the fluidizing sand and a freeboard section above the bed. The fluidized bed aids in the volatilization and combustion of the organic waste constituents. The solid particulate in the bed provides a sufficient heat capacity to volatilize organic constituents. The forced air used to fluidize the bed provides sufficient oxygen and turbulence to enhance the reactions of organics with oxygen to form carbon dioxide and water vapor. Additional time for conversion of the organic constituents is provided by the freeboard above the fluidized bed.

As a special example, the circulating fluidized bed uses high-velocity air to circulate and suspend the waste particles in a combustion loop, and operates at temperatures up to 870°C (1,600°F). Another example unit type—the infrared unit—uses electrical resistance heating elements or indirect-fired radiant U-tubes to heat material passing through the chamber on a conveyor belt and operates at temperatures up to 870°C (1,600°F).

12.2.8.3 Infrared Combustion

The infrared combustion technology is a mobile thermal processing system that uses electrically powered silicon carbide rods to heat organic wastes to combustion temperatures. Waste is fed into the primary chamber and exposed to infrared radiant heat (up to 1,850°F) provided by silicon carbide rods above the conveyor belt. A blower delivers air to selected locations along the belt to control the oxidation rate of the waste feed. Any remaining combustibles are incinerated in an afterburner.

12.2.8.4 Rotary Kilns

Rotary kiln (and fixed hearth) incineration technologies involve two chambers in the incineration process. Organic constituents in the waste are volatilized in the primary chamber. During this volatilization process, some of the organic constituents oxidize to form carbon dioxide and water vapor. High temperatures then cause the organic constituents to react with oxygen in the second chamber, forming carbon dioxide and water vapor.

The rotary kiln is a refractory-lined, slightly inclined, rotating cylinder that serves as a combustion chamber, and operates at temperatures up to 980°C (1,800°F). Typically, commercial incinerator designs consist of rotary kilns, equipped with an afterburner, a quench, and an air pollution control system.

12.2.9 *In Situ* Chemical Oxidation

In situ chemical oxidation (ISCO) refers to a general group of specialty remediation techniques or technologies in which chemical oxidants are delivered to the subsurface to rapidly degrade organic contaminants, with each variant technology representing unique combinations of oxidants and delivery techniques. Broadly speaking, it involves the application of a strong oxidizing agent in the ground via well injection or a specially designed injection tool. The oxidants degrade the target contaminants by converting them to benign compounds, usually H_2O, CO_2, and mineral salts. Specific primary oxidants commonly used for ISCO include hydrogen peroxide (H_2O_2); Fenton's reagent (i.e., an iron-catalyzed hydrogen peroxide—a liquid composed of Fe^{2+} + H_2O_2); permanganate $\left(MnO_4^-\right)$—typically potassium and sodium permanganate (usually, $KMnO_4$ in liquid form); and ozone (viz., O_3 gas). Each oxidant chemical is generally uniquely effective for different contaminants.

Some general examples of potential contaminants that are amenable to treatment by ISCO include chlorinated solvents, aromatic compounds (polyaromatic hydrocarbons [PAHs]), and petroleum products. Specific examples of target contaminants include benzene, toluene, ethylbenzene, and xylenes (BTEX), tetrachloroethylene (PCE), TCE, DCEs, VC, methyl-tert-butyl-ether, PAH compounds, and many other organic contaminants. It is notable that, historically, commonly used oxidants have not quite been effective with saturated aliphatic hydrocarbons (viz., octane and hexane) or chlorinated alkanes (viz., chloroform and carbon tetrachloride), and permanganate may have limited effectiveness against BTEX. In any case, the method is applicable to the remediation of both soil and groundwater contaminated with organic contaminants. Remediation of groundwater contamination using ISCO involves injecting oxidants and other amendments, as required, directly into the source zone and applicable downgradient plume. The oxidant chemicals then react with the contaminant, producing innocuous substances such as carbon dioxide (CO_2), water (H_2O), and inorganic chloride—albeit the full spectrum of reaction intermediates and by-products may not always be fully understood for all contaminants. Anyhow, an important attribute is the fact that the residual oxidizing agents are relatively benign in the environment.

ISCO can be used to treat dense nonaqueous-phase liquids (DNAPLs) as well as dissolved-phase contaminants. Indeed, ISCO is particularly suitable for contaminant source zone (e.g., DNAPL) treatment—albeit not quite effective for total plume management *per se*. In any case, migration of contaminants/by-products may have to be controlled as part of the overall remedial plan. In general, ISCO is most useful for source area mass reduction and intercepting of plumes to remove mobile contaminants. The appropriateness of ISCO technology at a site also depends on matching the oxidant and delivery system to the site contaminants and site conditions; for example, permanganate is not effective against BTEX, while peroxide and ozone are effective against BTEX, and this requires careful site characterization and screening to arrive at this decision point. Furthermore, oxidation is dependent on achieving adequate contact between oxidants and contaminants. Thus, failure to account for subsurface heterogeneities or preferential flow paths can result in extensive pockets of untreated contaminants. The applied reagents could also be consumed by natural organic matter or dissolved iron (rather than the contaminants), thereby compromising the effectiveness of the remediation system.

ISCO offers several advantages over conventional treatment technologies such as pump and treat. For instance, the technology does not generate large volumes of waste material that must be disposed of and/or treated, and indeed may not require aboveground treatment (as with pump-and-treat systems, thermal heating, and surfactant flushing). ISCO is also implemented over a much shorter time frame. Both of these advantages should result in savings on material, monitoring, and maintenance. Nonetheless, this technology also has various limitations and should not be considered a magic bullet for every site. Furthermore, application of ISCO may actually disrupt other remedies; for example, application of ISCO on a site that is benefiting from natural reductive dehalogenation may temporarily upset the geochemistry that facilitates the process. Also, it is noteworthy that application of chemical oxidation technology may be limited by site geology and geochemistry (e.g., pH, alkalinity/hardness, dissolved iron).

Other important advantages of ISCO include its relatively low cost and speed of reaction—albeit the design should appropriately account for the possible hazards of the chemicals and potential for vigorous uncontrolled reactions in the subsurface that could occur with Fenton's reagent. Indeed, volatile compounds may be released by even moderate changes in temperature. Thus, there could be a significant change in both the concentration and the distribution of flammable vapors and/or toxic nonflammable vapors when using an ISCO method. In any case, for chlorinated hydrocarbon remediation via chemical oxidation methods, the risk of a fire is reduced.

A key uncertainty about ISCO systems is that the radii of influence for different types of injections have not been adequately established for all soil types and hydrogeological conditions. Indeed, several recent case studies suggest that, for situations where the soil is tight, the number, geometry, and technique of injection are quite critical to the success or failure of an ISCO treatment. Thus, some experimentation and multiple attempts with injection configuration and injection method may be necessary for an effectual system design and implementation.

12.2.10 IN SITU SOIL FLUSHING

In situ soil flushing (or *in situ chemical leaching*) is a process by which in-place soils are flushed with water, usually mixed with a biodegradable nontoxic surfactant, in an effort to leach the compounds present in the soil into groundwater. The flushing agent is allowed to percolate into the soil and enhance the transport of contaminants to groundwater extraction wells for recovery. The groundwater is then collected downgradient or downstream of the leaching site for treatment, recycling, and/or disposal.

In general, high water solubility, a low soil–water partition coefficient (K_{oc}), and a porous soil matrix will aid in the effective removal of chemical contaminants from soils using the soil leaching technique. The K_{oc}, which is a measure of the equilibrium between the soil organic content and water, is indeed the leading factor controlling the effectiveness of soil flushing; a low K_{oc} value indicates a favorable leaching tendency of the constituent from the soil.

The *in situ* leaching technology is most applicable for soluble organics and metals at a low-to-medium concentration that are distributed over a wide area. In practice, soil flushing is primarily used to treat halogenated and nonhalogenated SVOCs, PCBs, and ordnance compounds from *in situ* materials; water-soluble inorganic contaminants may also be removed using soil flushing. The technology can also be used to treat various VOCs, SVOCs, fuels, and pesticides, but it may be less cost-effective than alternative technologies for these contaminant groups. On the whole, the addition of environmentally compatible surfactants may be used to increase the effective solubility of some organic compounds—albeit the flushing solution may alter the physical/chemical properties of the soil system. Additionally, residual flushing additives in the soil may be a concern and thus should be evaluated on a site-specific basis.

All in all, *in situ* soil flushing consists of the extraction of contaminants from the soil with water or other suitable aqueous solutions. Soil flushing is accomplished by passing the extraction fluid through in-place soils using an injection or infiltration process. Contaminants present in the soil partition into the flushing solution by mechanisms such as solubilization, emulsification, or chemical reaction. The contaminant-laden solution must then be recovered to prevent uncontrolled transport of contaminants. For biodegradable contaminants, it may be possible to add nutrients and distribute the flushing solution on the soil to promote contaminant bioremediation as well. In a typical application, inorganic and organic contaminants can be extracted from soil by flushing the soil with solvents; solvents are recovered, contaminants are extracted, and the solvents are recirculated through the soils (Figure 12.3). The *in situ* technology offers the potential for recovery of metals and can mobilize a wide range of organic and inorganic contaminants from coarse-grained soils.

12.2.10.1 Cosolvent Enhancement

Water is normally used as the flushing agent in the *in situ* leaching technique. However, other solvents may be used for contaminants that are tightly held or only slightly soluble in water. Chemically enhanced *in situ* soil flushing may have extensive applications, but such applications will generally require site-specific evaluation and system design. Indeed, for hydrophobic compounds such as most hydrocarbons, flushing with surfactants is likely to be more effective than flushing with water; on the other hand, flushing with water generally may suffice for hydrophilic compounds.

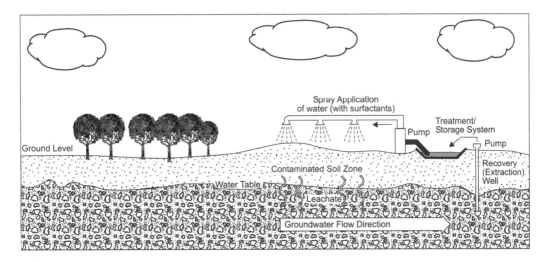

FIGURE 12.3 A general schematic of the soil leaching process.

Cosolvent flushing involves injecting a solvent mixture (e.g., water plus a miscible organic solvent such as alcohol) into either the vadose zone or saturated zone—or indeed into both—in order to extract organic contaminants. Solvents are selected on the basis of their ability to solubilize the contaminants and also on their environmental and human health effects. Thus, it is important to know the chemistry and toxicity of the surfactant. It is also important to understand the hydrogeology of the site to ensure that contaminants will be extracted once they are mobilized.

Cosolvent flushing can be applied to soils to dissolve either the source of contamination or the contaminant plume emanating from it. The cosolvent mixture is normally injected upgradient of the contaminated area, and the solvent with desorbed/dissolved contaminants is extracted downgradient and treated above ground. In most cases, treatment will be required to allow reuse of the fluid for continued flushing or to meet appropriate discharge standards prior to recycling or release to local, publicly owned wastewater treatment works or receiving streams. To the maximum extent practical, recovered fluids should be reused in the flushing process. Indeed, the separation of surfactants from recovered flushing fluids (for reuse in the process) could become a major factor in the cost of soil flushing techniques.

12.2.11 IN SITU VITRIFICATION

ISV is an innovative thermal treatment process that converts contaminated materials into a chemically inert, stable glass, and crystalline product. Indeed, the ISV technology is another *in situ* solidification/stabilization process that uses an electric current/energy to melt soil or other earthen materials at extremely high temperatures (1,600°C–2,000°C or 2,900°F–3,650°F) and thereby immobilize most inorganics and destroy organic pollutants by pyrolysis; inorganic pollutants are incorporated within the vitrified glass and crystalline mass. A vacuum-pressurized hood is typically placed over the vitrification zone to contain and process any contaminants emanating from the soil during vitrification. Water vapor and organic pyrolysis combustion products are captured in a hood, which draws the contaminants into an off-gas treatment system that removes particulates and other pollutants from the gas. The vitrification product is a chemically stable, leach-resistant, glass and crystalline material similar to obsidian or basalt rock. The process destroys and/or removes organic materials; radionuclides and heavy metals are retained within the molten soil.

The initial step in the vitrification process is to identify the boundaries of the area of contaminated soil to be treated. An array of four graphite electrodes is inserted into the ground to the desired treatment depth. The electrodes are placed vertically into the contaminated soil region, and an electrical current is applied; an electrical potential of over 12 kV is applied to the electrodes that establish the electrical current. The resultant power heats the path and surrounding soil to above fusion temperatures, and then, the soil is melted by the resulting high temperatures. When the melt cools and solidifies, the resulting material is stable and glass-like, with the contaminants bound in the solid. The molten soil zone grows downward and outward. Some newer designs can incorporate a moving electrode mechanism to achieve a greater process depth.

Overall, the ISV method may be used to provide solution to mixed wastes, including organic, inorganic, and radioactive wastes, in soils at a potentially contaminated site, up to about a 15 m (≈50 ft) depth. The ISV process can destroy or remove organics and immobilize most inorganics in contaminated soils, sludge, or other earthen materials. Indeed, the process has been tested on a broad range of VOCs and SVOCs, other organics including dioxins and PCBs, and on most priority pollutant metals and radionuclides. During the vitrification process, the major portion of the contaminants initially present in the soils are volatilized, with the remainder being worked in place in the hardened soil. The application of this technology can be rather expensive.

12.2.12 LANDFARMING

Landfarming (also known as *land treatment* or *land application*) is an aboveground remediation technology for soils that reduces contaminant concentrations predominantly through biodegradation.

It consists of a land treatment process by which contaminated materials are spread over an area to enhance naturally occurring processes such as volatilization, aeration, biodegradation, and photolysis. Standard earth-moving equipment is typically used to prepare the landfarm area and to apply the material for treatment. The treatment area usually has a low-permeability liner material to prevent or minimize leaching; it is usually also surrounded by berms to control potential runoff processes and erosion from the soil during rainstorm events.

General treatment requirements for the landfarming process include periodic applications of nutrients such as phosphorous and nitrogen, moisture control, and disking for oxygen exposure; biological treatments use bacteria and other microorganisms to degrade the contaminants. As such, landfarming may really be considered a full-scale bioremediation technology that usually incorporates liners and other methods to control leaching of contaminants and that also requires excavation and placement of contaminated soils, sediments, or sludges. Contaminated media are applied to lined beds and periodically turned over or tilled to aerate the waste. Soil conditions are often controlled to optimize the rate of contaminant degradation; conditions normally controlled include moisture content (usually by irrigation or spraying), aeration (by tilling the soil with a predetermined frequency, the soil is mixed and aerated), pH (buffered near neutral pH by adding crushed limestone or agricultural lime), and other amendments (e.g., soil bulking agents and nutrients).

The principle behind land application technology is simple. Waste that contains low concentrations of organic contaminants is spread over a large area and is allowed to interact with the soil and climate at the site. The soil is turned periodically using equipment such as a plow or disk to expose new surfaces and aerate the matrix. Waste, soil, climate, and biological agents interact dynamically as a system to degrade, transform, and immobilize waste constituents. In the typical process, contaminated media are usually treated in lifts that are up to 45 cm (or 18 in.) thick. When the desired level of treatment is achieved, the lift is removed and a new lift is constructed. It may be desirable to only remove the top of the remediated lift, and then construct the new lift by adding more contaminated media to the remaining material and mixing. This serves to inoculate the freshly added material with an actively degrading microbial culture, and can reduce treatment times.

Landfarming has been proven most successful in treating petroleum hydrocarbons. However, because the lighter, more volatile hydrocarbons such as gasoline are treated very successfully by processes that use their volatility (e.g., SVE), the use of aboveground bioremediation is usually limited to heavier hydrocarbons. Contaminants that have been successfully treated using landfarming include diesel fuel, fuel oils, jet fuels, oily sludge, wood-preserving wastes (pentachlorophenol, PCP and creosote), coke wastes, and certain pesticides. As a rule of thumb, the higher the molecular weight (and the more rings with a PAH), the slower the degradation rate. Also, the more chlorinated or nitrated the compound, the more difficult it is to degrade. On the whole, land application is primarily used to treat aerobically biodegradable compounds such as nonhalogenated VOCs and SVOCs.

12.2.13 PASSIVE REMEDIATION/NATURAL ATTENUATION

Passive remediation (or *natural attenuation* or *intrinsic remediation*) relies on natural processes (e.g., biodegradation, volatilization, photolysis, sorption, dispersion, and dilution) to remediate impacted environmental media, usually soils and groundwater. In fact, natural attenuation processes occur to varying degrees at virtually all sites with organic chemical contamination, and this state of affair or condition invariably contributes significantly to site remedial goals. For instance, biodegradation processes tend to be particularly important to the natural attenuation of some chlorinated solvents, under specific environmental conditions. Chlorinated solvents dissolved in water, sorbed on soil particles, or even in vapor form are the ones most readily amenable to natural attenuation processes, whereas bulk chlorinated solvents (viz., NAPLs) typically are not readily susceptible to natural attenuation in the short term. Consequently, the significance of natural attenuation processes to attain site remedial goals at any given site must be very carefully evaluated to ascertain its feasibility; oftentimes, extensive site characterization and monitoring tends to be necessary under such circumstances.

In fact, natural attenuation is not a "technology" *per se*, and there is significant debate among technical experts about its use at hazardous waste sites. Consideration of this option usually requires modeling and evaluation of contaminant degradation rates and pathways, and predicting contaminant concentration at downgradient receptor points, especially when plume is still expanding/migrating. The primary objective of site modeling is to demonstrate that natural processes of contaminant degradation will reduce contaminant concentrations below regulatory standards or risk-based levels in appropriate time frame before potential exposure pathways are completed. In addition, sampling and sample analysis is generally conducted throughout the process to confirm that degradation is proceeding at rates consistent with meeting cleanup objectives.

Broadly speaking, the natural attenuation processes typically include a variety of physical, chemical, and biological processes that, under favorable conditions, can act without direct human intervention to reduce the mass, toxicity, mobility, volume, and/or concentration of contaminants in various environmental matrices. The common forms of specific natural attenuation processes occurring under average field conditions include biodegradation, dispersion, dilution, sorption, volatilization, and abiotic degradation mechanisms. In practice, biodegradation processes are of special importance in natural attenuation because they reduce the mass of contaminants, as well as generally transform toxic contaminants to nontoxic by-products. Still, it is noteworthy that natural attenuation processes can also convert some contaminants to more toxic forms.

Typical target contaminants for natural attenuation are VOCs, SVOCs, and fuel hydrocarbons. Indeed, fuel and halogenated VOCs (chlorinated solvents) are thus far the most commonly evaluated for natural attenuation. Pesticides can also be allowed to naturally attenuate, but the process may be less effective and may be applicable to only some compounds within the group. In any case, passive remediation may be applicable at sites where contaminant migration is limited, where potential impact on the environment is minimal, and when health and safety considerations are insignificant. In the application of a passive remediation option, continued monitoring is usually used to demonstrate that contamination levels are being attenuated and exposure is not occurring. Thus, following site assessment, the only activity undertaken is a progressive monitoring program to evaluate the effectiveness of the "no-action" option in the management of a contaminated site problem.

12.2.13.1 Natural Attenuation via Intrinsic Bioremediation

In utilizing natural attenuation by intrinsic bioremediation, natural subsurface processes are generally allowed to reduce contaminant concentrations to "acceptable levels." Indeed, ambient conditions at many sites are suitable for microorganisms to degrade contaminants without human intervention (i.e., stimulation); for instance, naturally occurring microorganisms in soil and groundwater can degrade a range of common petroleum contaminants effectively. In these cases, intrinsic bioremediation can provide an attractive remedial alternative to more costly soil and groundwater cleanup methods. Essentially, the natural attenuation combines the effects of several natural processes that operate in the subsurface environment and can indeed reduce contamination to levels that are protective of human health and the environment.

It is noteworthy that the term "intrinsic bioremediation" has evolved to emphasize the essential role of indigenous microorganisms in achieving site remediation by natural attenuation; intrinsic biodegradation refers to the degradation of contaminants by indigenous microorganisms under the prevailing site conditions. Thus, in intrinsic bioremediation, no actions are taken to enhance the biodegradation of contaminants beyond the existing capacity of the system. In practice, intrinsic bioremediation is implemented only after demonstrating that the native microbial populations have the potential to reduce contaminant levels to meet remediation goals, and then monitoring the program to confirm that contaminants do not reach areas of potential concern at unacceptable concentrations. Indeed, before intrinsic bioremediation is fully implemented at a site, an evaluation is required to determine whether it is a viable remediation method. Preliminary screening to determine whether intrinsic bioremediation is applicable is followed by a detailed evaluation to determine whether it will be effective at meeting remediation goals. A detailed evaluation of intrinsic bioremediation should answer the following two key questions: (i) Is biodegradation of the contaminants already occurring? and (ii) is this occurring rapidly enough to remain protective of potential receptors? Subsequently, if intrinsic bioremediation is demonstrated to be a viable remedial option for a site, a long-term monitoring plan must be designed and implemented to verify ongoing effectiveness until remediation goals are met, and also to be able to detect unexpected contaminant migration away from the site that could impact potential receptors in the area. The monitoring strategy should consider appropriate sampling locations, frequency, and parameters to be measured. Semiannual or annual sampling may be sufficient if contaminant concentrations have been relatively stable during initial site monitoring. More frequent (e.g., monthly or quarterly) sampling may be required to resolve trends in the data when initial monitoring results fluctuate significantly. Parameters to measure should include, at a minimum, contaminant concentrations and water levels to track changes in the plume and groundwater flow direction. Electron acceptors, metabolic by-products, and general water quality parameters (e.g., temperature, pH, alkalinity, hardness, and redox potential) may be measured to monitor changes in ambient water quality, as well as to provide further evidence of ongoing remediation. Invariably, long-term monitoring is always a necessary part of any remediation system that includes natural attenuation as a component (USEPA, 1997c,g).

12.2.13.2 Monitored Natural Attenuation

Monitored natural attenuation (MNA) refers to the reliance on natural attenuation processes (within the context of a carefully controlled and monitored site cleanup approach) to achieve site-specific remedial action objectives within a time frame that is reasonable compared to that offered by other more active methods (USEPA, 1997c,g, 1999d,e,k). The "natural attenuation processes" that are at work in such a remediation approach include a variety of physical, chemical, or biological processes that, under favorable conditions, act without human intervention to reduce the mass, toxicity, mobility, volume, or concentration of contaminants in soil or groundwater. These *in situ* processes include biodegradation; dispersion; dilution; sorption; volatilization; and chemical or biological stabilization, transformation, or destruction of contaminants (USEPA, 1999d,e,k). It may be utilized as a complete remedy, or as part of a multi-remedy system, in order to address a site contamination problem.

In a typical contaminated site situation, the ultimate goals of MNA are for plume stabilization and mass reduction. The MNA can indeed be used alone or in concert with active/engineered remediation systems/technologies of a site remedy—as long as it can be proven effective in its role in the overall scheme of activities. It is noteworthy, however, that active remediation can alter natural attenuation by physically altering the plume and by changing its geochemistry. Congruently, whether remediation activities enhance or interfere with natural attenuation is to some extent dependent on the technology used. Thus, these interactions need to be taken into consideration when using MNA in concert with active remediation systems. In the long run, natural attenuation processes (viz., biodegradation, dispersion, dilution, sorption, volatilization, and abiotic degradation mechanisms) affect the fate and transport of a variety of contaminants in various environmental systems. When these processes are shown to produce outcomes that are protective of human health and the environment, and when an adequate monitoring program is put in place to document the effectiveness of these processes, they can be used as an exclusive remedy, or as a component of an engineered remedy at a site.

On the whole, MNA consists of the use of naturally occurring contaminant-degrading and contaminant-dispersing processes, combined with environmental monitoring, to remediate contaminated environmental matrices such as groundwater. In fact, natural attenuation processes occur in all environmental systems, but their effectiveness varies considerably from site to site and among different kinds of contaminants. Among other things, the efficiency of natural attenuation depends on particular hydrologic and geochemical characteristics of the environmental system into which the contaminants have been introduced. As such, it is possible to assess the efficiency of natural attenuation using a few general hydrologic, geochemical, and biological principles. Notwithstanding, it is often still quite difficult to identify those sites where MNA is appropriate, and just as importantly, those sites where natural attenuation is not appropriate, as a remedial strategy.

12.2.13.3   Natural Attenuation in Groundwater

In one of the most common scenarios for MNA, natural attenuation processes act to decrease concentrations of dissolved petroleum hydrocarbon and chlorinated solvent contaminants as they are transported by flowing groundwater. In this regard, the efficiency of natural attenuation processes refers to a quantitative comparison of contaminant transport rates to rates of biodegradation (USEPA, 1997c,g). If rates of biodegradation are fast relative to rates of transport, contaminant migration will be highly restricted and the efficiency of natural attenuation will be relatively high. Congruently, when natural attenuation efficiency is high, contaminant transport to sensitive receptor exposure points may be prevented and natural attenuation can then become an effective part of the overall site remediation program. On the other hand, if rates of biodegradation are slow relative to rates of transport, contaminants will not be as restricted and the efficiency of natural attenuation will be relatively low. Correspondingly, when natural attenuation efficiency is low, contaminants may move freely and reach receptor exposure points; in this case, natural attenuation may not be the sole remedy but could still be an appropriate component of a comprehensive remedial strategy.

In general, assessing the efficiency of natural attenuation requires information on at least the following types of factors (USEPA, 1997c,g):

- Concentrations of contaminants and daughter products in space and/or time
- Ambient geochemical conditions such as the reduction/oxidation (redox) state of groundwater
- Rates and directions of groundwater flow
- Rates of contaminant biodegradation
- Demographic considerations such as the presence of nearby receptor exposure points

Meanwhile, it is noteworthy that because the conditions that affect biodegradation of certain chemical compounds are reasonably well known (e.g., for petroleum hydrocarbons and chlorinated solvents), it is possible to assess the efficiency of natural attenuation with a high degree of confidence.

12.2.13.4   Natural Attenuation in Soil

In consideration of natural attenuation in soils, it is apparent that natural biotransformation processes such as dilution, dispersion, volatilization, biodegradation, adsorption, and chemical reactions with soil materials can reduce contaminant concentrations to acceptable levels. In general, natural attenuation may be considered for remediation of contaminants in soils if the following particularly important site-specific factors support its use:

- Protection of potential receptors during attenuation
- Favorable geological and geochemical conditions
- Documented reduction of degradable contaminant mass in a reasonable time frame in the surface and subsurface soils
- Confirmation in microcosm studies of contaminant cleanup
- For the persistent or conserved contaminants, indication of containment during and after natural attenuation

Meanwhile, it is noteworthy that surface and subsurface soils tend to have different characteristics in natural attenuation processes. Consequently, natural attenuation in both of contaminated surface soils and subsurface soils typically would require a very careful site-specific evaluation prior considering natural attenuation as a possible remedy.

As an example of contaminant behaviors in the soil matrix, it is evident that mobile organic contaminants in subsurface soils diffuse into soil vapor and aqueous phases and thus are relatively easily subject to natural subsurface processes that can attenuate these contaminants. On the other hand, most high-molecular-weight (persistent) organic and many inorganic contaminants may be immobilized in the subsurface soil matrix; these persistent organic contaminants are often difficult to degrade, and also, the inorganic metals are conserved. Without probable exposure routes, however, they do not represent significant risk unless unlikely events such as fresh solvent releases, chemical or biochemical transformation, or physical disturbances that increase their mobility or open exposure routes take place.

In contrast to the aforementioned exemplar, mobile organic contaminants in surface soils (typically, set at approximately 30–60 cm, or 1–2 ft of soil at a site) usually have already degraded, volatilized, or leached from the soil unless weathered free product has entrapped the contaminants. In addition, immobilized contaminants persist in surface soils for decades and thus, if undisturbed, are only slowly removed or not removed at all by natural attenuation processes. However, wind or water erosion may mobilize the persistent or conserved contaminants through soil transport into surface waters. Furthermore, the resulting contaminated sediments are often uncontrollable, since storms and natural events such as earthquakes can resuspend or distribute contaminants over wide areas.

12.2.14 PERMEABLE REACTIVE BARRIERS

Permeable reactive barriers (PRBs) (or *passive/reactive treatment walls*, or *permeable reactive treatment walls*, or *permeable reactive walls*) are structures installed underground across the flow path of a contaminated groundwater plume, in order to treat the contaminated groundwater passing through. Classically, they are comprised of a passive *in situ* treatment technology that uses natural groundwater flow conditions of a site for remediation.

In its simplest form, a treatment wall barrier consists of a trench installed across the flow path of a contaminant plume, allowing the water portion of the plume to passively move through the wall (Figure 12.4). This trench is filled with a reactive material, such as granular iron to reduce Cr (VI) or to dechlorinate halogenated organics; chelators to sequester selected metals; or other appropriate treatment material. Indeed, the barriers allow the passage of water while prohibiting the movement of contaminants by using such agents as zero-valent metals, chelators (ligands selected for their specificity for a given metal), sorbents, microbes, zeolite, and others. The "entrapped" contaminants will either be degraded or be retained in a concentrated form by the barrier material, which may require periodic replacement. As the groundwater passes through the treatment barrier, the contaminants react with the infilled media; for example, chlorinated organics that come in contact with an elemental iron treatment wall are degraded to potentially nontoxic dehalogenated organic compounds and inorganic chloride. In fact, the wall could provide permanent containment for relatively benign residues or provide a decreased volume of the more toxic contaminants for subsequent treatment.

In fact, several variations of PRB configurations are possible, including the more commonly used funnel-and-gate systems, which combine one or more permeable gate and impermeable funnel sections to capture wider contaminant plumes. Continuous reactive barriers may be beneficial when the plume width is smaller and/or if contaminant concentrations are lower. Pea gravel sections can also be installed bordering the reactive cell to facilitate uniform flow of contaminated groundwater through an iron cell. Depending on the site conditions, a funnel-and-gate system can be installed to handle large volumes of contaminated water without hydraulic control via pumping. In practice, methods of installation include constructing a trench across the contaminated groundwater flow

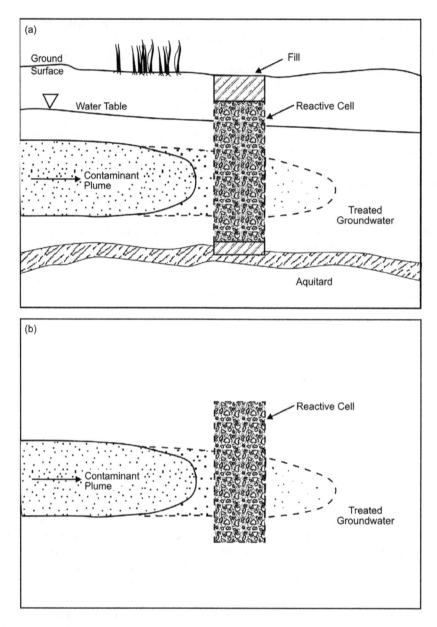

FIGURE 12.4 A general schematic of a PRB system. (a) Elevation view of a PRB; (b) plan view of a PRB configuration.

path by using either a funnel-and-gate system or a continuous reactive barrier. Indeed, a common treatment barrier configuration is the funnel-and-gate system; treatment walls or barriers are placed across the groundwater flow path typically using a funnel-and-gate system. The gate or reactive cell can be filled with various treatment media that are selected for their ability to clean up specific types of contaminants.

Although a variety of reactive media could be used to treat groundwater contaminants, the most commonly used media are zero-valent metals, particularly granular iron. As the zero-valent metal in the reactive cell corrodes, the resulting electron activity can reduce [chlorinated] compounds to potentially nontoxic products. Thus, typically, zero-valent iron particles are commonly used in the treatment cell, which react with the target contaminants such as chlorinated solvents (e.g., TCE,

PCE, DCE, and VC) in a strong reducing reaction to produce nontoxic and easily biodegradable by-products, such as light hydrocarbon chain C2–C5 compounds (viz., ethanes and ethenes), chloride and iron ions, and hydroxides. The reactive media are often bordered with sand or other porous materials (usually pea gravel), which encourage groundwater and contaminants to disperse and flow uniformly through the cell. The impermeable funnel walls, usually consisting of steel sheet piling or cement/bentonite slurry, direct the groundwater flow toward the reactive treatment gate.

The PRB technology is indeed an *in situ* approach for remediating groundwater contamination that combines subsurface fluid flow management with a passive chemical treatment zone. Removal of contaminants from a groundwater plume is achieved by altering chemical conditions in the plume as it moves through the reactive barrier. Because the reactive barrier approach is a passive treatment, a large plume of a variety of compounds of environmental concern can be treated in a cost-effective manner relative to traditional pump-and-treat systems. Typical target contaminant groups for passive treatment walls consist of VOCs, SVOCs, and inorganics; the technology can also be used in treating some fuel hydrocarbons—albeit it may be less effective for these types of contaminants. Indeed, reactive barriers are primarily used to treat halogenated VOCs and to reduce Cr (VI) to Cr (III). In any case, successful application of the PRB technology requires a proficient knowledge and understanding of the hydrogeologic characteristics of the site, specific groundwater flow conditions, and subsurface contaminants. Site-specific bench-scale studies and groundwater modeling are recommended prior to the design and implementation of reactive treatment walls.

Finally, it is worth mentioning here that a major advantage of a PRB system relates to the fact that no pumping or aboveground treatment is typically required; the contaminated water passively moves through the barrier. Because there are no aboveground installed structures, the affected property can generally be put to productive use even while it is being cleaned up.

12.2.15 Phytotechnologies and Phytoremediation

Phytotechnologies are a set of techniques that use plants to remediate or contain contaminants in soil, groundwater, surface water, or sediments; classically, the planted systems can degrade organic pollutants, extract heavy metals, and provide landfill covers. Some of these technologies have indeed become attractive alternatives to conventional cleanup technologies due to relatively low costs and the inherently aesthetic nature of planted sites. In general, the key to successfully applying any remediation technology is ensuring that the technology is applicable to the remediation objectives and site conditions.

Phytotechnology systems can be designed to provide control and containment, as well as remediation. If containment is the primary objective, the main focus should be on using rain interception and evapotranspiration or groundwater uptake and transpiration. An application for a phytotechnology project with the primary objective of containment should include modeling results regarding the effects of the plants on contaminant fate and transport. If a vegetative cover is proposed, hydrologic models to estimate infiltration or runoff should also be presented. For groundwater hydraulic control, this could include models that predict plume migration or stability; several models for these applications are available in the literature. If a containment strategy is not acceptable, the proposed application of phytotechnologies should include a mechanism for stabilizing, sequestering, reducing, degrading, metabolizing, or mineralizing the contaminants. The applications of phytotechnologies that combine containment with remediation include phytoremediation covers, groundcover systems for remediating surface soils, tree stands for remediating soil and groundwater, wetland treatment systems, riparian buffers, and aboveground hydroponic systems.

In general, the specific phytotechnology application to be used at a contaminated site depends on several mechanisms that can be used to address the different environmental conditions that may exist at a site; the specific mechanisms that are exploited depend on several factors, including the affected media, current site conditions, the specific contaminants or constituents of concern, the remedial objectives or goals, and the pertinent regulatory issues. In many cases, hydraulic control or containment is one of the remedial objectives for a site to ensure that contaminants do not migrate off-site or impact other receptors. To accomplish hydraulic control, vegetative groundcovers, tree hydraulic barriers, and wetland plant systems can be used to control surface water and groundwater movements as well as to physically stabilize the soil environment (i.e., reduce erosion, dust emissions, etc.). In addition to containment, another general objective in the remediation of a site is stabilization, accumulation, reduction, degradation, metabolism, or mineralization of specific contaminants in order to reduce the associated risks to human health and the environment. Therefore, the application of phytotechnologies is simply the logical and scientifically sound combination of a variety of feasible phytotechnology mechanisms. Depending on the chemical in the soil or groundwater, the specific species being used, soil conditions, and other factors, the specific mechanism that will be the primary method for achieving remedial goals may vary. Furthermore, several of these mechanisms may operate in series or differently, but in conjunction with each other. There also has been extensive scientific debate concerning which mechanisms are the predominant modes of action. A primary example of this is the use of deep-rooted trees to contain and remediate TCE. One school of thought has been that the TCE is unaffected as it passes through the rhizosphere but is taken up from the groundwater in its parent form and released into the atmosphere through phytovolatilization (Newman et al., 1997). A second school of thought has been that there is some degradation or transformation of the TCE that occurs while the contaminant is translocating through the trees (Anderson and Walton, 1991). Furthermore, these by-products are then released into the atmosphere through transpiration. Consequently, there is a combination of phytodegradation with phytovolatilization. A final school of thought for the remediation of TCE from groundwater has been that the parent compound is actually degraded in the rhizosphere with the subsequent uptake of the by-products into the trees to be eventually released during transpiration (Orchard et al., 1999). Therefore, the mechanisms involved in this phytotechnology application include rhizodegradation, followed by phytovolatilization. Anyhow, it seems to have been demonstrated that this application of phytotechnologies (regardless of which mechanisms are involved) is effective at treating TCE in groundwater (Chappell, 1998).

Meanwhile, it must be acknowledged here that phytotechnologies tend to be limited by plant growth rates, rooting depth, and duration of the growing season; because of these limitations, a longer restoration time may be required to achieve cleanup goals than with more conventional methods, such as excavation and landfilling or incineration. However, if the projected risks over time are shown to be minimal through a suitable risk analysis, phytotechnologies may be more cost-effective than other alternatives. Conversely, phytotechnologies are probably not the remediation technique of choice for sites that pose acute or chronic risks to humans and/or other ecological receptors. Furthermore, risks may change seasonally, depending on the growth cycle of the vegetation. Thus, phytotechnologies are not recommended for time-critical cleanups but may be most suitable for sites where restoration time requirements are less of an issue.

Finally, it is noteworthy that even areas with contaminated soils or groundwater can become revegetated through the establishment of naturally occurring plants; typically, a so-called forensic phytoremediation process can be used to identify plants that are capable of surviving in contaminated areas, some of which might also be capable of contributing to the degradation of the contaminants. The "forensic phytoremediation" process generally consists of the investigation of naturally revegetated contaminated areas to determine which plants have become established and why, as well as to determine the impact of these plants on the contamination (USEPA, 2000i). It is also notable that there are many factors that affect the performance of phytotechnologies, including the composition, concentration, solubility, toxicity, and other chemical properties of the contaminant. Furthermore, another important factor is the ability to bring the contaminated media in contact with the plant roots. For surface water remediation, constructing wetlands that bring the contaminated water into contact with the plants can extend the limits of the technology. Similarly, groundwater that is below the root zone of trees can be pumped to the surface and applied to the plants as irrigation water.

12.2.15.1 Phytoremediation in Practice

Phytoremediation uses plants known as hyperaccumulators and halophytes to remove, transfer, stabilize, or destroy contamination from various environmental matrices. These plants show preferential uptake of specific compounds, accumulating them in their tissues. For example, the high metal accumulation rate shown by some plants suggests that such plants may be used to clean up toxic metal-contaminated sites via a process of *phytoextraction* (see below). In fact, a small number of wild plants that grow on metal-contaminated soils accumulate large amounts of heavy metals in their roots and shoots; some of these species can accumulate unusually high concentrations of toxic metals to levels that far exceed the soil levels. By and large, this property may be exploited for soil reclamation if an easily cultivated, high biomass crop plant that can accumulate heavy metals is identified (Kumar et al., 1995). It is noteworthy that whereas very heavily contaminated soils may not support plant growth, sites with light to moderate contamination could be remediated by growing "specialty" plants. Each cleanup situation may indeed require a different plant species, or a number of plants in tandem.

Broadly speaking, phytoremediation may be applicable for the remediation of metals, pesticides, solvents, explosives, crude oil, PAHs, and landfill leachates. Indeed, phytoremediation can be used to treat a wide range of organic and inorganic contaminants. Specifically, phytoremediation may be most useful in the remediation of sites contaminated by heavy metals, and also for treating VOCs, CVOCs, and trinitrotoluene. It has also been quite effective for the removal of nitrates and ammonium from the groundwater; in fact, since all plants require a nitrogen source to grow, certain nitrogen-containing contaminants may be used by such plants. On the whole, reasonably significant advances have been made in the use of phytoremediation to address a variety of contaminated site problems. In the processes involved, phytoremediation is implemented by establishing a plant or community of plants that have been selected to provide the required remediation mechanisms. The technology exploits the natural hydraulic and metabolic processes of plants and thus is passive and solar driven.

The general types of phytoremediation technologies fall into two broad categories: (i) stabilization processes (e.g., phytostabilization) and (ii) removal processes (e.g., phytoaccumulation/phytoextraction, rhizofiltration, phytodegradation, rhizo(sphere) degradation, organic pumps, phytovolatilization) (see below). The technology can be used in combination with mechanical treatment methods, or as a "stand-alone" treatment method; typically, phytoremediation is indeed used in conjunction with other cleanup approaches in the site restoration effort.

12.2.15.2 Rhizodegradation

Rhizodegradation (also called *rhizosphere biodegradation*, *phytostimulation*, or *plant-assisted bioremediation/degradation*) involves the breakdown of contaminants in the soil through microbial activity that is enhanced by the presence of the rhizosphere consisting of the symbiotic relationship that occurs between plant root systems and microorganisms in the root zone. Generally speaking, natural substances released or excreted by the plant roots, such as sugars, alcohols, and acids, contain organic carbon that acts as nutrient/food sources for soil microorganisms, and the additional nutrients stimulate their biological activities. Essentially, microorganisms (viz., yeast, fungi, and/or bacteria) consume and degrade or transform organic substances for use as nutrient substances. Indeed, certain microorganisms can degrade organic substances such as fuels or solvents that are hazardous to humans and eco-receptors, and then convert them into harmless products through biodegradation. This provision enhances microbial activity in the root zone, resulting in a microbial contribution to soil contaminant degradation.

Rhizodegradation is aided by the way plants loosen the soil and transport oxygen and water to the area. The plants also enhance biodegradation by other mechanisms, such as breaking apart clods and transporting atmospheric oxygen to the root zone. Meanwhile, it is noteworthy that enhanced rhizosphere biodegradation takes place in the soil immediately surrounding plant roots. Plant roots also loosen the soil and then die, leaving paths for transport of water and aeration. This process tends to pull water to the surface zone and dry the lower saturated zones.

12.2.15.3 Phytoextraction

Phytoextraction (or *phytoaccumulation*) refers to the uptake of contaminants by plant roots and the translocation/accumulation (phytoextraction) of contaminants into plant tissues, usually the plant shoots and leaves. Typically, this is used for remediation of metal-contaminated soils, sediments, and groundwater. Invariably, the process of phytoextraction generally requires the translocation of the constituents of concern to the easily harvestable shoots and, in some cases, the roots as well. Subsequently, dried, ashed, or composted plant residues highly enriched in the constituents of concern may be isolated and managed as hazardous waste or recycled as metal ore (Kumar et al., 1995).

In the phytoaccumulation/phytoextraction process, specific species of plants (called hyperaccumulators) are used to absorb unusually large amounts of specific chemical constituents (in comparison with other plants and the ambient chemical concentration) from the environmental media and are subsequently harvested from the growing area. These plants are selected and planted at a site based on the type of contaminants present and other site conditions. After the plants have been allowed to grow for several weeks or months, they are harvested. The biomass is composted to recycle appropriate metals or incinerated, and the ash is sent to a landfill. Indeed, landfilling, incineration, and composting are options to dispose of or recycle the chemical constituents—albeit this may also depend on the results of a toxicity characteristic leaching procedure test as well costs. In any event, the planting and harvesting of plants may be repeated as necessary to bring contaminant levels down to allowable limits.

12.2.15.4 Rhizofiltration

Rhizofiltration is the adsorption or precipitation of contaminants onto plant roots or the absorption of contaminants into the roots when contaminants are in solution surrounding the root zone. Generally, the plants are raised in greenhouses hydroponically (with their roots in water, rather than in soil). Once a large root system has been developed, contaminated water is diverted and brought in contact with the plants, or the plants are moved and floated in the contaminated water. Ultimately, the plants are harvested and disposed as the roots become saturated with contaminants.

Rhizofiltration is indeed similar to phytoextraction; however, plant root systems are usually first developed to maturity in an aqueous environment within a greenhouse. When the root system is developed, contaminated water (usually with metal contamination) is brought to the plants and circulated through their water supply.

12.2.15.5 Phytodegradation

Phytodegradation (also called *phytotransformation*) is the process where plant enzymes completely mineralize or partially break down contaminant compounds. It consists of the metabolism of contaminants within plant tissues; plants produce enzymes, such as dehalogenase and oxygenase, that help catalyze degradation.

During the transpirational uptake of water, dissolved organic and inorganic contaminants in the subsurface can enter into the plant where they are subject to additional phytotechnology mechanisms. Specifically, once inside the plant, organic chemicals can be subject to various plant-produced enzymes that can break down the contaminants; this mechanism is known as phytodegradation. Under the circumstances, pollutants are degraded, used as nutrients, and incorporated into the plant tissues. In some cases, metabolic intermediate or end products are rereleased to the environment, depending on the contaminant and plant species.

12.2.15.6 Phytostabilization

Phytostabilization consists of the use of certain plant species to immobilize contaminants in the soil and groundwater through absorption and accumulation by roots, adsorption onto roots, or precipitation within the root zone and physical stabilization of soils. This process generally reduces the mobility of the contaminant and prevents migration to the groundwater or air.

Overall, the phytostabilization process involves the production of chemical compounds by a plant to immobilize contaminants at the interface between roots and soil. It consists of the use of plants to increase the sequestration of contaminants (usually metals) in the soil, that is, the ability of plant roots to sequester certain inorganic elements in the root zone. Similarly, the exudation of photosynthetic products into the rhizosphere can lead to the phytostabilization of organic compounds as well. Soil sequestration occurs as plants alter water flux and reduce contaminant mobility. Plants and microbial enzymes bind contaminants into soil (humification). Plants also incorporate free contaminants into plant roots (lignification) and prevent wind and water erosion.

12.2.15.7 Organic Pumps

Organic pumps consist of the use of plants to control the migration of contaminants in the groundwater by exploiting their natural hydraulic properties. Indeed, the ability of plants to take up and transpire large volumes of water from the subsurface has been used in phytotechnologies to provide hydraulic control at contaminated sites. This hydraulic control can be used to prevent the horizontal migration or vertical leaching of contaminants. Using trees for water control is estimated to generally cost approximately one-half the cost of traditional pump-and-treat systems.

12.2.15.8 Phytovolatilization

Phytovolatilization is the uptake and subsequent transpiration of a contaminant by a plant, with release of the contaminant or a modified form of the contaminant to the atmosphere from the plant occurring through the leaves. The process consists of the use of plants to remove contaminants from the subsurface, and then evaporating or volatilizing the contaminants from the leaf surface of the plant once it has traveled through the plant's system.

Essentially, phytovolatilization occurs as growing trees and other plants take up water and contaminants; some of these contaminants can pass through the plants to the leaves and volatilize into the atmosphere at comparatively low concentrations, with many organic compounds that are transpired by a plant being subject to photodegradation.

12.2.16 Precipitation/Coagulation/Flocculation

Metal precipitation from contaminated water involves the conversion of soluble heavy metal salts to insoluble salts that will precipitate. Precipitation is mainly used to convert dissolved ionic species into solid-phase particulates that can be removed from the aqueous phase by coagulation and filtration. The precipitate can then be removed from the treated water by physical methods such as clarification (settling) and/or filtration. The process usually uses pH adjustment, addition of a chemical precipitant, and flocculation. Typically, metals precipitate from the solution as hydroxides, sulfides, or carbonates.

In fact, precipitation of metals has long been the primary method of treating metal-laden industrial wastewaters. As a result of the success of metal precipitation in such applications, the technology is considered a candidate for use in the remediation of groundwater containing heavy metals, including their radioactive isotopes. Essentially, remedial application of this technology usually involves the removal of dissolved toxic metals and radionuclides. Depending on the process design, sludges may be amenable to metal recovery. Invariably, the solubilities of the specific metal contaminants and the required cleanup standards will dictate the specific process used. In some cases, process design will allow for the generation of sludges that can be sent to recyclers for metal recovery. At any rate, in groundwater treatment applications, the metal precipitation process is often used as a pretreatment for other treatment technologies, such as chemical oxidation or air stripping, where the presence of metals would interfere with the other treatment processes.

12.2.17 Soil Washing

Soil washing (also known as *solvent washing*) consists of excavating soils from contaminated areas and washing the contaminants from the soil using water or an aqueous solution. The contaminated effluent is then recovered, treated, and recycled, or otherwise disposed of, as appropriate. Water is normally used as the washing agent; however, other solvents may be used for contaminants that are tightly held or only slightly soluble in water. Solvents are selected on the basis of their ability to solubilize the contaminants and also on their environmental and human health effects. Thus, it is important to know the chemistry and toxicity of the surfactant of choice.

Basically, soil washing is accomplished by contacting soil with a wash solution, separating the soil and solution, and treating the solution. The solution is contacted with the soil and vigorously agitated to transfer contaminants into the wash solution. The process removes contaminants from soils by dissolving or suspending them in the wash solution (which is later treated by conventional wastewater treatment methods). Surfactants or similar mild solvents are often used to improve the removal of petroleum hydrocarbons and a wide variety of other organic contaminants from soils.

The target contaminant groups for soil washing are typically SVOCs, fuels, and inorganics; the technology can also be used on selected VOCs and pesticides. Specifically, the soil washing technique is most applicable for soluble organic chemicals and metals. Contaminant groups that may be effectively dealt with by soil washing include petroleum and fuel residues, heavy metals, PCBs, PAHs, pesticides, creosote, and cyanides (Pratt, 1993). Typically, inorganic and organic contaminants are extracted from the soil by washing the soil with solvents; the contaminated effluent is then recovered, contaminants are extracted, and the solvents are recirculated through the soils or properly disposed of. The technology tends to be most effective with coarse-grained soils with low clay content.

12.2.18 Thermal Desorption

Thermal desorption is a physical separation process and is not designed to destroy organic compounds *per se*. Essentially, wastes are heated to volatilize water and organic contaminants; a carrier gas or vacuum system transports the volatilized water and organics to a gas treatment system. In principle, thermal desorption is implemented by heating and agitating the contaminated material while it is exposed to a carrier gas or vacuum that transports volatilized water and organic contaminants to a gas treatment system. The bed temperatures and residence times designed into these systems will volatilize selected contaminants but typically will not oxidize or destroy them. Meanwhile, it is worth mentioning here that all thermal desorption systems require treatment of the off-gas to remove particulates and other contaminants. Particulates are generally removed by conventional particulate removal equipment, such as wet scrubbers or fabric filters; contaminants are removed through condensation followed by carbon adsorption, or they are destroyed in a secondary combustion chamber or a catalytic oxidizer. Most of these units are transportable.

Based on the operating temperature of the desorber, thermal desorption processes can be categorized into two groups: high-temperature thermal desorption (HTTD) and LTTD. The target contaminant groups for LTTD systems are nonhalogenated VOCs and fuels—albeit the technology can be used to treat SVOCs at reduced effectiveness. The target contaminants for HTTD are SVOCs, PAHs, PCBs, and pesticides; however, VOCs and fuels may also be treated, but treatment may be less cost-effective. Volatile metals may be removed by HTTD systems—albeit the presence of chlorine can affect the volatilization of some metals, such as lead. In practice, thermal desorbers are designed to heat soils to temperatures sufficient to cause constituents to volatilize and desorb (i.e., physically separate) from the soil. Although they are not designed to decompose organic constituents, thermal desorbers can cause some of the constituents to completely or partially decompose, depending on the specific organic chemicals present and on the temperature of the desorber system.

Thermal desorption is indeed a full-scale *ex situ* remedial technology that has been proven successful for remediating all types of contaminated material. Two common thermal desorption designs are the rotary dryer and the thermal screw. Rotary dryers are horizontal cylinders that can be indirect- or direct-fired; the dryer is normally inclined and rotated. For the indirect-fired, a direct-fired rotary dryer heats an airstream that, by direct contact, desorbs water and organic contaminants from the contaminated material. With the direct-fired, fire is applied directly upon the surface of contaminated media—the main purpose of the fire being to desorb contaminants from the contaminated material, although some contaminants may be thermally oxidized. For the thermal screw units, screw conveyors or hollow augers are used to transport the medium through an enclosed trough. Hot oil or steam circulates through the auger to indirectly heat the medium. In the indirect-heated system, an externally fired rotary dryer volatilizes the water and organics from the contaminated media into an inert carrier gas stream; the carrier gas is later treated to remove or recover the contaminants.

The thermal desorption process is applicable for the separation of organics from refinery wastes, coal tar wastes, wood-treating wastes, creosote-contaminated soils, hydrocarbon-contaminated soils, mixed (radioactive and hazardous) wastes, synthetic rubber processing waste, pesticides, and paint wastes. Generally speaking, thermal desorption systems have varying degrees of effectiveness against the full spectrum of organic contaminants.

12.2.19 Vapor Extraction Systems

Vapor extraction systems (VESs) may be applied to the removal of volatile organic chemicals (VOCs) present at contaminated sites. In principle, the VES is most applicable to the remediation of the higher volatile or lower molecular weight constituents of organic compounds only. As a general rule, heavier organic fractions, such as diesel fuel and fuel oils, are not candidates for remediation by vapor extraction. Also, vapor extraction is more effective at sites where the more volatile chemicals are still present, especially when the spill/release is reasonably recent. At most sites, the initial VOC recovery rates are relatively high, and then, it decreases asymptotically to zero with time.

The VES technology is a particularly economical and efficient means of removing VOCs from the subsurface environment. A very important advantage for using a VES relates to the fact that there is usually minimal site disruption during its implementation/operation. VESs can indeed be designed for many areas of site remediation such as contaminated soil piles and inaccessible locations (e.g., underneath buildings and similar structures). In its application, the VES may vary considerably in size and design, depending on site-specific requirements and/or conditions. In any case, a typically well-designed VES consists of a series of extraction/injection wells connected to a common manifold and a positive displacement air blower, among other surface equipment, such as emission controls, instrumentation, and electric motor. The most common applications of VESs include SVE, air or steam stripping, and air sparging; these systems are discussed in the sections that follow.

12.2.19.1 Soil Vapor Extraction

SVE (also known as *in situ soil venting, subsurface venting, vacuum extraction,* or *in situ soil stripping*) is an *in situ* remedial technique that uses soil aeration to treat subsurface zones of VOC contamination in soils. In the SVE technology, VOCs are extracted from soil by using a vacuum system. This soil cleanup technique uses vacuum blowers to pull large volumes of air through contaminated soil. The airflow flushes out the vapor-phase VOCs from the soil pore spaces, disrupting the equilibrium that exists between the contaminants on the soil and in the vapor. This causes volatilization of the contaminants and subsequent removal in the airstream. Treatment rate depends on airflow through the soils and how effectively the contaminants partition into the mobile air phase (Newman et al., 1993).

Figure 12.5 depicts the pertinent design elements of a typical SVE system. The general type of SVE system consists of extraction vents/wells, air inlets or injection vents/wells (optional), air

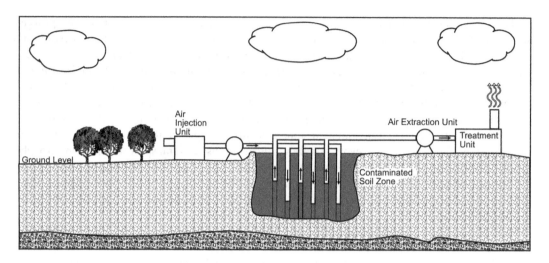

FIGURE 12.5   Illustrative example of a typical SVE system.

(piping) headers, vacuum pumps or air blowers, flow meters and controllers, vacuum gauges, sampling ports, air–water separator (optional), vapor treatment (optional), and an impermeable cap (optional) (Hutzler et al., 1990; USEPA, 1989q). The SVE system operation is based on the extraction of VOC-laden air from contaminated soils. Inlet or injection wells, usually located at the boundaries of the contaminated area, may be used to enhance VOC-laden airflow to the extraction wells. Inlet wells are passive, with ambient air being drawn into the ground at the well locations due to pressure differentials caused by the removal of air from the extraction wells; injection wells are active and force air into the ground at the well locations. The inlet air going to the injection well may be supplied by the VOC control treatment exhaust, or the vacuum pump (blower) exhaust at the site engineer's discretion. Insulation is occasionally used on the piping and headers, especially in colder climates, to prevent condensate freezing.

Designing an effectual SVE system: Fundamentally, the SVE process involves removing and venting VOCs from the vadose or unsaturated zone of contaminated soils by mechanically drawing or venting air through the soil matrix. Fresh air is injected or allowed to flow into the subsurface at locations in and around the contaminated soil to enhance the extraction process. This is carried out by connecting a vacuum pump or fan to one or more extraction wells, with the extraction wells typically being installed to penetrate the contaminant plume near the zone of highest VOC concentration. When suction is applied to the extraction wells, it induces a subsurface radial airflow toward perforations in the well casings. In addition, injection or ventilation wells (to facilitate the infiltration of clean air into the soil) may be placed at selected locations to help direct the flow of induced air toward the extraction wells. The VOC-laden air is withdrawn under vacuum from recovery or extraction wells that are placed in selected locations within the contaminated site. This air is then either vented directly to the atmosphere or vented to an aboveground level VOC treatment unit, such as a carbon adsorber or a catalytic incinerator, prior to being released to the atmosphere. The decision to use VOC control system treatment is largely dependent on VOC concentrations and applicable regulations. Also, the selection of a particular VOC treatment option may be based in part on individual site characteristics.

It is noteworthy that a SVE is generally effective under a wide range of site conditions. Extraction vents/wells are typically designed to fully penetrate the unsaturated soil zone or the geologic stratum to be cleaned. Spacing of extraction vents is usually based on an estimate of the radius of influence of an individual extraction vent; vent spacing typically ranges from about 4.5 to 30 m (≈15–100 ft) (Hutzler et al., 1990). Capping the entire site with plastic sheeting, clay, concrete, or asphalt enhances horizontal movements toward the extraction vent. In fact, impermeable caps extend the radius of influence around the extraction vent. The use of a ground surface cover will also prevent or minimize infiltration, which, in turn, reduces moisture content and further chemical migration. If water should be pulled from the extraction vents, an air–water separator is required to protect the blowers or pumps and to increase the overall efficiency of vapor treatment systems.

In practical terms, the complete design and operation of SVE systems depend on several factors, including the following (Hutzler et al., 1990):

- *Contamination volume*: The extent to which the contaminants are dispersed in the soil is important in deciding if vapor extraction will be cost-effective.
- *Groundwater depth*: SVE systems have been used in both shallow and deep unsaturated zones. However, where groundwater is more than 12 m ($\approx$40 ft) deep and contamination extends to the water table, a SVE system may be the only way to remove VOCs from the unsaturated zone.
- *Soil heterogeneity*: Inhomogeneities influence air movements, affecting the placement of extraction and inlet vents.
- *Site soil characteristics*: Higher contaminant removal efficiency can be anticipated in highly permeable soils. Soil moisture content or degree of saturation is also important because it is easier to draw air through drier soils.
- *Chemical properties*: In conjunction with site conditions and soil properties, chemical properties determine the feasibility of a SVE system. The system most effectively removes compounds that exhibit significant volatility at ambient subsurface temperatures. Generally, compounds with Henry's law constants >0.01, or vapor pressures >25 mm Hg (or >1 in. mercury), are removable by vapor extraction.
- *Contaminant location and area development*: If, in particular, contamination extends across property lines, beneath buildings, or beneath extensive utility trench network, then the applicability of VESs should be comprehensively evaluated in favor of intrusive remediation techniques. Since *in situ* vapor extraction can be generally conducted with minimal site disruption, this is a particularly favored remedial option in highly developed areas.
- *Operating variables*: Among several factors, operating variables such as higher airflow rates tend to increase vapor removal because the zone of influence is increased and air is forced through more of the air-filled pores. Also, water infiltration rate can be controlled by placing an impermeable cap over the site.
- *Response variables*: These consider system performance parameters that include air pressure gradients, VOC concentrations, and power usage; vapor removal rates are affected by chemicals' volatility, its sorptive capacity into soil, the airflow, initial distribution of chemical, and soil moisture content.

When properly designed and operated, SVE systems offer a relatively inexpensive means of removing chemicals without major disruption to the project site. Indeed, SVE is a cost-effective technique for VOC removal, and this is finding more widespread applications/uses. What is more, the method is a relatively simple concept and can be used in conjunction with other soil decontamination procedures such as biological degradation. Because contaminants must volatilize and partition into air undergoing removal, contaminated sites having significant amounts of low-volatility compounds (e.g., diesel fuel, fuel oils, or jet fuels) are often not targeted for remediation by soil venting. Sites that have soil heterogeneities that result in uneven air permeation can also prevent effective remediation by conventional soil venting.

In general, SVE is primarily used to treat halogenated and nonhalogenated VOCs. The technology is typically applicable only to volatile compounds with Henry's law constant greater than 0.01 or a vapor pressure greater than 0.5 mm Hg (0.02 in. Hg). Other factors such as moisture content, organic content, and air permeability of the soil will also influence the effectiveness of this treatment technology. In any case, SVE will generally not remove heavy oils, metals, PCBs, or dioxins.

Because the process involves the continuous flow of air through the soil, however, SVE often promotes the *in situ* biodegradation of low-volatility organic compounds that may be present.

Enhancing the performance of SVE systems: Removal of VOCs from vadose zone soils through vacuum extraction has become an important remedial alternative of choice for many contaminated sites. Thus, modifications that add to its effectiveness is always a welcome news; in this regard, the removal efficiency at contaminated sites can generally be enhanced as follows:

- By the use of temporary or permanent caps over the contaminated soil; in fact, if low-permeability strata are present, or if other nonvolatile contaminants are encountered with the VOCs, a permanent cap may be more cost-effective. In general, capping the entire site with plastic sheeting, clay, concrete, or asphalt enhances horizontal movements toward the extraction vent; impermeable caps extend the radius of influence around the extraction vent. The use of a ground surface cover will also prevent or minimize infiltration, which, in turn, reduces moisture content and further chemical migration.
- Overall costs of SVEs can be reduced and cleanup accelerated by applying heat to enhance vaporization. In fact, when a heat pump arrangement is installed, product recovery can be achieved along with better GAC adsorber performance (i.e., where GAC is used) and reduced operating costs. Furthermore, the restoration time frame can be reduced if vaporization and diffusion rates are increased by heating. An additional advantage of heating is the ability to remove less volatile chemicals.

In general, insulation is occasionally used on the piping and headers, especially in colder climates, to prevent condensate freezing.

12.2.19.2 Thermally Enhanced SVE

Thermally enhanced SVE is a full-scale technology that uses electrical resistance heating, electromagnetic/fiberoptic/radio frequency heating, or hot air/steam injection to increase the volatilization rate of semivolatile compounds and thus facilitate removal. Heating, especially RFH and electrical resistance heating, can improve airflow in high-moisture soils by evaporating water. Indeed, since high moisture content tends to be a limitation of standard SVE, the thermal enhancements may help overcome this shortcoming. The process is otherwise similar to standard SVE, except that it requires heat-resistant extraction wells. Extracted vapor can then be treated by a variety of existing technologies, such as GAC or incineration.

The system is usually designed to treat SVOCs but will consequentially treat VOCs as well. Depending on the temperatures achieved by the system, thermally enhanced SVE technologies can also be effective in treating some pesticides and fuels. It is noteworthy that after the application of this technique, subsurface conditions will usually be excellent for biodegradation of residual contaminants.

Electrical resistance heating: Electrical resistance heating uses an electrical current to heat less-permeable soils such as clays and fine-grained sediments, so that water and contaminants trapped in these relatively conductive regions are vaporized and ready for vacuum extraction. In a typical field application, electrodes are placed directly into the less-permeable soil matrix and activated so that electrical current passes through the soil, creating a resistance that then heats the soil. The heat dries out the soil causing it to fracture. These fractures make the soil more permeable, allowing the use of SVE to remove the contaminants. The heat created by electrical resistance heating also forces trapped liquids to vaporize and move to the "steam zone" for removal by SVE.

Of special mention here, the so-called six-phase soil heating (SPSH) is a distinctive electrical resistance heating that uses low-frequency electricity delivered to six electrodes in a circular array to heat soils; the SPSH system is elaborated further in Section 12.2.20. Essentially, the temperature of the soil and contaminant is increased during the SPSH, thereby increasing the contaminant's vapor pressure and its removal rate. SPSH also creates an *in situ* source of steam to strip contaminants from soil.

Radio frequency/electromagnetic heating: RFH is an *in situ* process that uses electromagnetic energy to heat soil in order to enhance SVE. The RFH technique heats a discrete volume of soil using rows of vertical electrodes embedded in soil (or other media). Heated soil volumes are bounded by two rows of ground electrodes with energy applied to a third row midway between the ground rows. The three rows act as a buried triplate capacitor. When energy is applied to the electrode array, heating begins at the top center and proceeds vertically downward and laterally outward through the soil volume. The technique can heat soils to over 300°C.

In general, RFH enhances SVE in four major ways: (i) contaminant vapor pressure and diffusivity are increased by heating, (ii) the soil permeability is increased by drying, (iii) an increase in the volatility of the contaminant from *in situ* steam stripping by the water vapor, and (iv) a decrease in the viscosity, which improves mobility. It is noteworthy, however, that the technology is self-limiting because, as the soil heats and dries, electrical current will likely stop flowing.

Hot air injection: Hot air or steam is injected below the contaminated zone to heat up contaminated soil. The heating enhances the release of contaminants from soil matrix. Ultimately, some VOCs and SVOCs are stripped from contaminated zone and brought to the surface through SVE.

12.2.19.3 Air Stripping

Air stripping is a remediation technique that comprises the physical removal of dissolved-phase contamination from a water stream. By bringing large volumes of air in contact with the contaminated water, a driving gradient from the water to the air can be created for the contaminants of concern. It is indeed a separation technology that takes advantage of the fact that certain chemicals are more soluble in air than in water (Nyer, 1993).

Fundamentally, the air stripping process involves pumping contaminated water containing VOCs from the ground and allowing it to trickle over packing material in an air stripping tower. At the same time, clean air is circulated past the packing material. When the contaminated water comes into contact with the clean air, the contaminants tend to volatilize from the water into the air. The contaminated air is then released into the atmosphere, or a GAC system.

In a typical air stripping process, contaminated water containing VOCs is countercurrently contacted with air in a packed tower (Figure 12.6). Contaminants, usually VOCs, are transferred from liquid phase to gaseous phase. By contacting contaminated water with clean air, dissolved

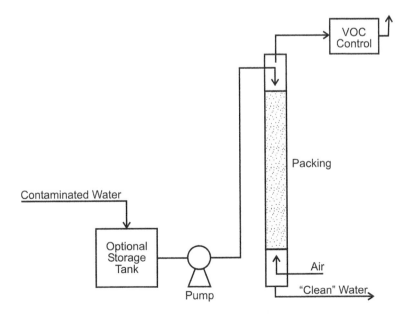

FIGURE 12.6 Schematic of a typical/representative air stripping process.

VOCs are transferred to the airstream to create equilibrium between the phases. The process takes place in a cylindrical tower, packed with inert material that allows sufficient air/water contact to remove volatiles from water. Contaminants are then removed from the airstream. An optional unit operation (e.g., catalytic oxidation or vapor-phase carbon adsorption) may be used for the control of air emissions. The VOCs are transferred to the gas phase during the intimate gas–liquid contact. The stripped water may further be treated in an optional carbon absorber polishing bed. The treated effluent water is either recycled as process water or properly discharged; in the case of a groundwater cleanup operation for site restoration programs, the treated water may be pumped back into the aquifer.

Air strippers are indeed designed to maximize the removal of VOCs from groundwater, leading to transfer of these contaminants into air that can be treated in a control device or discharged to the atmosphere. An air-stripping tower provides the air mechanism for, and water contact to, removing the volatiles from the water by countercurrent flow through a packing material. The removal efficiency of an air stripper depends upon the volatility of the contaminants. Undeniably, air stripping of chemical contaminants from contaminated water is an effective method of removing VOCs from the contaminated water. However, this method also transfers pollutants from the water to the gas phase, and the resulting air emissions may need to be appropriately controlled. It is noteworthy that air stripping is not a destruction technology; it merely moves contaminants from a liquid phase to an air phase, which is to be addressed differently.

Classically, air stripping is used to separate halogenated and nonhalogenated VOCs from water, but is ineffective for contaminants with low vapor pressure or high solubility such as inorganic salts. Henry's law constant is used to determine whether air stripping will be effective. Generally, organic compounds with Henry's law constants greater than 0.01 atm-m^3/mol are considered amenable to stripping. Some compounds that have been successfully separated from water using air stripping include BTEX, chloroethane, TCE, DCE, and perchloroethylene (PCE).

Practical design elements and considerations: The air stripping process involves the mass transfer of volatile contaminants from water to air. Air stripping is indeed a full-scale technology in which VOCs are partitioned from groundwater by greatly increasing the surface area of the contaminated water exposed to air. Archetypical forms of aeration methods used in the process include packed towers, diffused aeration, tray aeration, and spray aeration.

For groundwater remediation, the air stripping is typically conducted in a packed tower or low-profile aeration system. The typical packed tower air stripper includes a spray nozzle at the top of the tower to distribute contaminated water over the packing in the column, a fan to force air countercurrent to the water flow, and a sump at the bottom of the tower to collect decontaminated water. Auxiliary equipment that can be added to the basic air stripper includes an air heater to improve removal efficiencies; automated control systems with sump-level switches and safety features, such as differential pressure monitors, high sump-level switches, and explosion-proof components; and air emission control and treatment systems, such as activated carbon units, catalytic oxidizers, or thermal oxidizers. Packed tower air strippers are installed as permanent installations either on concrete pads or on a skid or a trailer. In general, aeration tanks strip volatile compounds by bubbling air into a tank through which contaminated water flows. A forced air blower and a distribution manifold are designed to ensure air–water contact without the need for any packing materials. The baffles and multiple units ensure adequate residence time for stripping to occur. Aeration tanks are typically sold as continuously operated skid-mounted units. The advantages offered by aeration tanks are considerably lower profiles (less than 2 m or 6 ft high) than those offered by packed towers (5–12 m, or 15–40 ft high) where height may be a problem, and the ability to modify performance or adapt to changing feed composition by adding or removing trays or chambers. The discharged air from aeration tanks can be treated using the same technology as for packed tower air discharge treatment.

Air strippers can be operated continuously or in a batch mode where the air stripper is intermittently fed from a collection tank. The batch mode ensures consistent air stripper performance and greater energy efficiency than continuously operated units because mixing in the storage tanks eliminates any inconsistencies in feed water composition.

Finally, it is noteworthy that modifying the packing configurations greatly increases removal efficiency. For instance, an innovative variant of the standard air stripper is the so-called low-profile air stripper that is offered by several commercial vendors; this unit packs a number of trays in a very small chamber to maximize air–water contact while minimizing space. Because of the significant vertical and horizontal space savings, these units can be more efficiently used for groundwater treatments.

12.2.19.4   Air Sparging

Air sparging is an *in situ* remedial technology used to reduce concentrations of volatile constituents that are adsorbed to soils and dissolved in groundwater. It consists of the highly controlled injection of air or oxygen into a contaminant plume in the soil *saturated* zone. In fact, air sparging is frequently used together with SVE, but it can also be used with other remedial technologies. In the processes involved, air pumped into contaminated groundwater is used to strip volatiles from the groundwater to the soil vadose zone for subsequent capture using SVE. Oxygen present in the air added to the groundwater and vadose zone soils can also enhance the biodegradation of contaminants below and above the water table. Basically, the injected air bubbles traverse horizontally and vertically in channels through the soil column, creating a transient air-filled porosity in which volatilization can occur, that is, by creating an underground stripper that removes volatile and semi-volatile organic contaminants by volatilization. Indeed, air sparging effectively creates a crude air stripper in the subsurface, with the soil acting as the packing; the injected air helps to flush the contaminants into the unsaturated zone.

In a typical design of an air sparging system (Figure 12.7), an array of vents (or shallow wells) penetrating the impacted area of the vadose zone are connected via manifolding to air blowers. The blowers create a partial vacuum in the vents and pull air, including VOCs, out of the soil.

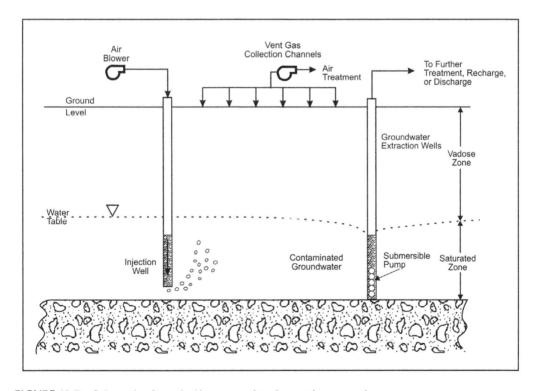

FIGURE 12.7   Schematic of a typical/representative air sparging system/process.

The air sparging treatment involves injecting the soils with air, which flows vertically and horizontally to form an oxygen-rich zone in which VOCs are volatilized. Air bubbles that contact dissolved- or adsorbed-phase contaminants in the aquifer cause the VOCs to volatilize. The volatilized organics are carried by the air bubbles into the vadose zone where they can be captured by a VES. Also, the sparged air maintains high DO content, which enhances natural biodegradation. A carbon treatment system is used to treat off-gases. Careful investigation for optimal system design will result in an efficient operating system.

The target contaminant groups for air sparging are VOCs and fuels. In fact, air sparging is primarily used to treat compounds with moderate to high Henry's law constants (i.e., high vapor pressure and low solubility), such as halogenated and nonhalogenated VOCs and nonhalogenated SVOCs. Methane can be used as an amendment to the sparged air to enhance cometabolism of chlorinated organics.

The overarching mechanism: In principle, air sparging is an *in situ* technology that is implemented by injecting pressurized air through a contaminated aquifer. Injected air traverses horizontally and vertically in channels through the soil column, creating an underground stripper that removes contaminants by volatilization. This injected air helps to flush (bubble) the contaminants up into the unsaturated zone where a VES is usually implemented in conjunction with air sparging to remove the generated vapor-phase contamination. In addition to the removal of contaminants by volatilization, oxygen added to contaminated groundwater and vadose zone soils can also enhance the biodegradation of contaminants below and above the water table. This technology is indeed designed to operate at high flow rates to maintain increased contact between groundwater and soil, and to strip more groundwater by sparging.

12.2.20 Miscellaneous/Integrated Remediation Techniques

There are several proven remediation technologies and processes that may be used in the management of contaminated site problems. However, no single technology is universally applicable with equal success to all contaminant types, and at all sites. Oftentimes, more than one remediation technique is needed to effectively address most contaminated site problems. In fact, treatment processes can be, and are usually, combined into process trains for more effective removal of contaminants and hazardous materials present at contaminated sites. For example, whereas biological treatment (with or without enhancement techniques) should result in the most desirable treatment scenario for petroleum-contaminated sites, the "hot spots" at such sites may best be handled by physical removal (i.e., excavation) and thermal treatment of the removed materials—rather than by biological methods. Consequently, several technologies (or combination of technologies) that can provide both efficient and cost-effective remediation or cleanup should normally be reviewed and explored as possible candidates in the remedy selection process.

Some reasonably common holistic techniques of possible general interest that are variations and/or combinations of some of the methods elaborated earlier on in this chapter are presented in the sections that follow. In general, multimechanism-type remediation technologies/techniques combine two or more biological, chemical, and/or physical processes into one integrated remediation system. These technologies apply a variety of treatment mechanisms in order to better destroy, separate, degrade, transform, or contain contamination that would otherwise be less effectively or efficiently remediated using a single technique. Meanwhile, it is worth emphasizing here that the contaminant treatment processes and techniques, as well as the remediation methods, processes, and technologies discussed earlier on in this chapter, are by no means complete and exhaustive, and neither is the additional discussions presented below.

12.2.20.1 Air Sparging-cum-SVE Combination Systems

A variety of *in situ* techniques have been used in attempts to remediate VOC-contaminated sites. However, most of the techniques seem to be of limited effectiveness when used to remediate saturated soils and groundwater. The use of an *in situ* air sparging system in conjunction with a SVE system has become increasingly popular as a more efficient and cost-effective method for the remediation of VOC-contaminated saturated soils and groundwater (Reddy et al., 1995); these systems (consisting of air injection wells, vapor extraction wells, and several other appurtenances) are generally most effective for localized contamination of known extent. Key to the processes involved here, air injected into the saturated zone transports the contaminants to the top of the saturated zone, and into the unsaturated zone, to be captured through the SVE extraction wells by an induced vacuum. Meanwhile, it is worth mentioning here that sparging systems generally require very sophisticated analysis of site hydrogeology and careful engineering before implementation; even so, overall, the system can be very cost-effective. Indeed, when designed and operated properly, this combination system can prove very cost-effective in the restoration of contaminated saturated soils and groundwater.

In practice, air sparging in combination with SVE has been used to remove and destroy organic solvents at difficult corrective action cleanup sites. In one such case in the State of New York, TCE and other VOCs were adsorbed onto soils below the water table at a facility located adjacent to a wetland area. Applying SVE alone would have required pumping out groundwater to lower the water table; however, this option was unacceptable because of the location of the wetlands. Consequently, an air sparging–vapor extraction design was selected after careful site evaluation and pilot studies. With air sparging, air injected into saturated soil travels vertically and horizontally to form an oxygen-rich zone in which adsorbed and dissolved VOCs are volatilized. As vapors rise from the saturated zone to the soil vadose zone above, VOCs are captured by the SVE system. In fact, used in conjunction with SVE, air sparging has emerged as a particularly effective treatment technology for soils and groundwater contaminated with VOCs.

12.2.20.2 Biosparging

Biosparging is an *in situ* remediation technology that uses indigenous microorganisms to biodegrade organic constituents in the saturated zone. During the biosparging process, air (or oxygen) and nutrients (if needed) are injected into the saturated zone to increase the biological activity of the indigenous microorganisms.

The biosparging process is indeed similar to air sparging; however, whereas air sparging removes contaminants primarily through volatilization, biosparging promotes the biodegradation of constituents rather than volatilization (generally by using lower flow rates than are used in air sparging). In practice, some degree of both volatilization and biodegradation occurs when either air sparging or biosparging is used.

Biosparging is most often used at sites with mid-weight petroleum products (e.g., diesel fuel and jet fuel); lighter petroleum products (e.g., gasoline) tend to volatilize readily and to be removed more rapidly using air sparging. Finally, it is noteworthy that heavier products (e.g., lubricating oils) generally take longer to biodegrade than lighter products, but biosparging can still be used at such sites.

12.2.20.3 Bioventing

Bioventing is a variation of the VES that comprises the delivery of oxygen to unsaturated soils by forced air movement for the purpose of enhancing biodegradation of organic contaminants. During bioventing, the oxygen concentration in the soil gas is increased by injecting air into the contaminated zone. Ultimately, increased microbial activity results in the degradation of contaminants that are less easily removed by volatilization using the VES. Unlike vacuum-enhanced vapor extraction, bioventing injects air into the contaminated media at a rate designed to maximize *in situ* biodegradation and minimize or eliminate off-gassing of volatilized contaminants to the atmosphere. Now, considering the fact that most local regulations often require that there should be permitting, monitoring, and/or treatment of soil-venting off-gases that discharge to the atmosphere, a modification of the conventional vapor extraction remediation process to allow contaminants to be biologically removed *in situ* will reduce or eliminate air emissions, and therefore significantly cut down remediation costs. Bioventing also biodegrades less volatile organic contaminants and allows treatment of less-permeable soils, because a reduced volume of air is required for treatment (Newman et al., 1993).

Bioventing is indeed a desirable *in situ* remediation technology that stimulates the natural *in situ* biodegradation of any aerobically degradable compounds in soil by providing oxygen to existing soil microorganisms. The process uses indigenous microorganisms to biodegrade organic constituents adsorbed to soils in the unsaturated soil zone. In contrast to soil vapor vacuum extraction, bioventing uses low airflow rates to provide only enough oxygen to sustain microbial activity. Oxygen that needs for bioventing is most commonly supplied by directing airflow (via direct air injection) through residual contamination in soil. It is noteworthy that gaseous substrates can be added with the air to promote cometabolic degradation of recalcitrant compounds, resulting in a process better referred to as *cometabolic bioventing*.

In general, conventional bioventing techniques are primarily used to treat aerobically biodegradable compounds such as nonhalogenated VOCs and SVOCs; compounds that have been successfully remediated include petroleum hydrocarbons, nonchlorinated solvents, some pesticides, and wood preservatives in contaminated soils. Correspondingly, *cometabolic bioventing* is applicable to contaminants such as TCE, trichloroethane, ethylene dibromide, and DCE that resist direct aerobic degradation. Indeed, in addition to the degradation of adsorbed fuel residuals, volatile compounds are biodegraded as vapors move slowly through biologically active soil. Additionally, some studies show that whereas bioremediation cannot degrade inorganic contaminants, bioremediation can be used to change the valence state of inorganics and consequentially cause adsorption, uptake, accumulation, and concentration of inorganics in micro- or macroorganisms; thus, this helps in stabilizing or removing inorganics from soil.

12.2.20.4　Confined Disposal Facilities

Confined disposal facilities (CDFs) are used to contain dredged sediments and can indeed be used to help contain halogenated and nonhalogenated VOCs and nonhalogenated SVOCs. The dikes needed to form CDF cells may be constructed at an upland location (above the water table), partially in the water near shore, or completely surrounded by water. A CDF may contain a large cell for material disposal, and adjoining cells for retention and decantation of turbid, supernatant water. A variety of linings have been used to reduce seepage through the dike walls; the most effective linings are clay or bentonite–cement slurries but sand, soil, and sediment linings have also been used.

Three key factors will normally influence the development and implementation of the CDF technology at a given site: location, design, and monitoring. First, the location of a CDF will depend on physical site parameters, such as the relative sizes of the site and the CDF technology required for remediation, and proximity to a navigable waterway; construction parameters, such as geology and hydrology of a site; and environmental effects of the CDF on the site. Second, the design of a CDF should be directed toward the goal of minimizing contaminant loss. Potential contaminant loss or escape pathways must therefore be identified, followed by the selection of controls and structures that will limit contaminant release through these migration/escape pathways. Common contaminant escape pathways include leaching through the bottom of a CDF, seepage through CDF dikes, volatilization to the air, and uptake by plants and animals. Caps are a generally effective method of minimizing contaminant loss through pathways associated with CDFs, but the selection of proper CDF liner material is also an important approach. Finally, continuous monitoring is required to ensure that the structural integrity of a CDF is maintained.

12.2.20.5　Constructed Wetlands

Wetland systems are those in which water is near enough to the soil surface to maintain saturated conditions year-round, and capable of supporting the associated wetland vegetation. Constructed wetlands are human-made wetlands designed to intercept and remove a wide range of contaminants from water; a constructed wetland treatment system incorporates principal ecosystem components found in wetlands, including organic materials (substrate), microbial fauna, and algae. As influent waters with organic contaminants, low pH, and/or contaminated with high metal concentrations flow through the aerobic and anaerobic zones of the wetland ecosystem, large hydrophobic organic compounds and metals are removed by ion exchange, adsorption, absorption, and precipitation through geochemical and microbial oxidation and reduction. Sorption occurs as metals in the water contact with humic or other organic substances in the soil medium. Oxidation and reduction reactions that occur in the aerobic and anaerobic zones, respectively, transform or degrade organic compounds and precipitate metals as hydroxides and sulfides. Precipitated and adsorbed metals settle in quiescent ponds or are filtered out as the water percolates through the soil or substrate.

Constructed wetlands are primarily used to treat halogenated and nonhalogenated VOCs, nonhalogenated SVOCs, and inorganics. Indeed, constructed wetlands have most commonly been used in wastewater treatment for the control of organic matter; nutrients, such as nitrogen and phosphorous; and suspended sediments (e.g., agricultural runoff). The wetland-based treatment process is also suitable for controlling trace metals and other toxic materials (e.g., acid mine drainage). Additionally, constructed wetlands have been used to treat acid mine drainage that has extreme acid conditions and high concentrations of iron, sulfate, and other trace metals. In fact, constructed wetlands not only show a reduction in iron and sulfate concentrations, but also show a recovery in pH levels. In the area of stormwater treatment, wetlands have shown the ability to remove fecal coliform bacteria, total petroleum hydrocarbons, and metals including lead, chromium, and zinc.

12.2.20.6 Electrical Resistive Heating (Six-Phase Heating)

The trademarked so-called *six-phase heating* (SPH) is an *in situ* thermal remediation technology that uses electrical resistive heating to remediate soil and groundwater. This process splits conventional electricity into six electrical phases for the electrical resistive heating of soil and groundwater. Each electrical phase is delivered to one of six electrodes placed in a hexagonal array. The voltage gradient between phases causes an electrical current to flow through the soil and groundwater. As the soil and groundwater are uniformly heated, steam strips volatile and semivolatile contaminants from the subsurface. Additionally, the SPH process increases the air permeability of the soils, which further enhances the rate of contaminant removal.

Essentially, SPH is a polyphase electrical technology that uses *in situ* resistive heating and steam stripping to accomplish subsurface remediation. A voltage control transformer converts conventional three-phase electricity into six electrical phases. These electrical phases are then delivered to the subsurface by vertical, angled, or horizontal electrodes installed using standard drilling techniques. Because the SPH electrodes are electrically out of phase with each other, electrical current flows from each electrode to all the other electrodes adjacent to it. It is the resistance of the subsurface to this current movement that causes heating. The result is a uniquely uniform subsurface heating pattern that can be generated in both the saturated and the vadose zones.

The SPH technique creates a uniform heating pattern by using the electrical resistance of the soil and groundwater within the target treatment volume to heat the volume internally. The soil and groundwater are analogous to a distributed matrix of series and parallel resisters that are continuous throughout the heated volume. Initially, the SPH current causes the soil and groundwater to heat to the boiling point of water. Steam is then generated throughout the heated volume as though thousands of individual heating elements existed in the heated volume. This integrated process of heating and internal steam generation has proven extremely effective and efficient for thermal remediation in both the vadose zone and groundwater. During the heating process, subsurface vapor extraction wells are used to remove steam and contaminant vapors as they are produced. A steam condenser separates the mixture of soil vapors, steam, and contaminants extracted from the subsurface into condensate and contaminant-laden vapor. If these waste streams require pretreatment before discharge, standard air abatement and water treatment technologies are used.

It is noteworthy that, in general, as electric current is conducted through the soil column and aquifer, the current flux is initially highest along paths of low electrical resistance where preferential heating occurs. As a practical example for addressing chlorinated hydrocarbon sites, the most heavily impacted portions of the subsurface are preferentially treated by SPH. Indeed, over time, chlorinated solvents undergo natural anaerobic dehalogenation, producing daughter compounds and free chloride ions. The resulting elevated ion content near high *in situ* concentrations of solvent has been found to be effective in producing low resistance pathways that are also heated preferentially. Silt or clay lenses in the vadose zone are also heated preferentially because they exhibit an elevated moisture content and relatively low electrical resistivity. Overall, these phenomena contribute to accelerated remediation by helping to focus heating where the contaminant is likely to reside and where diffusive processes would normally be rate limiting.

Six-phase soil heating: The removal of VOCs from tight, silty, clay, and clay-rich soils has proven difficult to remediate using conventional technologies. Vapor extraction, an *in situ* technique commonly used for removing VOCs from the subsurface, is limited in effectiveness for silty and clay-rich soils due to the very low permeability, high moisture content, and binding potential of these lithologies. Additionally, vapor extraction is not practicable under saturated conditions. However, the combination of *in situ* heating with SVE has proven effective for releasing DNAPL from tight soils. By heating the soil (and thereby increasing the contaminants vapor pressure), the removal efficiency of SVE can be improved, and the remediation time frame can be drastically shortened. Methods that have been used to heat soil *in situ* include radio frequency, hot air or steam injection, and resistive heating. *SPSH* has subsequently been developed (by Battelle Pacific Northwest Laboratory in the United States) as an improved process of resistive soil heating; SPSH uses commonly available equipment and so generally has a lower capital cost than RFH. Indeed, the six-phase power is generated by using standard utility transformers to split a three-phase current. By using a six-phase voltage pattern instead of a single-phase or three-phase voltage pattern, the voltage gradient between each electrode is kept constant, thus minimizing the power density variations in the soil and creating a more uniform heating pattern. Because the soils are heated internally, low-permeability clay soils and complex heterogeneous soil formations can be more effectively treated with SPSH than with hot air or steam injection, which are preferentially heat zones of higher permeability.

In SPSH, an array of six electrodes is spaced around a central extraction well. A six-phase electrical potential is applied to the electrode array, generating a voltage gradient throughout the zone of the array. As the electrical current generated by the voltage gradient passes through the soil, the resistance of the soil to the current flow causes the soil temperature to rise, thereby increasing the volatility of the contaminant VOCs. Furthermore, as the soils are heated to the boiling point of water, the water turns to steam, stripping the VOCs from the soil pore spaces. The vapor-phase VOCs and steam are collected by applying a vacuum to the central collection well. It is noteworthy that the current generated by the SPSH process will preferentially travel through low-permeability clay soils, as the higher water content and ionic potential of the clay soils provide a more favorable current path than sand or silt soils.

The SPSH process has been successfully demonstrated to effectively reduce the concentrations of chemicals such as PCE in clay soils, which also showed that the SPSH process preferentially heats clay soil over adjacent sandy soil. Indeed, direct *in situ* volatilization is an extremely important SPH remediation mechanism for contaminants with boiling points below that of water, such as TCE, the isomers of 1,2-dichloroethene, and VC. The ability to produce steam *in situ* represents the second significant mechanism for contaminant removal using SPH. Through preferential heating, SPH creates steam from within silt and clay stringers and lenses. The physical action of steam escaping these tight soil lenses drives contaminants out of these otherwise diffusion-limited portions of the soil matrix that tends to lock in contamination via low permeability or capillary forces. The released steam then acts as a carrier gas that, as it moves toward the surface, strips contaminants from both groundwater and the more permeable portions of the soil matrix. The presence of the steam also causes the boiling point of the DNAPL to become depressed due to the effects described by

"Dalton's law of partial pressure." Thus, the normal boiling point of PCE, for instance, is decreased from 121°C to 89°C in the presence of steam in the groundwater, causing free-phase PCE to be rapidly volatilized from the subsurface by the SPH process. Once in the vadose zone, rising steam and contaminant vapors are collected by conventional SVE wells. A condenser then separates the mixture into condensate and contaminant-laden vapor. If these waste streams require pretreatment before discharge, then standard air abatement and water treatment technologies are used.

Treating DNAPLS: When DNAPLs percolate downward through the subsurface, they usually become trapped below the groundwater table on low-permeability stringers and aquitards. As they migrate below the groundwater table, DNAPLs commonly form networks of pools and interlacing strings throughout the soil matrix. Once in an aquifer, DNAPLs represent distributed source terms that are extremely difficult to access and remediate.

To more effectively remove DNAPLs, it is important to heat the upper layer of the aquitard itself, not just the permeable zone, and SPH provides such opportunity (Figure 12.8). With SPH, simultaneous heating occurs within the aquitard as well as within other low-permeability units where the DNAPLs tend to pool. This means that steam is generated continuously along the top units of the aquitard and other low-permeability units where the DNAPL resides so that it can pass directly through the DNAPL pools. The physical action of steam escaping these tight soil lenses drives contaminants out of those portions of the soil matrix that tends to lock in contamination via low permeability or capillary forces. Released steam then acts as a carrier gas, sweeping contaminants out of the subsurface and into the SVE wells. As this steam moves toward the surface, it strips contaminants from both groundwater and the more permeable portions of the soil matrix.

In general, heating causes DNAPL to partition to the vapor phase, enabling a high degree of removal over relatively brief treatment periods. Indeed, Henry's law constant of typical DNAPL compounds (representing the equilibrium ratio of vapor phase to dissolved concentrations) typically increases roughly 15–20 times as temperatures increase from 10°C to 100°C (Heron et al., 1998). Heating further enhances vapor partitioning by increasing the aqueous contaminant solubility. Because of elevated vapor concentrations of the contaminant, it has been shown that the boiling point of DNAPL in the surface is reduced significantly below that of either the pure contaminant or water (de Voe and Udell, 1998). For example, PCE as a DNAPL boils at 88°C versus 100°C for pure water, or 121°C for pure PCE. Boiling generally results in a rapid release of DNAPL from the formation through pressure-induced advection and buoyant forces. It has also been suggested that continuous heating can cause pressure-driven fracturing in low-permeability soils, providing

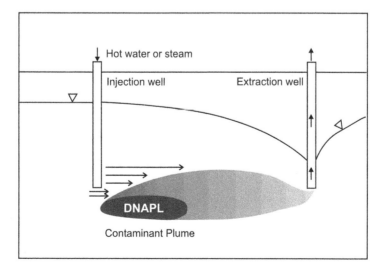

FIGURE 12.8 Schematic of a typical/representative DNAPL treatment system/process using SPH.

a potentially critical release mechanism for removing DNAPL from tight soil layers within the aquifer and soil column. The ability to form microfractures in low-permeability soils would indeed help explain why diffusive rebound of contaminant concentrations has not typically been observed following thermal remediation. Finally, by increasing subsurface temperatures to the boiling point of water, the SPH technology further speeds the removal of contaminants by *in situ* steam stripping. The ability to generate steam *in situ* represents a significant advantage of SPH over technologies that rely upon hydraulic transport and conductive transfer to deliver heat to the subsurface.

In summary, it is noteworthy that electricity takes pathways of least resistance when moving between electrodes, and these pathways are heated preferentially; examples of low resistance pathways include silt or clay lenses and areas of high free ion content. In an archetypical spill scenario, as chlorinated compounds, for instance, sink through the lithology, they become trapped on these same silt and clay lenses. Over time, trapped solvents undergo biological dehalogenation producing daughter compounds and free chloride ions. Thus, at DNAPL sites, the most impacted portions of the subsurface are also the low resistance electrical pathways that are preferentially treated by SPH. By increasing subsurface temperatures to the boiling point of water, SPH speeds the removal of contaminants by both increased volatilization and *in situ* steam stripping. As subsurface temperatures climb, contaminant vapor pressure, and the corresponding rate of contaminant extraction, increases.

12.2.20.7 Excavation and Treatment/Disposal

Excavation is the physical process of removing soil by digging and scooping it out for treatment and/or disposal. It is often an initial step in many of the site restoration technology options available for the treatment of contaminated soils. Soil excavation, transport, and disposal processes generally use mechanized equipment to move contaminated soil. During excavation actions, adequate precautions and measures are usually needed to minimize VOC emissions and fugitive dust generation.

Excavation and disposal at a regulated landfill should indeed be considered a reasonable or preferred option only when relatively small volumes of "hot spot" soils have to be removed from the contaminated site. This is because extensive soil excavation is often too costly and disruptive to normal operations. Excavation of contaminated soil also creates increased potential for exposure of site personnel and the general public to the constituents of concern. Furthermore, the potential for long-term liabilities is always a concern for landfill disposals since contributions of even small quantities to a disposal facility could make one a potentially responsible party at a future date. Other *in situ* remedial alternatives should therefore be carefully evaluated before excavating large volumes of soil from a contaminated site.

12.2.20.8 *Ex situ* SVE

Ex situ SVE is a full-scale technology in which soil is excavated and placed over a network of aboveground piping to which a vacuum is applied to encourage volatilization of organic compounds. Soil piles are generally covered with a geomembrane to prevent volatile emissions and to prevent the soil from becoming saturated by precipitation. The process also includes a system for handling off-gases.

In general, the target contaminant group for *ex situ* SVE is VOCs. Specific advantages over its *in situ* counterpart include the fact that (i) the excavation process forms an increased number of passageways, (ii) shallow groundwater no longer limits the process, (iii) leachate collection is possible, and (iv) treatment is more uniform and easily monitored. The major disadvantage over *in situ* SVE is the increased excavation costs.

12.2.20.9 Free Product Recovery

The recovery of free product hydrocarbon or solvents floating atop a groundwater table (i.e., light, NAPL [LNAPL]) is similar in concept to the pump-and-treat system. For example, LNAPLs can be removed using physical recovery techniques such as a single-pump system that produces water and free product, or a two-pump/two-well system that steepens the hydraulic gradient and recovers the accumulating free product. On the other hand, special circumstance problems call for greater innovation in the methods of choice. For example, there generally are no good methods for the recovery of DNAPLs. Remedial strategies for DNAPL sites typically involve containment, removal, or a combination of both. Indeed, whenever DNAPL pools are present, environmental regulations almost invariably mandate the removal of the free product (usually by-product pumpage). Still, removing DNAPL residual is more difficult, especially when it is below the water table.

12.2.20.10 Groundwater Pump-and-Treat Systems

By and large, the most common method of groundwater restoration has been the application of *pump-and-treat systems*—albeit the trend seems to moving away from this popular technology. Groundwater pump-and-treat systems involve contaminated groundwater being pumped out of the ground, treated by an appropriate treatment method to remove the contaminants of concern, and then finally reinjected into the ground, discharged into surface water bodies, or used otherwise. Technologies for groundwater extraction and treatment are generally used to address the treatment needs of site-specific conditions and regulatory requirements. A common approach is to combine technologies to achieve effective treatment and to meet requisite discharge criteria. The performance of these systems depends directly on site conditions and contaminant chemistry (NRC, 1994a).

Typical possible objectives of groundwater pumping include the removal of dissolved contaminants from the subsurface, and also containment of contaminated groundwater to prevent migration. The groundwater pumping actions can have a number of configurations and design objectives. Single pumping wells or a line of well points can be used to capture a plume. Single or multiple wells can be installed to divert groundwater by lowering the water table; they can also be used to prevent unconfined aquifers from contaminating lower aquifers separated by leaky formations. The water withdrawn by pumping may be treated and subsequently reinjected through one or more wells. The reinjection wells may be used to flush contaminants toward the pumping wells, or to create a hydraulic barrier to preclude further plume migration. In general, pump-and-treat systems may be more appropriate for containing contaminant plumes, or for use in initial emergency response actions at sites experiencing NAPL releases to groundwater. If free product hydrocarbons are present, then an oil/water separator may be required as part of the overall restoration program.

Invariably, groundwater extraction and treatment as a remedial action must address issues pertaining to the strategic and optimum design of the extraction-injection well network, as well as the selection of the proper treatment technology, such as air stripping, biodegradation, and GAC adsorption, for the extracted groundwater. It is noteworthy that several types of the treatment processes used may be affected by a number of extraneous factors. For example, iron in groundwater can precipitate out when oxidized, and may indeed have an adverse effect on air stripping and carbon adsorption treatment systems. Consequently, groundwater parameters such as total organic carbon, chemical oxygen demand, biochemical oxygen demand BOD, and iron content should be determined *a priori*, so as to provide input to the overall system design. Also, it must be recognized that treated water to be discharged may require compliance with certain pollutant discharge standards or criteria, such as compliance with the National Pollutant Discharge Elimination System standards in the United States, or indeed similar criteria. Additionally, the use of a pump-and-treat technology would generally require the use of an air pollution control device (APCD), such as GAC columns to remove contaminants from gases released into the atmosphere. For these reasons, the use of only pump-and-treat technology is often neither a cost-effective nor an efficient approach to the restoration of a contaminated aquifer.

12.2.20.11 Multiphase Extraction

Multiphase extraction (MPE) (also known as *dual-phase extraction* [DPE], *vacuum-enhanced recovery*, or sometimes *bioslurping*) is a full-scale *in situ* technology that uses a high vacuum system to remove various combinations of contaminated groundwater, immiscible contaminants (e.g., separate-phase petroleum product), and vapors from the vadose and saturated zones (e.g., hydrocarbon vapor from the subsurface). The system lowers the water table around the well, exposing more of the vadose zone; contaminants in the newly exposed vadose zone are then accessible to VESs. Extracted liquids and vapors are collected, separated, and treated aboveground; this is then stockpiled for disposal, or reinjected to the subsurface (where permissible under applicable local laws).

MPE systems maintain hydraulic control similar to groundwater pump-and-treat technologies; however, MPE also influences the movement of vapor in the unsaturated zone and is generally more effective than groundwater pump-and-treat technologies at removing contaminants from fine-grained soils. MPE may also stimulate aerobic biodegradation in the saturated and unsaturated zones by drawing uncontaminated soil gas and groundwater into oxygen-depleted zones. Thus, MPE may additionally help establish favorable conditions to implement MNA after the source of contamination has been removed.

Overall, MPE represents an enhancement of the SVE technology developed primarily for the remediation of VOCs in low- to moderate-permeability soils; it enhances VOC recovery by extracting soil vapor and groundwater simultaneously from the same well. The MPE system uses a vacuum pump to extract a mixture of fuel, vapor, and groundwater from a network of extraction wells, combining aspects of SVE and groundwater pump-and-treat technologies in a single remedial system. Unlike SVE, however, MPE can rapidly remove free product (petroleum) from the subsurface. In fact, the prime objective of a MPE system is to remove VOCs from the vadose zone before they migrate further to contaminate groundwater while simultaneously remediating impacted groundwater. During the process, the saturated zone is dewatered, allowing SVE to extract VOCs from previously saturated soils. In addition, the vacuum applied to a well that is also pumping groundwater increases the water yield of the well, thereby increasing the recovery of VOCs dissolved in groundwater.

Dual-phase (two-phase) extraction: Two-phase extraction provides airflow through the unsaturated zone to remediate VOCs and fuel contaminants by vapor extraction and/or bioventing. The airflow also extracts groundwater for treatment aboveground. The screen in the two-phase extraction well is positioned in both the unsaturated and the saturated zones. A vacuum applied to the well, using a drop tube near the water table, extracts soil vapor; the vapor movement entrains groundwater and carries it up the tube to the surface. Once above grade, the extracted vapors and/or groundwater is separated and treated. It is noteworthy that the drop tube is located below the static water level, and so the water-table elevation is lowered, exposing more contaminated soil to remediation by the airflow. When containment of vapors/liquids is necessary, the outcomes are generally better than those obtained through air sparging. Dual-phase vacuum extraction is indeed more effective than SVE for heterogeneous clays and fine sands. However, it is not recommended for lower permeability formations due to the potential to leave isolated lenses of undissolved product in the formation.

The target contaminant groups for DPE systems are VOCs and fuels (e.g., LNAPLs). Indeed, two-phase extraction is primarily used to treat halogenated and nonhalogenated VOCs and nonhalogenated SVOCs. In DPE systems used for liquid/vapor treatment, a high vacuum system is used to remove liquid and gas from low permeability or heterogeneous formations. The vacuum extraction well includes a screened section in the zone of contaminated soils and groundwater. It removes contaminants from above and below the water table. The system lowers the water table around the well, exposing more of the formation. Contaminants in the newly exposed vadose zone are then accessible to vapor extraction. Meanwhile, it is worth mentioning here that DPE for liquid/vapor treatment is generally combined with bioremediation, air sparging, or bioventing when the target contaminants include long-chain hydrocarbons. Indeed, the use of DPE with these technologies can shorten the cleanup time at a site. It can also be used with pump-and-treat technologies to recover groundwater in higher yielding aquifers.

The DPE process for undissolved liquid-phase organic compounds (also known as *free product recovery*, discussed earlier) is primarily used in cases where a fuel hydrocarbon lens more than 20 cm (8 in.) thick is floating on the water table. The free product is generally drawn up to the surface by a pumping system. Following recovery, it can be disposed of, reused directly in an operation not requiring high-purity materials, or purified prior to reuse. Recovery systems may be designed to recover only product, mixed product and water, or separate streams of product and water.

Vacuum-enhanced recovery (bioslurping): Bioslurping is an *in situ* technology that combines vacuum-assisted LNAPL recovery from contaminated aquifers with bioventing and SVE to simultaneously recover LNAPL, and bioremediate the vadose zone. Thus, the bioslurping process concurrently recovers free-product fuel from the water table and capillary fringe while also promoting aerobic bioremediation in the vadose zone of subsurface soils. In its utilization, a bioslurper system withdraws free-phase LNAPL from the water table, relatively small amounts of groundwater, and soil gas/vapor in the same process stream using the air lift created by a single aboveground vacuum pump. Groundwater is then separated from the free product and treated (as necessary) and then discharged or used otherwise; any free product recovered can be recycled. Soil gas/vapor is treated (as necessary) and then discharged. Invariably, the system is operated to cause very little drawdown of the water-table level, thus reducing the problem of free-product entrapment in soils when pumping is stopped.

Bioslurping is primarily used to recover LNAPL, and can indeed help remediate nonhalogenated VOCs and SVOCs in the unsaturated zone. The bioslurping process combines physical recovery of LNAPL, removal of LNAPL constituents by vaporization (SVE), and mineralization of LNAPL constituents by biological action (bioventing). The rate of mass removal due to each mechanism is site dependent. Indeed, the bioventing component can be a strong contributor to the total mass removal at sites with low-volatility fuel, whereas vaporization may be relatively high at sites with high-volatility fuel. The combined mechanisms applied in bioslurping allow an effective removal of LNAPL in the capillary fringe and LNAPL floating on the water table, while the residual organic in the unsaturated zone is mineralized by biological action and removed by vapor extraction.

In general, vacuum-enhanced recovery (or bioslurping) is performed using a tube positioned in a well so the end of the tube is near the water-table level in the hydrogeological formation. Vacuum is applied to the well using a single aboveground vacuum pump, and LNAPL and groundwater are removed from the well by air entrainment. The depth of the tube can be adjusted manually, if needed. The negative pressure established in the well depends on the air withdrawal rate and the permeability of the surrounding formation. Aeration of the unsaturated zone soils is achieved by withdrawing soil gas from the recovery well. The slurping action of the bioslurper system cycles between recovering liquid (free product and/or groundwater) and soil gas. The rate of soil–gas extraction is dependent on the recovery rate of liquid into the well. When free-product removal activities are complete, the bioslurper system can be converted to a conventional bioventing system to complete remediation of the unsaturated zone soils.

A prominent feature of the bioslurping process is the induced airflow, which in turn induces LNAPL flow toward the well. The pressure gradient created in the air phase results in a driving force on the LNAPL that can be significantly greater than the driving force that can be induced by pumping the LNAPL with no airflow. Also of significance is the fact that the vacuum extraction mechanism pulls LNAPL along more permeable horizontal zones. In addition, the continuity of the LNAPL phase is better maintained by eliminating the cone of depression formed during drawdown recovery, thus increasing the relative permeability for LNAPL. For these reasons, bioslurping has the potential for removing more LNAPL and at greater rates than do several other pumping mechanisms.

12.2.20.12 Skimming

Skimmer systems are effective at removing LNAPL from a well while withdrawing little or no water and producing little or no drawdown; thus, they have limited pressure head to move LNAPL toward the recovery point. Skimmers will generally recover LNAPL when the thickness of the floating layer is too thin to allow efficient recovery with a pump. Indeed, a skimmer can recover LNAPL even when the floating layer in the well is less than ¼ in. (or <7 mm) thick.

Skimming is typically done using a floating filter of oleophilic/hydrophobic mesh with a high affinity for nonpolar hydrocarbons, and the ability to reject polar molecules such as water. A mesh cylinder is designed to float in the LNAPL layer in a recovery well. LNAPL floating on the water surface in the well passes through the mesh while water is prevented from entering by the mesh. The LNAPL runs down into a collection pot and is periodically discharged by air pressure to a central holding tank on the surface. The pressurization cycle may be controlled by a timer, by high- and low-level switches, or manually.

Shallow wells with low recovery rates can use rope wick or belt skimmers. The rope wick or belt skimmer uses a continuous loop of rope or belt made up of an oleophilic/hydrophobic material. The rope or belt is strung through the LNAPL layer, and up through a pair of compression rollers. The rollers provide the motive force for the rope or belt while squeezing out any retained LNAPL into a small container. LNAPL collected in the container is pumped to a central holding tank periodically. Large trench recovery points can be fitted with drum or disk skimmers that are too large to fit into a well.

12.2.20.13 Subsurface Control Systems

The primary purpose of subsurface control systems is to prevent leachate migration and therefore reduce potential groundwater contamination via diversion, containment, or plume capture. Broadly speaking, subsurface control measures include capping and top liners, seepage basins and ditches, subsurface drains, ditches and bottom liners, impermeable barriers, groundwater pumping, and interceptor trenches (USEPA, 1985c).

Caps and *top liners* are generally used to reduce infiltration into a contaminated site, thereby reducing the amount of leachate that is generated. The primary objective for using *seepage basins* and *ditches* is to recharge site runoff or water withdrawn by wells or drains; it can also help improve the efficiency of plume capture by modifying groundwater flow patterns. *Subsurface drains*, *ditches*, and *bottom liners* are usually installed in the unsaturated zone to capture leachate before it reaches the saturated zone. *Impermeable barriers* are grout curtains, slurry walls, and sheet pilings installed in the saturated zone to divert uncontaminated groundwater around a site or to limit the migration of contaminated groundwater; barriers can indeed be placed in a number of locations relative to a contaminated site (e.g., upgradient, downgradient, or completely around). *Groundwater pumping* actions can have a number of configurations and design objectives. Single pumping wells or a line of well points can be used to capture a plume. Single or multiple wells can be installed to divert groundwater by lowering the water table; they can also be used to prevent unconfined aquifers from contaminating lower aquifers separated by leaky formations. The water withdrawn by pumping may be treated and subsequently reinjected through one or more wells. The reinjection wells may be used to flush contaminants toward the pumping wells, or to create a hydraulic barrier to preclude further plume migration. *Interceptor trenches* are drain systems that are installed in the saturated zone; they can be used to divert groundwater by lowering the water table, or to capture a plume. An illustrative example that combines several subsurface control measures in a site restoration program is shown in Figure 12.9.

Slurry walls: Slurry walls are used to generally contain contaminated groundwater, divert contaminated groundwater from a drinking water intake, divert uncontaminated groundwater flow, and/or provide a barrier for a groundwater treatment system. These subsurface barriers consist of a vertically excavated trench that is filled with slurry. The slurry hydraulically shores the trench to prevent collapse, and forms a filter cake to reduce groundwater flow. Slurry walls are often used where the waste mass is too large for treatment and where soluble and mobile constituents pose an imminent threat to a source of drinking water.

Most slurry walls are constructed of a soil, bentonite, and water mixture. The bentonite slurry is primarily used for wall stabilization during trench excavation. A soil–bentonite backfill material is then placed into the trench (displacing the slurry) to create the cutoff wall; walls of this composition provide a barrier with low permeability and chemical resistance at low cost. Other wall compositions, such as cement/bentonite, pozzolan/bentonite, attapulgite, organically modified bentonite, or slurry/geomembrane composite, may be used if greater structural strength is required or if chemical incompatibilities between bentonite and site contaminants exist.

Slurry walls are indeed a full-scale technology that has been used for decades as long-term solutions for controlling seepage. They are often used in conjunction with capping. Overall, the technology has demonstrated its effectiveness in containing more than 95% of the uncontaminated

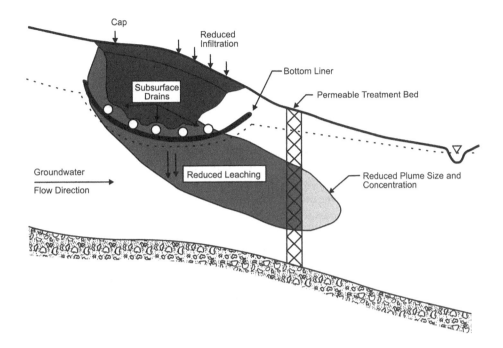

FIGURE 12.9 Illustrative representation of a combination subsurface control system.

groundwater issues; however, in contaminated groundwater applications, specific contaminant types may degrade the slurry wall components and reduce the long-term effectiveness.

Vertical cutoff walls: Vertical cutoff walls are installed to limit the migration of contaminated groundwater and can be applied to remediate halogenated and nonhalogentated VOCs and SVOCs, PCBs, ordnance compounds, inorganics, LNAPL, and DNAPL.

As a technique to control groundwater flow, sheet pile cutoff walls may be constructed by driving interlocking steel or high-density polyethylene (HDPE) into the ground. The joints between individual sheets are typically plugged with clay slurry (steel sheets) or an expanding gasket (HDPE sheets). The steel piles can be driven directly into the ground, whereas the synthetic piles need to be driven with steel backing that is removed once the synthetic sheet is in place. In many cases, barriers can be installed very quickly, thus potentially curtailing the spread of contamination during a remedial action implementation. It is noteworthy that vertical cutoff walls can be used in conjunction with pump-and-treat systems to reduce the volume of water that must be extracted to maintain hydraulic control around a contaminant source.

Grouting: Grouting is another direct method to control the migration of contaminated groundwater. A grout wall is constructed by injecting fluids under pressure into the ground. The grout moves away from the zone of injection, fills pores in the formation, and solidifies, which then reduces the hydraulic conductivity of the formation. Typical grouting compounds include cement, bentonite, and silicate.

Geomembranes: Geomembranes are synthetic sheets installed in open or slurry-supported trenches to control contaminant spread. Geomembranes can provide very low hydraulic conductivity. The sheets are typically constructed of either HDPE or polyvinyl chloride, or indeed other appropriate materials.

12.2.20.14 Surfactant-Enhanced Aquifer Remediation

Surfactant-enhanced aquifer remediation (SEAR), also known as *in situ surfactant flooding*, addresses the removal of residual NAPLs that have become trapped in the pore spaces of an aquifer. Recognize that NAPL contaminants exist as a separate organic phase, rather than in the dissolved (aqueous) phase, and they are commonly found in the vicinity of the original spill or leak, where they have become trapped by capillary and sorptive forces. Now, conventional pump-and-treat systems are mass-transfer-limited by the low aqueous solubilities of most NAPLs and are thus relatively ineffective in removing them. In fact, the NAPL contaminants that pose the greatest challenge for removal are the DNAPLs, for example, chlorinated solvents such as TCE and PCE; these compounds not only have low aqueous solubilities, but also biodegrade very slowly. Under the circumstances, the SEAR technology is unique in its ability to effectively remediate saturated zone contamination by DNAPLs.

SEAR is conducted by injecting a surfactant solution into the contaminated zone while simultaneously extracting water to maintain hydraulic control over the movement of the surfactant solution and the mobilized contaminants (Figure 12.10). In this process, a solution of surfactants (which are the primary ingredient of many soaps and detergents) is injected into the subsurface containing NAPLs. The surfactants increase the effective aqueous solubility of the NAPL contaminant; indeed, they can achieve a significant reduction in the interfacial tension between the NAPL and water phases, so that NAPL removal is greatly accelerated. Hydraulic control of the surfactant and NAPL is maintained by using higher pumping rates at the extraction wells than at the injection wells, and by the selective placement of water injection wells. (By the way, it is noteworthy that the surfactants used in SEAR are nontoxic, food grade, and biodegradable. Salts, and sometimes alcohols, may be added to adjust the surfactant properties that is, to improve the uptake of NAPL, or the hydraulic properties, such as the viscosity of the surfactant solution.) Surfactant flooding is followed by water flooding to remove residual contaminants and injected chemicals. Conventional wastewater treatment technologies may then be used to process the extracted effluent so long as surfactant foaming can be controlled, via the addition of an anti-foaming agent.

Overall, the removal of NAPL contaminants is frequently inefficient and expensive using conventional technologies such as pump-and-treat, due to the low solubilities and rates of dissolution of NAPL contaminants. If not addressed, however, NAPL will persist as a continuing source of contamination to surrounding soils and groundwater, prolonging attenuation of a groundwater plume. SEAR is designed to enhance the removal of NAPL from the subsurface by increasing the effective aqueous solubility of the NAPL and by reducing the interfacial tension between the water and NAPL phases. The technology has most often been applied at sites contaminated by chlorinated solvents existing as DNAPLs. In fact, DNAPLs present a unique remediation problem because they are fundamentally denser than water, and will characteristically migrate downward within an aquifer until retarded by a low permeability layer; therefore, they can reach depths that are not amenable

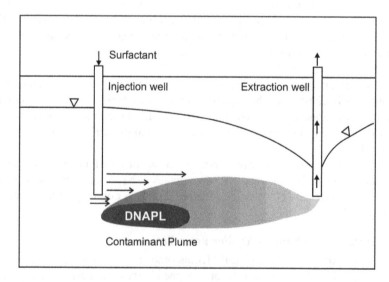

FIGURE 12.10   Schematic of a typical/representative DNAPL treatment system/process using surfactants.

to remediation by traditional methods such as excavation. To help abate such situations, SEAR has been applied to address various DNAPL scenarios with satisfactory outcomes, including for chlorinated solvents, creosote, PCBs from the subsurface, gasoline, jet fuels, and other special fuel oils.

Finally, the design of a surfactant flood requires comprehensive site data to identify the location and distribution of NAPL, to measure aquifer permeability and hydraulic gradients, and to determine the integrity and thickness of any underlying layer of low permeability (i.e., aquitard). If an aquitard is not present, SEAR must be specially designed to avoid downward movement of mobilized contaminants; this can be accomplished by adjusting the surfactant composition. Indeed, variations in aquifer permeability can reduce SEAR performance. In any case, because it is not possible to characterize all of the permeability heterogeneities in a system, it is advisable to use measures that will control the subsurface mobility of the surfactant solution when implementing a SEAR. For instance, by injecting polymer in the surfactant solution, or by creating surfactant foam in the higher permeability zones, it is possible to avoid short-circuiting of the surfactant solution through higher permeability zones, thus improving the sweep of surfactants through lower permeability zones. Meanwhile, it is noteworthy that the use of mobility control measures is subject to depth limitations because higher gradients are necessary to propagate a surfactant solution that contains polymer, or to sustain foam through an aquifer. In the end, surfactant floods in low-permeability systems must be carefully designed to avoid pore plugging.

12.3 DECIDING BETWEEN POSSIBLE CONTAMINATED SITE RESTORATION TECHNIQUES

The first and prime step of any remediation project consists of defining the remedial action objectives to be accomplished at the contaminated site; this typically involves gathering enough background site information and field data to make effectual assessments of remedial requirements and possible cleanup levels. Congruently, the first key determination then is whether "actual/true cleanup" or "simple containment" might be the most appropriate type of remedial action to undertake for the specific project. If direct cleanup is the chosen path forward, then the level of cleanup must be determined first; if containment is the chosen path, then the appropriate technique (e.g., groundwater pumping) is screened for possible use as a hydraulic or physical barrier to prevent off-site migration of contaminant plumes.

The next segment consists of the design and implementation of the appropriate remedial systems—generally based on data evaluated during the process of setting the project goals and objectives. In the case of groundwater pumping systems, for example, the criteria for well design, pumping system, and treatment are dependent on the physical site characteristics and contaminant type; actual treatment may include the design of a train of processes such as gravity segregation, air strippers, carbon systems tailored to remove specific contaminants. Another component of any groundwater extraction system typically would consist of a groundwater monitoring program to verify its effectiveness; indeed, monitoring the remedial action with wells and piezometers allows the operator to make iterative adjustments to the system in response to post-implementation changes in subsurface conditions potentially caused by the remediation, among other things.

The final component would generally consist of a mechanism to determine what may be viewed as the "termination requirements" or "exit strategy." Termination requirements are based on the cleanup objectives or remedial goals defined in the initial stage of the remedial process; the termination criteria are also dependent on the specific site aspects revealed during remedial operations. With the exit strategy clearly defined, well-formulated remedy decisions can be generated for the case-specific project site, resulting in the selection of the appropriate technology, a critical aspect in the design of a successful remedial program. To this end, a number of factors regarding a specific technology, such as its site applicability, performance efficiency, effectiveness, time requirements, public acceptance, and cost, must be prudently considered and/or evaluated throughout the selection process.

Table 12.1 provides a summary description of a selection process for some site restoration techniques that may be applied to the major environmental matrices that have become contaminated

TABLE 12.1

A Summary Description of Selected Site Restoration Techniques/Methods and Technologies Commonly Applied to Contaminated Site Problems

Contaminated Environmental Matrix	Remediation Option	Basic Technology & Process	Scope of Potential Applications	Limitations/Comments
Soils	Asphalt batching	Asphalt incorporation, involving the incorporation of petroleum-contaminated soils into hot asphalt mixes as a partial substitute for stone aggregate.	• Economical for larger volumes of contaminated materials. • Dependent on climate; most asphalt plants do not operate during cold weather.	• May be unsuitable for clays. • May require off-site transportation.
	Bioremediation/ biorestoration	Natural or enhanced biodegradation, involving a process to degrade organic compounds into innocuous materials. Needs shallow groundwater (<15 m [or 50 ft]) or an underlying impermeable silt or clay layer	• Most of the cost-effective for large volumes of contaminated material. • Most applicable when contamination extends into groundwater, and is of sufficient volume or depth below surface. • Minimal disruption to site operations, for in situ bioremediation.	• Labor-intensive, requires considerable maintenance. • Possibility for contaminant migration. • Loss of efficiency in soils containing certain chemicals and low pH.
	Encapsulation	*In situ* containment and isolation process that comprises isolating contaminated soils from the surrounding environment by use of clay caps, liners, slurry walls, and grout curtains	• Used only to prevent contaminant migration. • Applicable to most chemicals, provided such compound does not attack containment materials. • Open areas preferred. • Typical application scenario involves capping landfills to prevent leaching by recharge water.	• Long-term monitoring required. • Does not destroy contaminants; only prevents migration. • Solution may not be permanent.
	Chemical fixation	Solidification/stabilization treatment process, involving the addition of materials to decrease the mobility of the original waste constituents.	• Applicable to a wide variety of waste materials.	• Long-term monitoring required. • Future land use may be restricted.

(Continued)

TABLE 12.1 (*Continued*)

A Summary Description of Selected Site Restoration Techniques/Methods and Technologies Commonly Applied to Contaminated Site Problems

Contaminated Environmental Matrix	Remediation Option	Basic Technology & Process	Scope of Potential Applications	Limitations/Comments
	Excavation/landfill disposal	Process involves the removal of contaminated soils.	• Best for removal and disposal of limited volumes of "hot spot" materials from a vast area that may otherwise be "clean." • Common practice for shallow, highly contaminated soils.	• Potential for long-term liabilities. • Long-term monitoring required. • Trend is toward increasing disposal costs.
	Incineration	Thermal treatment	• Economy of scale for large volumes. • Applicable to a broad range of organic compounds.	• Generally high costs. • Potential disposal problems for residual materials; residual ash may require further treatment.
	In situ soil leaching/flushing	*In situ* soil leaching process that involves injecting or flushing in-place soils into groundwater to leach chemicals in soils into groundwater; surfactants may be added to facilitate the flushing process.	• Applicable to both organics and inorganics, but to different degrees. • Most applicable when contamination extends to groundwater table, requiring use in conjunction with extraction and treatment systems. • Site hydrogeology has strong influence.	• Leachate collection required. • Long-term monitoring required. • Potential problems with leaching fluid used. • Less feasible for complex mixtures of waste types. • Costs depends on site characteristics, contaminant constituents, and cleanup levels.
	Landfarming	Land treatment process, by which affected soils are removed and spread over a treatment area in a layer so as to enhance naturally occurring degradation processes.	• Best suited for lighter organic compounds. • Warmer temperatures are conductive to faster degradation rates, since temperature influences the rate of degradation.	• Emissions control is difficult. • Low cleanup levels may not be practical. • Requires relatively large treatment areas. • Certain chemicals may be toxic to native microbes that could be facilitating degradation process.

(Continued)

TABLE 12.1 (Continued)

A Summary Description of Selected Site Restoration Techniques/Methods and Technologies Commonly Applied to Contaminated Site Problems

Contaminated Environmental Matrix	Remediation Option	Basic Technology & Process	Scope of Potential Applications	Limitations/Comments
	Passive remediation	A "no action" option relies entirely on several natural processes to destroy the contaminants of concern.	• Site-specific; greater depth to groundwater, presence of aquitard, and low infiltration may minimize migration to groundwater. • Temperature may affect volatilization and natural degradation.	• Long-term monitoring required. • Long-term liabilities possible. • Low restoration levels may not be possible. • Effectiveness may be influenced by soil conditions. • Possible future land-use restrictions.
	VESs	*In situ* volatilization, involving the removal of volatile organic contaminants from subsurface soils by mechanically drawing air through the soil matrix.	• Applicable to VOCs. • Can be applied to wide areas. • Minimal disruption to site operations; conductive to both developed and undeveloped sites. • Most effective at higher concentrations. • Generally low costs.	• Venting and emissions difficult to control for very shallow areas. • Not effective below the water table. • Performance affected by soil conditions; less effective and/or longer time frame in fine-grained soils. • May not work for SVOCs.
Groundwater	Air stripping	Aeration process, effective for removal of volatile organic contaminants.	• Simple technology, usable in conjunction with other methods. • Low maintenance costs.	• Problems with air emissions; collection and treatment systems required for off-gas releases.
	Carbon absorption	Activated carbon absorption technology, based on the principle that certain organic contaminants preferentially absorb onto organic carbon.	• Simple technology, usable in conjunction with other methods. • Attractive as point-of-use treatment method. • More appropriate for aquifer restoration. • Applicable to a broad range of (organic and inorganic) contaminants.	• Generates "spent carbon" as a waste by-product. • High maintenance costs; carbon requires frequent regeneration.

(Continued)

TABLE 12.1 (*Continued*)

A Summary Description of Selected Site Restoration Techniques/Methods and Technologies Commonly Applied to Contaminated Site Problems

Contaminated Environmental Matrix	Remediation Option	Basic Technology & Process	Scope of Potential Applications	Limitations/Comments
	Groundwater extraction & treatment	Groundwater pump-and-treat system for the restoration of contaminated aquifers; it involves pumping groundwater for treatment at the surface.	• Very common approach for the restoration of contaminated groundwater. • Best used for containing contaminant plumes.	• Treated water to be discharged or reinjected may require compliance with certain pollutant discharge standards or criteria. • Generally requires use of APCDs to remove contaminants released into air. • Not considered cost-effective nor efficient for aquifer restoration.
	Passive remediation	A "no action" option that relies entirely on several natural processes to destroy the contaminants of concern.	• Several physical and chemical variables (e.g., temperature and alkalinity) may affect natural degradation and the other attenuation processes.	• Long-term monitoring required. • Long-term liabilities possible. • Low restoration levels may not be possible.

with a variety of chemical compounds; this partial listing consists of remedial action technologies, methods, and processes commonly used to address contaminated site problems. It is noteworthy that, in general, no single technology is universally applicable with equal success to all contaminant types, and at all sites. Thus, oftentimes, more than one remediation technique is needed to effectively address most contaminated site problems. In fact, treatment processes can be, and are usually, combined into process trains for more effective removal of contaminants and hazardous materials present at contaminated sites. Ultimately, the type of treatment or remedial technique selected for a given contaminated site problem situation will be dependent on the following key factors:

- The overall goal of the program
- The complexity of the problem
- The types of contaminants of concern
- The nature of the impacted and threatened media associated with the problem situation and
- The type of corrective actions being considered

With that in mind, it is notable that significant advances continue to be made on several fronts in the arena of contaminated site restoration in recent years, and this will essentially facilitate the screening and selection of the best-demonstrated and available technologies and processes for site cleanup efforts. In general, the remediation time frames, and also the costs associated with remediation programs, will vary significantly in accordance with site-specific conditions and the degree of cleanup attainment that is anticipated or desired.

13 Contaminated Site Risk Management Stratagems
The Design of Effectual Corrective Action Programs for Contaminated Site Problems

The effective management of contaminated site problems has become an important environmental priority that will remain a growing social challenge for years to come. This is due in part to the numerous complexities and inherent uncertainties involved in the evaluation of such problems vis-à-vis society's interest in reclaiming and redeveloping derelict lands. Indeed, whatever the cause of a contaminated site problem, the impacted media usually must be cleaned up. Under certain circumstances, however, active site cleanup may not be economically and/or technically feasible; in that case, risk assessment and monitoring of the situation, together with institutional control measures, may be an acceptable site management alternative in lieu of an active remedial action. Such action will generally ensure that at least the contamination is properly contained or controlled in some fashion—so as to minimize or even prevent the chance for its spreading or migrating off-site. This chapter expounds strategies and relevant protocols that can be used to aid the overall contaminated site restoration and management decision-making process.

13.1 THE GENERAL NATURE OF CONTAMINATED SITE RISK MANAGEMENT DECISIONS

Risk management is a decision-making process that entails weighing policy alternatives and then selecting the most appropriate regulatory action. This is accomplished by integrating the results of risk assessment with scientific data, as well as with social, economic, and political concerns—in order to arrive at an appropriate decision on a potential hazard situation (Cohrssen and Covello, 1989; NRC, 1994a–d; Seip and Heiberg, 1989; van Leeuwen and Hermens, 1995). Risk management may also include the design and implementation of policies and strategies that result from this decision-making process.

This section identifies important concepts, analytical tools, and key steps in the effectual design of archetypical contaminated site risk assessment and risk management programs. Such risk management programs are usually directed at risk reduction (i.e., taking measures to protect humans and/or the environment against previously identified risks), risk mitigation (i.e., implementing measures to remove risks), and/or risk prevention (i.e., instituting measures to completely prevent the occurrence of risks). Ultimately, the risk management action or program (viz., reduction, mitigation, and prevention) can generally help engender an increase in the level of protection to both ecological and human health, as well as enhance safety and assist in the reduction of liability.

13.1.1 The Role of Risk Assessment in Contaminated Site Risk Management and Related Decision-Making

Risk assessment is a systematic technique that can be used to generate estimates of significant and likely risk factors associated with environmental contamination and related contaminated site problems. Oftentimes, risk assessment is used as a management tool to facilitate effective decision-making on the control of environmental pollution and/or contaminated site problems. In fact, the chief purpose of risk assessment is to aid decision-making, and this focus should be maintained throughout any environmental or contaminated site risk management program. On the whole, the application of risk assessment to environmental contamination and/or contaminated site problems can likely remove some of the ambiguities in the decision-making process. It can also aid in the selection of prudent, technically feasible, and scientifically justifiable risk control or corrective actions that will help protect public health and the environment in a cost-effective manner. In the end, a risk management action is subsequently used to provide a context for balanced analysis and decision-making.

It is notable that risk assessments performed for environmental contamination and/or contaminated site problems usually will depend on an understanding of the fate and behavior of the chemical constituents of concern. Consequently, the fate and behavior issues in the various environmental compartments and exposure settings should be carefully analyzed with the best available scientific tools. For instance, the application of computer models (in a responsible manner) for the predictive simulation of environmental systems can be a very useful and cost-effective approach when evaluating potential risks and developing credible corrective action programs. In general, it is recommended that the procedures utilized in these efforts reflect current/state-of-the-art methods for conducting risk assessments (Asante-Duah, 2017).

Risk assessments do indeed provide decision-makers with scientifically defensible information for determining whether an environmental contamination and/or contaminated site problem poses a significant threat to the human health or the environment. Congruently, it would typically be conducted to assist in the development of cost-effective strategies for the management of environmental contamination and/or contaminated site problems. Among other things, the risk assessment process can be used to define the level of risk, which will in turn assist in determining the level of analysis, as well as the type of risk management actions to adopt for a given contaminated site and/or environmental management problem; the level of risk considered in such applications can be depicted in a risk-decision matrix (see, e.g., Asante-Duah, 2017)—in a manner that will help distinguish between imminent health hazards and risks. In general, this can be used as an aid for policy decisions, in order to develop variations in the scope of work necessary for case-specific contaminated site risk management programs. Ultimately, the risk assessment efforts can help minimize or eliminate potential long-term problems or liabilities that could result from hazards associated with environmental contamination and related contaminated site problems.

Overall, the benefits of risk assessment designed to facilitate contaminated site risk management decisions outweigh any possible disadvantages; still, it must be recognized that this process will not be without tribulations. Indeed, risk assessment is by no means a panacea. Its use, however, is an attempt to widen and extend the decision-maker's knowledge base, and thus improve the decision-making capability. In any case, the method deserves the effort required for its continual refinement as a contaminated site risk management tool.

13.1.1.1 Hazard Characterization as a Foundational Basis for Making Environmental and Contaminated Site Risk Management Decisions

Hazard accounting and characterization usually represents a very fundamental activity that needs to be undertaken before any credible risk management decisions and/or actions can take place. The general purpose of a hazard characterization is to make a qualitative judgment of the effect(s) caused by an agent or stressor under consideration and its relevance to a target population of interest. In translating hazard characterization into corresponding risk value or indicator, the processes involved need to consider, among other things, the severity of critical effects and the specific affected population groups (Asante-Duah, 2017). Overall, it is important to carefully consider the scenarios of interest (with respect to population, duration, exposure routes, etc.) in such characterization efforts—in order to arrive at realistic and pragmatic risk conclusions. Meanwhile, it is worth mentioning here that to ensure that risk assessments are maximally useful for risk management decisions, the questions that risk assessments need to address must be raised before the process begins—also recognizing that the more complex and multifaceted the problem to be dealt with, the more important the need to operate in this manner; indeed, by focusing on early and careful problem formulation, and on the options for managing the problem, implementation of this type of paradigm or structural framework can do much to improve the utility of risk assessment (NRC, 1996a,b, 2009).

In the final analysis, the levels and complexity of hazard and risk assessments (especially with regard to planning efforts and design elements) should generally be consistent with the goals of the overarching and/or anticipated decisions to be made in the long run. Indeed, one could argue that risk assessments should not be conducted unless it is clear that they are designed to answer very specific questions, and that the level of technical detail along with uncertainty and variability analysis is appropriate to the decision context; such attention to planning should probably assure the most efficient use of resources, as well as affirm the relevance of the risk assessment to decision-makers (NRC, 2009).

13.1.1.2 Risk-Based Decision-Making for Contaminated Site Problems

The assessment of health and environmental risks play an important role in site characterization activities, in corrective action plan (CAP), and also in risk mitigation and risk management strategies for contaminated site problems. A major objective of any site-specific risk assessment is to provide an estimate of the baseline risks posed by the existing conditions at a contaminated site and to further assist in the evaluation of site restoration options. Thus, appropriately applied, risk assessment techniques can be used to estimate the risks posed by site contaminants under various exposure scenarios and to further estimate the degree of risk reduction achievable by implementing various engineering remedies. Indeed, it is apparent that some form of risk assessment is inevitable if site characterization and corrective action response programs are to be conducted in a sensible and deliberate manner. This is because, the very process of performing risk assessment does lead to a better understanding and appreciation of the nature of the risks inherent in a study, and further helps develop steps that can be taken to reduce such risks. The application of the risk assessment process to contaminated site problems will generally serve to document the fact that risks to the human health and the environment have been evaluated and incorporated into the appropriate response actions. In fact, almost invariably, every process for developing corrective action response strategies should incorporate some concepts or principles of risk assessment. Thus, all decisions on site restoration plans for contaminated sites will include, implicitly or explicitly, some aspect of risk assessment.

On the whole, the application of risk assessment to contaminated site problems helps identify critical migration and exposure pathways, receptor exposure routes, and other extraneous factors contributing most to total risks. It also facilitates the determination of cost-effective risk reduction policies. Used in the CAP process, risk assessment generally provides a useful tool for evaluating the effectiveness of remedies at contaminated sites and also for determining acceptable cleanup levels. Indeed, risk-based corrective action programs do facilitate the selection of appropriate and cost-effective site restoration measures. Meanwhile, it is noteworthy that corrective action response programs for contaminated site problems may vary greatly—typically ranging from a "no-action" alternative to a variety of extensive and costly mitigative options. Regardless, the primary objective of every risk management program is to ensure public safety and welfare by protecting human health, environment, and public and private properties. The ability to select an appropriate and cost-effective corrective action response strategy that meets this goal will generally depend on a careful assessment of both short- and long-term risks associated with the contaminated site problem; it also depends on the case-specific mitigative measures. In the end, once quality-assured information has been compiled for a potential contaminated site problem, and a benchmark risk level has been established, an acceptable risk management strategy can then be determined that will be used to guide possible mitigative actions.

13.1.1.3 The Road to Effectual Risk Management and Corrective Action Decisions

Risk management and corrective action decisions generally are complex processes that involve a variety of technical, political, and socioeconomic considerations. In any event, to ensure public safety in all situations, contaminant migration beyond a compliance boundary into the public exposure domain must be below some stipulated risk-based maximum exposure level—classically established through a risk assessment process. Risk assessments are generally conducted to aid risk management and corrective action decisions. Notwithstanding the complexity and the fuzziness of issues involved, the ultimate goal of corrective actions at contaminated sites is to protect the public health and the environment. The application of risk assessment can remove some of the ambiguity in the decision-making process. It can also aid in the selection of prudent, technically feasible, and scientifically justifiable corrective actions that will help protect human and ecological health in a cost-effective manner.

As part of a risk management program, data generated in a risk assessment may help determine the need for and the degree of remediation at a contaminated site. However, to successfully apply the risk assessment process to a potentially contaminated site problem, the process must be tailored to the site-specific conditions and relevant regulatory constraints. In fact, it is almost imperative to make risk assessment an integral part of all contaminated site management programs, except that the level of detail will be case-specific—ranging from qualitative through semiquantitative to detailed quantitative analyses. Irrespective of the required level of detail, the continuum of acute to chronic hazards and exposures should be fully investigated in a comprehensive assessment; this will then allow the complete spectrum of risks to be defined for subsequent risk management decisions. The decision on the effort involved in the analysis (i.e., qualitative, quantitative, or combinations thereof) will usually be dependent on the complexity of the situation and the level of risk involved, anticipated, or predicted.

Ultimately, based on the results of a risk assessment, decisions can be made relating to the types of risk management actions needed for a given contaminated site problem. If unacceptable risk levels are identified, the risk assessment process can further be used in the evaluation of remedial alternatives. This will ensure that net risks to the human health and the environment are truly reduced to acceptable levels via the remedial action of choice.

13.1.2 A HOLISTIC LOOK AT THE GENERAL NATURE OF CONTAMINATED SITE RISK MANAGEMENT PROGRAMS

The management of contaminated site and chemical exposure problems usually involves competing and contradictory objectives—with the prime objective being to minimize both hazards and risk management action costs under multiple constraints. Usually, once a minimum acceptable and achievable level of protection has been established via risk assessment, alternative courses of action can be developed that weigh the magnitude of adverse consequences against the cost of risk management actions. In general, reducing hazards would require increasing costs, and cost minimization during hazard abatement will likely leave higher degrees of unmitigated hazards (Figure 13.1). In any event, a decision is commonly made on the basis of the alternative that accomplishes the desired objectives at the least total cost—total cost here being the sum of hazard cost and risk management cost.

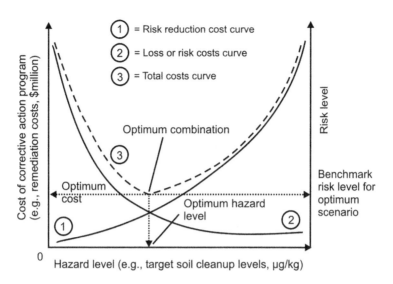

FIGURE 13.1 Risk reduction versus costs: a schematic of corrective action costs (e.g., cleanup or remediation costs) for varying hazard levels (e.g., chemical concentrations in environmental media or residual risk).

Holistically, risk management uses information from hazard analyses and/or risk assessment—along with information about technical resources; social, economic, and political values; and regulatory control or response options—to help determine what actions to take, in order to reduce or eliminate a risk. It is comprised of actions evaluated and implemented to help in risk reduction policies, and may include concepts for prioritizing the risks, as well as an evaluation of the costs and benefits of proposed risk reduction programs. Ultimately, risk assessment results, serving as input to risk management, generally help in the setting of priorities for a variety of contaminated site and chemical exposure problems—further to producing more efficient and consistent risk reduction policies. Risk management does indeed provide a context for balanced analysis and decision-making—with contaminated site risk management programs generally designed with the goal to minimize potential negative impacts associated with consequential chemical exposure problems.

13.1.2.1 A Framework for Designing Effectual Risk Management Programs

Contaminated site risk management programs are generally designed with the goal to minimize potential negative impacts associated with the prevailing contaminated site or chemical exposure problem. Typically, several pertinent questions relating to the nature and extent of contamination, exposure settings, migration and exposure pathways, populations potentially at risk, the nature and level of risks, environmental quality goals, regulatory policies, and availability of technically feasible remedial techniques are asked during the planning, development, and implementation of corrective action programs that are directed at mitigating contaminated site problems (Asante-Duah, 1996; BSI, 1988; Cairney, 1993; Jolley and Wang, 1993; USEPA, 1985e, 1987h,i, 1988e, 1989n,o, 1991a,h; WPCF, 1988). In the end, it is very important that the contaminated site management program helps address all relevant issues (Box 13.1)—which will normally affect the type of risk management decision accepted for a contaminated site problem.

To successfully apply the risk assessment process to a potential contaminated site or chemical exposure problem, the process must be tailored to the case-specific conditions and relevant regulatory constraints. Based on the results of a risk assessment, decisions can then be made relating to the types of risk management actions needed for a given contaminated site or chemical exposure problem. If unacceptable risk levels are identified, the risk assessment process can further be used in the evaluation of remedial or risk control action alternatives. This will ensure that net risks to

**BOX 13.1   KEY DECISION ELEMENTS ASSOCIATED WITH
THE DESIGN OF A RISK MANAGEMENT PROGRAM**

- Establish basis for contamination indicators.
- Gather and review background information for evidence of release.
- Determine area history, to help identify other possible sources of contamination.
- Identify potentially affected areas.
- Address health and safety issues associated with the case-specific situation—to include providing emergency response by mitigating release and potential hazards.
- Assess the fate and behavior characteristics of contaminants in the environment, including an identification of anticipated degradation, reaction, and/or decomposition by-products.
- Determine the critical environmental media of concern (such as air, surface water, groundwater, soils and sediments, and terrestrial and aquatic biota).
- Delineate potential migration pathways.
- Identify and characterize populations potentially at risk (i.e., potential human and ecological receptors).
- Determine receptor exposure pathways.
- Develop a conceptual representation or model for the problem situation.
- Evaluate potential exposure scenarios, and the possibility for human and ecosystem exposures.
- Characterize the general contaminated site problem—to include a general identification of the contaminant types and their characteristics, a delineation of the extent of contamination for affected matrices, and a mapping of areas where contaminants may impact human health and/or environment.
- Assess the environmental and health impacts of the contaminants, if they should reach critical human and ecological receptors.
- Determine the corrective action needs and formulate a risk management strategy—to include establishment of remedial action objectives (RAOs) and case-specific restoration goals; assessment of variables influencing selection of restoration goals and remedial systems; identification of remedial action alternatives; and development of general response actions.
- Design effective long-term monitoring and surveillance programs as a necessary part of an overall CAP.

human and ecological health are truly reduced to acceptable levels via the remedial or risk management action of choice. Figure 13.2 provides an illustrative framework that may be used or adapted to facilitate the environmental and contaminated site risk management decision-making process. The process will generally incorporate a consideration of the complex interactions existing between the environmental or exposure setting, regulatory policies, and technical feasibility of remedial technologies and/or risk management options. Ultimately, the tasks involved should help contaminated site risk analysts to identify, rank/categorize, and monitor the status of potential contaminated site or chemical exposure problems; identify field data needs and decide on the best investigation or sampling strategy; establish appropriate human and ecological health goals; and choose the risk management action that is most cost-effective in controlling or abating the risks associated with the contaminated site or chemical exposure problem.

In the arena of contaminated site and chemical exposure problems, it is noteworthy that risk management decisions should be based on a wide range of issues relevant to risk analysis—including various scientific opinions, contaminant toxicology, and professional judgment, along

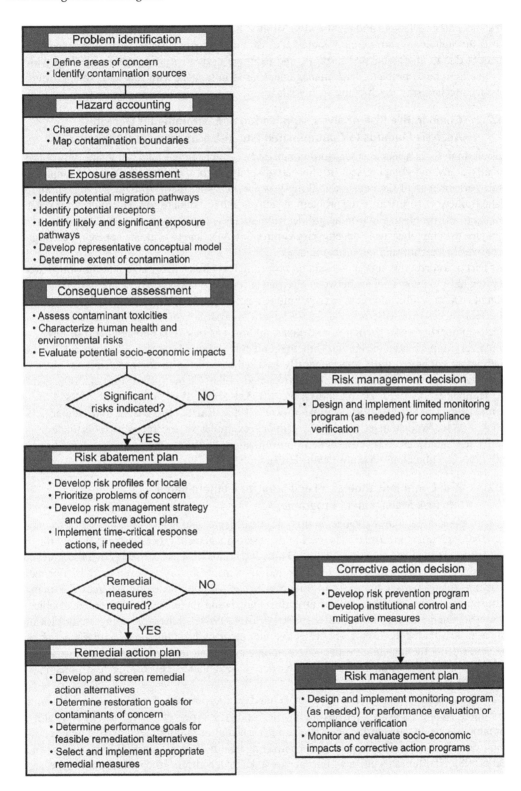

FIGURE 13.2 A risk-based decision framework for the management of contaminated site and related problems.

with socioeconomic factors and technical feasibility. It is also imperative to systematically identify hazards throughout an entire contaminated site risk management system, assess the potential consequences due to any associated hazards, and examine corrective measures for dealing with the case-specific type of problem. Risk management, used in tandem with risk assessment, offers the necessary mechanism for achieving such goals.

13.1.2.2 Comparative Risk Analysis: Application of Environmental Decision Analysis Methods to Contaminated Site Risk Management Programs

Decision analysis is a management tool comprised of a conceptual and systematic procedure for rationally analyzing complex sets of alternative solutions to a problem, in order to improve the overall performance of the decision-making process. Decision theory provides a logical and systematic framework to structure the problem objectives and to evaluate and rank alternative potential solutions to the problem. Environmental decision analyses typically involve the use of a series of techniques to comprehensively develop risk control or CAPs, and to evaluate appropriate mitigative alternatives in a technically defensible manner.

As part of a corrective action assessment program, it is almost inevitable that the policy analyst will often have to make choices between alternative remedial options. These are based on an evaluation of risk trade-offs and relative risks among available decision alternatives; evaluation of the cost-effectiveness of CAPs; or a risk–cost–benefit comparison of several management options. In fact, comparing risks, benefits, and costs among various risk management strategies can become very important in most environmental and contaminated site risk management programs. A number of analytical tools may generally be used to assist with the processes involved here (see, e.g., Ashford and Caldart, 2008; Bentkover et al., 1986; Clemen, 1991; Finkel, 2003; Haimes, 1981; Haimes et al., 1990; Hattis and Goble, 2003; Keeney, 1990; Lave et al., 1988; Lave and Omenn, 1986; Lind et al., 1991; Nathwani et al., 1990; Raiffa, 1968; Seip and Heiberg, 1989; USEPA, 1984c; Weinstein et al., 1980); primary examples of the relevant tools include cost-effectiveness analyses, risk–cost–benefit optimization, multi-attribute decision analysis, and utility theory applications (Asante-Duah, 2017).

13.1.2.3 Risk Communication as a Facilitator for Contaminated Site Risk Management Programs

Risk management combines socioeconomic, political, legal, and scientific approaches to manage risks. Risk assessment information is used in the risk management process to help in deciding how to best protect human and ecological health. Thus, essentially, risk assessment provides *information* on the risks—and risk management develops and implements an *action* based on that information. This means that risk assessment can in principle be carried out objectively, whereas risk management usually involves preferences and attitudes, and should therefore be considered a subjective activity (Seip and Heiberg, 1989; NRC, 1983b; USEPA, 1984c). The subjectivity of the risk management task calls for the use of very effective facilitator tools/techniques—with good risk communication being the logical choice; risk communication is an interactive process or exchange of information and opinions among interested parties or stakeholders concerning risk, potential risk, or perceived risk.

Risk communication has formally been defined as the process of conveying or transmitting information among interested parties about the following types of issues: levels of health and environmental risks; the significance or meaning of health or environmental risks; and decisions, actions, or policies aimed at managing or controlling health or environmental risks (Cohrssen and Covello, 1989). It offers a forum at which various stakeholders discuss the nature, magnitude, significance, or control of risks and related consequences with one another. Effective risk communication is indeed important for the implementation of an effectual risk management program. It is therefore quite important to give adequate consideration to risk communication issues when developing a risk management agenda. As a matter of fact, in many a situation, even credible risk assessment and risk management decisions may never get implemented unless they are effectively communicated to all interested stakeholders. Thus, risk communication should be viewed as rather vital to the risk assessment and risk management processes—and, ultimately, to the success of most risk management actions.

In practice, to be able to design an effectual risk management program, a variety of qualitative issues—such as relates to sound risk communication—become equally important in addition to any prior risk quantification. Risk communication may indeed dictate public perception and therefore public acceptance of risk management strategies and overall environmental and public health risk management decisions. One paramount goal of risk communication is to improve the agreement between the magnitude of a risk and the public's political and behavioral response to this risk—necessitating researchers to investigate a number of message characteristics and risk communication strategies (Weinstein et al., 1996; Weinstein and Sandman, 1993). The process involved provides information to a concerned public about potential human and ecological risks from exposure to toxic chemicals or similar environmental hazards. In fact, because the perception of risks often differs widely, risk communication typically requires a somehow perceptive approach and should involve genuine dialog (van Leeuwen and Hermens, 1995). Among several other factors, trust and credibility are believed to be key determinants in the realization of any risk communication goals. Apparently, defying a negative stereotype is crucial to improving perceptions of trust and credibility (Peters et al., 1997). Anyway, the literature available on the subject addresses several other important elements/issues—including checklists for improving both the process and the content of risk communication efforts. Finally, it is noteworthy that only limited presentation on the risk communication topic is given in this book—with more detailed elaboration/discussions to be found elsewhere in the literature (e.g., Cohrssen and Covello, 1989; Covello, 1992, 1993; Covello and Allen, 1988; Fisher and Johnson, 1989; Freudenburg and Pastor, 1992; Hance et al., 1990; Kasperson and Stallen, 1991; Laird, 1989; Leiss, 1989; Leiss and Chociolko, 1994; Lundgren, 1994; Morgan and Lave, 1990; NRC, 1989c; Pedersen, 1989; Peters et al., 1997; Renn, 1992; Silk and Kent, 1995; Slovic, 1993; van Leeuwen and Hermens, 1995; Vaughan, 1995; Weinstein et al., 1996; Weinstein and Sandman, 1993).

Designing an Effectual Risk Communication Program. Several rules and guidelines have been suggested/proposed to facilitate effective risk communication (e.g., Cohrssen and Covello, 1989; Covello and Allen, 1988)—albeit there are no easy prescriptions per se. In fact, it is very important that risk communication should consider and embrace several important elements (see, e.g., Asante-Duah, 2017)—in order to minimize or even prevent suspicion/outrage from a usually cynical public. Thoughtful consideration of the relevant elements should generally help move a potentially charged atmosphere to a responsible one, and one of cooperation and dialog. Ultimately, a proactive, planned program of risk communication will—at the very least—usually place the intended message in the public eye in advance of negative publicity and sensational media headlines. Under all circumstances, a reliable tool and channel of communication should be identified to ensure effective and timely transmittal of all relevant information. Overall, a systematic evaluation using structured decision methods—such as the use of the event tree approach (see, e.g., Asante-Duah, 1998)—can greatly help in this direction. The event tree illustrates the cause-and-effect ordering of event scenarios, with each event being shown by a branch of the event tree in the context of the decision problem. The event tree model structure can indeed aid risk communicators in improving the quality and effectiveness of their performance and presentations.

Ultimately, scientific information about health and environmental risks is generally communicated to the public through a variety of channels—ranging from warning signs to public meetings/forums involving representatives from government, industry, the media, the populations potentially at risk, and other sectors of the general public (Cohrssen and Covello, 1989). Important traditional techniques of risk communication usually consist of community/public education programs, "fact sheets," newsletters, public notices, workshops, focus groups, public meetings, and similar forum types—all of which seem to work very well if rightly implemented or utilized. Anyhow, irrespective of the approach or technique adopted, however, it must be acknowledged that hazard perception and risk thresholds tend to be quite different in different parts of the world—and maybe, at times, even between different communities within the same region, etc. In fact, there could also be variations within different sectors of a community or society within the same locale or region. Even so, such variances should not affect the general design principles when one is developing a risk communication program.

13.1.3 The Use of Contemporary Risk Mapping Tools: GIS in Contaminated Site Risk Management Applications

A Geographic Information System (GIS) is a computer-based tool used to capture, manipulate, process, and display spatial or geo-referenced data (Bernhardsen, 1992; Gattrell and Loytonen, 1998; Goodchild et al., 1993, 1996). It is a tool that can serve a wide range of research and surveillance purposes; it allows the layering of health, demographic, environmental, and other traditional data sources to be analyzed by their location on the earth's surface. Indeed, the GIS technology has become an important tool for environmental professionals—as this generally allows for the efficacious mapping and analysis of contaminated site and environmental management data.

Broadly speaking, understanding and communicating the association between environmental hazards and risks are essential requirements of an effective environmental protection policy. For instance, in the United States, routine public health surveillance programs have traditionally generated massive amounts of information. Environmental monitoring and simulation modeling projects also provide an equally large volume of data. But, due to the lack of a coordinated framework to organize, manage, analyze, and display the data, majority of this information had been poorly utilized in the past. This is one reason why the advances in the geospatial information technologies, such as GIS, have become so very useful in a whole wide range of contaminated site risk management functions.

It is apparent that recent advances in the application of GIS technology have significantly improved and will continue to revolutionize the spatial analysis of environmental contamination and related or concomitant issues. Indeed, the GIS of today provides a relatively easy tool for overlaying and analyzing disparate data sets that relate to each other by location on the earth's surface. The growing availability of health, demographic, and environmental databases containing local, regional, national, and international information are further propelling major advances in the use of GIS and computer mapping with spatial statistical analyses.

13.1.3.1 The General Role and Utilization of GIS Applications in Environmental Management

Every field of environmental and exposure modeling is increasingly using spatially distributed approaches, and the use of GIS methods will likely become even more widespread. Invariably, use of a GIS can allow for the processing of geo-referenced data and provide answers to such questions as relate to the particulars of a given location, the distribution of selected phenomena and their temporal changes, the impact of a specific event, or the relationships and systematic patterns of a region (Bernhardsen, 1992). In fact, it has been suggested that as a planning and policy tool, the GIS technology could be used to "regionalize" (or perhaps even "globalize") the risk analysis process—moving it from its traditional focus on a microscale (i.e., site-specific problems) to a true macroscale (e.g., urban or regional risk analysis, comparative risk analysis, risk equity analysis) (D. Rejeski, in Goodchild et al., 1993).

GIS is indeed a rapidly developing technology for handling, analyzing, and modeling geographic information. It is not a source of information per se, but only a way to manipulate information. Overall, the GIS technology provides a way to manage and analyze information—and perhaps even more importantly allows users to visualize information using combinations of different map layers.

Example application types for the utilization of GIS in environmental risk management decisions can be found in the literature elsewhere (e.g., Asante-Duah, 2017).

13.2 DESIGN ELEMENTS OF CORRECTIVE ACTION ASSESSMENT AND RESPONSE PROGRAMS

Corrective actions are generally designed with the goal of helping to minimize potential negative impacts associated with contaminated site problems. The processes involved in the design of an adequate strategy will commonly incorporate a consideration of the complex interactions existing between the hydrogeological environment, regulatory policies, and the technical feasibility of remedial technologies. Characteristically, answers would be sought for several pertinent questions relating to the nature and extent of contamination, exposure settings, migration and exposure pathways, populations potentially at risk, the nature and level of risks, and cleanup goals; this is best completed during the planning, development, and implementation of corrective action programs directed at restoring contaminated sites (BSI, 1988; Cairney, 1993; Jolley and Wang, 1993; USEPA, 1985e, 1987h,i, 1988e, 1989n,o, 1991a,h; WPCF, 1988). Indeed, it is very important that the corrective action assessment helps answer all relevant questions—which will ultimately affect the type of corrective action response decision accepted for a contaminated site problem.

Corrective action assessment programs are typically designed to facilitate the development of scientifically based methods necessary to support corrective action decisions. The process allows a multimedia approach to site characterization and the establishment of data quality objectives; identifies the specific parameters for which data must be collected; identifies the preferred data gathering, handling, and analytical techniques to be used; and allows project managers to develop site-specific and cost-effective cleanup solutions. In general, once quality-assured information has been compiled for a potentially contaminated site, and benchmark risk criteria have been established, acceptable cleanup criteria can be determined that will be used to guide further site restoration decisions. This will subsequently form an important basis for developing a CAP for the site.

Invariably, site assessments are a primary activity in the overall corrective action assessment process. The objective of the site assessment is to determine the nature and extent of potential impacts from the release or threat of release of hazardous substances. A site investigation effort, which is a major component of the site assessment activity, aims at collecting representative samples from the potentially contaminated site. Depending on the adequacy of historical data and the sufficiency of details available on the likely contaminants at a site, sampling programs can be designed to search for specific chemical constituents that become indicator parameters for the sample analyses. Where specific information about historical uses of the site is lacking, a more comprehensive sampling and analytical program will generally be required; in this case, the sampling and analysis program may be carried out in phases—moving from a more general scope to specificity as adequate information becomes available on the site. The results from these activities will facilitate a complete analysis of possible contaminants present at the site. The information thus obtained is used to determine current and potential future risks to the human health and the environment. Ultimately, corrective actions are developed and implemented with the principal objective to protect the public health and the environment.

All in all, the design of an effectual corrective action program for any contaminated site problem should consider several important decision elements (Box 13.2). These basic elements should be completely evaluated and the relevant information used to support appropriate corrective action decisions. In fact, an adequately conducted site assessment will allow a project manager to develop and select a final CAP or cleanup solution that best suits the conditions at a particular site. Invariably, an initial site characterization is typically conducted to gather adequate information for determining if particular remedial systems are suitable options for remediating the site. Although different levels of effort may be required for different sites, a comprehensive site assessment program is always

---

**BOX 13.2   ELEMENTS OF A CORRECTIVE ACTION
ASSESSMENT AND RESPONSE PROCESS**

- Characterization of the contaminated site, including the physical setting, site geology, topography, hydrogeology, and meteorological conditions
- Identification of contaminant types and their characteristics
- Assessment of the fate and transport characteristics of site contaminants, including the identification of anticipated degradation, reaction, and/or decomposition by-products
- Determination of the critical environmental media of concern (such as air, surface water, groundwater, soils and sediments, and terrestrial and aquatic biota)
- Delineation of potential migration pathways
- Identification and characterization of potential human and ecological receptors
- Development of a conceptual representation or model of the site
- Evaluation of potential exposure scenarios, and the potential for human and ecosystem exposures
- Assessment of the environmental and health impacts of the contaminants, if they should reach critical human and ecological receptors
- Decision-making on the corrective action needs for the site
- Development of a risk management strategy for the site
- Design of effective long-term monitoring and surveillance programs as a necessary part of an overall CAP

---

essential for making good corrective action decisions. The nature of this assessment will generally depend on the stage of investigation.

13.2.1   Diagnostic and Corrective Action Assessment of Contaminated Site Problems

The existence of contaminated sites may result in the release of chemical contaminants into various environmental compartments. Any contaminant released into the environment will be controlled by a complex set of processes (such as intermedia transfers, degradation, and biological uptake). In fact, many chemical contaminants are persistent in the environment and undergo complex interactions in more than one environmental medium. Contaminated sites should therefore be carefully and thoroughly investigated so that risks to potentially exposed populations can be determined with a reasonably high degree of accuracy. Typically, different levels of effort in the investigation will generally be required for different contaminated site problems. Ultimately, the application of a well-designed diagnostic assessment plan to a contaminated site problem will ensure that appropriate and cost-effective corrective measures are identified and implemented for the site.

A multimedia approach to site characterization is generally adopted for most contaminated site problems, so that the significance of possible air, water, soil, and biota contamination can be established through appropriate field sampling and analytical procedures; this is expected to yield high-quality environmental data needed to support the corrective action response decision. To accomplish this, samples are gathered and analyzed for the chemicals of potential concern in the appropriate media of interest. Proper analytical protocols in the field sampling and laboratory analysis procedures are used to minimize uncertainties associated with the data collection and evaluation activities. Ultimately, the information gathered during the remedial investigation (RI) or site characterization activities is used to map out the extent of contamination and to evaluate the potential risks associated with the subject site, so that an appropriate corrective action response decision can be made. Invariably, the health and environmental risk estimates become an important element in the corrective action decision process. This is because the risk assessment provides the decision-maker with technically defensible and scientifically valid procedures for determining whether or not a contaminated site represents significant adverse human health or ecological risks that warrants consideration as a candidate for mitigation.

Classically, the following general tasks are carried out as part of the overall diagnostic and corrective action programs designed to address contaminated site problems (Cairney, 1993; OBG, 1988; USEPA, 1988e):

- Characterize site as completely as possible—to include development of very good conceptual site model (CSM) for the site.
- Define and develop remediation objectives that are appropriate for the specific contamination problems.
- Identify remedial technologies that are capable of achieving the site restoration goal(s).
- Develop and screen remedial alternatives, and select only those that are superior based on engineering, environmental, and economic criteria, and which concurrently meet regulatory and community expectations.
- Perform detailed analyses of remedial alternatives by scrutinizing each of the initially selected alternatives.
- Choose, offering justification, the preferred remedial alternative(s).
- Implement appropriate remedy and/or risk management plan.

Information derived from the RI or site characterization will generally help determine the need for and the extent of cleanup necessary for a contaminated site. The early determination of restoration goals facilitates the development of a range of feasible site restoration decisions, which in turn helps focus remedy selection on the most effective remedial alternative(s) during a corrective measure appraisal.

Corrective measure evaluation typically is comprised of a feasibility study (FS) of remedial options, the purpose of which is to examine site characteristics, cleanup goals, and the performance of alternative remedial technologies so that the most effective approach for the restoration of a contaminated site can be identified. A well-designed FS addresses every contaminant migration pathway and environmental medium that poses, or could pose, unacceptable risks to the human health or the environment. To accomplish the tasks involved in this type of evaluation, several corrective measure evaluation tools will usually be used to assist in determining whether or not remediation is necessary for a contaminated site problem, and to further determine the cleanup goals and techniques appropriate for a given site.

13.2.1.1   The Role of Mathematical Models

Mathematical models can be used to perform several functions that will assist in the effective management of environmental contamination problems, as long as they are used properly and with an understanding of their limitations. In particular, mathematical models find several specific applications in corrective action assessments—ranging from problem definition and system conceptualization, to exposure and risk assessments, and to the development and evaluation of remedial options. In fact, since field data frequently are limited and insufficient to accurately and completely characterize a contaminated site and nearby conditions, models can be particularly useful for studying spatial and temporal variabilities, together with potential uncertainties. In addition, sensitivity analyses can be performed by varying specific parameters and then using models to explore the ramifications (as reflected by changes in model outputs).

Models can indeed be used for several purposes in the investigation, characterization, and/or management of contaminated site problems. The most common applications of mathematical models in corrective action assessment and response programs required for contaminated site management and restoration decisions relate to the following in particular:

- *To understand and predict contaminant fate and transport.* Models can be used to predict contaminant concentrations at receptor point locations or compliance boundaries (e.g., in the prediction of contaminant migration in various environmental compartments, or in the prediction of future concentrations of contaminants at a water supply or compliance boundary well). In addition, contaminant distribution as well as degradation with time can be determined via modeling. Models can also be used to predict flow paths and times of travel for contaminants, and therefore to delineate well-head protection areas.
- *To determine contaminant sources using back-tracking procedures.* Using back-tracking procedures from a given project location allows for the equitable allocation of cleanup costs among potentially responsible parties (PRPs) when several facilities collectively have contributed to contamination at a given project site (such as well locations beyond a compliance boundary).
- *To screen remedial alternatives.* Models can be used as a screening tool for ranking cleanup alternatives in feasibility studies (e.g., screening of alternatives is performed to eliminate those remedial actions deemed infeasible due to technical, public health, institutional, and/or cost reasons).
- *Conceptual design of optimal remediation systems.* Models can be used to refine and, in some cases, optimize conceptual designs prior to their implementation (e.g., the design of recharge well fields or basins for efficient injection/infiltration of treated wastewater for irrigation use; design of numbers, locations, and configurations of injection, extraction, and monitoring wells that will maximize system efficiency; and the design and evaluation of remedial schemes which use drains, barrier walls, and caps).
- *Analysis of remedial action alternatives.* Models can be used for the design of monitoring and CAPs or in the simulation of several scenarios during the design of groundwater extraction–injection well networks (e.g., the study of the interaction of surface water bodies (such as small streams, rivers, lakes) with groundwater when the aquifer system is stressed by extraction wells).
- *Performance evaluation of remedial alternatives.* Models can be used to evaluate the expected remedy performance during the FS, so that the effects of restoration or corrective action can be predicted. For example, models are used to estimate the effects of source-control actions on remediation.

Models (analytical or numerical) may also be used for the detailed analysis of corrective action alternatives. Appropriate models may be used to obtain information on the effectiveness, durability (i.e., design life), and expected exposures and risks as a result of implementing different remedial actions. In a corrective action evaluation program, models may be used to determine the general technical feasibility and any potential environmental impacts arising from implementation of different remedial actions.

Modeling can be used in the optimization of remedial action designs because a number of alternative designs can be evaluated rapidly and quantitatively. They can assist in the development of a conceptual design for the most cost-effective action—by simulating different configurations of the selected action. For example, a groundwater pumping action may be conceptually designed by evaluating pumping rates, number and spatial locations of wells, and location of screened intervals. Optimizing a conceptual design for the groundwater pumping action may involve evaluating alternative locations for wells, pumping rates, and remedial action configurations to identify which specific combination will be most effective (USEPA, 1985e). In addition, models are frequently used to help assess cleanup levels, determine the levels of required source removal, project the performance characteristics for remedial action designs, as well as formulate post-remediation and closure requirements.

In general, models usually simulate the response of a simplified version of a complex system. As such, their results are imperfect. Nonetheless, when used in a technically responsible manner, they can provide a very useful basis for making technically sound decisions about a contaminated site problem. They are particularly useful where several alternative scenarios are to be compared. In such cases, all the alternatives are compared on a similar basis; thus, whereas the numerical results of any single alternative may not be exact, the comparative results of showing that one alternative is superior to others will usually be valid. In planning for site characterization and restoration programs, one of the major benefits associated with using mathematical models is that environmental concentrations useful for exposure assessment and risk characterization can be estimated for several locations and time-periods of interest. In a corrective measure evaluation program, models may be used to determine the general technical feasibility as well as any potential environmental impacts arising from implementation of different remedial actions.

13.2.2 Integrating Land Reuse Plans into Anticipated Remedies

Identifying the reasonably anticipated future use of a given piece of land is an important consideration in the contaminated site risk management and/or cleanup process, and indeed is the first step for integrating reuse plans into a risk management and/or cleanup program. In general, the anticipated future land use helps determine the appropriate extent of remediation—because it affects the types and frequency of exposures that may result from any post-restoration residual contamination likely to remain at the site, which in turn affects the nature of the remedy chosen. As a corollary, the alternatives selected through the remedy selection may determine the extent to which hazardous constituents remain at the site and therefore affect subsequent available land uses. It is noteworthy that if there should be a decision at a future date to change the original land use in a way that makes further cleanup necessary to ensure adequate protectiveness, then such further actions must be carried out so that the relevant protection of all populations potentially at risk is not compromised under the new land-use scenario(s).

In general, land uses that will be available following completion of remedial action are determined as part of the remedy selection process; during this process, the goal of realizing the reasonably anticipated future land uses is considered along with several other factors. The outcome of this evaluation may be any combination of unrestricted uses, restricted uses, or other yet to be determined uses. Meanwhile, it must be recognized that in addition to analyzing human health exposure scenarios associated with certain land uses, ecological exposures may also need to be considered concurrently. In the final analysis, it is apparent a reuse assessment assists in developing the proper assumptions regarding the *general types* or *broad categories* of reuse that might reasonably occur at a contaminated site. Examples of land-use assumptions that oftentimes appear likely based on the conclusions of a typical reuse assessment include, but are not limited to, residential, commercial/industrial, recreational, and ecological preserve. More specific end uses (e.g., office complex, shopping center, and sports facility) are typically considered during the corrective action response evaluation—and during which time detailed planning information is more readily available.

13.2.2.1 The Contaminated Site Reuse Assessment Process

Typically, a reuse assessment involves collecting and evaluating a variety of information, in order to be able to develop assumptions about reasonably anticipated future land use(s) at contaminated sites. It generally may involve a review of available records, visual inspections of the site, and stakeholder discussions about reasonably anticipated future uses with local government officials, property owners, and community members. Information obtained from a reuse assessment can be particularly useful during the planning stages of a response action. Indeed, the resulting assumptions of reasonably anticipated future land uses would generally become part of the following activities in particular:

- The baseline risk assessment, when estimating potential future risks
- The development of remedial/removal action objectives, and the development and evaluation of corrective action response alternatives
- The selection of the appropriate response action that is required for the protection of human health and the environment.

Among other things, the RAOs developed during the corrective action assessment and response plan for a contaminated site problem should reflect the reasonably anticipated future land use or uses. The baseline risk assessment provides the basis for taking a remedial action at a contaminated site and supports the development of RAOs; land-use assumptions affect the exposure pathways that are evaluated in the baseline risk assessment. Current land use is critical in determining whether there is a current risk associated with a contaminated site, whereas future land use is important in estimating potential future threats. The results of the risk assessment aid in determining the degree of remediation necessary, in order to ensure long-term protection at the contaminated site. Also, future land-use assumptions allow the baseline risk assessment and the FS to become more focused with respect to developing practicable and cost-effective remedial alternatives—leading to site activities which are consistent with the reasonably anticipated future land use.

Broadly speaking, the baseline risk assessment generally needs only to consider the reasonably anticipated future land use; however, it may be valuable to evaluate risks associated with other land uses. Indeed, more than one future land-use assumption may be considered when decision-makers wish to understand the implications of unexpected exposures. In particular, when there is some uncertainty regarding the anticipated future land use, it may be useful to compare the potential risks associated with several land-use scenarios—in order to be able to gage the potential impact on human health and the environment, should the land use unexpectedly change. The magnitude of such potential impacts may be an important consideration in determining whether, and how, institutional controls should perhaps be used to restrict future uses.

13.2.2.2   The General Scope of Land Reuse Assessments

The scope and level of detail of the reuse assessment should be site-specific and tailored to the complexity of the site, the extent of the contamination, the level of redevelopment activity that has already occurred at the site, and the density of development in the vicinity of the site. To the extent possible, reuse assessments and the development of future land-use assumptions should rely on readily available information. For example, future industrial land use is likely to be a reasonable assumption where a site is currently used for industrial purposes or is located in an area where the surroundings are zoned for industrial uses, and the comprehensive redevelopment plan predicts that the site will continue to be used for industrial purposes. As another example, sites where the owner desires to maintain the current use may only require a limited assessment; and sites where future land-use decisions have already been determined and documented may simply require a review to confirm the information. However, large sites, or sites with potentially different future use scenarios, may benefit from multiple reuse assessments or an iterative approach to developing future land-use assumptions.

Essentially, in cases where the future land use is relatively certain, the RAOs generally should reflect this land use. In this case, alternative future land-use scenarios generally are not required unless it is impracticable to provide a protective remedy that allows for the desired use. On the other hand, in cases where the reasonably anticipated land use is uncertain, or where multiple uses are being considered, a range of potential future land-use options should generally be considered when developing RAOs . Thus, if, for example, information gathered for the reuse assessment suggests the site could be used either for recreational purposes or for commercial/light industrial activity, then in the process of identifying multiple potential reuse scenarios, the reuse assessment should consider input from potential stakeholders for the most likely scenarios. In other cases, alternative future land-use scenarios can be reflected in the reuse assessment by developing a range of remedial alternatives for detailed evaluation—with focus on those that could achieve different land-use potentials.

Although each reuse assessment will be different, a thorough evaluation should probably be performed for all reasonable ones—also recognizing that the scope and level of effort needed to complete a reuse assessment will be dependent on conditions at the site, and should therefore be tailored accordingly. Information supporting a reuse assessment should be obtained from existing and readily available sources to the extent possible. Ultimately, the reuse assessment should provide sufficient information to develop realistic assumptions of the reasonably anticipated future use(s) for a site. When the reuse assessment and the selected remedy results in categories of allowable future land use (e.g., commercial, industrial, and recreational), rather than an unrestricted use, appropriate institutional controls (that acknowledges the restricted land uses) should be identified in the decision document associated with the case-specific project. In fact, institutional controls should be used, whenever necessary and appropriate, in order to prevent exposure to contamination remaining on-site and/or to protect components of an ongoing remedy.

13.2.2.3 Land-Use Considerations in Remedy Selection

From the point of view of site-specific remedy selection, the remedy determines the cleanup levels, the volume of contaminated material to be treated, and the volume of contaminated material to be contained. Consequently, the remedy selection decision determines the size of the area that can be returned to productive use and the particular types of uses that will be possible following remediation. Indeed, the volume and concentration of contaminants left on-site, and thus the degree of residual risk at a site, will affect future land use. For example, a remedial alternative may include leaving in-place contaminants in soil at concentrations that would be protective for industrial exposures only, but not protective for residential exposures; in this case, institutional controls should be used to ensure that industrial use of the land is maintained, and thus to prevent its possible use for residential purposes, which could then result in elevated risks with respect to residential exposures. Conversely, a remedial alternative may result in no significant levels of contamination being left in place—and thus allow for unrestricted use (e.g., residential, industrial, and variety of other uses).

In general, if any remedial alternative developed during a FS requires a restricted land use in order to be protective, then it is essential that the alternative would include components that will ensure that it remains protective as such. In particular, institutional controls will generally be included with the selected alternative, in order to prevent an unanticipated change in land use, which could then result in unacceptable exposures to residual contamination—or at a minimum, this will alert future users to the residual risks and to monitor for any changes in land use. In such cases, institutional controls will play a key role in ensuring long-term protectiveness and should therefore be evaluated and implemented with the same degree of care as given to other elements of the remedy.

13.2.3 Developing RAOs

RAOs provide the foundation upon which site cleanup alternatives are developed. In general, RAOs should be such as to facilitate the development of remedial alternatives that would achieve cleanup levels associated with the reasonably anticipated future land use over as much of the contaminated site as possible. Still, it should be recognized that achieving either the reasonably anticipated land use or the land use preferred by the stakeholders may not be practicable across the entire site, or in some cases, at all. For example, as RI/FS data become available, this may indicate that the remedial alternatives under consideration for achieving a level of cleanup consistent with the reasonably anticipated future land use are neither cost-effective nor practicable. If this is the case, then the RAO may be revised in a manner that results in different, more reasonable, land use(s).

In cases where the future land use is relatively certain, the RAO generally should reflect this land use. In general, it need not include alternative land-use scenarios unless, as discussed above, it is impracticable to provide a protective remedy that allows for that use. A landfill site offers a classic example where it is highly likely that the future land use will remain unchanged (i.e., long-term waste management area); in such a case, a RAO could be established with a very high degree of certainty to reflect the reasonably anticipated future land use.

In cases where the reasonably anticipated future land use is highly uncertain, a range of the reasonably likely future land uses should be considered when developing RAOs. These likely future land uses can be displayed via developing a range of remedial alternatives that will achieve different land-use potentials. The remedy selection process will determine which alternative is most appropriate for the site and, consequently, the land use(s) available following remediation.

13.2.4 CORRECTIVE MEASURE EVALUATION

Corrective measure evaluation typically is comprised of a FS of the available remedial options, the purpose of which is to examine site characteristics, cleanup goals, and the performance of alternative remedial technologies so that the most effective approach for the restoration of a contaminated site can be identified. To accomplish the tasks involved in this type of evaluation, several corrective measure evaluation tools will usually be used to assist in determining whether or not remediation is necessary for a contaminated site problem, and to further determine the cleanup goals and techniques appropriate for a given site.

Corrective measure studies are generally designed to identify and evaluate remedial alternatives that are potentially suitable for addressing contaminated site problems. The ultimate goal of corrective action assessments is to select and implement the most cost-effective mitigation strategy that provides adequate protection to public health, welfare, and the environment. Oftentimes, a variety of decision tools may be used to assist the decision-maker with the choice of the optimum corrective measure; Appendix B discusses some example application tools appropriate for such purposes. Indeed, several other logistical tools can be used to support corrective action assessment programs, in order to arrive at informed decisions on potentially contaminated site problems; the use of mathematical models, decision analysis techniques, and statistical methods can be of particular interest in corrective measure evaluations.

In general, the characterization of contaminant migration and attenuation rates, as well as the successful evaluation, selection, and implementation of appropriate remedial option, requires completion of several tasks—typically to include the following key steps:

- Review available site data, and develop a preliminary conceptual model.
- Screen the site, and assess the potential for natural attenuation.
- Collect additional site characterization data, as necessary, and refine the conceptual model.
- Simulate remedial alternatives using analytical or numerical solute fate-and-transport models that allow incorporation of attenuation terms, as necessary.
- Identify current and future receptors, and conduct an exposure-pathway analysis.
- Determine whether source treatment will be remediation, removal, containment, or a combination of these.
- Implement best risk management and/or remedial action alternative.

It is noteworthy that a well-designed FS addresses every contaminant migration pathway and environmental medium that poses, or could pose, unacceptable risks to human health or the environment.

13.2.5 Consideration of Life-Cycle Design Concepts and Design Optimization

Life-cycle design (LCD) consists of a design that takes account of the complete life cycle of a given project—that is, in this case, with considerations given to changes in contaminant concentrations that occur over time, as well as technology limitations at different stages of system operation. In general, most remediation systems are not necessarily capable of quickly achieving the required cleanup goals determined for a target contaminant in a given environmental medium—at least not consistently; indeed, most systems may require a combination of technologies and/or operational procedures to be considered during the "life cycle" of the overall remedial action. The LCD concept allows for the selection and operation of remediation systems that account for anticipated changes in contaminant concentrations over time vis-à-vis the performance of system components for the different periods or segments.

Meanwhile, it is noteworthy here that during remediation, contaminant levels typically decrease until they reach an asymptotic level. Once asymptotic conditions are reached for several successive sampling periods, continuing remediation activities (using same remedial techniques) generally results only in little (perhaps miniscule) further decrease of contaminant concentrations. However, and as happens oftentimes, when active remediation is ceased, levels of contaminants recorded may show abrupt increase—a process described by the "rebound" effects. Consequently, when asymptotic behavior begins, then there may be a need to transition into alternative remedial strategies—especially if rebound effects are anticipated for the end of the initial or start-off remedial system.

13.3 THE CORRECTIVE ACTION ASSESSMENT AND RESPONSE PROCESS

Corrective action assessment and response programs are generally designed to facilitate the development of feasible and scientific-based methods necessary to support contaminated site restoration and management decisions. The typical tasks performed in this type of investigation are comprised of a site assessment, a risk analysis, an evaluation of remedial alternatives (in terms cost, probable effectiveness, etc.), the ranking of feasible remedial options, a recommendation of the overall best remedial technology, and the design and implementation of requisite remedial options, together with monitoring programs. Overall, the process allows a multimedia approach to site characterization; identifies the specific parameters for which data must be collected; identifies the preferred data gathering, handling, and analytical techniques to be used; and allows project managers to develop site-specific and cost-effective cleanup solutions.

Several formalized steps will normally be involved in the design of corrective action programs for contaminated site problems—with the major activities carried out typically comprised of the so-called RI and a FS. The RI is conducted to gather sufficient data in order to characterize conditions at a contaminated site. This may involve extensive fieldwork and site sampling used to identify site contaminants, determine their concentrations and distribution across the site, characterize migration pathways and routes of exposure to surrounding populations, and assess other site conditions that might affect cleanup options. Information derived from the RI will generally help determine the need for and the extent of cleanup necessary for a contaminated site. The early determination of remediation goals facilitates the development of a range of feasible site restoration decisions, which in turn helps focus remedy selection on the most effective remedial alternative(s). The FS is undertaken to identify and analyze potential site restoration alternatives. Typically, the FS uses a screening process to reduce the number of alternatives to a limited range of remedial options. The short-listed options are subjected to a detailed analysis in which various trade-offs are evaluated (such as in relation to cost-effectiveness, extent of cleanup, and permanence of a cleanup action). After review of the detailed analysis and upon selection of one of a number of cleanup options identified in the FS, the project then enters an engineering design phase in which plans and specification are developed for the selected remedy. Once the engineering design is completed, construction activities to deal with the actual cleanup of the site can be implemented. In some cases, such as is archetypical of groundwater remediation, cleanup activities may continue for several years.

In practice, the following general tasks would likely be carried out as part of the overall RI/FS program designed to address contaminated site problems (Cairney, 1993; OBG, 1988; USEPA, 1988e):

- Characterize site as completely as possible.
- Define and develop remediation objectives that are appropriate for the specific contamination problems.
- Identify remedial technologies that are capable of achieving the site restoration goal(s).
- Develop and screen remedial alternatives, and select only those that are superior based on engineering, environmental, and economic criteria, which concurrently meet regulatory and community expectations.
- Perform detailed analyses of remedial alternatives by scrutinizing each of the initially selected alternatives.
- Choose, offering justification, the preferred remedial alternative(s).

In general, once existing site information has been analyzed and a conceptual understanding of a site is obtained, potential RAOs should be defined for all impacted media at the contaminated site; a preliminary range of remedial action alternatives and associated technologies should then be identified. The identification of potential technologies at this stage will help ensure that data needed to evaluate them can be collected as early as possible in the corrective action process. Figure 13.3 summarizes the major tasks involved in a corrective action assessment and response program for

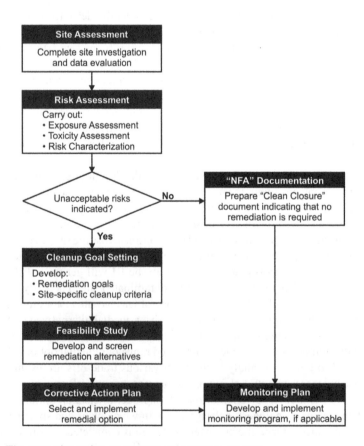

FIGURE 13.3 The corrective action assessment and response process.

potentially contaminated site problems. In the final analysis, the information gathered during RI activities is used to map out the extent of contamination and to evaluate the potential risks associated with the subject site so that an appropriate corrective action decision can be made.

13.3.1 SITE ASSESSMENT CONSIDERATIONS

When there is a source release, contaminants may be transported to different potential receptors via several environmental media (such as air, soils, groundwater, and surface water). In fact, a complexity of processes may affect contaminant migration at contaminated lands—resulting in human and ecological receptors outside the source area potentially being threatened. Consequently, it is imperative to adequately characterize a site and its surroundings through a well-designed site investigation program (in which all contaminant sources and impacted media are thoroughly investigated), in order to arrive at appropriate and cost-effective corrective action decisions. In the efforts involved on this path, a multimedia approach to site characterization is generally adopted, so that the significance of possible air, water, soil, and biota contamination can be established through appropriate field sampling and analytical procedures that will yield high-quality environmental data needed to support the corrective action decision. To accomplish this, samples are gathered and analyzed for the chemicals of potential concern in the appropriate media of interest. Effective analytical protocols in the sampling and laboratory procedures are used to minimize uncertainties associated with the data collection and evaluation activities. A quality assurance program (consisting of a system of documented checks which validate the reliability of acquired data sets) is an important part of this whole process. Indeed, the development and implementation of a firm quality assurance/quality control program during a sampling and analysis program is critical to obtaining reliable analytical results that becomes the basis for the corrective action decision process.

Site assessments are invariably a primary activity in the overall process involved in the management and restoration of contaminated site problems. The objective of the site assessment is to determine the nature and extent of potential impacts from the release or threat of release of hazardous substances. In all cases of contaminated site assessments, an initial qualitative evaluation of site conditions will help identify potential contaminant release source(s), determine the environmental media affected by each release, and broadly define the possible extent of the release(s); the questionnaire chart included in Box 13.7 (presented later on in Section 13.7) can be used to facilitate this process. Among other things, it is notable that for a confirmed release, the release source(s) must be stopped and hazards mitigated, and where applicable, any free product present should be removed immediately to prevent the development of health and safety hazards; removal will also help prevent further migration of free product into soils and/or groundwater. Subsequently, a site investigation effort, which is a major component of the site assessment activity, aims at collecting representative samples from the potentially contaminated site. Depending on the adequacy of historical data and the sufficiency of details about the likely contaminants at a site, sampling programs can be designed to search for specific chemical constituents that become indicator parameters for the sample analyses. Where specific information about historical uses of the site is lacking, a more comprehensive sampling and analytical program will generally be required; in this case, the sampling and analysis program may be carried out in phases—moving from a more general scope to one of specificity as adequate information becomes available about the site. The results from these activities will facilitate a complete analysis of possible contaminants present at the site. The information obtained is used to determine current and potential future risks to human health and the environment. Ultimately, corrective actions are developed and implemented with the principal objective to protect public health and the environment.

13.3.2 RISK ASSESSMENT CONSIDERATIONS

Invariably, the health and environmental risk estimates become an important element in the corrective action decision process. This is because, the risk assessment generally provides the decision-maker with technically defensible and scientifically valid procedures for determining whether or not a potentially contaminated site represents significant adverse human health or ecological risks that warrants consideration as a candidate for mitigation. Thus, risk assessment becomes an integral part of the RI process. This is used to facilitate decisions on whether or not remedial actions are needed to abate site-related risks and also in the enforcement of regulatory decisions. Meanwhile, it should be acknowledged here that remedial actions, by their nature, can alter or destroy aquatic and terrestrial habitats. The potential for the destruction or alteration of ecological habitats, and the consequences of ecosystem disturbances and other ecological effects, must therefore be given adequate consideration during the corrective action response process. Thus, remedial alternatives, in addition to being evaluated for the degree to which they protect human health, are also evaluated for their ability to protect ecological receptors and ecosystems. Indeed, it is very important to integrate ecological investigation results and general concerns into the overall site cleanup process.

It is worth mentioning here that as part of the overall corrective action response process, remedial alternatives are usually analyzed for both potentially long- and short-term health and environmental effects associated with the implementation of each remediation option. Indeed, a remedial action will often increase potential short-term exposures and risks at a contaminated site above baseline conditions. Nonetheless, escalated short-term risks should not become the principal determinant (or deterrent) of corrective action decisions; rather, this should be used to determine appropriate management practices or institutional controls during implementation of the selected remedial action. Consider, for instance, a remedial option at a site that involves excavation and removal of contaminated soil. In the absence of precautionary or mitigation measures, fugitive dust generation by heavy equipment and remedial activities may create unreasonably high short-term health hazards. On the other hand, these and other temporary sources of contaminant releases associated with construction and implementation of a remedy may not be adequate enough grounds for rejecting the particular remedial alternative. However, management practices, such as the temporary relocation of populations potentially at risk, should be considered in order to mitigate the health risks associated with such temporary source releases.

Finally, recognizing the fact that most contaminated sites cannot be cleaned to pristine conditions, a set of cleanup criteria are typically used to set the goals for the site restoration program to achieve. Of special interest, "preliminary remediation goals" (PRGs) are usually established as cleanup objectives early in the site characterization process. The early determination of remediation goals facilitates the development of a range of feasible corrective action decisions, which in turn helps focus remedy selection on the most effective remedial alternative(s). The development of PRGs requires site-specific data relating to the impacted media of interest, the chemicals of potential concern, and the probable future land uses. (Methods for developing the appropriate PRGs are discussed in greater detail in Chapter 11.) It is noteworthy that an initial list of PRGs may have to be revised when new data become available during the site characterization activities. In fact, PRGs are progressively refined into "final remediation goals" (FRGs) throughout the process leading up to the final remedy selection. Therefore, it is important to iteratively review and reevaluate the media and chemicals of potential concern, future land uses, and exposure assumptions originally identified at scooping—so that the PRGs can be appropriately modified to produce FRGs.

13.3.3 RAOs and Remedy Selection

RAOs provide the foundation upon which site cleanup alternatives are developed. As such, the RAOs should be as specific as possible—in order to help focus the remedy decision process; it should at least specify such relevant items as enumerated in Box 13.3. Examples of case-specific RAOs may include the following: protection of human and/or ecological receptors from direct contacting of contaminated soils, prevention of contaminants from migrating into groundwater, preventing contaminated groundwater migration to drinking water wells, reduction or elimination of contaminants in wetlands and creek sediments, and limiting the releases of contaminated materials

---

BOX 13.3   MAJOR CONSIDERATIONS FOR ESTABLISHING RAOs

- The contaminants of concern (CoCs)
- Potential receptors and populations-at-risk (PARs) (to include exposure routes)
- Cleanup goals (i.e., the target COC concentrations that serve as the acceptable remediation goals)
- Future land usage (e.g., industrial vs. commercial vs. residentialvs. "green space" or recreational usage)
- Timeframe before change in land use can be anticipated (i.e., life of any deed restrictions to cater for future land usage)
- Prevailing site accessibility (i.e., limited vs. unlimited access) and possible need for institutional controls

---

via runoff. In practice, development of RAOs that will properly support the selection of an appropriate corrective action response plan also depends on a careful assessment of both short- and long-term risks posed by the case-specific site, and as a general rule, corrective action response plans should provide for the removal and/or treatment of contaminants until a level necessary to protect the human health and the environment is achieved.

All in all, it is important to develop technically sound and cost-effective remedial action decisions for contaminated sites, in order to be able to achieve the desirable site closure goal. This is accomplished through the selection of appropriate remedy, the design of an optimal remedial action operational strategy, and a possible long-term site management program (that may include a long-term monitoring program that occurs after cleanup goals have been achieved to ensure continuing protectiveness of the human health and the environment).

13.4   CONTAMINATED SITE MANAGEMENT AND RESTORATION STRATEGIES

A variety of corrective action strategies may be used in the quest to restore contaminated sites into healthier conditions. The processes involved will generally incorporate a consideration of the complex interactions existing between the hydrogeological environment, regulatory policies, and the technical feasibility of remedial technologies. In any case, once RAOs have been defined, general response actions (such as treatment, containment, excavation, and pumping), or indeed other actions that may be taken to satisfy those objectives, can then be identified. In general, risks associated with contaminated site problems can usually be reduced by containment or removal (for treatment and/or disposal) of the CoCs. The chemical nature of the CoCs is a primary consideration in the identification of remedial alternatives that are capable of meeting the case-specific remediation goal(s). For instance, some technologies are designed primarily to control the release of heavy metals, whereas others are designed primarily to control the release of organic chemicals; thus, when both types of contaminants are present at a site, the possibility of combining two or more technologies to meet the remediation objectives should be carefully evaluated. Potential short-term risks that may arise as a result of remediation activities that are part of certain types of remedial actions should also be carefully evaluated and addressed as part of an implementation package. For example, vacuum extraction and soil excavation (used to remove contaminants from soils) and air stripping, in combination with granular activated carbon (used to remove contaminants from water) represent typical technologies that may result in the transfer of contaminants into air, and such releases would have to be treated before being released into the atmosphere; this requirement can indeed exert significant influence on the choice of cleanup technologies.

Overall, the site restoration goal and strategy selected for a contaminated site problem may vary significantly from one site to another due to the potential effects of several site-specific parameters.

A number of extraneous factors may also affect the selection of site restoration strategies. Ultimately, however, remedial action alternatives selected for contaminated site problems should be designed and operated in such a manner as to be fully protective of the human health and the environment.

13.4.1 GENERAL TYPES OF CORRECTIVE ACTION RESPONSES

The broad variety of remedial action alternatives associated with contaminated site problems may be categorized into the following general groups:

- No action or passive (which implicitly depends on natural processes such as volatilization, biodegradation, leaching, photolysis, and redox to destroy or attenuate the CoCs)
- Containment (i.e., in-place isolation/containment)
- Removal/disposal (i.e., contaminant removal and disposal elsewhere)
- Treatment/restoration (i.e., contaminant treatment on- or off-site for site restoration).

The first three groups will usually be limited to problems posing low-to-no risks, or when the fourth option—a generally preferred alternative—is not feasible. Long-term monitoring will invariably be required to supplement the use of these methods, with the extent of monitoring dependent on the category of remedial option. In fact, the principal remediation options that seem to have found widespread applications in the management of contaminated site problems have usually involved the removal of material from the case site for disposal elsewhere; retention and isolation of material on-site using an appropriate form of cover, barrier, or encapsulation system; physical, chemical, or biological treatment to eliminate or immobilize the contaminants; and/or lowering of the contaminant concentrations by diluting the contaminated material with clean material (Cairney, 1993).

Broadly speaking, the selection of a particular type of remedial option depends on the type of contaminants involved, cost-effectiveness, practicability, site conditions and accessibility, and applicable regulations that must be met in the site restoration process. For instance, specific characteristics of diesel fuel (such as its low volatility, relatively low health risk, and moderate viscosity) dictate that certain particular remediation options are better suited for diesel-contaminated soils, whereas the same remedial techniques may not be as effective in addressing other petroleum hydrocarbons. This is because the moderate viscosity of diesel makes remedial options such as containment or fixation reasonable candidate techniques to adopt for diesel-contaminated soils; on the other hand, the low diesel volatility makes vapor extraction systems (which may work reasonably well for the remediation of gasoline-contaminated soils, in most cases) ineffective for diesel fuels.

13.4.1.1 Containment versus Cleanup Strategies

Contaminated site restoration programs generally involve the use of containment or cleanup strategies. *Containment strategies* have the goal of preventing further migration of mobile contaminants, by controlling contaminant plume movements within a specified area and timeframe; a primary benefit of a containment strategy is that for some contaminants, given enough containment time, natural processes of attenuation will help reduce concentrations to acceptable levels. *Cleanup strategies* have the goal of removing contaminants or contaminated media in a specified area until acceptable concentration levels are attained.

13.4.1.2 Removal versus Remedial Action Programs

The cleanup of inactive hazardous waste sites may involve removal and/or remedial actions; a good understanding of the differences is important to a contaminated site project analyst. As an example, U.S. environmental regulations under the Superfund program (Section 104 of CERCLA) authorizes that a removal or remedial action be taken to protect public health, welfare, or the environment when there is a release or substantial threat of release of any hazardous substance, pollutant, or contaminant that may present an imminent and substantial danger to the public health or welfare.

The *removal action program* provides a mechanism to clean up acute, short-term, immediate risks to public health and/or the environment. The process usually involves immediate response to stabilize and contain a hazard situation; it is normally not a total site cleanup program. Such action is meant to reduce the threat sufficiently so that further studies can be conducted prior to any large-scale cleanup, without posing unacceptable short-term risks (Kunreuther and Gowda, 1990). The major factors considered in this case relate to the immediacy of the threat to human health and public safety. Consequently, the general criterion for completing a removal action is the determination of the level of contamination that will not pose short-term or acute risks to human health and/or the environment. The *remedial action program* provides the framework for an organized response to inactive hazardous waste sites, generally consisting of the discovery and appraisal, prioritization, investigation, and cleanup of the contaminated site (Kunreuther and Gowda, 1990). Once preliminary assessment and site investigation programs are completed, a quasi-risk model (such as the hazard ranking system, developed by the US EPA to determine whether or not a site qualifies for placement on the so-called 'National Priorities List', NPL) can be used for the prioritization; work may then proceed to fully characterize the site and to determine the appropriate cleanup strategies.

13.4.2 INTERIM CORRECTIVE ACTIONS

Interim corrective actions are measures used to address situations which pose imminent threat to the human health or the environment, or to prevent further environmental degradation or contaminant migration pending final decisions on the necessary long-term remedial or risk management activities. Thus, as an example, whenever excessive risks exist at a contaminated site, a decision should be made to implement interim corrective actions immediately, in order to protect public health and the environment. Common examples of interim corrective measures may include simply erecting a fence around a contaminated site in order to restrict/limit access to an impacted property; covering exposed contaminated soils with synthetic liner materials; applying dust suppressants to minimize emissions of contaminated fugitive dust; restricting use of contaminated water resources and/or providing alternative domestic/municipal water supplies; temporal displacement or relocation of nearby residents away from a hazardous waste site; and the implementation of such more elaborate control measures as installing a pump-and-treat system to prevent further migration of a groundwater contaminant plume. As an illustration of its application scenario, it is notable that the U.S. Superfund program uses the removal authority (provided under Section 104 of CERCLA) to accomplish this same objective where expedited response and/or emergency actions are needed.

Interim corrective actions will usually not represent the ultimate containment or treatment strategy implemented as solution for a contaminated site problem. These measures are developed primarily to minimize public exposure (to potentially acute risks/hazards) prior to developing a comprehensive corrective action program. Indeed, in many situations, it is possible to identify very early in the risk management and corrective action response process, measures that can, and should be taken to control potential receptor exposures to contamination, or to stop further environmental degradation from occurring. Typically, where it is obvious that the final remedy will require excavation and treatment or removal of contaminated "hot spots," such actions are best initiated as interim measures—rather than being deferred to a final remedy selection stage. In general, the interim measures may be relatively straightforward (such as erecting a fence or removing a small number of drums), or it may involve more elaborate control measures (such as installing a pump-and-treat system to prevent further migration of a groundwater contaminant plume).

13.4.3 CONDITIONS FOR DOCUMENTING *"NO-FURTHER-RESPONSE-ACTION"* DECISIONS

A major reason for conducting environmental investigation and characterization or corrective action assessments for potentially contaminated site problems is to be able to make informed decisions regarding site restoration and risk management programs. The fundamental purpose of a site restoration and risk management program is to protect the human health and the environment from the unintended consequences of contaminated site problems. In some situations, a decision that "no further action" (NFA) is required for a site or problem may be deemed appropriate for the case-specific contaminated site problem situation. The process involved in a NFA decision is intended to indicate that based on the best available information, no further response action is necessary to ensure that the case site does not pose significant risks to the human health or the environment. Under such circumstances, a NFA closure document will usually be prepared for the site or group of sites, in accordance with applicable regulatory requirements. The NFA document typically is a stand-alone report, containing sufficient information to support the "no-further-response-action" decision (NFRAD); it should therefore include site-specific evidence along with well-supported technical reasoning and justification for the NFRAD. Additionally, the NFA document should clearly address specific CSM hypotheses that have been tested to confirm that there are no likely and complete exposure pathways or scenarios associated with the site. It is noteworthy that contaminant cleanup standards specified by several regulations (such as media-specific action levels, site-specific risk-based criteria, and local or area background threshold levels) often form an important basis for the NFRAD.

In general, a broad categorization scheme for a variety of area designations may be used to facilitate the planning, development, and implementation of appropriate NFRADs. For instance, sites may be designated as belonging to a number of different groups, depending on the level of effort required to implement the appropriate response actions. Typical designations, which are by no means exhaustive, are elaborated below.

- *"Areas with No Confirmed Contamination" (ANCCs).* These consist of areas (e.g., suspected sources) where the results of records search and site investigations show that no hazardous substances were stored for any substantial period of time, released into the environment or site structures, or disposed of on the site property; or areas where the occurrence of such storage, release, or disposal is not considered to have been probable, so that no further response action is necessary. The ANCC determination can be made at any of several investigative decision stages. Consequently, NFA decisions under this category can be made if, after a preliminary site assessment (or an equivalent effort), it can be concluded that a source or suspected release of contamination does not exist and the site can be described as part of an ANCC.

- *"Areas below Action Levels" (ABALs).* These consist of impacted areas where no response or remedial action is required to ensure the protection of human health and the environment. ABALs include areas where an environmental investigation (such as a "Phase II"-type site assessment) has demonstrated that hazardous materials have been released, stored, or disposed of, but are present in quantities that require no response or remedial action to protect the human health and the environment. An ABAL designation means that levels of hazardous substances detected in a given area *do not* exceed media-specific action levels (e.g., chemical-specific risk-based concentrations), *do not* result in significant carcinogenic or noncarcinogenic risks, nor otherwise exceed applicable federal or state and local requirements. Consequently, NFA decisions under this category can be made if, after a baseline risk assessment, it can be concluded that the site is part of an ABAL or otherwise poses no significant risk to human health and the environment.

- *"Areas where Remedies Have Been Implemented/Completed" (ARICs).* These consist of areas where the site records indicate that hazardous materials are known to have been released or disposed of, but where all remedial actions necessary to protect the human health and the environment with respect to any hazardous substances remaining on the site have been taken. Consequently, NFA decisions under this category can be made if it can be demonstrated from the available evidence that the selected remedy is complete, that remediation goals have been met, and that the remaining contamination at the site does not pose significant threat to the human health or the environment.

Ultimately, the NFA determination rests on whether or not complete exposure pathways exist at a site, and whether or not any of the defined complete pathways are significant. It is, therefore, logical to infer that the NFA decision criteria are indeed linked to the use of the CSM as a decision tool. With that said, a careful evaluation of the elements or components of a CSM will usually serve as an important basis for many NFRADs. Such analysis involves a demonstration that the source–pathway–receptor linkage cannot be completed at the site or that even if the linkage can be completed, the risk posed by the contamination present does not exceed "acceptable" reference standards established for the site or area.

13.5   EVALUATION OF ALTERNATIVE SITE RESTORATION OPTIONS

In the process of developing contaminated site remediation alternatives, information on the nature and extent of contamination, applicable environmental regulations, contaminant fate and transport properties, and the toxicity of contaminants is used to guide decisions made about the potentially feasible and appropriate remedial options. Subsequently, the remedial action alternatives and associated technologies are screened to identify those that will likely be effective for the contaminants and media of interest at the specific project site. The broad groups of remedial alternatives generally screened during corrective measure assessments for contaminated site problems include containment, removal, and treatment of contaminated materials. Site characterization data are used to identify the general approach, or combination of approaches, that are likely to be most effective in addressing each impacted environmental matrix at a contaminated site.

Classically, once a list of remedial alternatives is developed, these alternatives are analyzed in detail so that the most appropriate for the site-specific problem could be selected for likely implementation. The analyses usually involve an initial screening, followed by a detailed evaluation. The initial screening of alternatives is designed to eliminate alternatives that are clearly inappropriate to the given situation or that are clearly inferior to other alternatives. Alternatives that remain after the initial screening are subjected to further detailed evaluation; additional data gathering may be required to complete this level of analysis. Furthermore, laboratory- or pilot-scale studies may be required at this stage—especially with respect to treatment technologies. Based on the results of the detailed analysis (that considers the type of contaminants involved, cost-effectiveness, practicability, general site conditions and accessibility, and applicable regulations that must be met in the site restoration process), the appropriate remedial alternative(s) can then be selected.

Overall, the selection of an appropriate CAP for a contaminated site problem depends on a careful assessment of both short- and long-term risks posed by the site. Risk assessment, among other tools, can be used to aid this process of selecting among several remedial options for contaminated site problems. Risk assessment techniques can indeed be used to quantify the human health risks and environmental hazards created by implementing specific remedial options at contaminated sites. As such, these procedures can help determine whether a particular remedial alternative will pose unacceptable risks following implementation and determine the specific remedial alternatives that will create the least risk with respect to the cleanup goals or RAOs for the site. By using such an approach, the remedy selection process ensures that remedies satisfy the following pertinent conditions:

- Be protective of human health and the environment.
- Attain media cleanup standards specified by regulatory requirements.
- Control the source(s) of releases so as to reduce or eliminate, to the extent practicable, further releases that may pose a threat to the human health or the environment.
- Comply with all relevant corrective action program specifications and standards.

The prominent criteria, as well as other extraneous factors, affecting the selection and implementation of a number of remediation options for contaminated sites are discussed in the proceeding sections; further details can be found elsewhere in the literature (e.g., Cairney, 1993; Calabrese and Kostecki, 1991; Nyer, 1993; OBG, 1988; Sims et al., 1986; USEPA, 1984c, 1985e, 1988e).

13.5.1 THE DEVELOPMENT AND SCREENING OF REMEDIAL ACTION ALTERNATIVES

The development of remedial alternatives for contaminated site problems involves compiling a limited number of site restoration options for source-control and/or remedial action. In general, remedial alternatives can be developed to address a specific contaminated medium (e.g., groundwater), specific areas of a contaminated site (e.g., a waste lagoon or "hot spots"), or an entire contaminated site problem. Typically, the development of a remedial action alternative will consist of the set of activities indicated in Box 13.4—with the RAOs made up of medium-specific or site-specific goals for protecting the human health and the environment (Pratt, 1993; USEPA, 1988e). These are developed upon completion of a site characterization program—in order to specify the area designated for site restoration, the restoration timeframe, the cleanup levels, and the feasible remedial techniques. Correspondingly, general response actions that describe those actions required in satisfying the RAOs are an important aspect of the remedial alternative development process. The general response actions may include treatment, containment, excavation, extraction, disposal, institutional controls, or a combination of these. Similar to RAOs, these are medium- and indeed site-specific. Process options are determined using effectiveness, implementability, and cost criteria. At this stage, however, the evaluation will usually focus on effectiveness factors with less effort directed at the implementability and cost evaluation.

In most situations, several potentially feasible remedial options are developed early on in the site restoration evaluation process. Consequently, it becomes necessary to screen out some of the available options, in order to reduce the number of alternatives that will be analyzed in detail.

**BOX 13.4   REQUIREMENTS FOR DEVELOPING
REMEDIATION ALTERNATIVES**

- Establish RAOs and cleanup goals.
- Identify potential treatment technologies and containment or disposal requirements for the CoCs (that will satisfy RAOs).
- Determine process options and general response actions (that will satisfy RAOs and cleanup goals).
- Identify volumes or areas of impacted media (to which general response actions might be required).
- Identify regulatory limits and compare with removal efficiencies of remedial techniques
- Prescreen remedial technologies and process options based on their effectiveness, implementability, and cost.
- Assembly technologies and their associated containment or disposal requirements into alternatives for the contaminated media.
- Establish PRGs and/or cleanup levels.
- Determine the area of attainment (i.e., area over which cleanup levels will be achieved for the contaminated site, encompassing the area outside the site boundary and up to the boundary of contaminant plume).
- Estimate the restoration timeframe (i.e., the period of time required to achieve selected cleanup levels at all locations within the area of attainment).
- Formulate remedial alternatives (i.e., compile and group technologies into appropriate remedial alternatives).

The screening process, usually done on a general basis and with limited effort, involves evaluating alternatives with respect to their effectiveness, implementability, and cost. In fact, because the screening process addresses methodologies to site restoration rather than specific remedial technologies, the evaluation is more qualitative (rather than being quantitative). However, the screening analysis uses the quantitative site characterization data to recommend an approach to the site restoration program. Altogether, it is important to consider as many alternatives as possible during the screening of remedial action measures. This will ensure that the most cost-effective technique is not excluded from consideration. On the other hand, it is impractical or uneconomical to conduct extensive and detailed evaluation of every alternative during the planning and preliminary design stages. Thus, the first step is to determine the potentially feasible alternatives that can be evaluated further, based on technical and economic factors.

When all is said and done, the development and screening of the preferred remedial action alternatives should be comprised of identifying a range of remediation options that will ensure adequate protection of public health and the environment. Depending on the site-specific circumstances, such remedial options may result in the complete elimination or destruction of the CoCs that are present at a project site, the reduction of contaminant concentrations to "acceptable" risk-based levels, and/or the prevention of exposure to the CoCs via engineering or institutional controls.

13.5.2 THE DETAILED ANALYSIS OF ALTERNATIVE SITE RESTORATION OPTIONS

After an initial screening of remedial alternatives, a detailed evaluation process is used to identify the remedial technology most likely to be successful from among the remedial approaches previously compiled during the screening analysis. The number of technologies selected for a detailed evaluation is determined based on knowledge of the alternatives that have proven to be successful under conditions similar to those at the case site; untested technologies and those known to perform inadequately under similar site conditions are generally not evaluated.

The detailed analysis of remedial alternatives is conducted with the principal objective of providing decision-makers with sufficient information to compare alternatives in a technically justifiable and socioeconomically acceptable manner. It follows the development and screening of feasible alternatives and precedes the actual selection of a remedy. The evaluation of both short-term and long-term risks is an important part of the detailed analyses. The detailed evaluation of the applicable alternative technologies being considered for a contaminated site problem will typically incorporate information on the following key issues:

- Successful application of the technology under similar site conditions—supported by identification of project locations, dates, and managing entity
- Total project cost—supported by an estimate itemizing technology testing, capital equipment, operating and maintenance labor, equipment, environmental testing and monitoring, and closure costs
- Risk reduction—supported by numeric estimates of risk posed to site workers or other receptors during remediation, and the risk posed by any contaminants remaining after remediation
- Project duration—supported by an estimated schedule showing major milestones, including any permitting activities that may be required
- Manageability of data gaps—supported by the identification of any environmental testing or treatability studies necessary to determine the effectiveness of a remedial technology under site conditions

In the detailed evaluation process, one or more of the screened remedial measures undergo detailed analyses that may also involve field/prototype investigations to identify the most cost-effective alternative. In fact, an important component of the detailed evaluation of remedial alternatives involves the assessment of design parameters for remedial technologies. It is noteworthy that the design of many remedial technologies requires data that may not generally be collected during routine site characterization or RI. On the other hand, it is important to consider data needs for design during scoping so as to minimize the amount of time needed to select and implement the remedy; the important data typically needed for the evaluation and design of various remedial technologies are enumerated elsewhere in the literature (e.g., USEPA, 1988e).

On the whole, the detailed analysis of remedial alternatives is completed with consideration given to several evaluation criteria (such as required by the National [Oil and Hazardous Substances Pollution] Contingency Plan (NCP) under the U.S. federal environmental programs); Box 13.5 contains the particularly important remedy selection decision factors that should be considered during the remedial action evaluation (USEPA, 1988e). Ultimately, the selected remedy will be the alternative found to provide the best balance of trade-offs among alternatives in terms of these evaluation criteria. This will generally satisfy several important requirements (such as the U.S. NCP requirements of providing the lowest cost alternative that is technologically feasible and reliable, which effectively mitigates and minimizes environmental damages, and provides adequate protection of public health, welfare, or environment (40 CFR 300.68(j))). Indeed, any remedies not meeting these criteria are likely to be eliminated from further consideration as a preferred alternative.

The outcome of the detailed evaluations will comprise a recommended technology or combination of technologies to restore each impacted medium posing "unacceptable" risks. If such a determination cannot be made with the available information, data gaps are identified and a program capable of providing the missing information is implemented. Generally speaking, the remedial alternative selected following the detailed evaluation should attain or exceed pertinent regulatory standards that pertain to the site, and should also realize sustained effectiveness. An alternative that does not meet the applicable standards may, however, be selected in any of the following types of situations (OBG, 1988):

- The selected alternative is not a final remedy, but is part of a more comprehensive remedy package.
- No alternative that attains or exceeds the standards is feasible.
- All alternatives that attain or exceed the requirements will result in other significantly adverse environmental effects.

That is, under certain circumstances, a "less-than-acceptable" remedy could become the technique of choice. Indeed, a number of other extraneous but important site-specific features may also affect the selection of the ultimate corrective measure; these include several site characteristics pertaining to surface features, subsurface conditions, populations-potentially-at-risk, climate, adjacent land uses, cultural and social situations, and regulatory climate.

BOX 13.5   REMEDY SELECTION DECISION CRITERIA

- Overall protection of human health and the environment—including consideration of future land uses
- Compliance with applicable laws and regulations
- Short- and long-term effectiveness, and permanence of corrective actions
- Potential short- and long-term liability issues
- Amount of contaminated media (such as soils and groundwater)
- Ease of implementation (i.e., the technical and administrative feasibility of a remedy)
- Cost-effectiveness and cost efficiency of plans
- Reduction of toxicity, mobility, or volume of contaminated materials
- Regulatory and community acceptance of the program

13.5.2.1 Treatability and Bench-Scale Investigations in a Focused FS

If remedial actions involving treatment are identified for a contaminated site problem, and if existing site and/or treatment data are insufficient to adequately evaluate such alternative, then the need for treatability or bench-scale studies should be established as early as possible in the corrective action assessment process. Treatability and bench-scale tests may be necessary to evaluate the effects of a particular technology on specific site contaminants. Such tests will typically involve pilot testing to gather information that will help assess the feasibility of selected technologies. In some situations, a pilot-scale study may be necessary to furnish performance data and to develop good cost estimates, so that a detailed analysis can be effectively carried out to aid the selection of a remedial option.

As an example of a typical application, the treatability study may consist of placing a small amount of a representative soil sample into a bowl or container, thereby creating a microcosm of the site and then dosing the microcosm with the selected site remedy. For instance, a measured amount of a contaminated soil system can be placed into a container and inoculated with the selected remedy. After a specified time period, a sample is taken from the microcosm and analyzed for the parameters of concern. This is continued until the soil has reached the desired cleanup criteria, or until the soil no longer exhibits a change in condition—indicating the end point of the lower limit of cleanup attainable (i.e., the asymptotic threshold level).

On the whole, treatability studies are conducted to provide sufficient data that will allow treatment alternatives to be fully developed and evaluated during a detailed analysis, and to support the remedial design of a selected alternative (USEPA, 1988e). Ultimately, it will help reduce cost and performance uncertainties for treatment alternatives to acceptable levels and, most importantly, result in an appropriate remedy being implemented in an effectual manner.

13.5.2.2 Comparative Risk Assessment of Alternative Remedial Measures

Risk assessment plays a very important role in the development of RAOs for contaminated sites, in the identification of feasible remedies that meet the remediation objectives, and in the selection of a protective and balanced remedial alternative. The risk evaluation of remedial alternatives involves the same general steps as a baseline risk assessment (see Part III)—albeit the baseline risk assessment typically is more refined and requires a greater degree of effort than the risk comparison of remedial alternatives. In any case, much of the data collected during the baseline risk assessment can also be used to calculate the long-term residual risk associated with remedial alternatives.

In analyzing exposures associated with remedial actions, it is noteworthy that treatment processes and technologies used as part of a remediation strategy may facilitate the transfer of contaminants into nonimpacted environmental matrices. On the other hand, well-engineered remedial alternatives are not expected in themselves to cause a net increase in exposures and impacts due to contaminant releases into the environment. Still, the risks associated with a remedial action should be evaluated so as to include the risks created by implementing a remedial alternative as well as the post-remediation risks associated with residual contamination that remains at the site. The difference between the site risks in the absence of remedial action (i.e., the baseline risk) and the risks associated with a remedial alternative will generally help define the net benefits associated with a given remedial option. In the end, it is crucial to ensure that the projected risks posed by a remedial option do not offset any benefits associated with reducing site contamination to achieve an established remediation or risk reduction goal.

13.5.3 Choosing between Remediation Options: Selection and Implementation of Remedial Options

The selection of a specific remedial alternative for a contaminated site problem depends, to a great extent, on the required cleanup criteria established for the site. Once the cleanup criteria have been determined, a variety of remedy techniques can be evaluated for containing and treating impacted media (such as soils and groundwater) associated with a contaminated site problem. The remediation approach may comprise the use of physical containment techniques and/or physical, chemical, and biological treatment processes. In a representative situation, the remedial action will include the use of techniques to contain the contamination plume, and to recover and treat the impacted matrices. Meanwhile, it is noteworthy that oftentimes, the best solution for remediation of potentially contaminated sites is not necessarily one specific technology—but rather a combination of several technologies or remedial options capable of addressing the site-specific concerns of the case site. For instance, soil vapor extraction can be successfully coupled with bioremediation and air sparging technologies in comprehensive site cleanup programs. Furthermore, treatment processes can be, and usually are, combined into process trains for more effective removal of contaminants.

Insofar as practicable, corrective action remedies must generally ensure, with a high level of confidence, that environmental damage and health impacts from the sources of contamination will not occur in the future. To this end, successful and long-term site restoration requires adequate mitigation strategies to remove the contaminant source(s). Source-control technologies that involve treatment of contaminated materials, or that otherwise do not rely on containment structures or systems to safeguard against future releases, should therefore be strongly preferred to those that offer temporary, or less reliable, controls.

13.5.3.1 Factors Affecting the Selection and Design of Remediation Techniques

Once media cleanup standards have been established, the potential remedies can be evaluated, and then, a selection made for the best demonstrated available technology (BDAT). Of utmost interest in this process is the nature of contamination; important contaminant properties that may affect the choice and design of remediation techniques include the following:

- Retardation (which is a fundamental factor in contaminant treatment design)
- Adsorption properties of compounds on solids
- Volatility of compounds
- Henry's law constant (usually for the design of air strippers)
- Solubility
- Octanol/water partition coefficient
- Specific gravity
- Degradation rate or half-life.

By using these types of information, several treatment methods can be selectively eliminated before significant resources are spent on a focused FS. For instance, by using the Henry's law constant data, a designer can estimate the removal efficiency that an air stripper will have on a particular compound; consequently, the designer may ascertain whether an air stripper should be studied further or eliminated as a possible candidate option. Carbon adsorption and biological treatment can be evaluated in a similar manner, using the appropriate chemical data. Indeed, several extraneous factors may also affect the choice of remedial options—all the while recognizing that in general, there is no single remedial technique that is best for all types of contamination and for all site conditions; thus, successful remediation efforts may have to rely on proper marriages between remediation technologies. A more comprehensive review of information necessary for these types of evaluation is provided elsewhere in the literature (e.g., Nyer, 1993; OBG, 1988; Sims et al., 1986; USEPA, 1985e).

13.5.3.2 Remediation of Contaminated Soils: In-Place
Remediation as a Method of Choice

The utility of any soil remediation option invariably depends on the nature of contamination and the level of risks posed by the contaminants. In any case, an in-place remedial technology or process will generally be given preference as a method of choice in the determination of soil remediation options for a contaminated site problem. In-place remedial techniques may include extraction (e.g., soil washing), immobilization (e.g., sorption, ion exchange, precipitation), chemical degradation (e.g., degradation involving oxidation, reduction, and polymerization reactions), biodegradation (using natural microorganisms or genetically engineered microbes), photolysis (that may include enhanced photodegradation achieved by the addition of proton donors in the form of polar solvents), attenuation (e.g., mixing of contaminated surface soil with clean soil), and reduction of volatilization effects (Sims et al., 1986). Typically, the in-place technique is used to contain the source of contamination or to remove contamination through treatment processes.

A primary consideration in the identification of appropriate site-specific in-place technologies for contaminated soils relates to the chemical nature of the CoCs. In fact, the design and implementation of an in-place treatment process requires information on characteristics of the contaminant/soil systems as a whole, with particular attention given to the following important variables (Charbeneau et al., 1992; Sims et al., 1986; USEPA, 1984c):

- *Depth to Contamination.* If contamination is limited to the upper 15–20 cm (≈6–8 in.) of the soil and is well above the water table, in-place treatment techniques may be much more easily applied than if the contamination extends well below the ground surface and into a seasonally high water table.
- *Contaminant Concentrations and Quantities.* The efficiency and effectiveness of an in-place process depends on both contaminant concentration levels and quantity of each contaminant present in a given area.

Contaminant and soil characteristics are used in the prescreening of in-place alternatives for their potential applicability to meet site restoration goals. Based on the contaminant, soil, and system characteristics, an analysis can be made of the pathways and rates of contaminant migration, as well as the potential for damage to the human health and the environment as a function of time, under conditions of "no action". The details of commonly used soil restoration techniques for addressing site contamination problems were previously discussed in Chapter 12.

13.5.3.3 Remediation of Groundwater Contamination Problems

Several remedial alternatives exist for the handling of groundwater contamination problems. Prominent among these are the following broad categories (Charbeneau et al., 1992): containment, source removal (including source excavation and pumped removal of product and/or contaminated water), *in situ* treatment (chemical or biological), and vacuum extraction. Depending on the site-specific situation, it often becomes necessary to use more than one method in order to achieve adequate remediation in the saturated and vadose zones.

Specific methods for mitigating contaminated groundwater plumes include the use of physical containment systems (such as impermeable barriers, hydraulic barriers, and subsurface collection systems) and leachate controls. The impermeable barriers may consist of slurry walls, grout curtains, and sheet piling; these can be used to contain, capture, or redirect groundwater flow for the mitigation of groundwater contamination problems. The hydraulic barriers (e.g., recovery wells and interceptor trenches) are used to modify hydraulic gradients around contained waters; this can be used to manipulate, through pumping/injection strategies, the movement and size of a contaminant plume—given the proper subsurface conditions. Leachate controls may include capping (to prevent or minimize rainwater from infiltrating through contaminated soil to groundwater) and/or by using subsurface drains (consisting of buried conduits that collect and convey leachate). The containment system acts to interrupt contaminant transport mechanisms, in order to prevent or minimize the continuing spread of contamination.

Groundwater pump-and-treat systems—consisting of extraction/injection well networks—have been a very commonly used remediation technique. In the remedial design for groundwater contamination problems, the recovery well systems are designed to intercept the contaminant plume so that no further degradation of the impacted aquifer occurs. Modeling is a very useful tool in the design of such systems (USEPA, 1985c). Typically, groundwater treatment technologies applicable to the recovered groundwater include physical treatment processes (such as phase-separated hydrocarbon recovery, air stripping, activated carbon adsorption, and filtration—which are all processes generally applied without the aid of chemical or biological agents); chemical processes (such as coagulation–precipitation, oxidation–reduction [redox], and neutralization); and biological methods (such as suspended-growth and fixed-film reactors, as well as *in situ* biodegradation).

13.5.3.4 The Challenges of Contaminated Aquifer Restoration Programs

It is apparent that the restoration of contaminated aquifer systems is one of the most challenging problems in CAP decisions. In attempts to get a good handle on this, far too much attention is often given to pump-and-treat remedial technologies. However, this technique tends to leave a great part of contaminant residues in the capillary fringe or vadose zone—oftentimes unaffected by groundwater pumping. In fact, a number of studies have been carried out to assess the feasibility and effectiveness of such a remedial strategy—generally with rather dismal outcomes reported. For instance, Yin (1988) uses a two-dimensional random-walk solute transport model to simulate aquifer restoration processes for groundwater that is contaminated by dissolved petroleum constituents. The model simulated proposed aquifer restorations by pumping out the contaminated groundwater. The aquifer restoration process consisted of a single pumping well located at the center of the contaminated area. The effectiveness of pumping was evaluated in terms of duration of pumping and concentration reduction. The simulation results were then used to evaluate the feasibility and effectiveness of aquifer restoration that can be attained by pumping out the contaminated water. The modeling results indicated that the pumping time/duration required for reducing the concentrations of benzene and xylene to acceptable levels was likely to extend over several years—to the point of affecting the efficacy of this site restoration technique.

Removal of the contaminated soil at the case/subject site, followed by treatment and/or disposal, will generally be a better strategy to adopt—in order to address source elimination. In fact, in most aquifer contamination problems, containment of the aquifer contaminants is the most immediate concern. This can be achieved through the use of a physical barrier (such as a grout curtain, slurry cutoff wall, or sheet piles) or by the creation of a hydraulic barrier resulting from a network of extraction (pumping) and injection (recharge) wells. The next requirement will be the removal of the mobile contaminants from the aquifer system. This may include free product recovery (e.g., of petroleum products floating on a water table), air stripping and vacuum extraction of the volatile organic contaminants, bioremediation, and a pump-and-treat technology for the soluble constituents. Insoluble constituents may remain adsorbed onto soil particles in the aquifer or vadose zone to be addressed differently. General details of technologies and processes used for the prevention and/or cleanup of groundwater contamination were previously discussed in Chapter 12.

13.5.4 Monitoring and Performance Evaluation of Remedial Options

The ultimate RAO for a contaminated site problem is to protect human health, environment, and public and private properties in the vicinity of the impacted site. Risks posed through contaminant migration pathways are generally eliminated, reduced, or controlled through treatment, engineering measures, or institutional controls. The amount of reduction in toxicity, mobility, or volume offered by treatment processes gives a measure of the anticipated performance of the treatment technologies that a remedy may use. This becomes an important factor in the decision regarding the remedial action alternative that should be selected and implemented for a contaminated site problem.

The adequacy of protection afforded by a remedial option to the human health and the environment is measured by whether or not the remedy meets the remediation goals. Thus, if after remedy implementation, it is determined that a remedial alternative does not (or will not) meet stipulated remediation or performance goals for all contaminants in the media of concern, then the residual risk remaining after implementation should be examined to determine whether other measures are necessary to assure protectiveness. The performance goal of remedial alternatives may be evaluated based on cleanup criteria and time period for the restoration of the contaminated site. The favored alternatives are compared based on the trade-offs between the time to attain an "acceptable" cleanup level and the costs associated with the remedial actions. However, the complexities in the fate and transport mechanisms at contaminated sites often make it difficult to predict with any degree of accuracy, the performance of site remedial actions. Consequently, there should be reasonable flexibility in the remediation process that allows for midcourse corrections—so that where necessary and possible, appropriate changes can be made to the remedial program to improve the performance of the selected remedial action alternative.

13.5.4.1 Monitoring Programs Network Design

In addition to adopting appropriate remedial actions, monitoring programs will normally be implemented to verify the long-term effectiveness of an overall corrective action program. Monitoring is indeed considered a very important component of corrective action response programs and can serve as a useful tool for evaluating the performance of remedial actions. Along this path, several monitoring parameters are important to the design of an effectual monitoring program. Of special interest is the selection of monitoring constituents/parameters; in general, the selection of monitoring constituents should consider the possibility for chemical constituents to be transformed over time and space. For instance, knowledge about the degradation of contaminants can be an extremely important factor in identifying monitoring constituents. Thus, specific monitoring constituents and indicator parameters may have to be modified as an investigation progresses in time, in order to account for likely transformation products. This is because physical, chemical, and biological degradation may transform certain constituents as the release ages and/or advances. In fact, despite the notion that most chemicals usually degrade into less toxic, more stable species, this is not universally true; for example, one of the degradation products of trichloroethylene (a carcinogen) is vinyl chloride, which is also a carcinogenic chemical. Consequently, the potential for physical, chemical, or biological transformations of constituents should be given adequate consideration in identifying monitoring constituents.

In general, several different monitoring programs may be implemented to evaluate the effectiveness of a remedy. For instance, an environmental monitoring program is normally used to ensure that chemical loadings from contaminated lands do not continually escape the influence of applicable remedial systems. Under a biomonitoring program, measurements of toxicity (through bioassays) and bioaccumulation can be used to assess the nature and extent of potential biological impacts in off-site areas. Where necessary, the remedial system is modified to ensure that the remedies are effective. Invariably, decisions regarding monitoring network design and operation are generally made in the light of available data. To a great extent, monitoring can be considered as an evolutionary process that should be refined as more relevant information is obtained. In fact, effective monitoring efforts are both dynamic and flexible, and this should be explicitly indicated in the site characterization plan. Ultimately, it is prudent to specify monitoring programs that will permit the collection of high-quality, representative data for the most sensitive chemical constituents of interest.

Finally, all other factors being equal, siting and placement of monitoring networks becomes a critical factor in monitoring programs. The success of the network would very much depend on whether or not representative samples or measurements can be obtained. Thus, it is important that the finite number of monitoring locations be placed in such a manner that will produce an accurate description of the prevailing site conditions. Meanwhile, it is worth mentioning here that most monitoring programs designed for contaminated site problems are directed at groundwater investigations—and perhaps surface water quality assessment to some extent. The practical elements of a viable long-term groundwater monitoring effort typically will consist of an evaluation of the hydrogeologic setting; proper well placement and construction; evaluation of well-performance and purging strategies; and the execution of effective sampling protocols (to include the selection of appropriate sampling mechanisms and materials, as well as sample collection and handling procedures) (USEPA, 1985e). Most of these elements, or variations thereof, are also applicable to monitoring programs in other environmental media.

13.5.4.2 Optimizing the Design of Environmental Monitoring Systems

Groundwater monitoring programs are generally designed to investigate the possibility of spread of contamination from a contaminated site. But the adequacy of groundwater monitoring systems that consist of upgradient and downgradient monitoring wells often comes into question. This is because groundwater contaminants often migrate along preferred pathways that are the result of heterogeneities within stratigraphic units (Osiensky, 1995). Several monitoring wells must, therefore, be completed "correctly" within individual heterogeneities to provide for the early detection of contaminant migration along these preferred flowpaths. Thus, the adequacy of a specific groundwater monitoring system is usually very much dependent on the number of available monitoring wells.

A more innovative approach, such as one involving the use of time series electrical potential field measurements in combination with groundwater monitoring wells, can greatly improve the effectiveness of groundwater monitoring systems required to detect the presence of site contamination (Osiensky, 1995). Typically, an electrical geophysical method is used that involves the measurement of electrical current flow through the earth materials under investigation. Changes in the electric potential field over time due to the presence of groundwater contaminants can provide early detection of leaks from chemical disposal facilities or contaminated sites (Osiensky, 1995). In fact, geophysical methods have historically been used to help define potential pathways for contaminant migration. For example, surface electrical resistivity methods can be used to reduce the number of monitoring wells needed to define the migration of certain contaminants away from contaminated sites. Time series electrical potential measurements can be particularly useful for the early detection of releases from contaminated sites. Data from this can then be used to evaluate the potential for groundwater contamination, prior to the drilling of groundwater monitoring wells. The data can therefore help in the determination of the optimal locations for groundwater monitoring wells.

13.6 PRESCRIPTIONS FOR COST-EFFECTIVE CORRECTIVE ACTION RESPONSE DECISIONS

In order to design an effectual corrective action response program for contaminated site problems, a number of relevant decision elements are usually evaluated as part of the corrective action assessment process for such sites (Box 13.6). The information derived from this evaluation will generally help define the following: concentrations of contaminants of interest at and near the impacted site; populations who might be exposed to site contamination; magnitude and frequency of potential receptor exposures; and the health and environmental effects that might result from site exposures. Subsequently, appropriate corrective action decisions being considered to learn more about the site, and/or to mitigate any risks, can be formulated to support the corrective action and risk management program.

13.6.1 A DECISION FRAMEWORK FOR RISK-BASED CORRECTIVE ACTION RESPONSE PROGRAMS

To reduce cleanup costs associated with contaminated site problems, it is important that the decision-making process involved be well defined. A systematic decision framework should therefore be used to develop the risk information necessary to support the corrective action decision-making process (Figure 13.4); such formulation will generally provide a rational protocol for determining the level of effort required in the evaluation and implementation of site restoration programs.

BOX 13.6 IMPORTANT DECISION ELEMENTS FOR DESIGNING CORRECTIVE ACTION PROGRAMS

- Establish basis for contamination indicators.
- Gather and review site background information for evidence of release.
- Determine site history, to help identify other possible sources of contamination.
- Identify potentially affected areas.
- Address health and safety issues associated with the case-specific situation.
- Address emergency response by mitigating release and potential hazards.
- Characterize site, to include an identification of the contaminants of potential concern and a delineation of the extent of contamination for affected matrices.
- Determine contaminant behavior in the environment, and develop working hypothesis about contaminant fate and transport.
- Determine contaminant migration and exposure pathways.
- Identify populations potentially at risk.
- Map areas where contaminants may impact human health and/or environment.
- Develop realistic exposure scenarios appropriate for the specific site.
- Conduct site-specific exposure assessments.
- Define magnitude/severity of health and environmental impacts.
- Establish remediation objectives and site-specific cleanup criteria.
- Assess variables influencing selection of cleanup criteria and remedial systems.
- Develop RAOs.
- Identify remedial action alternatives.
- Develop general response actions.

The approach will indeed allow for a comprehensive evaluation of risk mitigation measures needed for the corrective action assessment and response program. Overall, the use of a structured decision framework will generally facilitate rational decision-making in relation to site restoration and risk management efforts undertaken for potentially contaminated sites.

Conceptually, the decision framework shown in Figure 13.4 can be used (or adapted for use) to facilitate the decision-making process involved in contaminated site management problems. The principles and ideas of the site restoration framework can be used, on a site-specific basis, to help determine the extent of cleanup required at a particular contaminated site. The process will generally incorporate a consideration of the complex interactions existing between the hydrogeological environment, regulatory policies, and technical feasibility of remedial technologies. In the end, the decision processes involved should help environmental analysts identify, rank/categorize, and monitor the status of potentially contaminated sites; identify field data needs and decide on the best sampling strategy; establish appropriate remediation goals and/or cleanup levels; and choose the remedial action technique that is most cost-effective in managing site-related risks.

13.6.1.1 Elements of the Risk-Based Corrective Action Assessment and Response Process

Risk-based corrective action assessment and response programs are generally designed to facilitate the selection of appropriate and cost-effective site characterization and corrective action measures. With that goal in mind, it is apparent that the selection of an appropriate and cost-effective risk management plan for a contaminated site problem usually depends on a careful assessment of both short- and long-term risks posed by the case-specific problem. The key components of such an assessment, along with a representative site restoration decision process, are typically comprised of the tasks discussed below.

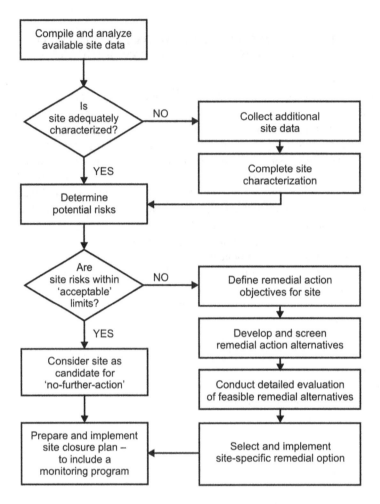

FIGURE 13.4 Risk management and corrective action decision framework for the management of contaminated site problems.

Task 1: *A Preliminary Site Appraisal*—that consists of the identification of possible source(s) of contaminant release. The purpose of the preliminary site appraisal is to quickly assess the potential for a site to adversely impact the environment and/or public health. This process typically involves: establishing basis for contamination indicators; gathering and reviewing background information for evidence of release; determining the case history, to help identify other possible sources of contamination; identifying potentially affected areas; addressing health and safety issues associated with the case-specific situation; and addressing emergency response by mitigating release and potential hazards. The site appraisal is initiated by the discovery of a potentially contaminated site. Conventional site reconnaissance procedures may be used in this qualitative site assessment that involves the collection and review of all available information (including an off-site reconnaissance to evaluate the source and nature of contamination, and the identification of any potential "outside" polluters or PRPs). Depending on the results of the preliminary survey, a site may or may not be referred for further action. In general, the preliminary site appraisal allows for site screening in order to establish a basis for more detailed site investigations.

Task 2: *A Site Assessment*—involving the characterization of site contamination, as well as a site categorization, where necessary. The objectives of the site assessment are to identify site contaminants and to determine site-specific characteristics that influence the migration of the contaminants. The process typically involves the following: characterizing problem locale (to include an identification of the contaminants of potential concern and a delineation of the extent of contamination for affected matrices), and determining contaminant behavior in the environment by developing a working hypothesis about contaminant fate and transport. Overall, site assessment activities are undertaken to better define the characteristics of a potentially contaminated site and its neighboring areas.

Task 3: *A Risk Appraisal*—that entails the determination of the migration and exposure pathways, as well as an assessment of the environmental fate and transport of the CoCs. The primary objective of the risk appraisal is to determine whether existing or potential receptors are presently, or may in the future, be at risk of adverse effects as a result of exposure to site contaminants. The process typically involves the following: determining contaminant migration and exposure pathways; identifying populations potentially at risk; mapping areas where contaminants may impact human health and/or environment; developing realistic exposure scenarios appropriate for the specific problem area; and conducting case-specific exposure assessments. To determine if potential receptors are at risk, it is necessary to identify potential migration and exposure pathways as well as contaminant exposure point concentrations in relation to acceptable threshold levels.

Task 4: *A Risk Determination*—to include an evaluation of the environmental and health impacts associated with contaminant releases, and also the development of site-specific cleanup criteria. The objective of the risk determination is to evaluate potential site risks, which can be compared against a benchmark risk; this then becomes the basis for developing cleanup or restoration goals and/or for risk management plans. The process typically involves defining the magnitude or severity of health and environmental impacts.

Task 5: *A Site Mitigation*—that is comprised of the development and implementation of a CAP that may include site remediation and/or monitoring programs. The process typically will involve the following: establishing remediation objectives and case-specific restoration goals; assessing variables that influence selection of restoration goals and remedial systems; developing RAOs; identifying remedial action alternatives; and developing general response actions. Site mitigation strategies and objectives are based on the protection of both current and potential future receptors that could become exposed to site contaminants. Consequently, the site restoration is carried out so as to leave the site in such a condition as to pose no significant risks to populations potentially at risk.

Each of the above tasks may also be comprised of a number of elements or subtasks. Although the tasks appear sequential in this presentation, it is noteworthy that certain aspects may be executed concurrently or indeed in an iterative manner.

In all cases of contaminated site assessments, a qualitative evaluation of site conditions identifies potential contaminant release source(s), determines the environmental media affected by each release, and broadly defines the possible extent of the release(s). For a confirmed release, the release source(s) should be stopped and hazards mitigated, insofar as possible. Where applicable, any free product present should be removed immediately to prevent the development of health and safety hazards. Removal will also help prevent further migration of free product into soil and/or groundwater. In fact, to meet environmental regulations and specifications for many a region, the appropriate regulatory agencies usually would have to be notified of any source release(s) and of steps taken in immediate response. In any event, the initiation of free product removal should *not* have to await approval by regulatory agencies. In fact, since free product removal, if necessary, is the first step in every cleanup, there is no justification to delay its commencement.

13.7 GENERAL ELEMENTS OF THE CONTAMINATED SITE RESTORATION AND MANAGEMENT DECISION-MAKING PROCESS

Corrective action programs for contaminated sites may vary greatly—ranging from a possible "no-action" alternative to a variety of extensive and costly remediation or site restoration options (such as those offered in Chapter 12). The primary objective of any corrective action response program is to ensure public safety and welfare by protecting human health, environment, and public and private properties. The ability to select an appropriate and cost-effective site restoration strategy that meets this goal will generally depend on a careful assessment of both short- and long-term risks potentially posed by the site; it also depends on the site-specific restoration goal that is established and accepted for the site (as determined by methods presented in Chapter 11), as well as the discretionary use of several corrective measure evaluation tools (such as those elaborated in Appendix B).

In general, the use of a systematic evaluation process will tend to facilitate rational decision-making on restoration efforts undertaken for contaminated site problems. Risk assessment is particularly useful in determining environmental restoration goals most appropriate for a contaminated site problem. It can also be used in the development of performance goals for various corrective action response alternatives. In fact, by utilizing methodologies that establishes restoration goals based on risk assessment principles, restoration programs can be carried out in a cost-effective and efficient manner. In the processes involved, risk assessment can be used to define the level of risk, which will in turn aid in determining the level of analysis and the type of risk management actions to adopt for a given contaminated site problem. The level of risk considered in such types of application may be depicted in a classical risk-decision matrix (see, e.g., Asante-Duah, 2017), and this may help distinguish between imminent health and/or environmental hazards and risks; this can then facilitate cleanup prioritization for multiple projects, as well as help with resource allocation strategies. Subsequently, this can be used as a facilitator for policy decisions, in order to be able to develop variations in the scope of work necessary for case-specific contaminated site management programs. Ultimately, information derived from the relevant evaluations generally helps formulate credible policies to support the risk management and/or site restoration actions.

In closing, it is noteworthy that several pertinent questions are typically explored during the planning, development, and implementation of corrective action programs that are directed at restoring contaminated sites (Box 13.7). These, and indeed several other issues, are relevant to making risk reduction and risk management decisions for potentially contaminated sites (Asante-Duah, 1996, 1998; BSI, 1988; Cairney, 1993; Jolley and Wang, 1993; USEPA, 1985e, 1987h,i, 1988e, 1989n,o, 1991a,h; WPCF, 1988). This chapter has enumerated several significant elements and factors that will help answer these important questions and will ultimately influence the type of corrective action decision accepted for a given contaminated site problem.

BOX 13.7   QUESTIONNAIRE CHART TO FACILITATE THE QUALITATIVE EVALUATION OF CORRECTIVE ACTION ASSESSMENT AND RESPONSE PROGRAMS

- Do contaminant sources exist?
- Are there visible sources of contamination?
- What are the sources of the contaminants?
- Is the site potentially contaminated with hazardous or toxic chemicals? What are the toxic agents involved?
- Are there any confining layers or porous layers in the soil horizon?
- Is there soil erosion or recent cuts or fills on site?

(Continued)

**BOX 13.7 (*Continued*)   QUESTIONNAIRE CHART TO FACILITATE
THE QUALITATIVE EVALUATION OF CORRECTIVE
ACTION ASSESSMENT AND RESPONSE PROGRAMS**

- What is the nature of drainage and surface flow patterns at the site and immediate vicinity?
- What are the site characteristics, hydrological features, meteorological or climatic factors, land-use patterns, and agricultural practices affecting the transport and distribution of site contaminants?
- What is the distribution of the chemicals over the site and vicinity?
- Are there known "hot spots" at the site?
- What is an appropriate background or control region to use for corrective action investigations?
- Is there any area that poses an immediate and life-threatening exposure?
- What are the important transport processes and migration pathways that contribute to contamination and/or exposure?
- Are there present or future potential receptors that could be adversely affected by the contaminant sources? In particular, are there sensitive ecosystems or residences located downgradient, downstream, or downwind from the site?
- Are there one or more pathways through which site contaminants might migrate from the source and reach potential receptors? What are the dominant routes of exposure at the site?
- Have populations already been impacted, and/or are populations potentially at risk? What are the potential risks posed to the human health and the environment if no further response action is taken at the site?
- Does the risk level exceed benchmark levels specified by environmental compliance regulations? What site-specific cleanup criteria will be appropriate for the site?
- At the indicated contaminant concentrations at the site, which areas are considered as posing risks to the environment or surrounding populations? Which areas must therefore be cleaned up in order to reduce risks to an "acceptable" level?
- Are estimated risk levels low enough—such that a "no-action" alternative is still protective of public health and the environment?
- What contaminants and environmental media should become the target for site remediation?
- How much contaminated material should be cleaned up to achieve an acceptable site restoration goal?
- Which remedial alternatives can be applied at the contaminated site in order to achieve adequate cleanup?
- Will exposure pathways be interrupted or will receptors be protected following a removal or remedial actions?
- What institutional control measures and risk management strategies are required in the overall corrective action decision?

14 Turning "Brownfields" into "Greenfields"

Illustrative Paradigms of Risk-Based Solutions to Contaminated Site Problems

Globally, it has come to be realized that drastic and dramatic risk control measures involving the cleanup of most contaminated sites to "pristine" conditions or even "desirably safe levels" is neither technically nor economically feasible—and may not even be warranted per se. Consequently, risk-based contaminated site management strategies seem to be gaining growing grounds and widespread acceptance—and this trend is expected to remain. Ultimately, the use of such decision-support tools and strategies should facilitate the design and implementation of policies and frameworks that will support sustainable management programs for contaminated site problems.

In general, an important strategy to adopt in the management of contaminated site problems usually relates to the use of a risk-based management protocol that can be brought in to address the several challenges faced in decisions to reutilize or redevelop such contaminated sites. The strategy should be such as to help minimize uncertainties that invariably surround the actual or perceived risks associated with these so-called "brownfield" sites. Indeed, risk assessment promises a way for developing appropriate strategies that can aid sustainable development and risk management decisions in this arena. All things considered, risk-based decision-making generally will result in the design of better risk management programs for contaminated site problems. This is because risk assessment can produce more efficient and consistent risk-reduction policies; it can also be used as a screening device for the setting of policy priorities. This chapter enumerates the general scope for the application of risk assessment and risk-based solutions, as pertains to the management of potentially contaminated site problems; it also discusses specific practical example situations for the utilization of the risk assessment paradigm.

14.1 GENERAL SCOPE OF THE PRACTICAL APPLICATION OF RISK-BASED SOLUTIONS TO ENVIRONMENTAL MANAGEMENT AND CONTAMINATED SITE PROBLEMS

Risk assessment has several specific applications that could affect the type of decisions to be made in relation to environmental and contaminated site risk management programs. The application of the risk assessment process to environmental risk management and contaminated site problems will generally serve to document the fact that risks to human health and the environment have been evaluated and incorporated into a set of appropriate response actions. Appropriately applied, risk assessment techniques can indeed be used to estimate the risks posed by environmental hazards under various exposure scenarios and to further estimate the degree of risk reduction achievable by implementing various scientific remedies. In fact, almost invariably, every process for developing effectual environmental restoration and risk management programs should incorporate some concepts or principles of risk assessment. In particular, all corrective action decisions and restoration plans for potentially contaminated site problems will include, implicitly or explicitly, some elements of risk assessment. A number of practical examples of the potential application of risk assessment principles, concepts, and techniques in environmental management practice—including the identification of key decision issues associated with specific problems—abound in the literature of risk analysis. Some of the broad applications often encountered in chemical exposure situations are annotated in Box 14.1—with further general discussion of some of the more prominent application scenarios offered later on. Anyhow, this listing and additional subsequent discussion is by no means complete and exhaustive, since variations or even completely different and unique problems may be resolved by use of one form of risk assessment methodology or another.

BOX 14.1 SELECT DECISION ISSUES AND SPECIFIC PROBLEMS TYPICALLY ADDRESSED BY THE USE OF RISK ASSESSMENT CONCEPTS AND TECHNIQUES IN ENVIRONMENTAL MANAGEMENT PROGRAMS

- Field sampling design and identification of data needs and/or data gaps. Risk assessment can aid in the development of cost-effective field sampling programs. This will then allow an adequate number of samples to be collected and analyzed, with an overall objective of determining the presence or absence of contamination, their extent and distribution, or of verifying the attainment of site mitigation criteria
- Determination of potential risks associated with industrial, commercial, and residential properties—to facilitate land-use decisions and/or restrictions
- Corrective measures' evaluation leading to the selection of remedial alternatives: risks posed by alternative remedial actions can be assessed before program implementation—that is, the risk assessment process allows performance goals of remedial alternatives to be established prior to their implementation.
- Facilitates the determination of cost-effective risk-reduction policies through the selection of feasible remedial alternatives protective of public health and the environment
- Evaluation of remedial alternatives to determine whether the proposed remedial action itself will have any deleterious environmental and human health effects
- Evaluation and ranking of potential liabilities from potentially contaminated properties
- Prioritization of potentially contaminated sites for remedial action; by providing consistent data for the rank ordering of potentially contaminated sites, risk assessment helps regulatory agencies and potentially responsible parties to prioritize cleanup actions
- Development of target cleanup criteria and guidelines for contaminated sites: helps define soil cleanup levels required for remediation decisions—that is, risk assessment provides the basis for determining the levels of chemicals that can remain at a site, or in environmental matrices, without impacting public health and the environment
- Facilitates decisions on the use of specific chemicals in manufacturing processes and industrial activities
- Preliminary screening for potential environmental impact problems. For instance, this typically may include an identification of factors contributing the most to overall risks of exposures from chemicals originating from a potentially contaminated site—incorporating an analysis of baseline risks, and a consistent process to document potential public health and environmental threats from potentially contaminated sites
- Design of monitoring programs (to identify chemicals present in various media, and their persistence)

(Continued)

BOX 14.1 (*Continued*)   SELECT DECISION ISSUES AND SPECIFIC PROBLEMS TYPICALLY ADDRESSED BY THE USE OF RISK ASSESSMENT CONCEPTS AND TECHNIQUES IN ENVIRONMENTAL MANAGEMENT PROGRAMS

- Ecological/environmental risk assessment for the identification of critical habitats and organisms exposed to environmental contaminants
- Probabilistic risk assessment for the evaluation of transportation risks associated with the movement of hazardous wastes
- Implementation of general risk management and risk-prevention programs for environmental and public health management planning
- Facilitation of property transactions by assisting developers, lenders, and buyers in the "safe" acquisition of both residential and commercial properties
- Evaluation of potential risks associated with the migration of contaminant vapors into building structures
- Streamlining of site remediation activities: facilitates more cost-effective site closures for contaminated sites
- Addressing the health and safety issues associated with environmental chemicals— that is, to determine "safe" exposure limits for toxic chemicals used or found in the workplace and residences
- Facilitation of decisions on alternative disposal methods for potentially hazardous waste materials
- Determination of discharge requirements for industrial facilities
- Hazardous waste facility design and operation—including evaluation and management of potential risks due to toxic air emissions from industrial facilities and incinerators
- Demonstration and justification for the acceptability of risks associated with "baseline" conditions, allowing regulatory agencies to accept a "no-action" recommendation in remedial action decisions
- Determination of potential risks anticipated from site remediation, allowing appropriate risk management and site control measures to be implemented. Remedial alternatives can be evaluated to determine whether the proposed remedial action itself will have any deleterious environmental and human health effects
- Evaluation of expected performance and effectiveness of alternative remediation techniques
- Analysis of human health impacts from chemical residues found in food products (such as contaminated fish and pesticide-treated produce)

14.1.1 DESIGN OF RISK-BASED SOLUTIONS TO CORRECTIVE ACTION RESPONSE AND ENVIRONMENTAL RESTORATION PROGRAMS

It is apparent that, some form of risk assessment is inevitable if environmental management programs and corrective actions are to be conducted in a sensible and deliberate manner. In fact, risk assessment seems to be gaining greater grounds in making public policy decisions in the control of risks associated with contaminated site problems. This situation may be attributed to the fact that the very process of performing a risk assessment does lead to a better understanding and appreciation of the nature of the risks inherent in a study and further helps develop steps that can be taken to reduce such risks. Among other things, the application of risk assessment to contaminated site and related environmental contamination problems helps identify critical migration and exposure pathways, key receptor exposure routes, and other extraneous factors contributing the most to the total risks. It also facilitates the determination of cost-effective risk reduction policies. Used in the corrective action planning process, risk assessment generally serves as a useful tool for evaluating the effectiveness of remedies associated with environmental contamination problems and also for determining acceptable public and ecological health restoration goals. By and large, the process will usually serve to document the fact that risks to human health and the environment have been evaluated and incorporated into the appropriate response actions. Inevitably, risk-based corrective action programs facilitate the selection of appropriate and cost-effective remedies or restoration measures. Ultimately, based on the results of a risk assessment, a more effectual decision can be made in relation to the types of risk management actions necessary to address a given contaminated site problem or a hazardous situation.

Indeed, the assessment of health and environmental risks plays an important role in site characterization activities, in corrective action planning, and also in risk mitigation and risk management strategies for potentially contaminated site problems. A major objective of any site-specific risk assessment is to provide an estimate of the baseline risks posed by the existing conditions at a contaminated site and to further assist in the evaluation of site-restoration options. Thus, appropriately applied, risk assessment techniques can be used to estimate the risks posed by site contaminants under various exposure scenarios and to further estimate the degree of risk reduction achievable by implementing various engineering remedies. Risk assessment is particularly useful in determining the level of cleanup most appropriate for potentially contaminated sites. By utilizing methodologies that establish cleanup criteria based on risk assessment principles, corrective action programs can be conducted in a cost-effective and efficient manner. Once realistic risk-reduction levels potentially achievable by various remedial alternatives are known, the decision maker can use other scientific criteria (such as implementability, reliability, operability, and cost) to select a final design alternative. Subsequently, an appropriate corrective action plan can be developed and implemented for the contaminated site.

Overall, a risk assessment will generally provide the decision maker with scientifically defensible procedures for determining whether or not a potential environmental contamination problem could represent a significant adverse health and environmental risk and if it should therefore be considered a candidate for mitigative actions. In fact, the use of health and environmental risk assessments in environmental management decisions in general and corrective action programs in particular is becoming an important regulatory requirement in several places. For instance, a number of environmental regulations and laws in various jurisdictions (e.g., regulations promulgated under the Comprehensive Environmental Response, Compensation, and Liability Act (CERCLA), Resource Conservation and Recovery Act (RCRA), Clean Air Act (CAA), and Clean Water Act in the United States) increasingly require risk-based approaches in determining cleanup goals and related decision parameters; indeed, risk-based solutions for addressing contaminated site problems seem to be a key ingredient in most contaminated site assessment and restoration programs in North America, particularly in the United States—and the trend is not much different in Canada. Similar trends and frameworks also seem to be found in much of Europe—as attested, for example, by the birth of activities and networks such as CLARINET (Contaminated Land Rehabilitation Network for Environmental Technologies) and CARACAS (Concerted Action on Risk Assessment for Contaminated Land) in the European contexts. CLARINET involves analyzing key issues in decision-making processes for the management of contaminated site problems, and CARACAS focuses on areas such as human toxicology, ecological risk assessment, and relevant methodologies to adopt for contaminated site risk assessments. Ultimately, the use of such decision-support tools and strategies should facilitate the design and implementation of policies and frameworks that will support sustainable management programs for contaminated site problems. Several issues that—directly or indirectly—affect public health and environmental management programs may truly be addressed by using some form of risk assessment or related tools, as illuminated by the variety of examples discussed in Section 14.2.

14.1.1.1 Risk-Based Evaluation of the Beneficial Reuse of Contaminated Materials

Environmental quality criteria for contaminated materials typically serve as benchmarks in the assessment of the degree of contamination, and also in the determination of the possible beneficial uses of the impacted materials, and whose uses are not expected to be adversarial to human health and the environment. By utilizing protocols that determine maximum acceptable or safe contaminant levels based on risk assessment methodologies, such impacted materials can be managed in a cost-effective manner. In fact, various analytical tools and models can generally be used in the development of appropriate action levels or criteria that will facilitate the implementation of effectual policy decisions on the impacted materials.

As an example, risk assessment principles can be incorporated into the development of statistical relationships between the nature of contamination from highway materials and the roadway types. For instance, by using the analytical results obtained from catch basin materials derived from roadways, hazards, and/or risks associated with cleanings from the different catch basins (associated with the different road types) can be determined. In addition, state-of-the-art risk models can be used to establish "safe" or "action" levels for the contaminants of concern found in the catch basin materials. This contaminant level can then serve as a reference or trigger criteria against which future catch basin cleanings can be compared, in order to determine if such materials are likely to present elevated risks to public health and/or the environment; the established criteria could also become an important determinant in the beneficial reuse analysis for the catch basin materials.

14.2 MANAGEMENT, REUSE, AND REDEVELOPMENT STRATEGIES FOR CONTAMINATED SITE PROBLEMS IN PRACTICE

Contaminated site risk management and site-restoration decisions may generally be formulated quite differently for varying situations or circumstances—such as one that is purely qualitative in nature, through a completely quantitative evaluation. In any event, in the applications of risk assessment to contaminated site problems, it is important to adequately characterize the exposure and physical settings for the problem situation, in order to allow for a proper application of appropriate risk assessment methods of approach. Unfortunately, there tends to be several unique complexities associated with real-life contaminated site problems and related chemical exposure scenarios—and this can seriously overburden the overall process. Also, the populations potentially at risk from contaminated site problems are usually heterogeneous—and this can greatly influence the anticipated impacts/consequences. Critical receptors should therefore be carefully identified with respect to numbers, locations (areal and temporal), sensitivities, etc., so that risks are neither underestimated nor conservatively overestimated. Indeed, the determination of potential risks associated with contaminated site problems invariably plays an important role in environmental and public health risk mitigation and/or risk management strategies—as demonstrated by the hypothetical example problems that follow.

14.2.1 RISK-BASED CORRECTIVE ACTION ASSESSMENTS: THE MANAGEMENT OF PETROLEUM-CONTAMINATED SITES AS AN EXAMPLE

The release of chemical substances from leaking underground storage tanks (USTs) is a common occurrence in several regions of the world. In particular, leakage of petroleum hydrocarbon products and other chemicals from USTs and pipelines is a frequent occurrence in commercial, industrial, and even domestic activities. Such leakages can result in the contamination of several environmental media—especially soils and groundwater. The subsequent impacts of such releases on the environment and public health are a particularly important environmental issue. Of an even greater interest/concern is groundwater contamination by petroleum products—because many communities depend on groundwater as their primary source of drinking water.

14.2.1.1 Composition of Petroleum Products

Specific petroleum products commonly found in USTs include gasoline, diesel, heating oil, aviation fuel, waste oils, and other related petroleum hydrocarbons. Typically, motor fuel alone is a mixture of over 200 petroleum-derived chemicals plus a few synthetic products that are added to improve the fuel performance (LUFT, 1989). Several of these chemical constituents can potentially affect human and ecological receptors if released into the environment.

Petroleum fuel contaminants of major health concern include benzene, toluene, ethylbenzene, and xylene—commonly referred by the acronym BTEX (benzene, toluene, ethylbenzene, and xylenes). These BTEX compounds have the potential to move through soils to contaminate groundwater. In fact, groundwater contamination from petroleum products, and particularly from leaking USTs, is a growing concern especially because of the potential carcinogenic and neurotoxic effects exhibited by some of the BTEX compounds. For instance, benzene is a known human carcinogen, whereas the others (i.e., toluene, ethylbenzene, and xylene) are noncarcinogens but are known to possess neurotoxicity effects. This means that the BTEX compounds generally constitute preselected target/indicator chemicals of concern for petroleum product releases—albeit some of the synthetic additives may present additional concerns. In any case, as BTEX oftentimes constitute the most toxic and environmentally mobile constituents, their selection as indicator chemicals assures that any corrective action cleanup criteria developed based on these may also adequately address the less toxic or the less mobile constituents.

In addition to an evaluation of BTEX, analysis for total petroleum hydrocarbons (TPHs) is commonly carried out for most petroleum-contaminated sites. This latter analysis detects aliphatic and aromatic constituents contained in the fuel. Detection is commonly reported as the sum total of all hydrocarbons in the sample—rather than as individual chemical constituents. Because the lighter fractions (such as BTEX) are more mobile, these constituents can migrate or dissipate away from the main body of contamination. Indeed, the less mobile hydrocarbons (such as those detected in TPH analysis) may give a more accurate indication of the actual contamination. As a consequence, soils are preferably analyzed for both the BTEX and TPH as indicators of contamination.

Further to BTEX and TPH analyses, and because of its extreme toxicity, the possible presence of organolead would generally be investigated where significant leaks of leaded motor fuel have occurred or where an investigator feels that there may be potential danger of exposure to organolead. Of course, other synthetic additives such as methyl-t-butylether (MTBE) may also become a major focal point in such evaluations.

It is noteworthy that, diesel fuels consist primarily of aliphatic hydrocarbons (though they may also contain some limited quantities of aromatic constituents [including benzene], depending on the source and the refining process). Consequently, TPH analysis is usually the only one initially required for leakage and spills from diesel storage.

14.2.1.2 The Subsurface Fate and Behavior of Petroleum Constituent Releases

Typically, in the event of contamination from a leaking UST used for petroleum products that have density less than that of water (e.g., motor fuel), the leaking product first enters the unsaturated soil below the tank. This eventually forms a floating layer on any underlying water table if the leaking quantity is large. Also, product vapor will enter the unsaturated soils around the tank and above the water table. Water-soluble fractions of the petroleum product, such as benzene and xylene, will eventually form a plume within the groundwater. The soluble plume will spread within the groundwater by diffusion and then move with the groundwater as it continues its flow downgradient.

In the unsaturated zone, petroleum constituents tend to move downward, under the influence of gravity, and laterally due to capillary forces and the heterogeneous characteristics of soil. The movement also depends on the viscosity of the product and the rate of product release. If the amount of petroleum constituents is large enough, it will pass through the capillary zone and reach the water table. For motor fuels that are immiscible with water and also less dense than water, a layer of petroleum constituents will lie on and slightly depress the water table. Only a small amount of hydrocarbons will dissolve in the water.

Flow in the saturated zone generally transports the contamination in the direction of decreasing hydraulic potential. Petroleum constituents that reach the groundwater table will dissolve in water and be transported by groundwater. Transport of dissolved constituents in the groundwater is governed by advection and dispersion of groundwater and by attenuation mechanisms (such as biodegradation and adsorption) of the soil media.

In general, the migration of contaminants such as petroleum constituents through the subsurface environment is governed by four major factors: the quantity or volume of release, the physical properties of the contaminants, the physical properties of the subsurface material (e.g., the adsorptive capacity of the earth materials), and the subsurface flow, such as the rates and directions of groundwater movement (Hunt et al., 1988; Yin, 1988). In addition, all processes that attenuate contaminant concentrations and/or limit the area of the contaminated zones will affect the fate of the source release.

In the end, the quantity of the contaminant determines whether it will reach the water table. Physical properties of the leaked substance that are important to migration include solubility, specific gravity, viscosity, and surface tension; biodegradability should also be considered important to long-term contaminant migration. Physical properties of subsurface materials important to the migration of petroleum constituents include the porosity, permeability, and homogeneity. Other important physical parameters would include soil organic carbon content available for partitioning and also the available oxygen to aid aerobic biodegradation. All these parameters do indeed affect the subsurface behavior of the contaminants of concern.

The Dominant Fate and Transport Processes for Source Releases. Releases of petroleum constituents from leaking USTs can contaminate soils and groundwater by the migration of free product through the unsaturated zone and by dissolution of certain constituents into groundwater. It is noteworthy that, although petroleum fuel odor may be reported during soil sampling in the unsaturated zone in a site investigation/appraisal, at some site locations, no hydrocarbons may be detected. This situation suggests that the mechanisms of groundwater contamination away from the gasoline spill are vaporization of gasoline components into the soil gas, migration of denser-than-air soil gas downward to the water table, followed by radial spreading and then component partitioning into the groundwater (Hunt et al., 1988). Indeed, several factors affect the fate and transport of the mobile liquid, vapor, and dissolved hydrocarbon phases.

In general, when petroleum products enter the soil, gravitational forces act to draw the fluid in a downward direction. Other forces act to retain it, which is either adsorbed to soil particles or trapped in soil pores. Petroleum hydrocarbon products can therefore become a major source of long-term contamination of soils and groundwater. The amount of product retained in the soil is of importance because this could determine both the degree of contamination and the likelihood of subsequent contaminant transport into other environmental compartments. With that said, a good understanding of the fate and transport processes is important in determining the likelihood of medium-specific releases and exposures that may arise from leaking USTs. A variety of environmental fate models may be used to predict the transport and fate of contaminants; these models may range from a simple mass-balance equation to multidimensional numerical solution of coupled differential equations. In any case, the basic equation governing fate and transport is based on the principle of conservation of mass for the contaminant.

In a typical situation when petroleum products have leaked from an UST, a nonaqueous-phase liquid is released that can move through soil pores. Some of this liquid is left behind as disconnected fluid, called ganglia because of strong capillary forces (Hunt et al., 1988). These ganglia make up the residual saturation in the soil and are the long-term source of contaminants released to the air and groundwater. Even where no regional groundwater flow may be evident, a far field transport of selective gasoline components may indicate groundwater contamination by subsurface vapor migration away from the spill. In the unsaturated zone, denser gas is produced by volatilization of liquid gasoline, and this gas sinks towards the water table and then spreads out over the capillary fringe. Air saturated with gasoline and in contact with capillary water and groundwater generally will lose the more water-soluble components as it spreads out radially. Thus, the more volatile, less water-soluble hydrocarbons are expected to move greater distances than the more water-soluble compounds such as benzene. Indeed, available investigation data strongly suggest that groundwater contaminants detected in wells away from a petroleum fuel spill may arrive mostly via the vapor phase and not necessarily by groundwater flow (Hunt et al., 1988). Thus, it is important, in several situations, to model subsurface vapor transport in order to adequately assess the potential for contamination from leaking USTs. Relevant information from this type of evaluation is also important in the choice of appropriate remediation technologies and/or cleanup strategies.

14.2.1.3 The Corrective Action and Risk Management Decision Process for UST Release Sites

It is almost a certainty that the result of most cleanup actions undertaken at petroleum-contaminated sites will leave some residual contamination. In fact, cleanup of all contaminated soil and dissolved product in groundwater is not always necessary to protect public health and the environment (LUFT, 1989). Consequently, it is important to develop cleanup goals attainable for appropriate remedial actions, based on estimates of health and environmental risks associated with the contaminants from UST releases. In general, however, generic cleanup levels for contaminated soil and dissolved product are undesirable, since conditions vary from one region to another; instead, site-specific cleanup levels are usually recommended for corrective action decisions (LUFT, 1989). Indeed, it is believed that the use of site-specific cleanup criteria could result in significant cost savings—because such approach can allow the use of corrective actions that are most cost-effective in risk reduction. As part of a UST release site investigation, therefore, one should determine appropriate cleanup levels and the attainable remediation goals prior to implementing any corrective action plans.

Classically, the corrective action and risk management decision process will involve the conduct of several important investigation tasks—with the most important ones noted as follows:

- *Site Categorization.* To facilitate a site-specific and phased approach to cost-effective corrective action decisions at petroleum-contaminated sites, a site categorization scheme may be adopted in the investigation of UST releases. The different categories will reflect the seriousness or hazardous nature of the potential release situation. In general, a site designation process may be used to categorize sites after observing the presence of one or more of a number of suspect conditions (such as tank closure, reported nuisance conditions, monitoring problems, and observed leakages). Existence of any of the suspect conditions provides justification to initiate a preliminary investigation that will confirm or disprove the suspected situation. Based on the preliminary site investigation and site history, the site can then be assigned to one of a number of categories; for the purpose of this discussion, sites are classified into the following three categories:
  - Category 1: Low risk sites (LrS) → no suspected soil contamination
  - Category 2: Medium risk sites (MrS) → suspected or known soil contamination
  - Category 3: High risk sites (HrS) → known soil and groundwater contamination

  Obviously, moving from the LrS category to the HrS sends one from a less serious to a more serious and hazardous scenario. For instance, the HrS category presents a case where both soil and groundwater contamination is confirmed. In general, if the field personnel suspect that more serious contamination has occurred than was anticipated, then a site may be reclassified from a lower risk category to an intermediary risk category, or to a higher risk category, as appropriate. Conversely, the discovery of a less serious situation may result in reclassifying a site from HrS to MrS or to LrS category. Relevant and standard fuel leak detection and screening methods can be used to aid field personnel in the classification/categorization tasks.
- *Contamination Assessment.* Typically, the LrS will require a field TPH test only. This method is recommended only when there is likelihood that a problem exists and a quick,

qualitative confirmation is desired. Thus, at an LrS, the field personnel may use a field TPH test to confirm the absence of soil contamination. In fact, only sites showing no evidence of possible soil contamination may use this somewhat qualitative form of analysis. The field TPH test can indeed prove helpful as an overall guidance tool.

If a field inspection or background check indicates that a site is suspected or known to have contaminated soil, or if a site has failed the field TPH test under category 1, then *quantitative* laboratory analysis of soil samples is usually needed. To ensure quality results, standard established procedures should be followed in the design of the sample collection and analysis procedures. For instance, soil samples should be quantitatively analyzed for BTEX using an appropriate or scientifically sanctioned analytical method. If any of these constituents is detected above the minimum practical quantitation limit, then the site investigation should proceed through a more detailed analysis that includes a general risk appraisal. If a general risk appraisal shows that groundwater is at risk, then further evaluation may be required, that will help verify an HrS category. The necessary procedures involve more detailed investigations and decisions above what is performed under an MrS. Among other things, groundwater gradient is determined using piezometers, which show the direction of groundwater and contaminant plume movements.

- *Corrective Action Decision Parameters.* Table 14.1 presents a comprehensive suite of decision elements typically used in the assessment and restoration of sites contaminated by petroleum hydrocarbon releases from leaking USTs.
- *The Site Management and Restoration Plan.* If the levels of contamination found at the spill site and impacted media are determined to exceed "acceptable" risk levels, then corrective actions—which may include site remediation—will generally be required. It must be recognized, however, that most cleanup actions cannot achieve a "zero" contamination level for petroleum-contaminated sites—or indeed for any contaminated site problem for that matter. The development of site-specific cleanup objectives necessary for the case-specific problem situation should therefore aid in the selection of appropriate site restoration strategies. Meanwhile, it is worth the mention here that, during the initial planning stages of the restoration program, it is generally recommended that soils at the contaminated site be screened to determine if it is possible for bioremediation to meet designated cleanup specifications for the site. As a general guide, cleanup levels of TPH ≤ 100 ppm and BTEX ≤ 10 ppb have been determined to be typical attainable cleanup limits for the bioremediation of gasoline- and diesel-contaminated sites (Barnhart, 1992). Bioremediation programs usually will require bench-scale tests and/or treatability studies to enhance the development of the design criteria and parameters. After the treatability study has demonstrated that the soil system will indeed support it, a full-scale program can then be implemented.

Generally speaking, cleanup of petroleum fuel spill sites may have to be carried out to a "realistic" level only. This is because once the major chemical constituents (such as benzene) have reached an approximately 100 ppb level in groundwater, for example, continued use of most groundwater remediation technologies may not yield any substantial improvement over that achieved through natural processes. In fact, trace residual hydrocarbons left in the subsurface at such "threshold" levels will degrade through natural processes in an equally timely manner.

As a whole, risk-assessment procedures can be used to establish cleanup objectives for most petroleum-contaminated sites. A site-specific risk assessment will guide the development of appropriate cleanup objectives with reference to site conditions, land uses, and exposure scenarios pertaining specifically to the case site and its vicinity. Such an assessment will provide directions to develop effectual corrective action and risk management plans.

TABLE 14.1

General Decision Elements Typically Used in the Corrective Action Assessment of UST Releases

Task (and Purpose)	Subtasks/Action Items	Major Decision Elements	Typical Investigation Rationale and/or Requirements
Preliminary site appraisal (to investigate indicated or suspected UST releases)	Establish basis for release indicators	Inventory variance/ discrepancy	Discrepancy of 0.5% of throughput over approximately 30-day period
		Tank closure	Cleaning of tank reveals corrosion (i.e., potential leak)
			Removal of tank reveals soil contamination or tank corrosion
		Nuisance conditions	Vapors in adjacent structures or utility conduits
			Non-aqueous phase liquid (NAPL) in adjacent structures or utility conduits
		Reported leakage	Documented release from tank or lines
		Piping problems	Loss of suction at dispensers
		Monitoring problems	Detection of NAPL in groundwater monitoring wells
			Detection of NAPL with interstitial monitoring devices
			Detection of vapors in tank pit monitoring wells
	Gather and review site background information for evidence of release	Geographic and geologic information	Topographic maps (for preliminary determination of groundwater flow direction, potential receiving water bodies, and residential communities)
			Geologic maps (for preliminary determination of subsurface conditions controlling the transport of contaminants)
			Soil survey (to determine characteristics of near surface soils with respect to contaminant transport)
			Local hydrogeologic records (for preliminary determination of groundwater depth and direction of groundwater flow)
			Well permits (for preliminary determination regarding the potential for impacting potable water wells)
		Engineering drawings	Delineate subsurface structures that may function to provide preferential contaminant transport pathways
			Determine the number, size, and location of subsurface storage tanks (gasoline, heating, and waste oil)

(Continued)

TABLE 14.1 (*Continued*)

General Decision Elements Typically Used in the Corrective Action Assessment of UST Releases

Task (and Purpose)	Subtasks/Action Items	Major Decision Elements	Typical Investigation Rationale and/or Requirements
		Interview of site personnel regarding storage tanks information, transportation mode, monitoring methods, etc.	Determine the method of product transfer from the tank to the dispenser
			Piping construction, location, and configuration
			Fill port location and configuration
			Determine if automated and/or manual monitoring devices are in use (i.e., interstitial monitoring [tanks and piping], tank pit monitoring wells, perimeter monitoring wells, and leak detectors)
			Determine property location with respect to environmentally sensitive areas (e.g., wildlife sanctuary)
		Site history	Determine previous land use to assess the potential for subsurface contamination from past site activities
			Assess the potential for contaminant transport from off-site sources
	Perform contamination assessment	Release dates	If a documented release(s) has occurred, defining the date is of importance when delineating the potential extent of contaminant migration within the subsurface
		Duration of release	If a leak rate is established during tank testing activities, than the duration of the release can be used to approximate the volume of product lost
		Quantities of liquid hydrocarbon lost	The volume of lost product dictates the severity of subsurface impact (i.e., large volumes indicate extensive soil degradation and most likely groundwater contamination; minor loss event may only impact a localized area of soil)
		Hydrocarbon type	Gasoline: vapor problems, high dissolved constituent (i.e., BTEX) concentrations in the groundwater, and rapid contaminant transport
			Diesel and/or fuel oils: minimal vapor problems, low dissolved constituent (i.e., BTEX) and high polynuclear aromatic hydrocarbon concentrations in the groundwater, and moderate to slow contaminant transport
		Origin of hydrocarbon	Source location to determine potential impact area
		Subsurface access points	Monitoring wells, storm drain catch basins, French drain systems, utility manholes tank access ports, dry wells, basements, etc.
	Identify potentially affected areas	Underground utilities	The presence of NAPL or hydrocarbon vapors in structures such as sanitary sewers, storm sewers, water lines, gas lines, dry wells, septic systems, power lines

(Continued)

TABLE 14.1 (*Continued*)

General Decision Elements Typically Used in the Corrective Action Assessment of UST Releases

Task (and Purpose)	Subtasks/Action Items	Major Decision Elements	Typical Investigation Rationale and/or Requirements
		Wells	Complaints of taste, odor, and sheen in the potable water of adjacent residences.
			The presence of NAPL and/or dissolved-phase contaminants (i.e., BTEX) in potable wells
		Surface water	The presence of NAPL on adjacent surface water bodies (i.e., streams, ponds, and marsh land)
		Residential properties	Presence of NAPL or hydrocarbon vapors in basement of adjacent residential structures
	Review, verify, and reconcile inventory records/data	Examination of overall physical facility (including calibration of dispenser meters, pressure testing of piping systems, and tightness test of storage systems)	Review record keeping procedures for potential discrepancy areas
			Look for obvious release characteristics (i.e., soil staining at fill ports, product in storm drains, etc.)
			Verify accuracy of dispenser meters
			Implement precision-testing procedures on piping system to determine integrity of pipe and fittings (e.g., leak rate > 0.05 gal/h indicates a release)
			Implement precision tests on individual subsurface storage tanks (e.g., leak rate > 0.05 gal/h indicates a release)
			Re-evaluate inventory records with tank/line test and dispenser calibration information
	Investigate other possible sources of hydrocarbon releases	Adjacent land uses	Review historic aerial photographs and records
		Other nearby tanks	Identify potential source areas based upon assumed direction of groundwater movement and product (contaminant) identification.
	Address health and safety issues	Health and safety requirements	Evaluate the potential adverse health and safety effects, associated with the release to the environment
		Regulatory requirements	Once a release is confirmed, identify all appropriate regulatory agencies/personnel (e.g., State regulatory agency and local fire marshal)
	Address emergency response	Release source control	If subsurface tank(s) are responsible for product discharge, evacuate the tank contents and remove the tank from service
		Free product removal	Implement emergency-response product-recovery efforts, gaining access to the subsurface through pit excavations and/or wells, to minimize the extent of environmental impact
		Imminent hazard mitigation	Address the potential for the development of an explosive environment
			Mitigate the presence of vapors in adjacent residential structures
			Provide alternate potable water supplies for impacted wells

(Continued)

TABLE 14.1 (*Continued*)

General Decision Elements Typically Used in the Corrective Action Assessment of UST Releases

Task (and Purpose)	Subtasks/Action Items	Major Decision Elements	Typical Investigation Rationale and/or Requirements
Site assessment (to confirm suspected UST releases)	Implement subsurface investigation to characterize site	Identify contaminated soils and determine extent of soil contamination	Establish area of contaminated soil utilizing a soil vapor survey and/or soil borings
			Determine the magnitude of soil contamination through the collection of depth-specific samples for TPH, BTEX, and gasoline additive analyses
		Identify free product on water table; identify dissolved contaminants in groundwater; and determine extent of groundwater contamination	Install groundwater monitoring wells that are screened across the water table to detect the presence of NAPL
			Conduct periodic liquid-level measurements to determine the depth of groundwater and the thickness of NAPL accumulations
			Conduct product recovery (numeric and/or field) tests to delineate the "true" product thickness
			Develop a groundwater monitoring well array designed to define the background water quality and the downgradient extent of dissolved contaminant transport.
			Collect groundwater samples to facilitate quantitative laboratory analysis for BTEX and gasoline additive constituents
		Determine presence and distribution of vapor accumulations	Implement a soil vapor investigation and/or conduct periodic vapor monitoring of adjacent utility conduits
		Determine sources and directions of contaminant migration	Utilize the monitoring well array to establish the direction of contaminant transport and define potential source areas
			Conduct periodic liquid level measurements to determine the hydraulic gradient and groundwater velocity
		Determine hydraulic properties controlling contaminant movement	Conduct hydraulic conductivity tests (slug and/or pump tests) to establish aquifer hydraulic properties
		Identify hydrogeologic conditions	Determine the following: depth to groundwater, type of aquifer (i.e., water table, confined, or perched), thickness of the water bearing strata, permeability, and porosity of the formation
		Determine regional recharge	Define the recharge area, typically upgradient topographically, for the aquifer of concern
		Determine discharge areas	Look for local streams, seeps, springs, and surface water bodies that function as groundwater discharge points.
		Determine outlet points where contamination can impact public health and safety	Withdrawal of groundwater from large volume production wells can have significant control over the direction of localized groundwater flow.
			Conduct a record search for potable water wells within the immediate area of the site.

(Continued)

TABLE 14.1 (*Continued*)

General Decision Elements Typically Used in the Corrective Action Assessment of UST Releases

Task (and Purpose)	Subtasks/Action Items	Major Decision Elements	Typical Investigation Rationale and/or Requirements
	Characterize petroleum hydrocarbons	Product type	Utilize visual observations, specific gravity, and gas chromatograph (GC) fingerprint analyses to characterize the type of product (such as gasoline; middle distillates [diesel, kerosene, jet fuels, and lighter fuel oils]; heavier fuel oils and lubricating oils; asphalts and tars)
		Physical properties	Viscosity which will control migration and adsorption of contaminants. Volatility which will control the production of hazardous vapors and the ratio of dissolved to NAPL-phase contamination within the groundwater.
		Chemical properties	Organic carbon partitioning coefficient, half-life, solubility, etc.
	Verify presence of hydrocarbons	Delineation of residual liquid hydrocarbons	Collect depth-specific soil samples above the water table for TPH and BTEX analysis to define the quantity of adsorbed contaminants. The percentage of organic carbon within the soil matrix must also be evaluated
		Delineation of liquid-phase hydrocarbons	Measure "apparent" product thickness from monitoring wells and conduct NAPL bail-down recovery tests to delineate the "true" product thickness and, therefore, provide more accurate estimates regarding the volume of product within the geologic formation.
		Delineation of dissolved-phase hydrocarbons in vadose zone	Conduct a soil gas survey utilizing a portable GC for constituent identification. Verify results and calibrate survey with quantitative laboratory analysis of soil samples
		Delineation of dissolved-phase hydrocarbons in groundwater zone	Conduct BTEX and gasoline additive analysis in all non-product-bearing wells.
		Delineation of vapor-phase	Conduct soil gas investigations
		Delineation of adsorbed-phase	Soil sampling to determine soil partitioning coefficient.
	Confirm presence of free phase hydrocarbons	Remove free-phase hydrocarbons	Implement manual (bailing) recovery of NAPL from all product-bearing wells until a centralized, automated, recovery program is initiated.
	Confirm presence of dissolved-phase contamination	Sample for indicator parameters	Sample potentially impacted potable wells and surface waters for indicator parameters (such as BTEX, MTBE, ethylene dibromide [EDB], lead, and other applicable additives)

(Continued)

TABLE 14.1 (*Continued*)

General Decision Elements Typically Used in the Corrective Action Assessment of UST Releases

Task (and Purpose)	Subtasks/Action Items	Major Decision Elements	Typical Investigation Rationale and/or Requirements
	Confirm presence of vapor-phase contamination	Delineate preferential vapor migration pathways	Review utility conduit drawings and engineering plans to delineate potential vapor migration pathways.
		Ventilate vapors to eliminate explosive environment	Evacuate hydrocarbon vapors from all structures, including utility conduits and residences, to reduce vapor concentrations below the lowest explosive limit.
		Institute remedial action program to reduce vapor concentrations to within acceptable risk-based levels	Establish either positive pressure within or negative pressure outside of adjacent structures possessing vapor contamination to mitigate the migration of vapors into the structure.
	Perform contamination pathways analysis	Develop working hypothesis about migration of liquid, vapor, and dissolved phases beneath and near releases site	Evaluate all information regarding indigenous stratigraphy and subsurface structures to assess preferential contaminant migration pathways, if any
		Predict timing and concentration levels of mobile phases reaching pathway outlets	Utilize aquifer hydraulic parameters and contaminant characteristics to implement a contaminant transport analysis (computer modeling)
		Determine populations potentially at risk	Having defined both spatial and temporal controls on contaminant transport, evaluate the potential impact area, and associated receptors
		Identify potential exposure routes	Map areas where hydrocarbon phases may impact human health and/or environment
			Map distribution of hydrocarbon phases and all potential pathway outlets
Risk appraisal (for contamination assessment of UST releases)	Determine factors affecting contaminant fate and transport	Adsorptive capacity of earth materials	Organic carbon content, soil/water partitioning coefficient, solubility of constituents
		Perching horizons and interconnected void spaces	Porosity and permeability
		Relative conductivity of earth materials to water and other hydrocarbon fluids	Hydraulic conductivity, grain size analysis, matrix sorting
		Rates and directions of groundwater movement	Hydraulic gradient, permeability, and porosity
		Processes that attenuate concentrations and limit area of contaminated zones	Dispersion, dilution, adsorption, biodegradation, and biotransformation.

(*Continued*)

TABLE 14.1 (*Continued*)

General Decision Elements Typically Used in the Corrective Action Assessment of UST Releases

Task (and Purpose)	Subtasks/Action Items	Major Decision Elements	Typical Investigation Rationale and/or Requirements
	Determine acceptable cleanup criteria to qualify risks	Establish liquid hydrocarbon criteria that confirms removal	Cessation of liquid hydrocarbon discharges into underground structures or openings. Attainment of practical limits for hydrocarbon recovery at wells and drains
		Establish dissolved hydrocarbon criteria (acceptable levels of BTEX), confirming removal	Presence of only traces of free hydrocarbons in monitoring wells. Background levels of offsite contaminant levels. Constituents reach asymptotic contaminant levels.
		Establish vapor hydrocarbon criteria	Explosion limit for total vapors in air. Offensive odors. Background levels for key constituents. Leaching potential of key constituents to groundwater or soil.
		Establish residual hydrocarbon criteria	Proximity of potential discharge points connecting leaching process.
	Perform exposure assessment	Indicator/target chemical selection	For gasoline, common indicators are BTEX.
		Select significant exposure pathways	Selection based on information on probable exposure scenarios.
		Address fate and transport modeling	Assess modeling needs and select appropriate model
	Define magnitude/ severity of health and environmental impacts	Potential for exposure, via contaminant mobility, and pathway analyses	Determine concentrations, toxicity, sensitive receptors, future land uses
Site remediation (consisting of an evaluation of the applicability of remedial alternatives)	Implement emergency response	Soil excavation	Remove or minimize contaminant source(s)
		Hydrocarbon vapor control	Eliminate explosive environment and detrimental health conditions
		NAPL recovery	Manual recovery utilizing hand bailing or vacuum truck without water table depression
		Dissolved hydrocarbon-groundwater control	Groundwater withdrawal creating a localized cone of depression to reverse contaminant migration
	Establish remediation objectives and cleanup criteria	Cleanup objectives and criteria	Negotiate cleanup standards with regulatory agency based on results of risk analysis
	Establish limits of applicability of corrective action program for vadose zone soil remediation	Soil excavation	Limited areal extent of contaminated soils Limited depth of contaminant migration Moderate to high concentrations of adsorbed contaminants Soils away from foundations and below ground structures
		Soil venting	Volatile contaminants Permeable soils Moderate to high concentrations of adsorbed contaminants Absence of NAPL

(Continued)

TABLE 14.1 (*Continued*)
General Decision Elements Typically Used in the Corrective Action Assessment of UST Releases

Task (and Purpose)	Subtasks/Action Items	Major Decision Elements	Typical Investigation Rationale and/or Requirements
		Enhanced biodegradation	Indigenous microbial population
			Nonreactive soil matrix
			Neutral pH groundwater
			Permeable soils
		Surfactant flushing	Permeable soils
			Low to moderate viscosity product
		Landfilling	Low to high concentration of adsorbed contaminants
			Low to high viscosity products
			Classified soils
		On-site aeration treatment	Volatile contaminants
			Limited depth of contaminant migration
			Low to high concentration of adsorbed contaminants
			Moderate to high permeable soils
			Air emission monitoring
		Asphalt batching/ incorporation	Low to high concentration of adsorbed contaminants
			Moderate to high viscosity product
			Moderate to low volatile content
	Establish limits of applicability of corrective action program to vapor-phase remediation	Soil venting	Volatile contaminants
			Permeable soils
			Moderate to high concentrations of adsorbed contaminants
			Absence of NAPL
	Establish limits of applicability of corrective action program to NAPL recovery	Interceptor trenches and drains	NAPL on perched water system
			Low permeability formations
			Shallow water table aquifers
			Limited volume of fugitive product
		Skimming systems	Shallow to moderate depth water table aquifers
			Perched water systems
			Minimal NAPL accumulations
			Low to moderate permeability formations
			Plume of large areal extent
			Low to moderate viscosity product
		Single-pump (total fluids) systems	Shallow to deep aquifers
			Low to moderate permeability formations
			Minimal NAPL accumulations
			Low to moderate viscosity product
			Requires phase separation (i.e., oil/water separator)
		Two-pump systems	Shallow to deep aquifers
			Substantial NAPL accumulations
			Moderate to high permeability formations
			Low to moderate viscosity products

(Continued)

TABLE 14.1 (*Continued*)

General Decision Elements Typically Used in the Corrective Action Assessment of UST Releases

Task (and Purpose)	Subtasks/Action Items	Major Decision Elements	Typical Investigation Rationale and/or Requirements
	Establish limits of applicability of corrective action program to dissolved hydrocarbon recovery/treatment	Air stripping	Volatile contaminants (i.e., BTEX)
			Low to high solubility constituents
			Large throughput volume
			Low to high VOC concentration
			Low ambient iron concentration
			Air emissions
		Activated carbon adsorption	Low to moderate solubility compounds
			Volatile contaminants
			Low to moderate throughput volume
			Low to moderate VOC concentration
			Low ambient iron and microbial concentration
			Carbon regeneration or disposal
		Chemical oxidation	Oxidizable contaminants
			Low to high throughput volume
			Low ambient iron concentration
		Combined air stripping and carbon adsorption	Moderate to highly volatile contaminants
			Stringent air emission and/or water discharge requirements
		Spray irrigation	Large surface area
			Liberal air emission and groundwater discharge requirements
			Indigenous microbial population
		In situ enhanced biodegradation	Indigenous microbial population
			Neutral pH groundwater
			Nonreactive aquifer matrix
			Permeable formation
		Natural (passive) biodegradation	Indigenous microbial population
			Low to moderate contaminant concentrations
			Plume of limited areal extent
			Low to moderate contaminant transport velocity

14.2.2 Design Protocol for a Corrective Action Investigation

This section illustrates the nature of decision elements typically used in the investigation of potentially contaminated sites. The example discusses a potential contamination problem at a small shopping center located in a rural township that depends almost exclusively on groundwater from a contiguous aquifer for its water supplies. The trigger for this investigation is the discovery that dry-cleaning solvents have spilled on concrete floor slabs at a dry-cleaning facility within this mini mall.

14.2.2.1 Background

A preliminary environmental site assessment carried out for the Village Shopping Center (VSC) indicated the potential for soil and groundwater contamination at this facility as a result of releases from dry-cleaning and laundry activities at one section of the mall. A follow-up Phase II-type assessment conducted for the site confirmed the presence of elevated levels of perchloroethylene (PCE) in the sampled soils. The soils beneath the facility location consist predominantly of dense clayey fine sands or fine sand, with occasional gravels. Some plastic liner materials (likely to have been used as plastic moisture barrier beneath the concrete floor slabs) were found in some of the exploratory borings for the site. It is believed that the plastic liner material found beneath the concrete floor may well have prevented extensive contamination of the subsurface environmental compartments.

14.2.2.2 Recommendations for the Corrective Action Investigation

Since PCE spills would have occurred on the concrete floors at the VSC facility, and because the plastic liner material may have been serving as a "barrier against further contaminant migration," it is quite possible that any PCE encountered in the exploratory soil borings could have been introduced into the soil after the barrier was broken during the soil sample coring activities. If this hypothesis is true, then the extent of soil contamination may be even less than suspected; in addition, the possibility of any extensive groundwater contamination can also be ruled out. Under such circumstances, a more detailed assessment may determine that no extensive and expensive remediation or cleanup program is necessary for the VSC facility. With the aforementioned in mind, this reasoning will also support the importance of studying complete building plans/layouts before any drilling activities that could actually facilitate the spreading of contaminants that would otherwise be sitting as a more easily removable free product.

To complete the investigations for the VSC facility that will allow for appropriate corrective action decisions, a number of issues must be fully explored and evaluated—including the following key attributes:

1. Groundwater beneath the site should be investigated to complete the site assessment. This is because soils beneath the site have already been impacted, and PCE is reasonably mobile in the type of soil formations at this site. That is, considering the mobility of PCE and the sandy nature of soils found at the VSC facility, the possibility of a contaminated aquifer beneath the facility cannot quite be ruled out. In particular, if it cannot be established that the liner materials may have prevented the PCE from migrating further into the subsurface environment, then the groundwater system beneath the site should be fully investigated. On the other hand, if it can be positively established that the plastic liner material did serve to prevent or minimize PCE migration into the subsurface environments, then a different set of exposure scenario may be developed for the corrective action program for this facility.

2. Site conditions should be adequately characterized, in order to help in fully defining the lateral and vertical extents of the PCE contamination. In this regard, it is notable that PCE has rather low adsorption to soil; consequently, if released to soils, it will generally be subject to an accelerated migration into the groundwater. In particular, PCE can move rapidly through sandy soils and may therefore reach groundwater more easily in such formations as found at the VSC facility.

3. Risk assessment procedures can be used to establish cleanup objectives for contaminated environmental media requiring corrective actions and/or for the implementation of risk management programs for contaminated sites. Such an assessment will help focus corrective action assessments and risk management plans for the VSC facility. The site-specific risk assessment will also include the development of appropriate site restoration goals with reference to site conditions, land uses, and exposure scenarios pertaining specifically to this shopping center and its vicinity. It is believed that the development of site-specific cleanup levels can result in significant cost savings in this investigation.

In the end, a complete characterization of the "contaminated zone" at the site should facilitate the screening and selection of the best available technology for a remedy evaluation program designed for the VSC facility. Prior to implementing any remediation plan for this site, appropriate risk-based cleanup criteria should be developed and compared with the levels of contamination presently to be found at the site (i.e., under the baseline conditions). Based on such criteria, it may well become apparent under the prevailing types of exposure scenarios that no cleanup is warranted after all. In fact, even if it is determined that some degree of cleanup is required, the cleanup criteria developed will aid in optimizing the efforts involved—so as to arrive at more cost-effective solutions than could otherwise have been achieved.

14.2.3 EVALUATION OF HUMAN HEALTH RISKS ASSOCIATED WITH AIRBORNE EXPOSURES TO ASBESTOS

There are two subdivisions of asbestos: the serpentine group containing only chrysotile (which consists of bundles of curly fibrils) and the amphibole group containing several minerals (which tend to be more straight and rigid). Asbestos is neither water soluble nor volatile, so that the form of concern is microscopic fibers (usually reported as, or measured in the environment in units of fibers/m^3 or fibers/cc). Now, processed asbestos has typically been fabricated into a wide variety of materials that have been used in consumer products (such as cigarette filters, wine filters, hair dryers, brake linings, vinyl floor tiles, and cement pipe) and also in a variety of construction materials (e.g., asbestos-cement pipe, flooring, friction products, roofing, sheeting, coating and papers, packaging and gaskets, thermal insulation, and electric insulation). Notwithstanding the apparent useful commercial attributes, asbestos has emerged as one of the most complex, alarming, costly, and tragic environmental health problems (Brooks et al., 1995). A case in point, asbestos materials are frequently removed and discarded during building renovations and demolitions. To ensure safe ambient conditions under such circumstances, it often becomes necessary to conduct an asbestos sampling and analysis—whose results can be used to support a risk assessment. This section presents a discussion of the investigation and assessment of the human health risks associated with worker exposures to asbestos in the ventilation systems of a commercial/office building.

14.2.3.1 Study Objective

The primary concern of the risk assessment for the ventilation systems in the case building is to determine the level of asbestos exposures that potential receptors (especially workers cleaning the ventilation systems) could experience and whether such exposure constitutes potential significant risks.

14.2.3.2 Summary Results of Environmental Sampling and Analysis

Standard air samples are usually collected on a filter paper, and fibers >5 μm long are counted with a phase contrast microscope; alternative approaches include both scanning and transmission electron microscopy (TEM) and X-ray diffraction. It is generally believed that fibers that are 5 μm or longer are of potential concern (USEPA, 1990a,b).

Following an asbestos identification survey of the case structure, air samples collected from suspect areas in the building's ventilation systems were analyzed using phase contrast microscopy (PCM), and highly suspect ones further analyzed by using TEM. The TEM analytical results are important because they serve as means/methods for distinguishing asbestos particles from other fibers or dust particles.

The PCM analysis produced concentration of asbestos fibers in the range of <0.002 to a maximum of 0.008 fibers/cm^3. From the TEM, chrysotile asbestos was determined to be at <0.004 structures per cm^3 (str/cm^3) in all the environmental air samples.

14.2.3.3 The Risk Estimation

For asbestos fibers to cause any disease in a potentially exposed population, they must gain access to the potential receptor's body. Since they do not pass through the intact skin, their main entry routes are by inhalation or ingestion of contaminated air or water (Brooks et al., 1995)—with the inhalation pathway apparently being the most critical in typical exposure scenarios. That is, for asbestos exposures, inhalation is expected to be the only significant exposure pathway. Consequently, intake is based on estimates of the asbestos concentration in air, the rate of contact with the contaminated air, and the duration of exposure. Subsequently, the intake is integrated with the toxicity index to determine the potential risks associated with any exposures.

Individual excess cancer risk is a function of the airborne contaminant concentration, the probability of an exposure causing risk, and the exposure duration (ED). By using the cancer risk equations presented earlier in Chapter 9, the cancer risk from asbestos exposures may be estimated in accordance with the following relationship:

$$\text{Cancer risk} = \left[\text{airborne fiber concentration} \left(\text{fibers}/\text{m}^3 \right) \right] \times \left[\text{exposure constant} \left(\text{unitless} \right) \right]$$
$$\times \left[\text{inhalation unit risk} \left(\left(100 \text{ PCM fibers}/\text{m}^3 \right)^{-1} \right) \right] \tag{14.1}$$

or

$$\text{Risk probability} = \text{Intake} \times \text{UR} = \left[C_a \times \text{INHf} \right] \times \text{UR} \tag{14.2}$$

The following exposure assumptions are used to facilitate the intake computation for this particular problem, as identified previously:

- It is assumed that workers cleaning the ventilation system will complete this task within 2 weeks for a 5-day workweek. Hence, the maximum ED is taken as, ED = 10 days—in comparison to a 70-year lifetime daily exposure.
- Assumed exposure time is 40 min per working hour, for an 8-h workday.
- Inhalation rate is 20 m^3/day (or 0.83 m^3/h).

The exposure evaluation utilizes the information obtained from the airborne fiber samples collected and analyzed for during the prior air sampling activities; to be conservative, the maximum concentrations measured from the analytical protocols are used in the risk estimation. Thence, the fraction of an individual's lifetime for which exposure occurs—represented by the inhalation factor—is estimated to be as follows:

$$\text{INHf} = \left(40/60 \right) \times \left(8/24 \right) \times \left(10/365 \right) \times \left(1/70 \right) = 8.7 \times 10^{-5}$$

Next, asbestos is considered carcinogenic with a unit risk of approximately 1.9×10^{-4} (100 PCM fibers/m^3)$^{-1}$ (see, e.g., DTSC/Cal EPA, 1994). Consequently, potential risk associated with the "possible" but unlikely (represented by an evaluation based on the PCM analysis results) and the reasonable/likely (represented by an evaluation based on TEM analysis results) asbestos concentrations are determined, respectively, as follows:

- Risk represented by results of the PCM analyses is estimated by integrating the following information,
  - PCM-based airborne fiber concentration (maximum) = 0.008 fibers/cc = 8×10^3 fibers/m^3
  - INHf = 8.7×10^{-5}
  - UR = 1.9×10^{-4} (100 PCM fibers/m^3)$^{-1}$ ≡ 1.9×10^{-6} per fibers/m^3

Hence,

$$\text{Cancer risk (based on PCM concentration)} = 1.32 \times 10^{-6}$$

- Risk represented by results of the TEM analyses is estimated by integrating the following information:
  - TEM-based airborne asbestos concentration (maximum) = 0.004 str/cc = 4×10^3 str/m^3
  - INHf = 8.7×10^{-5}
  - UR = 1.9×10^{-4} (100 PCM fibers/m^3)$^{-1}$ ≡ 1.9×10^{-6} per fibers/m^3

Hence,

$$\text{Cancer risk (based on TEM concentration)} = 6.6 \times 10^{-7}$$

14.2.3.4 A Risk Management Decision

All risk estimates indicated here are near the lower end of the generally acceptable risk range/spectrum (i.e., 10^{-4} to 10^{-6}). Thence, it may be concluded that asbestos in the case building should represent minimal potential risks of concern for workers entering the ventilation system to clean it up. Nonetheless, it is generally advisable to incorporate adequate worker protection through the use of appropriate respirators. In general, any asbestos abatement or removal program should indeed conform to strict health and safety requirements—with on-site enforcement of the specifications being carried out by a qualified health and safety officer or industrial hygienist.

14.2.4 A HUMAN HEALTH RISK ASSESSMENT FOR PCB RELEASE INTO THE ENVIRONMENT

Polychlorinated biphenyls (PCBs) are mixtures of synthetic organic chemicals. Different mixtures can take on forms ranging from oily liquids to waxy solids. Although their chemical properties vary widely, different mixtures can have many common components. Because of their non-inflammability, chemical stability, and insulating properties, commercial PCB mixtures had been used in many industrial applications, especially in capacitors, transformers, and other electrical equipment. These chemical properties, however, also contribute to the persistence of PCBs after they are released into the environment. In fact, because of evidence that PCBs persist in the environment and cause harmful effects, the manufacture of commercial mixtures was stopped in the late1970s—albeit existing PCBs continued in use.

What is more, PCBs persist in the body—providing a continuing source of internal exposure after external exposure stops. There may be greater-than-proportional effects from less-than-lifetime exposure, especially for persistent mixtures and for early-life exposures. PCBs are absorbed through ingestion, inhalation, and dermal exposure, after which they are transported similarly through the circulation. This provides a reasonable basis to expect similar internal effects from different routes of human exposure. Indeed, joint consideration of cancer studies and environmental processes leads to a conclusion that environmental PCB mixtures are highly likely to pose a risk of cancer to humans. Apart from the cancer effects, PCBs also have significant human health effects other than cancer—including neurotoxicity, reproductive and developmental toxicity, immune system suppression, liver damage, skin irritation, and endocrine disruption. Toxic effects have indeed been observed from acute and chronic exposures to PCB mixtures with varying chlorine content.

14.2.4.1 Problem Scenario

Consider a release of PCBs onto the ground near a lake. Potential pathways of human exposure have been determined to include vapor inhalation, drinking water, fish ingestion, and skin contact with ambient water and contaminated soil.

The population of interest includes anglers who consume an average of two 105 g portions of local fish each week; this translates into 30 g of fish ingestion per day (i.e., [2 × 105 g/week]/7 days/week = [210/7] = 30 g/day). They also spend most of their time in the area, on average, breathing 20 m^3 of air and drinking 2 L of water each day. Skin contact with ambient water and soil is negligible for this population. A 30-year human ED is assumed, with a representative life span of 70 years and an average body weight of 70 kg.

Environmental samples indicate long-term average concentrations of 0.01 µg/m³ in ambient air, 5 µg/L in drinking water, and 110 µg/kg in the edible portion of local fish. Issues pertaining to dust in ambient air and sediment in drinking water are considered negligible.

14.2.4.2   The Exposure Scenarios

Three different exposure pathways are assumed for this case problem—namely, vapor inhalation, water ingestion, and fish consumption. Because of partitioning, transformation, and bioaccumulation, different fractions of the original mixture are encountered through these pathways—and hence different potency values are appropriate. Vapor inhalation is associated with "low risk" (because evaporating congeners tend to have low chlorine content and be inclined to metabolism and elimination) so the low end of the range (upper-bound slope of 0.4 per mg/kg-day) is used for vapor inhalation (USEPA, 1996g). Similarly, ingestion of water-soluble congeners is associated with "low risk" (because dissolved congeners tend to have low chlorine content and be inclined to metabolism and elimination)—so the low end (of 0.07 per mg/kg-day) is also used for drinking water (USEPA, 1996g). (It is noteworthy that, if ambient air or drinking water had contained significant amounts of contaminated dust or sediment, the high-end potency values would be more appropriate, as adsorbed congeners tend to be of high chlorine content and persistence.) Finally, food chain exposure is more realistically associated with "high risk" (because aquatic organisms and fish selectively accumulate congeners of high chlorine content and persistence that are resistant to metabolism and elimination)—and thus, the high end of the range (upper-bound slope of 2 per mg/kg-day) is used for fish ingestion (USEPA, 1996g).

14.2.4.3   Risk Calculations

The lifetime average daily dose (LADD) is calculated as the product of concentration, C, intake rate, IR, and ED divided by body weight, BW, and lifetime, LT, as follows:

$$\text{Pathway exposure, LADD} = [C \times \text{IR} \times \text{ED}]/[\text{BW} \times \text{LT}] \qquad (14.3)$$

Thence,

$$\text{Vapor inhalation LADD} = \left[0.01\,\mu\text{g/m}^3 \times 20\,\text{m}^3/\text{day} \times 30\,\text{year}\right]/\left[70\,\text{kg} \times 70\,\text{year}\right]$$

$$= 1.2 \times 10^{-6}\,\text{mg/kg-day}$$

$$\text{Drinking water LADD} = \left[5.0\,\mu\text{g/L} \times 2\,\text{L/day} \times 30\,\text{year}\right]/\left[70\,\text{kg} \times 70\,\text{year}\right]$$

$$= 6.1 \times 10^{-5}\,\text{mg/kg-day}$$

$$\text{Fish ingestion LADD} = \left[110\,\mu\text{g/kg} \times 30\,\text{g/day} \times 30\,\text{year}\right]/\left[70\,\text{kg}/70\,\text{year}\right]$$

$$= 2.0 \times 10^{-5}\,\text{mg/kg-day}$$

Subsequently, for each pathway, the LADD is multiplied by the appropriate slope factor to arrive at the estimated risk, as follows:

$$\text{Pathway risk} = [\text{LADD}] \times [\text{cancer slope factor}] \qquad (14.4)$$

Thence,

$$\text{Vapor Inhalation Risk} = 1.2 \times 10^{-6}\,\text{mg/kg-day} \times 0.4\,\text{per mg/kg-day} = 4.8 \times 10^{-7}$$

$$\text{Drinking Water Risk} = 6.1 \times 10^{-5} \text{ mg/kg-day} \times 0.07 \text{ per mg/kg-day} = 4.3 \times 10^{-6}$$

$$\text{Fish Ingestion Risk} = 2.0 \times 10^{-5} \text{ mg/kg-day} \times 2 \text{ per mg/kg-day} = 4.0 \times 10^{-5}$$

Thus,

$$\text{Total LADD} = 8.2 \times 10^{-5} \text{ mg/kg-day}$$

and

$$\text{Total risk} = 4.5 \times 10^{-5}$$

14.2.4.4 A Risk Management Decision

The above-mentioned evaluation leads to a conclusion that fish ingestion is the principal pathway contributing to risk, and that drinking water and vapor inhalation are of lesser consequence. Indeed, it would be advisable to examine variability in fish consumption rates and fish tissue concentrations to determine whether some individuals are at much higher risk. In any case, it also is important to recognize that, this specific site exposure adds to a background level of exposure from other sources.

14.2.5 AN ILLUSTRATION OF A STATISTICAL EVALUATION USED TO FACILITATE A CORRECTIVE ACTION DECISION

This section consists of a simple statistical correlation analysis that is used to support a corrective action decision for a PCB-oil contaminated site located within the rural and predominantly agricultural community of NK4. This site has an area of about 30 acres (≈12 ha), about half of which is occupied by a 6-m deep lagoon. The lagoon contains oily liquids and PCBs in excess of 7,000 ppm in certain sediment pockets. A thick clay unit at the lagoon bottom provides protection of a water supply aquifer at the NK4 site. The lagoon water level is up to 3 m higher than the groundwater table, causing a groundwater mound in the vicinity of the lagoon. Nonetheless, there does not appear to be significant hydraulic flows of lagoon water into the aquifer, apparently because of possible sealing/barrier effects created by the oily bottom sludge together with the clay unit.

14.2.5.1 Study Objective

A simple statistical evaluation is being considered to facilitate cost-effective corrective action decisions for the NK4 site. It is generally believed that the PCBs will tend to be concentrated in the oils. The specific objective of this evaluation is to confirm if any reasonable correlation exists between the occurrence of oil/grease and PCB levels in the lagoon sediments. The rationale for this evaluation is that, if it is determined that high levels of oil/grease is associated with high levels of PCBs and vice versa, then a remedial design can be so tailored to extract PCB-laden oils first, followed by the dredging of PCB-contaminated "hot-spots" in the sediment zones. The oils and sediments can then be treated using distinct incineration processes.

14.2.5.2 Choice of Statistical Tests

Nonparametric statistical tests are used in this evaluation since these have fewer and less stringent assumptions in comparison with parametric analyses. The nonparametric correlational technique selected for this correlation analysis is the Spearman rank coefficient of correlation. This coefficient or number indicates the exact strength and direction of the relationship between the two sets of variables being compared, in this case, the PCB and oil/grease levels found in sediment samples taken from the lagoon at the NK4 site. A rough gage for interpreting the Spearman rank coefficient of correlation is given as follows (Sharp, 1979):

High : 0.85 to 1.0 (or, −0.85 to −1.0)

$$Moderate : 0.50 \text{ to } 0.84 \text{ (or, } -0.50 \text{ to } -0.84)$$

$$Low : 0 \text{ to } 0.49 \text{ (or, } 0 \text{ to } -0.49)$$

Subsequently, a two-tailed hypothesis test is performed to determine the level of significance of the estimated correlation coefficient.

14.2.5.3 Statistical Evaluation

The statistical evaluation process used in this study is described as follows:

- The Spearman rank correlation coefficient (or the Spearman rho), ρ, used in this evaluation is given by:

$$\rho = 1 - \frac{\left(6D^2\right)}{N\left(N^2 - 1\right)}$$

  where N is the number of individual points in a group and D represents the difference between the ranks for the groups. Results from the application of this technique to the PCB and oil/grease levels present in the lagoon sediment samples are presented in Table 14.2. A Spearman rho of 61% (i.e., $\rho = 0.61$) is indicated, which represents a moderate positive correlation between the levels of PCB and oil/grease found in the sediment samples taken from the lagoon.
- To test the level of significance of this correlation involves testing the null hypothesis, H_0: *No correlation exists between PCB and oil/grease levels in sediment* versus the corresponding alternative hypothesis, H_a: *PCB and oil/grease levels in lagoon sediments are related.* This is a two-tailed test and is performed at a significance level of $\alpha = 0.01$. Using standard statistical tables (e.g., Sharp, 1979), $\rho_{\alpha,n}$ from the tables is compared against the computed ρ value, and H_0 is rejected if $|\rho| > |\rho_{\alpha,n}|$, otherwise H_0 is accepted.
- At a level of significance of $\alpha = 0.01$ (which corresponds to 99% confidence level), H_0 is compared against H_a to arrive at a statistical decision. This helps determine whether or not the observed value of $\rho = 0.61$ differs from zero only by chance. The following steps are used to help arrive at this statistical decision (e.g., Sharp, 1979):

  i. Compute $Z = \rho\left\{(N-1)^{0.5}\right\}$

       i.e., $Z = 0.61\left\{(47-1)^{0.5}\right\} = 4.14$

  ii. Determine the probability of Z from standard statistical tables as follows:

$$Prob\{Z = 4.14\} = 2 \times (0.00003) = 0.00006$$

$$\text{i.e., } P = 0.00006$$

  iii. If the probability value, P, obtained from the tables is less than or equal to α, then H_0 is rejected—that is,

$$\text{Reject } H_0 \text{ if } P \leq \alpha$$

  Now, since P (=0.00006) < α (=0.01), H_0 is rejected and H_a accepted. This means that, there is a reasonable degree of correlation between the occurrence of oil/grease and PCB in the lagoon sediment samples.
- The statistical conclusion, made at a 99% level of confidence, is that high levels of PCBs are likely to occur where high levels of oil/grease exist in the lagoon at NK4.

TABLE 14.2

Statistical Evaluation Results: Statistical Correlation Analysis for PCB and Oil/Grease Contamination at the NK4 Site

Sampling Event/Group	PCB Concentration in Soils/Sediments (ppm)	Oil/Grease Concentration in Soils/Sediments (ppm)
1	14.6	9,630
2	1.8	507
3	8.3	2,230
4	3.3	2,870
5	67.7	98,900
6	20.9	71,600
7	25.6	71,600
8	31.1	43,700
9	46	81,600
10	194	188,000
11	1.5	2,020
12	3.9	91,200
13	3.8	16,600
14	3.8	14,300
15	951	230,000
16	826	25,000
17	22.7	73
18	7,241	130,000
19	5,639	1,700
20	840	110
21	2.5	620
22	0.7	100
23	5.2	110
24	57.6	150,000
25	6.1	270
26	13.4	7,400
27	8.9	15,000
28	0.6	1,900
29	9.0	4,800
30	3.0	15,000
31	13.1	85,000
32	2.0	1,300
33	2.3	2,080
34	78	130,000
35	16.9	77,000
36	203	109,000
37	119	56,000
38	25.6	25,000
39	88	107,000
40	7.2	6,800
41	114	107,000
42	142	60,000
43	99	170,000
44	34	3,700
45	9.9	12,000
46	6.3	7,200
47	4.3	750

Note: The Spearman rank correlation coefficient for the data set is 0.61.

14.2.5.4 The Decision Process

Results obtained from this statistical analysis indicate a good chance that high levels of PCBs will likely be found concentrated in the oil/grease. In fact, PCBs are not soluble in water and are generally immobile in the solid matrix. Consequently, if the oil can be extracted from the lagoon sediments and/or water, then PCB levels in these matrices may be greatly reduced. This would in turn reduce the volumes of high PCB-contaminated sediments and/or water that have to be treated. A recommended remedial strategy for the NK4 site will consist of the dredging and on-site incineration of the oil/PCB sludge and sediments, together with an on-site treatment of the lagoon water.

In general, it is expected that the costs to incinerate PCB-contaminated oil will be considerably less than the incineration of PCB-contaminated sediments. Thus, the knowledge gained from the statistical evaluation may be used to support a decision to initially extract PCB-laden oils, followed by a sediments' cleanup process. An effective restoration process may be dry excavation to remove sediments, after dewatering the lagoon. The lagoon can be dewatered using partially penetrating barrier walls in conjunction with a system of pumping wells. The barrier wall used in the lagoon dewatering system can also serve as a containment system that will minimize potential migration of contaminants; additionally, the wells may aid further groundwater pump and treat remedial actions likely required for groundwater remediation. The aqueous phase (to be treated and re-used or discharged) can then be separated from the PCB-laden oil phase (to be incinerated).

14.2.6 Illustration of the Site Restoration Assessment Process

A hypothetical problem is discussed here, to illustrate the site restoration requirements for a petroleum-contaminated site problem. This illustrative problem consists of the development of a corrective action plan in response to hydrocarbon releases from USTs at an operating facility.

14.2.6.1 Background Information

A limited environmental site assessment conducted for the Petro-X Fueling and Automotive Services facility located in East London indicated a high degree of soil and groundwater contamination within the site boundaries. This occurrence is the result of UST releases and also spills of organic solvents associated with the automotive service station. For the purpose of this illustration, it is assumed that analytical results from the site investigations indicate BTEX as the only chemical constituents of significant concern for this site.

14.2.6.2 Objective

The overall objective of the corrective action assessment for the Petro-X facility is to determine the type of remedial systems necessary to abate potential risks posed by the site.

14.2.6.3 Assessment of the Corrective Action Program Needs

Site conditions at the facility should be fully characterized, to help establish the vertical and lateral extents of hydrocarbon contamination from this site. Based on the information available from the previous site assessments for the Petro-X facility, it is apparent that petroleum product releases from the gasoline station have significantly impacted both soils and groundwater at the site. Additional sampling and characterization will provide the information necessary to properly characterize the site; to perform a risk assessment that will help establish cleanup criteria for the soil and groundwater matrices; and to develop cost-effective corrective action plans for the facility. Supplementary site information needed to complete a comprehensive site characterization for the Petro-X facility will, at a minimum, include the following:

- Sediment samples from drainage systems at and near the site
- Representative soil samples across the site, in particular, within areas with visually obvious contamination to help identify potential "hot-spots" at the facility.

- Groundwater monitoring wells that are placed in such a manner that upgradient and downgradient water quality relative to the potential "hot-spots" can be fully investigated.
- Limited background soil samples collected for the site vicinity that will become an important basis for comparison when developing cleanup levels for the site

The additional investigations should also include an assessment of the potential for hydrocarbon vapors to migrate along or within human-made conduits at and near the Petro-X facility location. This is because, among other things, such vapors may pose health hazards and threats of explosion or fire if concentrations reach explosive levels and an ignition source is present.

14.2.6.4 Identification of Feasible Remedial Options

Typically, the most frequently used treatment methods for groundwater that is contaminated with BTEX are air stripping and granular activated carbon (GAC) adsorption. For greater removal efficiency, air stripping is used in tandem with GAC (i.e., GAC/air stripping system). As necessary, remediation of the petroleum hydrocarbon plumes using a pump/treat/reinject method may be recommended for groundwater at the Petro-X facility. Reinjection of the treated water is considered to be necessary, especially to maintain the water supply capacity of the aquifer at this site. The preferred approach will be to pump from the edges of the mapped plume and reinject (after treatment) at the center of the plume. A remedial plan developed around four pumping wells and two injection wells is considered to be a reasonable starting point for an average-sized program for a contaminated gasoline fueling station.

An important site restoration technique for hydrocarbon-contaminated soils would be bioremediation; clearly identified "hot-spots" may however be removed for disposal or for treatment by other processes such as incineration or asphalt batching. Yet, other favored remediation techniques would consist of the use of in situ soil-venting systems.

14.2.6.5 The Corrective Action Plan

The corrective action plan for the Petro-X facility is divided into soil and groundwater remediation tasks. It has been decided to use vacuum extraction systems for the impacted soils and a pump-and-treat technology for the affected groundwater. The soil remediation tasks will consist of the design of a vapor extraction system (VES), the installation and operation of the VES (after obtaining applicable permits to operate the system), and laboratory analyses of verification borings (for monitoring purposes and performance evaluation). The groundwater remediation tasks will consist of the design of a groundwater treatment system (GTS), the installation and operation of the GTS, water sampling and analysis during system operation, and laboratory analyses of verification samples (for monitoring purposes and performance evaluation).

Since the Petro-X facility is still operational, the VES should be designed, installed, and operated in such a manner that will minimize potential disruption to normal activities at the facility and vicinity. The VES in this case will utilize an array of vapor extraction wells and air injection wells located in areas of impacted soils as determined from the prior site characterization and so designed to achieve maximum efficiency for contaminant removal; the use of the complementing air injection wells will generally improve the performance of the VES. The VES should be operated until the acceptable cleanup goals are achieved for the key contaminants of concern in the contaminated soils. Subsequently, verification samples (to ascertain the performance of the VES) should be collected and analyzed prior to closure.

The GTS for the Petro-X facility should also be designed, installed, and operated in such a manner that will minimize potential disruption to normal activities at the facility and vicinity. The GTS, a pump-and-treat system, will consist of both extraction and injection wells appropriately located so as to achieve maximum efficiency for contaminant removal from the groundwater. Groundwater pumped will be treated prior to reinjection or disposal. The GTS should be operated until the acceptable cleanup goals are achieved for the key contaminants of concern in groundwater.

Subsequently, verification samples (to ascertain the performance of the GTS) should be collected and analyzed prior to closure; furthermore, long-term monitoring wells should be installed down-gradient of the contaminated areas.

It is noteworthy that, in a number of situations, a detailed assessment may indicate that no extensive and expensive remediation or cleanup program is necessary for a contaminated site. Thus, prior to implementing any remediation plan for the Petro-X facility, it is recommended that appropriate risk-based cleanup criteria be developed and compared with the levels of contamination presently existing at the site (i.e., the baseline conditions). Based on such criteria, it may be determined from the types of exposure scenarios relevant to the site that only limited cleanup is warranted; this could save Petro-X substantial amounts of money as well as other potential problems and liabilities that the implementation of a remedial action program could carry. In fact, even if it is determined that some degree of cleanup is required, the cleanup criteria developed will aid in optimizing the efforts involved so as to arrive at more cost-effective solutions.

14.2.7 DEVELOPMENT OF A SITE CLOSURE PLAN FOR A CONTAMINATED SITE PROBLEM

The purpose of this section is to present the pertinent information relevant to the preparation of a decommissioning or closure plan for an abandoned industrial facility. This hypothetical facility has been used for a multitude of operations—including machine components cleaning, electroplating, sandblasting, painting, and vehicle maintenance. The site is located within an industrial estate in the State of California. Based on current zoning plans, it is anticipated that this land parcel will be used for residential developments in the near future.

14.2.7.1 Introduction and Background

The former industrial facility, located at A2Z in an industrially zoned area, operated for about two decades before being permanently closed. Site facilities include a main plant building, office buildings, storage tanks, and post-closure areas (that consist of surface impoundment for waste-water treatment operations and sludge ponds). Past operations at the plant required the storage of raw materials in aboveground tanks; the distribution of raw materials in pipelines; and the storage of chemicals, fuels, and waste materials in USTs. Historical uses of the site included component cleaning (in which acids, caustics, and chlorinated hydrocarbon-based solvents were used) and electroplating (for which major associated chemicals included cadmium, nickel, and chromium). Other significant activities included sandblasting of unpainted metal parts, painting, and vehicle maintenance.

Due to the sandblasting activities, incidental spillage during material handling and possible leakage of underground storage and transmission/distribution systems, soils and groundwater underlying the A2Z plant site are expected to be significantly impacted; this is the result of releases of chemical materials that were used in the industrial processes and related activities carried out at this facility. Preliminary remedial activities have been implemented to remove buried drums and storage tanks and to remove soil materials from some of the most heavily contaminated areas.

The soil materials at the A2Z site consist mostly of sand and silty sand, underlain by silts and clays that overlie some cherty shales and siltstones. An extensive program has been undertaken to define the nature and extent of the soil and groundwater contamination within the site boundary. The past sampling activities indicate that a number of chemicals of potential concern (COCs) associated with the site have impacted soils and groundwater, with the possibility to affect surface water in the vicinity of the site if timely corrective measures are not implemented.

14.2.7.2 Objective and Scope

The objective of this illustrative example is to develop a preliminary decommissioning/closure plan for the A2Z site. This will normally include the preparation of an initial documentation (consisting of a general description of the facilities to be decommissioned; a site plan drawn to scale, also showing surrounding land uses and natural features; the approximate time frame envisaged for the decommissioning or site cleanup; and the current official plan designation and zoning of the site), a preliminary inventory (which provides an initial understanding of the potential range and quantity of contaminants possibly present at the site), a sampling and analysis program, a risk assessment, and a corrective action program.

For the purpose of this illustrative example, only the sampling requirements, a risk assessment, and a proposed corrective action plan are discussed; general descriptive presentation of the facility layout and inventory data is not included here.

14.2.7.3 Sampling Requirements for the Decommissioning Plan

The general types of site data required as part of the decommissioning plan for the A2Z site include the following:

- Contaminant identities
- Chemical concentrations in the key sources and media of concern
- Characteristics of sources, in particular information on chemical release potentials
- Nature of the environmental setting (including groundwater flow direction) that may affect the fate, transport, and persistence of the COCs present at this site

Typically, a preliminary sampling program for soils, groundwater, surface water, and sediments is undertaken to identify on-site contamination. If it is apparent that contaminants may have or could migrate off-site, the initial sampling program should be extended to off-site locations for verification.

A preliminary sampling program is required for the A2Z site in order to completely characterize the site conditions, as well as determine the nature and extent of contamination present at the facility. This is achieved through sampling and analysis of all contaminant sources and all potentially impacted media. Since soil often is the major source of chemical releases to other media, the number, location, and type of soil samples collected could have significant impacts on the overall investigations. Consequently, a systematic grid pattern that covers the areal extent of this site is preferably used for the soil sampling program. Additionally, where higher degrees of contamination are visually apparent or suspected, a greater number of samples should be collected to better define any existing "hot spots" at the facility.

The sampling and analysis program for the A2Z site was so designed to produce representative samples to support the closure plan. Chemicals found in soils and groundwater at the A2Z site included both the organic and inorganic constituents—shown in Table 14.3.

14.2.7.4 Evaluation of Potential Risks

By virtue of the physical setting and the type of surface materials present at the site, the primary release mechanisms for the contaminants of concern at the A2Z facility have been determined to be via infiltration and migration into groundwater and also by wind erosion and fugitive dust transport. Surface water runoff and overland transport could become an important contributor to the spread of contamination if timely corrective actions are not taken. For the purpose of this discussion, the significant exposure pathways to potential receptors that are evaluated here relate to the air, groundwater, and soils exposure media only.

The following represents an evaluation of the potential impacts of the COCs present at the A2Z site. Residential exposure scenarios involving possible exposure of potential receptors to contaminated soils and to contaminated groundwater were evaluated using the methods of approach discussed in Part III; the results are presented in Tables 14.4 and 14.5, respectively. The 95% upper confidence level (UCL) concentrations of the COCs in the environmental samples were used as the exposure point concentrations in the evaluation of potential risks. Also, case-specific exposure parameters obtained from the literature (e.g., DTSC, 1994; OSA, 1993; USEPA, 1989c,f,g,k,n,o,p,r,

TABLE 14.3
Summary of the COCs at the A2Z Site

COCs in Soils	Important Synonyms or Trade Names or Chemical Formula	Chemical Abstracts Service Number (CAS No.)	95% UCL Soil Concentration (mg/kg)	COCs in Groundwater	Important Synonyms or Trade Names or Chemical Formula	Chemical Abstracts Service Number (CAS No.)	95% UCL Groundwater Concentration (µg/L)
Inorganic Chemicals							**Inorganic Chemicals**
Antimony	Sb	7440-36-0	4.0	Antimony	Sb	7440-36-0	3.4
Arsenic	As	7440-38-2	17.2	Arsenic	As	7440-38-2	7.6
Beryllium	Be	7440-41-7	1.4				
Cadmium	Cd	7440-43-9	4.5	Cadmium	Cd	7440-43-9	4.6
Chromium	Cr	16065-83-1	92.2	Chromium	Cr	16065-83-1	8,690
Chromium VI	Cr(VI)	7440-47-3	4.6	Chromium VI	Cr(VI)	7440-47-3	435
Cobalt	Co	7440-48-4	7.8	Manganese	Mn	7439-96-5	299
Manganese	Mn	7439-96-5	512	Molybdenum	Mo	7439-98-7	164
Mercury	Hg	7439-97-6	0.25	Nickel	Ni	7440-02-0	82.2
Molybdenum	Mo	7439-98-7	16.6	Vanadium	V	7440-62-2	54
Nickel	Ni	7440-02-0	90.0	Zinc	Zn	7440-66-6	64.6
Selenium	So	7782-49-2	1.4				
Vanadium	V	7440-62-2	62.9				
Zinc	Zn	7440-66-6	413				
Organic Compounds							**Organic Compounds**
Chloroform	Trichloromethane	67-66-3	0.008	Vinyl chloride	VC	75-01-4	0.729
Trichloroethene	TCE	79-01-6	32	1,1-Dichloroethene	1,1-DCE	75-35-4	3.31
1,1,2-Trichloroethane	1,1,2-TCA	79-00-5	0.020	trans-1,2-Dichloroethene	1,2-trans-DCE	156-60-5	11.8
Tetrachloroethene	PCE	127-18-4	0.033	cis-1,2-Dichloroethene	cis-1,2-DCE	540-59-0	1.220
Ethylbenzene	EB; phenylethane	100-41-4	0.009	Trichloroethene	TCE	79-01-6	548
Xylenes (mixed)	Dimethylbenzene	1330-20-7	0.044				

TABLE 14.4
Risk Screening for a Hypothetical Residential Population Exposure to Soils at the A2Z Site

Chemical of Potential Concern	95% UCL Soil Concentration (mg/kg)	Chemical-Specific Dermal Absorption (ABS_s)	Oral RfD (mg/kg-day)	Oral SF (1/mg/kg/day)	Inhalation RfD (mg/kg-day)	Inhalation SF (1/mg/kg/day)	Risk for Air	Hazard for Air	Risk for Soil	Hazard for Soil	Total Risk (Air + Soil)	Total Hazard (Air + Soil)	Risk-Based Soil Criteria (mg/kg)	Soil Criteria Exceeded?
Inorganic Chemicals														
Antimony	4.0	0.01	4.00E−04				0.00E+00	3.20E−04	0.00E+00	1.41E−01			28	No
Arsenic	17.2	0.03	3.00E−04	1.75E+00		1.20E+01	1.54E−06	1.83E−03	6.41E−05	9.54E−01			0.26	Yes
Beryllium	1.4	0.01	5.00E−03	4.30E+00		8.40E−00	8.76E−08	8.95E−06	1.06E−05	3.94E−03			0.13	Yes
Cadmium	4.5	0.001	5.00E−03			1.50E+02	5.03E−07	2.88E−04	0.00E+00	1.16E−01			9	No
Chromium(total)	92.2	0.01	1.00E+00				0.00E+00	2.95E−06	0.00E+00	1.30E−03			70,662	No
Chromium(vi)	4.6	0.00	5.00E−03	4.20E−01		5.10E+02	1.75E−06	2.94E−06	3.03E−05	1.18E−02			0.22	Yes
Cobalt	7.8	0.01	2.90E−04		2.90E−04		0.00E+00	8.59E−04	0.00E+00	3.79E−01			21	No
Manganese	512	0.01	1.40E−01		1.10E−04		0.00E+00	1.49E−01	0.00E+00	5.15E−02			2,557	No
Mercury	0.25	0.01	3.00E−04		8.60E−05		0.00E+00	9.29E−06	0.00E+00	1.17E−02			21	No
Molybdenum	16.6	0.01	5.00E−03		5.00E−03		0.00E+00	1.06E−04	0.00E+00	4.67E−02			354	No
Nickel	90.9	0.01	2.00E−02			9.10E−01	6.16E−07	1.45E−04	0.00E+00	6.40E−02			148	No
Selenium	1.4	0.01	5.00E−03				0.00E+00	8.95E−06	0.00E+00	3.94E−03			354	No
Vanadium	62.9	0.01	7.00E−03				0.00E+00	2.67E−04	0.00E+00	1.27E−01			496	No
Zinc	413	0.01	3.00E−01				0.00E+00	4.40E−06	0.00E+00	1.94E−02			21,259	No
Organic Compounds														
Chloroform	0.008	0.10	1.00E−02	3.10E−02	1.00E−02	1.00E−02	1.32E−09	2.98E−05	8.53E−10	2.05E−05			9	No
Trichloroethene	32	0.10	6.00E−03	1.50E−02	6.00E−03	1.00E−02	2.01E−06	1.44E−01	1.65E−06	1.37E−01			19	Yes
1,1,2-Trichloroethene	0.020	0.10	4.00E−03	5.70E−02	4.00E−03	5.00E−02	3.54E−09	6.77E−05	3.92E−09	1.28E−04			5	No
Tetrachloroethene	0.033	0.10	1.00E−02	5.10E−02	1.00E−02	5.10E−02	5.18E−09	4.36E−05	5.79E−09	0.45E−06			6	No
Ethylbenzene	0.009	0.10	1.00E−01		2.90E−01		0.00E+00	5.47E−07	0.00E+00	2.30E−06			3,901	No
Xylenes	0.044	0.10	2.00E+00		2.00E−01		0.00E+00	3.16E−06	0.00E+00	5.63E−07			78,028	No
							2.22E−06	0.30	7.94E−06	2.1	1.02E−04	2.4		

Notes:

1. The computational formulas and models used in this evaluation are discussed in Chapters 5 and 6 and further elaborated in Appendices F, I, and J.

2. Risk/hazard for air accounts for the volatilization effects (for volatile organic compounds) and airborne emissions of contaminated particulates (for nonvolatile chemicals) present at the site (see Section 5.8.1 and DTSC 1994).

3. Case-specific exposure parameters used in the calculations were obtained from the sources—DTSC (1994), OSA (1993) and USEPA (1989a,b, 1991a–h, 1992a,b,d,e,f,h,i,j).

TABLE 14.5
Risk Screening for a Hypothetical Residential Population Exposure to Groundwater at the A2Z Site

COCs	95% UCL Water Concentration (µg/L)	Chemical-Specific K_p (cm/h)	Toxicity Criteria				Risk of Water	Hazard of Water	Risk-Based Water Critereia (µg/L)	MCL Values as water Criteria (µg/L)	Water Criteria Exceeded?
			Oral RfD (mg/kg-day)	Oral SF (1/mg/kg-day)	Inhalation RfD (mg/kg-day)	Inhalation SF (1/mg/kg-day)					
Inorganic Chemicals											
Antimony	3.4	1.60E – 04	4.00E – 04				0.00E + 00	5.43E – 01	6.26	6	No
Arsenic	7.6	1.60E – 04	3.00E – 04	1.75E – 00		1.20E + 01	1.98E – 04	1.62E + 00	0.04	50	Within MCL
Cadmium	4.6	1.60E – 04	5.00E – 04			1.50E + 01	0.00E – 00	5.88E – 01	8	10	No
Chromium (total)	8,690	1.60E – 04	1.00E + 00				0.00E – 00	5.55E – 01	15,647	50	Within risk-based criteria
Chromium (vi)	435	1.60E – 04	5.00E – 03	4.20E – 01		5.10E + 02	2.72E – 03	5.56E – 00	0.16	na	Yes
Manganese	299	1.60E – 04	5.00E – 03		1.10E – 04		0.00E – 00	3.82E + 00	78	50	Yes
Molybdenum	164	1.60E – 04	5.00E – 03		5.00E – 03		0.00E – 00	2.10E + 01	78	na	Yes
Nickel	82.2	1.60E – 04	2.00E – 02				0.00E – 00	2.63E – 01	313	100	No
Vanadium	54	1.60E – 04	7.00E – 03				0.00E – 00	4.93E – 01	110	na	No
Zinc	64.6	1.60E – 04	3.00E – 01			9.10E – 01	0.00E – 00	1.38E – 02	4,694	5,000	No
Organic Compounds											
Vinyl Chloride	0.729	7.30E – 03		2.70E – 01		2.70E – 01	5.91E – 06	0.00E + 00	0.12	0.5	Yes
1,1-Dichloroethene	3.31	1.60E – 02	9.00E – 03	6.00E – 01	9.00E – 03	1.20E – 01	8.98E – 05	0.00E + 00	0.04	6.0	Within MCL
trans-1,2-Dichloroethene	11.8	1.00E – 02	2.00E – 02		2.00E – 02		0.00E + 00	7.58E – 02	156	10.0	Within risk-based criteria
cis-1,2-Dichloroethene	1,220	1.00E – 02	1.00E – 02		1.10E – 02		0.00E – 00	1.57E + 01	78	6.0	Yes
Trichloroethene	548	1.60E – 02	6.00E – 03	1.50E – 02	1.00E – 02	1.50E – 02	2.49E – 04	9.43E + 00	2.20	5.0	Yes
							3.72E – 03	40.8			

Notes: 1. The computational formulas and models used in this evaluation are discussed in Chapters 5 and 6, and further elaborated in appendices F, I and J.

2. Risk/hazard for water account for both volatile chemical emissions and nonvolatile chemical contributors present at the site (see Section 5.8.1 and DTSC, 1994).

3. Case-specific exposure parameters used in the calculations were obtained from the following sources—DTSC (1994, OSA (1993), and USEPA (1989c,f,g,k,n,o,p,r, 1991a–h, 1992a,b,d,e,f,h,i,j).

4. K_p, Chemical-specific dermal permeability coefficient for water; MCL, maximum contaminant level (for california); na, not available.

1991a–h, 1992a,b,d,e,f,h,i,j) were used in the modeling effort. Toxicity values used pertain to those promulgated into California regulations, since this facility is located in the State of California and in some cases that found within federal EPA guidance, as appropriate. It is assumed for the sake of simplicity that all the noncarcinogenic effects of the COCs are attributable to the same toxicological end point.

Table 14.4 consists of an evaluation of the potential risks associated with a hypothetical residential development at A2Z, assuming the contaminated soils present at the site remain in place. It is assumed that potential receptors may be exposed via inhalation of airborne contamination (consisting of particulates and/or volatile emissions), through the ingestion of contaminated soils, and by dermal contact with the contaminated soils at the site. Default exposure parameters indicated in the literature (e.g., DTSC, 1994; OSA, 1993; USEPA, 1989c,f,g,k,n,o,p,r, 1991a–h, 1992 a,b,d,e,f,h,i,j) were used for the calculations shown in this spreadsheet. Based on this scenario, it is determined that risks exceed the generally accepted benchmark of 10^{-6}. The noncarcinogenic hazard index also exceeds the reference level of unity. In particular, acceptable risk-based soil criteria are exceeded for arsenic, beryllium, hexavalent chromium, and trichloroethylene (TCE) at the 10^{-6} risk level.

Table 14.5 consists of an evaluation of the potential risks associated with a hypothetical population exposure to impacted groundwater originating from the A2Z site, assuming the contaminated water is not treated before going into a public water supply system. It is assumed that potential receptors may be exposed through the inhalation of volatiles during domestic usage of contaminated water, from the ingestion of contaminated water, and by dermal contact to contaminated waters. Default exposure parameters indicated in the literature (e.g., DTSC, 1994; OSA, 1993; USEPA, 1989c,f,g,k,n,o,p,rb, 1991a–h) were used for the calculations presented in this spreadsheet. Based on this scenario, it is apparent that risks exceed the generally accepted benchmark risk level of 10^{-6}. The reference hazard index of 1 is also exceeded for receptor exposures to raw/untreated groundwater from the A2Z site. The major contributors to the carcinogenic risk are from the general population exposure to arsenic, hexavalent chromium, 1,1-dichloroethylene (DCE), and TCE in groundwater. The most significant contributors to noncarcinogenic effects are hexavalent chromium, manganese, cis-1,2-DCE, and TCE in groundwater.

14.2.7.5 A Site Restoration Strategy

The risk evaluation presented previously indicates that the COCs present in soils at A2Z may pose significant risks to human receptors potentially exposed via the soil and groundwater media. Consequently, if the A2Z site is to be used for future residential developments or if raw/untreated groundwater from this site is to be used as a potable water supply source, then a comprehensive site restoration program is required to abate the imminent risks that the site poses. This should comprise of an integrated soil and groundwater remediation program. Thus, general response actions should be developed for each of the impacted environmental medium associated with the site (i.e., for both soils and groundwater). For other than a "no-action" situation, the site-specific risk-based criteria can be used to guide and support the development, screening, and selection of potentially feasible remedial alternatives.

14.2.7.6 The Corrective Action Plan

General response actions to consider and evaluate for the A2Z site include containment (e.g., capping and slurry walls), removal of "hot-spots" (e.g., excavation and disposal), treatment (e.g., by soil venting/vapor extraction, thermal treatment, biological treatment/bioremediation, soil washing, and immobilization) for the soils, and pump-and-treat systems (including air/steam stripping and GAC adsorption, etc.) for groundwater. Where necessary, treatability investigations should be conducted following the development and screening of the remedial alternatives for the site. The remedial alternatives selected as the most technically feasible will subsequently become the focus for more detailed analyses. The detailed evaluation ranks the potential response actions with particular attention given to several regulatory compliance requirements, community relations activities, effectiveness, implementability, and associated costs. Depending on what chemicals are determined to drive the remediation, one particular remedial technology/process may be preferred over another; a combination of various process options should be adopted if that is determined to be more cost effective and technically justifiable/defensible in its application.

14.2.8 Diagnostic Human Health Risk Assessment and Development of a Site Restoration Plan for a Contaminated Site Problem

The purpose of this section is to present a procedural illustration of the nature of human health risk evaluation required for the development of a site restoration program—often necessary as part of a decommissioning or closure plan for an abandoned industrial facility. A hypothetical facility that has been used for a multitude of operations is utilized in this illustrative example. The case site, owned by PLC Limited, is located within an industrial estate in the outskirts of London. Based on current zoning plans, it is anticipated that this land parcel could be used for a variety of commercial developments in the near future. A baseline risk assessment for this inactive site that previously housed the PLC facility is necessary, in order to help make the appropriate decision regarding utilization of this land parcel.

14.2.8.1 Introduction and Background

A former industrial facility, located in an industrially zoned area in the outskirts of metropolitan London, operated for over three decades before being permanently closed. Site facilities include a main plant building, office buildings, storage tanks, and post-closure areas (that consist of surface impoundment for wastewater treatment operations and sludge ponds). Past operations at the plant required the storage of raw materials in aboveground tanks; the distribution of raw materials in pipelines; and the storage of chemicals, fuels, and waste materials in USTs. Historical uses of the site included machine component cleaning (in which chlorinated hydrocarbon-based solvents were used) and electroplating (for which major associated chemicals included cadmium, nickel, and chromium). Other significant activities included sandblasting of unpainted metal parts, painting, and vehicle maintenance.

Due to the sandblasting activities, incidental spillage during materials' handling and possible leakage of underground storage and distribution systems, soils, and groundwater underlying the PLC plant site have been significantly impacted; this is the result of releases of chemical materials that were used in the industrial processes and related activities carried out at this facility. Preliminary remedial activities have already been implemented to remove buried drums and storage tanks, as well as to remove soil materials from some of the most heavily contaminated areas.

Under the current decommissioning program, the PLC facility could be zoned for a variety of commercial developments in the near future. The development of a site closure or redevelopment plan should therefore incorporate a diagnostic risk assessment that addresses potential impacts under all realistically feasible site uses and conditions.

14.2.8.2 Objective and Scope

The principal goal of any comprehensive corrective action program for the PLC site would be to prevent contaminant migration from the site to potential receptors and therefore prevent the endangerment of human health and the environment at and in the vicinity of the site. It is apparent, however, that releases at the PLC site have caused significant soil and groundwater contamination beneath this industrial facility. Several key environmental issues must therefore be addressed in the processes involved in the diagnostic assessment required of the decommissioning plan—including the following:

- Identification of the possible site related contaminants associated with past site activities
- Screening for the COCs to human health and the environment

- Estimation of the chemical concentrations in the impacted media of significant concern
- Determination of the populations potentially at risk from site contaminants
- Identification of representative and site-specific exposure scenarios
- Characterization of the potential risks associated with the site
- Development of site-specific cleanup criteria for the impacted matrices at the site

14.2.8.3   Technical Elements of the Diagnostic Risk Assessment Process

Figure 14.1 shows the major elements required of the process for developing a decommissioning plan for the PLC site. In particular, the following specific tasks are carried out, in order to accomplish the overall goal of the diagnostic risk assessment:

- Compile and characterize the list of contaminants present at the site.
- Compile the toxicological profiles of the CoCs.
- Investigate all possible contaminant migration pathways, and determine the pathways of concern.
- Identify targets in the vicinity of the site and all other populations potentially at risk (including possible sensitive receptors).

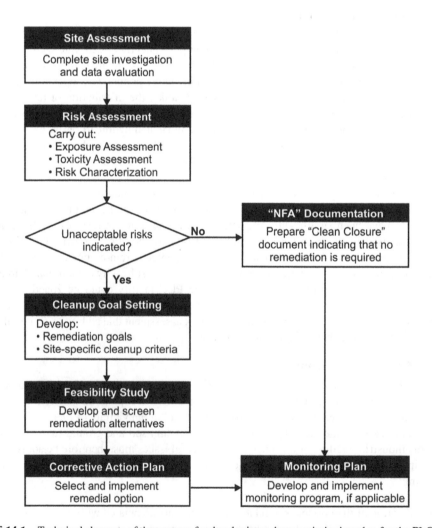

FIGURE 14.1   Technical elements of the process for developing a decommissioning plan for the PLC facility.

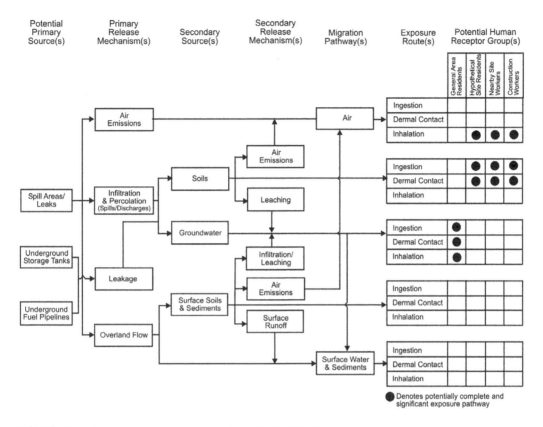

FIGURE 14.2 Conceptual exposure elements for the PLC facility.

- Develop a representative conceptual model for the site.
- Develop exposure scenarios, by integrating information on the populations potentially at risk with the likely and significant migration/exposure pathways (Figure 14.2).
- Calculate carcinogenic risks and noncarcinogenic hazard indices for the various receptor groups that have been determined to be potentially at risk.

Based on the type of exposure scenarios identified for this environmental setting, an environmental management strategy can be formulated such that the PLC site does not pose any significant risks.

14.2.8.4 Identification of Site Contaminants

Chemicals found in soils and groundwater at the PLC site consist of both organic and inorganic constituents, as summarized in Tables 14.6 and 14.7 (developed from the complete laboratory data package of site sampling results). The summary of analytical results is reported in these tables, together with the naturally occurring background threshold values, where background concentrations are available. The background levels are used as a screening indicator of possible media contamination that may be the result of past site activities.

14.2.8.5 Screening for COCs

A listed site contaminant is considered to be a chemical of potential concern if it is likely to have originated from past site activities and if it could potentially result in adverse effects to populations potentially at risk when such receptors are exposed to the particular compound. The site contaminants are properly screened (Figure 14.3) in order to arrive at the target chemicals listed in Table 14.8—as the COCs for the anticipated future use of the site. The maximum concentrations of

TABLE 14.6

Preliminary List of Possible Site Contaminants in Soils at the PLC Site

Possible Site Contaminant	Important Synonyms, Trade Names, or Chemical Formula	Naturally Occurring Background Level (mg/kg)	Maximum Soil Concentration (mg/kg)
Inorganic Chemicals			
Aluminum	Al	16,400	10,400
Antimony	Sb	0.3	1.1
Arsenic	As	10	12
Barium	Ba	98	81.1
Beryllium	Be	0.7	0.33
Cadmium	Cd	1.0	2.6
Calcium	Ca	3,350	8,350
Chromium	Cr	28	57.3
Cobalt	Co	4.0	64.6
Copper	Cu	46	47.3
Iron	Fe	13,500	22,900
Lead	Pb	22	15
Magnesium	Mg	3,900	2,420
Manganese	Mn	523	170
Mercury	Hg	0.2	1.1
Molybdenum	Mo	3.0	4.8
Nickel	Ni	13	19
Potassium	K	2,610	2,200
Selenium	Se	0.2	6.1
Silver	Ag	0.2	0.55
Sodium	Na	453	588
Thallium	Tl	0.4	0.415
Vanadium	V	58	34.1
Zinc	Zn	107	432
Organic Compounds			
Trichloroethene	TCE	not available	0.710
cis-1,2-Dichloroethene	cis-1,2-DCE	not available	0.007

the target chemicals in the environmental samples listed here will be used as the receptor exposure concentrations in this study.

14.2.8.6   Risk Characterization for Site-Specific Exposure Scenarios

Three different population groups—that is, onsite workers, site construction workers, and off-site residential populations—are considered in developing the requisite exposure scenarios anticipated for the PLC site. Table 14.9 provides a summary of the likely migration and exposure pathways that form a basis for estimating risks associated with this site.

Calculation of potential carcinogenic risks and noncarcinogenic hazards under the existing conditions at the PLC site is performed for the three different population groups identified in the conceptual site representation (shown in Table 14.9). The maximum concentrations of the target chemicals in the environmental samples (listed in Table 14.8) are used as the exposure point concentrations in this evaluation. Case-specific exposure parameters obtained from the technical literature (viz., DTSC, 1994; OSA, 1993; USEPA, 1989c,f,g,k,n,o,p,r, 1991a–h, 1992a,b,d,e,f,h,i,j) are used in the modeling efforts involved. Toxicity values used for the risk characterization pertain to those

TABLE 14.7
Preliminary List of Possible Site Contaminants in Groundwater at the PLC Site

Possible Site Contaminant	Important Synonyms, Trade Names, or Chemical Formula	Naturally Occurring Background Level (µg/L)	Maximum Water Concentration (µg/L)
Inorganic Chemicals			
Aluminum	Al	1,200	1,000
Antimony	Sb	10	5.9
Arsenic	As	7	5.0
Barium	Ba	276	133
Beryllium	Be	4.0	1.0
Calcium	Ca	197,000	54,300
Chromium	Cr	20	15
Cobalt	Co	13	6.3
Copper	Cu	58	22
Iron	Fe	3,530	3,000
Magnesium	Mg	119,000	37,800
Manganese	Mn	971	1,390
Molybdenum	Mo	6	6.5
Nickel	Ni	490	21
Potassium	K	13,300	6,550
Silver	Ag	12	4.0
Sodium	Na	420,000	124,000
Thallium	Tl	1.0	1.5
Vanadium	V	28	24
Zinc	Zn	80	70
Organic Compounds			
Trichloroethene	TCE	not available	939
cis-1,2-Dichloroethene	cis-1,2-DCE	not available	118

found in recent toxicological databases—in this case the Integrated Risk Information System (see, e.g., Appendix C). The computational processes involved are elaborated in the following:

- *Risk Characterization Associated with an On-site Worker.* Table 14.10 consists of an evaluation of the potential risks associated with a nearby and/or on-site worker (following the redevelopment of the site for commercial activities) being exposed to the COCs at the PLC site—assuming the contaminated soils remain in place. It is also assumed that potential receptors may be exposed via inhalation of airborne contamination (consisting predominantly of particulate emissions from fugitive dust), through the incidental ingestion of contaminated soils and by dermal contact with the contaminated soils at the site.

  Case-specific exposure parameters used in this evaluation conservatively assume that the on-site worker will be exposed at a frequency of 250 days/year over a 25-year period. Additional parameters include using a soil ingestion rate of 50 mg/day and airborne particulate emission rate of 50 µg/m^3. Other default exposure parameters indicated in the literature (e.g., DTSC, 1994; OSA, 1993; USEPA, 1989c,f,g,k,n,o,p,r, 1991a–h, 1992a,b,d,e,f,h,i,j) were used for the calculations shown in this spreadsheet.

  Based on this scenario, it is apparent that potential risks (of 1.7×10^{-5}) to an on-site worker at the PLC site could exceed a benchmark risk level of 10^{-6}. The noncarcinogenic

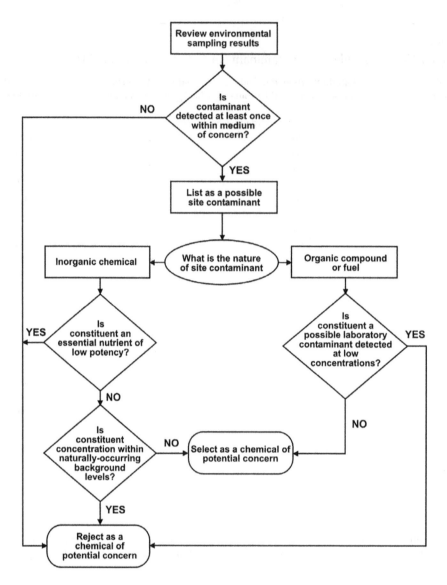

FIGURE 14.3   COPCs selection process for the PLC facility.

hazard index (of 0.4), however, is within the reference index of unity. The sole "risk driver"
in this case is arsenic.

• *Risk Characterization Associated with a Site Construction Worker.* Table 14.11 consists of
   an evaluation of the potential risks associated with a construction worker being exposed to
   the COCs at the PLC site—assuming no personal protection from the contaminated soils
   during site redevelopment activities. It is also assumed that potential receptors may be
   exposed via inhalation of airborne contamination (consisting predominantly of particulate
   emissions from fugitive dust), through the incidental ingestion of contaminated soils and
   by dermal contact with the contaminated soils at the site.

   Case-specific exposure parameters used in this evaluation conservatively assume that
   the construction worker will be exposed at a frequency of 250 days/year over a 1-year
   period. Additional parameters include using a soil ingestion rate of 480 mg/day and
   airborne particulate emission rate of 1,000 µg/m³. Other default exposure parameters

TABLE 14.8
Summary of the COCs at the PLC Site

Chemical of Potential Concern in Soils	Important Synonyms or Trade Names, or Chemical Formula	Maximum Soil Concentration (mg/kg)	Chemical of Potential Concern in Groundwater	Important Synonyms or Trade Names, or Chemical Formula	Maximum Groundwater Concentration (μg/L)
Inorganic Chemicals			**Inorganic Chemicals**		
Antimony	Sb	1.1	Manganese	Mn	1,390
Arsenic	As	12	Molybdenum	Mo	6.5
Cadmium	Cd	2.6	Thallium	Tl	1.5
Chromium	Cr	57.3			
Cobalt	Co	64.6			
Copper	Cu	47.3			
Mercury	Hg	1.1			
Molybdenum	Mo	4.8			
Nickel	Ni	19.0			
Selenium	Se	6.1			
Silver	Ag	0.55			
Thallium	Tl	0.415			
Zinc	Zn	432			
Organic Compounds			**Organic Compounds**		
Trichloroethene	TCE	0.710	Trichloroethene	TCE	939
cis-1,2-Dichloroethene	cis-1,2-DCE	0.007	cis-1,2-Dichloroethene	cis-1,2-DCE	118

indicated in the literature (viz., DTSC, 1994; OSA, 1993; USEPA, 1989c,f,g,k,n,o,p,r, 1991a–h, 1992a,b,d,e,f,h,i,j) were used for the calculations shown in this spreadsheet.

Based on this scenario, it is apparent that potential risks (of 2.5×10^{-6}) to a construction worker at the PLC site could marginally exceed an assumed benchmark risk level of 10^{-6}. The noncarcinogenic hazard index (of 1.6) also marginally exceeds the reference index of 1. The major "risk drivers" are arsenic and cobalt.

- *Risk Characterization for a Downgradient Residential Population Exposure to Groundwater.* Table 14.12 consists of an evaluation of the potential risks associated with a hypothetical downgradient population exposure to impacted groundwater that originates from the PLC site—assuming the contaminated water is not treated before going into a public water supply system. It is also assumed that potential receptors may be exposed through the inhalation of volatile constituents during domestic usage of contaminated water, from the ingestion of contaminated water and by dermal contact to contaminated waters.

Default exposure parameters indicated in the literature (viz., DTSC, 1994; OSA, 1993; USEPA, 1989c,f,g,k,n,o,p,r, 1991a–h, 1992a,b,d,e,f,h,i,j) were used for the calculations presented in this spreadsheet.

Based on this scenario, it is apparent that the generally acceptable benchmark risk level of 10^{-6} and the reference hazard index of 1 may both be exceeded by several orders of magnitude, as a result of receptor exposures to raw/untreated groundwater from the PLC site. The most significant contributors to site risks are from the general population exposure to manganese, thallium, TCE, and cis-1,2-DCE in groundwater.

For the noncarcinogenic effects, it is assumed for the sake of simplicity, that all the target chemicals have the same physiologic end point.

TABLE 14.9
Tabular Analysis Chart for the Exposure Scenarios Associated with the PLC Site

Contaminated Exposure Medium	Contaminant Release Source(s)	Contaminant Release Mechanism(s)	Potential Receptor Location	Receptor Groups Potentially at Risk	Potential Exposure Routes	Pathway Potentially Complete and Significant?
Air	Contaminated surface soils	Fugitive dust generation	On-site	Onsite facility worker	Inhalation	Yes
					Incidental ingestion	No
					Dermal absorption	No
				Construction worker	Inhalation	Yes
					Incidental ingestion	No
					Dermal absorption	No
			Off-site	Downwind worker	Inhalation	No
					Incidental ingestion	No
					Dermal absorption	No
				Downwind resident	Inhalation	No
					Incidental ingestion	No
					Dermal absorption	No
		Volatilization	On-site	Onsite facility worker	Inhalation	Yes
					Dermal absorption	No
				Construction workers	Inhalation	Yes
					Dermal absorption	No
			Off-site	Nearest downwind worker	Inhalation	No
					Dermal absorption	No
				Nearest downwind resident	Inhalation	No
					Dermal absorption	No
Soils	Contaminated soils and/or buried wastes	Direct contacting	On-site	Onsite facility worker	Incidental ingestion	Yes
					Dermal absorption	Yes
				Construction worker	Incidental ingestion	Yes
					Dermal absorption	Yes
			Off-site	Downwind worker	Incidental ingestion	No
					Dermal absorption	No
				Downwind resident	Incidental ingestion	No
					Dermal absorption	No
Surface water	Contaminated surface soils	Surface runoff into surface impoundments	On-site	Onsite facility worker	Inhalation	No
					Incidental ingestion	No
					Dermal absorption	No

(Continued)

TABLE 14.9 (*Continued*)
Tabular Analysis Chart for the Exposure Scenarios Associated with the PLC Site

Contaminated Exposure Medium	Contaminant Release Source(s)	Contaminant Release Mechanism(s)	Potential Receptor Location	Receptor Groups Potentially at Risk	Potential Exposure Routes	Pathway Potentially Complete and Significant?
				Construction worker	Inhalation	No
					Incidental ingestion	No
					Dermal absorption	No
		Erosional runoff	Off-site	Downslope resident	Inhalation	No
					Incidental ingestion	No
					Dermal absorption	No
				Recreational population	Inhalation	No
					Incidental ingestion	No
					Dermal absorption	No
	Contaminated groundwater	Groundwater discharge	Off-site	Downslope resident	Inhalation	No
					Incidental ingestion	No
					Dermal absorption	No
				Recreational population	Inhalation	No
					Incidental ingestion	No
					Dermal absorption	No
Groundwater	Contaminated soils	Infiltration/leaching	On-site	Onsite facility worker	Inhalation	No
					Incidental ingestion	No
					Dermal absorption	No
				Construction worker	Inhalation	No
					Incidental ingestion	No
					Dermal absorption	No
			Off-site	Downgradient resident	Inhalation	Yes
					Incidental ingestion	Yes
					Dermal absorption	Yes
Drainage sediments	Contaminated surface soils	Surface runoff/episodic overland flow	On-site	Onsite facility worker	Inhalation	No
					Incidental ingestion	No
					Dermal absorption	No
				Construction worker	Inhalation	No
					Incidental ingestion	No
					Dermal absorption	No
			Off-site	Nearest downgradient resident	Inhalation	No
					Incidental ingestion	No
					Dermal absorption	No

TABLE 14.10

Risk Screening for a Nearby and/or Onsite Worker Exposure to Soils at the PLC Site

Chemical of Potential Concern	Maximum Soil Concentration (mg/kg)	Chemical-Specific Dermal Absorption (ABS$_d$)	Toxicity Criteria				Risk for Air	Hazard for Air	Risk for Soil	Hazard for Soil	Total Risk (Air + Soil)	Total Hazard (Air + Soil)	Risk-Based Soil Criteria (mg/kg)	Soil Criteria Exceeded?
			Oral RfD (mg/kg-day)	Oral SF (1/mg/kg-day)	Inhalation RfD (mg/kg-day)	Inhalation SF (1/mg/kg-day)								
Inorganic Chemicals														
Antimony (Sb)	1.1	0.01	4.00E−04				0.00E+00	2.69E−05	0.00E+00	2.91E−03			375	No
Arsenic (As)	12	0.03	3.00E−04	1.75E+00		1.20E+01	5.03E−07	3.91E−04	1.64E−05	8.77E−02			0.71	Yes
Cadmium (Cd)	2.6	0.001	5.00E−04			1.50E+01	1.36E−07	5.09E−05	0.00E+00	2.84E−03			19	No
Chromium (Cr–total)	57.3	0.01	1.00E+00				0.00E+00	5.61E−07	0.00E+00	6.06E−05			9,37,615	No
Cobalt (Co)	64.6	0.01	2.90E−04		2.90E−04		0.00E+00	2.18E−03	0.00E+00	2.35E−01			272	No
Copper (Cu)	47.3	0.01	3.70E−02				0.00E+00	1.25E−05	0.00E+00	1.35E−03			34,692	No
Mercury (Hg)	1.1	0.01	3.00E−04		8.60E−05		0.00E+00	1.25E−04	0.00E+00	3.87E−03			275	No
Molybdenum (Mo)	4.8	0.01	5.00E−03		5.00E−03		0.00E+00	9.39E−06	0.00E+00	1.01E−03			4,688	No
Nickel (Ni)	19.0	0.01	2.00E−02			9.10E−01	6.04E−08	9.30E−06	0.00E+00	1.00E−03			314	No
Selenium (Se)	6.1	0.01	5.00E−03				0.00E+00	1.19E−05	0.00E+00	1.29E−03			4,688	No
Silver (Ag)	0.55	0.01	5.00E−03				0.00E+00	1.08E−06	0.00E+00	1.16E−04			4,688	No
Thallium (Tl)	0.415	0.01	8.00E−05				0.00E+00	5.08E−05	0.00E+00	5.48E−03			75	No
Zinc (Zn)	432	0.01	3.00E−01				0.00E+00	1.41E−05	0.00E+00	1.52E−03			2,81,284	No
Organic Compounds														
Trichloroethene (TCE)	0.710	0.1.0	6.00E−03	1.50E−02	6.00E−03	1.00E−02	2.48E−11	1.16E−06	2.34E−08	7.29E−04	2.34E−08	1.16E−06	30	No
cis-1,2-Dichloroethene (DCE)	0.007	0.1.0	1.00E−02		1.00E−02		0.00E+00	6.85E−09	0.00E+00	4.32E−06	0.00E+00	6.85E−09	1,620	No
							7.00E−07	*0.003*	*1.65E−05*	*0.35*	*1.72E−05*	*0.35*		

Notes:

1. The computational formulae and models used in this evaluation are discussed in Chapters 13 and 15.

2. Risk/hazard for air accounts for only the airborne emissions of contaminated particulates for all chemicals present at the site; strict volatilization effects are not included in this screening analysis.

3. Case-specific exposure parameters used in the calculations were obtained from the following sources—DTSC (1994), OSA (1993), and USEPA (1989c,f,g,k,n,o,p,r, 1991a–h, 1992a,b,d,e,f,h,i,j).

TABLE 14.11

Risk Screening for a Construction Worker Exposure to Soils at the PLC Site

Chemical of Potential Concern	Maximum Soil Concentration (mg/kg)	Chemical-Specific Dermal Absorption (ABS$_d$)	Toxicity Criteria				Risk for Air	Hazard for Air	Risk for Soil	Hazard for Soil	Total Risk (Air + Soil)	Total Hazard (Air + Soil)	Risk-Based Soil Criteria (mg/kg)	Soil Criteria Exceeded?
			Oral RfD (mg/kg-day)	Oral SF (1/mg/kg-day)	Inhalation RfD (mg/kg-day)	Inhalation SF (1/mg/kg-day)								
			Inorganic Chemicals											
Antimony (Sb)	1.1	0.01	4.00E − 04				0.00E + 00	5.38E − 04	0.00E + 00	1.45E − 02			73	No
Arsenic (As)	12	0.03	3.00E − 04	1.75E + 00		1.20E + 01	4.03E − 07	7.83E − 03	1.92E − 06	2.56E − 01			5	Yes
Cadmium (Cd)	2.6	0.001	5.00E − 04			1.50E + 01	1.09E − 07	1.02E − 03	0.00E + 00	2.47E − 02			24	No
Chromium (Cr)	57.3	0.01	1.00E + 00				0.00E + 00	1.12E − 05	0.00E + 00	3.02E − 04			1,83,154	No
Cobalt (Co)	64.6	0.01	2.90E − 04		2.90E − 04		0.00E + 00	4.36E − 02	0.00E + 00	1.17E + 00			53	Yes
Copper (Cu)	47.3	0.01	3.70E − 02				0.00E+00	2.50E − 04	0.00E + 00	6.73E − 03			6,777	No
Mercury (Hg)	4.8	0.01	3.00E − 04		8.60E − 05		0.00E + 00	2.50E − 03	0.00E + 00	1.93E − 02			50	No
Molybdenum (Mo)	19.0	0.01	5.00E − 03		5.00E − 03		0.00E + 00	1.88E − 04	0.00E + 00	5.05E − 03			916	No
Nickel (M)	19.0	0.01	2.00E − 02			9.10E − 01	4.83E − 08	1.86E − 04	0.00E + 00	5.00E − 03			393	No
Selenium (Se)	6.1	0.01	5.00E − 03				0.00E + 00	2.39E − 04	0.00E + 00	6.42E − 03			916	No
Silver (Ag)	0.55	0.01	5.00E − 03				0.00E + 00	2.15E − 05	0.00E + 00	5.79E − 04			916	No
Thallium (T)	0.415	0.01	8.00E − 05				0.00E + 00	1.02E − 03	0.00E + 00	2.73E − 02			15	No
Zinc (Zn)	432	0.01	3.00E − 01				0.00E + 00	2.82E − 04	0.00E + 00	7.58E − 03			54,946	No
			Organic Compounds											
Trichloroethene (TCE)	0.710	0.10	6.00E − 03	1.50E − 02	6.00E − 03	1.00E − 02	1.98E − 11	2.32E − 05	1.58E − 09	1.23E − 03	0.00E + 00	7.26E − 06	444	No
cis-1,2-Dichloroethene (DCE)	0.007	0.10	0.10		1.00E − 02	1.00E − 02	0.00E + 00	1.37E − 07	0.00E + 00	1.37E − 07	0.00E + 00		946	No
							5.60E − 07	*0.06*	*1.92E − 06*	*1.55*	*2.48E − 06*	*1.6*		

Note:

1. The computational formulae and models used in this evaluation are discussed in Chapters 13 and 15.

2. Risk/hazard for air accounts for only the airborne emissions of contaminated particulars for all chemicals present at the site; strict volatilization effects are not include in the screening analysis.

3. Case-specific exposure parameters used in the calculations were obtained from the following sources –DTSC (1994), OSA (1993), and USEPA (1989c,f,g,k,n,o,p,r, 1991a–h, 1992a,b,d,e,f,h,i,j).

TABLE 14.12

Risk Screening for a Downgradient Residential Population Exposure to Groundwater from the PLC Site

Chemical of Potential Concern	Maximum Water Concentration (µg/L)	Chemical-Specific K_p (cm/h)	Oral RfD (mg/kg-day)	Oral SF (1/mg/kg-day)	Inhalation RfD (mg/kg-day)	Inhalation SF (1/mg/kg-day)	Risk for Water	Hazard for Water	Risk Based Water Criteria (µg/L)	Water Criteria Exceeded?
Inorganic Chemicals										
Manganese (Mn)	1,390	1.60E − 04	5.00E − 03		1.40E − 05		0.00E + 00	1.78E + 01	78	Yes
Molybdenum (Mo)	6.5	1.60E − 04	5.00E − 03		5.00E − 03		0.00E + 00	8.31E − 02	78	No
Thallium (Tl)	1.5	1.60E − 04	8.00E − 05				0.00E + 00	1.20E + 00	1.3	Yes
Organic Compounds										
Trichloroethene (TCE)	939	1.60E − 02	6.00E − 03	1.50E − 02	6.00E − 03	1.00E − 02	3.57E − 04	2.02E + 01	2.6	Yes
cis-1,2-Dichloroethene (DCE)	118	1.00E − 02	1.00E − 02		1.00E − 02		0.00E + 00	1.52E + 00	78	Yes
							3.57E − 04	**40.7**		

Notes:

1. The computational formulae and models used in this evaluation are discussed in Chapters 13 and 15.

2. Risk/hazard for water account for both volatile chemical emissions and non-volatile chemical contributors present at the site (DTSC, 1994).

3. Case-specific exposure parameters used in the calculations were obtained from the following sources - DTSC (1994), OSA (1993), and USEPA (1989c,f,g,k,n,o,p,r, 1991a–h, 1992a,b,d,e,f,h,i,j).

4. K_p = chemical-specific dermal permeability coefficient for water (DTSC, 1994).

TABLE 14.13

Summary of the Risk Screening for the PLC Site

Receptor Group	Risk Parameter	Exposure Routes and Pathways				
		Inhalation Exposure to Soils	Oral Exposure to Soils	Total Exposure to Soils	Total Exposure to Groundwater	Overall Total Risk
Hypothetical	Cancer risk	–	–	–	3.6×10^{-4}	3.6×10^{-4}
downgradient resident	Hazard index	–	–	–	40.7	40.7
Nearby and/or onsite	Cancer risk	7.0×10^{-7}	1.6×10^{-5}	1.7×10^{-5}	–	1.7×10^{-5}
worker	Hazard index	0.003	0.4	0.4	–	0.4
Construction worker	Cancer risk	5.6×10^{-7}	1.9×10^{-6}	2.5×10^{-6}	–	2.5×10^{-6}
	Hazard index	0.06	1.6	1.6	–	1.6

Note: Risks due to oral exposure is the total contribution from ingestion and dermal absorption of chemicals present in the contaminated medium.

14.2.8.7 A Risk Management Decision

Table 14.13 summarizes the results of the diagnostic baseline risk assessment for the PLC site. In general, a cancer risk estimate greater that 10^{-6} or a noncarcinogenic hazard index greater than 1 indicate the presence of contamination that may pose a significant threat to human health. Overall, the levels of both carcinogenic and noncarcinogenic risks associated with the PLC site should *not* require time-critical removal action for contaminated soils present at this site. However, the use of untreated groundwater from aquifers underlying the site as a potable water supply source could pose significant risks to exposed individuals or a community that uses such water for culinary purposes. Some type of corrective action or institutional control measure may therefore be necessary for this site.

The risk evaluation presented earlier indicates that the COCs present at the PLC site may pose some degree of risks to human receptors potentially exposed via the soil and groundwater media. Consequently, if this site is to be used for future commercial redevelopment projects or if raw/ untreated groundwater from this site is to be used as a potable water supply source, then a comprehensive site restoration program may be necessary to abate the imminent risks that the site poses— especially from groundwater exposures.

Whereas, limited site control measures will probably be adequate to protect construction workers at the PLC site, a more extensive corrective action program will be needed for the impacted aquifer. Thus, general response actions (comprised of an integrated soil and groundwater remediation program) should be developed for each of the potentially impacted environmental media associated with the site. Ultimately, the redevelopment of the site for commercial purposes may require only limited restoration activities, which may indeed be accomplished through "incidental" site capping offered to the site following the construction of commercial buildings and pavements at the site.

14.2.8.8 A Site Restoration Strategy

This section consists of a brief review of remedial action alternatives considered for the contaminated site problem described previously. General response actions to consider and evaluate for the PLC site include containment (e.g., via capping and immobilization techniques) and treatment (e.g., by soil vapor extraction [SVE], thermal treatments, biological treatments, and soil washing) for the soils and pump-and-treat systems (including air or steam stripping and GAC adsorption) for groundwater. Table 14.14 presents a broader range of options considered and evaluated for this site. The remedial alternatives determined to be the most technically feasible subsequently become the focus for more detailed analyses. For other than a "no-action" situation, the site-specific risk-based

TABLE 14.14

Screening of a Range of Feasible Remedial Technologies and an Evaluation of Remediation Process Options for the PLC Site

Target Media for Remediation Program	General Response Action	Remedial Technology	Preferred Process Option	Description of Remediation Program	Screening Comments
Groundwater	No action	None	Not applicable	No action	Site remediation is required in order to achieve the remedial action objectives for this site.
	Pump-and-treat	Groundwater extraction	Extraction/injection wells	Series of wells to extract contaminated groundwater; injection wells inject uncontaminated or treated water to increase flow to extraction wells.	Preferred method of choice.
	Containment	Capping (using low permeability caps and liners)	Clay	Compacted clay, covered with soil over areas of contamination.	Potentially applicable as an effective alternative. Susceptible to cracking but has self-healing properties. Easily implemented. Low capital and maintenance costs.
			Multimedia cap	Clay and synthetic membrane, covered with soil over areas of contamination.	Potentially applicable as an effective alternative. Least susceptible to cracking. Easily implemented. Moderate capital and maintenance costs.
		Vertical barriers	Slurry wall	Trench around area of contamination is filled with bentonite slurry.	Potentially applicable.
			Grout curtain	Pressure injection of grout in a regular pattern of drilled holes.	Potentially applicable.
			Vibrating beam	Vibrating force used to advance beams into the ground, with injection of slurry as beam is withdrawn.	Potentially applicable.
		Horizontal barriers	Grout injection	Pressure injection of grout at depth through closely spaced drilled holes.	Potentially applicable.
			Block displacement	In conjunction with vertical barriers, injection of slurry in notched injection holes.	Potentially applicable.

(Continued)

TABLE 14.14 (*Continued*)
Screening of a Range of Feasible Remedial Technologies and an Evaluation of Remediation Process Options for the PLC Site

Target Media for Remediation Program	General Response Action	Remedial Technology	Preferred Process Option	Description of Remediation Program	Screening Comments
	Treatment	Gradient control	Hydraulic gradient manipulation	Use of hydraulic gradient to control flow.	Potentially applicable.
		Bioremediation	Aerobic digestion	Degradation of chemicals using microorganisms in an aerobic environment.	Potentially applicable.
			Anaerobic digestion	Degradation of chemicals using microorganisms in an anaerobic environment.	Potentially applicable.
		Physical treatment (e.g., physical seperation or destruction) and chemical	Air stripping	Mixing large volumes of air and water in a packed column to promote transfer of VOCs to air.	Potentially applicable to some contaminants.
			Carbon adsorption	Adsorption of contaminants onto activated carbon by passing water through carbon column.	Potentially applicable.
		treatment (e.g., chemical modification or destruction)	Reverse osmosis	Use of high pressure to force water through a membrane, leaving contaminants behind.	Potentially applicable.
			Incineration	Combustion.	Potentially applicable.
		Permeable treatment bed	Treatment bed	Treat shallow groundwater in place by constructing permeable treatment beds that can physically and chemically remove contaminants.	Potentially applicable as a temporary remedial measure.
Site soils	No action	None	Not applicable	No action.	Site remediation is required in order to achieve the remedial action objectives for this site.

(*Continued*)

TABLE 14.14 (*Continued*)

Screening of a Range of Feasible Remedial Technologies and an Evaluation of Remediation Process Options for the PLC Site

Target Media for Remediation Program	General Response Action	Remedial Technology	Preferred Process Option	Description of Remediation Program	Screening Comments
	Containment	Capping (using low permeability caps and liners)	Clay	Compacted clay, covered with soil over areas of contamination.	Potentially applicable as an effective alternative. Susceptible to cracking but has self-healing properties. Easily implemented. Low capital and maintenance costs.
			Multimedia cap	Clay and synthetic membrane, covered with soil over areas of contamination.	Potentially applicable as an effective alternative. Least susceptible to cracking. Easily implemented. Moderate capital and maintenance costs.
		Vertical barriers	Slurry wall	Trench around area of contamination is filled with bentonite slurry.	Potentially applicable.
			Grout curtain	Pressure injection of grout in a regular pattern of drilled holes.	Potentially applicable.
			Vibrating beam	Vibrating force used to advance beams into the ground, with injection of slurry as beam is withdrawn.	Potentially applicable.
		Horizontal barriers	Grout injection	Pressure injection of grout at depth through closely spaced drilled holes.	Potentially applicable.
			Block displacement	In conjunction with vertical barriers, injection of slurry in notched injection holes.	Potentially applicable to overburden.
		Gradient control	Hydraulic gradient manipulation	Use of hydraulic gradient to control leachate flow.	Potentially applicable.

(*Continued*)

TABLE 14.14 (Continued)
Screening of a Range of Feasible Remedial Technologies and an Evaluation of Remediation Process Options for the PLC Site

Target Media for Remediation Program	General Response Action	Remedial Technology	Preferred Process Option	Description of Remediation Program	Screening Comments
	Excavation/ Treatment	Excavation	Soils excavation/ removal	Mechanical removal of materials for treatment and/or redisposal.	Potentially applicable.
		Immobilization/ stabilization	Sorption and/or encapsulation	Use of pozzolanic agents to help immobilize site contaminants.	Potentially applicable.
		Physical treatment	Incineration and pyrolysis	Thermal treatment methods.	Potentially applicable, especially if NAPL should be encountered.
		Chemical treatment	Neutralization	Use of chemical reagents to attenuate contaminant toxicity effects.	Potentially applicable.
		Biological treatment	Bioremediation	Use of microorganisms to degrade organic contaminants.	Potentially applicable to some of the soil contaminants.
	Vacuum extraction	SVE	Extraction/injection wells	Series of wells to extract contaminated soil vapors; injection wells inject fresh air to increase flow to extraction wells.	Preferred method of choice.

criteria developed previously (shown in Tables 14.10–14.13) can be used to guide and support the development, screening, and selection of potentially feasible remedial alternatives.

Feasible Remediation Options. An important site restoration technique for organic contamination in soils would be bioremediation; clearly identified "hot-spots" may however be removed for disposal or for treatment by other processes such as incineration. Yet, other favored remediation techniques would consist of the use of in situ soil-venting systems. Inorganic contaminants that remain may be encapsulated or otherwise immobilized or may be removed through soil washing or leaching activities.

The most frequently used treatment methods for volatile organic contamination in groundwater are air stripping and GAC adsorption. For greater removal efficiency, air stripping is used in tandem with GAC (i.e., a GAC–air stripping remedial system). The overall remediation of the groundwater contamination at the PLC site may be accomplished by using a pump/treat/reinject system. The preferred approach will be to pump from the edges of the mapped plume and reinject (after treatment) at the center of the plume.

14.2.8.9   The Corrective Action Plan

The corrective action plan for the PLC facility is divided into soil and groundwater remediation tasks. A SVE system is tentatively recommended for the impacted soils and a pump-and-treat technology for the affected groundwater. The soil remediation will consist of the design, installation, and operation of a SVE system (to remove the VOCs from the soils in order to prevent further impacts on groundwater) and laboratory analyses of verification borings (for monitoring purposes and performance evaluation). The groundwater remediation will consist of the design, installation, and operation of a GTS, water sampling and analysis during operation, and laboratory analyses of verification samples (for monitoring purposes and performance evaluation).

The SVE System. The SVE system recommended for the PLC site restoration will utilize an array of vapor extraction wells and air injection wells located in areas of impacted soils, so designed to achieve maximum efficiency for contaminant removal; the use of the complementing air injection wells will generally improve the performance of the SVE system. The SVE system will be operated until the attainment of the acceptable cleanup levels for the key contaminants of concern in the contaminated soils. Subsequently, verification samples (to ascertain the performance of the SVE system) will be collected and analyzed prior to site closure or redevelopment.

The Groundwater Treatment System. The GTS recommended for the PLC site will be a pump-and-treat system that consists of both extraction and injection wells; the extraction-injection well network is so designed and placed as to achieve maximum efficiency for contaminant removal from the impacted groundwater. Groundwater pumped will be treated prior to reinjection or disposal. The GTS will be operated until the attainment of the acceptable cleanup goals established for the key contaminants of concern in groundwater. Subsequently, verification samples (to ascertain the performance of the GTS) will be collected and analyzed prior to closure; furthermore, long-term monitoring wells will be installed downgradient of the contaminated areas.

14.3   THE REUSE AND REVITALIZATION OF BROWNFIELD SITES: DEVELOPMENTS AND CONSTRUCTIONS ON OTHERWISE CONTAMINATED SITES

Several localities in just about every region around the world have some abandoned industrial or related commercial sites that have been lying dormant for several years. In fact, in the past, city planners and property developers have had the tendency to avoid this issue by going after "virgin" lands (or so-called "greenfields") located away from these old abandoned industrial or "brownfield" sites. In recent years, however, as it becomes more and more difficult to satisfy property demand in the urban areas, pressure has been mounting to redevelop some of the abandoned sites—and consequently the birth of the so-called "brownfields" economic redevelopment programs in some regions. Now, it is notable that slightly different definitions of the so-called "brownfields" may be encountered in the literature. For instance, the US EPA has defined a brownfield site as a vacant or abandoned, idled, or underutilized industrial or commercial facility where expansion, redevelopment, or reuse is complicated by real or perceived environmental contamination; this definition serves as a broad enough basis for the discussions offered here. Indeed, many of these sites lie in prime central and desirable locations where property value could be rather high. Anyhow, it is important to carefully deliberate on potential redevelopment plans from the outset of any brownfields project. After all, the redevelopment plan (or lack thereof) will govern most brownfields projects—from the identification of site investigation and cleanup standards and the ability to obtain financing, up to critical elements relating to the ultimate affordability or profitability of the project. Defining and understanding the overall long-term goals of the brownfields project and the decisions to be made throughout the project in support of those goals are indeed a crucial element in identifying appropriate technologies for site investigation and cleanup that typically precedes redevelopment or reuse.

Invariably, preparing brownfield sites for productive reuse requires the integration of numerous and variant elements—including financial issues, community involvement, liability considerations, environmental assessment and cleanup, regulatory requirements, as well as coordination among many stakeholder groups. In these efforts, the assessment and cleanup of a contaminated site must be carried out in a way that integrates all those factors into the overarching redevelopment process. In addition, it is noteworthy that the cleanup strategy will generally vary from site to site; at some sites, cleanup will typically be completed before the property is utilized or transferred to new owners—whereas at other sites, cleanup may take place simultaneously with construction and redevelopment activities. Regardless of when and how cleanup is accomplished, the challenge to any brownfields program is to clean up sites in accordance with redevelopment goals. Such goals may include cost-effectiveness, timeliness, and avoidance of adverse effects on structures on the site and on neighboring communities, as well as redevelopment of the land in a way that benefits communities and local economies.

As a whole, a brownfields' process is tailored to the specific end use, if that use is known. For example, if the redevelopment plan calls for the construction of a light industrial facility, it may be appropriate to apply industrial investigation and cleanup standards that are less stringent than those applicable to a property that is to be redeveloped for residential use. The standards required will affect every aspect of the project—from its overall cost (which is generally greater as the standards become more conservative), to the applicability of innovative characterization and cleanup technologies. It is noteworthy, however, that new information about contamination or cleanup may require that reuse plans be altered; thus, it is imperative to develop flexible plans so that revised cleanup needs can be incorporated into them. At any rate, if the end use is not known at the beginning of the project, then a general type of desirable development—be it industrial, commercial, residential, or indeed a mixed-use development of some sort—should be identified to keep the program in perspective and well focused. All in all, risk-based strategies allow for a cost-effective characterization, restoration, and/or management of most contaminated site problems. Indeed, since the return of contaminated sites to "pristine" or "virgin" conditions is almost impractical, the use of risk-based solutions to strike a fair balance between adequate protection of human health and the environment vis-à-vis implementation of economically and technically feasible remedies should be embraced as the best of strategies yet. In particular, risk assessment has become a very important tool in determining the level of remediation necessary to return so-called "brownfield" sites to productive use—and in such a manner that does not require extensive cleanups that will make the property unattractive to investors and/or developers. Among other things, the risk assessment approach allows for the establishment of target cleanup levels appropriate for the actual current or anticipated future land use; thus, a site that is expected to, for example, remain in industrial use and that is also located in a heavily industrial area need not be cleaned up to the more stringent levels typically required for a residential property or area. This mindset would tend to allow for an overall streamlined process for constructions on contaminated sites, as well as help minimize likely cleanup costs under a given redevelopment initiative—thus effectually tackling the usually vexing questions of "how safe is safe" as relates to pre-remediation land-use conditions and/or "how clean is clean" for establishing post-remediation land-use criteria.

Part V

Appendices

This section of the book contains a set of four appendices that is comprised of the following:

- Appendix A, *Glossary of Selected Terms and Definitions*—addressing scientific and environmental terminologies that are generally pertinent to the evaluation of contaminated site problems.
- Appendix B, *Selected Scientific Tools and Databases for Contaminated Site Risk Management and Site Restoration Decision-Making*—that describes a variety of decision-making tools and logistics, as well as computer databases and information libraries, that may find several useful applications in contaminated site risk assessment/management programs and related corrective action decision-making efforts.
- Appendix C, *Toxicological Parameters for Selected Environmental Chemicals*—consisting of some toxicological data that are generally pertinent to human health risk evaluations for chemical exposure problems.
- Appendix D, *Selected Units of Measurement and Noteworthy Expressions*—consisting of some selected units of measurements and noteworthy expressions of potential interest to the environmental professional, analyst, or decision-maker faced with environmental contamination and related contaminated site problems.

Appendix A
Glossary of Selected Terms and Definitions

Absorption: The movement of a substance through the outer boundary of a medium—or the process of one substance actually penetrating into the structure of another substance. It is generally used to refer to the uptake of a chemical by a cell or an organism following exposure through the skin, lungs, and/or gastrointestinal tract. *Systemic absorption*—refers to the flow of chemicals into the bloodstream. In general, chemicals can be absorbed through the skin into the bloodstream and then transported to other organs; chemicals can also be absorbed into the bloodstream after breathing or oral intake.

Absorption fraction: Refers to the percent or fraction of a chemical in contact with an organism that becomes absorbed into the receptor—that is, the relative amount of a substance at the exchange barrier that actually penetrates into the body of an organism. Typically, this is reported as the unitless fraction of the applied dose or as the percent absorbed—for example, relative amount of a substance on the skin that penetrates through the epidermis into the body. *Relative absorption fraction (RAF)*—the fraction obtained by dividing the absolute bioavailability from soil by the absolute bioavailability from the dosing medium used in the toxicity study from which the reference dose for human health risk assessment was determined.

Acceptable daily intake (ADI): An estimate of the maximum amount of a chemical/agent, expressed on a body mass basis (viz., in mg/kg body weight/day), to which a potential receptor (or individuals in a [sub]population) can be exposed to on a daily basis over an extended period of time (usually a lifetime) without suffering a deleterious effect, or without anticipating an adverse effect.

Acceptable risk: A risk level generally deemed by society to be acceptable or tolerable. This is commonly considered as a risk management term—with the acceptability of the risk being dependent on: available scientific data; social, economic, and political factors; and the perceived benefits arising from exposure to an agent/stressor/chemical.

Action level (AL): The limit of a chemical in selected media of concern above which there are potential adverse health and/or environmental effects. On the whole, this represents the environmental chemical concentration above which some corrective action (e.g., monitoring or remedial action) is typically required by regulation.

Activated carbon: A highly adsorbent form of carbon used to remove contaminants from fluid emissions or discharges. It is a special form of amorphous carbon, often derived from charcoal and treated to make it capable of adsorbing and retaining certain chemical substances; its high adsorptive capacity makes it useful for removing contaminants from liquid and vapor waste streams. *Activated carbon adsorption*—a treatment technology based on the principle that certain organic constituents preferentially adsorb to organic carbon.

Acute: Of short-term duration—i.e., occurring over a short time, usually a few minutes or hours. *Acute exposure*—refers to a single large exposure or dose to a chemical, generally occurring over a short period (usually lasting <24–96 h), in relation to the lifespan of the exposed organism. An acute exposure can result in short-term or long-term health effects. *Acute effect*—takes place a short time (up to 1 year) after exposure. *Acute toxicity*—refers to the development of symptoms of poisoning or the occurrence of adverse health effects after exposure to a single dose or multiple doses of a chemical within a short period of time. It represents the sudden onset of adverse health effects that are of short duration—generally resulting in cellular changes that are reversible.

Adsorption: The physical process occurring when liquids, gases, or suspended matter adhere to the surfaces of, or in the pores of, an adsorbent material. Adsorption is a physical process that occurs without chemical reaction. Generally refers to the removal of contaminants from a fluid stream by concentrating the constituents onto a solid material—and consists of the physical process of attracting and holding molecules of other chemical substances on the surface of a solid, usually by the formation of chemical bonds. A substance is considered *adsorbed* if the concentration in the boundary region of a solid (e.g., soil) particle is greater than in the interior of the contiguous phase. *Adsorption coefficient*—represents the ratio of a substance's total concentration in the sorbed phase to that in the solution.

Adverse effect: A biochemical change, functional impairment, or pathologic lesion that affects the performance of the whole organism, or reduces an organism's ability to respond to a future environmental challenge. It is often exhibited by a change in the morphology, physiology, growth, development, reproduction, or life span of an organism, system, or (sub)population that results in an impairment of functional capacity, an impairment of the capacity to compensate for additional stress, or an increase in susceptibility to other influences.

Aerobe: An organism that can grow in the presence of air or free oxygen.

Aerobic: An environment that has a partial pressure of oxygen similar to normal atmospheric conditions.

Aerosol: A suspension of liquid or solid particles in air.

Agent (also, stressor): Any physical, chemical, or biological entity that can induce an adverse response.

Air sparging: The blowing of air through a liquid for mixing purposes—in order to strip volatile materials (i.e., VOCs), or to add oxygen. Usually refers to the highly controlled injection of pressurized air into a contaminant plume in the soil's *saturated* zone or contaminated aquifer, in order to transfer the contaminants into the vapor phase and to aerate the groundwater. This consists of the injection of air below the water table to strip volatile contaminants from the saturated zone.

Air stripping: A remediation technique comprised of the physical removal of dissolved-phase contamination from a water stream. It generally consists of the mass transfer of compounds from an aqueous stream to a gaseous stream—and involves the pressurized injection of air below the water table to strip volatile contaminants from the saturated zone (or groundwater).

Aliphatic compounds: Organic compounds in which the carbon atoms exist as either straight or branched chains; examples include pentane, hexane, and octane.

Ambient: Pertaining to surrounding conditions or area. *Ambient medium*—one of the basic categories of material surrounding or contacting an organism (e.g., outdoor air, indoor air, water, or soil) through which chemicals or pollutants can move and reach the organism.

Anaerobic: An environment without oxygen or air. It consists of processes that require the absence of oxygen. *Anaerobe*—is an organism that grows in the absence of oxygen or air.

Analyte: A chemical component of a sample that is to be investigated or measured; for example, if the *analyte of interest* in an environmental sample is mercury, then the laboratory testing or analysis will determine the likely amount of mercury in the sample. The *analytical method* defines the sample preparation and instrumentation procedures or steps that must be performed to estimate the quantity of analyte in a given sample.

Anoxic: An atmosphere greatly deficient in oxygen—or more generally, condition characterized by low levels of free oxygen.

Antagonism (or, antagonistic chemical effect): A pharmacologic or toxicologic interaction in which the combined effect of two chemicals is less than the sum of the effect of each chemical acting alone. This phenomenon is the result of interference or inhibition of the effects of one chemical substance by the action of other chemicals—and reflects the counteracting effect of one chemical on another, thus diminishing their additive effects.

Anthropogenic: Caused or influenced by human activities or actions.

Applied dose: The amount of a substance in contact with the primary absorption boundaries of an organism (e.g., skin, lung, and gastrointestinal tract), and that is available for absorption. This actually is a measure of exposure—since it does *not* take absorption into consideration (See also, *administered dose*).

Aquifer: A geological formation, group of formation, or part of a formation, which is capable of yielding significant and usable quantities of groundwater to wells and/or springs.

Arithmetic mean (also, *average*): A statistical measure of central tendency for data from a normal distribution—defined, for a set of *n* values, as the sum of the values divided by *n*, as follows:

$$X_m = \frac{\sum_{i=1}^{n} X_i}{n}$$

Aromatic compounds: Organic compounds that contain carbon molecular ring structures (i.e., a benzene ring); examples include benzene, toluene, ethylbenzene, and xylenes (BTEX). These compounds are somewhat soluble, volatile, and mobile in the subsurface environment, and are a very useful indicator of contaminant migration.

Asphalt batching (also, *asphalt incorporation*): A method for treating hydrocarbon-contaminated soils—that involves the incorporation of petroleum-laden soils into hot asphalt mixes as a partial substitute for stone aggregate. This mixture can then be utilized for construction of pavements.

Attenuation: Any decrease in the amount or concentration of a pollutant in an environmental matrix as it moves in time and space. It represents the reduction or removal of contaminant constituents by a combination of physical, chemical, and/or biological factors acting upon the contaminated "parent" media.

Attributable risk (also, *incremental risk*): The difference between risk of exhibiting a certain adverse effect in the presence of a toxic substance in comparison with that risk to be expected in the absence of the substance (See also, *excess lifetime risk*).

Auger: A rotary drilling equipment used in soils or unconsolidated materials, that continuously removes cuttings from a borehole by mechanical means without the use of fluids.

Average concentration: A mathematical average of chemical concentration(s) from more than one sample—typically represented by the "arithmetic mean" or the "geometric mean" for environmental samples.

Average daily dose (*ADD*): The average dose calculated for the duration of receptor exposure, defined by:

$$\text{ADD}(\text{mg/kg} - \text{day}) = \frac{[\text{chemical concentration}] \times [\text{contact rate}]}{[\text{body weight}]}$$

This is used to estimate risks for chronic noncarcinogenic effects of environmental chemicals.

Averaging time: The time period over which a function (e.g., human exposure concentration of a chemical) is measured—yielding a time-weighted value.

Background (*threshold*) *level:* The normal, or typical, average ambient environmental concentration of a chemical constituent. It represents the amount of an agent in a medium (e.g., air, water, soil) that is not attributed to the source(s) under investigation in an exposure assessment. Overall, two types of background levels may exist for chemical substances—namely, naturally occurring concentrations and elevated anthropogenic levels resulting from non-site-related human activities. *Anthropogenic background levels*—refer to concentrations of chemicals that are present in the environment due to human-made, non-site sources (e.g., lead depositions from automobile exhaust and "neighboring" industry). *Naturally occurring background levels*—refer to ambient concentrations of chemicals that are present in the environment and yet have not been influenced by human activities (e.g., natural formations of aluminum, arsenic, and manganese).

Bacteria: A group of diverse and ubiquitous prokaryotic single-celled microorganisms.

Benchmark risk: A threshold level of risk, typically prescribed by regulations, and above which corrective measures will almost certainly have to be implemented to mitigate the risks.

Bioaccessibility: A term used in describing an event that relates to the absorption process upon exposure of an organism—and generally refers to the fraction of the administered substance that becomes solubilized in the gastrointestinal fluid. For the most part, solubility is a prerequisite of absorption, although small amounts of some chemicals in particulate or suspended/emulsified form may be absorbed by pinocytosis. Moreover, it is not simply the fraction dissolved that determines bioavailability, but also the rate of dissolution, which has physiological and geochemical influences. In and of itself, bioaccessibility is not a direct measure of the movement of a substance across a biological membrane (i.e., absorption or bioavailability). Indeed, the relationship of bioaccessibility to bioavailability is ancillary and the former need not be known in order to measure the latter. However, bioaccessibility (i.e., solubility) may serve as a surrogate for bioavailability if certain conditions are met.

Bioaccumulation: The progressive increase in amount of a chemical in an organism or part of an organism that occurs because the rate of intake exceeds the organism's ability to remove the substance from the body. This represents the retention and concentration of a chemical by an organism—that is the result of a buildup of the chemical in the organism as a consequence of the organism taking in more of the chemical than it can rid of in the same length of time, and therefore ends up storing the chemical in its tissue, etc. (See also, *bioconcentration*).

Bioassay: Measuring the effect(s) of environmental exposures by intentional exposure of living organisms to a chemical. It consists of tests used to evaluate the relative potency of a chemical by comparing its effects on a living organism with the effect of a standard preparation on the same type of organism.

Bioaugmentation: A process in which specially selected bacteria cultures that are predisposed to metabolize some target compound(s) are added to impacted media/soil, along with the nutrient materials, to encourage degradation of the contaminants of concern.

Bioconcentration: The accumulation of a chemical substance in tissues of organisms (such as fish) to levels greater than levels in the surrounding media (such as water) for the organism's habitat; this is often used synonymously with bioaccumulation. *Bioconcentration factor (BCF)*—is the ratio of the concentration of a chemical substance in an organism, at equilibrium to the concentration of the substance in the surrounding environmental medium. It is a measure of the amount of selected chemical substances that accumulate in humans or in biota (See also, *bioaccumulation*).

Biodegradation: The breakdown or decomposition of a substance into simpler substances by the action of microorganisms, usually in soil. The process may or may not detoxify the material that is decomposed. *Daughter product*—a compound that results directly from the biodegradation of another; for example, *cis*-1,2-dichloroethene (*cis*-1,2-DCE) is commonly a daughter product of trichloroethene (TCE). *Biodegradable*—capable of being metabolized by a biologic process or an organism.

Biomagnification: The serial accumulation of a chemical by organisms in the food chain, with higher concentrations occurring at each successive trophic level.

Bioremediation (also called *biorestoration*): The process by which biological/living organisms are used to degrade or transform (hazardous organic) contaminants. It is a viable and cost-effective remediation technique for treating a wide variety of contaminants (such as petroleum and aromatic hydrocarbons, chlorinated solvents, and pesticides). It relies on microorganisms to transform hazardous compounds found in environmental matrices into innocuous or less toxic metabolic products. *Natural in-situ bioremediation*—involves the attenuation of contaminants by indigenous (native) microorganisms without any manipulation.

Biostimulation: A process involving the addition of selected amounts of nutrient materials to stimulate or encourage the growth of the indigenous bacteria in the soil—ultimately resulting in the degradation of some target contaminant(s).

Biological uptake: The transfer of hazardous substances from the environment to plants, animals, and humans. This may be evaluated through environmental measurements, such as measurement of the amount of the substance in an organ known to be susceptible to that substance. More commonly, *biological dose measurements* are used to determine whether exposure has occurred. The presence of a chemical compound, or its metabolite, in human biologic specimens (such as blood, hair, or urine) is used to confirm exposure—and this can be an independent variable in evaluating the relationship between the exposure and any observed adverse health effects.

Biomagnification: The serial accumulation of a chemical by organisms in the food chain—with higher concentrations occurring at each successive trophic level.

Bioslurping: A remediation technology that combines vacuum-assisted free-product recovery with bioventing and soil vapor extraction to simultaneously recover free product from the water table as well as remediate contaminants in the vadose zone. Bioventing stimulates the aerobic bioremediation of hydrocarbon-contaminated soils, and vacuum-enhanced free-product recovery extracts light nonaqueous phase liquids (LNAPLs) from the capillary fringe and the water table.

Biota: All living organisms found within a prescribed volume or space.

Bioventing: A remediation technology designed to deliver electron acceptor to microorganisms in the vadose zone, in order to facilitate contaminant attenuation. It is a remediation process that supplies indigenous microorganisms with oxygen to support *in-situ* degradation of contaminants. An *electron acceptor* is a compound that receives electrons and is then reduced in an oxidation–reduction reaction; an *electron donor* is a compound that supplies electrons.

Borehole: A hole drilled into the earth and into which a well casing or screen can be installed to construct a well.

Brownfield: An abandoned, idled, or underused industrial or commercial facility where expansion or redevelopment is complicated by a real or perceived environmental contamination.

BTEX: Benzene, toluene, ethylbenzene, and xylenes.

Cancer: Refers to the development of a malignant tumor or abnormal formation of tissue. It is a disease characterized by malignant, uncontrolled invasive growth of body tissue cells.

Cancer slope factor (CSF) (also, *slope factor, SF, cancer potency factor, CPF, or cancer potency slope, CPS):* Health effect information factor commonly used to evaluate health hazard potentials for carcinogens. It is a plausible upper-bound estimate of the probability of a response per unit intake of a chemical over a lifetime—represented by the slope of the dose-response curve in the low-dose region. This parameter is used to estimate an upper-bound probability for an individual to develop cancer as a result of a lifetime of exposure to a particular level of a carcinogen. Generally, cancer slope factors are available from databases such as US EPA's Integrated Risk Information System (IRIS).

Capillary zone: The unsaturated area between ground surface and the water table. *Capillary fringe*—the porous material just above the water table which may hold water by capillarity (a property of surface tension that draws water upward) in the smaller soil void spaces.

Capture zone: The volume of aquifer from which groundwater is reasonably retrieved and/or contained as a result of the pumping of a well or network of wells.

Carcinogen: A cancer-producing chemical or substance. It represents any substance that is capable of inducing a cancer response in living organisms.

Carcinogenic: Capable of causing, and tending to produce or incite cancer in living organisms. That is, a substance able to produce malignant tumor growth.

Carcinogenicity: The power, ability, or tendency of a chemical, physical, or biological agent to produce cancerous tissues from normal tissue—in order to cause cancer in a living organism.

Chelates: The type of coordination compound in which a central metallic ion (Co_2+, Ni_2+, or Zn_2+) is attached by covalent bonds to two or more nonmetallic atoms in the same molecule, called ligands. *Chelating agents* are used to remove ions from solutions and soil.

Chemical (or, constituent) of potential concern (CoPC) (also, *Chemical of interest, CoI):* Chemical compound or substance present at elevated concentrations of concern, and that is likely to pose some threats to human health and/or environment; such chemicals become the focus of detailed characterization and/or remedial actions. *Chemical of potential ecological concern (CoPEC)*—often used to refer to chemical compound or substance present at elevated concentrations of concern, and that is likely to pose some threats to ecological health and/or environmental habitat. The CoPCs and CoPECs generally comprise the hazardous substances, pollutants, and contaminants that are investigated during the baseline risk assessment; the list of CoPCs/CoPECs may include all of the constituents whose data are of sufficient quality for use in the quantitative risk assessment, or a subset thereof. *Chemicals* (or *constituents*) *of concern (CoCs)*—represent the hazardous substances, pollutants, and contaminants that, at the end of a risk assessment, are found to be the "risk drivers" or those that may actually pose unacceptable human or ecological risks; the CoCs typically drive the need for a remedial action.

Chemical-specific adjustment factor (CSAF): A factor based on quantitative chemical-specific toxicokinetic or toxicodynamic data, which replaces the classical default uncertainty factor.

Chronic: Of long-term duration—i.e., occurring over a long period of time (usually more than 1 year). *Chronic daily intake (CDI)*—refers to the receptor exposure, expressed in mg/kg-day, averaged over a long period of time. *Chronic effect*—refers to an effect that is manifest after some time has elapsed from an initial exposure to a substance. *Chronic exposure*—refers to the long-term, low-level exposure to chemicals, i.e., the repeated exposure or doses to a chemical over a long period of time (usually lasting 6 months to a lifetime). It is generally used to define the continuous or intermittent long-term contact between an agent and a target. It is noteworthy that this may cause latent damage that does not appear until a later period in time. *Chronic toxicity*—refers to the occurrence of symptoms, diseases, or other adverse health effects that develop and persist over time, following exposure to a single dose or multiple doses of a chemical delivered over a relatively long period of time. This represents the adverse health effects that are of a long and continuous duration—generally resulting in cellular changes that are irreversible. Chronic toxicity usually consists of a prolonged health effect that may not become evident until many years after exposure.

Cleanup: Actions taken to abate a situation involving the release or threat of release of contaminants that could potentially affect human health and/or the environment. This typically involves a process to remove or attenuate contamination levels, in order to restore the impacted media to an "acceptable" or usable condition. *Cleanup level*—refers to the contaminant concentration goal of a remedial action, i.e., the concentration of media contaminant level to be attained through a remedial action.

Closure: All activities involved in taking a hazardous waste facility out of service and securing it for the duration required by applicable regulations and laws. *Site closures* typically follow the implementation of appropriate site restoration programs—with monitoring usually becoming part of the post-closure site activities.

Cometabolism: The process in which a compound is fortuitously degraded by an enzyme or cofactor produced during microbial metabolism of another compound.

Compliance: To conduct or implement an activity in accordance with stipulated legislative or regulatory requirements.

Concentration: Broadly refers to the amount/quantity of a material or agent dissolved or contained in unit quantity/volume in a given medium or system.

Conceptual site model (CSM): A conceptual physical representation of a site that summarizes general site characteristics, along with the nature and behaviors of contamination vis-à-vis likely receptors associated with the site. A CSM is a key element of corrective action assessment and implementation decisions—used to facilitate site characterization, risk assessment, remedial action design, and related efforts. *Conceptual model diagrams*—are visual representations of the CSMs, which serve as useful tools for communicating important pathways in a clear and concise way.

Confidence interval (CI): A statistical parameter used to specify a range, and the probability that an uncertain quantity falls within this range. *Confidence limits*—the upper and lower boundary values of a range of statistical probability numbers that define the CI. *Ninety five percent confidence limits (95% CL)*—refers to the limits of the range of values within which a single estimation will be included 95% of the time. For large samples sizes (i.e., $n > 30$),

$$95\%\mathrm{CL} = X_\mathrm{m} \pm \frac{1.96s}{\sqrt{n}}$$

where CL is the confidence level, and s is the estimate of the standard deviation of the mean (Xm). For a limited number of samples ($n < 30$), a confidence limit or confidence interval may be estimated from,

$$95\%\mathrm{CL} = X_\mathrm{m} \pm \frac{ts}{\sqrt{n}}$$

where t is the value of the Student t-distribution (refer to standard textbooks of statistics) for the desired confidence level and degrees of freedom, $(n - 1)$.

Consequence: The impacts resulting from the response associated with specified exposures, or loading or stress conditions.

Conservative assumption: Used in exposure and risk assessment, this expression refers to the selection of assumptions (when real-time data are absent) that are unlikely to lead to underestimation of exposure or risk. Conservative assumptions are those which tend to maximize estimates of exposure or dose—such as choosing a value near the high end of the concentration or intake rate range (See also, *worst case*).

Contact rate: Amount of an exposure or environmental medium (e.g., air, groundwater, surface water, soil, and cosmetics) contacted per unit time or per event (e.g., liters of groundwater ingested or milligrams of soil ingested per day).

Containment: Refers to systems used to prevent (or significantly reduce) the further spread of contamination. Such systems may consist of pumping (and/or injection) wells, and cut-off walls designed and placed at strategic locations. *Containment technology*—refers to those technologies designed to prevent the further migration of contaminants away from the current "areas of concern."

Contaminant (or, *pollutant*): Any substance or material that enters a system (such as the environment, human body, and food) where it is not normally found—as, e.g., any undesirable substance that is not naturally occurring and therefore not normally found in the environmental media of concern. This typically consists of any potentially harmful physical, chemical, biological, or radiological agent occurring in the environment, in consumer products, or at the workplace as a result of human activities. Such materials can potentially have adverse impacts upon exposure to an organism and/or could adversely impact public health and the environment simply by their presence in the ambient setting. *Contaminant release*—refers to the ability of a contaminant to enter into other environmental media/matrices (e.g., air, water, or soil) from its source(s) of origin. *Contaminant migration*—refers to the movement of a contaminant from its source through other matrices/media such as air, water, or soil. *Contaminant migration pathway*—is the path taken by the contaminants as they travel from the contaminated source through various environmental media.

Contaminant plume: A body of contaminated groundwater or vapor originating from a specific source and spreading out due to influences of factors such as local groundwater or soil vapor flow patterns and character of an aquifer. It represents the volume of groundwater or vapor containing the contaminants released from a pollution source.

Corrective action: Action taken to correct a problematic situation; it consists of an activity or set of activities undertaken to address an environmental contamination problem. A typical/common example involves the remediation of chemical contamination in soil and groundwater.

Cost-effective alternative: The *most cost-effective alternative* is the lowest cost alternative that is technologically feasible and reliable, and which effectively mitigates and minimizes environmental damage. It generally provides adequate protection of public health, welfare, and/or the environment.

Data quality assessment (DQA): Is the scientific and statistical evaluation of data to determine if data obtained from environmental data operations are of the right type, quality, and quantity to support their intended use. It is noteworthy that, DQA is built on a fundamental premise—that data *quality*, as a concept, is meaningful only when it relates to the *intended use* of the data.

Data quality objectives (DQOs): Qualitative and quantitative statements developed by analysts to specify the quality of data that, at a minimum, is needed and expected from a particular data collection activity (or hazard source characterization activity). It represents the full set of qualitative and quantitative constraints needed to specify the level of uncertainty that an analyst can accept when making a decision based on a particular set of data. This is determined based on the end use of the data to be collected. The *DQO process*—consists of a planning tool that enables an investigator to specify the quality of the data required to support the objectives of a given study. Overall, the DQO process results in a well thought out sampling and analysis plan. Despite the fact that the DQO process is considered flexible and iterative, it generally follows a well-defined sequence of stages that will allow effective and efficient data management.

Decision analysis: A process of systematic evaluation of alternative solutions to a problem where the decision is made under uncertainty. The approach is comprised of a conceptual and systematic procedure for analyzing complex sets of alternatives in a rational manner so as to improve the overall performance of a decision-making process.

Decision framework: Management tool designed to facilitate rational decision-making.

Degradation: The physical, chemical or biological breakdown of a complex compound into simpler compounds and byproducts.

de Minimus: A legal doctrine dealing with levels associated with insignificant versus significant issues relating to human exposures to chemicals that present very low risk. In general, this represents the level below which one need not be concerned—and, therefore, is of no public health consequence.

Dense nonaqueous phase liquids (DNAPLs): Organic liquids (with very low solubility), composed of one or more contaminants that are more dense than water, often coalescing in an immiscible layer at the bottom of a saturated geologic unit. Many chlorinated solvents are DNAPLs.

Dermal absorption: The absorption of materials/substances through the skin. *Dermally absorbed dose*—refers to the amount of the applied material (i.e., the "external dose") which becomes absorbed into the body.

Dermal adsorption: The process by which materials come into contact with the skin surface, but are then retained and adhered to the permeability barrier without being taken into the body.

Dermal exposure: Exposure of an organism or receptor through skin adsorption and possible absorption.

Detection limit (DL): The minimum concentration or weight of analyte that can be detected by a single measurement with a known confidence level. *Instrument detection limit (IDL)*—represents the lowest amount that can be distinguished from the normal "noise" of an analytical instrument (i.e., the smallest amount of a chemical detectable by an analytical instrument under ideal conditions). *Method detection limit (MDL)*—represents the lowest amount that can be distinguished from the normal "noise" of an analytical method (i.e., the smallest amount of a chemical detectable by a prescribed or specified method of analysis).

Deterministic model: A model that provides a single solution for a given set of stated variables. This type of model does not explicitly simulate the effects of uncertainty or variability, as changes in model outputs are due solely to changes in model components.

Diffusion: The migration of molecules, atoms, or ions from one fluid to another in a direction tending to equalize concentrations. It consists of a process whereby molecules move from a region of higher concentration to a region of lower concentration as a result of Brownian motion.

Dispersion: The overall mass transport process resulting from both molecular diffusion (which always occurs if there is a concentration gradient in the system) and the mixing of the constituent due to turbulence and velocity gradients within the system. It is the tendency for a solute to spread from the path that it would be expected to follow under advective transport. *Dispersivity*—a property that quantifies the physical dispersion of a solute that is being transported in a porous medium.

Dissolved product: The water-soluble fuel components of contaminant releases.

Domain (spatial and temporal): The limits of space and time that are specified in a risk assessment, or components thereof.

Dose: A stated quantity or concentration of a substance to which an organism is exposed over a continuous or intermittent duration of exposure; thus this provides a measure of the amount of a chemical substance received or taken in by potential receptors upon exposure—typically expressed as an amount of exposure (in mg) per unit body weight of the receptor (in kg). More specifically, it consists of the amount of agent that enters a target organ(ism) after crossing an exposure surface; if the exposure surface is an absorption barrier, the dose is an absorbed dose/uptake dose—otherwise it is considered an intake dose. *Total dose*—is the sum of chemical doses received by an individual from multiple exposure sources in a given interval as a result of interaction with all exposure or environmental media that contain the chemical substances of concern. Units of dose and total dose (mass) are often converted to units of mass per volume of physiological fluid or mass of tissue. *Exposure dose* (also referred to as *applied dose* or *potential dose*)—is the amount of an agent presented to an absorption barrier and available for absorption (i.e., the amount ingested, inhaled or applied to the skin); this amount may be the same as or greater than the absorbed dose. *Absorbed dose* (also called, *internal dose*)—is the amount or the concentration of a chemical substance (or its metabolites and adducts) actually entering an exposed organism (or pertinent biological matrices) via the lungs (for inhalation exposures), the gastrointestinal tract (for ingestion exposures), and/ or the skin (for dermal exposures). It represents the amount of chemical that, after contact with the exchange boundaries of an organism (viz., skin, lungs, gut), actually penetrates the exchange boundary and enters the circulatory system—i.e., the amount of a substance penetrating across an absorption barrier (represented by the exchange boundaries such as the skin, lung, and gastrointestinal tract) of an organism, via either physical or biological processes. The amount may be the same as or less than the applied dose. *Delivered dose*—denotes the amount of a substance available for biologically significant interactions in a target organ. *Biologically effective dose*—represents the amount of the chemical available for interaction with any particular organ, cell or macromolecular target. *Effective dose* (ED_{10})—refers to the dose corresponding to a 10% increase in an adverse effect, relative to the control response. *Lower limit on effective dose* (LED_{10})—the 95% lower confidence limit of the dose of a chemical needed to produce an adverse effect in 10% of those exposed to the chemical, relative to control.

Dose-response (or dose-effect): The quantitative relationship between the dose of a chemical substance and an effect caused by exposure to such substance. *Dose-response relationship*—refers to the relationship between a quantified exposure (dose), and the proportion of subjects demonstrating specific biological changes (response). *Dose-response curve*—refers to the graphical representation of the relationship between the degree of exposure to a chemical substance and the observed or predicted biological effects or response. *Dose-response assessment*—consists of a determination of the relationship between the magnitude of an administered, applied, or internal dose and a specific biological response. Response can be expressed as measured or observed incidence, percent response within groups of subjects (or populations), or as the probability of occurrence within a population. *Dose-response evaluation*—refers to the process of quantitatively evaluating toxicity information, and then characterizing the relationship between the dose of a chemical administered or received and the incidence of adverse health effects in the exposed population.

Dual-phase extraction (DPE): Also known as, *multiphase extraction (MPE), vacuum-enhanced recovery*, or sometimes *bioslurping*, is a full-scale technology that uses a high vacuum system to remove various combinations of contaminated ground water, separate-phase petroleum product, and hydrocarbon vapor from the subsurface. Extracted liquids and vapor are treated and collected for disposal, or reinjected to the subsurface (where permissible under applicable local laws).

Dump: A site used for the disposal of solid wastes without environmental controls or safeguards.

Ecological assessment endpoint: An explicit expression of the environmental value that is to be protected. An assessment endpoint includes both an ecological entity and specific attributes of that entity. *Ecological entity*—refers to a general term that may be used to reference a species, a group of species, an ecosystem function or characteristic, or a specific habitat.

Ecological measurement endpoint: A measurable ecological characteristic that is related to the valued characteristic chosen as the assessment endpoint.

Ecological risk assessment (ERA): The process that evaluates the likelihood that adverse ecological effects may occur or are occurring as a result of exposure to one or more stressors. The process typically involves defining and quantifying risks to nonhuman biota and determining the acceptability of the estimated risks. It consists of determining the probability and magnitude of adverse effects of environmental hazards on nonhuman biota.

Ecosystem: The interacting system of a biological community and its abiotic (i.e., nonliving) environment.

Ecotoxicity assessment: The measurement of effects of environmental toxicants on indigenous populations of organisms.

Effect: The response arising from a chemical-contacting episode. In general, this represents the change in the state or dynamics of an organism, system, or (sub)population caused by the exposure to an agent. *Local effect*—refers to the response that occurs at the site of first contact. *Systemic effect*—refers to the response that requires absorption and distribution of the chemical, and this tends to affect the receptor at sites farther away from the entry point(s).

Effect assessment: Consists of the combination of analysis and inference of possible consequences of the exposure to a particular stressor/agent based on knowledge of the dose-effect relationship associated with that particular stressor/agent in a specific target organism, system, or (sub) population.

Effective porosity: The ratio of the volume of inter-connected voids through which fluid can flow to the total volume of material.

Empirical model: A model with a structure that is based on experience or experimentation, and does not necessarily have a structure informed by a causal theory of the modeled process. This type of model can be used to develop relationships that are useful for forecasting and describing trends in behavior but may not necessarily be mechanistically relevant. Empirical dose-response models can be derived from experimental or epidemiologic observations.

Endangerment assessment: A case-specific risk assessment of the actual or potential danger to human health and welfare, and also the environment, that is associated with the release of hazardous chemicals into various exposure or environmental media or matrices.

Endpoint (toxic): An observable or measurable biological or biochemical effect (e.g., metabolite concentration in a target tissue) used as an index of the impacts of a chemical on a cell, tissue, organ, organism, etc. This is usually referred to as *toxicological endpoint* or *physiological endpoint* in the context of chemical toxicity assessments.

Enhanced rhizospere biodegradation: Enhanced biodegradation of contaminants near plant roots where compounds exuded by the roots increase microbial biodegradation activity. Other plant processes such as water uptake by the plant roots can enhance biodegradation by drawing contaminants to the root zone.

Environmental fate: The "destination" or "destiny" of a chemical after release or escape from a given source into the environment, and following transport through various environmental compartments. For example, in a contaminated land situation, it may consist of the movement of a chemical through the environment by transport in air, water, sediment, and soil—culminating in exposures to living organisms. It represents the disposition of a material in the various environmental compartments (e.g., soil, sediment, water, air, and biota) as result of transport, transformation, and degradation.

Environmental medium: A part of the environment for which reasonably distinct boundaries can be specified. Typical environmental media addressed in chemical risk assessments may include air, surface water, groundwater, soil, sediment, fruits, vegetables, meat, dairy, and fish—or indeed any other parts of the environment that could contain contaminants of concern.

Environmental pollutant: Any entity that contaminates any ambient media, including surface water, groundwater, soil, or air.

Environmental toxicant: Agents present in the surroundings of an organism that are harmful to the health of such organisms.

Event-tree analysis: A procedure, utilizing deductive logic, often used to evaluate a series of events that lead to an upset or accident scenario. It offers a systematic approach for analyzing the types of exposure scenarios that can result from a chemical exposure problem.

Excess (or *incremental*) *lifetime risk:* The additional or extra risk (above normal background rate) incurred over the lifetime of an individual as a result of exposure to a toxic substance (See also, *attributable risk*).

Exposure: The situation of receiving a dose of a substance, or coming in contact with a hazard. It represents the contact of an organism with a chemical, biological, or physical agent available at the exchange boundary (e.g., lungs, gut, or skin) during a specified time period. This is typically expressed by the concentration or amount of a particular agent that reaches a target organism, system, or (sub)population at a specific frequency for a defined duration. *Exposure conditions*—refer to factors (such as location and time) that may have significant effects on an exposed population's response to a hazard situation. *Exposure period*—refers to the time of continuous contact between an agent and a target receptor. *Exposure duration*—refers to the length of time over which continuous or intermittent contacts occur between an agent and a target receptor; it is generally represented by the length of time that a potential receptor is exposed to the hazards or contaminants of concern in a defined exposure scenario. *Exposure event*—refers to an incident or occurrence of continuous contact between a chemical or physical agent and a target receptor, usually defined by time (e.g., number of days or hours of contact). *Exposure frequency*—refers to the number of exposure events within a given exposure duration; it is generally represented by the number of times (per year or per event) that a potential receptor would be exposed to contaminants of concern in a defined exposure scenario. *Exposure parameters* (*or factors*)—refer to the variables used in the calculation of intake (e.g., exposure duration, breathing rate, food ingestion rate, and average body weight); these may consist of standard exposure factors that may be needed to calculate a potential receptor's exposure to toxic chemicals (in the environment).

Exposure assessment: The qualitative or quantitative estimation, or the measurement, of the dose or amount of a chemical to which potential receptors have been exposed, or could potentially be exposed to. This process comprises of the determination of the magnitude, frequency, duration, route, and extent of exposure (to the chemicals or hazards of potential concern). This generally corresponds to the process of estimating or measuring the magnitude, frequency, and duration of exposure to an agent, along with the number and characteristics of the population exposed; ideally, this should describe the sources, pathways, routes, and the uncertainties in the assessment. *Exposure activity pattern data*—relates to information on human activities used in exposure assessments. These may include a description of the activity, frequency of activity, duration of time spent performing the activity, and the microenvironment in which the activity occurs.

Exposure (*point*) *concentration* (*EPC*): The concentration of a chemical (in its transport or carrier medium) at the point of receptor contact. *Exposure point*—refers to a location of potential contact between an organism and a hazardous (viz., biological, chemical or physical) agent.

Exposure investigation: The collection and analysis of site-specific information to determine if human populations have been exposed to hazardous substances. The site-specific information may include environmental sampling, exposure-dose reconstruction, biologic or biomedical testing, and evaluation of medical information. The information from an exposure investigation can be used to support a complete public health risk assessment and subsequent risk management programs.

Exposure model: A conceptual or mathematical representation of the exposure process.

Exposure pathway: The course a chemical, biological, or physical agent takes from a source to an exposed population or organism. It describes a unique mechanism by which an individual or population is exposed to chemical, biological, or physical agents at or originating from a contaminant release source.

Exposure route: The avenue or path (such as inhalation, ingestion, and dermal contact) by which a chemical/agent enters a target receptor or organism after contact. It represents the way in which a potential receptor may contact a chemical substance; for example, drinking (ingestion) and bathing (skin contact) are two different routes of exposure to contaminants that may be found in water.

Exposure scenario: A set of conditions or assumptions about hazard sources, exposure pathways, concentrations of chemicals, and potential receptors or exposed organisms that facilitates the evaluation and quantification of exposure(s) in a given situation. Broadly, it represents the combination of facts, assumptions, and inferences that define a discrete situation where potential exposures may occur; these may include the source, the exposed population(s), the timeframe of exposure occurrence, the microenvironment(s), and related activities. *Potentially exposed*—refers to the situation where valid information, usually analytical environmental data, indicates the presence of chemical(s) of a public health concern in one or more environmental media contacting humans (e.g., air, drinking water, soil, food chain, and surface water), and where there is evidence that some of the target populations have well-defined route(s) of exposure (e.g., drinking contaminated water, breathing contaminated air, contacting contaminated soil, or eating contaminated food) associated with them. Although actual exposure is generally not confirmed for a "potentially exposed" receptor, this type of exposure scenario would typically have to be adequately evaluated during the exposure assessment.

Ex situ: Out of the original position—i.e., excavated. It refers to a technology or process in which the contaminated material is removed from the site or location of contamination for treatment.

Extraction well: A well specifically designed to remove groundwater and/or soil vapor.

Extrapolation: An estimate of response or quantity at a point outside the range of the experimental data, usually via the use of a mathematical model. This consists of the estimation of unknown numerical values of an empirical (measured) function by extending or projecting from known values/observations to points outside the range of data that were used to calibrate the function. In chemical exposure situations, this may comprise of the estimation of a measured response in a different species, or by a different route than that used in the experimental study of interest—i.e., species-to-species; route-to-route; acute-to-chronic; high-to-low dose; etc. For instance, the quantitative risk estimates for carcinogens are generally low-dose extrapolations based on observations made at higher doses.

Exudates: Release of soluble organic matter from the roots of plants to enhance availability of nutrients or as a by-product of fine root degradation.

Fate: The pattern of distribution of an agent/chemical, its derivatives, or metabolites in an organism, system, compartment, or (sub)population of concern as a result of transport, partitioning, transformation, and/or degradation.

Fault-tree analysis: A procedure often used to evaluate series of events that lead to an upset or accident scenario, using inductive logic.

Feasibility Study (FS): The analysis and selection of alternative remedial or corrective actions for hazardous waste or contaminated sites. The process identifies and evaluates remedial alternatives by utilizing a variety of appropriate environmental, engineering, and economic criteria.

Field Sampling Plan (FSP): A documentation that defines in detail, the sampling and data gathering activities to be used in the investigation of a potentially contaminated land.

Free product: Chemical constituents that float on groundwater, or that remains "unadulterated" as a contaminant pool in the environment, following a product release or spill.

Fugitive dust: Atmospheric dust arising from disturbances of particulate matter exposed to the air. Fugitive dust emissions typically consist of the release of chemicals from contaminated surface soil into the air, attached to dust particles.

Geographic Information System (GIS): Computer-based tool used to capture, manipulate, process, and display spatial or geo-referenced data for solving complex resource, environmental, and social problems. GIS is indeed a rapidly developing technology for handling, analyzing, and modeling geographic information. It is not a source of information *per se*, but only a way to manipulate information. When the manipulation and presentation of data relates to geographic locations of interest, then our understanding of the real world is enhanced.

Geometric mean: A statistical measure of the central tendency for data from a positively skewed distribution (viz., lognormal), given by:

$$X_{gm} = \left[(X_1)(X_2)(X_3)...(X_n) \right]^{1/n}$$

or,

$$X_{gm} = \text{anti} \log \left\{ \frac{\sum_{i=1}^{n} \ln[X_i]}{n} \right\}$$

Ground-level concentration (GLC): The air pollutant concentration at ground level—generally used in monitoring or modeling assessments.

Groundwater: Water found beneath the ground surface; it represents underground waters, whether present in a well-defined aquifer, or present temporarily in the vadose (unsaturated soil) zone. Groundwater is primarily water that has seeped down from the surface by migrating through the interstitial spaces in soils and geologic formations. *Groundwater monitoring well*—a well designed and installed specifically to provide groundwater samples for contaminant analyses.

Hazard: That innate character or property which has the potential for creating adverse and/or undesirable consequences. It represents the inherent adverse effect that a chemical or other object poses—and defines the chance that a particular substance will have an adverse effect on human health or the environment under a particular set of circumstances that creates an exposure to that substance. Thus, hazard is simply a source of risk that does not necessarily imply actual potential for occurrence; a hazard produces risk only if an exposure pathway exists, and if exposures create the possibility of adverse consequences.

Hazard assessment: The process of determining whether exposure to an agent can cause an increase in the incidence of a particular adverse health effect (e.g., cancer, and birth defects), and whether the adverse health effect is likely to occur in the target receptor populations potentially at risk. This involves gathering and evaluating data on types of injury or consequences that may be produced by a hazardous situation or substance. The process includes hazard identification and hazard characterization.

Hazard characterization: The qualitative and, wherever possible, quantitative (or semiquantitative) description of the inherent property of an agent/stressor or situation having the potential to cause adverse effects. Wherever possible, this would tend to include a dose-response assessment and its attendant uncertainties.

Hazard identification: The systematic identification of potential accidents, upset conditions, etc.—consisting of a process of determining whether or not, for instance, a particular substance or chemical is causally linked to particular health effects. This is generally comprised of the identification of the type and nature of adverse effects that an agent has with respect to its inherent capacity to affect an organism, system, or (sub)population. Thus, the process involves determining whether exposure to an agent or hazard can cause an increase in the incidence of a particular adverse response or health effect in receptors of interest.

Hazard quotient (HQ): The ratio of a single substance exposure level for a specified time period to the "allowable" or "acceptable" intake limit/level of that substance derived from a similar exposure period. For a particular chemical and mechanism of intake (e.g., oral, dermal, and inhalation), the hazard quotient is defined by the ratio of the average daily dose (ADD) of the chemical to the reference dose (RfD) for that chemical; or the ratio of the exposure concentration to the reference concentration (RfC). A value of less than 1.0 indicates the risk of exposure is likely insignificant; a value greater than 1.0 indicates a potentially significant risk. *Hazard index (HI)*—is the sum of several hazard quotients (HQs) for multiple substances and/or multiple exposure pathways.

Hazard Ranking System (HRS): A scoring system used (by the US EPA) to assess the relative threat associated with actual or potential releases of hazardous substances at contaminated sites.

Hazardous substance: Any substance that can cause harm to human health or the environment whenever excessive exposure occurs. *Hazardous waste*—is that byproduct which has the potential to cause detrimental effects on human health and/or the environment if it is not managed in an efficient manner. Typically, this refers to wastes that are ignitable, explosive, corrosive, reactive, toxic, radioactive, pathological, or has some other property that produces substantial risk to life.

Heavy metals: Members of a group of metallic elements that are recognized as toxic, and are generally bioaccumulative. The term arises from the relatively high atomic weights of these elements.

"Hot-spot": Term often used to denote zones where contaminants are present at much higher concentrations than the immediate surrounding areas. It tends to represent a relatively small area that is highly contaminated within a study area.

Human-equivalent concentration or dose: The human concentration (for inhalation exposure) or dose (for oral exposure) of an agent that is believed to induce the same magnitude of toxic effect as the exposure concentration or dose in experimental animal species. Generally speaking, the *human equivalent dose* represents the dose that, when administered to humans, produces effects comparable to that produced by a dose in experimental animals. This adjustment may incorporate toxicokinetic information on the particular agent, if available, or use a default procedure.

Human health risk: The likelihood (or probability) that a given exposure or series of exposures to a hazardous substance will cause adverse health impacts on individual receptors experiencing the exposures.

Hydraulic conductivity: A measure of the relative ability of earth materials (viz., soil, sediment, or rock) to transmit fluid—and that is dependent on the type of fluid passing through the material.

Hydrocarbon: Organic chemicals/compounds, such as benzene, that contain atoms of both hydrogen and carbon.

Hydrophilic: Having greater affinity for water—or "water-loving." Hydrophilic compounds tend to become dissolved in water.

Hydrophobic: Tending *not* to combine with water—or less affinity for water. Hydrophobic compounds tend to avoid dissolving in water and are more attracted to nonpolar liquids (e.g., oils) or solids.

Hydroponic: The cultivation of plants by placing the roots in liquid nutrient solutions rather than soil.

Immunoassay: The identification of a substance (or contaminants in a sample) based on its capacity for selected antibodies to stimulate an immune response in the substance (or sample). In environmental assessments, it consists of a site investigation method used to determine whether or not contaminants are present at a project site, based on their ability to bind to antibodies produced by a living organism in response to the contaminant.

Incineration: A thermal treatment/degradation process by which contaminated materials are exposed to excessive heat in a variety of incinerator types. The incineration process typically involves the thermal destruction of contaminants by burning under controlled conditions. Depending on the intensity of the heat, the contaminants of concern are volatilized and/or destroyed during the incineration process.

Incompatible wastes: Wastes, which when mixed with other materials without controls, may create fire, explosion, or other severe hazards.

Indicator species: A species that is surveyed or sampled for analysis (usually in ecological risk assessments) because it is believed to represent the biotic community, some functional or taxonomic group, or some population that cannot be easily sampled or surveyed.

Individual excess lifetime cancer risk: An upper-bound estimate of the increased cancer risk, expressed as a probability that an individual receptor could expect from exposure over a lifetime.

Ingestion exposure: An exposure type whereby chemical substances enter the body through the mouth, and into the gastrointestinal system. After ingestion, chemicals can be absorbed into the blood and distributed throughout the body.

Inhalation exposure: The intake of a substance by receptors through the respiratory tract system. Exposure may occur from inhaling contaminants—which become deposited in the lungs, taken into the blood, or both.

Initiating event: A specific trigger action that could potentially give rise to some degree of risk being incurred.

In-situ: Refers to in-place, without excavation—i.e., within the contaminated material or matrix itself. Usually used to refer to a technology or treatment process that is carried out in place, without removal of the contaminated material or matrix.

In-situ vitrification: The heating of the subsurface to extremely high temperatures in order to destroy organic contaminants and to "entomb" inorganic constituents. Upon cooling, the treated materials solidify, incorporating inorganic contaminants and ash.

Institutional control: A legal or institutional mechanism that is implemented to limit access to or use of a property, or at least to provide warning of potential hazards. This may consist of land use restrictions that are imposed (by property deed restrictions) on a potentially contaminated property.

Intake: The amount of material inhaled, ingested, or dermally absorbed during a specified time period. It is a measure of exposure—often expressed in units of mg/kg-day. More broadly, it may be viewed as the process by which an agent crosses an outer exposure surface of a target without passing an absorption barrier—as typically happens through ingestion or inhalation.

Integrated Risk Information System (IRIS): A US EPA database containing verified toxicity parameters (e.g., reference doses [RfDs] and slope factors [SFs]), and also up-to-date health risk and EPA regulatory information for numerous chemicals. It does indeed serve as a very important source of toxicity information for health and environmental risk assessment (See, Appendix B).

Interim Action: A preliminary action that initiates remediation of a contaminated land, but may also constitute part of the final remedy.

Investigation-generated wastes (IGWs) (also, *Investigation-derived wastes, IDWs):* Wastes generated in the process of collecting samples during a remedial investigation or an environmental characterization activity. Such wastes must be handled according to all relevant and applicable regulatory requirements. The wastes may include soil, groundwater, used personal protective equipment, decontamination fluids, and disposable sampling equipment.

*K*d *(soil/water partition coefficient):* Provides a soil- or sediment-specific measure of the extent of chemical partitioning between soil or sediment and water, unadjusted for the dependence on organic carbon.

*K*oc *(organic carbon adsorption coefficient):* Provides a measure of the extent of chemical partitioning between organic carbon and water at equilibrium. It is a measure that indicates the extent to which a compound will sorb to the solid organic content of geologic media in the subsurface. The *K*oc is computed as the ratio of the amount of chemical sorbed per unit weight of organic carbon in the soil, to the concentration of the chemical in solution at equilibrium.

*K*ow *(octanol/water partition coefficient):* Provides a measure of the extent of chemical partitioning between water and octanol at equilibrium. It is a dimensionless constant that provides a measure of how an organic compound will partition between an organic phase and water. This measure indicates the extent to which a compound is attracted to an organic phase (for which octanol is a proxy) and hence the compound's tendency to sorb to subsurface materials. The *K*ow is computed by dividing the amount that will dissolve in octanol by the amount that will dissolve in water; the greater the value, the greater the tendency to sorb in the subsurface. Thus, a low Kow (or log-Kow) indicates that a chemical readily partitions into a water phase, while a high Kow (or log-Kow) indicates that the chemical prefers to stay in the organic phase. It can be used to provide an indication of the quantity of the chemical that will be taken up by the plants.

Kw (*water/air partition coefficient*)*:* Provides a measure of the distribution of a chemical between water and air at equilibrium.

Landfarming: The application of biodegradable organic wastes onto a land surface and their incorporation into the surface soil so that they degrade more readily.

Landfill: A controlled site for the disposal of wastes on land, generally operated in accordance with regulatory safety and environmental compliance requirements.

Latency period: A seemingly inactive period—such as the time between the initial induction of a health effect from first exposures to a chemical agent and the manifestation or detection of actual health effects of interest (Often used to identify the period between exposure to a carcinogen and development of a tumor).

LC_{50} (*Mean lethal concentration*)*:* The lowest concentration of a chemical in air or water that will be fatal to 50% of test organisms living in that media, under specified conditions.

LD_{50} (*Mean lethal dose*)*:* The single dose (ingested or dermally absorbed) that is required to kill 50% of a test animal group. Also represents the median lethal dose value.

Leachate: Aqueous, often-contaminated, liquid generated when water percolates, or trickles, through waste materials or contaminated sites and then collects components of those wastes. Leaching usually occurs at landfills as a result of infiltration of rainwater or snowmelt, and may result in hazardous chemicals entering soils, surface water, or groundwater.

Lifetime average daily dose (LADD): The exposure, expressed as mass of a substance contacted and absorbed per unit body weight per unit time, averaged over a lifetime. It is usually used to calculate carcinogenic risks—and takes into account the fact that, whereas carcinogenic risks are determined with an assumption of lifetime exposure, actual exposures may be for a shorter period of time. Indeed, the LADD may be derived from the ADD—to reflect the difference between the length of the exposure period and the exposed person's lifetime, as follows:

$$\text{LADD} = \text{ADD} \times \frac{\text{Exposure period}}{\text{Lifetime}}$$

Lifetime exposure: The total amount of exposure to a substance or hazard that a potential receptor would be subjected to in a lifetime.

Lifetime risk: Risk that arises from lifetime exposure to a chemical substance or hazard.

Light nonaqueous phase liquids (LNAPLs): Organic liquids that are lighter than water, and capable of forming an immiscible layer that floats on the water table; also referred to as "floaters" (e.g., gasoline and fuel oil).

Lipophilic: The property of a chemical/substance to have a strong affinity for lipid, fats, or oils— i.e., being highly soluble in nonpolar organic solvents. Also, refers to a physicochemical property that describes a partitioning equilibrium of solute molecules between water and an immiscible organic solvent that favors the latter.

Lipophobic: The property of a chemical to be antagonistic to lipid—i.e., incapable of dissolving in or dispersing uniformly in fats, oils, or nonpolar organic solvents.

LOAEC (Lowest-observed-adverse-effect-concentration): The lowest concentration in an exposure medium in a study that is associated with an adverse effect on the test organisms. It represents the lowest concentration of a substance, found by experiment or observation, that causes an adverse alteration of morphology, functional capacity, growth, development or lifespan of the target organisms distinguishable from normal (control) organisms of the same species and strain under the same defined conditions of exposure.

LOAEL (Lowest-observed-adverse-effect level): The lowest dose or exposure level, expressed in mg/kg body weight/day, at which adverse effects are noted in the exposed population. It represents the chemical dose rate or exposure level causing statistically or biologically significant increases in frequency or severity of adverse effects between the exposed and control groups. In practice, this consists of the lowest amount of a substance, found by experiment or observation, that causes an adverse alteration of morphology, functional capacity, growth, development or lifespan of the target organisms distinguishable from normal (control) organisms of the same species and strain under the same defined conditions of exposure. *LOAEL*$_a$—refers to the LOAEL values adjusted by dividing by one or more safety factors.

Local effect: A biological response occurring at the site of first contact between the toxic substance and the organism.

LOEL (Lowest-observed-effect-level): The lowest exposure or dose level to a substance at which effects are observed in the exposed population; the effects may or may not be serious. In a given study, it is the lowest dose or exposure level at which a statistically or biologically significant effect is observed in the exposed population compared with an appropriate unexposed control group.

Matrix (or, medium): The predominant material (e.g., food, consumer products—such as cosmetics, soils, water, and air) surrounding or containing a constituent or agent of interest—and generally comprising the environmental or exposure sample being investigated.

Maximum daily dose (MDD): The maximum dose calculated for the duration of receptor exposure—and used to estimate risks for subchronic or acute noncarcinogenic effects from chemical exposures.

Microbe: A microscopic or ultramicroscopic organism (e.g., bacterium or virus).

Microorganisms: Includes bacteria, algae, fungi and viruses.

Mineralization: The breakdown of organic matter to inorganic materials (such as carbon dioxide and water) by bacteria and fungi.

Mitigation: The process of reducing or alleviating a hazard or problem situation.

Model: A simplification of reality that is constructed to gain insights into select attributes of a particular physical, biologic, economic, or social system. *Mathematical models* express the simplification in quantitative terms—and generally would consist of mathematical function(s) with parameters that can be adjusted so that the function closely describes a set of empirical data.

Modeling: Refers to the use of mathematical equations to simulate and predict real events and processes.

Monitoring: Process involving the measurement of concentrations of chemicals in environmental media, or in tissues of human receptors and other biological organisms. *Biological monitoring*—consists of measuring chemicals in biological materials (e.g., blood, urine, and breath) to determine whether chemical exposure in living organisms (e.g., humans, animals, or plants) has occurred.

Monte Carlo simulation: A process in which outcomes of events or variables are determined by selecting random numbers—subject to a defined probability law.

Natural Attenuation: Naturally occurring processes (usually in soil and groundwater environments) that act without human intervention to reduce the mass, toxicity, mobility, volume, and/or concentration of environmental contaminants. It typically is facilitated by various actions—consisting mostly of physical, chemical, and biological activities or processes such as dilution, dispersion, sorption, volatilization, hydrolysis, direct oxidation, reductive dehalogenation, co-metabolism, and biotransformation. *Monitored Natural Attenuation (MNA)*—refers to a remediation approach that systematically monitors the natural processes responsible for reducing contaminant concentration in order to achieve site-specific remedial action objectives without human intervention that involves active remedial activities.

NOAEC (No-observed-adverse-effect-concentration): The highest concentration of a substance, found by experiment or observation, that causes no detectable adverse alteration of morphology, functional capacity, growth, development or lifespan of the target organisms under defined conditions of exposure. This generally represents the highest concentration in an exposure medium in a study that is *not* associated with an adverse effect on the test organisms. Meanwhile, alterations may be detected that are judged not to be adverse.

NOAEL (No-observed-adverse-effect level): The highest amount of a substance, found by experiment or observation, that causes no detectable adverse alteration of morphology, functional capacity, growth, development or lifespan of the target organisms under defined conditions of exposure. That is, the highest level at which a chemical causes no observable adverse effect in the species being tested or the exposed population. It represents chemical intakes or exposure levels at which there are no statistically or biologically significant increases in frequency or severity of adverse effects between the exposed and control groups—meaning statistically significant effects are observed at this level, but they are not considered to be adverse nor precursors to adverse effects. Meanwhile, alterations may be detected that are judged not to be adverse. $NOAEL_a$—refers to NOAEL values adjusted by dividing by one or more safety factors.

NOEL (No-observed-effect level): The highest level at which a chemical causes no observable changes in the species or exposed populations under investigation. In a given study, this represents the dose rate or exposure level of chemical at which there are no statistically or biologically significant increases in frequency or severity of any effects between the exposed and control groups.

Nonaqueous phase liquid (NAPL): Organic compounds in the liquid phase, and that is not completely miscible in water. It consists of a liquid that partitions to a phase other than the aqueous phase. If the NAPL has a specific gravity of less than 1.0, it is referred to as light NAPL (LNAPL); and if the NAPL has a specific gravity of greater than 1.0, it is referred to as dense NAPL (DNAPL).

Nonparametric statistics: Statistical techniques whose application is independent of the actual distribution of the underlying population from which the data were collected.

Non-threshold toxicant: A chemical for which there is no dose or exposure concentration below which the critical effect will not be observed or expected to occur.

Nutrients: Elements or compounds essential as raw materials for organism growth and development. Nitrogen, phosphorous, potassium, and numerous other mineral elements are essential plant nutrients.

Off-site: Areas outside the boundaries or limits of a presumed contaminated land.

On-site: The boundaries or limits of a presumed contaminated land.

Organic carbon content of soils or sediments (f_{oc}, %): This reflects the amount of organic matter present—and generally correlates with the tendency of chemicals to accumulate in the soil or sediment. The accumulation of chemicals in soils or sediments is frequently the result of adsorption onto organic matter. Thus, in general, the higher the organic carbon content of the soil or sediment, the more a contaminant will be adsorbed to the soil particles, rather than be dissolved in the water or gasses permeating the soil or sediment.

Oxidation–reduction potential (ORP) (also, *redox potential, Eh):* A measure of the relative tendency of the environmental medium (such as groundwater) to accept or transfer electrons.

Oxygen release compounds (ORCs): Chemical additives used to increase the dissolved oxygen concentration in an environmental medium (such as groundwater) so as to promote the mineralization of aerobically biodegradable compounds.

Parameters: Terms in a model that determine the specific model form. For computational models, these terms are fixed during a model run or simulation, and they define the model output. They can be changed in different runs as a method of conducting sensitivity analysis or to achieve calibration goals.

Particulate matter: Small/fine, discrete, solid or liquid particles/bodies, especially those suspended in a liquid or gaseous medium—such as dust, smoke, mist, fumes, or smog suspended in air or atmospheric emissions.

Partitioning: Refers to the state of separation or division of a given substance into two or more compartments. This consists of a chemical equilibrium condition in which a chemical's concentration is apportioned between two different phases, according to the partition coefficient. *Partition coefficient*—is a term used to describe the relative amount of a substance partitioned between two different phases, such as a solid and a liquid or a liquid and a gas. It is the ratio of the chemical's concentration in one phase to its concentration in the other phase. For instance, a *blood-to-air partition coefficient* is the ratio of a chemical's concentration between blood and air when at equilibrium.

Pathway: Any specific route via which environmental chemicals or stressors take in order to travel away from the source in order to reach potential receptors or individuals.

PEL (Permissible exposure limit): A maximum (legally enforceable) allowable level for a chemical in workplace air.

Permeable reactive barrier (PRB): A trench or cell created in the subsurface and filled with transmissive media that allows or even enhances groundwater flow through the barrier, and also destroys contamination in the passing groundwater or transforms the contaminants into less toxic or nontoxic state.

Permeability: A measure of a material's ability to transmit fluid. It is the capacity of a porous medium to transmit a fluid subjected to an energy gradient (the hydraulic gradient, in the case of water).

Persistence: Attribute of a chemical substance which describes the length of time that such substance remains in a particular environmental compartment before it is physically removed, chemically modified, or biologically transformed.

pH: A measure of the acidity or alkalinity of a material or medium.

Photodegradation: Any chemical breakdown reaction that is initiated by ultraviolet rays from sunlight.

Phytoaccumulation: The build-up of contaminants (usually metals) that occurs as the contaminants are transferred from the contaminated media (usually soil and/or groundwater) to roots and other plant parts.

Phytodegradation: A process in which plants degrade (break down) organic pollutants through their metabolic processes.

Phytoextraction: The use of plants to extract contaminants (such as metals) from the environment (usually contaminated soils); the "saturated" plants are subsequently harvested and disposed of and/or appropriately managed.

Phytoremediation: The use of plants to remediate contaminated environmental matrices such as soil, sediments, surface water, or groundwater.

Phytostabilization: The use of plants to reduce bioavailability and offsite migration of contaminants.

Phytovolatilization: The use of plants to volatilize contaminants (usually organic solvents, etc.) from contaminated soil or water.

Pica: The behavior in children and toddlers (usually under age 6 years) involving the intentional eating/mouthing of large quantities of dirt and other objects. More broadly stated, it may be seen as a behavior characterized by deliberate ingestion of nonnutritive substances (such as contaminated soil) by anyone in a (sub)population.

Plume (or contaminant plume): A zone containing predominantly dissolved (or vapor phase) and sorbed contaminants that usually originates from a given contaminant or pollution source areas. It refers to an area of chemicals in a particular medium (such as air or groundwater), moving away from its source in a long band or column. A plume can be a column of smoke from a chimney, or chemicals moving with groundwater; common examples may consist of a body of contaminated groundwater or vapor originating from a specific source and spreading out due to influences of environmental factors such as local groundwater conditions or soil vapor flow patterns, or wind directions.

PM-10, PM_{10}: Particulate matter with physical/aerodynamic diameter <10 μm. It represents the respirable particulate emissions. *Aerodynamic diameter*—is the diameter of a particle with the same settling velocity as a spherical particle with unit density (1 g/cm^3); this parameter, which depends on particle density, is often used to describe particle size. *Respirable fraction (of dust)* (also, *respirable particulate matter*)—is the fraction of dust particles that enter the respiratory system because of their size distribution; generally, the size of these particles correspond to aerodynamic diameter of ≤10 μm.

Pollutant: A potentially harmful physical, chemical, or biological agent occurring in the environment, in consumer products, or at the workplace as a result of human activities.

Pollution (or *contamination*)*:* Refers to the release of a physical, chemical, or biological agent into an environment; this typically has the potential to impact human and/or ecological health.

Population-at-risk (*PAR*)*:* A population group or subgroup that is more susceptible or sensitive to a hazard or chemical exposure than is the general population.

Population excess cancer burden: An upper-bound estimate of the increase in cancer cases in a population as a result of exposure to a carcinogen.

Porosity: The ratio of the volume of void space in earth materials, to the total volume of the material. The wider the range of grain sizes the lower the porosity. *Effective porosity*—is the percentage of void space that is available for fluid flow.

Potency: A measure of the relative toxicity of a chemical.

Potentiation (*of chemical effects*)*:* The effect of a chemical that enhances the toxicity of another chemical.

Potentially responsible party (*PRP*)*:* Generally, refers to those identified as potentially liable for the cleanup at specified waste sites.

ppb (*parts per billion*)*:* An amount of substance in a billion parts of another material—also expressed by μg/kg or μg/liter, and equivalent to (1×10^{-9}) (NB: A billion is often used to represent a thousand millions, i.e., 10^{-9}, in some places, such as in the USA and France; whereas a billion represents a million millions, i.e., 10^{-12}, in some other places, such as in the UK and Germany).

ppm (*parts per million*)*:* An amount of substance in a million parts of another material—also expressed by mg/kg or mg/liter, and equivalent to (1×10^{-6}).

ppt (*parts per trillion*)*:* An amount of substance in a trillion parts of another material— also expressed by *ng*/kg or *ng*/liter and equivalent to (1×10^{-12}) (NB: A trillion is often used to represent a million times a million or a thousand billions, i.e., 10^{-12}, in some places, such as in the USA and France; whereas a trillion represents a million billions, i.e., 10^{-18}, in some other places, such as in the UK and Germany).

Preliminary Assessment (*PA*)*:* A survey and evaluation whereby sites are characterized with respect to their potential to release significant amounts of contaminants into the environment.

Preliminary site appraisal: Process used for quick assessment of site's potential to adversely affect the environment and/or public health.

Probability: The likelihood of an event occurring—numerically represented by a value between 0 and 1; a probability of 1 means an event is certain to happen, whereas a probability of 0 means an event is certain *not* to happen.

Probability distributions: Mathematical equations or graphical representations of the relationship between all the possible values (or outcomes) a variable can have and the likelihood (expressed as a number between zero and unity) that the variable will have a particular value. Probability distributions can be discrete or continuous. *Discrete distributions* can be represented as bar charts that describe the probabilities of a specific finite number of values a variable can have. *Continuous distributions* (also referred to by *probability density function [pdf]*) are represented as smooth curves, and describe probabilities for variables that can have a continuous range and infinite number of possible values. The area under the smooth curve between two points represents the probability that the true value of the variable lies between those points. Another type of probability distribution of significant interest is the *cumulative distribution*—which shows the probability of a variable being equal to or less than each value within the appropriate range for that variable.

Proxy concentration: Assigned chemical concentration value for situations where sample data may not be available, or when it is impossible to quantify accurately.

Pump-and-treat technology: A class of remedial technologies that require contaminated groundwater to be pumped to the surface for above-ground treatment.

Qualitative: Description of a situation without numerical specifications.

Quality assurance (QA): A system of activities designed to assure that the quality control system in a study or investigation is performing adequately. It typically would consist of the management of information and data sets from an investigation, to ensure that they meet the data quality objectives. *Quality control (QC)*—a system of specific efforts designed to test and control the quality of data obtained in an investigation. This comprises of the management of activities involved in the collection and analysis of data to assure they meet the data quality objectives—and it also represents the system of activities required to provide information as to whether the quality assurance system is performing adequately.

Quantitation limit (QL): The lowest level at which a chemical can be accurately and reproducibly quantitated. It usually is equal to the instrument detection limit (IDL) multiplied by a factor of 3 to 5, but varies for different chemicals and different samples.

Quantitative: Description of a situation that is presented in reasonably exact numerical terms.

Radio frequency (RF) heating: An *in-situ* remediation process for removing contaminants (usually VOCs and SVOCs) from soil, in which microwave energy is used to heat soil and vaporize contaminants for subsequent collection and treatment.

Reasonable maximum exposure (RME): A concept that attempts to identify the highest exposure (and, therefore, the greatest risk) that could reasonably be expected to occur in a given population.

Receptor: Members of a potentially exposed population, such as persons or organisms that are potentially exposed to concentrations of a particular chemical compound of concern. *Sensitive receptor*—individual in a population who is particularly susceptible to health impacts due to exposure to a chemical substance.

Reduction/Oxidation (Redox) reactions: A chemical or biological reaction wherein an electron is transferred from an electron donor to an electron acceptor. *Electron acceptor*—is a compound capable of accepting electrons during oxidation–reduction reactions. Microorganisms obtain energy by transferring electrons from electron donors such as organic compounds (or sometimes, reduced inorganic compounds such as sulfide) to an electron acceptor. Generally speaking, electron acceptors are compounds that are relatively oxidized—and include oxygen, nitrate, Fe(III), Mn(IV), sulfate, carbon dioxide, or in some cases the chlorinated aliphatic hydrocarbons such as PCE, TCE, DCE, and vinyl chloride. *Electron donor*—is a compound capable of supplying (i.e., giving up) electrons during oxidation–reduction reactions. Microorganisms obtain energy by transferring electrons from electron donors such as organic compounds (or sometimes, reduced inorganic compounds such as sulfide) to an electron acceptor. Generally speaking, electron donors are compounds that are relatively reduced—and include fuel hydrocarbons and native organic carbon. *Metabolic byproduct*—is a product of the reaction between an electron donor and an electron acceptor. Metabolic byproducts include volatile fatty acids, daughter products of chlorinated aliphatic hydrocarbons, methane, and chloride.

Reductive dechlorination: Reduction of a chlorine-containing organic compound via the replacement of chlorine with hydrogen. It is a process in which chlorinated hydrocarbon compounds serve as electron acceptors, resulting in the sequential removal of chlorine atoms and transformation into nontoxic ions, ethenes, and ethanes.

Reference concentration (RfC): An estimate of a daily inhalation exposure to the human population (including sensitive subgroups) that is likely to be without an appreciable risk of deleterious non-cancer effects during a lifetime. It represents a concentration of a noncarcinogenic chemical substance in an environmental medium to which exposure can occur over a prolonged period without expected adverse effect; the medium in this case is usually air—with the concentration expressed in mg of chemical per m^3 of air.

Reference dose (RfD): The estimate of lifetime daily oral exposure of a noncarcinogenic substance for the general human population (including sensitive receptors) which appears to be without an appreciable risk of deleterious effects, consistent with the threshold concept. This constitutes the maximum amount of a chemical that the human body can absorb without experiencing chronic health effects, expressed in mg of chemical per kg body weight per day. Generally, RfDs are available from databases such as US EPA's Integrated Risk Information System (IRIS)—and serves as the toxicity value for a chemical in human health risk assessment used for evaluating the noncarcinogenic effects that could result from exposures to chemicals of concern (See, Appendix B).

Remedial action: Actions consistent with a permanent remedy in the event of a release of a hazardous substance into the environment. This is meant to prevent or minimize the effects of such releases so that they do not migrate farther away and/or cause substantial danger to present or future public health or welfare, or the environment.

Remedial action objective (RAO): Cleanup objectives that specify the level of cleanup, area of cleanup (or area of attainment), and the time required to achieve required cleanup (i.e., the restoration time-frame).

Remedial alternative: An action considered in the feasibility study, that is intended to reduce or eliminate significant risks to human health and/or the environment at a contaminated land.

Remedial investigation (RI): The field investigations of hazardous waste sites to determine pathways, nature, and extent of contamination, as well as to identify preliminary alternative remedial actions. It addresses data collection and site characterization used to identify and assess threats or potential threats to human health and the environment posed by a potentially contaminated land.

Remediation: An activity directed towards the reduction of toxicity, mobility, and/or volumes of contaminants at an impacted site. The process consists of the cleaning up of a potentially contaminated land, in order to prevent or minimize the potential release and migration of hazardous substances from the impacted media that could cause adverse impacts to present or future public health and welfare, or the environment.

Removal action: Generally refers to a response action that is implemented to address a direct threat to human health or the environment.

Representative sample: A sample that is assumed *not* to be significantly different from the population of samples available.

Response (toxic): The reaction of a body or organ to a chemical substance or other physical, chemical, or biological agent. This is generally reflected in the change(s) developed in the state or dynamics of an organism, system, or (sub)population in reaction to exposure to an agent.

Restoration time frame: The length of time required to achieve requisite cleanup levels or site restoration goals.

Retardation coefficient: A measure of how quickly a contaminant moves through the ground compared to water. It is computed as the ratio of the total contaminant mass in a unit aquifer volume to the contaminant mass in solution.

Rhizosphere: Soil in the area surrounding plant roots that is influenced by the plant root—typically a few millimeters or at most centimeters from the plant root. It is noteworthy that, this area is higher in nutrients and thus has a higher and more active microbial population (that could be important to biodegradation processes). *Rhizosphere bioremediation*—Using the bacteria, fungi and protozoans that occur in the biologically rich zone of the immediate vicinity of plant roots to treat or degrade organic contaminants. *Rhizofiltration*—the uptake of contaminants by the roots of plants immersed in water; when the roots become "saturated" with the target contaminants they are harvested and appropriately managed.

Risk: The probability or likelihood of an adverse consequence/effect from a hazardous situation or hazard, or the potential for the realization of undesirable adverse consequences from impending events. In chemical exposure situations, it is generally used to provide a measure of the probability and severity of an adverse effect to health, property, or the environment under specific circumstances of exposure to a chemical agent or a mixture of chemicals. In quantitative probability terms, risk may be expressed in values ranging from zero (representing the certainty that harm will not occur) to one (representing the certainty that harm will occur). In risk assessment practice, the following represent examples of how risk is typically expressed: 1E-4 or 10^{-4} = a risk of 1/10,000; 1E-5 or

$10^{-5} = 1/100,000$; 1E-6 or $10^{-6} = 1/1,000,000$; 1.3E-3 or $1.3 \times 10^{-3} = $ a risk of $1.3/1,000 = 1/770$; 8E-3 or $8 \times 10^{-3} = $ a risk of $1/125$; and 1.2E-5 or $1.2 \times 10^{-5} = $ a risk of $1/83,000$. *Individual risk*—refers to the probability that an individual person in a population will experience an adverse effect from exposures to hazards. It is used to define the frequency at which an individual may be expected to sustain a given level of harm from the realization of specified hazards. In general, this is identical to "population" or "societal" risk—unless if specific population subgroups can be identified that have different (i.e., higher or lower) risks. *Societal* (or *population*) *risk*—refers to the relationship between the frequency and the number of people suffering from a specified level of harm in a given population, as a result of the realization of specified hazards. *Relative risk*—refers to the ratio of incidence or risk among exposed individuals to incidence or risk among nonexposed individuals. *Residual risk*—refers to the risk of adverse consequences that remains after corrective actions have been implemented. *Cumulative risk*—refers to the total added risks from all sources and exposure routes that an individual or group is exposed to. *Aggregate risk*—refers to the sum total of individual increased risks of an adverse health effect in an exposed population.

Risk acceptability/acceptance: Refers to the willingness of an individual, group, or society to accept a specific level of risk in order to obtain some reward or benefit.

Risk analysis: A process for controlling situations where an organism, system, or (sub)population could be exposed to a hazard. The risk analysis process is generally viewed as consisting of three key components—namely, risk assessment, risk management, and risk communication.

Risk appraisal: A review of whether existing or potential biologic receptors are presently, or may in the future, be at risk of adverse effects as a result of exposures to chemicals originating or found in the human environments.

Risk assessment: The determination of the type and degree of hazard posed by an agent; the extent to which a particular group of receptors have been or may become exposed to the agent; and the present or potential future health risk that exists due to the agent. It generally comprises of a methodology that combines exposure assessment with health and environmental effect data to estimate risks to human or environmental target organisms that may arise from exposure to various hazardous substances. The process is intended to calculate or estimate the risk to a given target organism, system, or (sub)population—including the identification of attendant uncertainties, following exposure to a particular agent, and taking into account the inherent characteristics of the agent of concern as well as the characteristics of the specific target system. In the context of human exposure to chemical substances, risk assessment involves the determination of potential adverse health effects from exposure to the chemicals of potential concern—including both quantitative and qualitative expressions of risk. Overall, the process of risk assessment involves four key steps, namely: hazard identification, dose-response assessment (or hazard characterization), exposure assessment, and risk characterization. *Baseline risk assessment*—is an assessment conducted to identify and evaluate threats to human health and the environment prior to the implementation of any active corrective or remedial action to abate the potential hazards.

Risk-based concentration: A chemical concentration determined based on an evaluation of the compound's overall risk to health upon exposure.

Risk characterization: The qualitative and, wherever possible, quantitative determination (including attendant uncertainties) of the probability of occurrence of known and potential adverse effects of an agent in a given organism, system, or (sub)population, under defined exposure conditions. It generally would consist of the estimation of the incidence and severity of the adverse effects likely to occur in a human population or ecological group due to actual or predicted exposure to a substance or hazard.

Risk communication: Activities carried out to ensure that messages and strategies designed to prevent exposure, adverse health effects, and diminished quality of life are effectively communicated to the public and/or stakeholders. As part of a broader risk prevention strategy, risk communication supports education efforts by promoting public awareness, increasing knowledge, and motivating individuals to take action to reduce their exposure to hazardous substances.

Risk control: The process of managing risks associated with a hazard situation. It may involve the implementation, enforcement, and re-evaluation of the effectiveness of corrective measures from time to time.

Risk decision: The complex public policy decision relating to the control of risks associated with hazardous situations.

Risk determination: An evaluation of the environmental and health impacts associated with chemical releases and/or exposures.

Risk estimation: The process of quantifying the probability and consequence values for a hazard situation. In general, it is comprised of the quantification of the probability, including attendant uncertainties, that specific adverse effects will occur in an organism, system, or (sub)population due to actual or predicted exposure. The process is used to determine the extent and probability of adverse effects of the hazards identified, and to produce a measure of the level of health, property, or environmental risks being assessed. A *risk estimate* is comprised of a description of the probability that a potential receptor exposed to a specified dose of a chemical will develop an adverse response.

Risk evaluation: This generally refers to the establishment of a qualitative or quantitative relationship between risks and benefits of exposure to an agent—involving the complex process of determining the significance of the identified hazards and estimated risks to the system concerned or affected by the exposure, as well as the significance of the benefits brought about by the agent. The effort is made up of the complex process of developing acceptable levels of risk to individuals or society. It is the stage at which values and judgments enter into the decision-making process.

Risk group: A real or hypothetical exposure group composed of the general or specific population groups.

Risk management: The steps and processes taken to reduce, abate, or eliminate the risk that has been revealed by a risk assessment. For chemical exposure situations, it consists of measures or actions taken to ensure that the level of risk to human health or the environment as a result of possible exposure to the chemicals of concern does not exceed the pre-established acceptable limit (e.g., 1E-06). The process focuses on decisions about whether an assessed risk is sufficiently high to present a public health concern, and also about the appropriate means for controlling the risks that are judged to be significant. The decision-making process involved takes account of political, social, economic, and engineering constraints, together with risk-related information, in order to develop, analyze, and compare management options and then select the appropriate managerial or regulatory response to a potential hazard situation.

Risk perception: Refers to the magnitude of a risk as is perceived by an individual or population. It consists of a convolution of the measured risk together with the preconceptions of the observer.

Risk reduction: The action of lowering the probability of occurrence and/or the value of a risk consequence, thereby reducing the magnitude of the risk.

Risk-specific dose (RSD): An estimate of the daily dose of a carcinogen which, over a lifetime, will result in an incidence of cancer equal to a given specified (usually the acceptable) risk level.

Sample blank: Blanks are samples considered to be the same as the environmental samples of interest except with regards to one factor whose influence on the samples is being evaluated. Sample blanks are used to ensure that contaminant concentrations actually reflect site conditions, and are not artifacts of the sampling and sample handling processes. The blank consists of laboratory distilled, deionized water that accompanies the empty sample bottles to the field as well as the samples returning to the laboratory, where it is not opened until both blank and the actual site samples are ready to be analyzed.

Sample duplicate: Two samples taken from the same source at the same time and analyzed under identical conditions.

Sample quantitation limit (SQL) (also called *practical quantitation limit, PQL*)*:* The lowest level that can be reliably achieved within specified limits of precision and accuracy during routine laboratory operating conditions. It represents a detection limit that has been corrected for sample characteristics, sample preparation, and analytical adjustments such as dilution. Typically, the PQL or SQL will be about 5–10 times the chemical-specific detection limit.

Sampling and analysis plan (SAP): Documentation that consists of a quality assurance project plan (QAPP) and a field sampling plan (FSP). The *QAPP*—contains documentation of all relevant QA and QC programs for the case-specific project. The *FSP*—is a documentation that defines in detail, the sampling and data gathering activities to be used in the investigation of a potential environmental contamination or chemical exposure problem.

Saturated zone: An underground geologic formation in which the pore spaces or interstitial spaces in the formation are completely filled with water under pressure, equal to or greater than atmospheric pressure.

Sediment: Soil that is normally covered with water. It generally is considered to provide a direct exposure pathway to aquatic life.

Seepage velocity: The average velocity of ground water in a porous medium.

Sensitivity: The degree to which the outputs of a quantitative assessment are affected by changes in selected input parameters or assumptions.

Sensitivity analysis: A method used to examine the operation of a system by measuring the deviation of its nominal behavior due to perturbations in the performance of its components from their nominal values. In risk assessment, this may involve an analysis of the relationship of individual factors (such as chemical concentration, population parameter, exposure parameter, and environmental medium) to variability in the resulting estimates of exposure and risk. Typically, it consists of a quantitative evaluation of how input parameters influence the model output (e.g. dose metrics).

Site assessment: Process used to identify toxic substances that may be present at a site and to present site-specific characteristics that influence the migration of contaminants.

Site categorization: A classification of sites to reflect the uniqueness of each site.

Site characterization: A process that attempts to identify the types and sources of contaminants present at a site and the site's hydrogeologic characteristics.

Site cleanup: The decontamination of a site initiated as a result of the discovery of contamination at a site or property.

Site mitigation: The process of cleaning up a contaminated land in order to return it to an environmentally acceptable state.

Slurry wall: An underground barrier designed to stop groundwater flow. This is usually constructed by digging a trench and then backfilling it with a slurry rich bentonite clay or equivalent.

Soil gas: The vapor or gas found in the unsaturated soil zone.

Soil vapor extraction (SVE): Also known as *in-situ soil venting, subsurface venting, vacuum extraction,* or *in-situ soil-stripping,* is a technique that uses soil aeration to treat subsurface zones of VOC contamination in soils.

Solubility: A measure of the ability of a substance to dissolve in a fluid—i.e., the amount of a substance that can be dissolved in water or other solvent or fluid medium.

Sorption: The processes that remove solutes from the fluid phase and concentrate them on the solid phase of a medium.

Source area (or, *Source zone*): Sections at a contaminated land where contaminants originated, and/or where contamination remains in place. The source area may stretch beyond the original contaminant spill site—and may also include regions along the contaminant flow path where contaminants are present in precipitated or nonaqueous-phase liquid form.

Source removal: Extraction of contamination from its place of origin or accumulation in the subsurface.

Stabilization: The conversion of a substance to a form that does not readily change its physical or chemical characteristics.

Standard: A general term used to describe legally established values above which regulatory action will be required.

Standard deviation: The most widely used statistical measure to describe the dispersion of a data set—defined for a set of *n* values as follows:

$$s = \sqrt{\frac{\sum_{i=1}^{n}(X_i - X_m)^2}{(n-1)}}$$

where X_m is the arithmetic mean for the data set of *n* values. The higher the value of this descriptor, the broader is the dispersion of data set about the mean.

Statistical Significance: The probability that a result likely to be due to chance alone. By convention, a difference between two groups is usually considered statistically significant if chance could explain it only 5% of the time or less. Study design considerations may influence the *a priori* choice of a different statistical significance level.

Stochastic model: A model that involves random variables (See also, *variable*).

Stochasticity: Variability in parameters (or in models containing such parameters) that may be attributed to the inherent variability of the system under consideration.

Stressor (also, *Agent*): Any physical, chemical, or biological entity that can induce an adverse response in an organism. *Stressor-response profile*—summarizes the data on the effects of a stressor and the relationship of the data to the assessment endpoint. This is typically a product of characterization of ecological effects in the analysis phase of an ERA.

Subchronic: Relating to intermediate duration, and usually used to describe studies or exposure levels spanning 5–90 days duration. *Subchronic daily intake (SDI)*—refers to the exposure, expressed in mg/kg-day, averaged over a portion of a lifetime. *Subchronic exposure*—refers to the short-term, high-level exposure to chemicals, i.e., the maximum exposure or doses to a chemical over a portion (approximately 10%) of a lifetime of an organism. In general, this represents a contact between an agent and a target of intermediate duration between acute and chronic.

Substrate: A compound, substance, or nutrient used by microorganisms to derive energy for growth. The term can refer to either an electron acceptor or an electron donor. In remediation technologies, various substrates may be injected into the contaminated media to facilitate and/or instigate bioremediation of contaminants in the impacted media (usually soil and groundwater).

Surfactant: A surface-active chemical agent (usually made up of phosphates) used in detergents to produce lathering; it is a substance that lowers the surface tension of a liquid. It bonds with oils and other immiscible compounds to aid their transport in water.

Surfactant-enhanced aquifer remediation (SEAR): The injection of surfactants into a targeted source zone to recover NAPL contaminants, either by enhanced solubilization or by mobilization due to the reduction of interfacial tension.

Surrogate data: A substitute data or measurement on a given substance or agent that is used to estimate analogous or corresponding values of another substance/agent.

Synergism (of chemical effects): A pharmacologic or toxicologic interaction in which the combined effect of two or more chemicals is greater than the sum of the effect of each chemical acting alone. In other words, it is the aspect of two or more agents interacting to produce an effect greater than the sum of the agents' individual effects. More generally, this represents the effects arising from a combination of two or more events, efforts, or substances that are greater than would be expected from adding the individual effects.

Systemic: Pertaining to, or affecting, the body as a whole—or acting in a portion of the body other than the site of entry; generally used to refer to non-cancer effects. *Systemic effect*—relates to those effects that require absorption and distribution of the toxicant to a site distant from its original entry point, and at which distant point any effects are produced. Most chemical substances that produce systemic toxicity do not cause a similar degree of toxicity in all organs, but usually demonstrate major toxicity to one or two organs; these are referred to as the *target organs* of toxicity for that chemical. *Systemic toxicity*—relates to toxic effects as a result of absorption and distribution of a toxicant to a site distant from its entry point, at which distant point any effects are produced. It is noteworthy that not all chemicals that produce systemic effects cause the same degree of toxicity in all organs.

Target organ (or, Tissue): The biological organ(s) that are most adversely affected by exposure to a chemical substance or physical agent. That is, the organ affected by a specific chemical in a particular species. *Target organ toxicity*—the adverse effects or disease states manifested in specific organs of the body of an organism.

Terrestrial: That relating to land—as distinct from air or water.

Threshold: The lowest dose or exposure of a chemical at which a specified measurable/deleterious effect is observed and below which such effect is not observed. *Threshold dose*—is the minimum exposure dose of a chemical that will evoke a stipulated toxicological response. *Toxicological threshold*—refers to the concentration at which a compound begins to exhibit toxic effects. *Threshold limit*—a chemical concentration above which adverse health and/or environmental effects may occur. *Threshold hypothesis*—refers to the assumption that no chemical injury occurs below a specified level of exposure or dose.

Threshold chemical/toxicant (also, *nonzero threshold chemical*): Refers to a substance that is known or assumed to have no adverse effects below a certain dose—i.e., a chemical for which the critical effect is observed or expected to occur only above a certain dose or exposure concentration. *Non-threshold chemical* (also called, *zero threshold chemical*)—refers to a substance that is known, suspected, or assumed to potentially cause some adverse response or toxic effect at any dose above zero. Thus, any level of exposure is deemed to involve some risk—and this has traditionally been used only in regard to carcinogenesis.

Tolerable intake is the estimated maximum amount of an agent, expressed on a body mass basis, to which each individual in a (sub)population may be exposed over a specified period without appreciable risk.

Tolerance limit: The level or concentration of a chemical residue in a media of concern above which adverse health effects are possible, and above which levels corrective action should therefore be undertaken.

Toxic: Harmful or deleterious with respect to the effects produced by exposure to a chemical substance. *Toxic substance*—refers to any material or mixture that is capable of causing an unreasonable threat or adverse effects to human health and/or the environment. These include chemical elements and compounds such as lead, benzene, dioxin, and others that have toxic (poisonous) properties when exposure via ingestion, inhalation, or absorption into the organism occurs. There is a large variation in the degree of toxicity among toxic substances and in the exposure levels that induce toxicity.

Toxicant: Any synthetic or natural chemical with an ability to produce adverse health effects. It represents a poisonous contaminant that may injure an exposed organism.

Toxicity: The inherent property of a substance to cause any adverse physiological effects (on living organisms). It represents the degree to which a chemical substance elicits a deleterious or adverse effect upon the biological system of an organism exposed to the substance over a designated time period. This generally indicates the harmful effects produced by a chemical substance—and reflects on the quality or degree of being poisonous or harmful to human or ecological receptors. *Delayed toxicity*—refers to the development of disease states or symptoms long (usually several months or years) after exposure to a toxicant. *Immediate toxicity*—refers to the rapid development or onset of disease states or symptoms following exposure to a toxicant.

Toxicity assessment: Evaluation of the toxicity of a chemical based on all available human and animal data. It consists of the characterization of the toxicological properties and effects of a chemical substance, with special emphasis on the establishment of dose-response characteristics.

Toxicity reference value (TRV): An estimate of an "acceptable" chemical dose to a wildlife species used in ecological risk assessments. Toxicity reference values are similar to reference doses used in human health risk assessment but are determined for ecological receptors rather than humans.

Toxicological profile: A documentation about a specific substance in which scientific interpretation is provided from all known information on the substance; this also includes specifying the levels at which individuals or populations may be harmed if exposed. The toxicological profile may also identify significant gaps in knowledge on the substance, and serves to initiate further research, where needed.

Treatment: A change in the composition or concentration of a waste substance so as to make it less hazardous, or to make it acceptable at disposal and reuse facilities. It involves the application of technological process(es) or means to a contaminant or waste in order to render it nonhazardous, or less hazardous, or more suitable for resource recovery.

Trip blank: A trip blank is transported just like actual samples, but does not contain the chemicals to be analyzed. The purpose of this blank is to evaluate the possibility that a chemical could seep into samples (to adulterate them) during transportation to the laboratory.

Turbidity: A property of water imparted by suspended solids, and measured by interference with the passage of light.

Uncertainty: The lack of confidence in the estimate of a variable's magnitude or probability of occurrence. It refers to lack of knowledge—and is generally attributable to the lack or incompleteness of information. Thus, uncertainty can often be reduced with greater knowledge of the system, or by collecting more and better experimental or simulation data. Quantitative uncertainty analysis attempts to analyze and describe the degree to which a calculated value may differ from the true value; it sometimes uses probability distributions. Uncertainty depends on the quality, quantity, and relevance of data—as well as on the reliability and relevance of models and assumptions.

Uncertainty factor (UF) (also called, *safety factor*): In toxicological evaluations, this refers to a factor that is used to provide a margin of error when extrapolating from experimental animals to estimate human health risks.

Underground Storage Tank (UST): A tank fully or partially located below the ground surface, and that is designed to hold gasoline or other petroleum products, or indeed other chemical products.

Unit cancer risk (UCR): The excess lifetime risk of cancer due to a continuous lifetime exposure/ dose of one unit of carcinogenic chemical concentration (caused by one unit of exposure in the low exposure region). It is a measure of the probability of an individual developing cancer as a result of exposure to a specified unit ambient concentration.

Unit risk (UR): The upper-bound (plausible upper limit) estimate of the probability of contracting cancer as a result of constant/continuous exposure to an agent at a concentration of 1 µg/L in water, or 1 µg/m^3 in air over the individual lifetime. The interpretation of unit risk would be as follows: if unit risk = 5.5×10^{-6} µg/L, 5.5 excess tumors are expected to develop per 1,000,000 people, if exposed daily for a lifetime to 1 µg of the chemical in 1 liter of drinking water.

Upper-bound estimate: The estimate not likely to be lower than the true (risk) value. That is, an estimate of the plausible upper limit to the true value of the quantity.

Upper confidence limit, 95% (95% UCL): The upper limit on a normal distribution curve below which the observed mean of a data set will occur 95% of the time. This is also equivalent to stating that, there is at most a 5% chance of the true mean being greater than the observed value. In other words, it is a value that equals or exceeds the true mean 95% of the time. Thus, assuming a random and normal distribution, this is the range of values below which a given value will fall 95% of the time (See also, *confidence interval*).

Utility: As used in decision analysis, refers to a quantitative measure of the strength of a preferred outcome.

Vadose zone: Also called the *unsaturated soil zone*, this is the zone between the ground surface and the top of the groundwater table. It is represented by the unsaturated zone of soil above the groundwater, extending from the bottom of the capillary fringe all the way to the soil surface.

Validation: Process by which the reliability and relevance of a particular approach (or model) is established for a defined purpose.

Variability: Variability refers to true differences in attributes due to heterogeneity or diversity. Differences among individuals in a population are referred to as *inter-individual variability*; differences for one individual over time are referred to as *intra-individual variability*. Variability is usually not reducible by further measurement or study, although it can be better characterized.

Variable: In mathematics, a variable is used to represent a quantity that has the potential to change. In the physical sciences and engineering, a variable is a quantity whose value may vary over the course of an experiment (including simulations), across samples, or during the operation of a system. In statistics, a *random variable* is that whose observed outcomes may be considered products of a stochastic or random experiment; their probability distributions can be estimated from observations. Generally, when a variable is fixed to take on a particular value for a computation, it is referred to as a *parameter*.

Volatile organic compound (VOC): Any organic compound that has a great tendency to vaporize, and is susceptible to atmospheric photochemical reactions. Such chemical volatilizes (evaporates) relatively easily when exposed to air—i.e., capable of becoming vapor at relatively low temperatures. VOCs generally consist of substances containing carbon and different proportions of other elements such as hydrogen, oxygen, fluorine, chlorine, bromine, sulfur, or nitrogen—and these substances easily become vapors or gases. A significant number of the VOCs found in human environments are commonly used as solvents (e.g., paint thinners, lacquer thinner, degreasers, and dry cleaning fluids). In general, volatile compounds are amenable to analysis by the purge and trap techniques.

Volatilization: The transfer of a chemical from the liquid or solid into the gaseous phase. *Volatility*—is a measure of the tendency of a compound to vaporize or evaporate (usually from a liquid state) at standard temperature and pressure.

Water table: The top of the saturated zone where confined groundwater is under atmospheric pressure—i.e., the level at the top of the zone of groundwater saturation. *Water table depression*—refers to a drop in water table level, generally caused by mechanical or natural groundwater pumping.

Wetland: An area of land (usually of significant ecological interest or concern) covered by marsh, swamp, or moisture-rich tidal areas.

Worst case: A semi-qualitative term that refers to the maximum possible exposure, dose, or risk to an exposed person or group, that could conceivably occur—i.e., regardless of whether or not this exposure, dose, or risk actually occurs or is observed in a specific population. Typically, this should refer to a hypothetical situation in which everything that can plausibly happen to maximize exposure, dose, or risk actually takes place. This worst case may indeed occur (or may even be observed) in a given population; however, since this is usually a very unlikely set of circumstances, in most cases, a worst-case estimate will be somewhat higher than actually occurs in a specific population. In most health risk assessments, a worst-case scenario is essentially a type of bounding estimate (See also, *conservative assumption*).

Xenobiotics: Substances noticeably foreign to an organism—i.e., substances that are not naturally produced within the organism. Also, substances not normally present in the environment—such as a pesticide or other environmental pollutant. Most xenobiotics are considered pollutants.

Zone of saturation: Refers to the soil layer in the ground in which all available interstitial voids (cracks, crevices, holes, etc.) are filled with water. The level of the top of this zone is the *water table*.

Appendix B
Selected Scientific Tools and Databases for Contaminated Site Risk Management and Site Restoration Decision-Making

Oftentimes, a variety of scientific and analytical tools are used to assist decision-makers with various issues associated with the management of contaminated site problems. Indeed, a variety of decision-making tools and logistics, as well as computer databases and information libraries containing important records on numerous chemical substances exist within the scientific community that may find extensive useful applications in environmental and risk management programs designed to address contaminated site and related problems. A limited number and select examples of such logistical application tools, computer software, scientific models, and database systems of general interest to environmental assessments and risk management programs are featured below in this appendix; this select list may be of interest, especially because of their international appeal and/ or their wealth of relevant support information—albeit it is not at all meant to be wholly representative of the comprehensive and diverse number of resources containing important environmental and risk management information that are available in practice. Really, the presentation here is only meant to demonstrate the overall wealth of scientific information that already exists—and which should therefore be consulted whenever possible, in order to obtain the relevant environmental and risk management support information necessary for making risk management and/or site corrective action decisions about an environmental contamination or contaminated site problem.

On the whole, a diversity of information sources are available to facilitate various types of contaminated site risk management tasks—with only a partial listing provided in the following sections. Meanwhile, it must be emphasized that the list provided here is by no means complete and exhaustive—and neither does it cover the broad spectrum of what is available to the scientific communities and/or the general public. Indeed, several other similar logistical and scientific tools are available that can be used to support environmental and risk management programs, in order to arrive at informed decisions on contaminated site and related problems. Further listings and/or information may generally be obtainable on the *Internet*—which serves as a very important and contemporary international network search system. Also, traditional libraries and directories of environment- and risk-related professional groups/associations may provide the necessary up-to-date contacts.

Finally, it is noteworthy that the choice of one particular model or tool over another—some of which are proprietary or a registered trademark—will generally be problem specific. Furthermore, the mention of any particular database, model, or software in this title does not necessarily constitute an endorsement of such product as being the most preferred, since each has its own merits and limitations. Anyhow, in the end, it is quite important that extra care be exercised in the choice of an appropriate tool for specific problems.

B.1 EXAMPLE DECISION SUPPORT TOOLS AND LOGISTICAL COMPUTER SOFTWARE

A select number of scientific application tools (consisting of scientific models and software) potentially appropriate for the management of contaminated site problems is enumerated in the following subsections. This listing is by no means complete and exhaustive; in fact, recent years have seen a proliferation of software systems for a variety of environmental chemical release and fate, contaminant behaviors, chemical exposure, risk assessment, and risk management studies. Care must therefore be exercised in the choice of an appropriate tool for case-specific problems.

B.1.1   The *BIOCHLOR* Model

BIOCHLOR is an easy-to-use screening model that simulates remediation by natural attenuation of dissolved solvents at chlorinated solvent release sites. It can be used to simulate solute transport without decay, solute transport with biodegredation/ biotransformation modeled as a sequential first-order process, and solute transport with biotransformation modeled as a sequential first-order decay process with two different reaction zones (i.e., each zone has a different set of rate coefficient values).

The BIOCHLOR software, programmed in the Microsoft Excel spreadsheet environment and based on the Domenico analytical solute transport model, has the ability to simulate 1-D advection, 3-D dispersion, linear adsorption, and biotransformation via reductive dechlorination (the dominant biotransformation process at most chlorinated solvent sites). Reductive dechlorination is assumed to occur under anaerobic conditions and dissolved solvent degradation is assumed to follow a sequential first-order decay process.

The BIOCHLOR Program is written as an Excel Spreadsheet and therefore requires Microsoft Excel to run. The program includes a Natural Attenuation Screening Protocol scoring system to help the user determine the potential for reductive dechlorination from site data.

B.1.2   The *CALTOX* Model

CalTOX is a multimedia, multi-pathway risk assessment model that allows stochastic simulation to be carried out. The CalTOX spreadsheet encompasses a multimedia transport and transformation model that uses equations based on conservation of mass and chemical equilibrium; it calculates the gains and losses in each environmental compartment over time, by accounting for both transport from one compartment into another and also chemical biodegradation and transformation. Overall, it is an innovative spreadsheet model that relates the concentration of a chemical in soil to the risk of an adverse health effect for a person living or working on or near a contaminated soil source. The model computes site-specific health-based soil cleanup concentrations for specified target risk levels and/or estimates human health risks for given soil concentrations at the site. It is a fugacity model for evaluating the time-dependent movement of contaminants in various environmental media. A noteworthy feature is that the model makes the distinction between the environmental concentration and the exposure concentration.

On the whole, the CalTOX model predicts the time-dependent concentrations of a chemical in seven environmental compartments—comprised of air, water, three soil layers, sediment, and plants at a site. After partitioning the concentration of the chemical to these environmental compartments, CalTOX determines the chemical concentration in the exposure media of breathing zone air, drinking water, food, and soil that people inhale, ingest, and contact dermally. CalTOX then uses the standard equations (such as found in the US Environmental Protection Agency (EPA) Risk Assessment Guidance for Superfund [USEPA, 1989n]) to estimate exposure and risk.

CalTOX has the capability to carry out Monte Carlo simulations with a spreadsheet add-in program. It quantitatively addresses both uncertainty and variability by allowing the presentation of both the risks and the calculated cleanup goals as probability distributions—allowing for a clearer distinction between the risk assessment and risk management steps in site remediation decisions. Used in this manner, CalTOX will produce a range of risks and/or health-based soil target cleanup levels that reflect the uncertainty/variability of the estimates.

CalTOX was developed by the California EPA and is available for free downloading from their website. In addition to the site-specific risk assessments, results from CalTOX can be exported to other programs (such as *Crystal Ball*) for Monte Carlo simulation.

Further information on CalTOX may be obtained from the following: California EPA (Cal EPA), Department of Toxic Substances Control (DTSC), Sacramento, California, United States and Lawrence Livermore National Laboratory, Berkeley, California, United States.

B.1.3 *CatReg* Software for Categorical Regression Analysis

CatReg is a computer program, written in the R-programming language, to support the conduct of exposure-response analyses by toxicologists and health scientists. It can be used to perform categorical regression analyses on toxicity data after effects have been assigned to ordinal severity categories (e.g., no effect, adverse effect, and severe effect) and bracketed with up to two independent variables corresponding to the exposure conditions (e.g., concentration and duration) under which the effects occurred. *CatReg* calculates the probabilities of the different severity categories over the continuum of the variables describing exposure conditions. The categorization of observed responses allows expression of dichotomous, continuous, and descriptive data in terms of effect severity—and supports the analysis of data from single studies or a combination of similar studies.

CatReg reads data from ordinary text files in which data are separated by commas. A query-based interface guides the user through the modeling process. Simple commands provide model summary statistics, parameter estimates, diagnostics, and graphical displays. The special features offered by *CatReg* include options for the following:

- Stratifying the analysis by user-specified covariates (e.g., species and sex)
- Choosing among several basic forms of the exposure-response curve
- Using effects assigned to a range of severity categories, rather than a single category
- Using cluster-correlated data
- Incorporating user-specified weights
- Using aggregate data
- Query-based exclusion of user-specified data (i.e., filtering) for sensitivity analysis

Indeed, there are many potential applications of the *CatReg* program in the analysis of health effects studies and other types of data. Although the software was developed to support toxicity assessment for acute inhalation exposures, the US EPA encourages a broader application of this software. The user manual, which contains illustrated examples, provides ideas on adapting *CatReg* for situation-specific applications.

Further information on, and access to, *CatReg* is obtainable from US EPA, Office of Research and Development (ORD), National Center for Environmental Assessment (NCEA), Research Triangle Park, North Carolina, United States.

B.1.4 The *IEUBK* Model

Lead (Pb) poisoning seems to present potentially significant risks to the health and welfare of children all over the world in this day and age. The Integrated Exposure Uptake Biokinetic (IEUBK) model for lead in children attempts to predict blood-lead concentrations (PbBs) for children exposed to Pb in their environment. Meanwhile, it is worth the mention here that measured PbB concentration is not only an indication of exposure but is a widely used index to discern potential future health problems.

The IEUBK model for lead in children is a menu-driven, user-friendly model designed to predict the probable PbB concentrations (via pharmacokinetic modeling) for children aged between

6 months and 7 years who have been exposed to Pb in various environmental media (e.g., air, water, soil, dust, paint, diet, and other sources). The model has the following four key functional components:

- *Exposure component*—compares Pb concentrations in environmental media with the amount of Pb entering a child's body.
- *Uptake component*—compares Pb intake into the lungs or digestive tract with the amount of Pb absorbed into the child's blood.
- *Biokinetic component*—shows the transfer of Pb between blood and other body tissues or the elimination of Pb from the body altogether.
- *Probability distribution component*—shows a probability of a certain outcome (e.g., a PbB concentration greater than 10 µgPb/dL in an exposed child based on the parameters used in the model).

It is noteworthy that, in the United States, the US EPA and the Centers for Disease Control and Prevention (CDC) have determined that childhood PbB concentrations at or above 5 micrograms of Pb per deciliter of blood (i.e., ≥5 µgPb/dL) present risks to children's health. The IEUBK model can calculate the probability of children's PbB concentrations exceeding 5 µgPb/dL (or other user-entered value). By varying the data entered into the model, the user can evaluate how changes in environmental conditions may affect PbB levels in exposed children.

The IEUBK model allows the user to input relevant absorption parameters, (e.g., the fraction of Pb absorbed from water) as well as rates for intake and exposure. Using these inputs, the IEUBK model then swiftly calculates and recalculates likely outcomes using a complex set of equations to estimate the potential concentration of Pb in the blood for a hypothetical child or population of children (6 months to 7 years). Overall, the model is intended to the following:

- Estimate a typical child's long-term exposure to Pb in and around his/her residence
- Provide an accurate estimate of the geometric average PbB concentration for a typical child aged 6 months to 7 years
- Provide a basis for estimating the risk of elevated PbB concentration for a hypothetical child
- Predict likely changes in the risk of elevated PbB concentration from exposure to soil, dust, water, or air following rigorous efforts/actions to reduce such exposure
- Provide assistance in determining target cleanup levels at specific residential sites for soil or dust containing high amounts of Pb
- Provide assistance in estimating PbB levels associated with the Pb concentration of soil or dust at undeveloped sites that may be developed at a later date

A major advantage of the IEUBK model is the fact that it takes into consideration the several different media through which children can be exposed.

Further information on IEUBK may be obtained from the US EPA's Office of Emergency and Remedial Response, Washington, DC, United States.

B.1.5 THE *LEADSPREAD* MODEL

LEADSPREAD, the CalEPA/DTSC Lead Risk Assessment Spreadsheet, is a tool for evaluating exposure and the potential for adverse health effects that could result from exposure to lead in the environment. Basically, it consists of a mathematical model for estimating blood lead concentrations as a result of contacts with lead-contaminated environmental media. The model can be used to determine blood levels associated with multiple pathway exposures to lead. A distributional approach is used with this model—allowing estimation of various percentiles of blood-lead concentration associated with a given set of inputs.

Overall, the LEADSPREAD model provides a computer spreadsheet methodology for evaluating exposure and the potential for adverse health effects resulting from multi-pathway exposure to inorganic lead via dietary intake, drinking water, soil and dust ingestion, inhalation, and dermal contact. Each of these pathways is represented by an equation relating incremental blood lead increase to a concentration in an environmental medium, using contact rates and empirically determined ratios. The contributions via all pathways are added to arrive at an estimate of median blood-lead concentration resulting from the multi-pathway exposure. The 90th, 95th, 98th, and 99th percentile concentrations are estimated from the median by assuming a log-normal distribution with a geometric standard deviation (GSD) of 1.6.

LeadSpread 8 is the most current version (as of the time of this writing) of the DTSC Lead Risk Assessment Spreadsheet. Among other things, the risk-based soil concentration developed in "LeadSpread 8" is generally based on the CalEPA Office of Environmental Health Hazard Assessment's more recent developments—consisting of a new toxicity evaluation of lead that replaces the previous 10 µg/dL threshold blood concentration with a source-specific "benchmark change" of 1 µg/dL incremental blood lead criterion; this is meant to be implemented as an estimate of the "Exposure Point Concentration" (EPC), usually based on the 95% confidence limit on the arithmetic mean—not as a "not to exceed" soil concentration.

Further information on LEADSPREAD may be obtained from the Office of Scientific Affairs, DTSC, California EPA, Sacramento, California, United States.

B.1.6   THE RISK-BASED CORRECTIVE ACTION TOOL KIT

The risk-based corrective action (RBCA) spreadsheet system/tool kit is a complete step-by-step package for the calculation of site-specific risk-based soil and groundwater cleanup goals, which will then facilitate the development of site remediation plans. The system includes fate and transport models for major and significant exposure pathways (i.e. air, groundwater, and soil), together with an integrated chemical/toxicological library of several chemical compounds (i.e. over 600 and also expandable by the user).

RBCA is indeed a standardized approach to designing remediation strategies for contaminated sites. It was developed by the American Society for Testing and Materials (ASTM) to help prioritize sites according to the urgency and type of corrective action needed to protect human health and the environment. The RBCA process allows for the calculation of baseline risks and cleanup standards, as well as for remedy selection and compliance monitoring at petroleum release sites. The user simply provides site-specific data to determine exposure concentrations, average daily intakes, baseline risk levels, and risk-based cleanup levels.

Further information on the RBCA tool kit/spreadsheet system may be obtained from the following source: ASTM, Philadelphia, Pennsylvania, United States.

B.1.7   THE *ReOpt* (REMEDIATION OPTIONS) SOFTWARE

ReOpt is a remediation software that can be used in the selection of suitable technologies for the cleanup of contaminated sites. It speeds up site cleanup decisions because the ReOpt software enables a user to quickly and easily review a variety of remediation options and determine their effectiveness for the particular site under investigation. ReOpt contains information about technologies that might potentially be used for cleanup at contaminated sites, auxiliary information about possible hazardous or radioactive contaminants at such sites, and selected pertinent regulations that govern disposal of wastes containing these contaminants. The user specifies a series of conditions, and ReOpt provides a short list of cleanup technology choices specific to the particular situation. The technology selection is based on site characteristics and cleanup strategy.

ReOpt enables engineers and planners involved in environmental restoration efforts to quickly identify potentially applicable environmental restoration technologies and access corresponding information that is required to select cleanup activities for contaminated sites. The analyst can automatically select potentially appropriate technologies by simply specifying the contaminants or contaminated medium of interest. The analyst can also select any technology and review the technical description of the process with accompanying schematic diagrams, as well as examine the technical and regulatory constraints that govern the technology.

Information on ReOpt may be obtained from the following sources: Sierra Geophysics, Inc., Seattle, Washington, United States and Battelle Pacific Northwest Laboratory, Richland, Washington, United States.

B.1.8   THE *RISK*ASSISTANT* SOFTWARE

The RISK*ASSISTANT software is designed to assist the user in rapidly evaluating exposures and human health risks from chemicals in the environment at a particular site. It is designed to evaluate human health risks associated with *chronic* exposures to chemicals. The user need only provide measurements or estimates of the concentrations of chemicals in the air, surface water, groundwater, soil, sediment, and/or biota.

RISK*ASSISTANT generates two types of risk estimate: (1) for potential cancer-causing chemicals, the increased probability of individuals getting cancer from a particular exposure; and (2) for potential non-cancer toxic effects, a comparison of the expected level of exposure to an exposure that is assumed to be essentially without risk. Indeed, the model provides an array of analytical tools, databases, and information-handling capabilities for human health risk assessment—and it has the ability to tailor exposure and risk assessments to local conditions. In fact, RISK*ASSISTANT uses standard approaches to generate estimates of exposure and risk using detailed, locally relevant information, and then test them against many alternative assumptions.

Further information on RISK*ASSISTANT may be obtained from the following: Hampshire Research Institute, Alexandria, Virginia; US EPA, Research Triangle Park, North Carolina; California EPA, Sacramento, California; New Jersey Department of Environmental Protection, Trenton, New Jersey; and Delaware Department of Natural Resources and Environmental Control, Dover, Delaware, United States.

B.1.9   THE SITE CONCEPTUAL EXPOSURE MODEL BUILDER

The Site Conceptual Exposure Model (SCEM) Builder is a user-friendly computer application created by the US EPA to assist environmental restoration program managers (ERPMs) in preparing SCEMs. It consists of a computer graphics tool that considerably shortens the time required to generate SCEM diagrams and associated documentation. Once a SCEM has been constructed, the ERPM can quickly modify any of the variables. The SCEM Builder can then generate a range of SCEMs that reflect various "what-if" scenarios and allow for data uncertainties at a case site to be bounded by the individual SCEMs.

SCEMs are generally used as a planning tool during the environmental site investigation phase to allocate finite financial and personnel resources to address data gaps, identify sources of contamination, release mechanisms, exposure pathways, and human or ecological receptors. The SCEMs include a visual presentation of site conditions—and provide a narrative description of the assumptions used in the model. The ERPMs can use the information in SCEMs to develop risk assessment data quality objectives (DQOs) and prioritize field sampling activities, thereby reducing the uncertainty associated with risk characterization.

The purpose of a SCEM is to provide a conceptual understanding of the potential for exposure to hazardous and/or radiological contaminants at a site based on the source of contamination, the release mechanism, the exposure pathway, and the receptor. A SCEM includes a graphical presentation that relates the source of contamination to human and ecological receptors. Based on a SCEM, a data collection strategy can be developed to prioritize field sampling activities and reduce uncertainty in risk characterization (e.g., contaminant release/transport mechanisms, and receptor profiles). A SCEM may also provide sufficient information to allow for development of a strategy for early response actions to address exposure pathways that are considered complete and pose an imminent risk to public health.

The SCEM Builder conceptually connects the source of contamination to human and ecological receptors by means of graphical boxes and lines. Sources of contamination and receptors are represented by boxes; and release and exposure mechanisms are represented by lines. It is noteworthy that, contaminant fate and transport are implicitly treated by the SCEM Builder by means of a release mechanism from one source to another and by means of an exposure mechanism from a source to a receptor.

The SCEM Builder was developed by the US EPA's Office of Environmental Policy and Guidance as a planning tool for public health and ecological risk assessments.

B.1.10   THE STOCHASTIC HUMAN EXPOSURE AND DOSE SIMULATION MODELS

The Stochastic Human Exposure and Dose Simulation (SHEDS) models are considered probabilistic models that can estimate the exposures that people typically confront from chemicals encountered in everyday activities. The models are able to generate predictions of aggregate and cumulative exposures over time—in order to engender risk assessments that are protective of human health.

SHEDS can estimate the range of total chemical exposures in a population from different exposure pathways over different time periods, given a set of demographic characteristics. SHEDS can also help identify critical exposure pathways, factors, and uncertainties. Overall, the SHEDS models estimate the range of total chemical exposures in a population from four exposure pathways, viz. inhalation, skin contact, and dietary and non-dietary ingestion. These estimates are calculated using available data—such as dietary consumption surveys; human activity information; and observed or modeled levels in food, water, air, and on surfaces such as counters and floors. The data on chemical concentrations and exposure factors used in SHEDS are typically based on measurements collected in field studies and published literature.

Among other things, SHEDS models have been successfully used by the US EPA to help in the following:

- Improve pesticide-related risk assessments
- Evaluate risks to children posed by chemically treated play sets
- Improve risk assessment for chemicals in food
- Prioritize chemicals for further study on the basis of risk
- Prioritize data needs

In the end, SHEDS enhances estimates of exposure in many contexts; for instance, it has been used to better inform EPA human health risk assessments and risk management decisions. Indeed, the SHEDS models are generally used by representatives from academia, industry, government, and consulting firms globally.

The US EPA has developed a number of different SHEDS models and modules that address specific research questions about chemical exposure. Further information on SHEDS may be obtained from ORD, US EPA, United States.

B.1.11 OTHER MISCELLANEOUS COMPUTER SOFTWARE

A variety of other decision support tools and computer software are available to facilitate various risk assessment and/or environmental management tasks—including the following specific ones identified as follows:

- *AIR3D.* The American Petroleum Institute (API)'s AIR3D model is a software package that has been designed to simulate airflow in the vadose zone during the use of soil vapor extraction (SVE) or soil venting as a site restoration measure. It is a powerful tool to assist site engineers in the efficient design of vapor extraction systems.

   AIR3D is both a deterministic model and an optimization model. Linear programming techniques are incorporated to determine the optimum number and location of venting wells. Optimization can be performed based upon minimizing the number of wells or on minimizing the cost of installing the system.

   Further information on AIR3D may be obtained from the following: API, Washington, DC, United States.

- *AIRFLOW/SVE.* It is a comprehensive SVE model capable of simulating soil vapor pressure distributions, vapor flow velocities, and multicomponent soil vapor concentrations. It can be used to generate graphical displays of contaminant concentration versus time and also contaminant mass remaining versus time for multicomponent soil vapors. In addition, the program also contains an online database of chemical properties for more than 60 commonly encountered organic contaminants.

   AIRFLOW/SVE consists of a combined finite element-finite difference-particle tracking model, developed as a practical tool for the design of SVE systems for site restoration programs. The finite difference formulation of the numerical model allows the flexibility to a reasonably accurate representation of heterogeneous sites, complex boundary conditions, well dimensions, and to some extent residual nonaqueous phase liquid (NAPL) contaminant sources. Typical model applications include the simulation of radial-symmetric SVE systems; calculation of effective radius of a SVE system; design of optimal screen position of extraction well; calculation of vapor flow rate for a given vacuum that is created; estimation of cleanup times; and determination of vapor concentrations in vacuum well airstreams.

   Further information on AIRFLOW/SVE may be obtained from the following: Waterloo Hydrogeologic, Inc., Waterloo, Ontario, Canada.

- *API DSS (Decision Support System for Exposure and Risk Assessment).* The API's exposure and risk assessment DSS is a software system designed to assist environmental professionals in the estimation of human exposures and risks from sites contaminated with petroleum products. The DSS is a user-friendly tool that can be used to estimate site-specific exposures and risks; identify the need for site remediation; develop and negotiate site-specific cleanup levels with regulatory agencies; and efficiently and effectively evaluate the effect of parameter uncertainty and variability in the input parameters on estimated risks, using Monte Carlo techniques. It estimates receptor point concentrations by executing fully incorporated unsaturated zone, saturated zone, air emission, air dispersion, and particulate emission models.

   The computational modules of the DSS can be implemented in either a deterministic or Monte Carlo mode; the latter is used to quantify the uncertainty in the exposure and risk values that could result from uncertainties in the input parameters.

   From physical, chemical, and toxicological property data provided in the DSS databases, risk assessments can be conducted for 16 hydrocarbons, 6 petroleum additives, and 3 metals. The databases can also be expanded to include up to 100 other constituents.

   Further information on API DSS may be obtained from the following: API, Washington, DC, United States.

- *API PRDF (Petroleum Release Decision Framework).* The API's PRDF has been designed to provide a logical approach for site characterization, collecting and archiving field data, focusing on key decision parameters, and developing and evaluating potential corrective action plans.

  The PRDF is structured to provide consistent, comprehensive, and systematic way to identify possible contaminant sources, to characterize the site and contaminants potentially present, and to help assess actual or potential exposures.

  Further information on API PRDF may be obtained from the following: API Washington, DC, United States.

- *ASTER (ASsessment Tools for the Evaluation of Risk).* The ASTER was developed by the US EPA to assist regulators in performing ecological risk assessments. The objective was to devise a software system that could assist regulators in hazard ranking and the development of comprehensive risk assessments.

  ASTER is a Unix-based computer program that integrates database information and quantitative structure activity relationships (QSARs) to assess the environmental risk of discrete chemicals. It integrates the AQUIRE (AQUatic toxicity Information REtrieval) toxic effects database and the QSAR system, a structure-activity based expert system.

  When empirical data are not available, mechanistically based predictive models are used to estimate ecotoxicology end points, chemical properties, biodegradation, and environmental partitioning. ASTER is designed to provide high quality data for discrete chemicals, when available in the associated databases and mechanistically based QSAR estimates when data are lacking. The QSAR system includes a database of measured physicochemical properties such as melting point, boiling point, vapor pressure, and water solubility as well as more than 56,000 molecular structures stored as SMILES (Simplified Molecular Input Line Entry System) strings for specific chemicals. Searches are conducted by identifying a chemical of interest using the Chemical Abstract Service registry number, a chemical name, or a SMILES string.

- *BIOSCREEN.* Three-dimensional contaminant transport for dissolved-phase hydrocarbons in saturated zone under the influences of oxygen, nitrate, iron, sulfate, and methane limited biodegradation. Advection, dispersion, adsorption, first-order decay, and instantaneous reactions under aerobic and anaerobic conditions. It is an easy-to-use screening tool which is based on the Domenico analytical solute transport model.

- *CRYSTAL BALL.* It is a user-friendly, graphical-oriented forecasting and risk analysis program that helps diminish the uncertainty associated with decision-making. The system uses standard spreadsheet models in its application.

  Through the use of Monte Carlo simulation techniques, *Crystal Ball* forecasts the entire range of results possible for a given situation. It also shows the confidence levels, in order that the analyst will know the likelihood for any specific event to take place. The tool further allows sensitivity evaluations to be carried out in a very effective manner—by the use of sensitivity charts; *Crystal Ball* calculates sensitivity by computing rank correlation coefficients between every assumption and every forecast cell during simulation. With an intuitive graphical interface, *Crystal Ball* gives users powerful capabilities to perform uncertainty analyses based on Monte Carlo simulations.

  Further information on *Crystal Ball* may be obtained from Decisioneering, Inc., Boulder, Colorado, United States.

- *EPA-Expo-Box.* The US EPA's "EXPOsure toolBOX" (EPA-Expo-Box) was developed by the ORD, as a compendium of exposure assessment tools that links to exposure assessment guidance documents, databases, models, key references materials, and other related resources. Overall, the toolbox provides a variety of exposure assessment resources organized into six "Tool Sets"—each containing a series of modules.

The EPA's EPA-Expo-Box is indeed a toolbox created to assist individuals from within government, industry, academia, and the general public with assessing exposure. The toolbox allows the user to navigate from different starting points, depending on the problem-specific needs. Further information on EPA-Expo-Box may be obtained from ORD, US EPA, United States.

*   *GEOEAS.* The GEOEAS geostatistical software is an interactive software package for two-dimensional geostatistical analyses of spatially distributed data. Programs are provided in this package for data file management, data transformations, univariate statistics, variogram analysis, cross-validation, kriging, contour mapping, post plots, and line/scatter graphs.

    The GEOEAS system features include hierarchical menus, full-screen data entry, and graphical displays to provide a high degree of user interaction. Users can also very easily alter parameter values to recalculate results or reproduce graphical displays—thus providing a very useful "what if"-type of analysis capabilities.

    Further information on GEOEAS may be obtained from the US EPA Environmental Laboratory, Las Vegas, Nevada, United States.

*   *GEOTOX.* It is a multimedia environmental transport and transformation model. It consists of a computer program designed to calculate time-varying chemical concentrations in air, soil, groundwater, and surface water. The model can be applied to constant or time-varying chemical sources.

    The GEOTOX model uses two sets of input data—one providing the properties of the environment or landscape receiving the contaminants and the other describing the properties of the contaminants. Model output is in the form of environmental concentrations, intake by various exposure pathways, as well as total intake.

    GEOTOX can be used to predict/derive contaminant concentrations in various multimedia compartments, which are subsequently combined with appropriate human inhalation and ingestion rates and absorption factors to calculate exposure. Relative health risk for a number of chemicals can be calculated.

    Further information on GEOTOX may be obtained from the following: Lawrence Livermore National Laboratory, Berkeley, California, United States.

*   *HELP (Hydrologic Evaluation of Landfill Performance model).* HELP is a quasi-two-dimensional deterministic numerical, finite-difference model that computes a daily water budget for a landfill represented as a series of horizontal layers. It models leaching from landfills into the unsaturated soils beneath. HELP is used for water balance computation, and for the estimation of chemical emissions, and also for leachate quality assessment. It models both organic and inorganic compounds, using rainfall and waste solubility to model the leachate concentrations leaving the landfill.

    Further information on HELP may be obtained from the following: US EPA's National Computer Center, Research Triangle Park, North Carolina, United States; and U.S. Corps of Engineers, Sacramento, California, United States.

*   *HSSM (the Hydrocarbon Spill Screening Model).* HSSM is a screening tool for light non-aqueous phase liquid (LNAPL) impacts to the groundwater table. The model consists of separate modules for addressing LNAPL flow through the vadose zone, LNAPL spreading in the capillary fringe—at the water table and dissolved LNAPL groundwater transport to potential receptor exposure locations. These modules are based on simplified conceptualizations of the flow and transport phenomena which were used so that the resulting model would be practical, though approximate, tool.

    The HSSM model is intended for the simulation of subsurface releases of LNAPLs and is used to estimate the impacts of this type of pollutant on water table aquifers. HSSM offers a simplified approximate analysis for emergency response, initial phases of site investigations, facilities siting and permitting, and underground storage tank programs.

Further information on HSSM may be obtained from the following: US EPA's Robert S. Kerr Environmental Research Laboratory, Ada, Oklahoma, United States.

• *MEPAS (Multimedia Environmental Pollutant Assessment System).* MEPAS is an analytical model that has been developed to address problems at hazardous waste sites. It is a versatile tool that can handle a diversity of different types of source terms. MEPAS couples contaminant release, migration and fate for environmental media (groundwater, surface water, air) with exposure routes (inhalation, ingestion, dermal contact, external dose), and risk/health consequences for radiological and non-radiological carcinogens and noncarcinogens.

Overall, MEPAS develops an integrated, site-specific, multimedia environmental assessment. It can simulate the transport and distribution of contaminants over time and space within air, water, soil, and food chain pathways. MEPAS incorporates a sector-averaged Gaussian plume algorithm to simulate the atmospheric transport of contaminants, simulates groundwater transport using a three-dimensional algorithm, uses a simplistic approach to modeling the surface water pathway, and includes food chains as an integral part of its exposure-dose component. It can model both on-site and off-site contaminant exposures. It estimates long-term health effects at receptor locations, as well as normalized maximum hourly concentrations for determining acute effects.

In the end, MEPAS integrates and evaluates transport and exposure pathways for chemicals and radioactive releases according to their potential human health impacts. It takes the nontraditional approach of combining all major exposure pathways into a multimedia computational tool for public health impact.

Further information on MEPAS may be obtained from the following: Battelle—Pacific Northwest National Laboratory, Richland, Washington, United States.

• *MMSOIL.* MMSOIL (The Multimedia Contaminant Fate, Transport, and Exposure Model) is a multimedia screening tool developed for the "relative comparison" of hazardous waste sites. The model performs a mass balance for the air and groundwater pathways separately, relative to the initial source term.

MMSOIL includes a sector-averaged Gaussian plume algorithm to simulate the atmospheric transport of contaminants, has a complex groundwater modeling component, includes a simplistic approach to modeling the surface water pathway, and considers food chain modeling "externally"/separately. It can model both on-site and off-site chemical exposures.

MMSOIL was designed specifically to simulate the release of toxic chemicals from underground storage tanks, surface impoundments, waste piles, and landfills. It can model the fate and transport of chemicals and calculates human exposure and health risk, as well as concentration in all the important media.

Further information on MMSOIL may be obtained from the following: ORD, US EPA, Washington, DC.

• *MULTIMED (MULTIMEDia exposure assessment model).* It is a computer model for simulating the transport and transformation of contaminants released from a hazardous waste disposal facility into the multimedia environment. The MULTIMED model simulates releases into air and soil—including the unsaturated (vadose) and saturated zones and possible interception of the subsurface contaminant plume by a surface stream. It further simulates movement through the air, soil, groundwater, and surface water media to contact humans and other potentially affected receptors. Uncertainties in parameter values used in the model are quantified using Monte Carlo simulation techniques.

MULTIMED is typically used to simulate the movement of contaminants leaching from a waste disposal facility. It is intended for general exposure and risk assessments of waste facilities and for the analyses of the impacts of engineering and management controls.

Further information on MULTIMED may be obtained from the following: Environmental Research Laboratory, Office of Research and Development, US EPA, Athens, Georgia, United States.

- *NAPL Simulator.* The *NAPL Simulator* is used for the simulation of the contamination of soils and aquifers that results from the release of organic liquids commonly referred to as NAPL.

  The *NAPL Simulator* is applicable to three interrelated zones: a vadose zone that is in contact with the atmosphere, a capillary zone, and a water-table aquifer zone. Three mobile phases are accommodated: water, NAPL, and gas. A 3-phase "k-S-P" sub-model accommodates capillary and fluid entrapment hysteresis. NAPL dissolution and volatilization are accounted through rate-limited mass transfer sub-models. The numerical solution is based on a Hermite collocation finite element discretization. The simulator provides an accurate solution of a coupled set of nonlinear partial differential equations that are generated by combining fundamental balance equations with constitutive thermodynamic relationships.

- *RAAS (Remedial Action Assessment System).* It is a sophisticated, Windows-based software package that serves as a tool for analyzing and evaluating the trade-offs necessary to select a preferred approach for restoring a contaminated site. It helps the user to develop a detailed site description; estimate baseline and residual risks to public health posed by the site; identify and screen applicable environmental restoration technologies; formulate and evaluate technically feasible, complete remedial response alternatives; and assess and compare a remedial response alternative across some set criteria to provide a basis for final remedy selection.

  RAAS can be used to quantitatively evaluate the effectiveness of several remedial alternatives in terms of concentration, risk reduction, and effect on media properties. In addition, the methodology enables the user to assess and cross-compare those remedial response alternatives across some set critera (e.g., cost, human health, risk reduction, and time) to provide a basis for comparisons among them.

  Further information on RAAS may be obtained from the following: Battelle—Pacific Northwest National Laboratory, Richland, Washington, United States.

- *RACER (Remedial Action Cost Engineering and Requirements).* The RACER is a cost-estimating tool that may be used in any of the phases of a contaminated site remediation project.

  RACER is a parametric cost modeling system that uses a patented methodology for estimating costs, particularly as pertains to the environmental and construction industries in the United States. An estimate in RACER is created by tailoring the generic engineering solutions embedded in the software to suite the project-specific needs; this is done by adding site-specific parameters to reflect project-specific conditions and requirements. The tailored design is then translated into specific quantities of work, and the quantities are priced using current [US] price data.

  Further information on the RACER cost estimating software/tool may be obtained from the following: Talisman Partners, Ltd., Englewood, Colorado, United States.

- *Regional Screening Levels (RSLs).* The RSL tables provide comparison values for residential and commercial/industrial exposures to soil, air, and tap water (drinking water). Included here are tables of risk-based screening levels—calculated using the latest toxicity values, default exposure assumptions, and physical and chemical properties; also included is a calculator where default parameters can be changed to reflect site-specific risks.

  Overall, this tool presents standardized risk-based screening levels and variable risk-based screening level calculation equations for chemical contaminants. The risk-based screening levels for chemicals are based on the carcinogenicity and systemic toxicity of the analytes of interest. In the end, screening levels are presented in the default tables for residential soil, outdoor worker soil, residential indoor air, worker indoor air, and tap water. In addition, the calculator provides a fish ingestion equation. The standardized or default screening levels used in the tables are based on default exposure parameters and incorporate exposure factors that present RME conditions.

Further information on the RSLs is obtainable from US EPA, ORD, NCEA, Arlington, Virginia, United States.

- *RISC (Risk Identification of Soil Contamination).* It is a knowledge-based framework for risk identification and evaluation of sites with contaminated soils. It consists of computer modules that facilitate site investigations, risk analyses, and priority ranking for former industrial facilities.

  The RISC framework uses expert information on the fate and behavior of contaminants in soil systems to predict potential risks to human health and the environment that could result from contaminated site problems. Dutch, English, and German versions of this computer model system are available.

  Further information on the RISC computer model system may be obtained from the following: Van Hall Institute, Groningen, the Netherlands.

- *RISKPRO.* It is a complete software system designed to predict the environmental risks and effects of a wide range of human health-threatening situations. It consists of a multi-media/multi-pathway environmental pollution modeling system—providing for modeling tools to predict exposure from pollutants in the air, soil, and water.

  RISKPRO is used to evaluate receptor exposures and risks from environmental contaminants. It graphically represents its results through maps, bar charts, wind rose diagrams, isopleth diagrams, pie charts, and distributional charts. Its mapping capabilities can also allow the user to create custom maps showing data and locations of environmental contaminant plumes.

  Further information on RISKPRO may be obtained from General Sciences Corporation (GSC), Laurel, Maryland, United States.

- *SITES (The Contaminated Sites Risk Management System).* It is a flexible interactive PC-computerized decision-support tool for organizing several relevant information needed to conduct risk management analyses for contaminated sites. It has the dimensionality to model multiple chemicals, pathways, population groups, health effects, and remedial actions. The model uses information from diverse sources, such as site investigations, transport and fate modeling, behavioral and exposure estimates, and toxicology.

  SITES is indeed a computer-based integrating framework used to help evaluate and compare site investigation and remedial action alternatives in terms of health and environmental effects and total economic costs/impacts. The user completely defines the scope of the analyses. Both deterministic and probabilistic analyses are possible.

  The decision-tree structure in SITES allows for explicit examination of key uncertainties and the efficient evaluation of numerous scenarios. The model's design and computer implementation facilitates quick and extensive sensitivity analyses.

  Further information on SITES may be obtained from the following: EPRI (Electric Power Research Institute), Palo Alto, California, United States.

- *TOXIC.* It is a microcomputer program that calculates the incremental risk to the hypothetical maximum exposed individual from hazardous waste incineration. It calculates exposure to each pollutant individually, using a specified dispersion coefficient (which is the ratio of pollutant concentration in air [in $\mu g/m^3$] to pollutant emission rate [in gm/s]).

  TOXIC is used in hazardous waste facility risk analysis. It is a flexible and convenient tool for performing inhalation risk assessments for hazardous waste incinerators—and produces point estimates of inhalation risks.

  Further information on TOXIC may be obtained from Rowe Research and Engineering Associates, Alexandria, Virginia, United States.

B.2   SELECTED DATABASES AND INFORMATION LIBRARY FOR
CONTAMINATED SITE RISK MANAGEMENT AND DECISION-MAKING

Several databases containing information on numerous chemical substances exist within the scientific community that may find extensive useful applications in the management of various types of chemical exposure and contaminated site problems. Example databases of general interest to contaminated site assessments and risk management programs are presented in the following subsections; this select list is of interest, especially because of their international appeal and/or their wealth of risk management support information. Indeed, the presentation is meant to simply demonstrate the overall wealth of scientific information that already exists—and that should generally be consulted to provide the relevant support information essential for risk determination and/or contaminated site risk management actions.

B.2.1   The "UNEP Chemicals"/International Register
of Potentially Toxic Chemicals Database

In 1972, the United Nations Conference on the Human Environment, held in Stockholm, recommended the setting up of an international registry of data on chemicals likely to enter and damage the environment. Subsequently, in 1974, the Governing Council of the United Nations Environment Programme (UNEP) decided to establish both a chemical's register and a global network for the exchange of information that the register would contain. The definition of the register's ultimate objectives was subsequently expounded to address the following:

- Make data on chemicals readily available to those who need it, by facilitating access to existing data on the production, distribution, release, and disposal of chemicals and their effects on humans and their environments—and thereby contribute to a more efficient use of national and international resources available for the evaluation of the effects of the chemicals and their control.
- On the basis of information in the register, identify and draw attention to the major/ important gaps in existing knowledge on the effects of chemicals and related available information—and then encourage research to fill those gaps.
- Identify or help identify the potential hazards originating from chemicals and waste materials—and then improve people awareness of such hazards.
- Assemble information on existing policies for control and regulation of hazardous chemicals at national, regional, and global levels—by ultimately providing information about national, regional, and global policies; regulatory measures and standards; and recommendations for the control of potentially toxic chemicals.
- Facilitate the implementation of policies necessary for the exchange of information on chemicals in possible international trades.

In 1976, a central unit for the register—named the International Register of Potentially Toxic Chemicals (IRPTC)—was created in Geneva, Switzerland, with the main function of collecting, storing, and disseminating data on chemicals and also to operate a global network for information exchange. IRPTC network partners (i.e., the designation assigned to participants outside the central unit) consisted of National Correspondents appointed by governments, national and international institutions, national academies (of sciences), industrial research centers, and specialized research institutions.

Chemicals examined by the IRPTC have generally been chosen from national and international priority lists. The key selection criteria used include the quantity of production and use, the toxicity to humans and ecosystems, persistence in the environment, and the rate of accumulation in living organisms.

As a final point, it is noteworthy that the IRPTC database has essentially been "transformed" into what is also known as the *UNEP Chemicals*—which has more or less become the focus for all activities undertaken by UNEP to ensure the globally sound management of hazardous chemicals; indeed, it is built upon the same solid technical foundation of the IRPTC—and aims to promote chemicals safety by providing countries with access to information on toxic chemicals; facilitate or catalyze global actions to reduce or eliminate chemicals risks; and to assist countries in building their capacities for safe production, use, and disposal of hazardous chemicals. At any rate, for all intents and purposes, the discussion provided here for UNEP's original IRPTC may reasonably be used interchangeably with *UNEP Chemicals*.

B.2.1.1   General Types of Information in the (Original) IRPTC Databases

IRPTC stores information that would aid in the assessment of the risks and hazards posed by a chemical substance to human health and environment. The major types of information collected include that relating to the behavior of chemicals and information on chemical regulations. Information on the behavior of chemicals is obtained from various sources such as national and international institutions, industries, universities, private databanks, libraries, academic institutions, scientific journals, and United Nations bodies such as the International Programme on Chemical Safety (IPCS). Regulatory information on chemicals is largely contributed by IRPTC National Correspondents. Specific criteria are used in the selection of information for entry into the databases. Whenever possible, IRPTC uses data sources cited in the secondary literature produced by national and international panels of experts to maximize reliability and quality. The data are then extracted from the primary literature. Validation is performed prior to data entry and storage on a computer at the United Nations International Computing Centre.

Overall, the complete IRPTC file structure consists of databases relating to the following key subject matter and areas of interest: Legal; Mammalian and Special Toxicity Studies; Chemobiokinetics and Effects on Organisms in the Environment; Environmental Fate Tests, and Environmental Fate and Pathways into the Environment; and Identifiers, Production, Processes and Waste.

The IRPTC *Legal* database contains national and international recommendations and legal mechanisms related to chemical substances control in environmental media such as air, water, wastes, soils, sediments, biota, foods, drugs, and consumer products. This setup allows for rapid access to the regulatory mechanisms of several nations and to international recommendations for safe handling and use of chemicals.

The *Mammalian Toxicity* database provides information on the toxic behavior of chemical substances in humans; toxicity studies on laboratory animals are included as a means of predicting potential human effects. The *Special Toxicity* databases contain information on particular effects of chemicals on mammals, such as mutagenicity and carcinogenicity, as well as data on nonmammalian species when relevant for the description of a particular effect.

The *Chemobiokinetics and Effects on Organisms in the Environment* databases provide data that will permit the reliable assessment of the hazard of chemicals present in the environment to man. The absorption, distribution, metabolism and excretion of drugs; chemicals; and endogenous substances are described in the *Chemobiokinetics* databases. *The Effects on Organisms in the Environment* databases contain toxicological information regarding chemicals in relation to ecosystems and to aquatic and terrestrial organisms at various nutritional levels.

The Environmental Fate Tests, and Environmental Fate and Pathways into the Environment databases assess the risk presented by chemicals to the environment.

The *Identifiers, Production, Processes and Waste* databases contain miscellaneous information about chemicals—including physical and chemical properties, hazard classification for chemical production and trade statistics of chemicals on worldwide or regional basis, information on production methods, information on uses and quantities of use for chemicals, data on persistence of chemicals in various environmental compartments or media, information on the intake of chemicals by humans in different geographical areas, and sampling methods for various media and species, as well as analytical protocols for obtaining reliable data; recommendable methods for the treatment and disposal of chemicals.

B.2.1.2 The Role of the (Original) IRPTC in Risk Assessment and Environmental Management

The IPRTC, with its carefully designed database structure, serves as a sound model for national and regional data systems. More importantly, it brings consistency to information exchange procedures within the international community. Indeed, the IPRTC serves as an essential international tool for chemicals hazard assessment, as well as a mechanism for information exchange on several chemicals. The wealth of scientific information contained in the IRPTC can serve as an invaluable database for a variety of environmental and public health risk management programs.

Further information on the IRPTC (and related tools or databases) may be obtained from the National Correspondent to the IRPTC and scientific bodies/institutions (such as a country's National Academy of Sciences)—as well as via UNEP and related Internet websites. [By the way, it is noteworthy that, following the successful implementation of the IRPTC databases, a number of countries created National Registers of Potentially Toxic Chemicals that is completely compatible with the IRPTC system.]

B.2.2 THE INTEGRATED RISK INFORMATION SYSTEM DATABASE

The Integrated Risk Information System (IRIS), that has been prepared and maintained by the Office of Health and Environmental Assessment of the US EPA, is an electronic database containing health risk and regulatory information on several specific chemicals. The IRIS database was created by the US EPA in 1985 (and made publicly available in 1988) as a mechanism for developing consistent consensus positions on potential health effects of chemical substances. Indeed, IRIS was originally developed for the US EPA staff—in response to a growing demand for consistent risk information on chemical substances for use in decision-making and regulatory activities. On the whole, it serves as an online database of chemical-specific risk information; it is also a primary source of EPA health hazard assessment and related information on several chemicals of broad environmental concern. It is noteworthy that each IRIS assessment can cover a chemical, a group of related chemicals, or a complex mixture. It is also notable that the information in IRIS is generally accessible to even those without extensive training in toxicology but with some rudimentary knowledge of health and related sciences.

Broadly speaking, the IRIS database provides information on how chemicals affect human health and is a primary source of EPA risk assessment information on chemicals of environmental and public health concern. It serves as a guide for the hazard identification and dose–response assessment steps of EPA risk assessments. More importantly, IRIS makes chemical-specific risk information readily available to those who must perform risk assessments—and also increases consistency in risk management decisions. The information in IRIS generally represents expert Agency consensus; in fact, this Agency-wide agreement on risk information is one of the most valuable aspects of IRIS. Chemicals are added to IRIS on a regular basis—with chemical file sections in the system being updated as new information is made available to the responsible review groups.

B.2.2.1 General Types of Information in IRIS

The IRIS database consists of a collection of computer files covering several individual chemicals. To aid users in accessing and understanding the data in the IRIS chemical files, the following key supportive documentation is provided as an important component of the system:

- Alphabetical list of the chemical files in IRIS and list of chemicals by Chemical Abstracts Service (CAS) number
- Background documents describing the rationales and methods used in arriving at the results shown in the chemical files

- A user's guide that presents step-by-step procedures for using IRIS to retrieve chemical information
- An example exercise in which the use of IRIS is demonstrated
- Glossaries in which definitions are provided for the acronyms, abbreviations, and specialized risk assessment terms used in the chemical files and in the background documents

The chemical files contain descriptive and numerical information on several subjects—including oral and inhalation reference doses (RfDs) for chronic noncarcinogenic health effects, as well as oral and inhalation cancer slope factors (SFs) and unit cancer risks (UCRs) for chronic exposures to carcinogens. It also contains supplementary data on acute health hazards and physical/chemical properties of the chemicals. In fact, the primary types of health assessment information in IRIS are oral RfDs and inhalation reference concentrations (RfCs) for noncarcinogens, as well as oral and inhalation carcinogen assessment parameters. RfDs and concentrations are estimated human chemical exposures over a lifetime and that are just below the expected thresholds for adverse health effects. The carcinogen assessments include a weight-of-evidence classification, oral and inhalation quantitative risk information, including SFs, along with unit risks calculated from those SFs. [An SF is the estimated lifetime cancer risk per unit of the chemical absorbed, assuming lifetime exposure.]

Overall, summary information in IRIS consists of three components: derivation of oral chronic RfD and inhalation chronic RfC, for non-cancer critical effects, cancer classification (and cancer hazard narrative for the more recent assessments), and quantitative cancer risk estimates. Indeed, the IRIS information has generally focused on the documentation of toxicity values (i.e., RfD, RfC, cancer unit risk, and SF) and cancer classification; the bases for these numerical values and evaluative outcomes are typically provided in an abbreviated and succinct manner. Anyhow, details for the scientific rationale can be found in supporting documents, and references for these assessment documents and key studies are provided in the bibliography sections. Meanwhile, it is notable that since 1997, IRIS summaries and accompanying support documents, including a summary and response to external peer-review comments, have been publicly available in full text on the IRIS website—with the Internet site now being EPA's primary repository for IRIS (comprising the "IRIS assessment" for a given chemical substance as a whole). By and large, prevailing information on IRIS at any one time generally represents the state-of-the-science and state-of-the-practice in risk assessment—that is, as existed when each assessment was prepared.

Finally, it is noteworthy that because exposure assessment pertains to exposure at a particular place, IRIS cannot provide situation-specific information on exposure. However, IRIS can be used with an exposure assessment to characterize the risk of chemical exposure. This risk characterization can then be used to decide on what actions to take to protect human health.

B.2.2.2   The Role of IRIS in Risk Assessment and Environmental Management

IRIS is a tool that provides hazard identification and dose–response assessment information but does not provide situation- or problem-specific information on individual instances of exposure. It is a computerized library of current information that is updated periodically. Combined with site-specific or national exposure information, the summary health information in IRIS could thenceforth be used by risk assessors/analysts and others to evaluate potential public health risks from environmental chemicals. Also, combined with specific exposure information, the data in IRIS can be used to characterize the public health risks of a chemical of potential concern under specific scenarios, which can then facilitate the development of effectual corrective action decisions designed to protect public health. The information in IRIS can indeed be used to develop corrective action and risk management decision for chemical exposure problems—such as achieved via the application of risk assessment and risk management procedures.

The IRIS Program is located within EPA's NCEA in the ORD. Further information on and access to the IRIS database may be obtained via the US EPA Internet website. Alternatively, the following groups may be contacted for pertinent needs: IRIS User Support, US EPA, Environmental Criteria and Assessment Office, Cincinnati, Ohio, United States; and National Library of Medicine (NLM), Bethesda, Maryland, United States.

B.2.3   The International Toxicity Estimates for Risks Database

International Toxicity Estimates for Risk (ITER) is a database of human health risk values and supporting information—generally comprised of data to aid human health risk assessments. Overall, the database consists of chemical files containing data with information that come mostly from: the US EPA, the US Agency for Toxic Substances and Disease Registry (ATSDR)/CDC, Rijksinstituut Voor Volksgezondheid en Miliouhygiene (RIVM) (National Institute of Public Health and the Environment) of the Netherlands, Health Canada, the International Agency for Research on Cancer (IARC) and indeed a number of independent parties offering peer-reviewed risk values. Among other things, this includes direct links to EPA's IRIS, and to ATSDR's "Toxicological Profiles" for each chemical file—and also has the ability to print reports. Meanwhile, it is worth the mention here that the data in the ITER database are presented in a comparative fashion—allowing the user to view what conclusions each organization has reached; in addition, a brief explanation of differences is provided. The database is typically updated several times a year.

In general, the values and text in the ITER database would have been extracted from credible published documents and data systems of the original author organizations. Independently derived values, which have undergone external peer review at a TERA-sponsored peer-review meeting, are also listed in the ITER database. The risk values are compiled into a consistent format, so that comparisons can be made readily by informed users. The necessary conversions are performed so that direct comparisons can be made—and the synopsis text is written so as to help the user better understand the similarities and differences between the values of the different organizations. At the end of each so-identified "Level 3" summary in the ITER database, the user will find listed a source and/or link for further information about that particular assessment.

The ITER database is compiled by *Toxicology Excellence for Risk Assessment (TERA)*, a non-profit corporation "with a mission dedicated to the best use of toxicity data for the development of risk values" (according to the organization). It is noteworthy that TERA is said to prevent conflicts of interest in part through its nonprofit status, as well as policy of informed and neutral guidance. Consequently, TERA generally helps environmental, industry, and government groups find common ground through the application of good science to risk assessment. Apparently, the general motivation has been that, in fostering successful partnerships, improvements in the science and practice of risk assessment will follow.

B.2.3.1   General Types of Information in ITER

ITER is considered a toxicology data file on the NLM's *Toxicology Data Network* (TOXNET)— containing data in support of human health risk assessments. It is compiled by TERA and contains over 650 chemical records—with key data coming from a number of international establishments (such as ATSDR/CDC; Health Canada; RIVM—the Netherlands; US EPA; IARC; NSF International; and independent parties whose risk values have undergone peer review). As an example, RIVM develops human-toxicological risk limits (i.e., maximum permissible risk levels, MPRLs) for a variety of chemicals based on chemical assessments that are compiled in the framework of the Dutch government program on risks in relation to soil quality. As another example, information on the toxic effects of chemical exposure in humans and experimental animals is contained in the ATSDR "Toxicological Profiles"; these documents also contain dose–response information for different routes of exposure—and when information is available, the Toxicological Profiles also contain a discussion of toxic interactive effects with other chemicals, as well as a description of potentially sensitive human populations. All these sources may indeed serve as a basis for some of the values reported in the ITER database. All in all, the ITER data, focusing on hazard identification and dose–response assessment, is extracted from each agency's assessment—and contains links to the source documentation.

Among the key data provided in ITER are ATSDR's minimal risk levels; Health Canada's tolerable intakes/concentrations and tumorigenic doses/concentrations; EPA's carcinogen classifications, unit risks, SFs, oral RfDs, and inhalation RfCs; RIVM's maximum permissible risk levels; NSF International's RfDs and carcinogen risk levels; IARC's cancer classifications; and non-cancer and/or cancer risk values (that have undergone peer review) derived by independent parties. Finally, it is notable that ITER provides comparison charts of international risk assessment information in a side-by-side format and explains differences in risk values derived by different organizations.

B.2.3.2 The Role of ITER in Risk Assessment and Environmental Management

ITER consists of a compilation of human health risk values for chemicals of environmental and/or public health concern from several health organizations worldwide. These values are developed for multiple purposes depending on the particular organization's function. They are principally used as guidance or regulatory levels against which human exposures from chemicals in the air, food, soil, and water can be compared.

As of the early part of the year 2016, ITER contained values mostly from the following major organizations: Health Canada, RIVM, US ATSDR/CDC, US EPA, IARC, NSF International, and other independent parties offering peer-reviewed risk values. Anyhow, in the future, it is expected that the ITER database will include additional chemicals and health information from various other organizations such as the World Health Organization (WHO)/IPCS. On the whole, the information in the ITER database is useful to risk assessors and risk managers needing human health toxicity values to make risk-based decisions. ITER allows the user to compare a number of key organizations' values and to determine the best value to use for the human exposure situation being evaluated.

Further information on, and access to, the ITER database may be obtained via the ITER and/or TERA Internet websites—as well as the NLM and TOXNET websites, among others.

B.2.4 *RAMAS* Library of Ecological Software

RAMAS is a software library for building ecological models. RAMAS programs incorporate species-specific data to predict the future changes in the population and assess the risk of population extinction or explosion and chances of recovery from a disturbance. All programs have user-friendly menu systems and context-sensitive, online help facilities.

The family of population simulators provided by the RAMAS Library consists of models that cover almost any type of circumstance and provides a very good demonstration for the principles of population dynamics.

B.2.4.1 General Types of Information in the RAMAS Library

Some of the major ecological software contained in the RAMAS Library includes the following:

- *RAMAS/age, for modeling fluctuations in age-structured populations.* It is an interactive simulator for age-structured population dynamics. The program predicts how many individuals there will be in an age class in future years and estimates the probability that the population will exceed or fall below a specified level of abundance.
- *RAMAS/ecoBound for ecological boundary delineation.* It provides a selection of traditional and contemporary methods to draw ecological boundaries in a single software package.
- *RAMAS/ecotoxicology, for population-level risk assessment.* It translates individual-level impacts to population-level risk assessment. The program imports data from standard laboratory bioassay experiments, incorporates these data into the parameters of a population

model, and performs a risk assessment by analyzing population-level differences between control and impacted samples.

- *RAMAS/GIS, linking landscape data with population viability analysis* (i.e., linking Geographical Information System, GIS with metapopulation dynamics). It is a comprehensive extinction risk assessment system, in which the user can run the metapopulation model to predict the risk of species extinction, expected occupancy rates, and metapopulation abundance. The program combines geographic and demographic data for risk assessment.
- *RAMAS/metapop, for viability analysis for stage-structured metapopulations.* It is an interactive program that allows the user to build models for species that live in multiple patches and incorporates the spatial aspects of metapopulation dynamics. The program can be used to predict extinction risks and explore management options such as reserve design, translocations, and to assess human impact on fragmented populations.
- *RAMAS/space, consisting of spatially structured models for conservation biology.* It is an interactive program of metapopulation modeling, which incorporates the spatial aspects of metapopulation dynamics. It is used in conservation biology to predict extinction risks and explore management options such as reserve design, translocations, and reintroductions. The program predicts how many individuals there will be in the metapopulation and in each of its populations in future years, estimates the probability that the species will go extinct, will fall below exceed a specified total abundance, and calculates the distribution of extinction times.
- *RAMAS/stage, for analyzing stage-structured populations.* It allows a user to build, run, and analyze discrete-time models for species with virtually any life history. It is especially useful for modeling species with complex life histories. The program estimates the chance that a population will go extinct or suffer a decline; it also estimates the chance that the population will grow to some level and/or the risks that an abundance will fall below, or above, any threshold.
- *RAMAS/time, for ecological time series analysis.* It is designed to analyze time series and to model immediate and delayed density-dependent feedback that affect organism numbers. It also determines the qualitative type of dynamical behavior that characterizes the studied population.
- *Risk Calc, for calculating bounds on point estimates.* It allows the user to determine the uncertainties in the risk estimates, so that reliability of any predictions can be determined. The program uses intervals and fuzzy numbers to represent uncertainty and provides an environment to compute with these numbers in which all uncertainties are carried forward automatically.

B.2.4.2   Sources of Information on the RAMAS Library of Software

Further information on RAMAS library of software may be obtained from the following source: Applied Biomathematics (Ecological Research and Software), Setauket, New York, United States.

B.2.5   THE ECOTOX (*ECOTOX*ICOLOGY) DATABASE

The ECOTOXicology database (ECOTOX) is a source for locating single chemical toxicity information/data for aquatic and terrestrial life (including both plants and wildlife). ECOTOX is a useful tool for examining impacts of chemicals on the environment. Peer-reviewed literature is the primary source of information encoded in the database. Pertinent information on the species, chemical, test methods, and results presented by the author(s) are abstracted and entered into the database. Another source of test results is independently compiled data files provided by various United States and international government agencies.

Indeed, there has been an increased emphasis in ecotoxicology to improve and enhance the process by which ecological and human health hazard assessments are performed. This has resulted in a range of activity from the hazard evaluation of new and existing chemicals and pesticides to the evaluation of impact assessments associated with effluent, leachate, and environmental monitoring data. Additionally, research programs within the US EPA that are associated with these efforts are expanding to further support the scientific basis of ecological and human health risk assessments for chemical pollutants. The US EPA and other government agencies have a need for ready access to toxicity information for use in characterizing, diagnosing, and predicting effects associated with chemical stressors and to support the development of eco-criteria and thresholds for natural resource use.

Literature acquisition, data entry, and data abstraction for ECOTOX are continually conducted. As data are entered and quality assured, the aquatic database system is updated—usually on a quarterly basis. It is noteworthy that the time lag from conducting a literature search, acquiring the publication, and encoding it into the ECOTOX database can be up to or exceed 6 months. For this reason, it is suggested that the user should conduct additional searches of the most recent publication year to ensure the capture of data that may not have been entered into the ECOTOX database. This will also help catch any relevant literature that could have been missed during ECOTOX's attempted comprehensive searches.

ECOTOX was created, and is maintained, by the US EPA, ORD, and the National Health and Environmental Effects Research Laboratory's Mid-Continent Ecology Division. It is recommended that, prior to using ECOTOX data for analysis or summary projects, you should consult the original scientific paper to ensure an understanding of the context of the data retrieved from ECOTOX.

B.2.5.1   General Types of Data Included in ECOTOX

ECOTOX is a source for locating single chemical toxicity data from three US EPA ecological effects databases—namely, AQUIRE, TERRETOX, and PHYTOTOX. The AQUIRE and TERRETOX databases contain information on lethal, sublethal, and residue effects.

Aquatic data in AQUIRE are limited to test organisms that are exclusively aquatic (saltwater and freshwater). Species that are associated with the water but do not have gills, such as ducks and geese, are included in the terrestrial database. Amphibians are included in both AQUIRE and TERRETOX databases, with the life stages that exist exclusively in the water (e.g., tadpole) located in AQUIRE and the terrestrial life stage (e.g., adult) in TERRETOX. Bacteria and virus are not included in the ECOTOX database. TERRETOX is the terrestrial animal database. Its primary focus is wildlife species, but when data gaps exist for a particular chemical, data for domestic species are included. PHYTOTOX is a terrestrial plant database. In the development of PHYTOTOX, the tests represent mostly agricultural chemicals and predominantly agricultural species.

The AQUIRE (AQUatic toxicity Information REtrieval) database was initially developed to hold acute laboratory data but underwent major changes in the early 1990s, encoding additional parameters to better present data from outdoor field exposures and chronic exposures. The AQUIRE database includes toxic effects data on all aquatic species including plants and animals and freshwater and saltwater species. TERRETOX is the terrestrial animal database. Its primary focus is wildlife species, but the database does include information on domestic species. PHYTOTOX is a terrestrial plant database that includes lethal and sublethal toxic effects data.

Relevant literature for ECOTOX is retrieved using a comprehensive search strategy. The strategy is designed to locate worldwide aquatic and terrestrial ecological effects literature. Chemical-specific literature search strategies have been used in the past for the TERRETOX and PHYTOTOX databases, but in general, the literature searches intersect general habitat and test condition terms with general effects and end point terms. Exclusion terms are used to eliminate non-applicable studies, but these are examined extensively to ensure that they do not inadvertently eliminate applicable literature. Furthermore, publications are continually and systematically acquired and reviewed. In addition to the published literature, computerized data files from the public sector and available unpublished reports are also acquired and critiqued.

B.2.5.2 Sources of Information on the ECOTOX Database

Further information on ECOTOX may be obtained from the following source: "US Environmental Protection Agency, 2002. ECOTOX User Guide: ECOTOXicology Database System, Version 3.0. [Available at: www.epa.gov/ecotox/]."

B.2.6 OTHER MISCELLANEOUS INFORMATION SOURCES

A variety of other information sources are available to facilitate various risk assessment and/or contaminated site risk management tasks—such as the additional listing provided in the following.

- *ATSDR Toxicological Profiles.* The US ATSDR "Toxicological Profiles" contain information on the toxic effects of chemical exposure in humans and experimental animals. These documents also contain dose–response information for different routes of exposure. When information is available, the Toxicological Profiles also contain a discussion of toxic interactive effects with other chemicals, as well as a description of potentially sensitive human populations. Overall, the ATSDR toxicological profile succinctly characterizes the toxicologic and adverse health effects information for the hazardous substance of interest or concern. Each peer-reviewed profile identifies and reviews the key literature that describes a hazardous substance's toxicologic properties; other pertinent literature is also presented but is described in less detail than the key studies. All in all, the focus of the profile is on health and toxicologic information; thus, each profile begins with a *Public Health Statement* that summarizes in nontechnical language, a substance's relevant properties.

  Further information on and access to the ATSDR toxicological profiles may be obtained via the ATSDR Internet website, or by contacting the ATSDR, US Department of Health and Human Service, Atlanta, GA, United States.
- *The Chemical Substances Information Network (CSIN).* The CSIN is not a database but rather an interactive network system that links together a number of databases relating to several chemical substances. The CSIN accesses data on chemical nomenclature, composition, structure, properties, toxicological information, health and environmental effects, production and uses, regulations, etc.

  The CSIN and the databases it accesses are in the public domain; however, users may have to make independent arrangements with vendors of those databases in the network that needs to be used for specific assignments. Further information on CSIN may be obtained from CIS, Baltimore, Maryland, United States.
- *Health Effects Assessment Summary Tables (HEAST).* The HEAST is a comprehensive listing consisting almost entirely of provisional risk assessment information relative to oral and inhalation routes for various chemical compounds. These entries in the HEAST are limited to analytes that have undergone review and have the concurrence of individual US EPA program offices, and each is supported by an Agency reference. This risk assessment information has not, however, had enough review to be recognized as high quality, Agency-wide consensus information. Chemicals listed as "HEAST Table 2" (HEAST2) were derived from alternate methods that are not currently practiced by the RfD/RfC Work Group. "HEAST Table 2" values consist primarily of inhalation RfC values determined from a methodology that does not follow the interim inhalation methods adopted by the US EPA and RfC or RfD values based on route-to-route extrapolation with inadequate pharmacokinetic and toxicity data.
- *MMEDE (Multimedia-Modeling Environmental Database and Editor).* The MMEDE is a user-friendly database interface for physical, chemical, and toxicological parameters typically associated with environmental assessments. The parameters for evaluation of impacts of hazardous and radioactive materials can be viewed, estimated, modified,

printed, deleted, and exported. The database parameters include physical parameters, dose factors, toxicity factors, environmental transfer factors, environmental decay half times, and other parameters of relevance. A source-of-information citation is used for every parameter value.

Further information on MMEDE may be obtained from Battelle—Pacific Northwest National Laboratory, Richland, Washington, United States.

Further listings of variant tools of potential interest to environmental and contaminated site risk management programs may generally be accessed on the *Internet.*

Appendix C
Toxicological Parameters for Selected Environmental Chemicals

Carcinogenic and noncarcinogenic toxicity indices relevant to the estimation of human health risks—generally represented by the cancer slope factor (SF)-cum-"inhalation unit risk" (IUR) factor and reference dose (RfD)-cum-"reference concentration" (RfC), respectively—are presented in Table C.1 for selected chemical constituents that may be encountered in the human environments. A more complete and up-to-date listing may be obtained from a variety of toxicological databases—such as the Integrated Risk Information System (IRIS), developed and maintained by the US EPA (see Appendix B.2.2).

It is noteworthy that ecological toxicity parameters are purposively excluded in this presentation—especially because they tend to be more unique to the widely varying receptors addressed in ecological risk assessments. In general, information on ecological toxicity parameters—often represented by the toxicity reference values (TRVs)—may be obtained from sources such as the ECOTOX database (see Appendix B.2.5).

TABLE C.1
Toxicological Parameters of Selected Environmental Chemicals

Chemical Name	Toxicity Index			
	Oral *SF* (mg/kg-day)$^{-1}$	Inhalation *UR* (μg/m^3)$^{-1}$	Oral *RfD* (mg/ kg-day)	Inhalation *RfC* (mg/m^3)
Inorganic Chemicals				
Aluminum (Al)			1.0E + 00	5.0E − 03
Antimony (Sb)			4.0E − 04	
Arsenic (As)	1.5E + 00	4.3E − 03	3.0E − 04	1.5E − 05
Barium (Ba)			2.0E − 01	5.0E − 04
Beryllium (Be)		2.4E − 03	2.0E − 03	2.0E − 05
Cadmium (Cd)		1.8E − 03	5.0E − 04	1.0E − 05
Chromium (Cr—total)			1.5E + 00	
Chromium VI (Cr^{+6})[a]	5.0E − 01	1.2E − 02	3.0E − 03	1.0E − 04
Cobalt (Co)		9.0E − 03	3.0E − 04	6.0E − 06
Cyanide (CN)—free			6.0E − 04	8.0E − 04
Manganese (Mn)			1.4E − 01	5.0E − 05
Mercury (Hg)			3.0E − 04	3.0E − 04
Molybdenum (Mo)			5.0E − 03	
Nickel (Ni)		2.6E − 04	2.0E − 02	9.0E − 05
Selenium (Se)			5.0E − 03	2.0E − 02
Silver (Ag)			5.0E − 03	
Thallium (Tl)			1.0E − 05	
Vanadium (V)			5.0E − 03	1.0E − 04
Zinc (Zn)			3.0E − 01	

(Continued)

TABLE C.1 (*Continued*)
Toxicological Parameters of Selected Environmental Chemicals

Chemical Name	Oral *SF* (mg/kg-day)$^{-1}$	Inhalation *UR* (μg/m³)$^{-1}$	Oral *RfD* (mg/kg-day)	Inhalation *RfC* (mg/m³)
		Toxicity Index		
Organic Compounds				
Acetone			9.0E − 01	3.1E + 01
Alachlor	5.2E − 02		1.0E − 02	
Aldicarb			1.0E − 03	
Anthracene			3.0E − 01	
Atrazine	2.3E − 01		3.5E − 02	
Benzene	5.5E − 02	7.8E − 06	4.0E − 03	3.0E − 02
Benzo(*a*)anthracene	7.3E − 01	1.1E − 04		
Benzo(*a*)pyrene [BaP]	7.3E + 00	1.1E − 03		
Benzo(*b*)fluoranthene	7.3E − 01	1.1E − 04		
Benzo(*k*)fluoranthene	7.3E − 02	1.1E − 04		
Benzoic acid			4.0E + 00	
Bis(2-ethylhexyl)phthalate	1.4E − 02	2.4E − 06	2.0E − 02	
Bromodichloromethane	6.2E − 02	3.7E − 05	2.0E − 02	
Bromoform	7.9E − 03	1.1E − 06	2.0E − 02	
Carbon disulfide			1.0E − 01	7.0E − 01
Carbon tetrachloride	7.0E − 02	6.0E − 06	4.0E − 03	1.0E − 01
Chlordane	3.5E − 01	1.0E − 04	5.0E − 04	7.0E − 04
Chlorobenzene			2.0E − 02	5.0E − 02
Chloroform	3.1E − 02	2.3E − 05	1.0E − 02	9.8E − 02
2-Chlorophenol			5.0E − 03	
Chrysene	7.3E − 03	1.1E − 05		
o-Cresol [2-Methylphenol]			5.0E − 02	6.0E − 01
m-Cresol [3-Methylphenol]			5.0E − 02	6.0E − 01
Cyclohexanone			5.0E + 00	7.0E − 01
1,4-Dibromobenzene			1.0E − 02	
Dibromochloromethane	8.4E − 02		2.0E − 02	
1,2-Dibromomethane[EDB]	2.0E + 00	6.0E − 04	9.0E − 03	9.0E − 03
1,2-Dichlorobenzene			9.0E − 02	2.0E − 01
Dichlorodifluoromethane			2.0E − 01	1.0E − 01
p,p′-Dichlorodiphenyl-dichloroethane [DDD]	2.4E − 01	6.9E − 05		
p,p′-Dichlorodiphenyl-dichloroethylene [DDE]	3.4E − 01	9.7E − 05		
p,p′-Dichlorodiphenyl-trichloroethane [DDT]	3.4E − 01	9.7E − 05	5.0E − 04	
1,1-Dichloroethane	5.7E − 03	1.6E − 06	2.0E − 01	
1,2-Dichloroethane	9.1E − 02	2.6E − 05	6.0E − 03	7.0E − 03
1,1-Dichloroethene			5.0E − 02	2.0E − 01
cis-1,2-Dichloroethene			2.0E − 03	

(*Continued*)

TABLE C.1 (*Continued*)
Toxicological Parameters of Selected Environmental Chemicals

		Toxicity Index		
Chemical Name	Oral *SF* (mg/kg-day)$^{-1}$	Inhalation *UR* (μg/m^3)$^{-1}$	Oral *RfD* (mg/kg-day)	Inhalation *RfC* (mg/m^3)
trans-1,2-Dichloroethene			2.0E − 02	
2,4-Dichlorophenol			3.0E − 03	
Dieldrin	1.6E + 01	4.6E − 03	5.0E − 05	
Di(2-ethylhexyl)phthalate [DEHP]	1.4E − 02	2.4E − 06	2.0E − 02	
Diethyl phthalate			8.0E − 01	
2,4-Dimethylphenol			2.0E − 02	
2,6-Dimethylphenol			6.0E − 04	
3,4-Dimethylphenol			1.0E − 03	
m-Dinitrobenzene			1.0E − 04	
1,4-Dioxane	1.1E − 01	5.0E − 06	3.0E − 02	3.0E − 02
Endosulfan			6.0E − 03	
Endrin			3.0E − 04	
Ethylbenzene	1.1E − 02	2.5E − 06	1.0E − 01	1.0E + 00
Ethyl chloride (chloroethane)				1.0E + 01
Ethyl ether			2.0E − 01	
Ethylene glycol			2.0E + 00	
Fluoranthene			4.0E − 02	
Fluorene			4.0E − 02	
Formaldehyde		1.3E − 05	2.0E − 01	9.8E − 03
Furan			1.0E − 03	
Heptachlor	4.5E + 00	1.3E − 03	5.0E − 04	
Hexachlorobenzene	1.6E + 00	4.6E − 04	8.0E − 04	
Hexachlorodibenzo-*p*-dioxin [HxCDD]	6.2E + 03	1.3E + 00		
Hexachloroethane	4.0E − 02	1.1E − 05	7.0E − 04	3.0E − 02
n-Hexane				7.0E − 01
Indeno(1,2,3-*c,d*)pyrene	7.E − 01	1.1E − 04		
Isobutyl alcohol			3.0E − 01	
Lindane [gamma-HCH]	1.1E + 00	3.1E − 04	3.0E − 04	
Malathion			2.0E − 02	
Methanol			2.0E + 00	2.0E + 01
Methyl mercury			1.0E − 04	
Methyl parathion			2.5E − 04	
Methylene chloride [dichloromethane]	2.0E − 03	1.0E − 08	6.0E − 03	6.0E − 01
Methyl ethyl ketone [MEK]			6.0E − 01	5.0E + 00
Methyl isobutyl ketone [MIBK]				3.0E + 00
Mirex	1.8E + 01	5.1E − 03	2.0E − 04	
Nitrobenzene		4.5E − 05	2.0E − 03	9.0E − 03
n-Nitroso-di-*n*-butylamine	5.4E + 00	1.6E − 03		
n-Nitroso-di-*n*-methylethylamine	2.2E + 01	6.3E − 03		
n-Nitroso-di-*n*-propylamine	7.0E + 00	2.0E − 03		

(Continued)

TABLE C.1 (*Continued*)
Toxicological Parameters of Selected Environmental Chemicals

Chemical Name	Oral *SF* (mg/kg-day)$^{-1}$	Inhalation *UR* (μg/m^3)$^{-1}$	Oral *RfD* (mg/kg-day)	Inhalation *RfC* (mg/m^3)
		Toxicity Index		
n-Nitrosodiethanolamine	2.8E + 00	8.0E − 04		
n-Nitrosodiethylamine	1.5E + 02	4.3E − 02		
n-Nitrosodimethylamine	5.1E + 01	1.4E − 02		
n-Nitrosodiphenylamine	4.9E − 03	2.6E − 06		
Pentachlorobenzene			8.0E − 04	
Pentachlorophenol	4.0E − 01	5.1E − 06	5.0E − 03	
Phenol			3.0E − 01	2.0E − 01
Polychlorinated biphenyls [PCBs][b]	7.0E − 02 to 2.0E + 00	2.0E − 05 to 5.7E − 04		
Pyrene			3.0E − 02	
Styrene			2.0E − 01	1.0E + 00
1,2,4,5-Tetrachlorobenzene			3.0E − 04	
1,1,1,2-Tetrachloroethane	2.6E − 02	7.4E − 06	3.0E − 02	
1,1,2,2-Tetrachloroethane	2.0E − 01	5.8E − 05	2.0E − 02	
Tetrachloroethene	2.1E − 03	2.6E − 07	6.0E − 03	4.0E − 02
2,3,4,6-Tetrachlorophenol			3.0E − 02	
Toluene			8.0E − 02	5.0E + 00
Toxaphene	1.1E + 00	3.2E − 04		
1,2,4-Trichlorobenzene	2.9E − 02		1.0E − 02	2.0E − 03
1,1,1-Trichloroethane			2.0E + 00	5.0E + 00
1,1,2-Trichloroethane	5.7E − 02	1.6E − 05	4.0E − 03	2.0E − 04
Trichloroeth[yl]ene	4.6E − 02	4.1E − 06	5.0 E − 04	2.0E − 03
1,1,2-Trichloro-1,2,2- trifluoroethane [CFC-113]			3.0E + 01	
Trichlorofluoromethane			3.0E − 01	
2,4,5-Trichlorophenol			1.0E − 01	
2,4,6-Trichlorophenol	1.1E − 02	3.1E − 06	1.0E − 03	
1,1,2-Trichloropropane			5.0E − 03	
1,2,3-Trichloropropane	3.0E + 01		4.0E − 03	3.0E − 04
Triethylamine				7.0E − 03
1,3,5-Trinitrobenzene			3.0E − 02	
2,4,6-Trinitrotoluene [TNT]	3.0E − 02		5.0E − 04	
o-Xylene			2.0E − 01	1.0E − 01
Xylenes (mixed)			2.0E − 01	1.0E − 01
Others				
Asbestos (units of per fibers/mL)[c]		2.3E − 01		
Hydrazine	3.0E + 00	4.9E − 03		3.0E − 05

(*Continued*)

TABLE C.1 (*Continued*)
Toxicological Parameters of Selected Environmental Chemicals

| | | Toxicity Index | | |
| | | | | |
Chemical Name	Oral *SF* (mg/kg-day)$^{-1}$	Inhalation UR (µg/m^3)$^{-1}$	Oral *RfD* (mg/kg-day)	Inhalation *RfC* (mg/m^3)
Hydrogen chloride				2.0E − 02
Hydrogen cyanide			6.0E − 04	8.0E − 04
Hydrogen sulfide				2.0E − 03

Notes:

[a] IUR = 8×10^{-6} mg/m^3 (for exposure to Cr^{+6} acid mists and dissolved aerosols); and IUR = 1×10^{-4} mg/m^3 (for exposure to Cr^{+6} particulate matter).

[b] Tiers of human potency and slope estimates exist for environmental mixtures of PCBs. For instance, for high risk and persistent PCB congeners or isomers, an upper-bound slope of 2.0E + 00 per mg/kg-day and a central slope of 1.0E + 00 per mg/kg-day may be used for the oral SF, etc.; for low risk and persistent PCBs, an upper-bound slope of 4E − 01 per mg/kg-day and a central slope of 3E − 01 per mg/kg-day may be used for the oral SF, etc.; and for the lowest risk and persistent PCBs, an upper-bound slope of 7E − 02 per mg/kg-day and a central slope of 4E − 02 per mg/kg-day may be used for the oral SF, etc.

[c] Note the different set of units applied here; also, 2.3E − 01 per fibers/mL ≡ 2.3E − 07 per fibers/m^3. It is also noteworthy that, regulatory agencies (such as the California EPA) have used a significantly more restrictive value of 1.9 per fibers/mL (≡1.9E − 06 per fibers/m^3) as the inhalation SF for asbestos in some risk management and remedy decisions.

Appendix D
Selected Units of Measurement and Noteworthy Expressions

Some selected units of measurements and noteworthy expressions (of potential interest to the environmental professional, analyst, or decision maker) are provided in the following:

D.1 MASS/WEIGHT UNITS

g	gram(s)
ton (metric)	tonne = 1×10^6 g
Mg	Megagram(s), metric ton(s) = 10^6 g
kg	kilogram(s) = 10^3 g
mg	milligram(s) = 10^{-3} g
μg	microgram(s) = 10^{-6} g
ng	nanogram(s) = 10^{-9} g
pg	picogram(s) = 10^{-12} g
mol	mole, molecular weight (mol. wt.) in grams

D.2 VOLUMETRIC UNITS

cc or cm^3	cubic centimeter(s) $\approx$ 1 mL = 10^{-3} L
mL	milliliter(s) = 10^{-3} L
L	liter(s) = 10^3 cm^3
m^3	cubic meter(s) = 10^3 L

D.3 ENVIRONMENTAL/CHEMICAL CONCENTRATION UNITS

ppm	parts per million
ppb	parts per billion
ppt	parts per trillion

These are used for expressing/specifying the relative masses of contaminant within a given exposure matrix/medium. It is worth mentioning here that because water is necessarily assigned a mass of 1 kilogram per liter, mass-to-mass and mass-to-volume measurements are interchangeable for this particular medium.

NB: (1) A billion is often used to represent a thousand millions [i.e., 10^9] in some regions, such as in the United States and France; whereas a billion represents a million millions [i.e., 10^{12}] in some other jurisdictions, such as in the United Kingdom and Germany. (2) A trillion is often used to represent a million times a million or a thousand billions [i.e., 10^{12}] in some regions, such as in the United States and France; whereas a trillion represents a million billions [i.e., 10^{18}] in some other jurisdictions, such as in the United Kingdom and Germany.

D.4 CONCENTRATION EQUIVALENTS

1 ppm	$\equiv$	mg/kg or mg/L $\equiv 10^{-6}$
1 ppb	$\equiv$	μg/kg or μg/L $\equiv 10^{-9}$
1 ppt	$\equiv$	ng/kg or ng/L $\equiv 10^{-12}$

D.4.1 CONCENTRATIONS IN SOLID MEDIA [E.G., SOILS]

Mg/kg	mg chemical per kg weight of sampled solid medium
μg/kg	μg chemical per kg weight of sampled solid medium

D.4.2 CONCENTRATIONS IN WATER OR OTHER LIQUID MEDIA

Mg/L	mg chemical per liter of total liquid volume
μg/L	μg chemical per liter of total liquid volume

D.4.3 CONCENTRATIONS IN AIR MEDIA

Mg/m^3	mg chemical per m^3 of total fluid volume
μg/m^3	μg chemical per m^3 of total fluid volume

To convert from ppm to mg/m^3 use the following conversion relationship:

$$\left[mg/m^3 \right] = \left[ppm \right] \times \frac{\left[molecular\,weight\,of\,substance, in\,g/mol \right]}{24.45}$$

To convert from ppm to μg/m^3, use the following conversion relationship:

$$\left[μg/m^3 \right]\left[ppm \right] \times \left[molecular\,weight\,of\,substance, in\,g/mol \right] \times 40.9$$

Note: The above conversion relationships assume standard temperature and pressure (STP), that is, temperature of 25°C and barometric pressure of 760 mm Hg (or 1 atm). More generally, to convert from ppm to mg/m^3, the following equation can be used:

$$mg/m^3 = \frac{\left[ppm \times MW \right]}{V}$$

where MW is the molecular weight of the gas, and V is the volume of 1 gram molecular weight of the airborne contaminant under review. This is further derived by using the formula: V = RT/P, where R is the ideal gas constant, T is the temperature in Kelvin (K = 273.16 + T°C), and P is the pressure in mm Hg. The value of R is 62.4 when T is in Kelvin, the pressure is expressed in units of mm Hg, and the volume is in liters. The value of R differs if the temperature is expressed in degrees Fahrenheit (°F) or if other units of pressure are used (e.g., atmospheres, kilopascals).

D.5 UNITS OF CHEMICAL INTAKE AND DOSE

mg/kg-day = milligrams of chemical exposure per unit body weight of exposed receptor per day

D.6 TYPICAL EXPRESSIONS COMMONLY USED IN RISK ASSESSMENT AND ENVIRONMENTAL MANAGEMENT PROGRAMS

D.6.1 ORDER OF MAGNITUDE

Reference to an "order of magnitude" means a tenfold difference or a multiplicative factor of ten—that is, the base parameter may vary by a factor of 10. Hence, "two orders of magnitude" means a factor of about 100; "three orders of magnitude" implies a factor of about 1,000; and so on. For example, "three orders of magnitude" may be used to describe the difference between 3 and 3,000 ($=3 \times 10^3$). The expression is often used in reference to the calculation of environmental quantities or risk probabilities.

D.6.2 EXPONENTIALS DENOTED BY 10^K

Superscript refers to the number of times 10 that is multiplied by itself. For example, $10^2 = 10 \times 10 = 100$; $10^3 = 10 \times 10 \times 10 = 1,000$; $10^6 = 10 \times 10 \times 10 \times 10 \times 10 \times 10 = 1,000,000$. [*NB: It is notable that 10^0 is equivalent to 1.*]

D.6.3 EXPONENTIALS DENOTED BY 10^{-K}

Negative superscript is equivalent to the reciprocal of the positive term, that is, $10^{-\kappa}$ is equivalent to $1/10^\kappa$. For example, $10^{-1} = 1/10^1 = 1/10 = 0.1$; $10^{-2} = 1/10^2 = 1/(10 \times 10) = 0.01$; $10^{-3} = 1/10^3 = 1/(10 \times 10 \times 10) = 0.001$; $10^{-6} = 1/10^6 = 1/(10 \times 10 \times 10 \times 10 \times 10 \times 10) = 0.000001$.

D.6.4 EXPONENTIALS DENOTED BY X.YZ E+K

Number after the E indicates the power to which 10 is raised and then multiplied by the preceding term (i.e., the number of times 10^κ is multiplied by preceding term, or $X.YZ \times 10^\kappa$). For example, $1.00E-01 = 1.00 \times 10^{-1} = 0.1$; $1.23E+04 = 1.23 \times 10^{+4} = 12,300$; $4.44E+05 = 4.44 \times 10^5 = 444,000$.

D.6.5 CONSERVATIVE ASSUMPTION

Used in exposure and risk assessment, this expression refers to the selection of assumptions (when real-time data are absent) that are unlikely to lead to underestimation of exposure or risk. Conservative assumptions are those which tend to maximize estimates of exposure or dose—such as choosing a value near the high end of the concentration or intake rate range.

D.6.6 WORST CASE

A semi-qualitative term that refers to the maximum possible exposure, dose, or risk to an exposed person or group that could conceivably occur—that is, regardless of whether or not this exposure, dose, or risk actually occurs or is observed in a specific population. Typically, this would refer to a hypothetical situation in which everything that can plausibly happen to maximize exposure, dose, or risk actually takes place. Under some circumstances, this worst case may indeed occur (or may even be observed) in a given population; however, since this is usually a very unlikely set of circumstances, in most cases, a worst-case estimate will tend to be somewhat higher than what actually occurs in a specific population. In most health risk assessments, a "worst-case scenario" is essentially a type of "upper-bounding" estimate.

D.6.7   RISK OF 1×10^{-6} (OR SIMPLY, 10^{-6})

Also written as 0.000001, or one in a million, means that one additional case of cancer risk is projected in a population of one million people exposed to a certain level of chemical X over their lifetimes. Similarly, a risk of 5×10^{-3} corresponds to 5 in 1,000 or 1 in 200 persons; and a risk of 2×10^{-6} means two chances in a million of the exposure causing cancer.

Bibliography

Abbaspour, KC, R Schulin, E Schläppi, and H Flühler, 1996. A Bayesian approach for incorporating uncertainty and data worth in environmental projects, *Environmental Modelling and Assessment*, 1(3/4): 151–158.

Acar, YB and DE Daniel (eds.), 1995. *Geoenvironment 2000: Characterization, Containment, Remediation, and Performance in Environmental Geotechnics, Volumes 1 and 2*, ASCE Geotechnical Special Publication No. 46, American Society of Civil Engineers, New York.

ACS (American Chemical Society), 1999. Innovative subsurface remediation: field testing of physical, chemical, and characterization technologies. In ACS Symposium Series 725, Brusseau, ML, Sabatini, DA, Gierke, JS, and Annable, MD (eds.), American Chemical Society, Washington, DC.

ACS, RFF, and Boroush, M, 1998. *Understanding Risk Analysis (A Short Guide for Health, Safety and Environmental Policy Making)*, A Publication of the American Chemical Society (ACS) and Resources for the Future (RFF), Washington, DC.

Adeel, Z, RG Luthy, and DA Edwards, 1995. Modeling transport of multiphase organic compounds: segregated transport-sorption/solubilization numerical techniques, *Water Resources Research*, 31(8): 2035–2045.

Ahmad, YJ, SE El Serafy, and E Lutz, 1989. *Environmental Accounting for Sustainable Development*, The World Bank, Washington, DC.

AIHC, 1994. *Exposure Factors Sourcebook*, American Industrial Health Council (AIHC), Washington, DC.

Aitchison, J and JAC Brown, 1969. *The Lognormal Distribution*, Cambridge University Press, Cambridge, UK.

Alberta Environment and Alberta Labor, 1989. *Subsurface Remediation Guidelines for Underground Storage Tanks*, Draft (August, 1989), Alberta MUST (Management of Underground Storage Tanks) Project, A Joint Project of the Departments of Environment and Labour, Edmonton, AB.

Alder, T, 1996. Botanical cleanup crews, *Science News*, 150: 42–43.

Allan, M and GM Richardson, 1998. Probability density functions describing 24-hour inhalation rates for use in human health risk assessments, *Human and Ecological Risk Assessment*, 4(2): 379–408.

Allen, AO, 1990. *Probability, Statistics and Queueing Theory with Computer Science Applications*, 2nd edition, Academic Press, Boston, MA.

Allen, D, 1996. *Pollution Prevention for Chemical Processes*, John Wiley & Sons, New York.

Allen, RG, ME Jensen, JL Wright, and RD Burman, 1989. Operational estimates of reference evapotranspiration, *Agronomy Journal*, 81: 650–662.

Alloway, BJ and DC Ayres, 1993. *Chemical Principles of Environmental Pollution*, Blackie Academic & Professional/Chapman & Hall, London, UK.

Al-Saleh, IA and L Coate, 1995. Lead exposure in Saudi Arabia from the use of traditional cosmetics and medical remedies, *Environmental Geochemistry and Health*, 17: 29–31.

Ananichev, K, 1976. *Environment: International Aspects*, Progress Publishers, Moscow, Russia.

Andelman, JB and DW Underhill, 1988. *Health Effects From Hazardous Waste Sites*, Lewis Publishers, Chelsea, MI.

Anderson, MP and WW Woessner, 1992. *Applied Groundwater Modeling: Simulation of Flow and Advective Transport*, Academic Press, Cambridge, MA.

Anderson, PS and AI Yuhas, 1996. Improving risk management by characterizing reality: a benefit of probabilistic risk assessment, *Human and Ecological Risk Assessment*, 2: 55–58.

Anderson, TA and BT Walton, 1991. Fate of trichloroethylene in soil-plant systems, *American Chemical Society Extended Abstract, Division of Environmental Chemistry*, pp. 197–200.

Anderson, TA and BT Walton, 1995. Comparative fate of ^{14}C trichloroethylene in the root zone of plants from a former solvent disposal site, *Environmental Toxicology and Chemistry*, 14: 2041–2047.

Ankley, GT, DM Di Toro, DJ Hansen, and WJ Berry, 1996. Technical basis and proposal for deriving sediment quality criteria for metals, *Environmental Toxicology and Chemistry*, 15(2): 2056–2066.

Annable, MD, PSC Rao, DP Dai, K Hatfield, W Graham, AL Wood, and CG Entfield, 1998. Partitioning Tracers for Measuring Residual NAPL: field scale test results, *Journal of Environmental Engineering*, 124(6): 498–502.

ARB (Air Resources Board), 1991. *Soil Decontamination*, Compliance Assistance Program, Air Resources Board, Compliance Division, California.

Aris, R, 1994. *Mathematical Modelling Techniques*, Dover Publications, Inc., New York.

Asante-Duah, DK and IV Nagy, 1997. *International Trade in Hazardous Waste*, E&FN Spon/Chapman & Hall, London, UK.

Asante-Duah, DK, 1990. Quantitative Risk Assessment as a Decision Tool for Hazardous Waste Management, In *Proceedings of 44th Purdue Industrial Waste Conference, May, 1989*, Lewis Publishers, Chelsea, MI, pp. 111–123.

Asante-Duah, DK, 1996a. *Management of Contaminated Site Problems*, CRC Press, Boca Raton, FL.

Asante-Duah, DK, 1996b. *Managing Contaminated Sites: Problem Diagnosis and Development of Site Restoration*, John Wiley & Sons, Chichester, UK.

Asante-Duah, DK, 1998. *Risk Assessment in Environmental Management: A Guide for Managing Chemical Contamination Problems*, John Wiley & Sons, Chichester, UK.

Asante-Duah, DK, DS Bowles, and LR Anderson, 1991. Framework for the Risk Analysis of Hazardous Waste Facilities, In *Proceedings of Sixth International Conference on Applications of Statistics and Probability in Civil Engineering, CERRA/ICASP 6*, Mexico.

Asante-Duah, K, 2017. *Public Health Risk Assessment for Human Exposure to Chemicals*, 2nd edition, Springer Publishers, Dordrecht, The Netherlands.

Ashford, NA and CS Miller, 1998. *Chemical Exposures: Low Levels and High Stakes*, John Wiley & Sons, New York.

Ashford, NA, and CC Caldart, 2008. *Environmental Law, Policy, and Economics: Reclaiming the Environmental Agenda*. MIT Press, Cambridge.

ASTM, 1994a. *Risk-based Corrective Action Guidance*, American Society for Testing and Materials (ASTM), Philadelphia, PA.

ASTM, 1994b. *Standard Guide for Comparing Ground-Water Flow Model Simulations to Site-Specific Information Standard*, ASTM D5490-93, American Society for Testing and Materials (ASTM), Philadelphia, PA.

ASTM, 1994c. *Standard Guide for Conducting a Sensitivity Analysis for a Ground-Water Flow Model Application*, ASTM D5611-94, American Society for Testing and Materials (ASTM), Philadelphia, PA.

ASTM, 1994d. *Standard Guide for Defining Boundary Conditions in Ground-Water Flow Modeling*, Standard, ASTM D5609-94, American Society for Testing and Materials (ASTM), Philadelphia, PA.

ASTM, 1995a. *Standard Guide for Developing Conceptual Site Models for Contaminated Sites*, Designation E1689-95. 1995 Annual Book of ASTM Standards, American Society for Testing and Materials (ASTM), Philadelphia, PA.

ASTM, 1995b. *Standard Guide for Risk-based Corrective Action Applied at Petroleum-Release Sites*, ASTM (E1739-95), American Society for Testing and Materials (ASTM), Philadelphia, PA.

ASTM, 1995c. *Standard Guide for Subsurface Flow and Transport Modeling*, ASTM D5880-95, American Society for Testing and Materials (ASTM), Philadelphia, PA.

ASTM, 1995d. *Standard Guide for Documenting a Ground-Water Flow Model Application*, ASTM D5718-95, American Society for Testing and Materials (ASTM), Philadelphia, PA.

ASTM, 1996a. *Standard Practice for Generation of Environmental Data Related to Waste Management Activities: Development of Data Quality Objectives*, ASTM Standards D5792-95, American Society for Testing and Materials (ASTM), Philadelphia, PA.

ASTM, 1996b. *Standard Guide for Calibrating a Ground-Water Flow Model Application*, ASTM D5981-96, American Society for Testing and Materials (ASTM), Philadelphia, PA.

ASTM, 1997a. *ASTM Standards on Environmental Sampling*, 2nd edition, American Society for Testing and Materials (ASTM), Philadelphia, PA.

ASTM, 1997b. *ASTM Standards Related to Environmental Site Characterization*, American Society for Testing and Materials (ASTM), Philadelphia, PA.

ASTM, 1998a. *Standard Guide for Remediation of Ground Water by Natural Attenuation at Petroleum Release Sites*, ASTM Standards E1943-98, American Society for Testing and Materials (ASTM), Philadelphia, PA.

ASTM, 1998b. *Standard Guide for Selecting a Ground-Water Modeling Code. Standard*, ASTM D6170-97, American Society for Testing and Materials (ASTM), Philadelphia, PA.

ASTM, 1999. *Standard Guide for Developing Appropriate Statistical Approaches for Ground-Water Detection Monitoring Programs*, ASTM D6312-98, American Society for Testing and Materials (ASTM), Philadelphia, PA.

Aswathanarayana, U, 1995. *Geoenvironment: An Introduction*, A.A. Balkema Publishers, Rotterdam, The Netherlands.

ATSDR, 1990–Present. *ATSDR Case Studies in Environmental Medicine*, Agency for Toxic Substances and Disease Registry (ATSDR), US Department of Health and Human Service, Atlanta, GA.

ATSDR, 1994. *Priority Health Conditions: An Integrated Strategy to Evaluate the Relationship between Illness and Exposure to Hazardous Substances*, Agency for Toxic Substances and Disease Registry (ATSDR), US Department of Health and Human Service, Atlanta, GA.

ATSDR, 1997. *Technical Support Document for ATSDR Interim Policy Guideline: Dioxin and Dioxin-like Compounds in Soil*, Agency for Toxic Substances and Disease Registry (ATSDR), US Department of Health and Human Services, Atlanta, GA.

ATSDR, 1999a. *Toxicological Profile for Lead*, Agency for Toxic Substances and Disease Registry (ATSDR), US Department of Health and Human Services, Atlanta, GA.

ATSDR, 1999b. *Toxicological Profile for Mercury*, Agency for Toxic Substances and Disease Registry, US Department of Health and Human Service, Atlanta, GA.

Awasthi, S, HA Glick, RH Fletcher, and N Ahmed, 1996. Ambient air pollution and respiratory symptoms complex in preschool children, *Indian Journal of Medical Research*, 104: 257–262.

Baasel, WD, 1985. *Economic Methods for Multipollutant Analysis and Evaluation*, Marcel Dekker, New York.

Bailar III, JC, J Needleman, BL Berney, and JM McGinnis (eds.), 1993. *Assessing Risks to Health—Methodological Approaches*, Auburn House, Westport, CT.

Bailar, J, 1991. How dangerous is dioxin? *New England Journal of Medicine*, 324: 260–262.

Baker AJM, RR Brooks and RD Reeves, 1998. Growing for gold and copper and zinc, *New Science*, 177: 44–48.

Bansal, RC and M Goyal, 2005. *Activated Carbon Adsorption*, CRC Press, Boca Raton, FL.

Baran, JR, Jr., GA Pope, C Schultz, WH Wade, V Weerasooriya, and A Yapa, 1996. Toxic spill remediation of chlorinated hydrocarbons via microemulsion formation. In *Surfactants in Solution*, AK Chattopadhyay and KL Mittal (eds.), pp. 393–411, Marcel Dekker, New York.

Baran, JR, Jr., GA Pope, WH Wade, and V Weerasooriya, 1996. Surfactant systems for soil and aquifer remediation of JP4 jet fuel, *Journal of Dispersion Science Technology*, 17(2): 131–138.

Baran, JR, Jr., GA Pope, WH Wade, V Weerasooriya, and A Yapa, 1994. Microemulsion formation with chlorinated hydrocarbon liquids, *Journal of Colloid Interface Science*, 168(1): 67.

Baran, JR, Jr., GA Pope, WH Wade, V Weerasooriya, and A Yapa, 1994. Microemulsion formation with chlorinated hydrocarbons of different polarity liquids, *Environmental Science Technology*, 28: 1361–1366.

Barnard, R and G Olivetti, 1990. Rapid assessment of industrial waste production based on available employment statistics, *Waste Management and Research*, 8(2): 139–144.

Barnes, DG and M Dourson, 1988. Reference dose (RfD): description and use in health risk assessments, *Regulatory Toxicology and Pharmacology*, 8: 471–486.

Barnhart, M, 1992. Bioremediation: do it yourself, *Soils*, August–September 1992, pp. 14–19.

Barnthouse, LW and GW Suter II (eds.), 1986. *User's Manual for Ecological Risk Assessment*, ORNL-6251, Oak Ridge National Lab., Oak Ridge, TN.

Barnthouse, LW, 1994. Issues in ecological risk assessment: the Committee on Risk Assessment Methodology (CRAM) perspective, *Risk Analysis*, 14(3): 251–256.

Barrie, LA, D Gregor, B Hargrave, R Lake, D Muir, R Shearer, B Tracey, and T Biddleman, 1992. Arctic contaminants: sources, occurrence and pathways, *Science of the Total Environment*, 122: 1–74.

Barry, PSI, 1975. A comparison of concentrations of lead in human tissue, *British Journal of Industrial Medicine*, 32: 119–139.

Barry, PSI, 1981. Concentrations of lead in the tissues of children, *British Journal of Industrial Medicine*, 38: 61–71.

Bartell, SM, RH Gardner, and RV O'Neill, 1992. *Ecological Risk Estimation*, Lewis Publishers, Chelsea, MI.

Bartram, J and R Balance (eds.), 1996. *Water Quality Monitoring*, E & FN Spon/Chapman & Hall, London, UK.

Batchelor, B, J Valdés, and V Araganth, 1998. Stochastic risk assessment of sites contaminated by hazardous wastes, *Journal of Environmental Engineering*, 124(4): 380–388.

Bate, R (ed.), 1997. *What Risk? (Science, Politics & Public Health)*, Butterworth-Heinemann, Oxford, UK.

Bates, DV, 1992. Health indices of the adverse effects of air pollution: the question of coherence, *Environmental Research*, 59: 336–349.

Bates, DV, 1994. *Environmental Health Risks and Public Policy*, University of Washington Press, Seattle, WA.

Bates, MN, A Smith, and K Cantor, 1995. Case-control study of bladder cancer and arsenic in drinking water, *American Journal of Epidemiology*, 141: 523–530.

Batstone, R, JE Smith, Jr., and D Wilson (eds.), 1989. The Safe Disposal of Hazardous Wastes—The Special Needs and Problems of Developing Countries, *Volumes 1–3*, A Joint Study Sponsored by the World Bank, the World Health Organization (WHO), and the United Nations Environment Programme (UNEP), World Bank Technical Paper 0253-7494, No. 93, The World Bank, Washington, DC.

Battelle, 1993. *The Second International In Situ and On-Site Bioremediation Symposium [Five Voulumes]*, Battelle Press: Columbus, OH.

Battelle, 1995. *The Third International In Situ and On-Site Bioremediation Symposium [Eleven Voulumes]*, Battelle Press: Columbus, OH.

Battelle, 1997. *The Fourth International In Situ and On-Site Bioremediation Symposium [Five Voulumes]*, Battelle Press: Columbus, OH.

Battelle, 1998. *The First International Conference on Remediation of Chlorinated and Recalcitrant Compounds [Six Voulumes]*, Battelle Press: Columbus, OH.

Battelle, 2000. Recommendations for Modeling the Geochemical and Hydraulic Performance of a Permeable Reactive Barrier, Prepared for Naval Facilities Engineering Service Center, Port Hueneme, CA.

Baum, EJ, 1998. *Chemical Property Estimation: Theory and Application*, Lewis Publishers, Boca Raton, FL.

Bean, MC, 1988. Speaking of risk, *ASCE Civil Engineers*, 58(2): 59–61.

Bear, J, 1972. *Flow through Porous Media*, Elsevier, New York.

Beck, LW, AW Maki, NR Artman, and ER Wilson, 1981. Outline and criteria for evaluating the safety of new chemicals, *Regulatory Toxicology Pharmacology*, 1: 19–58.

Begon, M, JL Harper, and CR Townsend, 1986. *Ecology: Individuals Populations, and Communities*, Sinauer Associates, Sunderland, MA.

Belzer, RB, 2002. Exposure assessment at a crossroads: the risk of success, *Journal of Exposure Analysis and Environmental Epidemiology*, 12(2): 96–103.

Benedetti, M, I Lavarone, and P Comba, 2001. Cancer risk associated with residential proximity to industrial sites: a review, *Archives of Environmental Health*, 56(4): 342–349.

Bennett, DH, TE McKone, JS Evans, WW Nazaroff, MD Margni, O Jolliet, and KR Smith, 2002. Defining intake fraction, *Environmental Science and Technology*, 36(9): 206A–211A.

Bentkover, JD, VT Covello, and J Mumpower, 1986. *Benefits Assessment: The State of the Art*, Riedel Publishing, Boston, MA.

Berlow, PP, DJ Burton, and JI Routh, 1982. *Introduction to the Chemistry of Life*, Saunders College Publishing, Philadelphia, PA.

Berne, RM and MN Levy, 1993. *Physiology*, 3rd edition, Mosby Year Book, St. Louis, MO.

Bernhardsen, T, 1992. *Geographic Information Systems*, Viak IT, Arendal, Norway.

Bertazzi, PA, L Riboldi, A Pesatori, L Radice, and C Zocchetti, 1987. Cancer mortality of capacitor manufacturing workers, *American Journal of Industrial Medicine*, 11: 165–176.

Berthouex, PM and LC Brown, 1994. *Statistics for Environmental Engineers*, CRC Press, Boca Raton, FL.

Berthouex, PM and LC Brown, 2002. *Statistics for Environmental Engineers*, 2nd edition, CRC Press, Boca Raton, FL.

Bertollini, R, MD Lebowitz, R Saracci, and DA Savitz (eds.), 1996. *Environmental Epidemiology (Exposure and Disease)*, CRC Press, Boca Raton, FL.

Bharagava, RN, 2017. *Environmental Pollutants and their Bioremediation Approaches*, CRC Press, Boca Raton, FL.

Bhatt, HG, RM Sykes, and TL Sweeney (eds.), 1986. *Management of Toxic and Hazardous Wastes*, Lewis Publishers, Chelsea, MI.

Binder, S, D Sokal, and D Maughan, 1986. Estimating the amount of soil ingested by young children through tracer elements, *Archives of Environmental Health*, 41(6): 341–345.

Blancato, JN, RN Brown, CC Dary, and MA Saleh (eds.), 1996. *Biomarkers for Agrochemicals and Toxic Substances: Applications and Risk Assessment*, ACS Symposium Series No. 643, American Chemical Society, Washington, DC.

Blumenthal, DS (ed.), 1985. *Introduction to Environmental Health*, Springer Publishing Co., New York.

BMA (British Medical Association), 1991. *Hazardous Waste and Human Health*, Oxford University Press, Oxford, UK.

BMA, 1998. *Health & Environmental Impact Assessment (An Integrated Approach)*, Earthscan Publications, London, UK.

BNL, 1999. *Striking Results from Brookhaven Ecology Facility: Trees Grow Faster in Simulated Future CO_2-Rich Atmosphere*, Brookhaven National Laboratory (BNL) Report 99-43. [Available at: www.bnl.gov/bnlweb/pubaf/pr/bnlpr051399.html].

Bogen, KT, 1990. *Uncertainty in Environmental Health Risk Assessment*, Garland Publishing, New York.

Bogen, KT, 1994. A note on compounded conservatism, *Risk Analysis*, 14: 379–381.

Bogen, KT, 1995. Methods to approximate joint uncertainty and variability in risk, *Risk Analysis*, 15(3): 411–419.

Boulding, JR and JS Ginn, 2003. *Practical Handbook of Soil, Vadose Zone, and Ground-Water Contamination: Assessment, Prevention, and Remediation*, 2nd edition, CRC Press, Boca Raton, FL.

Boulding, JR, 1994. *Description and Sampling of Contaminated Soils (A Field Manual)*, 2nd edition, CRC Press, Boca Raton, FL.

Bourrel, M and RS Schechter, 1988. *Microemulsions and Related Systems*, Marcel Dekker, New York.

Bowers, TS, NS Shifrin, and BL Murphy, 1996. Statistical approach to meeting soil cleanup goals, *Environmental Science and Technology*, 30(5): 1437–1444.

Bowles, DS, LR Anderson, and TF Glover, 1987. Design Level Risk Assessment for Dams, In *Proceedings of Structures Congress, ASCE*, Orlando, FL, pp. 210–225.

Boyce, CP, 1998. Comparison of approaches for developing distributions for carcinogenic slope factors, *Human and Ecological Risk Assessment*, 4(2): 527–577.

Bradley, PM and FH Chapelle, 1996. Anaerobic mineralization of vinyl chloride in Fe(Ill)-reducing aquifer sediments, *Environmental Science and Technology*, 40: 2084–2086.

Brady, PV, MV Brady, and DJ Borns, 1998. *Natural Attenuation: CERCLA, RBCA's, and the Future of Environmental Remediation*, Lewis Publishers, Boca Raton, FL.

Bregman, JI and KM Mackenthun, 1992. *Environmental Impact Statements*, Lewis Publishers, Chelsea, MI.

Bretherick, L, 1979. *Handbook of Reactive Chemical Hazards*, 2nd edition, Butterworth Publishers, Wolburn, MA.

Briggs, GG, RH Bromilow, and AA Evans, 1982. Relationship between lipophicity and root uptake and translocation of non-ionized chemicals by barley, *Pesticide Science*, 13: 495–504.

Bromley, DW and K Segerson (eds.), 1992. *The Social Response to Environmental Risk: Policy Formulation in an Age of Uncertainty*, Kluwer Academic Publishers, Boston, MA.

Brooks, SM, M Gochfield, J Herzstein, RJ Jackson, and MB Schenker, 1995. *Environmental Medicine*, Mosby, Mosby-Year Book, St. Louis, MO.

Brown, C, 1978. Statistical aspects of extrapolation of dichotomous dose response data, *Journal of the National Cancer Institute*, 60: 101–108.

Brown, DP, 1987. Mortality of workers exposed to polychlorinated biphenyls—An update, *Archives of Environmental Health*, 42(6): 333–339.

Brown, HS, 1986. A critical review of current approaches to determining "How Clean is Clean" at hazardous waste sites, *Hazardous Wastes and Hazardous Materials*, 3(3): 233–260.

Brown, HS, R Guble, and S Tatelbaum, 1988. Methodology for assessing hazards of contaminants to seafood, *Regulatory Toxicology and Pharmacology*, 8: 76–100.

Brown, JF, Jr. and RE Wagner, 1990. PCB movement, dechlorination, and detoxication in the Acushnet Estuary, *Environmental Toxicology and Chemistry*, 9: 1215–1233.

Brown, JF, Jr., 1994. Determination of PCB metabolic, excretion, and accumulation rates for use as indicators of biological response and relative risk, *Environmental Scince Technology*, 28(13): 2295–2305.

Brown, KW, GB Evans, Jr., and BD Frentrup (eds.), 1983. *Hazardous Waste and Treatment*, Butterworth Publishers: Boston, MA.

Brum, G, L McKane, and G Karp, 1994. *Biology: Exploring Life*, 2nd edition, John Wiley & Sons, New York.

Brunekreef, B, 1997. Air pollution and life expectancy: is there a relation? *Occupational and Environmental Medicine*, 54: 781–784.

Brusseau, ML, DA Sabatini, JS Gierke, and MD Annable (eds.), 1999. *Innovative Subsurface Remediation*, ACS Symposium Series, American Chemical Society, Washington, DC.

BSI, 1988. *Draft for Development, DD175: 1988 Code of Practice for the Identification of Potentially Contaminated Land and its Investigation*, British Standards Institution (BSI), London, UK.

Buchel, KH, 1983. *Chemistry of Pesticides*, John Wiley & Sons, New York.

Buchmann, MF, 1998. NOAA Screening Quick Reference Tables (SQuiRT), NOAA HAZMAT Report 97-2, Hazardous Materials Response and Assessment Division, National Oceanic and Atmospheric Administration, Seattle, WA [Available at: http://response.restoration.noaa.gov/living/SQuiRT/SQuiRT.html].

Budd, P, J Montgomery, A Cox, P Krause, B Barreiro, and RG Thomas, 1998. The distribution of lead within ancient and modern human teeth: implications for long-term and historical exposure monitoring, *Science of the Total Environment*, 220(2–3): 121–136.

Burmaster, DE and K von Stackelberg, 1991. Using Monte Carlo simulations in public health risk assessments: estimating and presenting full distributions of risk, *Journal of Exposure Analysis and Environmental Epidemiology*, 1: 491–512.

Burmaster, DE and RH Harris, 1993. The magnitude of compounding conservatisms in Superfund risk assessments, *Risk Analysis*, 13: 131–134.

Burmaster, DE, 1996. Benefits and costs of using probabilistic techniques in human health risk assessments—With emphasis on site-specific risk assessments, *Human and Ecological Risk Assessment*, 2: 35–43.

Burns, LA, CG Ingersoll, and GA Pascoe, 1994. Ecological risk assessment: application of new approaches and uncertainty analysis, *Environmental Toxicology and Chemistry*, 13(12): 1873–1874.

Burrough, PA and AU Frank, 1995. Concepts and paradigms in spatial information: are current geographical information systems truly generic? *International Journal of Geographical Information Systems*, 9(2): 101–116.

Byrnes, ME, 1994. *Field Sampling Methods for Remedial Investigations*, CRC Press, Boca Raton, FL.

Byrnes, ME, 2000. *Sampling and Surveying Radiological Environments*, CRC Press, Boca Raton, FL.

Byrnes, ME, 2008. *Field Sampling Methods for Remedial Investigations*, 2nd edition, CRC Press, Boca Raton, FL.

Cai, XH, C Brown, J Adhiya, SJ Traina, and RT Sayre, 1999. Growth and heavy metal binding properties of transgenic *chlamydomonas* expressing a foreign metallothionen gene, *International Journal of Phytoremediation*, 1: 53–66.

Cairney, T (ed.), 1987. *Reclaiming Contaminated Land*, Blackie Academic & Professional, Glasgow, UK.

Cairney, T (ed.), 1993. *Contaminated Land (Problems and Solutions)*, Blackie Academic & Professional, Glasgow/Chapman and Hall, London/Lewis Publishers, Boca Raton, FL.

Cairney, T, 1995. *The Re-Use of Contaminated Land: A Handbook of Risk Assessment*, John Wiley & Sons, Chichester, UK.

Cairns, J, Jr. and TV Crawford (eds.), 1991. *Integrated Environmental Management*, Lewis Publishers, Chelsea, MI.

Calabrese, EJ and EJ Stanek, 1992. Distinguishing outdoor soil ingestion from indoor dust ingestion in a soil pica child, *Regulatory Toxicology and Pharmacology*, 15: 83–85.

Calabrese, EJ and EJ Stanek, 1995. Resolving intertracer inconsistencies in soil ingestion estimation, *Environmental Health Perspectives*, 103(5): 454–456.

Calabrese, EJ and PT Kostecki (ed.), 1991. *Hydrocarbon Contaminated Soils, Volume 1*, Lewis Publishers, Chelsea, MI.

Calabrese, EJ and PT Kostecki, 1988. *Soils Contaminated by Petroleum: Environment and Public Health Effects*, John Wiley & Sons, New York.

Calabrese, EJ and PT Kostecki, 1989. *Petroleum Contaminated Soils, Volume 2*, Lewis Publishers, Chelsea, MI.

Calabrese, EJ and PT Kostecki, 1992. *Risk Assessment and Environmental Fate Methodologies*, CRC Press, Boca Raton, FL.

Calabrese, EJ, 1984. *Principles of Animal Extrapolation*, John Wiley & Sons, New York.

Calabrese, EJ, EJ Stanek, and CE Gilbert, 1991. Evidence of soil-pica behavior and quantification of soil ingested, *Human Exposure and Toxicology*, 10: 245–249.

Calabrese, EJ, EJ Stanek, CE Gilbert, and RM Barnes, 1990. Preliminary adult soil ingestion estimates: results of a pilot study, *Regulatory Toxicology and Pharmacology*, 12: 88–95.

Calabrese, EJ, PT Kostecki, and M Bonazountas, 1996. *Contaminated Soils, Volume 1*, Amherst Scientific Publishers (ASP), Amherst, MA.

Calabrese, EJ, R Barnes, EJ Stanek III, H Pastides, CE Gilbert, P Veneman, X Wang, A Lasztity, and PT Kostecki, 1989. How much soil do young children ingest: an epidemiologic study, *Regulatory Toxicology and Pharmacology*, 10: 123–137.

Calmano, W and U Forstner (eds.), 1996. *Sediments and Toxic Substances: Environmental Effects and Ecotoxicity*, Springer-Verlag, Berlin, Germany.

Calow, P (ed.), 1993. *Handbook of Ecotoxicology*, Blackwell Scientific Publications, London, UK.

Calow, P (ed.), 1998. *Handbook of Environmental Risk Assessment and Management*, Blackwell Science, London, UK.

Campbell, R and MP Greaves, 1990. *Anatomy and Community Structure of the Rhizosphere*, John Wiley & Sons, West Sussex, UK.

Canter, LW, RC Knox, and DM Fairchild, 1988. *Ground Water Quality Protection*, Lewis Publishers, Chelsea, MI.

CAPCOA, 1989. Air Toxics Assessment Manual, California Air Pollution Control Officers Association (CAPCOA), Draft Manual, August, 1987 (ammended, 1989), California.

CAPCOA, 1990. Air Toxics "Hot Spots" Program, Risk Assessment Guidelines, California Air Pollution Control Officers Association (CAPCOA), California.

Carlsen, TM, 1996. Ecological risk to fossorial vertebrates from volatile organic compounds in soil, *Risk Analysis*, 16(2): 211–219.

Carrington, CD and PM Bolger, 1998. Uncertainty and risk assessment, *Human and Ecological Risk Assessment*, 4(2): 253–257.

Carson, R, 1962. *Silent Spring*, Houghton Mifflin Co., New York.

Carson, R, 1994. *Silent Spring—With an Introduction by Vice President Al Gore*, Houghton Mifflin Co., New York.

Carson, WH (ed.), 1990. *The Global Ecology Handbook—What You Can Do about the Environmental Crisis, The Global Tommorrow Coalition*, Beacon Press, Boston, MA.

Casarett, LJ and J Doull, 1975. *Toxicology: The Basic Science of Poisons*, MacMillan Publishing Co., New York.

Cassidy, K, 1996. Approaches to the risk assessment and control of major industrial chemical and related hazards in the United Kingdom, *International Journal of Environment and Pollution*, 6(4–6): 361–387.

CCME (Canadian Council of Ministers of the Environment), 1991. Interim Canadian Environmental Quality Criteria for Contaminated Sites, Report CCME EPC-CS34, The National Contaminated Sites Remediation Program, Canadian Council of Ministers of the Environment (CCME), Winnipeg, MB.

CCME, 1993. Guidance Manual on Sampling, Analysis, and Data Management for Contaminated Sites. Volume I: Main Report (Report CCME EPC-NCS62E), and Volume II: Analytical Method Summaries (Report CCME EPC-NCS66E), Canadian Council of Ministers of the Environment, The National Contaminated Sites Remediation Program, Canadian Council of Ministers of the Environment (CCME), Winnipeg, MB.

CCME, 1994. Subsurface Assessment Handbook for Contaminated Sites, Canadian Council of Ministers of the Environment (CCME), The National Contaminated Sites Remediation Program (NCSRP), Report No. CCME-EPC-NCSRP-48E (March, 1994), Ottawa, ON.

CDHS, 1986. The California Site Mitigation Decision Tree Manual, California Department of Health Services (CDHS), Toxic Substances Control Division, Sacremento, CA.

CDHS, 1990. Scientific and Technical Standards for Hazardous Waste Sites, California Department of Health Services (CDHS), Toxic Substances Control Program, Technical Services Branch, Sacramento, CA.

CEC (Commission on the European Communities), 1986. *Risk Assessment for Hazardous Installations*, Pergamon Press, Oxford, UK.

CEQ, 1989. *Risk Analysis: A Guide to Principles and Methods for Analyzing Health and Environmental Risks*, NTIS: PB89-137772, Council on Environmental Quality, Washington, DC.

Chandra, R, 2015. *Advances in Biodegradation and Bioremediation of Industrial Waste*, CRC Press, Boca Raton, FL.

Chandra, R, NK Dubey, and V Kumar, 2017. *Phytoremediation of Environmental Pollutants*, CRC Press, Boca Raton, FL.

Chapelle, FH, PB McMahon, NM Dubrovsky, RF Fujii, ET Oaksford, and DA Vroblesky, 1995. Deducing the distribution of terminal electron-accepting processes in hydrologically diverse groundwater systems, *Water Resources Research*, 31: 359–371.

Chapman, SW and BL Parker, 2005. Plume persistence due to aquitard back diffusion following dense non-aqueous phase liquid removal or isolation, *Water Resource Research*, 41(12): 1–17.

Chappell, J, 1998. Phytoremediation of TCE in Groundwater Using Populus, Status report prepared for USEPA, Technology Innovation Office, February [Available at: http://clu-in.org/products/phytotce.htm].

Charbeneau, RJ, PB Bedient, and RC Loehr (eds.), 1992. *Groundwater Remediation*, Water Quality Management Library, *Volume 8*, Technomic Publishing Co., Inc., Lancaster, PA.

Chatterji, M (ed.), 1987. *Hazardous Materials Disposal: Siting and Management*, Gower Publishing Co. Ltd., Avebury, UK.

Chen, JJ, DW Gaylor, and RL Kodell, 1990. Estimation of the joint risk from multiple-compound exposure based on single-compound experiments, *Risk Analysis*, 10: 285–290.

Chen, JP, LK Wang, MS Wang, Y Hung, and NK Shammas (eds.), 2016. *Remediation of Heavy Metals in the Environment*, CRC Press, Boca Raton, FL.

Chen, L, 1995. Testing the mean of skewed distributions, *Journal of the American Statistical Association*, 90: 767–772.

Cheremisinoff, NP and ML Graffia, 1995. *Environmental and Health & Safety Management: A Guide to Compliance*, Noyes Publications, Park Ridge, NJ.

Chiu, HS and KL Tsang, 1990. Reduction of treatment cost by using communal treatment facilities, *Waste Mgmnt and Research*, 8(2): 165–167.

Cho, HJ, RJ Fiacco, and M Daly, 2002. Soil vapor extraction and chemical oxidation to remediate chlorinated solvents in fractured crystalline bedrock: pilot study results and lessons learned, *Remediation*, 12(2): 35–50.

Chouaniere, D, P Wild, JM Fontana, M Héry, M Fournier, V Baudin, I Subra, D Rousselle, JP Toamain, S Saurin, and MR Ardiot, 2002. Neurobehavioral disturbances arising from occupational toluene exposure, *American Journal of Industrial Medicine*, 41(2): 77–88.

Christakos, G and DT Hristopulos, 1998. *Spatiotemporal Environmental Health Modelling: A Tractatus Stochasticus*, Kluwer Academic Publishers, Dordrecht, The Netherlands.

Christen, K, 2001. The arsenic threat worsens, *Environmental Science and Technology*, 35(13): 286A–291A.

Christian, B and T Griffith (eds.), 2016. *Algorithms to Live By: The Computer Science of Human Decisions*, Picador, New York.

Chrostowski, PC, LJ Pearsall, and C Shaw, 1985. Risk assessment as a management tool for inactive hazardous materials disposal sites, *Environmental Management*, 9(5): 433–442.

Churchill, JE and WE Kaye, 2001. Recent chemical exposures and blood volatile organic compound levels in a large population-based sample, *Archives of Environmental Health*, 56(2): 157–166.

Cipollini, ML and JL Pickering, 1986. Determination of the phytotoxicity of barium in leach-field disposal of gas well brines, *Plant and Soil*, 92: 159–169.

Clark, M, 1996. *Transport Modeling for Environmental Engineers and Scientists*, John Wiley & Sons, New York.

Clausing, O, AB Brunekreef, and JH van Wijnen, 1987. A method for estimating soil ingestion by children, *International Archives of Occupational and Environmental Health*, 59(1): 73–82.

Clayson, DB, Krewski, D, and Munro, I (eds.), 1985. *Toxicological Risk Assessment, Volumes 1 and 2*, CRC Press, Boca Raton, FL.

Clayton, CA, ED Pellizari, and JJ Quackenboss, 2002. National human exposure assessment survey: analysis of exposure pathways and routes for arsenic and lead in EPA Region 5, *Journal of Exposure Analysis and Environmental Epidemiology*, 12(1): 29–43.

Cleek, RL and AL Bunge, 1993. A new method for estimating dermal absorption from chemical exposure, 1: general approach, *Pharmaceutical Research*, 10: 497–506.

Clemen, RT, 1996. *Making Hard Decisions: An Introduction to Decision Analysis* (2nd ed.), Duxbury Press, Boston, MA.

Clement Associates, Inc., 1988. *Multi-Pathway Health Risk Assessment Impact Guidance Document*, South Coast Air Quality Management District, Los Angeles, CA.

Clewell, HJ and ME Andersen, 1985. Risk assessment extrapolations and physiological modeling, *Toxicology and Industrial Health*, 1: 111–132.

CMA (Chemical Manufacturers Association), 1984. *Risk Management of Existing Chemicals*, Chemical Manufacturers Association (CMA), Washington, DC.

CMA, 1985. *Risk Analysis in the Chemical Industry*, Government Institutes, Rockville, MD.

Cogliano, VJ, 1997. Plausible upper bounds: are their sums plausible? *Risk Analysis*, 17(1): 77–84.

Cohen, RM and JW Mercer, 1993. *DNAPL Site Evaluation*, C.K. Smoley Press, Boca Raton, FL.

Cohen, Y, 1986. Organic pollutant transport, *Environmental Science and Technology*, 20(6): 538–545.

Cohrssen, JJ and VT Covello, 1989. *Risk Analysis: A Guide to Principles and Methods for Analyzing Health and Environmental Risks*, National Technical Information Service (NTIS), US Department of Commerce, Springfield, VA.

Colborn, T, FS Vom Saal, and AM Soto, 1993. Developmental effects of endocrine-disrupting chemicals in wild-life and humans, *Environmental Health Perspectives*, 101: 378–384.

Cole, GM, 1994. *Assessment and Remediation of Petroleum-Contaminated Sites*, CRC Press, Boca Raton, FL.

Colten, CE and PN Skinner, 1996. *The Road to Love Canal: Managing Industrial Waste before EPA*, University of Texas Press, Austin, TX.

Compton, HR, DM Haroski, SR Hirsh, and JG Wrobel, 1998. Pilot-scale use of trees to address VOC contamination. In *Bioremediation and Phytoremediation, Chlorinated and Recalcitrant Compounds*, GB Wickramanayake and RE Hinchee (eds.), Battelle Press, Columbus, OH, pp. 245–250.

Conover, WJ, 1980. *Practical Nonparametric Statistics*, 2nd edition, John Wiley & Sons, New York.

The Conservation Foundation, 1985. *Risk Assessment and Risk Control*, The Conservation Foundation, Washington, DC.

Conway, MF and SH Boutwell, 1987. The Use of Risk Assessment to Define a Corrective Action Plan for Leaking Underground Storage Tanks, In *Proceeding of the NWWA/API Conference on Petroleum Hydrocarbons & Organic Chemicals in Ground Water-Prevention, Detection & Restoration*, November, 1987, Water Well Publication Co., OH, pp. 19–40.

Conway, RA (ed.), 1982. *Environmental Risk Analysis of Chemicals*, Van Nostrand Reinhold Co., New York.

Corn, M. (ed.), 1993. *Handbook of Hazardous Materials*, Academic Press, San Diego, CA.

Cothern, CR (ed.), 1993. *Comparative Environmental Risk Assessment*, CRC Press, Boca Raton, FL.

Cothern, CR and NP Ross (eds.), 1994. *Environmental Statistics, Assessment, and Forecasting*, CRC Press, Boca Raton, FL.

Counter, SA, LH Buchanan, F Ortega, and G Laurell, 2002. Elevated blood mercury and neuro-otological observations in children of the Ecuadorian gold mines, *Journal of Toxicology and Environmental Health A*, 65(2): 149–163.

Couture, LA, Elwell, MR, and Birnbaum, LS, 1988. Dioxin-like effects observed in male rats following exposure to octachlorodibenzo-p-dioxin (OCDD) during a 13-week study, *Toxicology and Applied Pharmacology*, 93: 31–46.

Covello VT, McCallum DB, and Pavlova MT, 1989. Principles and guidelines for improving risk communication. In *Effective Risk Communication: The Role and Responsibility of Government and Nongovernment Organizations*, VT Covello, DB McCallum, and MT Pavlova (eds.), pp. 3–16, Plenum Press, New York.

Covello, VT and F Allen, 1988. *Seven Cardinal Rules of Risk Communication*, US EPA Office of Policy Analysis, Washington, DC.

Covello, VT and J Mumpower, 1985. Risk analysis and risk management: an historical perspective, *Risk Analysis*, 5: 103–120.

Covello, VT and MW Merkhofer, 1993. *Risk Assessment Methods: Approaches for Assessing Health and Environmental Risks*, Plenum Press, New York.

Covello, VT, 1992. Trust and credibility in risk communication, *Health Environment Digest*, 6(1): 1–3.

Covello, VT, 1993. Risk communication and occupational medicine, *Journal of Occupational Medicine*, 35(1): 18–19.

Covello, VT, J Menkes, and J Mumpower (eds.), 1986. *Risk Evaluation and Management*, Contemporary Issues in Risk Analysis, *Volume 1*, Plenum Press, New York.

Covello, VT, LB Lave, AA Moghissi, and VRR Uppuluri, 1987. *Uncertainty in Risk Assessment, Risk Management, and Decision Making*, Advances in Risk Analysis, *Volume 4*, Plenum Press, New York.

Cowherd, CM, GE Muleski, PJ Engelhart, and DA Gillette, 1985. Rapid Assessment of Exposure to Particulate Emissions From Surface Contamination Sites, Prepared for US EPA, Office of Health and Environmental Assessment, Washington, DC, EPA/600/8-85/002, Midwest Research Institute, Kansas City, MO.

Cox, DC and PC Baybutt, 1981. Methods for uncertainty analysis: a comparative survey, *Risk Analysis*, 1(4): 251–258.

Cox, LA and PF Ricci, 1992. Dealing with uncertainty—From health risk assessment to environmental decision making, *Journal of Energy Engineering*, 118(2): 77–94.

Cox, SJ and NRS Tait, 1991. *Reliability, Safety & Risk Management: An Integrated Approach*, Butterworth-Heinemann, Oxford, UK.

Crandall, RW and BL Lave (eds.), 1981. *The Scientific Basis of Risk Assessment*, Brookings Institution, Washington, DC.

Crane, M, MC Newman, PF Chapman, and J Fenlon (eds.), 2002. *Risk Assessment with Time to Event Models*, CRC Press, Boca Raton, FL.

Cranor, CF, 1993. *Regulating Toxic Substances: A Philosophy of Science and the Law*, Oxford University Press, New York.

Crawford-Brown, DJ, 1999. *Risk-Based Environmental Decisions: Methods and Culture*, Kluwer Academic Publishers, Dordrecht, The Netherlands.

Cressie, NA, 1994. *Statistics for Spatial Data*, Revised Edition, John Wiley & Sons, New York.

Crockett, AB, 1998. Background levels of metals in soils, McMurdo Station, Antarctica, *Environmental Monitoring and Assessment*, 50: 289–296.

Crosby, DG, 1998. *Environmental Toxicology and Chemistry*, Oxford University Press, New York.

Crouch, EAC and R Wilson, 1982. *Risk/Benefit Analysis*, Ballinger, Boston, MA.

Crouch, EAC, R Wilson, and L Zeise, 1983. The risks of drinking water, *Water Resources Research*, 19(6): 1359–1375.

Crump, KS and RB Howe, 1984. The multistage model with time-dependent dose pattern: applications of carcinogenic risk assessment, *Risk Analysis*, 4: 163–176.

Crump, KS, 1981. An improved procedure for low-dose carcinogenic risk assessment from animal data, *Journal of Environmental Toxicology*, 5: 339–346.

Crump, KS, 1984. A new method for determining allowable daily intakes, *Fundamentals of Applied Toxicology*, 4: 854–871.

Crump, KS, 1995. Calculation of benchmark doses from continuous data, *Risk Analysis*, 15, 79–89.

CRWQCB (California Regional Water Quality Control Board), 1989. The Designated Level Methodology for Waste Classification and Cleanup Level Determination, Staff Report, Central Coast Region, California Regional Water Quality Control Board (June 1989).

CSA (Canadian Standards Association), 1991. *Risk Analysis Requirents and Guidelines*, CAN/CSA-Q634-91, Canadian Standards Association, Rexdale, ON.

Csuros, M and C Csuros, 2002. *Environmental Sampling and Analysis for Metals*, CRC Press, Boca Raton, FL.

Csuros, M, 1994. *Environmental Sampling and Analysis for Technicians*, CRC Press, Boca Raton, FL.

Cullen, AC, 1994. Measures of compounding conservatism in probabilistic risk assessment, *Risk Analysis*, 14(4): 389–393.

Currado, GM and S Harrad, 1998. Comparison of polychlorinated biphenyl concentrations in indoor and outdoor air and the potential significance of inhalation as a human exposure pathway, *Environmental Science and Technology*, 32(20): 3043–3047.

D'Agostino, RB and MA Stephens, 1986. *Goodness-of-Fit Techniques*, Marcel Dekker, New York.

D'Agostino, RB, 1971. An omnibus test of normality for moderate and large size samples, *Biometrika*, 58: 341–348.

Dakins, ME, JE Toll, and MJ Small, 1994. Risk-based environmental remediation: decision framework and role of uncertainty, *Environmental Toxicology and Chemistry*, 13(12): 1907–1915.

Dakins, ME, JE Toll, MJ Small, and KP Brand, 1996. Risk-based environmental remediation: Bayesian Monte Carlo analysis and the expected value of sample information, *Risk Analysis*, 16(1): 67–79.

Daniel, DE (ed.), 1993. *Geotechnical Practice for Waste Disposal*, Chapman & Hall, London, UK.

Daniels, SL, 1978. Environmental Evaluation and Regulatory Assessment of Industrial Chemicals, In *51st Annual Conference, Water Pollution Control Federation*, Anaheim, CA.

Daugherty, J, 1998. *Assessment of Chemical Exposures—Calculation Methods for Environmental Professionals*, Lewis Publishers, Boca Raton, FL.

Davey, B and T Halliday (eds.), 1994. *Human Biology and Health: An Evolutionary Approach*, Open University Press, Buckingham, UK.

Davies, JC (ed.), 1996. *Comparing Environmental Risks: Tools for Setting Government Priorities*, Resources for the Future, Washington, DC.

Davis, CE, 1993. *The Politics of Hazardous Waste*, Prentice-Hall, Englewood Cliffs, NJ.

Davis, DL, HL Bradlow, M Wolff, T Wooddruff, DG Hoel, and H Anton-Culver, 1993. Medical hypothesis: xenoestrogens as preventable causes of breast cancer, *Environmental Health Perspectives*, 101: 372–377.

Dawson, GW and D Sanning, 1982. Exposure-response analysis for setting site restoration criteria. *Proceedings of National Conference on Management of Uncontrolled Hazardous Waste Sites*, Washington, DC.

Day, RW, 2001. *Soil Testing Manual: Procedures, Classification Data, and Sampling Practices*, McGraw-Hill, New York.

de Serres, FJ and AD Bloom (eds.), 1996. *Ecotoxicity and Human Health (A Biological Approach to Environmental Remediation)*, CRC Press, Boca Raton, FL.

de Voe, C and Udell, KS, 1998. Thermodynamic and hydrodynamic behavior of water and DNAPLs during heating. In *Remediation of Chlorinated and Recalcitrant Compounds: Nonaqueous-Phase Liquids*, GB Wickramanayake, AR Gavaskar and N Gupta (eds.), pp. 61–66, Battelle Press, Columbus, OH.

Decisioneering, Inc., 1996. *Crystal Ball, Version 4.0 User Manual*, Decisioneering, Inc., Denver, CO.

Dennison, MS, 1997. *Brownfields Redevelopment: Programs and Strategies for Rehabilitating Contaminated Real Estate*, Government Institutes/ABS Consulting, Rockville, MD.

Derelanko, MJ and MA Hollinger (eds.), 1995. *CRC Handbook of Toxicology*, CRC Press, Boca Raton, FL.

Desu, MM and D Raghavarao, 1990. *Sample Size Methodology*, Academic Press, San Diego, CA.

Devinny, JS, LG Everett, JCS Lu, and RL Stollar, 1990. *Subsurface Migration of Hazardous Wastes*, Van Nostrand Reinhold Co., New York.

Dewailly, É, JJ Ryan, C Laliberté, S Bruneau, J-P Weber, S Gingras, and G Carrier, 1994. Exposure of remote maritime populations to coplanar PCBs, *Environmental Health Perspectives*, 102(Suppl 1): 205–209.

Dewailly, É, J-P Weber, S Gingras, and C Laliberté, 1991. Coplanar PCBs in human milk in the province of Québec, Canada: are they more toxic than dioxin for breast-fed infants? *Bulletin of Environmental Contamination and Toxicology*, 47: 491–498.

Dewailly, E, P Ayotte, C Laliberte, J-P Webber, S Gingras, and AJ Nantel, 1996. Polychlorinated biphenyl (PCB) and dichlorodiphenyl dichloroethylene (DDE) concentrations in the breast milk of women in Quebec, *American Journal of Public Health*, 86(9): 1241–1246.

Dewailly, E, P Ayotte, S Bruneau, C Laliberte, DCG Muir, and RJ Norstrom, 1993. Inuit exposure to organo-chlorines through the aquatic food chain in Arctic Quebec, *Environmental Health Perspectives*, 101(7): 618–620.

Dewailly, E, S Dodin, R Verreault, P Ayotte, L Sauvé, J Morin, and J Brisson, 1994. High organochlorine body burden in women with estrogen receptor-positive breast cancer, *Journal of National Cancer Institute*, 86: 232–234.

Dienhart, CM, 1973. *Basic Human Anatomy and Physiology*, W.B. Saunders and Company, Philadelphia, PA.

Diesler, PF (ed.), 1984. *Reducing the Carcinogenic Risks in Industry*, Marcel Dekker, New York.

Dodge, CJ and AJ Francis, 1997. Biotransformation of binary and ternary citric acid complexes of iron and uranium, *Environmental Science and Technology*, 31: 3062–3067.

DoE (Department of the Environment), 1994. *Sampling Strategies for Contaminated Land*, CLR Report No.4, Department of the Environment, London, UK.

DOE (U.S. Department of Energy), 1987. *The Remedial Action Priority System (RAPS): Mathematical Formulations*, US Department of Energy, Office of Environment, Safety & Health, Washington, DC.

DoE, 1995. *A Guide to Risk Assessment and Risk Management for Environmental Protection*, UK Department of the Environment, HMSO, London, UK.

Doherty, N, P Kleindorfer, and H Kunreuther, 1990. An insurance perspective on an integrated waste management strategy. In *Integrating Insurance and Risk Management for Hazardous Wastes*, H Kunreuther and MV Rajeev Gowda, Kluwer Academic Publishers, Boston, MA, pp. 271–302.

Domencio, PA and FW Schwartz, 1998. *Physical and Chemical Hydrogeology*, 2nd edition, John Wiley & Sons, New York.

Donaldson, RM and RF Barreras, 1996. Intestinal absorption of trace quantities of chromium, *Journal of Laboratory and Clinical Medicine*, 68: 484–493.

Donnelly, PK and JS Fletcher, 1994. Potential use of mycorrhizal fungi as bioremediation agents. *Bioremediation Through Rhizosphere Technology*, T.A. Anderson and J.R. Coats (eds.), American Chemical Society, Washington, DC.

Douben, PET, 1998. *Pollution Risk Assessment and Management*, John Wiley & Sons, Chichester, UK.

Dourson, ML and FC Lu, 1995. Safety/Risk assessment of chemicals compared for different expert groups, *Biomedical and Environmental Sciences*, 8: 1–13.

Dourson, ML and JF Stara, 1983. Regulatory history and experimental support of uncertainty (safety) factors, *Regulatory Toxicology and Pharmacology*, 3: 224–238.

Dourson, ML and SP Felter, 1997. Route-to-route extrapolation of the toxic potency of MTBE, *Risk Analysis*, 25: 43–57.

Dowdy, D, TE McKone, and DPH Hsieh, 1996. The use of molecular connectivity index for estimating bio-transfer factors, *Environmental Science and Technology*, 30: 984–989.

Dragan, ES (ed.), 2014. *Advanced Separations by Specialized Sorbents*, CRC Press, Boca Raton, FL.

Dragun, J, 1998. *The Soil Chemistry of Hazardous Materials*, 2nd edition, Amherst Scientific Publishers, Amherst, MA.

Draper, NR and H Smith, 1998. *Applied Regression Analysis*, 3rd edition, John Wiley & Sons, New York.

Drever, JI, 1997. *The Geochemistry of Natural Waters: Surface and Groundwater Environments*, 3rd edition, Prentice-Hall, New York.

Driscoll, FG (ed.), 1986. *Groundwater and Wells*, 2nd edition, Johnson Division, St. Paul, MN.

Driver, J, SR Baker, and D McCallum, 2001. *Residential Exposure Assessment: A Sourcebook*, Kluwer Academic Publishers, Dordrecht, The Netherlands.

Driver, JH, JJ Konz, and GK Whitmyre, 1989. Soil adherence to human skin, *Bulletin Environmental Contamination Toxicology*, 43: 814–820.

Driver, JH, ME Ginevan, and GK Whitmyre, 1996. Estimation of dietary exposure to chemicals: a case study illustrating methods of distributional analyses for food consumption data, *Risk Analysis*, 16(6): 763–771.

DTSC, 1994a. *CalTOX, A Multimedia Exposure Model for Hazardous Waste Sites*, Office of Scientific Affairs, CalEPA/DTSC, Sacramento, CA.

DTSC, 1994b. *Preliminary Endangerment Assessment Guidance Manual (A Guidance Manual for Evaluating Hazardous Substance Release Sites)*, California Environmental Protection Agency, Department of Toxic Substances Control (DTSC), Sacramento, CA.

DTSC/Cal-EPA, 1992. *Supplemental Guidance for Human Health Multimedia Risk Assessments of Hazardous Waste Sites and Permitted Facilities*, Department of Toxic Substances Control (DTSC), Cal-EPA, Sacramento, CA.

Duff, RM and JC Kissel, 1996. Effect of soil loading on dermal absorption efficiency from contaminated soils, *Journal of Toxicology and Environmental Health*, 48: 93–106.

Dunster, HJ and W Vinck, 1979. The Assessment of Risk, Its Value and Limitations. In *European Nuclear Conference*, Hamburg, FRG.

Dushenkov, S, D Vasudev, Y Kapulnik, D Gleba, D Fleisher, KC Ting, and B Ensley, 1997b. Removal of uranium from water using terrestrial plants, *Environmental Science and Technology*, 31: 3468–3474.

Dushenkov, S, Y Kapulnik, M Blaylock, B Sorochinsky, I Raskin, and B Ensley, 1997a. Phytoremediation: a novel approach to an old problem. In *Global Environmental Biotechnology*, DL Wise (ed.), Elsevier Science BV, Amsterdam, The Netherlands, pp. 563–572.

Dwarakanath, V and GA Pope, 1998. A new approach for estimating alcohol partition coefficients between nonaqueous phase liquids and water, *Environmental Science and Technology*, 32(11): 1662–1666.

Dwarakanath, V and GA Pope, 1999. Surfactant enhanced aquifer remediation. In *Surfactants: Fundamentals and Applications in the Petroleum Industry*, LL Schramm (ed.), Ch. 11, pp. 433–460, Cambridge University Press, Cambridge, UK.

Dwarakanath, V, K Kostarelos, GA Pope, D Shotts, and WH Wade, 1999. Anionic surfactant remediation of soil columns contaminated by nonaqueous phase liquids, *Journal of Contaminant Hydrology*, 38(4): 465–488.

Dwyer, JP, 1990. The pathology of symbolic legislation, *Ecology Law Quarterly*, 17: 233–316.

Dykeman, R, G Aguilar-Madrid, T Smith, CA Juárez-Pérez, GM Piacitelli, H Hu, and M Hernandez-Avila, 2002. Lead exposure in Mexican radiator repair workers, *American Journal of Industrial Medicine*, 41(3): 179–187.

Earle, TC and G Cvetkovich, 1997. Culture, cosmopolitanism, and risk management, *Risk Analysis*, 17(1): 55–65.

Efron, B and R Tibshirani, 1993. *An Introduction to the Bootstrap*, Chapman & Hall, New York.

Efron, B, 1982. *The Jackknife, the Bootstrap and Other Resampling Plans*, SIAM, Philadelphia, PA.

Einarson, MD and DM Mackay, 2001. Predicting impacts of groundwater contamination, *Environmental Science and Technology*, 35(3): 66A–73A.

Eisenberg, JNS and TE McKone, 1998. Decision tree method for the classification of chemical pollutants: incorporation of across-chemical variability and within-chemical uncertainty, *Environmental Science and Technology*, 32(21): 3396–3404.

Ellis, B and JF Rees, 1995. Contaminated land remediation in the UK with reference to risk assessment: two case studies, *Journal of the Institute of Water and Environmental Management*, 9(1): 27–36.

Environment Canada, 1997. Environmental Assessments of Priority Substances Under the Canadian Environmental Protection Act. Guidance Manual Version 1.0, EPS/2/CC/3E, Chemicals Evaluation Division, Commercial Chemicals Evaluation Branch, Environment Canada, Ottawa, ON, Canada.

Erickson, AJ, 1977. *Aids for estimating soil erodibility – "K" Value Class and Tolerance*. US Dept. of Agriculture, Soil Conservation Service, Salt Lake City, UT.

Erickson, BE, 2002. Analyzing the ignored environmental contaminants, *Environmental Science and Technology*, 30(7): 140A–145A.

Erickson, MD, 1993. *Remediation of PCB Spills*, CRC Press, Boca Raton, FL.

Erickson, RL and RD Morrison, 1995. *Environmental Reports and Remediation Plans: Forensic and Legal Review*, John Wiley & Sons, New York.

Eschenroeder, A, RJ Jaeger, JJ Ospital, and C Doyle, 1986. Health risk assessment of human exposure to soil amended with sewage sludge contaminated with polychlorinated dibenzodioxins and dibenzofurans, *Veterinary and Human Toxicology*, 28: 356–442.

Ess, TH, 1981a. Risk Acceptability, In *Proceedings of National Conference on Risk & Decision Analysis for Hazardous Waste Disposal*, August 24–27, Baltimore, MD, pp. 164–174.

Ess, TH, 1981b. Risk Estimation, In *Proceedings of National Conference on Risk & Decision Analysis for Hazardous Waste Disposal*, August 24–27, Baltimore, MD, pp. 155–163.

Evans, JS and SJS Baird, 1998. Accounting for missing data in noncancer risk assessment, *Human and Ecological Risk Assessment*, 4(2): 291–317.

Evans, LJ, 1989. Chemistry of metal retention by soils, *Environmental Science and Technology*, 23(9): 1047–1056.

Faes, C, M Aerts, H Geys, and G Molenberghs, 2007. Model averaging using fractional polynomials to estimate a safe level of exposure, *Risk Analysis*, 27: 111–123.

Fagerlin, A, PA Ubel, DM Smith, and BJ Zikmund–Fisher, 2007. Making numbers matter: Present and future research in risk communication, *American Journal of Health Behavior*, 31(Suppl. 1): S47–S56.

Falck, F, A Ricci, MS Wolff, J Godbold, and P Deckers, 1992. Pesticides and polychlorinated biphenyl residues in human breast lipids and their relation to breast cancer, *Archives of Environmental Health*, 47: 143–146.

Farmer, A, 1997. *Managing Environmental Pollution*, Routledge, London, UK.

FDA, 1994. *Action Levels for Poisonous or Deleterious Substances in Human Food and Animal Feed*, US Food and Drug Administration (FDA), Washington, DC.

Feenstra, S, DM Mackay, and JA Cherry, 1991. A method for assessing residual NAPL based on organic chemical concentrations in soil samples, *Ground Water Monitoring Review*, 11(2): 128–136.

Felter, S and M Dourson, 1998. The inexact science of risk assessment (and implications for risk management), *Human and Ecological Risk Assessment*, 4(2): 245–251.

Felter, SP and ML Dourson, 1997. Hexavalent chromium contaminated soils: options for risk assessment and risk management, *Regulatory Toxicology and Pharmacology*, 25: 43–59.

Fenner, K, C Kooijman, M Scheringer, and K Hungerbuhler, 2002. Including transformation products into the risk assessment for chemicals: the case of nonylphenol ethoxylate usage in Switzerland, *Environmental Science and Technology*, 36(6): 1147–1154.

Ferro, AM, 1998. Biological Pump and Treat Systems Using Poplar Trees, In *Presentation at IBC's 3rd Annual International Conference on Phytoremediation*, June 22–25, 1998, Houston, TX.

Ferro, AM, J Kennedy, and D Knight, 1997. Greenhouse Scale Evaluation of Phytoremediation for Soil Contaminated with Wood Preservatives, In *4th International In-situ and On-Site Bioremediation Symposium*, 3, April 28–May 1 1997, New Orleans, LA, pp. 309–314.

Ferro, AM, RC Sims and B Bugbee, 1994. Hycrest crested wheatgrass accelerates the degradation of pentachlorophenol, *Journal of Environmental Quality*, 23: 272–279.

Ferson, S, 2002. *RAMAS Risk Calc 4.0 Software: Risk Assessment with Uncertain Numbers*, CRC Press, Boca Raton, FL.

Fetter, CW, 1993. *Contaminant Hydrogeology*, Macmillan Publishing Co., New York.

Filliben, JJ, 1975. The probability plot correlation coefficient test for normality, *Technometrics*, 17: 111–117.

Finkel, A, 1990. *Confronting Uncertainty in Risk Management*, Resources for the Future, Washington, DC.

Finkel, AM and D Golding (eds.), 1994. *Worst Things First? (The Debate over Risk-Based National Environmental Priorities)*, Resources for the Future, Washington, DC.

Finkel, AM and JS Evans, 1987. Evaluating the benefits of uncertainty reduction in environmental health risk management, *Journal of the Air Pollution Control Association*, 37: 1164–1171.

Finkel, AM, 1995. Toward less misleading comparisons of uncertain risks: the example of aflatoxin and alar, *Environmental Health Perspectives*, 103: 376–385.

Finkel, AM, 2003. Too much of the [National Research Council's] 'Red Book' is still ahead of its time, *Human and Ecological Risk Assessment*, 9(5): 1253–1271.

Finley, B and DP Paustenbach, 1994. The benefits of probabilistic exposure assessment: three case studies involving contaminated air, water, and soil, *Risk Analysis*, 14: 53–73.

Finley, B, D Proctor, P Scott, N Harrington, D Paustenbach, and P Price, 1994. Recommended distributions for exposure factors frequently used in health risk assessment, *Risk Analysis*, 14: 533–553.

Finley, B, PK Scott, and DA Mayhall, 1994. Development of a standard soil-to-skin adherence probability density function for use in Monte Carlo analyses of dermal exposures, *Risk Analysis*, 14: 555–569.

Fischhoff, B, S Lichtenstein, P Slovic, S Derby, and R Keeney, 1981. *Acceptable Risk*, Cambridge University Press, New York.

Fisher, A and FR Johnson, 1989. Conventional wisdom on risk communication and evidence from a field experiment, *Risk Analysis*, 9(2): 209–213.

Fitchko, J, 1989. *Criteria for Contaminated Soil/Sediment CleanDup*, Pudvan Publishing Co., Northbrook, IL.

Flathman, PE and GR Lanza, 1998. Phytoremediation: current views on an emerging green technology, *Journal of Soil Contamination*, 7: 415–432.

Flegal, AR and DR Smith, 1992. Blood lead concentrations in preindustrial humans, *New England Journal of Medicine*, 326: 1293–1294.

Flegal, AR and DR Smith, 1995. Measurement of environmental lead contamination and human exposure, *Reviews of Environmental Contamination and Toxicology*, 143: 1–45.

Fletcher JS, 2000. Biosystem Treatment of Recalcitrant Soil Contaminants, In *Presentation at the USEPA Phytoremediation State of the Science Conference*, May 1–2, Boston, MA.

Forester, WS and JH Skinner (ed.), 1987. *International Perspectives on Hazardous Waste Management--A Report from the International Solid Wastes and Public Cleansing Association (ISWA) Working Group on Hazardous Wastes*, Academic Press, London, UK.

Fox, JF, KA Hogan, and A Davis, 2017. Dose-response modeling with summary data from developmental toxicity studies, *Risk Analysis*, 37(5): 905–917.

FPC (Florida Petroleum Council), 1986. *Benzene in Florida Groundwater: An Assessment of the Significance to Human Health*, Florida Petroleum Council, Tallahassee, FL.

Francis, BM, 1994. *Toxic Substances in the Environment*, John Wiley & Sons, New York.

Fredrickson, JK, H Bolton Jr., and FJ Brockman, 1993. In situ and on-site bioreclamation, *Environmental Science and Technology*, 27(9): 1711–1716.

Freeze RA and DB McWhorter, 1997. A framework for assessing risk reduction due to DNAPL mass removal from low-permeability soils, *Ground Water*, 35(1): 111–123.

Freeze, RA and JA Cherry, 1979. *Groundwater*, Prentice-Hall, Englewood Cliffs, NJ.

Freeze, RA, 2000. *The Environmental Pendulum*, University of California Press, Berkeley, CA.

Freudenburg WR and SR Pastor, 1992. NIMBYs and LULUs: stalking the syndromes, *Journal of Social Issues*, 48(4): 39–61.

Freund, JE and RE Walpole (eds.), 1987. *Mathematical Statistics*, Prentice-Hall, Englewood Cliffs, NJ.

Frey, SE, H Destaillats, S Cohn, S Ahrentzen, and MP Fraser, 2014. Characterization of indoor air quality and resident health in an Arizona senior housing apartment building, *Journal of the Air and Waste Management Association*, 64(11): 1251–1259.

Frick, C, R Farrell, and J Germida, 1999. *Assessment of Phytoremediation as an In-Situ Technique for Cleaning Oil-Contaminated Sites*, Petroleum Technology Alliance of Canada (PTAC), Calgary, AB.

Frohse, F, M Brodel, and L Schlossberg, 1961. *Atlas of Human Anatomy*, Barnes & Noble Books/Harper & Row Publishers, New York.

Frosig, A, H Bendixen, and D Sherson, 2001. Pulmonary deposition of particles in welders: on-site measurements, *Archives of Environmental Health*, 56(6): 513–521.

Furst, A, 1990. Yes, but is it a human carcinogen? *Journal of the American College of Toxicology*, 9: 1–18.

Garbutt, J, 1995a. *Environmental Law*, 2nd edition, John Wiley & Sons, Chichester, UK.

Garbutt, J, 1995b. *Waste Management Law*, 2nd edition, John Wiley & Sons, Chichester, UK.

Gardels, MC and TJ Sorg, 1989. A laboratory study of the leaching of lead from water faucets, *Journal of the American Water Works Association*, 81(7): 101–113.

Garrett, P, 1988. *How to Sample Groundwater and Soils*, National Water Well Association (NWWA), Dublin, OH.

Gatliff, EG, 1994. Vegetative remediation process offers advantages over traditional pump-and-treat technologies, *Remediation*, 4(Summer): 343–352.

Gattrell, A and M Loytonen (eds.), 1998. *GIS and Health*, Taylor & Francis Group, London, UK.

Gauglitz, PA, JS Roberts, TM Bergsman, SM Caley, WO Heath, MC Miller, RW Moss, and R Schalla, 1994. Six-Phase Soil Heating Accelerates VOC Extraction from Clay Soil, In *Presented at Spectrum '94: International Nuclear and Hazardous Waste Management*, August 14–18 1994, Atlanta, GA.

Gavaskar, AR, Gupta, N, Sass, BM, Jansoy, RJ, and O'Sullivan, D, 1998. *Permeable Barriers for Groundwater Remediation: Design, Construction, and Monitoring*, Battelle Press, Columbus, OH.

Gaylor, DW and JJ Chen, 1996. A simple upper limit for the sum of the risks of the components in a mixture, *Risk Analysis*, 16(3): 395–398.

Gaylor, DW and Kodell, RL, 1980. Linear interpolation algorithm for low dose risk assessment of toxic substances, *Journal of Environmental Pathology and Toxicology*, 4: 305–312.

Gaylor, DW and Shapiro, RE, 1979. Extrapolation and risk estimation for carcinogenesis. In *Advances in Modern Toxicology Volume 1, New Concepts in Safety Evaluation Part 2*, MA Mehlman, RE Shapiro, and H Blumenthal (eds.), Hemisphere, New York, pp. 65–85.

Gaylor, DW and W Slikker, 1990. Risk assessment for neurotoxic effects, *Neurotoxicology*, 11: 211–218.

Gaylor, DW, FF Kadlubar, and FA Beland, 1992. Application of biomarkers to risk assessment, *Environmental Health Perspectives*, 98: 139–141.

Gefell, MJ, Hamilton, LA, and Stout, DJ, 1999, A Comparison between Low-Flow and Passive-Diffusion Bag Sampling Results for Dissolved Volatile Organics in Fractured Sedimentary Bedrock, In *Proceedings of the Petroleum and Organic Chemicals in Ground Water—Prevention, Detection, and Remediation Conference*, November 17–19, 1999, Houston, TX, pp. 304–315.

Gerber, GB, IA Leonard, and P Jacquet, 1980. Toxicity, mutagenicity and teratogenicity of lead, *Mutation Research*, 76: 115–141.

Gerity, TR and CJ Henry, 1990. *Principles of Route-to-Route Extrapolation for Risk Assessment*, Elsevier Publishing, Amsterdam, The Netherlands.

Ghadiri, H and CW Rose (eds.), 1992. *Modeling Chemical Transport in Soils: Natural and Applied Contaminants*, CRC Press, Boca Raton, FL.

Gheorghe, AV and M Nicolet-Monnier, 1995. *Integrated Regional Risk Assessment, Volumes 1 and 2*, Kluwer Academic Publishers, Dordrecht, The Netherlands.

Gibbons, RD and DE Coleman, 2001. *Statistical Methods for Detection and Quantification of Environmental Contamination*, John Wiley & Sons, New York.

Gibbons, RD, 1994. *Statistical Methods for Groundwater Monitoring*, John Wiley & Sons, New York.

Gibbs, LM, 1982. *Love Canal: My Story*, SUNY Press, Albany, NY.

Gierthy, JF, KF Arcaro, and M Floyd, 1995. Assessment and implications of PCB estrogenicity, *Organo-Halogen Compounds*, 25: 419–423.

Gilbert, RO, 1987. *Statistical Methods for Environmental Pollution Monitoring*, Van Nostrand Reinhold Co., New York.

Gilliland, FD, KT Berhane, YF Li, DH Kim, and HG Margolis, 2002. Dietary magnesium, potassium, sodium, and children's lung function, *American Journal of Epidemiology*, 155(2): 125–131.

Gilpin, A, 1995. *Environmental Impact Assessment (EIA): Cutting edge for the twenty-first century*, Cambridge University Press, Cambridge, UK.

Ginevan, ME and DE Splitstone, 1997. Improving remediation decisions at hazardous waste sites with risk-based geostatistical analysis, *Environmental Science and Technology*, 31(2): 92A–96A.

Glasson, J, R Therivel, and A Chadwick, 1994. *Introduction to Environmental Impact Assessment*, UCL Press, London, UK.

Glaze, WH and JW Kang, 1988. Advanced oxidation processes for treating groundwater contaminated with TCE and PCE: laboratory studies, *Journal AWWA*, 80(5): 57–63.

Gleit, A, 1985. Estimation for small normal data sets with detection limits, *Environmental Science and Technology*, 19: 1201–1206.

Glickman, TS and M Gough (eds.), 1990, *Readings in Risk*, Resources for the Future, Washington, DC.

Glowa, JR, 1991. Dose-effect approaches to risk assessment, *Neuroscience and Biobehavioral Reviews*, 15: 153–158.

Gochfeld, M, 1997. Factors influencing susceptibility to metals, *Environmental Health Perspectives*, 105(Suppl 4): 817–822.

Goldman, M, 1996. Cancer risk of low-level exposure, *Science*, 271(5257): 1821–1822.

Goldman, SJ, K Jackson, and TA Bursztynsky, 1986. *Erosion and Sediment Control Handbook*, McGraw-Hill, New York.

Goldstein, BD, 1989. The maximally exposed individual: an inappropriate basis for public health decision-making, *Environmental Forum*, 6: 13–16.

Goldstein, BD, 1995. The who, what, when, where, and why of risk characterization, *Policy Studies Journal*, 23(1): 70–75.

Goldstein, KG, AR Vitolins, D Navon, BL Parker, S Chapman, and GA Anderson, 2004. Characterization and pilot-scale studies for chemical oxidation remediation of fractured shale, *Remediation*, 14 (Autumn 4): 19–37.

Goodchild, MF, BO Parks, and LT Steyaert (eds.), 1993. *Environmental Modeling with GIS*, Oxford University Press, New York.

Goodchild, MF, LT Steyaert, BO Parks, C Johnston, D Maidment, M Crane, and S Glendinning (eds.), 1996. *GIS and Environmental Modeling: Progress and Research Issues*, GIS World Books, Fort Collins, CO.

Goodrich, MT and JT McCord, 1995. Quantification of uncertainty in exposure assessment at hazardous waste sites, *Ground Water*, 33(5): 727–732.

Gordon, MP, S Strand, and L Newman, 1998. Final Report: Degradation of Environmental Pollutants by Plants. [Available at: http://es.epa.gov/ncerqa/final/gordon.html] USEPA, National Center for Environmental Research, Office of Research and Development, Washington, DC.

Gordon, SI, 1985. *Computer Models in Environmental Planning*, Van Nostrand Reinhold Co., New York.

Gorelick, SM, RA Freeze, D Donohue, and JF Keely, 1993. *Groundwater Contamination (Optimal Capture and Containment)*, CRC Press, Boca Raton, FL.

Gots, RE, 1993. *Toxic Risks*, CRC Press, Boca Raton, FL.

Gould, JR (ed.), 1975, *Adsorption at Interfaces*, ACS Symposium Series 8, American Chemical Society, Washington, DC.

Grasso, D, 1993. *Hazardous Waste Site Remediation (Source Control)*, CRC Press, Boca Raton, FL.

Gratt, LB, 1996. *Air Toxic Risk Assessment and Management: Public Health Risk from Normal Operations*, Van Nostrand Reinhold Co., New York.

Greenberg, R, T Andrews, P Kakarla, and R Watts, 1998. In situ fenton-like oxidation of volatile organics: laboratory-, pilot-, and full-scale demonstrations, *Remediation*, 8(2): 29–42.

Gregory, R and H Kunreuther, 1990. Successful siting incentives, *Civil Engineering*, 60(4): 73–75.

Grimberg, SJ, CT Miller, and AD Aitken, 1996. Surfactant-enhanced dissolution of phenanthrene into water for laminar flow conditions, *Environmental Science Technology*, 30(10); 2967–2974.

Grisham, JW (ed.), 1986. *Health Aspects of the Disposal of Waste Chemicals*, Pergamon Press, Oxford, UK.

Guswa, JH, WJ Lyman, AS Donigian Jr., TYR Lo, and EW Shanahan, 1984. *Groundwater Contamination and Emergency Response Guide*, Noyes Publications, Park Ridge, NJ.

Guyton, AC, 1968. *Textbook of Medical Physiology*, W.B. Saunders, Philadelphia, PA.

Guyton, AC, 1971. *Basic Human Physiology: Normal Functions and Mechanisms of Disease*, W.B. Saunders, Philadelphia, PA.

Guyton, AC, 1982. *Human Physiology and Mechanisms of Disease* (3rd ed.), W.B. Saunders, Philadelphia, PA.

Guyton, AC, 1986. *Textbook of Medical Physiology* (7th ed.), W.B. Saunders, Philadelphia, PA.

Haas, CN and PA Scheff, 1990. Estimation of averages in truncated samples, *Environmental Science and Technology*, 24: 912–919.

Haas, CN, 1997. Importance of distributional form in characterizing inputs to Monte Carlo risk assessments, *Risk Analysis*, 17(1): 107–113.

Hadley, PW and RM Sedman, 1990. A health-based approach for sampling shallow soils at hazardous waste sites using the AALsoil contact criterion, *Environmental Health Perspectives*, 18: 203–207.

Haimes, YY (ed.), 1981. *Risk/Benefit Analysis in Water Resources Planning and Management*, Plenum Press, New York.

Haimes, YY, L Duan, and V Tulsiani, 1990. Multiobjective decision-tree analysis, *Risk Analysis*, 10(1): 111–129.

Haines AH, 1985. *Methods for the Oxidation of Organic Compounds*, Academic Press, Orlando, FL.

Haith, DA, 1980. A mathematical model for estimating pesticide losses in runoff, *Journal of Environmental Quality*, 9(3): 428–433.

Hall, P, 1988. Theoretical comparison of bootstrap confidence intervals, *Annals of Statistics*, 16: 927–953.

Hall, P, 1992. On the removal of skewness by transformation, *Journal of the Royal Statistical Society B*, 54: 221–228.

Hallenbeck, WH and Cunningham, KM, 1988. *Quantitative Risk Assessment for Environmental and Occupational Health*, 4th Printing, Lewis Publishers, Chelsea, MI.

Hallenbeck, WH and KM Cunningham-Burns, 1985. *Pesticides and Human Health*, Springer-Verlag, New York.

Hamed, MM and PB Bedient, 1997a. On the effect of probability distributions of input variables in public health risk assessment, *Risk Analysis*, 17(1): 97–105.

Hamed, MM and PB Bedient, 1997b. On the performance of computational methods for the assessment of risk from ground-water contamination, *Ground Water*, 35(4): 638–646.

Hamed, MM, 1999. Probabilistic sensitivity analysis of public health risk assessment from contaminated soil, *Journal of Soil Contamination*, 8(3): 285–306.

Hamed, MM, 2000. Impact of random variables probability distribution on public health risk assessment from contaminated soil, *Journal of Soil Contamination*, 9(2): 99–117.

Hammitt, JK and AI Shlyakhter, 1999. The expected value of information and the probability of surprise, *Risk Analysis*, 19(1): 135–152.

Hammitt, JK, 1995. Can more information increase uncertainty? *Chance*, 8(3): 15–17, 36.

Hance, BJ, C Chess, and PM Sandman, 1990. *Industry Risk Communication Manual: Improving Dialogue with Communities*, Lewis Publishers, Boca Raton, FL.

Hanley, N and CL Spash, 1995. *Cost-benefit Analysis and the Environment*, Second reprint, Edward Elgar Publishing Ltd., Hampshire, UK.

Hansen, PE and SE Jorgensen (eds.), 1991. *Introduction to Environmental Management, Developments in Environmental Modelling, 18*, Elsevier, Amsterdam, The Netherlands.

Hansson, S-O, 1989. Dimensions of risk, *Risk Analysis*, 9(1): 107–112.

Hansson, S-O, 1996a. Decision making under great uncertainty, *Philosophy of the Social Sciences*, 26(3): 369–386.

Hansson, S-O, 1996b. What is philosophy of risk? *Theoria*, 62: 169–186.

Haque, MI, 2017. *Mechanics of Groundwater in Porous Media*, CRC Press, Boca Raton, FL.

Harrad, S, 2000. *Persistent Organic Pollutants: Environmental Behavior and Pathways of Human Exposure*, Kluwer Academic Publishers, Dordrecht, The Netherlands.

Harris, SA, AM Sass-Kortsak, PN Corey, and JT Purdham, 2002. Development of models to predict dose of pesticides in professional turf applicators, *Journal of Exposure Analysis and Environmental Epidemiology*, 12(2): 130–144.

Harrison, RM and DPH Laxen (eds.), 1981. *Lead Pollution, Causes and Control*, Chapman & Hall, London, UK.

Hasan, SE, 1996. *Geology and Hazardous Waste Management*, Prentice-Hall, Englewood Cliffs, NJ.

Hathaway, GJ, NH Proctor, JP Hughes, and ML Fischman, 1991. *Proctor and Hughes' Chemical Hazards of the Workplace*, 3rd edition, Van Nostrand Reinhold Co., New York.

Hattis, D and DE Burmaster, 1994. Assessment of variability and uncertainty distributions for practical risk analyses, *Risk Analysis*, 14(5): 713–729.

Hattis, D and R Goble, 2003. The red book, risk assessment, and policy analysis: The road not taken, *Human and Ecological Risk Assessment*, 9(5): 1297–1306.

Haun, JW, 1991. *Guide to the Management of Hazardous Waste*, Fulcrum Publishing, Golden, CO.

Hauptmann, M, H Pohlabeln, JH Lubin, KH Jöckel, W Ahrens, I Brüske-Hohlfeld, and H- Wichmann, 2002. The exposure-time-response relationship between occupational asbestos exposure and lung cancer in two German case-control studies, *American Journal of Industrial Medicine*, 41(2): 89–97.

Hauser, VL, 2008. *Evapotranspiration Covers for Landfills and Waste Sites*, CRC Press, Boca Raton, FL.

Hawken, P, 1993. *The Ecology of Commerce: A Declaration of Sustainability*, Harper Business: New York.

Hawley, GG, 1981. *The Condensed Chemical Dictionary*, 10th edition, Van Nostrand Reinhold Co., New York.

Hawley, JK, 1985. Assessment of health risks from exposure to contaminated soil, *Risk Analysis*, 5(4): 289–302.

Health Canada, 1994. *Human Health Risk Assessment for Priority Substances*, Environmental Health Directorate, Canadian Environmental Protection Act, Health Canada, Ottawa, ON.

Heaton, CP, CL Rugh, N-J Wang, and RB Meagher, 1998. Phytoremediation of mercury- and methylmercury-polluted soils using genetically engineered plants, *Journal of Soil Contamination*, 7: 497–509.

Hee, SSQ, 1999. *Hazardous Waste Analysis*, Government Institutes/ABS Consulting, Rockville, MD.

Heikkila, P, R Riala, M Hämeilä, E Nykyri, and P Pfäffli, 2002. Occupational exposure to bitumen during road paving, *AIHA Journal*, 63(2): 156–165.

Helsel, DR and RM Hirsch, 1992. *Statistical Methods in Water Resources*, Elsevier, New York.

Helsel, DR, 1990. Less than obvious: statistical treatment of data below the detection limit, *Environmental Science and Technology*, 24: 1766–1774.

Helton, JC, 1993. Risk, uncertainty in risk, and the EPA release limits for radioactive waste disposal, *Nuclear Technology*, 101: 18–39.

Hemond, HF and EJ Fechner, 1994. *Chemical Fate and Transport in the Environment*, Academic Press, San Diego, CA.

Henderson, M, 1987. *Living with Risk: The Choices, The Decisions*, The British Medical Association Guide, John Wiley & Sons, New York.

Henke, KR, V Kühnel, DJ Stepan, RH Fraley, CM Robinson, DS Charlton, HM Gust, and NS Bloom, 1993. *Critical Review of Mercury Contamination Issues Relevant to Manometers at Natural Gas Industry Sites*, Gas Research Institute, Chicago, IL.

Hernández-Avila M, Smith D, Meneses F, Sanin LH, and H Hu, 1998. The influence of bone and blood lead on plasma lead levels in environmentally exposed adults, *Environmental Health Perspectives*, 106: 473–477.

Heron, G, Christensen, TH, Heron, T, and TH Larsen, 1998. Thermally enhanced remediation at DNAPL sites: the competition between downward mobilization and upward volatilization. In *Remediation of Chlorinated and Recalcitrant Compounds: Nonaqueous-Phase Liquids*, Battelle Press, Columbus, OH.

Hertwich, EG, WS Pease, and TE McKone, 1998. Evaluating toxic impact assessment methods: what works best? *Environmental Science and Technology*, 32(5): 138A–144A.

Hertz, DB and H Thomas, 1983. *Risk Analysis and Its Applications*, John Wiley & Sons, New York.

Hester, RE and RM Harrison (eds.), 1995. *Waste Treatment and Disposal*, The Royal Society of Chemistry, Herts, UK.

Hewitt, AD, TF Jenkins, and CL Grant, 1995. Collection, handling, and storage: keys to improved data quality for volatile organic compounds in soil, *American Environmental Laboratory*, 7: 25–28.

Hill, AB, 1965. The environment and disease: association or causation? *Proceedings of the Royal Society of Medicine*, 58: 295–300.

Hilts, SR, 1996. A co-operative approach to risk management in an active lead/zinc smelter community, *Environmental Geochemistry and Health*, 18: 17–24.

Hinchee, HJ, HJ Reisinger, D Burris, B Marks, and J Stepek, 1986. Underground Fuel Contamination, Investigation, and Remediation: A Risk Assessment Approach to How Clean Is Clean, In *Proceedings of the NWWA/API Conference on Petroleum Hydrocarbons & Organic Chemicals in Ground Water-Prevention, Detection & Restoration*, November, 1986, Water Well Publication Co., OH, pp. 539–564.

Hinchman, RR, MC Negri, and EG Gatliff, 1997. Phytoremediation: Using Green Plants to Clean Up Contaminated Soil, Groundwater, and Wastewater, Report submitted to the US Department of Energy, assistant secretary for Energy Efficient and Renewable Energy under Contract W-31-109-Eng-38.

Hipel, KW, 1988. Nonparametric approaches to environmental impact assessment, *Water Resources Bulletin, AWRA*, 24(3): 487–491.

Hirvonen, A, T Tuhkanen, and P Kalliokoski, 1996. Treatment of TCE- and PCE- contaminated groundwater using UV/H_2O_2 and O_3/H_2O_2 oxidation processes, *Water Science Technology*, 33(6): 67–73.

HMTRI (The Hazardous Materials Training and Research Institute), 1997. *Site Characterization: Sampling and Analysis*, John Wiley & Sons, New York.

Hodas, N, Q Meng, MM Lunden, and BJ Turpin, 2014. Toward refined estimates of ambient $PM_{2.5}$ exposure: evaluation of a physical outdoor-to-indoor transport model, *Atmospheric Environment*, 83: 229–236.

Hoddinott, KB (ed.), 1992. *Superfund Risk Assessment in Soil Contamination Studies*, ASTM Publication STP 1158, American Society for Testing and Materials, Philadelphia, PA.

Hoel, DG, Gaylor, DW, Kirschstein, RL, Saffiotti, U, and Schneiderman, MA, 1975. Estimation of risks of irreversible, delayed toxicity, *Journal of Toxicology and Environmental Health*, 1, 133–151.

Hoffman, DJ, BA Rattner, GA Burton Jr., and J Cairns Jr., 1995. *Handbook of Ecotoxicology*, CRC Press, Boca Raton, FL.

Hoffman, DJ, Rattner, BA, and RJ Hall, 1990. Wildlife toxicology, *Environmental Science and Technology*, 24: 276.

Hoffmann, FO and JS Hammonds, 1992. *An Introductory Guide to Uncertainty Analysis in Environmental and Health Risk Assessment*, Environmental Sciences Division, Oak Ridge National Laboratory, TN, ESD Publication 3920, Prepared for the US Department of Energy, Washington, DC.

Hogan, MD, 1983. Extrapolation of animal carcinogenicity data: limitations and pitfalls, *Environmental Health Perspectives*, 47, 333–337.

Holman, E, R Francis, and G Gray, 2017. Part I—Comparing noncancer chronic human health reference values: an analysis of science policy choices, *Risk Analysis*, 37(5): 861–878.

Holman, E, R Francis, and G Gray, 2017. Part II—Quantitative evaluation of choices used in setting noncancer chronic human health reference values across organizations, *Risk Analysis*, 37(5): 879–892.

Holmes, G, BR Singh, and L Theodore, 1993. *Handbook of Environmental Management and Technology*, John Wiley & Sons, New York.

Holmes, KK, Jr., JH Shirai, KY Richter, and JC Kissel, 1999. Field measurement of dermal soil loadings in occupational and recreational activities, *Journal of Environmental Research*, 80(2): 148–157.

Holtta, P, H Kiviranta, A Leppäniemi, T Vartiainen, PL Lukinmaa, and S Alaluusua, 2001. Developmental dental defects in children who reside by a river polluted by dioxins and furans, *Archives of Environmental Health*, 56(6): 522–528.

Homburger, F, JA Hayes, and EW Pelikan (eds.), 1983. *A Guide to General Toxicology*, Karger Publishers, New York.

Honeycutt, RC and DJ Schabacker (eds.), 1994. *Mechanisms of Pesticide Movement into Groundwater*, CRC Press, Boca Raton, FL.

HRI (Hampshire Research Institute), 1995. *Risk*Assistant for Windows*, The Hampshire Research Institute, Alexandria, VA.

Hrudey, SE, W Chen, and CG Rousseaux, 1996. *Bioavailability in Environmental Risk Assessment*, CRC Press, Boca Raton, FL.

HSE (Health and Safety Executive), 1989a. *Risk Criteria for Land-Use Planning in the Vicinity of Major Industrial Hazards*, Health and Safety Executive, HMSO, London, UK.

HSE, 1989b. *Quantified Risk Assessment—Its Input to Decision Making*, Health and Safety Executive, HMSO, London, UK.

Huckle, KR, 1991. *Risk Assessment—Regulatory Need or Nightmare*, Shell publications, Shell Center, London, UK.

Hudak, PF, 1998. Groundwater monitoring strategies for variable versus constant contaminant loading functions, *Environmental Monitoring and Assessment*, 50: 271–288.

Hudson, R, R Tucker, and M Haegeli, 1984. *Handbook of Toxicity of Pesticides to Wildlife*, 2nd edition, USFWS Resources Publication No. 153, U.S. Fish and Wildlife Service, Washington, DC.

Huff, JE, 1993. Chemicals and cancer in humans: first evidence in experimental animals, *Environmental Health Perspectives*, 100: 201–210.

Hughes, WW, 1996. *Essentials of Environmental Toxicology: The Effects of Environmentally Hazardous Substances on Human Health*, Taylor & Francis Group, Washington, DC.

Huh, C, 1979. Interfacial tensions and solubilizing ability of a microemulsion phase that coexists with oil and brine, *Journal of Colloid Interface Science*, 71(2): 408.

Hunt, JR, JT Geller, N Sitar, and KS Udell, 1988. Subsurface Transport Processes for Gasoline Components, In *Specialty Conference Proceedings, Joint CSCE-ASCE National Conference on Environmental Engineeringr*, July 13–15, 1988, Vancouver, BC, Canada, pp. 536–543.

Hutzler, NJ, JS Gierke, and BE Murphy, 1990. Vaporizing VOCs, *Civil Engineering, ASCE*, 60(4): 57–60.

Hwang, J-S and C-C Chan, 2002. Effects of air pollution on daily clinic visits for lower respiratory tract illness, *American Journal of Epidemiology*, 155(1): 1–10.

Hwang, ST and JW Falco, 1986. Estimation of multimedia exposures related to hazardous waste facilities. In *Pollutants in a Multimedia Environment*, Y Cohen (ed.), pp. 577–586, Plenum Press, New York.

Hyman, M and RR Dupont, 2001. *Groundwater and Soil Remediation: Process Design and Cost Estimating of Proven Technologies*, American Society of Civil Engineers (ASCE), Reston, VA.

IARC (International Agency for Research on Cancer), 1972–1985. *IARC Monographs on the Evaluation of the Carcinogenic Risk of Chemicals to Man, (Multi-volume Work)*, International Agency for Research on Cancer (IARC), World Health Organization, Geneva, Switzerland.

IARC, 1982. *IARC Monographs on the Evaluation of the Carcinogenic Risk of Chemicals to Humans: Chemicals, Industrial Processes and Industries Associated with Cancer in Humans*, Supplement 4, International Agency for Research on Cancer (IARC), Lyon, France.

IARC, 1984. *IARC Monographs on the Evaluation of the Carcinogenic Risk of Chemicals to Humans, Volume 33*, International Agency for Research on Cancer (IARC), Lyon, France.

IARC, 1987. *IARC Monographs on the Evaluation of Carcinogenic Risks of Chemicals to Humans: Overall Evaluations of Carcinogenicity*, Supplement 7, International Agency for Research on Cancer (IARC), Lyon, France.

IARC, 1988. *IARC Monographs on Evaluation of Carcinogenic Risks to Humans, Volume 43*, International Agency for Research on Cancer (IARC), Lyon, France.

IARC, 2006. IARC Monographs on the evaluation of the carcinogenic risk of chemicals to man, (Multi-volume work). International Agency for Research on Cancer (IARC), World Health Organization, Geneva, Switzerland.

ICRCL, 1987. *Guidance on the Assessment and Redevelopment of Contaminated Land, ICRCL 59/83*, 2nd edition, Interdepartmental Committee on the Redevelopment of Contaminated Land (ICRCL), Department of the Environment, Central Directorate on Environmental Protection, London, UK.

ICRP, 1975–[current]. *Publication Series of the International Commission on Radiological Protection (ICRP)*, Pergamon Press, Oxford, UK.

IJC (International Joint Commission), 1986. Literature Review of the Effects of Persistent Toxic Substances on Great Lakes Biota, Report of the Health of Aquatic Communities Task Force, International Joint Commission.

Ikegami, M, M Yoneda, T Tsuji, O Bannai, and S Morisawa, 2014. Effect of particle size on risk assessment of direct soil ingestion and metals adhered to children's hands at playgrounds, *Risk Analysis*, 34(9), 1677–1687.

Illing, HP, 1999. Are societal judgments being incorporated into the uncertainty factors used in toxicological risk assessment, *Regulatory Toxicology and Pharmacology*, 29(3): 300–308.

Iman, RL and JC Helton, 1988. An investigation of uncertainty and sensitivity analysis techniques for computer models, *Risk Analysis*, 8: 71–90.

IOM, 2013. *Environmental Decisions in the Face of Uncertainty*, The National Academies Press, Washington, DC.

IRPTC, 1978. *Attributes for the Chemical Data Register of the International Register of Potentially Toxic Chemicals, Register Attribute Series, No. 1*, International Register of Potentially Toxic Chemicals, UNEP, Geneva, Switzerland.

IRPTC, 1985a. *International Register of Potentially Toxic Chemicals, Part A*, International Register of Potentially Toxic Chemicals, UNEP, Geneva, Switzerland.

IRPTC, 1985b. *Industrial Hazardous Waste Management*, Industry and Environment Office and the International Register of Potentially Toxic Chemicals, UNEP, Geneva, Switzerland.

IRPTC, 1985c. *Treatment and Disposal Methods for Waste Chemicals: IRPTC File*, Data Profile Series No. 5, International Register of Potentially Toxic Chemicals, UNEP, Geneva, Switzerland.

Isaaks, EH and RM Srivastava, 1989. *Applied Geostatistics*, Oxford University Press, Cambridge, UK.

Isaxon, C, A Gudmundsson, E Nordin, L Lönnblad, A Dahl, G Wieslander, M Bohgard, and A Wierzbicka, 2015. Contribution of indoor-generated particles to residential exposure, *Atmospheric Environment*, 106: 458–466.

Iskandar, IK (ed.), 2001. *Environmental Restoration of Metals-Contaminated Soils*, CRC Press, Boca Raton, FL.

István P and JB Jones Jr., 1997. *The Handbook of Trace Elements*, St. Lucie Press, Boca Raton, FL.

ITRC, 2002. *A Systematic Approach to In Situ Bioremediation in Groundwater Including Decision Trees on In Situ Bioremediation for Nitrates, Carbon Tetrachloride, and Perchlorate*, Guidance Document, The Interstate Technology & Regulatory Council (ITRC), Washington, DC.

ITRC, 2003a. *An Introduction to Characterizing Sites Contaminated with DNAPLs*, Guidance Document, The Interstate Technology & Regulatory Council (ITRC), Washington, DC.

ITRC, 2003b. *Technical and Regulatory Guidance for Design, Installation, and Monitoring of Alternative Final Landfill Covers*, Guidance Document, The Interstate Technology & Regulatory Council (ITRC), Washington, DC.

ITRC, 2003c. *Technology Overview Using Case Studies of Alternative Landfill Technologies and Associated Regulatory Topics*, Guidance Document, The Interstate Technology & Regulatory Council (ITRC), Washington, DC.

ITRC, 2003d. *Technical and Regulatory Guidance for Surfactant/Cosolvent Flushing of DNAPL Source Zones*, Guidance Document, The Interstate Technology & Regulatory Council (ITRC), Washington, DC.

ITRC, 2003e. *Technical and Regulatory Guidance for the Triad Approach: A New Paradigm for Environmental Project Management*, Guidance Document, The Interstate Technology & Regulatory Council (ITRC), Washington, DC.

ITRC, 2003f. *DNAPL Guidance—An Introduction to Characterizing Sites with DNAPL*, Guidance Document, The Interstate Technology & Regulatory Council (ITRC), Washington, DC.

ITRC, 2003g. *Technical and Regulatory Guidance Document for Constructed Treatment Wetlands*, Guidance Document, The Interstate Technology & Regulatory Council (ITRC), Washington, DC.

ITRC, 2003h. *Technical and Regulatory Guidance for the Triad Approach: A New Paradigm for Environmental Project Management*, Guidance Document, The Interstate Technology & Regulatory Council (ITRC), Washington, DC.

ITRC, 2004. *Technical and Regulatory Guidance for Using Polyethylene Diffusion Bag Samplers to Monitor Volatile Organic Compounds in Groundwater*, Guidance Document, The Interstate Technology & Regulatory Council (ITRC), Washington, DC.

ITRC, 2005a. *Permeable Reactive Barriers: Lessons Learned/New Directions*, Guidance Document, The Interstate Technology & Regulatory Council (ITRC), Washington, DC.

ITRC, 2005b. *Examination of Risk-Based Screening Values and Approaches of Selected States*, Guidance Document, The Interstate Technology & Regulatory Council (ITRC), Washington, DC.

ITRC, 2005c. *Technical and Regulatory Guidance for In-situ Chemical Oxidation of Contaminated Soil and Groundwater*, Guidance Document, The Interstate Technology & Regulatory Council (ITRC), Washington, DC.

ITRC, 2007a. *Protocol for Use of Five Passive Samplers to Sample for a Variety of Contaminants in Groundwater*, Guidance Document, The Interstate Technology & Regulatory Council (ITRC), Washington, DC.

ITRC, 2007b. *Triad Implementation Guide*, Guidance Document, The Interstate Technology & Regulatory Council (ITRC), Washington, DC.

ITRC, 2007c. *Vapor Intrusion Pathway: A Practical Guideline*, Guidance Document, The Interstate Technology & Regulatory Council (ITRC), Washington, DC.

ITRC, 2007d. *Vapor Intrusion Pathway: Investigative Approaches for Typical Scenarios (A Supplement to Vapor Intrusion Pathway: A Practical Guideline)*, Guidance Document, The Interstate Technology & Regulatory Council (ITRC), Washington, DC.

ITRC, 2008a. *In Situ Bioremediation of Chlorinated Ethene: DNAPL Source Zones*, BioDNAPL-3, Guidance Document, The Interstate Technology & Regulatory Council (ITRC), Washington, DC.

ITRC, 2008b. *Enhanced Attenuation: Chlorinated Organics*, Guidance Document, The Interstate Technology & Regulatory Council (ITRC), Washington, DC.

ITRC, 2008c. *Use of Risk Assessment in Management of Contaminated Sites*, Guidance Document, The Interstate Technology & Regulatory Council (ITRC), Washington, DC.

ITRC, 2009a. *Phytotechnology Technical and Regulatory Guidance and Decision Trees*, Revised. Guidance Document, The Interstate Technology & Regulatory Council (ITRC), Washington, DC.

ITRC, 2009b. *Evaluating Natural Source Zone Depletion at Sites with LNAPL*, Guidance Document, The Interstate Technology & Regulatory Council (ITRC), Washington, DC.

ITRC, 2009c. *Evaluating LNAPL Remedial Technologies for Achieving Project Goals*, Guidance Document, The Interstate Technology & Regulatory Council (ITRC), Washington, DC.

ITRC, 2010a. *Use and Measurement of Mass Flux and Mass Discharge*, Guidance Document, The Interstate Technology & Regulatory Council (ITRC), Washington, DC.

ITRC, 2010b. *Use and Measurement of Mass Flux and Mass Discharge*, Guidance Document, The Interstate Technology & Regulatory Council (ITRC), Washington, DC.

ITRC, 2011a. *Permeable Reactive Barrier: Technology Update*, Guidance Document, The Interstate Technology & Regulatory Council (ITRC), Washington, DC.

ITRC, 2011b. *Project Risk Management for Site Remediation*, Guidance Document, The Interstate Technology & Regulatory Council (ITRC), Washington, DC.

ITRC, 2011c. *Development of Performance Specifications for Solidification/Stabilization*, Guidance Document, The Interstate Technology & Regulatory Council (ITRC), Washington, DC.

ITRC, 2011d. *Green and Sustainable Remediation: A Practical Framework*, Guidance Document, The Interstate Technology & Regulatory Council (ITRC), Washington, DC.

ITRC, 2011e. *Green and Sustainable Remediation: State of the Science and Practice*, Guidance Document, The Interstate Technology & Regulatory Council (ITRC), Washington, DC.

ITRC, 2011f. *Integrated DNAPL Site Strategy*, Guidance Document, The Interstate Technology & Regulatory Council (ITRC), Washington, DC.

ITRC, 2012a. *Incremental Sampling Methodology*, Guidance Document, The Interstate Technology & Regulatory Council (ITRC), Washington, DC.

ITRC, 2012b. *Using Remediation Risk Management to Address Groundwater Cleanup Challenges at Complex Sites*, Guidance Document, The Interstate Technology & Regulatory Council (ITRC), Washington, DC.

ITRC, 2013a. *Environmental Molecular Diagnostics New Site Characterization and Remediation Enhancement Tools*, Guidance Document, The Interstate Technology & Regulatory Council (ITRC), Washington, DC.

ITRC, 2013b. *Groundwater Statistics and Monitoring Compliance: Statistical Tools for the Project Life Cycle*, Guidance Document, The Interstate Technology & Regulatory Council (ITRC), Washington, DC.

ITRC, 2014. *Petroleum Vapor Intrusion: Fundamentals of Screening, Investigation, and Management*, Guidance Document, The Interstate Technology & Regulatory Council (ITRC), Washington, DC.

ITRC, 2015a. *Decision Making at Contaminated Sites: Issues and Options in Human Health Risk Assessment*, Guidance Document, The Interstate Technology & Regulatory Council (ITRC), Washington, DC.

ITRC, 2015b. *Integrated DNAPL Site Characterization and Tools Selection*, Guidance Document, The Interstate Technology & Regulatory Council (ITRC), Washington, DC.

ITRC, 2016. *Geospatial Analysis for Optimization at Environmental Sites*, Guidance Document, The Interstate Technology & Regulatory Council (ITRC), Washington, DC.

ITRC, 2017. *Light Nonaqueous Phase Liquids (LNAPL) Update*, Guidance Document, The Interstate Technology & Regulatory Council (ITRC), Washington, DC.

Jackson, RE and P Mariner, 1995. Estimating DNAPL composition and VOC dilution from extraction well data, *Ground Water*, 33(3): 407–414.

Jain, RK, LV Urban, GS Stacey, and HE Balbach, 1993. *Environmental Assessment*, McGraw-Hill, New York.

Jarabek, AM, MG Menache, JH Overton, ML Dourson, and FJ Miller, 1990. The US Environmental Protection Agency's inhalation RfD methodology: risk assessment in air toxics, *Toxicol Ind Health*, 6: 279–301.

Jasonoff, S, 1993. Bridging the two cultures of risk analysis, *Risk Analysis*, 13: 122–128.

Javandel, I and C-F Tsang, 1986. Capture-zone type curves: a tool for aquifer cleanup, *Groundwater*, 24: 616–625.

JCC (J.C. Consltancy Ltd, London), 1986. *Risk Assessment for Hazardous Installations*, Pergamon Press, Oxford, UK.

Jennings, AA and P Suresh, 1986. Risk penalty functions for hazardous waste management, *ASCE, Journal of Environmental Engineering*, 112(1): 105–122.

Jewell, T and J Steele, 1996. UK regulatory reform and the pursuit of 'sustainable development': the Environment Act 1995, *Journal of Environmental Law*, 8(2): 283–300.

Ji, W and ML Brusseau, 1998. A general mathematical model for chemical-enhanced flushing of soil contaminated by organic compounds, *Water Resource Research*, 34(7): 1635–1648.

Jin, M, GW Butler, RE Jackson, PE Mariner, JF Pickens, GA Pope, CL Brown, and DC McKinney, 1997. Sensitivity models and design protocol for partitioning tracer tests in alluvial aquifers, *Ground Water*, 36(6): 964–972.

Jin, M, M Delshad, V Dwarakanath, DC McKinney, GA Pope, K Sepehrnoori, CE Tilburg, and RE Jackson, 1995. Partitioning tracer test for detection, estimation and remediation performance assessment of subsurface nonaqueous phase liquids, *Water Resources Research*, 31(5): 1201–1211.

Jo, WK, CP Weisel, and PJ Lioy, 1990a. Chloroform exposure and the health risk associated with multiple uses of chlorinated tap water, *Risk Analysis*, 10: 581–585.

Jo, WK, CP Weisel, and PJ Lioy, 1990b. Routes of chloroform exposure and body burden from showering with chlorinated tap water, *Risk Analysis*, 10: 575–580.

Joffe, M and J Mindell, 2002. A framework for the evidence base to support health impact assessment, *Journal of Epidemiology and Community Health*, 56(2): 132–132.

Johnson BL and Jones DE, 1992. ATSDR's activities and views on exposure assessment, *Journal of Exposure Analysis and Environmental Epidemiology*, (Suppl 1): 1–17.

Johnson, BB and VT Covello, 1987. *Social and Cultural Construction of Risk: Essays on Risk Selection and Perception*, Kluwer Academic Publishers, Norwell, MA.

Johnson, NJ, 1978. Modified *t*-tests and confidence intervals for asymmetrical populations, *The American Statistician*, 73: 536–544.

Johnson, PC and RA Ettinger, 1991. A heuristic model for predicting the intrusion rate of contaminant vapours into buildings, *Environmental Science and Technology*, 25(8): 1445–1452.

Jolley, RL and RGM Wang (eds.), 1993. *Effective and Safe Waste Management: Interfacing Sciences and Engineering with Monitoring and Risk Analysis*, Lewis Publishers, Boca Raton, FL.

Jones, I and DN Lerner, 1995, Level-determined sampling in an uncased borehole, *Journal of Hydrology*, 171: 291–317.

Jones, I, Lerner, DN, and OP Baines, 1999. Multiport sock samplers: a low cost technology for effective multilevel ground water sampling, *Ground Water Monitoring and Remediation*, 19(1): 134–142.

Jones, N, C Thornton, D Mark, and R Harrison, 2000. Indoor/outdoor relationships of particulate matter in domestic homes with roadside, urban and rural locations, *Atmospheric Environment*, 34(16): 2603–2612.

Jones, RB, 1995. *Risk-Based Management: A Reliability-Centered Approach*, Gulf Publishing Co., Houston, TX.

Jorgensen, EP (ed.), 1989. *The Poisoned Well—New Strategies for Groundwater Protection*, Sierra Club Legal Defense Fund, Island Press, Washington, DC.

Jury, WA, WJ Farmer, and WF Spencer, 1984. Behavior assessment model for trace organics in soil: II. Chemical classification and parameter sensitivity, *Journal of Environmental Quality*, 13(4): 567–572.

Kabata-Pendias, A and H Pendias, 1984. *Trace Elements in Soils and Plants*, CRC Press, Boca Raton, FL.

Kabata-Pendias, A, 2001. *Trace Elements in Soils and Plants*, 3rd editon, CRC Press, Boca Raton, FL.

Kadlec, RH and RL Knight, 1996. *Treatment Wetlands*, CRC Press, Boca Raton, FL.

Kaplan, A, 1964. *The Conduct of Enquiry: Methodology for Behavioral Science*, Chandler Publishing Co., San Francisco, CA.

Kara, J, 1992. Geopolitics and the environment: the Case of Central Europe, *Environmental Politics*, 1(2): 18–36.

Karol, RH, 2003. *Chemical Grouting And Soil Stabilization*, 3rd edition, CRC Press, Boca Raton, FL.

Kasperson, RE and PJM Stallen (eds.), 1991. *Communicating Risks to the Public: International Perspectives*, Kluwer Academic Press, Boston, MA.

Kastenberg, WE and HC Yeh, 1993. Assessing public exposure to pesticides—Contaminated ground water, *Journal of Ground Water*, 31(5): 746–752.

Kastenberg, WE, TE McKone, and D Okrent, 1976. On Risk Assessment in the Absence of Complete Data, UCLA Report No. UCLA-ENG-677, School of Engineering and Applied Science, Los Angeles, CA.

Kasting, GB and PJ Robinson, 1993. Can we assign an upper limit to skin permeability? *Pharmaceutical Research*, 10: 930–939.

Kates, RW, 1978. *Risk Assessment of Environmental Hazard*, SCOPE Report 8, John Wiley & Sons, New York.

Katz, M and D Thornton, 1997. *Environmental Management Tools on the Internet: Accessing the World of Environmental Information*, St. Lucie Press, Delray Beach, FL.

Kearney, J, L Wallace, M MacNeill, ME Héroux, W Kindzierski, and AJ Wheeler, 2014. Residential infiltration of fine and ultrafine particles in Edmonton, *Atmospheric Environment*, 94: 793–805.

Keeney, RD and H Raiffa, 1976. *Decisions with Multiple Objectives: Preferences and Value Tradeoffs*, John Wiley & Sons, New York.

Keeney, RL, 1984. Ethics, decision analysis, and public risk, *Risk Analysis*, 4: 117–129.

Keeney, RL, 1990. Mortality risks induced by economic expenditures, *Risk Analysis*, 10(1): 147–159.

Keith, LH (ed.), 1988. *Principles of Environmental Sampling*, American Chemical Society (ACS), Washington, DC.

Keith, LH (ed.), 1992. *Compilation of E.P.A.'s Sampling and Analysis Methods*, CRC Press, Boca Raton, FL.

Keith, LH, 1991. *Environmental Sampling and Analysis—A Practical Guide*, Lewis Publishers, Boca Raton, FL.

Kempa, ES (ed.), 1991. *Environmental Impact of Hazardous Wastes*, PZITS Publishing Department, Warsaw, Poland.

Kent, C, 1998. *Basics of Toxicology*, John Wiley & Sons, New York.

Kim, SB, RL Kodell, and H Moon, 2014. A diversity index for model space selection in the estimation of benchmark and infectious doses via model averaging, *Risk Analysis*, 34(3), 453–464.

Kimmel, CA, and DW Gaylor, 1988. Issues in qualitative and quantitative risk analysis for developmental toxicology, *Risk Analysis*, 8, 15–20.

King, JJ, 1995. *The Environmental Dictionary, and Regulatory Cross-Reference*, 3rd edition, John Wiley & Sons, New York.

Kirtland, BC and CM Aelion, 2000. Petroleum mass removal from low permeability sediment using air sparging/soil vapor extraction: impact of continuous or pulsed operation, *Journal of Contaminant Hydrology*, (41): 367–383.

Kissel, J, JH Shirai, KY Richter, and RA Fenske, 1998. Investigation of dermal contact with soil in controlled trials, *Journal of Soil Contamination*, 7: 737–752.

Kissel, J, KY Richter, and RA Fenske, 1996. Field measurement of dermal soil loading attributed to various activities: implications for exposure assessment, *Risk Analysis*, 16(1): 115–125.

Klaassen, CD, 2002. Xenobiotic transporters: another protective mechanism for chemicals, *International Journal of Toxicology*, 21(1): 7–12.

Klaassen, CD, Amdur, MO, and Doull, J (eds.), 1986. *Casarett and Doull's Toxicology: The Basic Science of Poisons*, 3rd edition, Macmillan Publishing Co., New York.

Klaassen, CD, Amdur, MO, and Doull, J (eds.), 1996. *Casarett and Doull's Toxicology: The Basic Science of Poisons*, 5th edition, McGraw-Hill, New York.

Kleindorfer, PR and HC Kunreuther (ed.), 1987. *Insuring and Managing Hazardous Risks: From Seveso to Bhopal and Beyond*, Springer-Verlag, Berlin, Germany.

Klemm, DE, S Lummus, and S Eaton, 1997. Air Sparging in Various Lithologies: 3 Case Studies, In *Proceedings from the Fourth International In Situ and On-Site Bioremediation Symposium*, April 28–May 1 (In Situ and On Site Bioremediation: *Volume 1*), New Orleans, LA, pp. 193–198.

Kletz, T, 1994. *Learning from Accidents*, 2nd edition, Butterwort-Heinemann, Oxford, UK.

Koch, GS, Jr. and RF Link, 1980. *Statistical Analyses of Geological Data, Volumes 1 and 2*, Dover, New York.

Koch, I, L Wang, CA Ollson, WR Cullen, and KJ Reimer, 2000. The predominance of inorganic arsenic species in plants from Yellowknife, Northwest Territories, Canada, *Environmental Science and Technology*, 34: 22–26.

Kocher, DC and FO Hoffman, 1991. Regulating environmental carcinogens: where do we draw the line? *Environmental Science and Technology*, 25: 1986–1989.

Kodell, RL and JJ Chen, 1994. Reducing conservatism in risk estimation for mixtures of carcinogens, *Risk Analysis*, 14: 327–332.

Koenig, JQ, 1999. *Health Effects of Ambient Air Pollution: How Safe is the Air we Breathe?* Kluwer Academic Publishers, Dordrecht, The Netherlands.

Kolluru, R (ed.), 1994. *Environmental Strategies Handbook: A Guide to Effective Policies and Practices*, McGraw-Hill, New York.

Kolluru, RV, SM Bartell, RM Pitblado, and RS Stricoff (eds.), 1996. *Risk Assessment and Management Handbook (for Environmental, Health, and Safety Professionals)*, McGraw-Hill, New York.

Kostarelos, K, GA Pope, BA Rouse, and GM Shook, 1998. A new concept: the use of neutrally-buoyant microemulsions for DNAPL remediation, *Journal of Contaminant Hydrology*, 34: 383–397.

Kostecki, PT and EJ Calabrese (eds.), 1989. *Petroleum Contaminated Soils, Volumes 1–3*, Lewis Publishers, Chelsea, MI.

Kostecki, PT and EJ Calabrese (eds.), 1991. *Hydrocarbon Contaminated Soils and Groundwater, Volume 1*, Lewis Publishers, Chelsea, MI.

Kostecki, PT and EJ Calabrese (eds.), 1992. *Contaminated Soils (Diesel Fuel Contamination)*, CRC Press, Boca Raton, FL.

Krantzberg, G, JH Hartig, and MA Zarull, 2000. Sediment management: deciding when to intervene, *Environmental Science and Technology*, 34: 22A–27A.

Kraus, J, S Nelson, P Boersma, and A Maciey, 1997. Comparison of Pre-/Post-Sparging VOC Concentrations in Soil and Groundwater, In *Proceedings from the Fourth International In Situ and On-Site Bioremediation Symposium*, April 28–May 1 (In Situ and On-Site Bioremediation: *Volume 1*), New Orleans, LA, p. 128.

Krauskopf, KB and DK Bird, 1995. *Introduction to Geochemistry*, 3rd edition, McGraw-Hill, New York.

Kreith, F (ed.), 1994. *Handbook of Solid Waste Management*, McGraw-Hill, New York.

Krewski, D and J van Ryzin, 1981. Dose response models for quantal response toxicity data. In *Statistics and Related Topics*, M Csorgo, DA Dawson, JNK Rao, and AKE Saleh (eds.), North Holland, New York, pp. 201–231.

Krewski, D, C Brown, and D Murdoch, 1984. Determining safe levels of exposure: safety factors for mathematical models, *Fundamental and Applied Toxicology*, 4: S383–S394.

Kruseman, GP and NA de Ridder, 1990. *Analysis and Evaluation of Pumping Test Data*. International Institute for Land Reclamation and Improvement, Wageningen, The Netherlands.

Kueper, BH and Davies, KL, 2009. Assessment and Delineation of DNAPL Source Zones at Hazardous Waste Sites. *EPA Ground Water Issue Paper*, EPA/600/R-09/119.

Kumar, PBAN, V Dushenkov, H Motto, and I Raskin, 1995. Phytoextraction: the use of plants to remove heavy metals from soils, *Environmental Science and Technology*, 29: 1232–1238.

Kunkel DB, 1986. The toxic emergency, *Emergency Medicine*, 18: 207–217.

Kunreuther, H and MV Rajeev Gowda (ed.), 1990. *Integrating Insurance and Risk Management for Hazardous Wastes*, Kluwer Academic Publishers, Boston, MA.

Kunreuther, H and P Slovic, 1996. Science, values, and risk, *Annals of the American Academy of Political and Social Science*, 545: 116–125.

Kuo, HW, TF Chiang, II Lo, JS Lai, CC Chan, and JD Wang, 1998. Estimates of cancer risk from chloroform exposure during showering in Taiwan, *The Science of the Total Environment*, 218: 1–7.

Kuo, J, 2014. *Practical Design Calculations for Groundwater and Soil Remediation*, 2nd edition, CRC Press, Boca Raton, FL.

Kushner, EJ, 1976. On determining the statistical parameters for pollution concentration from a truncated data set, *Atmospheric Environment*, 10: 975–979.

LaGoy, PK and CO Schulz, 1993. Background sampling: an example of the need for reasonableness in risk assessment, *Risk Analysis*, 13(5): 483–484.

LaGoy, PK, 1987. Estimated soil ingestion rates for use in risk assessment, *Risk Analysis*, 7(3): 355–359.

LaGoy, PK, 1994. *Risk Assessment, Principles and Applications for Hazardous Waste and Related Sites*, Noyes Data Corp., Park Ridge, NJ.

Laib, RJ, Rose, N, and Brunn, H, 1991. Hepatocarcinogenicity of PCB congeners, *Toxicological Environmental Chemistry*, 34: 19–22.

Laird, FN, 1989. The decline of deference: the political context of risk communication, *Risk Analysis*, 9(2): 543–550.

Lake, LW, 1989. *Enhanced Oil Recovery*, Prentice Hall, Englewood Cliff, NJ.

Land, CE, 1971. Confidence intervals for linear functions of the normal mean and variance, *Annuals of Mathematical Statistics*, 42: 1187–1205.

Landmeyer, J, 2000. Effects of Woody Plants on Ground-Water Hydrology and Contaminant Concentrations, In *Presentation at USEPA Phytoremediation State of the Science Conference*, May 1–2, Boston, MA.

Lanphear, B and R Byrd, 1998. Community characteristics associated with elevated blood lead levels in children, *Pediatrics*, 101(2): 264–271.

Larsen, RJ and ML Marx, 1985. *An Introduction to Probability and its Applications*, Prentice-Hall, Englewood Cliffs, NJ.

Lave, LB (ed.), 1982. *Quantitative Risk Assessment in Regulation*, The Brooking Institute, Washington, DC.

Lave, LB and AC Upton (eds.), 1987. *Toxic Chemicals, Health, and the Environment*, The Johns Hopkins University Press, Baltimore, MD.

Lave, LB and GS Omenn, 1986. Cost-effectiveness of short-term tests for carcinogenicity, *Nature*, 324(6092), 29–34.

Lave, LB, FK Ennever, HS Rosenkranz, and GS Omenn, 1988. Information value of rodent bioassay, *Nature*, 336(6200), 631–633.

Law, AM and WD Kelton, 1991. *Simulation Modeling and Analysis*, McGraw-Hill, New York.

Layton, DW, 1993. Metabolically consistent breathing rates for use in dose assessments, *Health Physics*, 64(1): 23–36.

Lee, BM, SD Yoo, and S Kim, 2002. A proposed methodology of cancer risk assessment modeling using biomarkers, *Journal of Toxicology and Environmental Health A*, 65(5–6): 341–354.

Lee, C and S Lin, 2000. *Handbook of Environmental Engineering Calculations*, McGraw-Hill, New York.

Lee, DG, 1980. *The Oxidation of Organic Compounds by Permanganate Ion and Hexavalent Chromium*, Open Court Pub. Co., La Salle, IL.

Lee, JA, 1985. *The Environment, Public Health, and Human Ecology—Considerations for Economic Development*, A World Bank Publication, The Johns Hopkins University Press, Baltimore, MD.

Lee, RC and JC Kissel, 1995. Probabilistic prediction of exposures to arsenic contaminated residential soil, *Environmental Geochemistry and Health*, 17: 159–168.

Lee, RC, JR Fricke, WE Wright, and W Haerer, 1995. Development of a probabilistic blood lead prediction model, *Environmental Geochemistry and Health*, 17: 169–181.

Lee, RG, WC Becker, and DW Collins, 1989. Lead at the tap: sources and control, *Journal of the American Water Works Association*, 81(7): 52–62.

Lee, YW, MF Dahab, and I Bogardi, 1995. Nitrate-risk assessment using fuzzy-set approach, *Journal of Environmental Engineering*, 121(3): 245–256.

Lees, FB, 1980. *Loss Prevention in the Process Industries, Volume 1*, Butterworths, Boston, MA.

Leeson, A and RE Hinchee, 1997. *Soil Bioventing: Principles and Practice*, Battelle Press: Columbus, OH.

Leidel, N and KA Busch, 1985. Statistical design and data analysis requirements. In *Patty's Industrial Hygiene and Toxicology*, LJ Cralley and LV Cralley (eds.), Ch. 3, pp. 43–98, *Volume 3a*, 2nd edition, John Wiley & Sons, New York.

Leiss, W and C Chociolko, 1994. *Risk and Responsibility*, McGill-Queen's University Press, Montreal, QC.

Leiss, W, 1989. *Prospects and Problems in Risk Communication*, Institute for Risk Research, University of Waterloo, Waterloo, ON.

Lepow, ML, L Bruckman, M Gillette, S Markowitz, R Robino, and J Kapish, 1975. Investigations into sources of lead in the environment of urban children, *Environmental Research*, 10: 415–426.

Lepow, ML, M Bruckman, L Robino, S Markowitz, M Gillette, and J Kapish, 1974. Role of Airborne lead in increased body burden of lead in hartford children, *Environmental Health Perspectives*, 6: 99–101.

Lesage, S and RE Jackson (eds.), 1992. *Groundwater Contamination and Analysis at Hazardous Waste Sites*, Marcel Dekker, New York.

Levallois, P, M Lavoie, L Goulet, AJ Nantel, and S Gingra, 1991. Blood lead levels in children and pregnant women living near a lead-reclamation plant, *Canadian Medical Association Journal*, 144(7): 877–885.

Levesque, B, P Ayotte, R Tardif, L Ferron, S Gingras, E Schlouch, G Gingras, P Levallois, and É Dewailly, 2002. Cancer risk associated with household exposure to chloroform, *Journal of Toxicology and Environmental Health A*, 65(7): 489–502.

Levine, AG, 1982. *Love Canal: Science, Politics/, and People*, Lexington Books, Lexington, MA.

Levine, DG and AC Upton (eds.), 1992. *Management of Hazardous Agents, Volumes 1 and 2*, Praeger, Westport, CT.

Li, YH, 2000. *A Compendium of Geochemistry: From Solar Nebula to the Human Brain*, Princeton University Press, Princeton, NY.

Lifson, MW, 1972. *Decision and Risk Analysis for Practicing Engineers*, Barnes and Noble, Cahners Books, Boston, MA.

Lima, DB, B Parker, J Meyer, 2012. Dechlorinating microorganisms in a sedimentary rock matrix contaminated with a mixture of VOCs. *Environmental Science and Technology*, 46(11): 5756–5763.

Lind, NC, JS Nathwani, and E Siddall, 1991. *Managing Risks in the Public Interest*, Institute for Risk Research, University of Waterloo, Waterloo, ON.

Linders, JBHJ, 2000. *Modelling of Environmental Chemical Exposure and Risk*, Kluwer Academic Publishers, Dordrecht, The Netherlands.

Lindsay, WL, 1979. *Chemical Equilibria in Soils*, John Wiley & Sons: New York.

Linthurst, RA, P Bourdeau, and RG Tardiff (eds.), 1995. *Methods to Assess the Effects of Chemicals on Ecosystems*, SCOPE 53/IPCS Joint Activity 23/SGOMSEC 10, John Wiley & Sons, Chichester, UK.

Lipkus, IM, 2007. Numeric, verbal, and visual formats of conveying health risks: Suggested best practices and future recommendations, *Medical Decision Making*, 27(5), 696–713.

Lippit, J, J Walsh, M Scott, and A DiPuccio, 1986. Cost of Remedial Actions at Uncontrolled Hazardous Waste Sites: Worker Health and Safety Considerations, Project Summary Report No. EPA/600/S2-86/037, US EPA, Washington, DC.

Lippmann, M (ed.), 1992. *Environmental Toxicants: Human Exposures and Their Health Effects*, Van Nostrand Reinhold Co., New York.

Liptak, JF and G Lombardo, 1996. The development of chemical-specific, risk-based soil cleanup guidelines results in timely and cost-effective remediation, *Journal of Soil Contamination*, 5(1): 83–94.

Liptak, SC, JW Atwater, and DS Mavinic (eds.), 1988. In *Proceedings of the 1988 Joint CSCE-ASCE National Conference on Environmental Engineering*, July 13–15, 1988, Pan Pacific Hotel, Vancouver, BC, Canada.

Lister, C, 1996. *European Union Environmental Law: A Guide for Industry*, John Wiley & Sons, Chichester, UK.

Little, JC, JM Daisey, and WW Nazaroff, 1992. Transport of subsurface contaminants in buildings: an expo-
    sure pathway for volatile organics, *Environmental Science and Technology*, 26(11): 2058–2066.

Lockhart, WL, R Wagemann, B Tracey, D Sutherland, and DJ Thomas, 1992. Presence and implications of
    chemical contaminants in the freshwaters of the Canadian Arctic, *Science of the Total Environment*,
    122: 165–243.

Loehr, RC (ed.), 1976. *Land as a Waste Management Alternative*, Ann Arbor Science Publ. Inc., Ann
    Arbor, MI.

Long, FA and GE Schweitzer (eds.), 1982. *Risk Assessment at Hazardous Waste Sites*, American Chemical
    Society, Washington, DC.

Long, WL, 1990. Economic aspects of transport and disposal of hazardous wastes, *Marine Policy International
    Journal*, 14(3): 198–204.

Louvar, JF and BD Louvar, 1998. *Health and Environmental Risk Analysis: Fundamentals with Applications*,
    Prentice-Hall, Upper Saddle River, NJ.

Lowe, DF, CL Oubre, and CH Ward, 1998. *Surfactants and Cosolvents for NAPL Remediation—A Technology
    Practices Manual*, Lewis Publishers, Boca Raton, FL.

Lowe, DF, CL Oubre, and CH Ward, 1999a. *Surfactants and Cosolvents for DNAPL Remediation—A
    Technology Practices Manual*, CRC Press, Boca Raton, FL.

Lowe, DF, CL Oubre, and CH Ward, 1999b. *Reuse of Surfactants and Cosolvents for DNAPL Remediation*,
    Lewis Publishers, Boca Raton, FL.

Lowe, DF, CL Oubre, and CH Ward, 1999c. *Soil Vapor Extraction Using Radio Frequency Heating: Resource
    Manual and Technology Demonstration*, CRC Press, Boca Raton, FL.

Lowe, KS, FG Gardner, and RL Siegrist, 2002. Field evaluation of in situ chemical oxidation through vertical
    well-to-well recirculation of NaMnO4, *GWMR*, 22(Winter 1): 106–115.

Lowrance, WW, 1976. *Of Acceptable Risk: Science and the Determination of Safety*, William Kaufman, Inc.,
    Los Altos, CA.

Lu, FC, 1985a. *Basic Toxicology*, Hemisphere: Washington, DC.

Lu, FC, 1985b. Safety assessments of chemicals with threshold effects, *Regulatory Toxicology and
    Pharmacology*, 5: 121–132.

Lu, FC, 1988. Acceptable daily intake: inception, evolution, and application, *Regulatory Toxicology and
    Pharmacology*, 8: 45–60.

Lu, FC, 1996. *Basic Toxicology: Fundamentals, Target Organs, and Risk Assessment*, 3rd edition, Taylor &
    Francis Group, Washington, DC.

LUFT, 1989. *Leaking Underground Fuel Tank Field Manual: Guidelines for Site Assessment, Cleanup, and
    Underground Storage Tank Closure, State of California, Leaking Underground Fuel Tank Task Force*,
    The Task Force, Sacramento, CA.

Lum M, 1991. Benefits to Conducting Midcourse Reviews. In *Evaluation and Effective Risk Communications
    Workshop Proceedings*, A Fisher, M Pavolva, and V Covello, (eds.), Pub. No. EPA/600/9-90/054, US
    Environmental Protection Agency, Washington, DC.

Lundgren, R, 1994. *Risk Communication: A Handbook for Communicating Environmental, Safety, and
    Health Risks*, Battelle Press, Columbus, OH.

Lurakis MF and JM Pitone, 1984. Occupational lead exposure, acute intoxication, and chronic nephropathy:
    report of a case and review of the literature, *Journal of the American Osteopathic Association*, 83:
    361–366.

Lyman, WJ, WF Reehl, and DH Rosenblatt, 1990. *Handbook of Chemical Property Estimation Methods:
    Environmental Behavior of Organic Compounds*, American Chemical Society, Washington, DC.

Ma, L, KM Komar, C Tu, W Zhang, Y Cai, and ED Kennelley, 2001. A fern that hyperaccumulates arsenic,
    Nature, 409: 579.

Maas, RP and SC Patch, 1990. *Dynamics of Lead Contamination of Residential Tapwater: Implications for
    Effective Control*, Technical Report #90-004, University of North Carolina-Ashville Environmental
    Quality Institute, Ashville, NC.

Macari, EJ, JD Frost, and LF Pumarada (eds.), 1995. *Geo-Environmental Issues Facing the Americas*, ASCE
    Geotechnical Special Publication No.47, American Society of Civil Engineers, New York.

MacCarthy, LS and D Mackay, 1993. Enhancing ecotoxicological modeling and assessment, *Environmental
    Science and Technology*, 27(9): 1719–1728.

MacDonald, JA and BE Rittmann, 1993. Performance standards for in situ bioremediation, *Environmental
    Science and Technology*, 27(10): 1974–1979.

Macintosh, DL, GW Suter II, and FO Hoffman, 1994. Uses of probabilistic exposure models in ecological risk
    assessments of contaminated sites, *Risk Analysis*, 14: 405–419.

Mackay, D and ATK Yeun, 1983. Mass transfer coefficient correlations for volatilization of organic solutes from water, *Environmental Science and Technology*, 17: 211–217.

Mackay, D and PJ Leinonen, 1975. Rate of evaporation of low-solubility contaminants from water bodies, *Environmental Science and Technology*, 9: 1178–1180.

Mackay, D and RS Boethling, 2000. *Handbook of Property Estimation Methods for Environmental and Health Science*, CRC Press, Boca Raton, FL.

Mackay, D, 2001. *Multimedia Environmental Models: The Fugacity Approach*, 2nd edition, CRC Press, Boca Raton, FL.

Mackenzie, PD, DP Horney, and TM Sivavec, 1999. Mineral precipitation and porosity losses in granular iron columns, *Journal of Hazardous Materials*, 68: 1–17.

MacNeill, M, J Kearney, L Wallace, M Gibson, ME Héroux, J Kuchta, JR Guernsey, and AJ Wheeler, 2014. Quantifying the contribution of ambient and indoor-generated fine particles to indoor air in residential environments, *Indoor Air*, 24(4): 362–375.

Mahaffey KR, 1995. Nutrition and lead: strategies for public health, *Environmental Health Perspectives*, 103(Suppl 5): 191–196.

Maheux, PJ and RCE McKee, 1997. An In Situ Air Sparging Method: Contiguous Belled Sand Columns, In *Proceedings from the Fourth International In Situ and On-Site Bioremediation Symposium*, April 28–May 1 (In Situ and On Site Bioremediation: *Volume 1*), New Orleans, LA, pp. 221–226.

Mahmood, RJ and Sims, RC, 1986. Mobility of organics in land treatment systems, *Journal of Environmental Engineering*, 112(2): 236–245.

Malina, Jr., JF (ed.), 1989. Environmental Engineering, In *Proceedings of the 1989 Specialty Conference*, July 10–12, 1989, Austin, TX, ASCE, New York.

Manahan, S, 1992. *Toxicological Chemistry*, CRC Press, Boca Raton, FL.

Manahan, SE, 1993. *Fundamentals of Environmental Chemistry*, CRC Press, Boca Raton, FL.

Manly, BFJ, 1997. *Randomization, Bootstrap, and Monte Carlo Methods in Biology*, 2nd edition, Chapman and Hall, London, UK.

Mansour, M (ed.), 1993. *Fate and Prediction of Environmental Chemicals in Soils, Plants, and Aquatic Systems*, CRC Press, Boca Raton, FL.

Marcus WL, 1986. Lead health effects in drinking water, *Toxicology and Industrial Health*, 2: 363–400.

Mariner, PE, M Jin, and RE Jackson, 1997. An algorithm for the estimation of DNAPL saturation and composition from typical soil chemical analyses. *Ground Water Monitoring and Remediation*, 17(2): 122–129.

Markowitz ME and Rosen JF, 1991. Need for the lead mobilization test in children with lead poisoning, *Journal of Pediatrics*, 119(2): 305–310.

Martin, EJ and JH Johnson Jr. (eds.), 1987. *Hazardous Waste Management Engineering*, Van Nostrand Reinhold Co., New York.

Martin, HW, TR Young, DI Kaplan, L Simon, and DC Adriano, 1996. Evaluation of three herbaceous index plant species for bioavailability of soil cadmium, chromium, nickel, and vanadium, *Plant and Soil*, 182: 199–207.

Martin, WF, JM Lippitt, and TG Prothero, 1992. *Hazardous Waste Handbook for Health and Safety*, 2nd edition, Butterworth-Heinemann, London, UK.

Mason, A and BH Kueper, 1996. Numerical simulation of surfactant-enhanced solubilization of pooled DNAPL, *Environmental Science Technology*, 30(11): 3205–3215.

Massaro, EJ, (ed.), 1998. *Handbook of Human Toxicology*, CRC Press, Boca Raton, FL.

Massmann, J and RA Freeze, 1987. Groundwater contamination from waste management sites: the interaction between risk-based engineering design and regulatory policy 1. Methodology 2. Results, *Water Resources Research*, 23(2): 351–380.

Masters, GM, 1998. *Introduction to Environmental Engineering and Science*, 2nd edition, Prentice Hall, Upper Saddle River, NJ.

Mathews, JT, 1991. *Preserving the Global Environment: The Challenge of Shared Leadership*, W.W. Norton & Co., New York.

Matso, K, 1995. Mother nature's pump and treat, *Civil Engineering*, 65: 46–49.

Matz, CJ, DM Stieb, and O Brion, 2015. Urban–rural differences in daily time-activity patterns, occupational activity and housing characteristics, *Environmental Health*, 14(1): 1–11.

Maughan, JT, 1993. *Ecological Assessment of Hazardous Waste Sites*, Van Nostrand Reinhold Co., New York.

Maurits la Riviere, JW, 1989. Threats to the World's Water, *Scientific American*, Managing Planet Earth, September 1989, Special Issue.

Maxwell, RM and WE Kastenberg, 1999. Stochastic environmental risk analysis: an integrated methodology for predicting cancer risk from contaminated groundwater, *Stochastic Environmental Research and Risk Assessment*, 13: 27–47.

Maxwell, RM, SD Pelmulder, AFB Tompson, and WE Kastenberg, 1998. On the development of a new methodology for groundwater driven health risk assessment, *Water Resources Research*, 34(4): 833–847.

Mayer, AS and CT Miller, 1992. The influence of porous media characteristics and measurement scale on pore-scale distributions of residual nonaqueous phase liquids, *Journal of Contamination Hydrology*, 11: 189–213.

Mayer, AS, L Zhong, and GA Pope, 1999. Measurement of mass-transfer rates for surfactant-enhanced solubilization of nonaqueous phase liquids, *Environmental Science Technology*, 33: 2965–2972.

Mayer, FL and MR Ellersick, 1986. Manual of Acute Toxicity: Interpretation and Database for 410 Chemicals and 66 Species of Freshwater Animals, US Department of Interior, Fish and Wildlife Service, resource Publ. 160, Washington, DC.

McColl, RS (ed.), 1987. *Environmental Health Risks: Assessment and Management*, Institute for Risk Research, University of Waterloo, Waterloo, ON.

McDonagh, A and MA Byrne, 2014. The influence of human physical activity and contaminated clothing type on particle resuspension, *Journal of Environmental Radioactivity*, 127: 119–126.

McIntyre, T, 2001. *PhytoRem: A Global CD-ROM Database of Aquatic and Terrestrial Plants that Sequester, Accumulate, or Hyperaccumulate Heavy Metals*, Environment Canada, Hull, QC.

McKone, TE and JI Daniels, 1991. Estimating human exposure through multiple pathways from air, water, and soil, *Regulatory Toxicology and Pharmacology*, 13: 36–61.

McKone, TE and JP Knezovich, 1991. The transfer of trichloroethylene (TCE) from a shower to indoor air: experimental measurements and their implications, *Journal of Air and Waste Management*, 41: 832–837.

McKone, TE and KT Bogen, 1991. Predicting the uncertainties in risk assessment, *Environmental Science and Technology*, 25: 1674–1681.

McKone, TE and RA Howd, 1992. Estimating dermal uptake of nonionic organic chemicals from water and soil: part 1, unified fugacity-based models for risk assessments, *Risk Analysis*, 12: 543–557.

McKone, TE, 1987. Human exposure to volatile organic compounds in household tap water: the indoor inhalation pathway, *Environmental Science and Technology*, 21(12): 1194–1201.

McKone, TE, 1989. Household exposure models, *Toxicology Letters*, 49: 321–339.

McKone, TE, 1993. Linking a PBPK model for chloroform with measured breath concentrations in showers: implications for dermal exposure models, *Journal of Exposure Analysis and Environmental Epidemiology*, 3: 339–365.

McKone, TE, 1994. Uncertainty and variability in human exposures to soil contaminants through home-grown food: a Monte Carlo assessment, *Risk Analysis*, 14: 449–463.

McKone, TE, WE Kastenberg, and D Okrent, 1983. The use of landscape chemical cycles for indexing the health risks of toxic elements and radionuclides, *Risk Analysis*, 3(3): 189–205.

McTernan, WF and E Kaplan (eds.), 1990. *Risk Assessment for Groundwater Pollution Control*, ASCE Monograph, American Society of Civil Engineers, New York.

Meagher RB and C Rugh, 1996. Phytoremediation of Mercury Pollution Using a Modified Bacterial Mercuric-ion Reductase Gene, In *International Phytoremediation Conference*, May 8–10, 1996, Arlington, VA.

Means, 2000. *Environmental Remediation Cost Data—Unit Price*. R.S. Means Company, Kingston, MA.

Medina, VF, SL Larson, W Perez, and LE Agwaramgbo, 2000. Evaluation of Minced and Pureed Plants for Phytotreatment of Munitions, In *Presentation at the Second International Conference on Remediation of Chlorinated and Recalcitrant Compounds*, May 22–25, Monterey, CA.

Meek, ME, 2001. Categorical default uncertainty factors—Interspecies variation and adequacy of database, *Human and Ecological Risk Assessment*, 7: 157–163.

Menzie-Cura & Associates, 1996. An Assessment of the Risk Assessment Paradigm for Ecological Risk Assessment, Report Prepared for the Presidential/Congressional Commission on Risk Assessment and Risk Management, Washington, DC.

Merck, 1989. *The Merck Index: An Encyclopedia of Chemicals, Drugs and Biologicals*, 11th (Centennial) edition, Merck & Co., Rockway, NJ.

Meyer, CR, 1983. Liver dysfunction in residents exposed to leachate from a toxic waste dump, *Environmental Health Perspectives*, 48: 9–13.

Meyer, PB, RH Williams, and KR Yount, 1995. *Contaminated Land: Reclamation, Redevelopment and Reuse in the United States and the European Union*, Edward Elgar Publishing Ltd., Aldershot, UK.

Michigan Department of Environmental Quality, 2014. Groundwater Modeling, Remediation and Redevelopment Division Resource Materials, *RRD-Resource Material* 25-2013-01.

Mielke HW, D Dugas, PW Mielke Jr., KS Smith, SL Smith, and CR Gonzales, 1997. Associations between soil lead and childhood blood lead in urban New Orleans and rural Lafourche Parish of Louisiana, *Environmental Health Perspectives*, 105: 950–954.

Millard, SP, 1997. *Environmental Stats for S-Plus User's Manual, Version 1.0*, Probability, Statistics, and Information, Seattle, WA.

Miller, DW (ed.), 1980. *Waste Disposal Effects on Ground Water*, Premier Press, Berkeley, CA.

Miller, GT, 1991. *Environmental Science: Sustaining the Earth*, Wadsworth Publishing Co., Belmont, CA.

Miller, I and JE Freund, 1985. *Probability and Statistics for Engineers*, 3rd edition, Prentice-Hall, Englewood Cliffs, NJ.

Millette, JR and SM Hays, 1994. *Settled Asbsestos Dust Sampling and Analysis*, CRC Press, Boca Raton, FL.

Millner, GC, RC James, and AC Nye, 1992. Human health-based soil cleanup guidelines for diesel fuel No. 2, *Journal of Soil Contamination*, 1(2): 103–157.

Mills, WB, Dean, JD, Porcella, DB, MJ Ungs, SA Gherini, KV Summers, M Lingfung, GL Rupp, and GL Bowie. 1982. Water quality assessment: a screening procedure for toxic and conventional pollutants: Parts 1, 2, and 3. *Athens, GA: U.S. Environmental Protection Agency. Environmental Research Laboratory. Office of Research and Development. EPA600/6-821004 a,b,c.*

Mitchell JW (ed.), 1987. Occupational medicine forum: lead toxicity and reproduction, *Journal of Occupational Medicine*, 29: 397–399.

Mitchell, JK and GD Bubenzer, 1980. Soil loss estimation. In *Soil Erosion*, pp. 17–62, MJ Kirkby and PC Morgan (eds.), John Wiley & Sons, London, UK.

Mitchell, JK, 1993. *Fundamentals of Soil Behavior*, 2nd edition, John Wiley & Sons, New York.

Mitchell, P and D Barr, 1995. The nature and significance of public exposure to arsenic: a review of its relevance to South West England, *Environmental Geochemistry and Health*, 17: 57–82.

Mitsch, WJ and SE Jorgesen (eds.), 1989. *Ecological Engineering, An Introduction into Ecotechnology*, John Wiley & Sons, New York.

Mockus, J, 1972. Estimation of direct runoff from storm rainfall. In *National Engineering Handbook. Section 4: Hydrology*, AT Hjelmfelt (ed.), U.S. Department of Agriculture. Soil Conservation Service, Washington, DC.

Moeller, DW, 1992. *Environmental Health*, Harvard University Press, Cambridge, MA.

Moeller, DW, 1997. *Environmental Health*, Revised Edition, Harvard University Press, Cambridge, MA.

Mohamed, AMO and K Côté, 1999. Decision analysis of polluted sites—A fuzzy set approach, *Waste Management*, 19: 519–533.

Mohanty NR and IW Wei, 1992. Oxidation of 2,4-dinitrotoluene using fenton's reagent: reaction mechanisms and their practical applications, *Journal of Hazardous Wastes and Hazardous Material*, 10: 71–183.

Monahan, DJ, 1990. Estimation of hazardous wastes from employment statistics—Victoria, Australia, *Waste Management and Research*, 8(2): 145–149.

Moore, AO, 1987. *Making Polluters Pay—A Citizen's Guide to Legal Action and Organizing*, Environmental Action Foundation, Washington, DC.

Moore, MR, A Goldberg, and AC Yeung-Laiwah, 1987. Lead effects on the hemebiosynthetic pathway, *Annals of the New York Academy of Sciences*, 514: 191–202.

Morgan, MG and L Lave, 1990. Ethical considerations in risk communication practice and research, *Risk Analysis*, 10(3): 355–358.

Morgan, MG and M Henrion, 1991. *Uncertainty: A Guide to Dealing with Uncertainty in Quantitative Risk and Policy Analysis*, Oxford University Press, Cambridge, UK.

Morgan, MG, 2009. *Best Practice Approaches for Characterizing, Communicating and Incorporating Scientific Uncertainty in Climate Decision Making*, National Oceanic and Atmospheric Administration, Washington, DC.

Morris P and R Therivel (eds.), 1995. *Methods of Environmental Impact Assessment*, UCL Press, London, UK.

Morrison, RD, 1999. *Environmental Forensics: Principles and Applications*, CRC Press, Boca Raton, FL.

Muehlberger, EW, P Hicks, and K Harris, 1997. Biosparging of a Dissolved Petroleum Hydrocarbon Plume, In *Proceedings from the Fourth International In Situ and On-Site Bioremediation Symposium*, April 28–May 1 (In Situ and On Site Bioremediation: *Volume 1*), New Orleans, LA, p. 209.

Muir DCG, R Wagemann, BT Hargrave, DJ Thomas, DB Peakall, and RJ Norstrom, 1992. Arctic marine ecosystem contamination, *Science of the Total Environment*, 122: 75–134.

Mulkey, LA, 1984. Multimedia fate and transport models: an overview, *Journal of Toxicology-Clinical Toxicology*, 21(1–2): 65–95.

Munro, IC and Krewski, DR, 1981. Risk assessment and regulatory decision making, food cosmet, *Toxicology*, 19: 549–560.

Murphy, F and WN Herkelrath, 1996. A sample-freezing drive shoe for a wire line piston sampler, *Ground Water Monitoring and Review*, 16(Summer 3): 86–90.

Mushak, P and AF Crocetti, 1995. Risk and revisionism in arsenic cancer risk assessment, *Environmental Health Perspectives*, 103: 684–688.

NAE (National Academy of Engineering), 1993. *Keeping Pace with Science and Engineering: Case Studies in Environmental Regulation*, National Academy Press, Washington, DC.

Naidu, R and V Birke (eds.), 2014. *Permeable Reactive Barrier: Sustainable Groundwater Remediation*, CRC Press, Boca Raton, FL.

Nathwani, J, NC Lind, and E Siddall, 1990. Risk-Benefit Balancing in Risk Management: Measures of Benefits and Detriments, In *Presented at the Annual Meetingg of the Society for Risk Analysis*, 29th October–1 November, 1989, San Francisco, CA, Institute for Risk Research Paper No.18, Waterloo, ON.

NATO/CCMS, 1988a. Pilot Study on International Information Exchange on Dioxins and Related Compounds, North Atlantic Treaty Organization, Committee on the Challenges of Modern Society, Report 176, August 1988.

NATO/CCMS, 1988b. Scientific Basis for the Development of International Toxicity Equivalency Factor (I-TEF), Method of Risk Assessment for Complex Mixtures of Dioxins and Related Compounds, North Atlantic Treaty Organization, Committee on the Challenges of Modern Society, Report No. 178, December 1988.

NCI (National Cancer Institute), 1992. *Making Health Communication Programs Work: A Planner's Guide*, NIH Publication No. 92(1493): 64–65, National Cancer Institute, Washington, DC.

Needleman HL, C Gunnoe, A Leviton, R Reed, H Peresie, C Maher, and P Barrett, 1979. Deficits in psychologic and classroom performance of children with elevated dentine lead levels, *New England Journal of Medicine*, 300: 689–695.

Needleman, HL and CA Gatsonis, 1990. Low-level lead exposure and the IQ of children: a meta-analysis of modern studies, *Journal of the American Medical Association*, 263(5): 673–678.

Neely, WB, 1980. *Chemicals in the Environment (Distribution, Transport, Fate, Analysis)*, Marcel Dekker, New York.

Neely, WB, 1994. *Introduction to Chemical Exposure and Risk Assessment*, CRC Press, Boca Raton, FL.

Negri, MC, RR Hinchman, and JB Wozniak, 2000. Capturing a Mixed Contaminant Plume: Tritium Phytoevaporation at Argonne National Laboratory's Area 319, In *Presentation at USEPA Phytoremediation State of the Science Conference*, May 1–2, Boston, MA.

Nelson, DE, BW, Hesse, and RT Croyle, 2009. *Making Data Talk: Communicating Public Health Data to the Public, Policy Makers, and the Press*, Oxford University Press, New York.

Nendza, M, 1997. *Structure-Activity Relationships in Environmental Sciences*, Kluwer Academic Publishers, Dordrecht, The Netherlands.

Neubacher, FP, 1988. *Policy Recommendations for the Prevention of Hazardous Waste*, Elsevier Science Publishers, Amsterdam, The Netherlands.

Neuman, SP, 2005. Trends, prospects and challenges in qualifying flow and transport through fractured rocks, *Hydrogeology Journal*, 13(1): 124–147.

Neumann, DA and CA Kimmel (eds.), 1999. *Human Variability in Response to Chemical Exposures*, CRC Press, Boca Raton, FL.

Newell, CJ, HS Rifai, JT Wilson, JA Connor, JA Aziz, and MP Suarez, 2002. *Calculation and Use of First-Order Rate Constants for Monitored Natural Attenuation Studies, Ground Water Issues*, EPA/540/S-02/500, US Environmental Protection Agency (USEPA), National Risk Management Research Laboratory, Cincinnati, OH.

Newman, A, 1995. Plant enzymes set for bioremediation field study, *Environment Science and Technology*, 29: 18a.

Newman, B, M Martinson, G Smith, and L McCain, 1993. 'Dig-and-Mix' bioventing enhances hydrocarbon degradation at service station site, *Hazmat World*, December 1993: 34–40.

Newman, LA, MP Gordon, P Heilman, DL Cannon, E Lory, K Miller, J Osgood, and SE Strand, 1999. Phytoremediation of MTBE at a California Naval Site, *Soil & Groundwater Cleanup*, February/March, 42–45.

Newman, LA, SE Strand, N Choe, J Duffy, G Ekuan, M Ruszaj, BB Shurtleff, J Wilmoth, P Heilman, and MP Gordon, 1997. Uptake and biotransformation of trichloroethylene by hybrid poplars, *Environment Science and Technology*, 31: 1062–1067.

Ng, KL and DM Hamby, 1997. Fundamentals for establishing a risk communication program, *Health Physics*, 73(3): 473–482.

Nielsen, DM (ed.), 1991. *Practical Handbook of Groundwater Monitoring*, Lewis Publishers, Chelsea, MI.

Niessen, WR, 2010. *Combustion and Incineration Processes: Applications in Environmental Engineering*, 4th edition, CRC Press, Boca Raton, FL.

NIOSH (National Institute for Occupational Safety and Health), 1982. Registry of Toxic Effects of Chemical Substances, RL Tatken and RJ Lewis (eds.), US Department of Health and Human Services (DDSH), NIOSH, Cincinnati, OH DHHS (NIOSH) Publication No. 83-107.

Norris, RD (ed.), 1994. *Handbook of Bioremediation*, CRC Press, Boca Raton, FL.

Norrman, J, 2001. *Decision Analysis under Risk and Uncertainty at Contaminated Sites—A Literature Review*, SGI Varia 501, Swedish Geotechnical Institute (SGI), Linkoping, Sweden.

NRC (National Research Council), 1972. *Specifications and Criteria for Biochemical Compounds*, 3rd edition, National Academy of Sciences, Washington, DC.

NRC, 1977a. *Drinking Water and Health, Volume 1*, National Academy Press, Washington, DC.

NRC, 1977b. *Drinking Water and Health, Safe Drinking Water Committee, Advisory Center on Toxicology*, National Academy of Sciences, Washington, DC.

NRC, 1977c. *Environmental Monitoring, Analytical Studies for the U.S. Environmental Protection Agency, Volume 4*, National Academy Press, Washington, DC.

NRC, 1980a. *Drinking Water and Health, Volume 2*, National Academy Press, Washington, DC.

NRC, 1980b. *Drinking Water and Health, Volume 3*, National Academy Press, Washington, DC.

NRC, 1981a. *Prudent Practices for Handling Hazardous Chemicals in Laboratories*, National Academy Press: Washington, DC.

NRC, 1981b. *Testing for Effects of Chemicals on Ecosystems*, NAS Press, Washington, DC.

NRC, 1982a. *Risk and Decision-Making: Perspective and Research*, NRC Committee on Risk and Decision-Making, National Academy Press, Washington, DC.

NRC, 1982b. *Drinking Water and Health, Volume 4*, National Academy Press, Washington, DC.

NRC, 1983a. *Drinking Water and Health, Volume 5*, National Academy Press, Washington, DC.

NRC, 1983b. *Risk Assessment in the Federal Government: Managing the Process*, National Research Council, Committee on the Institutional Means for Assessment of Risks to Public Health, National Academy Press, Washington, DC.

NRC, 1988. *Hazardous Waste Site Management: Water Quality Issues*, National Academy Press, Washington, DC.

NRC, 1989a. *Biological Markers of Air-Pollution Stress and Damage in Forests*, NAS Press, Washington, DC.

NRC, 1989b, *Ground Water Models: Scientific and Regulatory Applications*, National Research Council (NRC), National Academy Press, Washington, DC.

NRC, 1989c, *Improving Risk Communication*, National Research Council, Committee on Risk Perception and Communication, National Academy Press, Washington, DC.

NRC, 1991a, *Frontiers in Assessing Human Exposure to Environmental Toxicants*, National Academy Press, Washington, DC.

NRC, 1991b, *Human Exposure Assessment for Airborne Pollutants: Advances and Opportunities*, National Academy Press, Washington, DC.

NRC, 1993a, *Ground Water Vulnerability Assessment: Predicting Relative Contamination Potential Under Conditions of Uncertainty*, National Academy Press, Washington, DC.

NRC, 1993b. *In Situ Bioremediation: When Does It Work?* National Academy Press, Washington, DC.

NRC, 1993c. *Issues in Risk Assessment*, National Academy Press, Washington, DC.

NRC, 1993d. *Pesticides in the Diets of Infants and Children*, National Academy Press, Washington, DC.

NRC, 1994a. *Alternatives for Ground Water Cleanup*, Committee on Ground Water Cleanup Alternatives, National Academy Press, Washington, DC.

NRC, 1994b. *Building Consensus through Risk Assessment and Risk Management*, National Academy Press, Washington, DC.

NRC, 1994c. *Ranking Hazardous Waste Sites for Remedial Action*, National Academy Press, Washington, DC.

NRC, 1994d, *Science and Judgment in Risk Assessment*, National Research Council, Committee on Risk Assessment of Hazardous Air Pollutants, National Academy Press, Washington, DC.

NRC, 1995. *Improving the Environment: An Evaluation of DOE's Environmental Management Program*, Washington, DC, National Academy Press.

NRC, 1996a. *Linking Science and Technology to Society's Environmental Goals*, National Forum on Science and Technology Goals: Environment, National Research Council (NRC), National Academy Press, Washington, DC.

NRC, 1996b. *Understanding Risk: Informing Decisions in a Democratic Society*, National Academy Press, Washington, DC.

NRC, 1996c. *Rock Fractures and Fluid Flow: Contemporary Understanding and Applications*, National Academies of Sciences, Engineering, and Medicine, The National Academies Press, Washington, DC.

NRC, 1997. *Innovations in Ground Water and Soil Cleanup: From Concept to Commercialization*, National Academy Press, Washington, DC.

NRC, 2000. *Copper in Drinking Water*, National Academy Press, Washington, DC.

NRC, 2001. *Conceptual Models of Flow and Transport in the Fractured Vadose Zone*, National Academies of Sciences, Engineering, and Medicine, The National Academies Press, Washington, DC.

NRC, 2002a. *Estimating the Public Health Benefits of Proposed Air Pollution Regulations*, The National Academies Press, Washington, DC.

NRC, 2002b. *Scientific Frontiers in Developmental Toxicology and Risk Assessment*, National Academy Press, Washington, DC.

NRC, 2005. *Health Implications of Perchlorate Ingestion*, The National Academies Press, Washington, DC.

NRC, 2006a. *Assessing the Human Risks of Trichloroethylene*, The National Academies Press, Washington, DC.

NRC, 2006b. *Health Risks for Dioxin and Related Compounds/Evaluation of the EPA Reassessment*, The National Academies Press, Washington, DC.

NRC, 2006c. *Health Risks from Exposures to Low Levels of Ionizing Radiation: BEIR VII*, The National Academies Press, Washington, DC.

NRC, 2006d. *Human Biomonitoring for Environmental Chemicals*, Committee on Human Biomonitoring for Environmental Toxicants, The National Academies Press, Washington, DC.

NRC, 2006e. *Toxicity Testing for Assessment of Environmental Agents: Interim Report*, The National Academies Press, Washington, DC.

NRC, 2007a. *Toxicity Testing in the 21st Century: A Vision and a Strategy*, The National Academies Press, Washington, DC.

NRC, 2008a. *Phthalates and Cumulative Risk Assessment: The Tasks Ahead*, National Academies of Sciences, Engineering, and Medicine, The National Academies Press, Washington, DC.

NRC, 2008b. *Public Participation in Environmental Assessment and Decision Making*, The National Academies Press, Washington, DC.

NRC, 2009. *Science and Decisions: Advancing Risk Assessment*, Committee on Improving Risk Analysis Approaches Used by the U.S. EPA The National Academies Press, Washington, DC.

NRC, 2011. *A Risk-Characterization Framework for Decision Making at the Food and Drug Administration*, The National Academies Press, Washington, DC.

NRC, 2012a. *Exposure Science in the 21st Century: A Vision and a Strategy*, National Academies of Sciences, Engineering, and Medicine, The National Academies Press, Washington, DC.

NRC, 2012b. *Science for Environmental Protection: The Road Ahead*, National Academies of Sciences, Engineering, and Medicine, The National Academies Press, Washington, DC.

NRC, 2015. *Characterization, Modeling, Monitoring, and Remediation of Fractured Rock*, National Academies of Sciences, Engineering, and Medicine, The National Academies Press, Washington, DC.

NRC, 2016. *Health Risks of Indoor Exposure to Particulate Matter: Workshop Summary*, National Academies of Sciences, Engineering, and Medicine, The National Academies Press, Washington, DC.

NTP, 1991. *Sixth Annual Report on Carcinogens*, National Toxicology Program (NTP), US Department of Health and Human Services, Public Health Service, Washington, DC.

Nyer EK and EG Gatliff, 1996. Phytoremediation, *Ground Water Monitoring and Remediation*, 16: 58–62.

Nyer, EK, 1992. *Groundwater Treatment Technology*, 2nd edition, Van Nostrand Reinhold Co., New York.

Nyer, EK, 1993. *Practical Techniques for Groundwater and Soil Remediation*, Lewis Publishers, Boca Raton, FL.

O'Hare, M, L Bacow, and D Sanderson, 1983. *Facility Siting and Public Opposition*, Van Nostrand Reinhold Co., New York.

O'Shay, TA and KB Hoddinott (eds.), 1994. *Analysis of Soils Contaminated with Petroleum Constituents*, ASTM Publication, STP 1221, ASTM, Philadelphia, PA.

O'Flaherty, EJ, 1998. A physiologically based kinetic model for lead in children and adults, *Environmental Health Perspectives*, 106(Suppl 6): 1495–1503.

OBG (O'Brien & Gere Engineers, Inc.), 1988. *Hazardous Waste Site Remediation: The Engineer's Perspective*, Van Nostrand Reinhold Co., New York.

Odermatt, JR and JA Menatti, 1996. Methodology for using contaminated soil leachability testing to determine soil cleanup levels at contaminated petroleum underground storage tank (UST) sites, *Journal of Soil Contamination*, 5(2): 157–169.

Odziemkowski, MS, TT Schuhmacher, RW Gillham, and EJ Reardon, 1998. Mechanism of oxide film formation on iron in simulating groundwater solutions: raman spectral studies, *Corrosion Studies*, 40(2/3): 371–389.

OECD (Organization for Economic Cooperation and Development), 1986a. *Existing Chemicals—Systematic Investigation, Priority Setting and Chemical Review*, Organization for Economic Cooperation and Development, Paris, France.

OECD, 1986b. Report of the OECD Workshop on Practical Approaches to the Assessment of Environmental Exposure, April 14–18, 1986, Vienna, Austria.

OECD, 1989. *Compendium of Environmental Exposure Assessment Methods for Chemicals*, Environmental Monographs, No.27, OECD, Paris, France.

OECD, 1993. *Occupational and Consumer Exposure Assessment*, OECD Environment Monograph 69, Organization for Economic Cooperation and Development (OECD), Paris, France.

OECD, 1994. *Environmental Indicators: OECD Core Set*, Organization for Economic Cooperation and Development (OECD), Paris, France.

Olin, SS (ed.), 1999. *Exposure to Contaminants in Drinking Water: Estimating Uptake through the Skin and by Inhalation*, CRC Press, Boca Raton, FL.

Olsen, PE and JS Fletcher, 1999. Field evaluation of mulberry root structure with regard to phytoremediation, *Bioremediation Journal*, 3(1): 27–33.

Onalaja, AO and L Claudio, 2000. Genetic susceptibility to lead poisoning, *Environmental Health Perspectives*, 108(Suppl 1): 23–28.

Onishi, Y, AR Olsen, MA Parkhurst, and G Whelan, 1985. Computer-based environmental exposure and risk assessment methodology for hazardous materials, *Journal of Hazardous Materials*, 10(1985): 389–417.

Orchard, BJ, JK Chard, WJ Doucette, and B Bugbee, 1999. Laboratory Studies on Plant Uptake of TCE, In *Presentation at the 5th International In-Situ and On-Site Bioremediation Symposium*, Battelle, San Diego, CA.

OSA, 1992. *Supplemental Guidance for Human Health Multimedia Risk Assessments of Hazardous Waste Sites and Permitted Facilities*, Office of Scientific Affairs (OSA), Cal-EPA, DTSC, Sacramento, CA.

OSA, 1993. Supplemental Guidance for Human Health Multimedia Risk Assessments of Hazardous Waste Sites and Permitted Facilities, Cal EPA, DTSC, 1993.

OSHA, 1980. Identification, classification, and regulation of potential occupational carcinogens, Occupational Safety and Health Administration (OSHA), US Department of Labor, *Federal Register*, 45(13): 5001–5296.

Osiensky, JL, 1995. Time series electrical potential field measurements for early detection of groundwater contamination, *Journal of Environmental Science and Health*, A30(7): 1601–1626.

OSTP (Office of Science and Technology Policy), 1985. Chemical carcinogens: a review of the science and its associated principles, *Federal Register*, 50: 10372–10442.

OTA (Office of Technology Assessment), 1981. *Assessment of Technologies for Determining Cancer Risks from the Environment*, Office of Technology Assessment, Washington, DC.

OTA, 1983. *Technologies and Management Strategies for Hazardous Waste Control*, US Congress, Office of Technology Assessment, Washington, DC.

OTA, 1993. *Researching Health Risks, Office of Technology Assessment (OTA)*, US Congress, US Government Printing Office, Washington, DC.

Ott, WR, 1995. *Environmental Statistics and Data Analysis*, CRC Press, Boca Raton, FL.

Ottoboni, MA, 1997. *The Dose Makes the Poison (A Plain Language Guide to Toxicology)*, 2nd edition, Van Nostrand Reinhold Co., ITP, New York.

Overcash, MR and D Pal, 1979. *Design of Land Treatment Systems for Industrial Wastes—Theory and Practice*, Ann Arbor Science Publ. Inc., Ann Arbor, MI.

Paasivirta, J, 1991. *Chemical Ecotoxicology*, Lewis Publishers, Chelsea, MI.

Page, GW and M Greenberg, 1982. Maximum contaminant levels for toxic substances in water: a statistical approach, *Water Resources Bulletin*, 18(6 December), 955–962.

Painter, HA and TF Zabel, 1988. Review of the Environmental Safety of LAS, Report to the European Centre of Studies on Linear Alkylbenzene and Derivatives (ECOSOL), p. 140.

Pankow, JF and JA Cherry, 1996. *Dense Chlorinated Solvents and Other DNAPLs in Groundwater: History, Behavior, and Remediation*, Waterloo Press, Portland, OR.

Park, CN and RD Snee, 1983. Quantitative risk assessment: state of the art for carcinogenesis, *The American Statistician*, 37(4): 427–441.

Parkhurst, DF, 1998. Arithmetic versus geometric means for environmental concentration data, *Environmental Science and Technology*, 32(3): 92A–98A.

Pastorok, RA, SM Bartell, S Ferson, and LR Ginzburg (eds.), 2002. *Ecological Modeling in Risk Assessment*, CRC Press, Boca Raton, FL.

Patch, SC, RP Maas, and JP Pope, 1998. Lead leaching from faucet fixtures under residential conditions, *Environmental Health*, 61(3): 18–21.

Patil, GP and CR Rao (eds.), 1994. *Handbook of Statistics, Volume 12*, Environmental Statistics, North-Holland, NY.

Patnaik, P, 1992. *A Comprehensive Guide to the Hazardous Properties of Chemical Substances*, Van Nostrand Reinhold Co., New York.

Patrick, DR (ed.), 1996. *Toxic Air Pollution Handbook*, Van Nostrand Reinhold Co., New York.

Patton, DE, 1993. The ABCs of risk assessment, *EPA Journal*, 19: 10–15.

Paustenbach, DJ (ed.), 1988. *The Risk Assessment of Environmental Hazards: A Textbook of Case Studies*, John Wiley & Sons, New York.

Paustenbach, DJ, GM Bruce, and P Chrostowski, 1997. Current views on the oral bioavailability of inorganic mercury in soil: implications for health risk assessments, *Risk Analysis*, 17(5): 533–544.

Payne, E, M Gallagher, S Pinizzotto, and E Nobles-Harris, 1997. Air Sparging Below Hydrocarbon Free Product Without Vapor Control. In *Proceedings from the Fourth International In Situ and On-Site Bioremediation Symposium*, April 28–May 1 (In Situ and on Site Bioremediation: *Volume 1*), New Orleans, LA, pp. 181–186.

Payne, FC, JA Quinnan, and ST Potter, 2008. *Remediation Hydraulics*, CRC Press, Boca Raton, FL.

Peck, DL (ed.), 1989. *Psychosocial Effects of Hazardous Toxic Waste Disposal on Communities*, Charles C. Thomas Publishers, Springfield, IL.

Pedersen, J, 1989. *Public Perception of Risk Associated with the Siting of Hazardous Waste Treatment Facilities*, European Foundation for the Improvement of Living and Working Conditions, Dublin, Eire.

Pennell, KD, GA Pope, and LM Abriola, 1996. Influence of viscous and buoyancy forces on the mobilization of residual tetrachloroethylene during surfactant flushing, *Environmental Science Technology*, 30(4): 1328–1335.

Pennell, KD, LM Abriola, and WJ Weber Jr., 1993. Surfactant enhanced solubilization of residual dodecane in soil columns 1. Experimental investigation, *Environmental Science Technology*, 27(12): 2332–2340.

Peres, MM, M Onur, and A Reynolds, 1989. A new analysis procedure for determining aquifer properties from slug test data, *Water Resources Research*, 25(7): 1591–1602.

Perket, CL (ed.), 1986. *Quality Control in Remedial Site Investigation: Hazardous and Industrial Solid Waste Testing, Volume 5*, ASTM STP 925, American Society for Testing and Materials (ASTM), Philadelphia, PA.

Petak, WJ and AA Atkisson, 1982. *Natural Hazard Risk Assessment and Public Policy: Anticipating the Unexpected*, Springer-Verlag, New York.

Peters, RG, VT Covello, and DB McCallum, 1997. The determinants of trust and credibility in environmental risk communication: an empirical study, *Risk Analysis*, 17(1): 43–54.

Petito Boyce, C and MR Garry, 2003. Developing risk-based target concentrations for carcinogenic polycyclic aromatic hydrocarbon compounds assuming human consumption of aquatic biota, *Journal of Toxicology and Environmental Health B Critical Reviews*, 6: 497–520.

Petito Boyce, C, AS Lewis, SN Sax, M Eldan, SM Cohen, and BD Beck, 2008. Probabilistic analysis of human health risks associated with background concentrations of inorganic arsenic: use of a margin of exposure approach, *Human and Ecological Risk Assessment*, 14: 1159–1201.

Petts, J, T Cairney, and M Smith, 1997. *Risk-Based Contaminated Land Investigation and Assessment*, John Wiley & Sons, Chichester, UK.

Peurrung, LM and Schalla, R, 1998. *Six-Phase Soil Heating of the Saturated Zone. Remediation of Chlorinated and Recalcitrant Compounds: Physical, Chemical, and Thermal Technologies*, Battelle Press, Columbus, OH.

Phelan, MJ, 1998. Environmental health policy decisions: the role of uncertainty in economic analysis, *Journal of Environmental Health*, 61(5): 8–13.

Pichtel, J, 2014. *Waste Management Practices: Municipal, Hazardous, and Industrial*, 2nd edition, CRC Press, Boca Raton, FL.

Pickering, QH and C Henderson, 1966. The acute toxicity of some heavy metals to different species of warm water fish, *International Journal of Air and Water Pollution*, 10: 453–463.

Piddington KW, 1989. Sovereignty and the environment, *Environment*, 31(7): 18–20, 35–39.

Pierzynski, GM, JT Sims, and GF Vance, 1994. *Soils and Environmental Quality*, CRC Press, Boca Raton, FL.

Pierzynski, GM, JT Sims, and GF Vance, 2000. *Soils and Environmental Quality*, 2nd edition, CRC Press, Boca Raton, FL.

Piomelli S, Rosen JF, Chisolm JJ Jr., and Graef JW, 1984. Management of childhood lead poisoning, *Journal of Pediatrics*, 105(4): 523–532.

Pirkle JL, Kaufmann RB, Brody DJ, Hickman T, Gunter EW, and Paschal DC, 1998. Exposure of the U.S. population to lead, 1991–1994, *Environ Health Perspect*, 106: 745–750.

Pirkle, JL, WR Harlan, JR Landis, and J Schwartz, 1985. The relationship between blood lead levels and blood pressure and its cardiovascular risk implications, *American Journal of Epidemiology*, 121: 246–258.

Plummer, LN and W Back, 1980. The mass balance approach: applications to interpreting the chemical evolution of hydrologic systems, *American Journal of Science*, 280: 130–142.

Pocock, SJ, M Smith, and P Baghurst, 1994. Environmental lead and children's intelligence: a systematic review of the epidemiological evidence, *British Medical Journal*, 309: 1189–1197.

Pollard, SJ, R Yearsley, N Reynard, IC Meadowcroft, R Duarte-Davidson, and SL Duerden, 2002. Current directions in the practice of environmental risk assessment in the United Kingdom, *Environmental Science and Technology*, 36(4): 530–538.

Pope, GA and M Baviere, 1991. Reduction of capillary forces by surfactants. In *Basic Concepts in EOR Processes*, M Baviere (ed.), pp. 89–122, Elsevier, London, UK.

Porcella, DB, 1994. Mercury in the environment: geochemistry. In *Mercury Pollution: Integration and Synthesis*, CJ Watras and JW Huckabee (eds.), CRC Press, Boca Raton, FL, pp. 3–19.

Postel, S, 1988. *Controlling Toxic Chemicals. State of the World*, Worldwatch Institute, New York.

Posthuma, L, GW Suter II, and TP Traas (eds.), 2002. *Species Sensitivity Distributions in Ecotoxicology*, CRC Press, Boca Raton, FL.

Potts, RO and RH Guy, 1992. Predicting skin permeability, *Pharmaceutical Research*, 9: 663–669.

Power, M and LS McCarty, 1996. Probabilistic risk assessment: betting on its future, *Human and Ecological Risk Assessment*, 2: 30–34.

Power, M and LS McCarty, 1998. A comparative analysis of environmental risk assessment/risk management frameworks, *Environmental Science and Technology*, 32(9): 224A–231A.

Prager, JC, 1995. *Environmental Contaminant Reference Databook*, Van Nostrand Reinhold Co., New York.

Pratt, M (ed.), 1993. *Remedial Processes for Contaminated Land*, Institution of Chemical Engineers, Warwickshire, UK.

Price, PS, CL Curry, PE Goodrum, MN Gray, JI McCrodden, NW Harrington, H Carlson-Lynch, and RE Keenan, 1996. Monte Carlo modeling of time-dependent exposures using a microexposure event approach, *Risk Analysis*, 16(3): 339–348.

Prins, G and R Stamp, 1991. *Top Guns & Toxic Whales: The Environment & Global Security*, Earthscan Publications Ltd., London, UK.

Prosser, CL and FA Brown, 1961. *Comparative Animal Physiology*, 2nd edition, WB Saunders Co., Philadelphia, PA.

Pugh, DM and JV Tarazona, 1998. *Regulation for Chemical Safety in Europe: Analysis, Comment and Criticism*, Kluwer Academic Publishers, Dordrecht, The Netherlands.

Putnam, RD, 1986. Review of toxicology of inorganic lead, *AIHA Journal*, 47: 700–703.

Rabinowitz, M, 1998. Historical perspective on lead biokinetic models, *Environmental Health Perspectives*, 106(Suppl 6): 1461–1465.

Rabl, A, JV Spadaro, and PD McGavran, 1998. Health risks of air pollution from incinerators: a perspective, *Waste Management and Research*, 16(4): 365–388.

Rai, PK, 2018. *Phytoremediation of Emerging Contaminants in Wetlands*, CRC Press, Boca Raton, FL.

Raiffa, H, 1968. *Decision Analysis: Introductory Lectures on Choices Under Uncertainty*, Random House, New York.

Rail, CC, 1989. *Groundwater Contamination: Sources, Control and Preventive Measures*, Technomic Publishing Co., Inc., Lancaster, PA.

Raloff, J, 1996. Tap water's toxic touch and vapors, *Science News*, 149(6): 84.

Ram, NM, M Leahy, E Carey, and J Cawley Jr., 1999. Environmental sleuth at work, *Environmental Science and Technology*, 33(21): 465A–469A.

Ramamoorthy, S and E Baddaloo, 1991. *Evaluation of Environmental Data for Regulatory and Impact Assessment*. Studies in Environmental Science 41, Elsevier Science Publishers B.V., Amsterdam, The Netherlands.

Rappaport, SM and J Selvin, 1987. A method for evaluating the mean exposure from a lognormal distribution, *AIHA Journal*, 48: 374–379.

Rappe, C, Buser, HR, and Bosshardt, H-P, 1979. Dioxins, dibenzofurans and other polyhalogenated aromatics: production, use, formation, and destruction, *Annals of the New York Academy of Sciences*, 320: 1–18.

Ray, DL, 1990. *Trashing the Planet*, Regnery Gateway, Washington, DC.

Raymond, C, 1995. New state programs encourage industrial development: balancing environmental protection and economic development, *Soil & Groundwater Cleanup*, November 1995, 10–12.

Reardon, EJ, 1995. Anaerobic corrosion of granular iron: measurement and interpretation of hydrogen evolution rates, *Environmental Science and Technology*, 29(12): 2936–2945.

Rebovich, DJ, 1992. *Dangerous Ground: The World of Hazardous Waste Crime*, Transaction Publishers, London, UK.

Reddy, KR, S Kosgi, and J Zhou, 1995. A review of in-situ air sparging for the remediation of VOC-contaminated saturated soils and groundwater, *Hazardous Waste and Hazardous Materials*, 12(2): 97–112.

Reddy, MB, RH Guy, and AL Bunge, 2000. Does epidermal turnover reduce percutaneous penetration? *Pharmaceutical Research*, 17: 1414–1419.

Reed, SC, RW Crites, and EJ Middlebrooks, 1995. *Natural Systems for Waste Management and Treatment*, 2nd edition, McGraw-Hill, New York.

Reeve, RN, 1994. *Environmental Analysis*, John Wiley & Sons, New York.

Regan MJ and Desvousges WH, 1990. *Communicating Environmental Risks: A Guide to Practical Evaluations*, Publication No. 230-01-91-001, US Environmental Protection Agency, Washington, DC.

Renn, O, 1992. Risk communication: towards a rational dialogue with the public, *Journal of Hazardous Materials*, 29(3): 465–519.

Renn, O, 1999. A model for an analytic-deliberative process in risk management, *Environmental Science and Technology*, 33(18): 3049–3055.

Renshaw CE, GD Zynda, and JC Fountain, 1997. Permeability reductions induced by sorption of surfactant, *Water Resources Research*, 33(3): 371–378.

Renwick, AG, JL Dorne, and K Walton, 2001. Pathway-related factors: The potential for human data to improve the scientific basis of risk, *Human and Ecological Risk Assessment*, 7(1): 165–180.

Ricci, PF (ed.), 1985. *Principles of Health Risk Assessment*, Prentice-Hall, Englewood Cliffs, NJ.

Ricci, PF and MD Rowe (ed.), 1985. *Health and Environmental Risk Assessment*, Pergamon Press, New York.

Rice, DC, 1996. Behavioral effects of lead: commonalities between experimental and epidemiologic data, *Environmental Health Perspectives*, 104(Suppl 2): 337–351.

Richards, D and WD Rowe, 1999. Decision-making with heterogeneous sources of information, *Risk Analysis*, 19(1): 69–81.

Richardson, GM, 1996. Deterministic versus probabilistic risk assessment: strengths and weaknesses in a regulatory context, *Human and Ecological Risk Assessment*, 2: 44–54.

Richardson, M (ed.), 1995. *Environmental Toxicology Assessment*, Taylor & Francis Group, London, UK.

Richardson, ML (ed.), 1986. *Toxic Hazard Assessment of Chemicals*, Royal Society of Chemistry, London, UK.

Richardson, ML (ed.), 1990. *Risk Assessment of Chemicals in the Environment*, Royal Society of Chemistry, Cambridge, UK.

Richardson, ML (ed.), 1992. *Risk Management of Chemicals*, Royal Society of Chemicals, Cambridge, UK.

Richardson, SD, JE Simmons, and G Rice, 2002. Disinfection byproducts: The next generation, *Environmental Science & Technology*, 36(9): 198A–205A.

Roberts, A (ed.), 2014. *Human Anatomy and Coloring Book*, DK Publishing, New York.

Roberts, MS and KA Walters (eds.), 1998. *Dermal Absorption and Toxicity Assessment*, Marcel Dekker, New York.

Rock, SA and P Sayre, 1999. Phytoremediation of hazardous wastes: potential regulatory acceptability, *Environmental Regulations and Permitting*, 8(3): 33–42.

Rodier, PM, 1995. Developing brain as a target of toxicity, *Environmental Health Perspectives*, 103(Suppl 6): 73–76.

Rodricks, J and Taylor, MR, 1983. Application of risk assessment to food safety decision making, *Regulatory Toxicology and Pharmacology*, 3: 275–307.

Rodricks, JV and RG Tardiff (eds.), 1984. *Assessment and Management of Chemical Risks*, ACS Symposium Series 239, American Chemical Society, Washington, DC.

Rodricks, JV, 1984. Risk assessment at hazardous waste disposal sites, *Hazardous Waste and Hazardous Materials*, 1(3): 333–362.

Romieu I, Lacasana M, and McConnell R, 1997. Lead exposure in Latin America and the Caribbean, *Environmental Health Perspectives*, 105: 398–405.

Romieu I, Palazuelos E, Hernandez Avila M, Rios C, Muñoz I, Jimenez C, and G Cahero, 1994. Sources of lead exposure in Mexico City, *Environmental Health Perspectives*, 102: 384–389.

Romieu I, T Carreon, L Lopez, E Palazuelos, C Rios, Y Manuel, and M Hernandez-Avila, 1995. Environmental urban lead exposure and blood lead levels in children of Mexico City, *Environmental Health Perspectives*, 103: 1036–1040.

Rong, Y (ed.), 2018, *Fundamentals of Environmental Site Assessment and Remediation*, CRC Press, Boca Raton, FL.

Rosén, L and HE LeGrand, 1997. An outline of a guidance framework for assessing hydrogeological risks at early stages, *Ground Water*, 35(2): 195–204.

Rousselle, C, V Pernelet-Joly, C Mourton-Gilles, JP Lepoittevin, R Vincent, A Lefranc, and R Garnier, 2014. Risk assessment of dimethylfumarate residues in dwellings following contamination by treated furniture, *Risk Analysis*, 34(5): 879–888.

Rowe, WD, 1977. *An Anatomy of Risk*, John Wiley & Sons, New York.

Rowe, WD, 1983. *Evaluation Methods for Environmental Standards*, CRC Press, Boca Raton, FL.

Rowe, WD, 1994. Understanding uncertainty, *Risk Analysis*, 14(5): 743–750.

Rowland, AJ and P Cooper, 1983. *Environment and Health*, Edward Arnold, London, UK.

Royal Society of London, 1983. *Risk Assessment: A Study Group Report*, The Royal Society, London, UK.

Royston, P, 1995. A remark on algorithm AS 181: the W-test for normality, *Applied Statistics*, 44: 547–551.

Rubin, E, 2000. Potential for Phytoremediation of Methyl-Tert-Butyl-Ether (MTBE), In *Presentation at the 10th Annual West Coast Conference on Contaminated Soils and Water*, March 20–23, San Diego, CA.

Ruby, MV, A Davis, R Schoof, S Eberle, and CM Sellstone, 1996. Estimation of lead and arsenic bioavailability using a physiologically-based extraction test, *Environmental Science and Technology*, 30: 422–430.

Ruby, MV, R Schoof, W Brattin, M Goldade, G Post, M Harnois, DE Mosby, SW Casteel, W Berti, M Carpenter, and D Edwards, 1999. Advances in evaluating the oral bioavailability of inorganics in soil for use in human health risk assessment, *Environmental Science and Technology*, 33(21): 3697–3705.

Ruckelshaus, WD, 1985. Risk, science, and democracy, *Issues in Science and Technology*, 1(Spring 3): 19–38.

Russell, DL, 2011. *Remediation Manual for Contaminated Sites*, CRC Press, Boca Raton, FL.

Russell, M and M Gruber, 1987. Risk assessment in environmental policy-making, *Science*, 236: 286–290.

Ryan, PB, 1991. An overview of human exposure modeling, *Journal of Exposure Analysis and Environmental Epidemiology*, 1(4): 453–474.

Sachs, L, 1984. *Applied Statistics—A Handbook of Techniques*, Springer-Verlag, New York.

Sáez, AE and JC Baygents, 2014. *Environmental Transport Phenomena*, CRC Press, Boca Raton, FL.

Safe, S, 1994. Polychlorinated biphenyls (PCBs): environmental impact, biochemical and toxic responses, and implications for risk assessment, *Critical Reviews in Toxicology*, 24(2): 87–149.

Safe, SH, 1998. Hazard and risk assessment of chemical mixtures using the toxic equivalency factor approach, *Environmental Health Perspectives*, 106 (Suppl 4): 1051–1058.

Sale, T, C Newell, H Stroo, R Hinchee, and P Johnson, 2008. Frequently Asked Questions Regarding Management of Chlorinated Solvent in Soils and Groundwater, ESTCP ER-0530.

Sale, TC and C Newell, 2011. Guide for Selecting Remedies for Subsurface Releases of Chlorinated Solvents, ESTCP ER-200530.

Saleh, MA, JN Blancato, and CH Nauman (eds.), 1994. *Biomarkers of Human Exposure to Pesticides*, ACS Symposium Series, American Chemical Society (ACS), Washington, DC.

Salt, DE, M Blaylock, NPBA Kumar, V Dushenkov, BD Ensley, I Chet, and I Raskin, 1995. Phytoremediation: a novel strategy for the removal of toxic metals from the environment using plants, *Biotechnology*, 13: 468–474.

Saltzman, BE, 1997. Health risk assessment of fluctuating concentrations using lognormal models, *Journal of the Air and Waste Management Association*, 47: 1152–1160.

Salvato, JA, Jr., 1982. *Environmental Engineering and Sanitation*, 3rd edition, John Wiley & Sons, New York.

Samiullah, Y, 1990. *Prediction of the Environmental Fate of Chemicals*, Elsevier Applied Science (in association with BP), London, UK.

Samuels, ER and JC Meranger, 1984. Preliminary studies on the leaching of some trace metals from kitchen faucets, *Water Research*, 18(1): 75–80.

Sandman, PM, 1993. *Responding to Community Outrage: Strategies for Effective Risk Communication*, American Industrial Hygiene Association, Fairfax, VA.

Santos, SL and J Sullivan, 1988. The use of risk assessment for establishing corrective action levels at RCRA sites. Issue Paper/Report by Focus Group Risk Communication and Environmental Management Consultants, Medford, MA.

Sara, MN, 1993. *Standard Handbook of Site Assessment for Solid and Hazardous Waste Facilities*, CRC Press, Boca Raton, FL.

Sara, MN, 2003. *Site Assessment and Remediation Handbook*, CRC Press, Boca Raton, FL.

Sarnat, SE, JA Sarnat, J Mulholland, V Isakov, H Özkaynak, HH Chang, M Klein, and PE Tolbert, 2013. Application of alternative spatiotemporal metrics of ambient air pollution exposure in a time-series epidemiological study in Atlanta, *Journal of Exposure Science and Environmental Epidemiology*, 23(6): 593–605.

Satkin, RL and PB Bedient, 1988. Effectiveness of various aquifer restoration schemes under variable hydrogeologic conditions, *Ground Water*, 26(4): 488–498.

Sax, NI and RJ Lewis Sr., 1987. *Hawley's Condensed Chemical Dictionary*, Van Nostrad Reinhold Co., New York.

Sax, NI, 1979. *Dangerous Properties of Industrial Materials*, 5th edition, Van Nostrand Reinhold Co., New York.

Saxena, J and F Fisher (eds.), 1981. *Hazard Assessment of Chemicals*, Academic Press, New York.

Sayetta, RB, 1986. Pica: an overview, *American Family Physician*, 33(5): 181–185.

Scanlon, VC and T Sanders, 1995. *Essentials of Anatomy and Physiology*, 2nd edition, F.A. Davis, New York.

Schaefer, CE, EB White, GM Lavorgna, and MD Annable, 2016. Dense nonaqueous-phase liquid architecture in fractured bedrock: implications for treatment and plume longevity, *Environmental Science and Technology*, 50: 207–213.

Schaefer, CE, RM Towne, D Root, and JE McCray, 2012. Assessment of chemical oxidation for treatment of DNAPL in fractured sandstone blocks, *Journal of Environmental Engineering*, 257: 174–188.

Schecter, A, 1994. *Dioxins and Health*, Kluwer Academic Publishers, Dordrecht, The Netherlands.

Schecter, A, J Startin, C Wright, M Kelly, O Papke, A Lis, M Ball, and Olson, JR, 1994. Congener-specific levels of dioxins and dibenzofurans in U.S. food and estimated daily dioxin equivalent intake, *Environmental Health Perspectives*, 102: 962–966.

Scheer, D, C Benighaus, L Benighaus, O Renn, S Gold, B Röder, GF Böl. 2014. The distinction between risk and hazard: Understanding and use in stakeholder communication, *Risk Analysis*, 34(7): 1270–1285.

Schleicher, K (ed.), 1992. *Pollution Knows No Frontiers: A Reader*, Paragon House Publishers, New York.

Schnoor JL, LA Light, SC McCutcheon, NL Wolfe, and LH Carriera, 1995. Phytoremediation of organic and nutrient contaminants, *Environment Science and Technology*, 29: 318–323.

Schnoor, JL, 1996. *Environmental Modeling—Fate and Transport of Pollutants in Water, Air, and Soil*, John Wiley & Sons, New York.

Schnoor, JL, 1997. *Phytoremediation*, TE-80-01, Ground-Water Remediation Technologies Analysis Center, Pittsburgh PA.

Schock, MR and CH Neff, 1988. Trace metal contamination from brass faucets, *Journal of the American Water Works Association*, 80(11): 47–56.

Schrader-Frechette, KS, 1991. *Risk and Rationality*, University of California Press, Berkeley, CA.

Schramm, G and JJ Warford (eds.), 1989. *Contaminated Site Management and Economic Development*, A World Bank Publication, The Johns Hopkins University Press, Baltimore, MD.

Schroeder, RL, 1985. *Habitat Suitability Index Models: Northern Bobwhite*, US Department of Interior, Fish and Wildlife Service, Washington, DC.

Schroll, R and I Scheunert, 1993. Uptake pathways of Octachlorodibenzo-*p*-dioxin from soil by carrots, *Chemosphere*, 26(9): 1631–1640.

Schulin, R, A Desaules, R Webster, and B Von Steiger (eds.), 1993. *Soil Monitoring: Early Detection and Surveying of Soil Contamination and Degradation*, Birkhäuser Verlag, Basel, Switzerland.

Schulz, TW and S Griffin, 1999. Estimating risk assessment exposure point concentrations when data are not normal or lognormal, *Risk Analysis*, 19(4): 577–584.

Schuurmann, G and B Markert, 1966. *Ecotoxicology*, John Wiley & Sons, New York.

Schwab, GO, RK Frevert, TW Edminster, and KK Barnes, 1966. *Soil and Water Conservation Engineering* (2nd ed.), John Wiley & Sons, New York.

Schwab, PA, LE Newman, M Banks, and K Nedunari, 2000. *Dewatering, Remediation, and Evaluation of Dredged Sediments*, Center for Integrated Remediation Using Managed Natural Systems (CIRUMNS), USEPA Hazardous Substance Research Center, West Lafayette, IN.

Schwartz, J, 1994. Low-level lead exposure and children's IQ: a meta-analysis and search for a threshold, *Environmental Research*, 65: 42–55.

Schwartz, SI and WB Pratt, 1990. *Hazardous Waste from Small Quantity Generators—Strategies and Solutions for Business and Government*, Island Press, Washington, DC.

Schwarzenbach, RC, RW Scholz, A Heitzer, B Staubli, and B Grossmann, 1999. A regional perspective on contaminated site remediation—fate of materials and pollutants, *Environmental Science and Technology*, 33(14): 2305–2310.

Schwarzenbach, RC, RW Scholz, A Heitzer, B Staubli, and B Grossmann, 1999. A regional perspective on contaminated site remediation—Fate of materials and pollutants, *Environmental Science and Technology*, 33(14): 2305–2310.

Schwarzenbach, RP, PM Gschwend, and DM Imboden, 1993. *Environmental Organic Chemistry*, John Wiley & Sons, New York.

Schwing, RC and WA Albers Jr. (eds.), 1980. *Societal Risk Assessment: How Safe is Safe Enough*, Plenum Press, New York.

Searle, CE (ed.), 1976. *Chemical Carcinogens*, ACS Monograph 173, American Chemical Society, Washington, DC.

Seaton, A, 1996. Particles in the air: the enigma of urban air pollution, *Journal of Royal Society of Medicine*, 89: 604–607.

Sebek, V (ed.), 1990. Maritime Transport, Control and Disposal of Hazardous Waste, *Marine Policy International Journal*, Special Issue, 14(3).

Sedman, R and RS Mahmood, 1994. Soil ingestion by children and adults reconsidered using the results of recent tracer studies, *Air and Waste*, 44: 141–144.

Sedman, RM, 1989. The development of applied action levels for soil contact: a scenario for the exposure of humans to soil in a residential setting, *Environmental Health Perspectives*, 79: 291–313.

Seip HM and AB Heiberg (eds.), 1989. *Risk Management of Chemicals in the Environment*, NATO, Challenges of Modern Society, *Volume 12*, Plenum Press, New York.

Selim, HM (ed.), 2017. *Competitive Sorption and Transport of Heavy Metals in Soils and Geological Media*, CRC Press, Boca Raton, FL.

Selim, HM, 2017. *Dynamics and Bioavailability of Heavy Metals in the Rootzone*, CRC Press, Boca Raton, FL.

Selim, HM, 2017. *Transport & Fate of Chemicals in Soils: Principles & Applications*, CRC Press, Boca Raton, FL.

Selroos, J-O, DD Walker, A Ström, B Gylling, and S Follin, 2002. Comparison of alternative modeling approaches for groundwater flow in fractured rock, *Journal of Hydrology*, 257: 174–188.

Shacklette, HT and JG Boerngen, 1984. Element Concentrations in Soils and Other Surficial Materials of the Conterminous United States, USGS Professional Paper No. 1270.

Shalat, SL, True, LD, Fleming, LE, and Pace, PE, 1989. Kidney cancer in utility workers exposed to polychlorinated biphenyls (PCBs), *British Journal of Industrial Medicine*, 46(11): 823–824.

Shapiro, SS and MB Wilk, 1965. An analysis of variance test for normality (complete samples), *Biometrika*, 52: 591–611.

Sharp, VF, 1979. *Statistics for the Social Sciences*, Little, Brown & Co., Boston, MA.

Sheppard, SC, 1995. Parameter values to model the soil ingestion pathway, *Environmental Monitoring and Assessment*, 34: 27–44.

Shere, ME, 1995. The myth of meaningful environmental risk assessment, *Harvard Environmental Law Review*, 19(2): 409–492.

Sherman JD, 1994. *Chemical Exposure and Disease*, Princeton Scientific Publishing Co., Princeton, NJ.

Shiau, B, JD Rouse, DA Sabatini, and JH Harwell, 1995. Surfactant selection for optimizing surfactant enhanced subsurface remediation. In *Surfactant Enhanced Subsurface Remediation Emerging Technologies*, DA Sabatini, RC Knox, and JH Harwell (eds.), ACS symposium series 594, American Chemical Society, Washington, DC, pp. 65–81.

Shimp, JF, JC Tracey, LC Davis, E Lee, W Huang, LE Erickson, and JL Schnoor, 1993. Beneficial effects of plants in the remediation of soil and groundwater contaminated with organic materials, *Critical Review in Environmental Science and Technology*, 23: 41–77.

Shook, GM, GA Pope, and K Kostarelos, 1998. Prediction and minimization of vertical migration of DNAPLs using surfactant enhanced aquifer remediation at neutral buoyancy, *Journal of Contaminant Hydrology*, 34: 363–382.

Shrader-Frechette, KS, 1985. *Risk Analysis and Scientific Method*, D. Reidel Publishing Co., Boston, MA.

Sidall, JN, 1983. *Probabilistic Engineering Design: Principles and Applications*, Marcel Dekker, New York.

Siegrist, R, M Urynowicz, O West, M Crimi, and K Lowe, 2001. *Principles and Practices of In Situ Chemical Oxidation Using Permanganate*, Battelle Press, Columbus, OH.

Siegrist, RL, KS Lowe, LC Murdoch, TL Case, and DA Pickering, 1999. In situ oxidation by fracture emplaced reactive solids, *Journal of Environmental Engineering*, 125(5): 429–440.

Siegrist, RL, M Crimi, and TJ Simpkin (eds.), 2011. *In Situ Chemical Oxidation for Groundwater Remediation*, Springer Science & Business Media, New York: 678.

Siegrist, RL, MA Urynowicz, OR West, ML Crimi, and KS Lowe, 2001. *Guidance for In Situ Chemical Oxidation at Contaminated Sites: Technology Overview with a Focus on Permanganate Systems*, Battelle Press, Columbus, OH.

Silberhorn, EM, H Glauert, and LW Robertson, 1990. Carcinogenicity of polyhalogenated biphenyls: PCBs and PBBs, *Critical Reviews in Toxicology*, 20(6): 439–496.

Silk, JC and MB Kent (eds.), 1995. *Hazard Communication Compliance Manual*, Society for Chemical Hazard Communication/BNA Books, Washington, DC.

Silkworth, JB and JF Brown Jr., 1996. Evaluating the impact of exposure to environmental contaminants on human health, *Clinical Chemistry*, 42: 1345–1349.

Sims, R, D Sorenson, J Sims, J McLean, R Mahmood, R Dupont, J Jurinak, and K Wagner, 1986. *Contaminated Surface Soils In-Place Treatment Techniques*, Noyes Publications, Park Ridge, NJ.

Sims, RC and Sims, JL, 1986. Cleanup of contaminated soils. In *Utilization, Treatment, and Disposal of Waste on Land*, ECA Runge (ed.) Soil Science Society of America, Madison, WI, pp. 257–278.

Sims, RC, 1990. Soil remediation techniques at uncontrolled hazardous waste sites, a critical review, *Journal of the Air and Waste Management Association*, 40(5): 704–732.

Singhroy VH, DD Nebert, and AI Johnson (eds.), 1996. *Remote Sensing and GIS for Site Characterization: Applications and Standards*, ASTM Publication No. STP 1279, ASTM, Philadelphia, PA.

Sinks, T, G Steele, AB Smith, K Watkins, and RA Shults, 1992. Mortality among workers exposed to poly-chlorinated biphenyls, *American Journal of Epidemiology*, 136(4): 389–398.

Sitnig, M, 1985. *Handbook of Toxic and Hazardous Chemicals and Carcinogens*, Noyes Data Corp., Park Ridge, NJ.

Sittig, M, 1994. *World-Wide Limits for Toxic and Hazardous Chemicals in Air, Water and Soil*, Noyes Publications, Park Ridge, NJ.

Slob, W, 1994. Uncertainty analysis in multiplicative models, *Risk Analysis*, 14: 571–576.

Slovic, P, 1993. Perceived risk, trust, and democracy, *Risk Analysis*, 13(6): 675–682.

Slovic, P, 1997. Public perception of risk, *Journal of Environmental Health*, 59(9): 22–24.

Smith AE, PB Ryan, and JS Evans, 1992. The effect of neglecting correlations when propagating uncertainty and estimating population distribution of risk, *Risk Analysis*, 12: 457–474.

Smith, AH, 1987. Infant exposure assessment for breast milk dioxins and furans derived from waste incineration emissions, *Risk Analysis*, 7(3): 347–353.

Smith, AH, S Sciortino, H Goeden, and CC Wright, 1996. Consideration of background exposures in the management of hazardous waste sites: a new approach to risk assessment, *Risk Analysis*, 16(5): 619–625.

Smith, CM, DC Christiani, and KT Kelsey (eds.), 1994. *Chemical Risk Assessment and Occupational Health*, Auburn House, Westport, CT.

Smith, JE, 1995. Generalized Chebychev inequality: theory and applications in decision analysis, *Operations Research*, 43: 807–825.

Smith, RL, 1994. Use of Monte Carlo simulation for human exposure assessment at a Superfund site, *Risk Analysis*, 14(4): 433–439.

Smith, RP, 1992. *A Primer of Environmental Toxicology*, Lea & Febiger, Philadelphia, PA.

Smith, TT, Jr., 1996. Regulatory reform in the USA and Europe, *Journal of Environmental Law*, 8(2): 257–282.

Sole, M, 1997. *In Situ Chemical Oxidation Safety Advisory*, Bureau of Petroleum Storage Systems, Florida Department of Environmental Protection, Tallahassee, FL.

Solomon, KR, 1996. Overview of recent developments in ecotoxicological risk assessment, *Risk Analysis*, 16(5): 627–633.

Speir, T, JA August, and CW Feltham, 1992. Assessment of the feasibility of using CCA (copper, chromium and arsenic)-treated and boric acid-treated sawdust as soil amendments. I. Plant growth and element uptake, *Plant and Soil*, 142: 235–248.

Spellman, FR, 2015. *Handbook of Environmental Engineering*, CRC Press, Boca Raton, FL.

Spencer, EY, 1982. *Guide to the Chemicals Used in Crop Protection*, 7th edition, Publication No. 1093, Research Institute, Agriculture Canada. Information Canada, Ottawa, ON.

Spiegelhalter, D, M Pearson, and I Short, 2011. Visualizing uncertainty about the future, *Science*, 333(6048): 1393–1400.

Splitstone, DE, 1991. How clean is clean…statistically? *Pollution Engineering*, 23: 90–96.

Sposito, G, CS LeVesque, JP LeClaire, and N Sensi, 1984. *Methodologies to Predict the Mobility and Availability of Hazardous Metals in Sludge-Amended Soils*, Contribution No. 189, California Water Resource Center, University of California, Berkeley, CA.

Stanek, EJ and EJ Calabrese, 1990. A guide to interpreting soil ingestion studies, *Regulatory Toxicology and Pharmacology*, 13: 263–292.

Stanek, EJ and EJ Calabrese, 1995a. Daily estimates of soil ingestion in children, *Environmental Health Perspectives*, 103(3): 276–285.

Stanek, EJ and EJ Calabrese, 1995b. Soil ingestion estimates for use in site evaluations based on the best tracer method, *Human and Ecological Risk Assessment*, 1: 133–156.

Starr C and C Whipple, 1980. Risks of risk decisions, *Science*, 208: 1114.

Starr, C, R Rudman, and C Whipple, 1976. Philosophical basis for risk analysis, *Annual Review of Energy*, 1: 629–662.

States, JB, PT Hang, TB Shoemaker, LW Reed, and EB Reed, 1978. *A System Approach to Ecological Baseline Studies, FQS/DBS-78/21*, USFWS, Washington, DC.

Steele, G, P Stehr-Green, and E Welty, 1986. Estimates of the biologic half-life of polychlorinated biphenyls in human serum, *New England Journal of Medicine*, 314(14): 926–927.

Stenesh, J, 1989. *Dictionary of Biochemistry and Molecular Biology*, John Wiley & Sons, New York.

Stepan, DJ, RH Fraley, and DS Charlton, 1995. *Remediation of Mercury-Contaminated Soils: Development and Testing of Technologies*, Gas Research Institute, Chicago, IL.

Sterling, SN, BL Parker, JA Cherry, JH Williams, JW Lane, and FP Haeni, 2005. Vertical cross contamination of trichloroethylene in a borehole in fractured sandstone, *Ground Water*, 43(4): 557–573.

Stout, SA, AD Uhler, TG Naymik, and KJ McCarthy, 1998. Environmental forensics: unraveling site liability, *Environmental Science and Technology*, 32(11): 260A–264A.

Strbak, L, 2000. In Situ Flushing with Surfactants and Cosolvents, Prepared for US Environmental Protection Agency, Office of Solid Waste and Emergency Response, Washington, DC.

Suarez, MP and HS Rifai, 1999. Biodegradation rates for fuel hydrocarbons and chlorinated solvents in groundwater, *Bioremediation*, 3(4): 337–362.

Suess, MJ and JW Huismans (eds.), 1983. *Management of Hazardous Wastes: Policy Guidelines and Code of Practice*, WHO Regional Publication, European Series No.14, World Health Organization, Regional Office for Europe, Copenhagen, Denmark.

Susarla, S, ST Bacchus, NL Wolfe, and SC McCutcheon, 1999. Phytotransformation of Perchlorate Using Parrot-Feather, *Soil & Groundwater Cleanup*, February/March: 20–23.

Suter II, GW, 1996. Toxicological benchmarks for screening contaminants of potential concern for effects on freshwater biota, *Environmental Toxicology and Chemistry*, 15(7): 1232–1241.

Suter II, GW, BW Cornaby, CT Hadden, RN Hull, M Stack, and FA Zafran, 1995. An approach for balancing health and ecological risks at hazardous waste sites, *Risk Analysis*, 15: 221–231.

Suter II, GW, RA Efroymson, BE Sample, and DS Jones, 2000. *Ecological Risk Assessment for Contaminated Sites*, CRC Press, Boca Raton, FL.

Suter, GW, 1993. *Ecological Risk Assessment*, Lewis Publishers, Boca Raton, FL.

Suthersan, SS and FC Payne, 2004. *In Situ Remediation Engineering*, CRC Press, Boca Raton, FL.

Sutherson, SS, 2001. *Natural and Enhanced Remediation Systems*, CRC Press, Boca Raton, FL.

Sutherson, SS, 2005. *In Situ Bioremediation*, CRC Press, Boca Raton, FL.

Sutherson, SS, J Horst, M Schnobrich, N Welty, and J McDonough, 2016. *Remediation Engineering: Design Concepts*, 2nd edition, CRC Press, Boca Raton, FL.

Swann, RL and A Eschenroeder (eds.), 1983. *Fate of Chemicals in the Environment*, ACS Symposium Series 225, American Chemical Society, Washington, DC.

Swett, GH, 1992. Bioremediation: Myths vs. Realities, *Environmental Protection*, May 1992.

Tabbaa, AA and JA Stegemann, 2005. Stabilisation/Solidification Treatment and Remediation, In *Proceedings of the International Conference on Stabilisation/Solidification Treatment and Remediation*, April 12–13, 2005, Cambridge, UK. CRC Press, Boca Raton, FL.

Taiz, L and E Zeiger, 1991. *Plant Physiology*, The Benjamin/Cummings Publishing Company, Inc., Redwood City, CA.

Talbot, EO and GF Craun (eds.), 1995. *Introduction to Environmental Epidemiology*, CRC Press, Boca Raton, FL.

Talley, J, 2005. *Bioremediation of Recalcitrant Compounds*, CRC Press, Boca Raton, FL.

Talmage, SS and BT Walton, 1993. Food chain transfer and potential renal toxicity to small mammals at a contaminated terrestrial field site, *Ecotoxicology*, 2: 243–256.

Tan, WY, 1991. *Stochastic Models of Carcinogenesis*, Statistics: Textbooks and Monographs, *Volume 116*, Marcel Dekker, New York.

Tang, WZ, 2003. *Physicochemical Treatment of Hazardous Wastes*, CRC Press, Boca Raton, FL.

Tardiff, RG and JV Rodricks (eds.), 1987. *Toxic Substances and Human Risk*, Plenum Press, New York.

Tasca, JJ, Saunders, MF, and RS Prann, 1989. Terrestrial Food-Chain Model for Risk Assessment. In *Superfund '89: Proceedings 10th National Conference.*

Taylor, AC, 1993. Using objective and subjective information to develop distributions for probabilistic exposure assessment, *Journal of Exposure Analysis and Environmental Epidemiology*, 3(3): 285–298.

Telford, WM, LP Geldart, RE Sheriff, and DA Keys, 1976. *Applied Geophysics*, Cambridge University Press, New York.

Tellez-Rojo, MM, M Hernandez-Avila, T González-Cossío, I Romieu, A Aro, E Palazuelos, J Schwartz, and H Hu, 2002. Impact of breastfeeding on the mobilization of lead from bone, *American Journal of Epidemiology*, 155(5): 420–428.

Tessier, A and PGC Campbell, 1987. Partitioning of trace metals in sediments: relationships with bioavailability, *Hydrobiologia*, 149: 43–52.

Testa, SM and DL Winegardner, 1990. *Restoration of Petroleum-Contaminated Aquifers*, CRC Press, Boca Raton, FL.

Testa, SM, 1994. *Geological Aspects of Hazardous Waste Management*, CRC Press, Boca Raton, FL.

Theiss, JC, 1983. The ranking of chemicals for carcinogenic potency, *Regulatory Toxicology and Pharmacology*, 3: 320–328.

Theodore, L and MK Theodore, 2009. *Introduction to Environmental Management*, CRC Press, Boca Raton, FL.

Theodore, L, JP Reynolds, and FB Taylor, 1989. *Accident and Emergency Management*, John Wiley & Sons, New York.

Thibodeaux, LJ and ST Hwang, 1982. Landfarming of petroleum wastes—Modeling the air emission problem, *Environmental Progress*, 1: 42–46.

Thibodeaux, LJ, 1979. *Chemodynamics: Environmental Movement of Chemicals in Air, Water and Soil*, John Wiley & Sons, New York.

Thibodeaux, LJ, 1996. *Environmental Chemodynamics*, 2nd edition, John Wiley & Sons, New York.

Thomas, DJ, B Tracey, H Marshall, and RJ Norstrom, 1992. Arctic terrestrial ecosystem contamination, *Science of the Total Environment*, 122: 135–164.

Thomas, F, J Reider, A Ferro, and D Tsao, 1998. Using Trees as a Barrier Strip to Metals-Contaminated Saline Groundwater, In *Presentation at the 3rd Annual International Conference on Phytoremediation*, June 22–25, Houston, TX.

Thompson, KM and DE Burmaster, 1991. Parametric distributions for soil ingestion in children, *Risk Analysis*, 11: 339–342.

Thompson, KM, DE Burmaster, and AC Crouch, 1992. Monte Carlo techniques for quantitative uncertainty analysis in public health risk assessments, *Risk Analysis*, 12: 53–63.

Thompson, SK, 1992. *Sampling*, John Wiley & Sons, New York.

Tian, B, 2016. *GIS Technology Applications in Environmental and Earth Sciences*, CRC Press, Boca Raton, FL.

Timbrell, JA, 1995. *Introduction to Toxicology*, 2nd edition, Taylor & Francis Group, London, UK.

Todd AC, Wetmur JC, Moline JM, Godbold JH, Levin SM, and Landrigan PJ, 1996. Unraveling the chronic toxicity of lead: an essential priority for environmental health, *Environmental Health Perspectives*, 104(Suppl 1): 141–146.

Tohn E, Dixon S, Rupp R, and Clark S, 2000. A pilot study examining changes in dust lead loading on walls and ceilings after lead hazard control interventions, *Environ Health Perspectives*, 108: 453–456.

Tolba, MK (ed.), 1988. *Evolving Environmental Perceptions: From Stockholm to Nairobi*, United Nations Environment Programme, UNEP, Nairobi, Kenya.

Tolba, MK, 1990. The global agenda and the hazardous waste challenge, *Marine Policy International Journal*, 14(3): 205–209.

Tomatis, L, J Huff, I Hertz-Picciotto, D Dandler, J Bucher, P Boffetta, O Axelson, A Blair, J Taylor, L Stayner, and JC Barrett, 1997. Avoided and avoidable risks of cancer, *Carcinogenesis*, 18(1): 97–105.

Tossell, RW, 2000. Uptake of Arsenic by Tamarisk and Eucalyptus Under Saline Conditions, In *Presented at the 2nd International Conference on Remediation of Chlorinated and Recalcitrant Compounds*, May 22–25, Monterey, CA.

Travis, CC and AD Arms, 1988. Bioconcentration of organics in beef, milk, and vegetation, *Environmental Science and Technology*, 22: 271–274.

Travis, CC and HA Hattemer-Frey, 1988. Determining an acceptable level of risk, *Environmental Science and Technology*, 22(8): 873–876.

Travis, CC, SA Richter, EAC Crouch, R Wilson, and ED Klema, 1987. Cancer risk management, *Environmental Science and Technology*, 21: 415–420.

Tsao, D, 1998. *Phytoremediation Technologies*, Guidance Document, BP Amoco Technology, Naperville, IL.

Tsuji, JS and KM Serl, 1996. Current uses of the EPA lead model to assess health risk and action levels for soil, *Environmental Geochemistry and Health*, 18: 25–33.

Tukey, JW, 1977. *Exploratory Data Analysis*, Addison Wesley, Reading, MA.

Turnberg, WL, 1996. *Biohazardous Waste: Risk Assessment, Policy, and Management*, John Wiley & Sons, New York.

UNICRI, 1993. *Environmental Crime, Sanctioning Strategies and Sustainable Development*, UNICRI Publication No. 50, United Nations Interregional Crime and Justice Research Institute/Australian Institute of Criminology, Rome, Italy/Canberra, NSW.

US Army Corps of Engineers, 1998. *Engineering and Design- Monitoring Well Design, Installation, and Documentation at Hazardous, Toxic, and Radioactive Waste Sites*. EM 1110-1-4000. November.

US EPA, 1992b. *Dermal Exposure Assessment: Principles and Applications*, EPA/600/8-91/011B, Office of Health and Environmental Assessment, USEPA, Washington, DC.

US EPA, 1994g. *Radiation Site Cleanup Regulations: Technical Support Document for the Development of Radionuclide Cleanup Levels for Soil*, EPA 402-R-96-011A (September 1994).

US EPA, 1997c. *Establishment of Cleanup Levels for CERCLA Sites with Radioactive Contamination*, OSWER Directive 9200.4-18 (August 1997).

US EPA, 1997e. *Exposure Factors Handbook, Volumes 1–5*, EPA/600/0-95-002Fa, Fb, Fc, Office of Research and Development, USEPA, Washington, DC.

USBR (US Bureau of Reclamation), 1977. *Design of Small Dams. A Water Resources Technical Publication* (2nd ed.), 797p, U. S. Government Printing Office, Washington, DC.

USBR, 1986. *Guidelines to Decision Analysis*, Memo. No.7, ACER Techn, Denver, CO.

USDHS, 1989. *Public Health Service, Fifth Annual Report on Carcinogens, Summary*, US Department of Health and Human Services (USDHS), Washington, DC.

USDHS, 2002. *Report on Carcinogens*, 10th edition, US Department of Health and Human Services, Public Health Service, National Toxicology Program, Washington, DC.

USDOE (US Department of Energy), 1995. In Situ Air Stripping Using Horizontal Wells: Innovative Technology Summary Report, Report No.:DOE/EM-0269; NTIS Number: DE96003564, Prepared by Stone and Webster Environmental Technology and Services, Boston, MA.

USEPA (US Environmental Protection Agency), 1973. *Biological Field and Laboratory Methods for Measuring the Quality of Surface Waters and Effluents*, EPA/670/4-73/001, US Environmental Protection Agency, National Environmental Research Center, Cincinnati, OH.

USEPA, 1974. Safe Drinking Water Act, US Environmental Protection Agency, Public Law 93–523.

USEPA, 1982. *Test Methods for Evaluating Solid Waste: Physical/Chemical Methods*, 1st edition, SW-846, USEPA, Washington, DC.

USEPA, 1983. *Hazardous Waste Land Treatment*, Revised edition, SW-8974, USEPA, Cincinnati, OH.

USEPA, 1984a. *Approaches to Risk Assessment for Multiple Chemical Exposures*, EPA-600/9-84-008, Environmental Criteria and Assessment Office, US Environmental Protection Agency (USEPA), Cincinnati, OH.

USEPA, 1984b. *Review of In-Place Treatment Techniques for Contaminated Surface Soils, Volumes 1 and 2*, EPA-540/2-84-003a and b, US Environmental Protection Agency, Hazardous Waste Engineering Research Laboratory, Cincinnati, OH.

USEPA, 1984c. *Risk Assessment and Management: Framework for Decision Making*, EPA 600/9-85-002, US Environmental Protection Agency (USEPA), Washington, DC.

USEPA, 1985a. *Characterization of Hazardous Waste Sites: A Methods Manual, Volume 1*, EPA-600/4-84-075, *site investigations*, US Environmental protection Agency, Environmental Monitoring Systems Laboratory, Las Vegas, NV.

USEPA, 1985b. *Development of Statistical Distribution or Ranges of Standard Factors Used in Exposure Assessments*, US Environmental protection Agency, Office of Health and Environmental Assessment, Washington, DC.

USEPA, 1985c. *Modeling Remedial Actions at Uncontrolled Hazardous Waste Sites*, EPA/540/2-85/001 (April, 1985), Office of Emergency and Remedial Response, Washington, DC.

USEPA, 1985d. *Practical Guide to Ground-Water Sampling*, EPA/600/2-85/104 (September, 1985), Robert S. Kerr Environmental Research Laboratory, Office of Research and Development, USEPA, Ada, OK.

USEPA, 1985e. *Principles of Risk Assessment: A Nontechnical Review*, Office of Policy Analysis, US Environmental Protection Agency (USEPA), Washington, DC.

USEPA, 1985f. *Rapid Assessment of Exposure to Particulate Emissions From Surface Contamination Sites*, EPA/600/8-85/002, NTIS PB85-192219, Office of Health and Environmental Assessment, Washington, DC.

USEPA, 1985g. *Toxicology Handbook*, Office of Waste Programs Enforcement, Washington, DC.

USEPA, 1986a. *Ecological Risk Assessment*, Hazard Evaluation Division Standard Evaluation Procedure, Washington, DC.

USEPA, 1986b. *Guidelines for the Health Risk Assessment of Chemical Mixtures*, EPA/630/R-98/002, Risk Assessment Forum, Office of Research and Development, US Environmental Protection Agency (USEPA), Washington, DC.

USEPA, 1986c. Guidelines for carcinogen risk assessment, *Federal Register*, 51(185): 33992–34003, CFR 2984, September 24.

USEPA, 1986d. *Guidelines for Mutagenicity Risk Assessment*, EPA/630/R-98/003, Risk Assessment Forum, Office of Research and Development, USEPA, Washington, DC.

USEPA, 1986e. *Methods for Assessing Exposure to Chemical Substances, Volume 8: Methods for Assessing Environmental Pathways of Food Contamination*, EPA/560/5-85-008, Exposure Evaluation Division, Office of Toxic Substances, Washington, DC.

USEPA, 1986f. *Registry of Toxic Effects of Chemical Substances*, USEPA, Research Triangle Park, NC.

USEPA, 1986g. *Superfund Public Health Evaluation Manual*, EPA/540/1-86/060, Office of Emergency and Remedial Response, Washington, DC.

USEPA, 1986h. *Superfund Risk Assessment Information Directory*, EPA/540/1-86/061, Office of Emergency and Remedial Response, Washington, DC.

USEPA, 1987a. *Alternate Concentration Limit Guidance*, Report No. EPA/530-SW-87-017, OSWER Directive 9481-00-6C, USEPA, Office of Solid Waste, Waste Management Division, Washington, DC.

USEPA, 1987b. *Data Quality Objectives for Remedial Response Activities*, EPA/540/G-87/003, Office of Emergency and Remedial Response, Washington, DC.

USEPA, 1987c. *Data Quality Objectives for Remedial Response Activities: Example Scenario*, EPA/540/G-87/004 (March, 1987), USEPA, Washington, DC.

USEPA, 1987e. *Quality Assurance Program Plan*, EPA/600/X-87/241, Quality Assurance Management Staff, USEPA, Las Vegas, NV.

USEPA, 1987f. *RCRA Facility Investigation (RFI) Guidance*, EPA/530/SW-87/001, Office of Emergency and Remedial Response, Washington, DC.

USEPA, 1987g. *Selection Criteria for Mathematical Models Used in Exposure Assessments: Surface Water Models*, EPA-600/8-87/042, Office of Health and Environmental Assessment, US Environmental Protection Agency, Washington, DC.

USEPA, 1987h. *Technical Guidance for Hazard Analysis*, Office of Emergency and Remedial Response, Washington, DC.

USEPA, 1987i. *The New Superfund: What It Is, How It Works*, US Environmental Protection Agency, Washington, DC.

USEPA, 1988a. *A Workbook of Screening Techniques for Assessing Impacts of Toxic Air Pollutants*, EPA-450/4-88-009, Office of Air Quality Planning and Standards, Research Triangle Park, NC.

USEPA, 1988b. *CERCLA Compliance with Other Laws Manual (Inetrim Final)*, EPA/540/G-89/006, Office of Solid Waste and Emergency Response, Washington, DC.

USEPA, 1988c. *Estimating Toxicity of Industrial Chemicals to Aquatic Organisms Using Structure Activity Relationships*, EPA/560/6-88/001, Office of Toxic Substances, Washington, DC.

USEPA, 1988d. *GEO-EAS (Geostatistical Environmental Assessment Software) User's Guide*, EPA/600/4-88/033a, Environmental Monitoring Systems Laboratory, Office of R&D, Las Vegas, NV.

USEPA, 1988e. *Guidance for Conducting Remedial Investigations and Feasibility Studies Under CERCLA*, EPA/540/G-89/004, OSWER Directive 9355.3-01, Office of Emergency and Remedial Response, Washington, DC.

USEPA, 1988f. *Guidance on Remedial Actions for Contaminated Ground Water at Superfund Sites*, EPA/540/G-88/003, Office of Emergency and Remedial Response, Washington, DC.

USEPA, 1988g. *Interim Report on Sampling Design Methodology*, EPA/600/X-88/408, Environmental Monitoring Support Laboratory, Las Vegas, NV.

USEPA, 1988h. *Review of Ecological Risk Assessment Methods*, Office of Policy, Planning and Evaluation, Washington, DC.

USEPA, 1988i. *Selection Criteria for Mathematical Models Used in Exposure Assessments: Ground-Water Models*, EPA-600/8-88/075, Office of Health and Environmental Assessment, Washington, DC.

USEPA, 1988j. *Superfund Exposure Assessment Manual*, Report No. EPA/540/1-88/001, OSWER Directive 9285.5-1, USEPA, Office of Remedial Response, Washington, DC.

USEPA, 1988k. *Superfund Exposure Assessment Manual*, EPA/540/1-88/001, OSWER Directive 9285.5-1, USEPA, Office of Remedial Response, Washington, DC.

USEPA, 1989a. *Application of Air Pathway Analyses for Superfund Activities, Air/Superfund National Technical Guidance Study Series, Procedures for Conducting Air Pathway Analyses for Superfund Applications, Volume 1*, EPA-450/1-89-001, Interim Final, Office of Air Quality Planning and Standards, Research Triangle Park, NC.

USEPA, 1989b. *CERCLA Compliance with Other Laws Manual: Part II--Clean Air Act and Other Environmental Statutes and State Requirements*, EPA/540/G-89/009. OSWER Directive 9234.1-02, Office of Emergency and Remedial Response, Washington, DC.

USEPA, 1989c. *Ecological Assessments of Hazardous Waste Sites: A Field and Laboratory Reference Document*, EPA/600/3-89/013, Office of Research and Development—Corvallis Environmental Research Laboratory.

USEPA, 1989d. *Estimating Air Emissions from Petroleum UST Cleanups*, Office of Underground Storage Tanks, Washington, DC.

USEPA, 1989e. *Estimation of Air Emissions from Cleanup Activities at Superfund Sites*, Air/Superfund National Technical Guidance Study Series, *Volume 3*, EPA-450/1-89-003, Interim Final, Office of Air Quality Planning and Standards, Research Triangle Park, NC.

USEPA, 1989f. *Exposure Assessment Methods Handbook*, Office of Health and Environmental Assessment, USEPA, Cincinnati, OH.

USEPA, 1989g. *Exposure Factors Handbook*, EPA/600/8-89/043, Office of Health and Environmental Assessment, Washington, DC.

USEPA, 1989h. *Ground-water Sampling for Metals Analyses*, EPA/540/4-89-001, Office of Solid Waste and Emergency Response, Washington, DC.

USEPA, 1989i. *Interim Methods for Development of Inhalation Reference Doses*, EPA/600/8-88/066F, Office of Health and Environmental Assessment, Washington, DC.

USEPA, 1989j. *Interim Procedures for Estimating Risks Associated with Exposures to Mixtures of Chlorinated dibenzo-p-dioxins and -dibenzofurans (CDDs and CDFs)*, EPA/625/3-89/016, Risk Assessment Forum, Washington, DC.

USEPA, 1989k. *Methods for Evaluating the Attainment of Cleanup Standards. Volume I: Soils and Solid Media*, EPA/230/2-89/042, Office of Policy, Planning and Evaluation, Washington, DC.

USEPA, 1989l. *Procedures for Conducting Air Pathway Analyses for Superfund Applications, Volume IV—Procedures for Dispersion Modeling and Air Monitoring for Superfund Air Pathway Analysis*, Air/Superfund National Technical Guidance Study Series, EPA-450/1-89-004, Interim Final, Office of Air Quality Planning and Standards, Research Triangle Park, NC.

USEPA, 1989m. *Review and Evaluation of Area Source Dispersion Algorithms for Emission Sources at Superfund Sites*, EPA-450/4-89-020, Office of Air Quality Planning and Standards, Research Triangle Park, NC.

USEPA, 1989n. *Risk Assessment Guidance for Superfund, Volume I—Human Health Evaluation Manual (Part A)*, Interim Final. [Available at: www.epa.gov/superfund/programs/risk/ragsa/], EPA/540/1-89/002, Office of Emergency and Remedial Response, US Environmental Protection Agency, Washington, DC.

USEPA, 1989o. *Risk Assessment Guidance for Superfund. Volume II--Environmental Evaluation Manual*, EPA/540/1-89/001, Office of Emergency and Remedial Response, Washington, DC.

USEPA, 1989p. *Soil Sampling Quality Assurance User's Guide*, 2nd edition, EPA/600/8-89/046, Experimental Monitoring Support Laboratory (EMSL), ORD, USEPA, Las Vegas, NV.

USEPA, 1989q. *Soil Vapor Extraction VOC Control Technology Assessment*, EPA-450/4-89-017 (September 1989), Office of Air Quality Planning and Standards, Research Triangle Park, NC.

USEPA, 1989r. *User's Guide to the Contract Laboratory Program*, OSWER Dir. 9240.0-1, Office of Emergency and Remedial Response.

USEPA, 1990a. *Air Stripper Design Manual*, Air/Superfund National Technical Guidance Study Series, EPA-450/4-90-003 (May 1990), Office of Air Quality Planning and Standards, Research Triangle Park, NC.

USEPA, 1990b. *Air/Superfund National Technical Guidance Study Series, Development of Example Procedures for Evaluating the Air Impacts of Soil Excavation Associated with Superfund Remedial Actions*, EPA-450/4-90-014, Office of Air Quality Planning and Standards, Research Triangle Park, NC.

USEPA, 1990c. *Development of Risk Assessment Methodology for Surface Disposal of Municipal Sludge*, EPA/600/6-90/001, Office of Research and Development, Washington, DC.

USEPA, 1990d. *Emission Factors for Superfund Remediation Technologies*, Draft, Office of Air Quality Planning and Standards, Research Triangle Park, NC.

USEPA, 1990e. *Environmental Asbestos Assessment Manual, Superfund Method for the Determination of Asbestos in Ambient Air, Part 1: Method (EPA/540/2-90/005a) & Part 2: Technical Background Document (EPA/540/2-90/005b)*, US Environmental Protection Agency (USEPA), Washington, DC.

USEPA, 1990f. *Estimation of Baseline Air Emissions at Superfund Sites. Air/Superfund National Technical Guidance Study Series, Procedures for Conducting Air Pathway Analyses for Superfund Applications, Volume 2*, EPA-450/1-89-002a, Office of Air Quality Planning and Standards, Research Triangle Park, NC.

USEPA, 1990g. *Guidance for Data Useability in Risk Assessment*, EPA/540/G-90/008, Interim Final, Office of Emergency and Remedial Response, Washington, DC.

USEPA, 1990h. Hazard ranking system, *Federal Register*, 55(241), 51532–51666, EPA, 40 CFR Part 300, December 14, 1990.

USEPA, 1990i. *State of the Practice of Ecological Risk Assessment Document*, USEPA draft report, Office of Pesticides and Toxic Substances, Washington, DC.

USEPA, 1991a. *Conducting Remedial Investigations/Feasibility Studies for CERCLA Municipal Landfill Sites*, EPA/540/P-91/001 (OSWER Directive 9355.3-11), Office of Emergency and Remedial Response, Washington, DC.

USEPA, 1991b. *Guidance for Performing Site Inspections Under CERCLA—Interim Version*, OSWER Directive 9345.1-06, Office of Emergency and Remedial Response, Washington, DC.

USEPA, 1991c. *Guidelines for Developmental Toxicity Risk Assessment*, EPA/600/FR-91/001, Risk Assessment Forum, Office of Research and Development, USEPA, Washington, DC.

USEPA, 1991d. *Management of Investigation-Derived Wastes During Site Inspections*, EPA/540/G-91/009 (May 1991), Office of Emergency and Remedial Response, USEPA, Washington, DC.

USEPA, 1991e. *Risk Assessment Guidance for Superfund, Volume 1: Human Health Evaluation Manual (Part B, Development of Risk-Based Preliminary Remediation Goals)*, EPA/540/R-92/003, Office of Emergency and Remedial Response, Washington, DC.

USEPA, 1991f. *Risk Assessment Guidance for Superfund, Volume 1: Human Health Evaluation Manual (Part C, Risk Evaluation of Remedial Alternatives)*, EPA/540/R-92/004, Office of Emergency and Remedial Response, Washington, DC.

USEPA, 1991g. *Risk Assessment Guidance for Superfund, Volume I: Human Health Evaluation Manual. Supplemental Guidance, "Standard Default Exposure Factors"*, OSWER Directive: 9285.6-03 (March, 1991), Interim Final, US Environmental Protection Agency, Office of Emergency and Remedial Response, Washington, DC.

USEPA, 1991h. *The Role of Baseline Risk Assessment in Superfund Remedy Selection Decisions*, OSWER Directive: 9355.0-30, Office of Solid Waste and Emergency Response, Washington, DC.

USEPA, 1992a. *A Supplemental Guidance to RAGS: Calculating the Concentration Term*. Publication 9285.7-081, Office of Solid Waste and Emergency Response, US Environmental Protection Agency, Washington, DC.

USEPA, 1992c. *Estimating Potential for Occurrence of DNAPL at Superfund Sites*, Publication 9355.4-07FS, Office of Solid Waste and Emergency Response (OSWER), US Environmental Protection Agency, Washington, DC.

USEPA, 1992d. *Framework for Ecological Risk Assessment*, EPA/630/R-92/001 (February, 1992), Risk Assessment Forum, Washington, DC.

USEPA, 1992e. *Guidance Document on the Statistical Analysis of Ground-Water Monitoring Data at RCRA Facilities*, EPA/530/R-93/003, Office of Solid Waste, US Environmental Protection Agency, Washington, DC.

USEPA, 1992f. *Guidance for Data Useability in Risk Assessment (Parts A & B)*, Publication No. 9285.7-09A&B, Office of Emergency and Remedial Response, USEPA, Washington, DC.

USEPA, 1992g. *Guideline for Predictive Baseline Emissions Estimation Procedures for Superfund Sites*, Air/Superfund National Technical Guidance Study Series, EPA 450/I-92-002, Interim Final, Office of Health and Environmental Assessment, USEPA, Washington, DC.

USEPA, 1992h. *Guidelines for Exposure Assessment*, EPA/600/Z-92/001, Risk Assessment Forum, Office of Research and Development, Office of Health and Environmental Assessment, USEPA, Washington, DC.

USEPA, 1992i. *Methods for Evaluating the Attainments of Cleanup Standards: Volume 2: Ground Water*, EPA/230/R-92/014, Office of Policy, Planning, and Evaluation, US Environmental Protection Agency, Washington, DC.

USEPA, 1992j. *Supplemental Guidance to RAGS: Calculating the Concentration Term*, Publication No. 9285.7-08I, Office of Emergency and Remedial Response, USEPA, Washington, DC.

USEPA, 1993a. *National Ambient Air Quality Standard for Particulate Matter*, 40 CFR, Part 50.6, Office of Emergency and Remedial Response, Washington, DC.

USEPA, 1993b. *Selection Criteria for Mathematical Models Used in Exposure Assessments: Atmospheric Dispersion Models*, EPA-600/8-91/038, Office of Research and Development, US Environmental Protection Agency, Washington, DC.

USEPA, 1993c. *Supplemental Guidance to RAGS: Estimating Risk from Groundwater Contamination*, Office of Solid Waste and Emergency Response, USEPA, Washington, DC.

USEPA, 1993d. *Wildlife Exposure Factors Handbook, Volumes 1 and 2*, EPA/600/R-93/187a & EPA/600/R-93/187b, Office of Research and Development, US Environmental Protection Agency, Washington, DC.

USEPA, 1994a. *Ecological Risk Assessment Issue Papers*, EPA/630/R-94/009, Risk Assessment Forum, Washington, DC.

USEPA, 1994b. *Emerging Technology Report—Cross-Flow Pervaporation System for Removal of VOCs from Contaminated Wastewater*, EPA/540/R-94/512, Office of Research and Development, US Environmental Protection Agency, Cincinnati, OH.

USEPA, 1994c. *Estimating Exposures to Dioxin-like Compounds*, EPA/600/6-88/005Cb, Office of Research and Development, USEPA, Washington, DC.

USEPA, 1994d. *Estimating Radiogenic Cancer Risks*, EPA 402-R-93-076, USEPA, Washington, DC.

USEPA, 1994e. *Guidance for the Data Quality Objectives Process*, EPA/600/R-96/055, Office of Research and Development, Washington, DC.

USEPA, 1994f. *Guidance Manual for the Integrated Exposure Uptake Biokinetic Model for Lead in Children*, EPA/540/R-93/081, Office of Emergency and Remedial Response, US Environmental Protection Agency (USEPA), Washington, DC.

USEPA, 1994h. *Statistical Methods for Evaluating the Attainment of Cleanup Standards, Volume 3: Reference-Based Standards for Soils and Solid Media*, EPA 230-R-94-004, US EPA Office of Policy, Planning, and Evaluation, Washington, DC.

USEPA, 1995–1996. *Guidance for Assessing Chemical Contaminant Data for Use in Fish Advisories, Volumes 1–5*, Office of Water, USEPA, Washington, DC.

USEPA, 1995a. *A Guide to the Biosolids Risk Assessments for the EPA Part 503 Rule*, EPA/832-B-93-005, Office of Wastewater Management, USEPA, Washington, DC.

USEPA, 1995b. *Ecological Risk: A Primer for Risk Managers*, EPA/734/R-95/001, Washington, DC.

USEPA, 1995c. *Guidance for Risk Characterization*, Science Policy Council, USEPA, Washington, DC.

USEPA, 1995d. *Surfactant Injection for Ground Water Remedi-ation: State Regulators' Perspectives and Experiences*, EPA/542/R-95/011, Technology Innovation Office, US Environmental Protection Agency, Washington, DC.

USEPA, 1995e. *The Use of the Benchmark Dose Approach in Health Risk Assessment*, EPA/630/R-94/007, Risk Assessment Forum, Office of Research and Development, US Environmental Protection Agency (USEPA), Washington, DC.

USEPA, 1996a. *A Citizen's Guide to Treatment Walls*, EPA/542/F-96/016, Office of Solid Waste and Emergency Response, US Environmental Protection Agency, Washington, DC.

USEPA, 1996b. *Data Quality Evaluation Statistical Toolbox (DataQUEST)*, EPA QA/G-9D, EPA/600/R-96/084, Office of Research and Development, US Environmental Protection Agency, Washington, DC.

USEPA, 1996c. *Guidance for Data Quality Assessment, Practical Methods for Data Analysis*, EPA QA/G-9, QA96 Version, EPA/600/R-96/084, US EPA Office of Research and Development, Washington, DC.

USEPA, 1996d. *Guidelines for Developmental Toxicity Risk Assessment*, 51 FR 34028-34040, October1996, US Environmental Protection Agency, Washington, DC.

USEPA, 1996e. *Guidelines for Reproductive Toxicity Risk Assessment*, EPA/630-96/009, USEPA, Washington, DC.

USEPA, 1996f. *Interim Approach to Assessing Risks Associated with Adult Exposures to Lead in Soil*, US Environmental Protection Agency (USEPA), Washington, DC.

USEPA, 1996g. *PCBs: Cancer Dose-Response Assessment and Application to Environmental Mixtures*, EPA/600/P-96/001F (September 1996), National Center for Environmental Assessment, Office of Research and Development, US Environmental Protection Agency, Washington, DC.

USEPA, 1996h. *Proposed Guidelines for Carcinogenic Risk Assessment*, EPA/600/P-92/003C (April 1996), US Environmental Protection Agency, Washington, DC.

USEPA, 1996i. *Radiation Exposure and Risk Assessment Manual*, EPA 402-R-96-016, US Environmental Protection Agency, Washington, DC.

USEPA, 1996j. *Soil Screening Guidance: Technical Background Document*, EPA/540/R-95/128, Office of Emergency and Remedial Response, Washington, DC.

USEPA, 1996k. *Soil Screening Guidance: User's Guide*, EPA/540/R-96/018, Office of Emergency and Remedial Response, Washington, DC.

USEPA, 1996l. *State Policies Concerning the Use of Injectants for In Situ Groundwater Remediation*, EPA/542/R-96/001, Technology Innovation Office, US Environmental Protection Agency, Washington, DC.

USEPA, 1997a. *Data Quality Evaluation Statistical Toolbox (DataQUEST) User's Guide*, EPA QA/G-9D QA96 Version, EPA/600/R-96/085. [Available at: www.epa.gov/quality/qs-docs/g9d-final.pdf; Software available at www.epa.gov/quality/qs-docs/dquest96.exe], Office of Research and Development, US Environmental Protection Agency, Washington, DC.

USEPA, 1997b. *Ecological Risk Assessment Guidance for Superfund: Process for Designing and Conducting Ecological Risk Assessments*, ERAGS, EPA 540-R-97-006, OSWER Directive # 9285.7-25, June 1997, US Environmental Protection Agency, Washington, DC.

USEPA, 1997d. *Estimating Radiogenic Cancer Risks*, USEPA, Washington, DC.

USEPA, 1997f. *Guiding Principles for Monte Carlo Analysis*, EPA/630/R-97/001, Office of Research and Development, USEPA, Washington, DC.

USEPA, 1997g. *Groundwater Currents*, EPA/542-N-97-006 (December 1997), US Environmental Protection Agency, Washington, DC.

USEPA, 1997h. *Policy for Use of Probabilistic Analysis in Risk Assessment*, EPA/630/R-97/001, Office of Research and Development, USEPA, Washington, DC.

USEPA, 1997i. *Site Conceptual Exposure Model Builder*, User Manual, U.S. Department of Energy, Office of Environmental Policy and Assistance, RCRA/CERCLA Division, EH-413, Washington, DC.

USEPA, 1997j. *The Brownfields Economic Redevelopment Initiative*, EPA 500-F-97-156, Office of Solid Waste and Emergency Response, US Environmental Protection Agency, Washington, DC.

USEPA, 1997k. *The Lognormal Distribution in Environmental Applications*, EPA/600/R-97/006, Office of Research and Development/Office of Solid Waste and Emergency Response, USEPA, Washington, DC.

USEPA, 1997l. *Users Guide for the Johnson and Ettinger (1991) Model for Subsurface Vapor Intrusion into Buildings*, Office of Emergency and Remedial Response, USEPA, Washington, DC.

USEPA, 1998a. *A Citizen's Guide to Phytoremediation*, EPA/542/F-98/011, Office of Solid Waste and Emergency Response, US Environmental Protection Agency, Washington, DC.

USEPA, 1998b. *Guidance for Data Quality Assessment, Practical Methods for Data Analysis*, EPA QA/G-9, QA97 Update, EPA/600/R-96/084, US EPA Office of Research and Development, Washington, DC.

USEPA, 1998c. *Guidelines for Ecological Risk Assessment*, EPA 630/R-95/002F (April 1998), US Environmental Protection Agency, Washington, DC.

USEPA, 1998d. *Guidelines for Neurotoxicity Risk Assessment*, EPA/630/R-95/001F, Risk Assessment Forum, Office of Research and Development, USEPA, Washington, DC.

USEPA, 1998e. *Human Health Risk Assessment Protocol for Hazardous Waste Combustion Facilities, Volumes 1–3*, EPA530-D-98-001A/-001B/-001C, Office of Solid Waste and Emergency Response, USEPA, Washington, DC.

USEPA, 1998f. *Risk Assessment Guidance for Superfund, Volume 1- Human Health Evaluation Manual, Part D*. [Available at www.epa.gov/superfund/programs/risk/ragsd/], Publication 9285:7-01D, Office of Solid Waste and Emergency Response, US Environmental Protection Agency, Washington, DC.

USEPA, 1998g. *Supplemental Guidance to RAGS: The Use of Probability Analysis in Risk Assessment (Part E), Draft Guidance*, Office of Solid Waste and Emergency Response, USEPA, Washington, DC.

USEPA, 1998h. *Technical Protocol for Evaluating Natural Attenuation of Chlorinated Solvents in Ground Water*, EPA/600/R-98/128 (September, 1998).

USEPA, 1999a. *Air Quality Criteria for Particulate Matter, Volumes 1–3*, EPA/600/P-99/002a, /002b, /002c, Office of Research and Development, USEPA, Washington, DC.

USEPA, 1999b. *Cancer Risk Coefficients for Environmental Exposure to Radionuclides*, Federal Guidance Report No. 13, EPA 402-R-99-001, Office of Radiation and Indoor Air, USEPA, Washington, DC.

USEPA, 1999c. *Estimating Radiogenic Cancer Risks, Addendum: Uncertainty Analysis*, EPA 402-R-99-003, USEPA, Washington, DC.

USEPA, 1999d. *Field Applications of In Situ Remediation Technologies: Permeable Reactive Barriers*, EPA/542/R-99/002, Office of Solid Waste and Emergency Response, US Environmental Protection Agency, Washington, DC.

USEPA, 1999e. *Phytoremediation Resource Guide*, EPA/542/B-99/003, US Environmental Protection Agency, USEPA, Washington, DC.

USEPA, 1999f. *Report of the Workshop on Selecting Input Distributions for Probabilistic Assessments*, EPA/630/R-98/004, Risk Assessment Forum, Office of Research and Development, USEPA, Washington, DC.

USEPA, 1999g. *Screening Level Ecological Risk Assessment Protocol, Appendix C: Media-to-Receptor BCF Values and Appendix D: Wildlife Measurement Assessment Receptor BCF Values*, US Environmental Protection Agency, USEPA, Washington, DC.

USEPA, 1999h. *Surfactant-Enhanced Subsurface Remediation to Remove DNAPL: Groundwater Currents*, EPA/542/N-99/006, Solid Waste and Emergency Response, US Environmental Protection Agency, Washington, DC.

USEPA, 1999i. *Three Dimensional DNAPL Fate and Transport Model*, Publication EPA/600/R-99/011 (February 1999), Office of Research and Development (ORD), US Environmental Protection Agency, Washington, DC.

USEPA, 1999j. *Understanding Variation in Partition Coefficient, K_d, Values, Volume 1 and 2*, EPA402-R-99-004A&B, Office of Radiation and Indoor Air, US Environmental Protection Agency, Washington, DC.

USEPA, 1999k. *Use of Monitored Natural Attenuation at Superfund, RCRA Corrective Action, and Underground Storage Tank Sites*, OSWER Directive 9200.4-17P (April 1999), EPA 540-F-99-009, USEPA, Washington, DC.

USEPA, 2000a. *CatReg Software Documentation*, EPA/600/R-98/053F, Office of Research and Development, National Center for Environmental Assessment, US Environmental Protection Agency, Research Triangle Park, NC.

USEPA, 2000b. *CatReg Software User Manual*, EPA EPA/600/R-98/052F, Office of Research and Development, National Center for Environmental Assessment, US Environmental Protection Agency, Research Triangle Park, NC.

USEPA, 2000c. *Data Quality Objectives Process for Hazardous Waste Site Investigations.* [Available at: www.epa.gov/quality/qs-docs/g4hw-final.pdf], EPA QA/G-4HW, EPA/600/R-00/007, Final. Office of Environmental Information, US Environmental Protection Agency, Washington, DC.

USEPA, 2000d. *Guidance for Data Quality Assessment: Practical Methods for Data Analysis.* [Available at: www.epa.gov/r10earth/offices/oea/epaqag9b.pdf], EPA QA/G-9, EPA/600/R-96/084, QA00 Update. Office of Environmental Information, US Environmental Protection Agency, Washington, DC.

USEPA, 2000e. *Guidance for the Data Quality Objectives Process*, EPA QA/G-4, EPA/600/R-96/055, Office of Environmental Information, US Environmental Protection Agency, Washington, DC.

USEPA, 2000f. *Innovations in Site Characterization: Geophysical Investigation at Hazardous Waste Sites*, EPA-542-R-00-003, Office of Solid Waste and Emergency Response, US Environmental Protection Agency, Washington, DC.

USEPA, 2000g. *Soil Screening Guidance for Radionuclides: Technical Background Document*, EPA/540-R-00-006, Office of Radiation and Indoor Air, US Environmental Protection Agency, Washington, DC.

USEPA, 2000h. *Soil Screening Guidance for Radionuclides: User's Guide*, EPA/540-R-00-007, Office of Radiation and Indoor Air, US Environmental Protection Agency, Washington, DC.

USEPA, 2000i. *Introduction to Phytoremediation*, EPA/600/R-99/107, Office of Research and Development, US Environmental Protection Agency, Washington, DC.

USEPA, 2000j. *A Resource for MGP Site Characterization and Remediation*, EPA-542-R-00-005, Office of Solid Waste and Emergency Response, US Environmental Protection Agency, Washington, DC.

USEPA, 2001a. *EPA Requirements for Quality Assurance Project Plans*, EPA QA/R-5, EPA/240/B-01/003, Office of Environmental Information, US Environmental Protection Agency, Washington, DC.

USEPA, 2001b. *EPA Requirements for Quality Management Plans*, EPA QA/R-2, EPA/240/B-01/002, Office of Environmental Information, US Environmental Protection Agency, Washington, DC.

USEPA, 2001c. *Guidance for Preparing Standard Operating Procedures (SOPs)*, EPA QA/G-6, EPA/240/B-01/004, Office of Environmental Information, US Environmental Protection Agency, Washington, DC.

USEPA, 2001d. *Providing Solutions for a Better Tomorrow—Reducing the Risks Associated with Lead in Soil*, EPA/600/F-01/014, Office of Research and Development, US Environmental Protection Agency (USEPA), Washington, DC.

USEPA, 2001e. *Reusing Cleaned Up Superfund Sites: Recreational Use of Land Above Hazardous Waste Containment Areas*, EPA 540-K-01-002, Office of Emergency and Remedial Response, US Environmental Protection Agency, Washington, DC.

USEPA, 2001f. *Risk Assessment Guidance for Superfund: Volume I, Human Health Evaluation Manual (Part D, Standardized Planning, Reporting, and Review of Superfund Risk Assessments)*, Publication 9285.7-47, Office of Emergency and Remedial Response, US Environmental Protection Agency, Washington, DC.

USEPA, 2001g. *Risk Assessment Guidance for Superfund, Volume 3- Part A, Process for Conducting a Probabilistic Risk Assessment Draft.* Office of Solid Waste and Emergency Response, US Environmental Protection Agency, Washington, DC.

USEPA, 2001h. *The Data Quality Objectives Decision Error Feasibility Trials (DEFT) Software*, EPA QA/G-4D, EPA/240/B-01/007, Office of Environmental Information, US Environmental Protection Agency, Washington, DC.

USEPA, 2002a. *Calculating Upper Confidence Limits for Exposure Point Concentrations at Hazardous Waste Sites*, OSWER 9285.6-10 (December 2002), Office of Emergency and Remedial Response, US Environmental Protection Agency, Washington, DC.

USEPA, 2002b. *Elements for Effective Management of Operating Pump and Treat Systems*, EPA-542-R-02-009, Office of Solid Waste and Emergency Response, US Environmental Protection Agency, Washington, DC.

USEPA, 2002c. *Groundwater Remedies Selected at Superfund Sites*, EPA-542-R-01-022, Office of Solid Waste and Emergency Response, US Environmental Protection Agency, Washington, DC.

USEPA, 2002d. *Innovations in Site Characterization: Geophysical Investigation at Hazardous Waste Sites*, EPA-542-R-00-003, Office of Solid Waste and Emergency Response, US Environmental Protection Agency, Washington, DC.

USEPA, 2002e. *Risk Assessment Guidance for Superfund, Volume I: Human Health Evaluation Manual (Part E, Supplemental Guidance for Dermal Risk Assessment)*, EPA/540/R/99/005, Office of Emergency and Remedial Response, USEPA, Washington, DC.

USEPA, 2002f. *Role of Background in the CERCLA Cleanup Program*, OSWER 9285.6-07P, Office of Emergency and Remedial Response, US Environmental Protection Agency, Washington, DC.

USEPA, 2003a. *Framework for Cumulative Risk Assessment*, EPA/600/P-02/001F, National Center for Environmental Assessment, Risk Assessment Forum, US Environmental Protection Agency, Washington, DC.

USEPA, 2003b. *Human Health Toxicity Values in Superfund Risk Assessments*, OSWER Directive 9285.8-53 (December 5, 2003), Office of Superfund Remediation and Technology Innovation, Washington, DC.

USEPA, 2003c. *Recommendations of the Technical Review Workgroup for Lead for an Approach to Assessing Risks Associated with Adult Exposures to Lead in Soil*, Final, EPA-540-R-03-001 (January).

USEPA, 2004a. *Community Air Screening How-To Manual, A Step-by-Step Guide to Using Risk-Based Screening to Identify Priorities for Improving Outdoor Air Quality*, EPA 744-B-04-001, US Environmental Protection Agency, Washington, DC.

USEPA, 2004b. *Risk Assessment Guidance for Superfund: Volume 1—Human Health Evaluation Manual (Part E, Supplemental Guidance for Dermal Risk Assessment)*, Final, EPA/540/R/99/005 (July 2004), Office of Emergency and Remedial Response, US Environmental Protection Agency, Washington, DC.

USEPA, 2004c. *Example Exposure Scenarios*, EPA/600/R-03/036, Center for Environmental Assessment, US Environmental Protection Agency, Washington, DC.

USEPA, 2004d. *Risk Assessment Principles and Practices*, Staff Paper, EPA/100/B-04/001, Office of the Science Advisor, US Environmental Protection Agency, Washington, DC.

USEPA, 2004e. *User's Guide for Evaluating Subsurface Vapor Intrusion Into Buildings*, Office of Emergency and Remedial Response, Washington, DC.

USEPA, 2004f. *Air Quality Criteria for Particulate Matter: Volume II*, EPA/600/P-95/001bF, Office of Research and Development, Washington, DC.

USEPA, 2004g. *An Examination of EPA Risk Assessment Principles and Practices*, Office of the Science Advisor, Environmental Protection Agency, Washington, DC.

USEPA, 2005a. *Guidelines for Carcinogen Risk Assessment*, EPA/630/P-03/001B (March 2005), US EPA, Office of Research and Development, National Center for Environmental Assessment, Washington, DC.

USEPA, 2005b. *Supplemental Guidance for Assessing Susceptibility from Early-Life Exposure to Carcinogens*, EPA/630/R-03/003F (March 2005), US EPA, Office of Research and Development, National Center for Environmental Assessment, Washington, DC.

USEPA, 2005c. *Guidance on Selecting Age Groups for Monitoring and Assessing Childhood Exposures to Environmental Contaminants*, EPA/630/P-03/003F, Risk Assessment Forum, Washington, DC.

USEPA, 2005d. *Contaminated Sediment Remediation Guidance for Hazardous Waste Sites*, OSWER 9355.0-85, EPA-540-R-05-012, US Environmental Protection Agency, Office of Solid Waste and Emergency Response, Washington, DC.

USEPA, 2006a. *Approaches to the Application of Physiologically Based Pharmacokinetic (PBPK) Models and Supporting Data in Risk Assessment (Final Report)*, EPA/600/R-05/043F, National Center for Environmental Assessment, NCEA, US Environmental Protection Agency, Washington, DC.

USEPA, 2006b. *A Framework for Assessing Health Risk of Environmental Exposures to Children (Final)*, EPA/600/R-05/093F, 2006, National Center for Environmental Assessment, US Environmental Protection Agency, Washington, DC.

USEPA, 2007a. *Framework for Metals Risk Assessment*, EPA/120/R-07/001, Environmental Protection Agency, Washington, DC.

USEPA, 2007b. *Concepts, Methods and Data Sources for Cumulative Health Risk Assessment of Multiple Chemicals, Exposures and Effects: A Resource Document*, EPA/600/R-06/013F, ORD, NCEA, US Environmental Protection Agency, Cincinnati, OH.

USEPA, 2007c. *Risk Communication in Action: The Risk Communication Workbook*, EPA, Washington, DC.

USEPA, 2008. *A Systematic Approach for Evaluation of Capture Zones at Pump and Treat Systems Final Project Report*, EPA 600/R-08/003, US Environmental Protection Agency, Washington, DC.

USEPA, 2009. *Risk Assessment Guidance for Superfund Volume I: Human Health Evaluation Manual (Part F, Supplemental Guidance for Inhalation Risk Assessment)*, EPA-540-R-070-002, OSWER 9285.7-82 (January 2009), Office of Superfund Remediation and Technology Innovation, Environmental Protection Agency, Washington, DC.

USEPA, 2010a. *Quantitative Health Risk Assessment for Particulate Matter*, EPA-452/R-10-005, US EPA, Research Triangle Park, NC.

USEPA, 2010b. *Recommended Toxicity Equivalence Factors (TEFs) for Human Health Risk Assessments of 2,3,7,8-Tetrachlorodibenzo-p-dioxin and Dioxin-Like Compounds*, EPA/600/R-10/005 (December), Risk Assessment Forum, Washington, DC.

USEPA, 2011a. *Exposure Factors Handbook: 2011 Edition*, EPA/600/R-09/052F, US EPA, Office of Research and Development, National Center for Environmental Assessment, Washington, DC.

USEPA, 2011b. *Environmental Cleanup Best Management Practices: Effective Use of the Project Life cycle Conceptual Site Model*, US Environmental Protection Agency, Washington, DC.

USEPA, 2012. Benchmark dose technical guidance, EPA/100/R-12/001, Risk Assessment Forum, U.S. Environmental Protection Agency, Washington, DC.

USEPA, 2013a. *ProUCL Version 5.0.00. Statistical Software for Environmental Applications for Data Sets with and without Nondetect Observations*, Technical Support Center for Monitoring and Site Characterization, US Environmental Protection Agency, Washington, DC, Updated September 19, 2013.

USEPA, 2013b. *ProUCL Version 5.0.00. Technical Guide (draft)*, EPA/600/R-07/041 (September 2013), Office of Research and Development, US Environmental Protection Agency, Washington, DC.

USEPA, 2013c. *ProUCL Version 5.0.00 User Guide*, EPA/600/R-07/041 (September 2013), Office of Research and Development, US Environmental Protection Agency, Washington, DC.

USEPA, 2014a. *Human Health Evaluation Manual, Supplemental Guidance: Update of Standard Default Exposure Factors*, OSWER Directive 9200.1-120 (February 6, 2014), Assessment and Remediation Division, Office of Superfund Remediation and Technology Innovation, US Environmental Protection Agency, Washington, DC.

USEPA, 2014b. *Groundwater Remedy Completion Strategy. Moving Forward with an End in Mind*, OSWER Directive 9200.2-144, US Environmental Protection Agency, Washington, DC.

USEPA, 1987d. *Handbook for Conducting Endangerment Assessments*, USEPA, Research Triangle Park, NC.

USEPA-NWWA, 1989. *Handbook of Suggested Practices for the Design and Installation of Ground Water Monitoring Wells*, National Water Well Association, Dublin, OH.

Van Bavel, CHM, 1996. Potential evapotranspiration: the combination concept and its experimental verification, *Water Resources Research*, 2: 445–467.

van Emden, HF and DB Peakall, 1996. *Beyond Silent Spring*, Kluwer Academic Publishers, Dordrecht, The Netherlands.

Van Emon, JM and CL Gerlach, 1995. A status report on field-portable immunoassay, *Journal of Environmental Science and Technology*, 29(7): 312A–317A.

van Leeuwen, CJ and JLM Hermens (eds.), 1995. *Risk Assessment of Chemicals: An Introduction*, Kluwer Academic Publishers, Dordrecht, The Netherlands.

van Leeuwen, FXR, 1997. Derivation of toxic equivalency factors (TEFs) for dioxin-like compounds in humans and wildlife, *Organohalogen Componds*, 34: 237–269.

van Leeuwen, K, 1990. Ecotoxicological effects assessment in the Netherlands, *Environmental Management*, 14: 779–792.

Van Ryzin, J, 1980. Quantitative risk assessment, *Journal of Occupational Medicine*, 22: 321–326.

van Veen, MP, 1996. A general model for exposure and uptake from consumer products, *Risk Analysis*, 16(3): 331–338.

van Wijnen, JH, P Clausing, and B Brunekreef, 1990. Estimated soil ingestion by children, *Environmental Research*, 51: 147–162.

Vandegrift, GF, DT Reed, and IR Tasker (eds.), 1992. *Environmental Remediation: Removing Organic and Metal Ion Pollutants*, ACS Symposium Series 509, American Chemical Society, Washington, DC.

Vane, LM, L Hitchens, FR Alvarez, EL Giroux, 2000. Field demonstration of pervaporation for the separation of volatile organic compounds from a surfactant-based soil remediation fluid, *Journal of Hazardous Materials*, B81: 141–166.

Vasudev, D, T Ledder, S Dushenkov, A Epstein, N Kumar, Y Kapulnik, B Ensley, G Huddleston, J Cornish, I Raskin, B Sorochinsky, M Ruchko, A Prokhnevsky, A Mikheev, and D Grodzinsky, 1996. Removal of Radionuclide Contamination from Water by Metal-Accumulating Terrestrial Plants. In *Presentation at the In-Situ Soil and Sediment Remediation Conference*, New Orleans, LA.

Vaughan, E, 1995. The significance of socioeconomic and ethnic diversity for the risk communication process, *Risk Analysis*, 15(2): 169–180.

Vermeire, TG, P van der Poel, R van de Laar, and H Roelfzema, 1993. Estimation of consumer exposure to chemicals: application of simple models, *Science of the Total Environment*, 135: 155–176.

Verschueren, K, 1983. *Handbook of Environmental Data on Organic Chemicals*, 2nd edition, Van Nostrand Reinhold Co., New York.

Vesley, D, 1999. *Human Health and the Environment: A Turn of the Century Perspective*, Kluwer Academic Publishers, Dordrecht, The Netherlands.

Viessman, W, GL Lewis, and JW Knapp, 1989. *Introduction to Hydrology*, 3rd edition, Harper & Row Publishers, New York.

Visschers, VHM, RM Meertens, WWF Passchier, and NNK De Vries, 2009. Probability information in risk communication: A review of the research literature, *Risk Analysis*, 29(2): 267–287.

Vogel, TM and PL McCarty, 1985. Biotransformation of tetrachloroethylene to trichloroethylene, dichloroethylene, vinyl chloride, and carbon dioxide under methanogenic conditions, *Applied and Environmental Microbiology*, 49(5): 1080–1083.

Volpp, C, 1988. "Is It Safe or Isn't It?": An Overview of Risk Assessment, *Water Resource News*, New Jersey Department of Environmental Protection, Division of Water Resources, 4(1), New Jersey.

Vorhees, DJ, AC Cullen, and LM Altshul, 1997. Exposure to polychlorinated biphenyls in residential indoor air and outdoor air near a Superfund site, *Environmental Science and Technology*, 31: 3612–3618.

Vroblesky, DA and Campbell, TR, 2001, Equilibration times, stability, and compound selectivity of diffusion samplers for collection of ground-water VOC concentrations, *Advances in Environmental Research*, 5(1): 1–12.

Vroblesky, DA and Hyde, WT, 1997, Diffusion samplers as an inexpensive approach to monitoring VOCs in ground water, *Ground Water Monitoring and Remediation*, 17(3): 177–184.

Vroblesky, DA, 2000, Simple, Inexpensive Diffusion Samplers for Monitoring VOCs in Ground Water, In *The Second International Conference on Remediation of Chlorinated and Recalcitrant Compounds*, May 22–25, 2000, Monterey, CA.

Vukovic, M and A Soro, 1992. *Determination of Hydraulic Conductivity of Porous Media from Grain-Size Composition*, Water Resources Publications, Littleton, CO.

Wade, A, GW Wallace, SF Siegwald, WA Lee, and KC McKinney, 1996. Performance Comparison Between a Horizontal and a Vertical Air Sparging Well: A Full-Scale One-Year Pilot Study, In *Proceedings of 10th National Outdoor Action Conference Expo*, pp. 189–206.

Walker, CH, 2008. *Organic Pollutants: An Ecotoxicological Perspective* (2nd ed.), CRC Press, Boca Raton, FL.

Wallace, L, 1996. Indoor particles: a review, *Journal of the Air and Waste Management Association*, 46: 98–126.

Walton, WC, 1984. *Practical Aspects of Ground Water Modeling*, National Water Well Association, Westerville, OH.

Washburn, ST and KG Edelmann, 1999. Development of risk-based remediation strategies, *Practice Periodical of Hazardous, Toxic, and Radioactive Waste Management*, 3(2): 77–82.

Watts, RJ, 1998. *Hazardous Wastes: Sources, Pathways, Receptors*, John Wiley & Sons, New York.

Weast, RC (ed.), 1984. *Handbook of Chemistry and Physics*, 65th edition, CRC Press, Boca Raton, FL.

Weeks, KR, CJ Bruell, and NR Mohanty, 2000. Use of Fenton's reagent for the degradation of TCE in aqueous systems and soil slurries, *Journal of Soil and Sediment Contamination*, 9(4): 331–345.

Weerasooriya, V, SL Yeh, GA Pope, and WH Wade, 2000. *Integrated Demonstration of Surfactant-Enhanced Aquifer Remediation with Surfactant Regeneration and Reuse*, ACS Symposium Series 740, Chapter 3, American Chemical Society, Washington, DC.

Weinstein, MC, HV Fineberg, AS Elstein, HS Frazier, D Neuhauser, RR Neutra, and BJ McNeil, 1980. *Clinical Decision Analysis*. W.B. Saunders, Philadelphia, PA.

Weinstein, ND and PM Sandman, 1993. Some criteria for evaluating risk messages, *Risk Analysis*, 13(1): 103–114.

Weinstein, ND, K Kolb, and BD Goldstein, 1996. Using time intervals between expected events to communicate risk magnitudes, *Risk Analysis*, 16(3): 305–308.

Weisman, J, 1996. AMS adds realism to chemical risk assessment, *Science*, 271(5247): 286–287.

Weiss, EB (ed.), 1992. *Environmental Change and International Law: New Challenges and Dimensions*, United Nations University Press, Tokyo, Japan.

Wentsel, RS, TW LaPoint, M Simini, RT Checkai, D Ludwig, and LW Brewer, 1994. *Tri-Service Procedural Guidelines for Ecological Risk Assessments*, ADA297968. U.S. Army ERDEC, Aberdeen Proving Ground, MD.

Weschler, CJ and WW Nazaroff, 2008. Semivolatile organic compounds in indoor environments, *Atmospheric Environment*, 42(40): 9018–9040.

Weschler, CJ and WW Nazaroff, 2010. SVOC partitioning between the gas phase and settled dust indoors, *Atmospheric Environment*, 44(30): 3609–3620.

West, GB, JH Brown, and BJ Enquist, 1997. A general model of the origin of allometric scaling laws in biology, *Science*, 276: 122–126.

Wheeler, MW and AJ Bailer, 2007. Properties of model-averaged BMDLs: A study of model averaging in dichotomous response risk estimation, *Risk Analysis*, 27, 659–670.

Wheeler, MW and AJ Bailer, 2009. Comparing model averaging with other model selection strategies for benchmark dose estimation, *Environmental and Ecological Statistics*, 16, 37–51.

Whipple, C, 1987. *De Minimis Risk, Contemporary Issues in Risk Analysis, Volume 2*, Plenum Press, New York.

Whipple, W, Jr., 1994. *New Perspectives in Water Supply*, Lewis Publishers, Boca Raton, FL.

WHO (World Health Organization), 1983. Management of Hazardous Waste, WHO Regional European Series No.14.

WHO, 1990. *Public Health Impact of Pesticides Used in Agriculture*, World Health Organization and United Nations Environment Programme, WHO, Geneva, Switzerland.

WHO, 2010. WHO human health risk assessment toolkit: Chemical hazards. IPCS Harmonization Project Document No. 8, The International Programme on Chemical Safety (IPCS). WHO Press, World Health Organization, Geneva, Switzerland.

Whyte, AV and I Burton (eds.), 1980. *Environmental Risk Assessment, SCOPE Report 15*, John Wiley & Sons, New York.

Wiedemeirer, TH, HS Rifai, CJ Newell, and JT Wilson, 1999. *Natural Attenuation of Fuels and Chlorinated Solvents in the Subsurface*, John Wiley & Sons, New York.

Wiens, JA and KR Parker, 1995. Analyzing the effects of accidental environmental impacts: approaches and assumptions, *Ecological Applications*, 5(4): 1069–1083.

Wilkes, CR, MJ Small, CI Davidson, and JB Andelman, 1996. Modeling the effects of water usage and co-behavior on inhalation exposures to contaminant volatilized from household water, *Journal of Exposure and Environmental Epidemiology*, 6: 393–412.

Williams, DR, JC Paslawski, and GM Richardson, 1996. Development of a screening relationship to describe migration of contaminant vapors into buildings, *Journal of Soil Contamination*, 5(2): 141–156.

Williams, FLR and SA Ogston, 2002. Identifying populations at risk from environmental contamination from point sources, *Occupational and Environmental Medicine*, 59(1): 2–8.

Williams, JD, IH Coulson, P Susitaival, and SM Winhoven, 2008. An outbreak of furniture dermatitis in the UK. *British Journal of Dermatology*, 159(1): 233–234.

Williams, JR, 1975. Sediment-yield prediction with the universal equation using runoff energy factor. In Present and prospective technology for predicting sediment yields and sources. U.S. Department of Agriculture. ARS-S-40.

Williams, PL and JL Burson (eds.), 1985. *Industrial Toxicology*, Van Nostrand Reinhold Co., New York.

Willis, MC, 1996. *Medical Terminology: The Language of Health Care*, Williams & Wilkins, Baltimore, MD.

Wilson, DJ, 1995. *Modeling of In Situ Techniques for Treatment of Contaminated Soils*, Technomic Publishing Co., Inc., Lancaster, PA.

Wilson, JD, 1996. Threshold for carcinogens: a review of the relevant science and its implications for regulatory policy, *Discussion Paper No. 96-21*, Resources for the Future, Washington, DC.

Wilson, JD, 1997. So carcinogens have thresholds: how do we decide what exposure levels should be considered safe? *Risk Analysis*, 17(1): 1–3.

Wilson, JT and R Kolhatkar, 2002. Role of natural attenuation in life cycle of MTBE plumes, *Journal of Environmental Engineering*, 128(9): 876–882.

Wilson, LG, LG Everett, and SJ Cullen (eds.), 1995. *Handbook of Vadose Zone Characterization and Monitoring*, CRC Press, Boca Raton, FL.

Wilson, R and EAC Crouch, 1987. Risk assessment and comparisons: an introduction, *Science*, 236: 267–270.

Winegardner, DL and SM Testa, 2000. *Restoration of Contaminated Aquifers: Petroleum Hydrocarbons and Organic Compounds*, 2nd edition, CRC Press, Boca Raton, FL.

Winter, CK, 1992. Dietary pesticide risk assessment, *Reviews of Environmental Contamination and Toxicology*, 127: 23–67.

Wong, A, 1993. A note on inference for the mean parameter of the gamma distribution, *Statistics and Probability Letters*, 17: 61–66.

Wong, MH (ed.), 2012. *Environmental Contamination: Health Risks and Ecological Restoration*, CRC Press, Boca Raton, FL.

Wonnacott, TH and RJ Wonnacott, 1972. *Introductory Statistics*, 2nd edition, John Wiley & Sons, New York.

Wood, TK, H Shim, D Ryoo, JS Gibbons, and JG Burken, 2000. Root-Colonizing Genetically-Engineered Bacteria for Trichloroethylene Phytoremediation, In *Presentation at the 2 nd International Conference on Remediation of Chlorinated and Recalcitrant Compounds*, May 22–25, Monterey, CA.

Woodside, G, 1993. *Hazardous Materials and Hazardous Waste Management: A Technical Guide*, John Wiley & Sons, New York.

The World Bank, 1985. *Manual of Industrial Hazard Assessment Techniques*, Office of Environment and Scientific Affairs, Washington, DC.

The World Bank, 1989. *Striking a Balance—The Environmental Challenge of Development*, IBRD/The World Bank, Washington, DC.

Worster, D, 1993. *The Wealth of Nature: Environmental History and the Ecological Imagination*, Oxford University Press, New York.

WPCF (Water Pollution Control Federation), 1988. *Hazardous Waste Site Remediation: Assessment and Characterization*, A Special Publication of the WPCF, Technical Practice Committee, Alexandria, VA.

WPCF, 1990. *Hazardous Waste Site Remediation Management*, A Special Publication of the WPCF, Technical Practice Committee, Alexandria, VA.

Yakowitz, H, 1989. *Monitoring and Control of Transfrontier Movements of Hazardous Wastes: An International Overview*, Technical Paper W/0587M, OECD, Paris, France.

Yakowitz, H, 1990. Monitoring and control of transfrontier movements of hazardous wastes: an international overview. In *The Management of Hazardous Substances in the Environment*, KL Zirm and J Mayer (eds.), Elsevier Applied Science, Elsevier Science Publishers Ltd., pp. 139–162, London, UK.

Yan, YE and FW Schwartz, 1999. Oxidative degradation and kinetics of chlorinated ethylenes by potassium permanganate, *Journal of Contaminant Hydrology*, 37: 343–365.

Yang, JT and WE Bye, 1979. *A Guidance for Protection of Ground Water Resources from the Effects of Accidental Spills of Hydrocarbons and Other Hazardous Substances*, EPA-570/9-79-017, USEPA, Washington, DC.

Yanosky, JD, CJ Paciorek, F Laden, JE Hart, RC Puett, D Liao, and HH Suh, 2014. Spatio-temporal modeling of particulate air pollution in the conterminous United States using geographic and meteorological predictors. *Environmental Health*, 13(1): 63.

Yaws, CL, 1999. *Chemical Properties Handbook*, McGraw-Hill, New York.

Yin, SCL, 1988. Modeling Groundwater Transport of Dissolved Gasoline, In *Specialty Conference Proceedings Joint CSCE-ASCE National Conference on Environmental Engineering*, Vancouver, BC, Canada, July 13–15, 1988, p. 544–551.

Yong, RN and CN Mulligan, 2003. *Natural Attenuation of Contaminants in Soils*, CRC Press, Boca Raton, FL.

Yong, RN, 2000. *Geoenvironmental Engineering: Contaminated Soils, Pollutant Fate, and Mitigation*, CRC Press, Boca Raton, FL.

Yong, RN, 2002. *Geoenvironmental Engineering: Contaminated Soils, Pollutant Fate, and Mitigation*, CRC Press, Boca Raton, FL.

Yong, RN, AMO Mohamed, and BP Warkentin, 1992. *Principles of Contaminant Transport in Soils*, Developments in Geotechnical Engineering, 73, Elsevier Scientific Publishers B.V., Amsterdam, The Netherlands.

Zakrzewski, SF, 1991. *Principles of Environmental Toxicology*, American Chemical Society (ACS), Washington, DC.

Zakrzewski, SF, 1997. *Principles of Environmental Toxicology*, 2nd edition, ACS Monograph 190, American Chemical Society, Washington, DC.

Zartarian, VG and JO Leckie, 1998. Dermal exposure: the missing link, *Environmental Science and Technology*, 32(5): 134A–137A.

Zauli Sajani, S, I Ricciardelli, A Trentini, D Bacco, C Maccone, S Castellazzi, P Lauriola, V Poluzzi, and RM Harrison, 2015. Spatial and indoor/outdoor gradients in urban concentrations of ultrafine particles and $PM_{2.5}$ mass and chemical components, *Atmospheric Environment*, 103: 307–320.

Zehnder, AJB (ed.), 1995. *Soil and Groundwater Pollution: Fundamentals, Risk Assessment and Legislation*, Kluwer Academic Publishers, Dordrecht, The Netherlands.

Zemba, SG, LC Green, EAC Crouch, and RR Lester, 1996. Quantitative risk assessment of stack emissions from municipal waste combusters, *Journal of Hazardous Materials*, 47: 229–275.

Zhang, Q, L Davis, and L Erickson, 2000. Transport of Methyl Tert-Butyl Ether (MTBE) Through Alfalfa Plants, In *Presentation at USEPA Phytoremediation State of the Science Conference*, May 1–2, Boston, MA.

Zhou, X-H and S Gao, 2000. One-sided confidence intervals for means of positively skewed distributions, *The American Statistician*, 54: 100–104.

Ziegler, J, 1993. Toxicity tests in animals: extrapolating to human risks, *Environmental Health Perspectives*, 101: 396–406.

Zirm, KL and J Mayer (eds.), 1990. *The Management of Hazardous Substances in the Environment*, Elsevier Applied Science, Elsevier Science Publishers Ltd., London, UK.

Zirschy, JH and DJ Harris, 1986. Geostatistical analysis of hazardous waste site data, *ASCE Journal of Environmental Engineering*, 112(4): 770–784.

Zogg, HA, 1987. *"Zurich" Hazard Analysis*, Zurich Insurance Group, Risk Engineering, Zurich, Switzerland.

Zoller, U (ed.), 1994. *Groundwater Contamination and Control*, Marcel Dekker, New York.

Index

A

Absorbed dose, 192, 203, 204, 232, 236, 237, 474, 475
Absorption
    fraction, 232, 467
    factor, 192, 196, 198–204, 206, 237, 267, 291, 292, 302, 303, 305, 307, 506
Acceptable chemical exposure level (ACEL)
    for carcinogenic chemicals, 290–291
    cleanup criteria, 296, 308, 309
    for noncarcinogenic chemicals, 291–292
    risk-specific concentrations, air, 293
    risk-specific concentrations, water, 293–294
    soil chemical limits, 300–301
    tolerable chemical concentrations, 294–295
Acceptable daily intake (ADI), 181, 221, 294–296, 467
Acceptable risk, 72, 161, 281, 295, 298, 301, 303, 305, 307, 313, 372, 397, 419, 425, 432, 444, 467, 490
Acceptable soil concentration (ASC), 148, 149, 300, 301, 303
Action level, 25, 94, 95, 112, 394, 415, 467
Activated carbon, 318, 319, 348, 391, 459, 467
Activated carbon adsorption
    defined, 318, 467
    technology, restoration methods/technologies, 318
Acute exposure, 86, 90–91, 209, 467
Acute toxicity, 21, 467
Adsorption, 122, 125, 126, 129, 132, 139, 318, 334, 340, 352, 353, 400, 417, 425, 429, 468, 474, 484, 505; *see also* Sorption
    activated carbon, 316, 318, 402, 467
    coefficients, 122, 124, 468, 481
Aggregate effects, chemical mixtures, 237–238
AIR3D, 504
Airborne pollutants
    airborne dust/particulate concentrations, 139–140
    vapor concentrations, VOCs, 140–142
    vapor intrusion (VI), 167
Air contamination
    background samples, 65
    water/air partition coefficient (K_w), 124
Air emissions
    categories, 138
    classification, 137–138
    from contaminated sites, 137
    releases, 137, 142
    sources, 137–138
AIRFLOW/SVE, 504
Air pathway exposure assessment (APEA)
    components of, 138–139
    design of, 139
Air sparging, 322, 343, 349–351, 358, 400, 468
Air stripping, 341, 347–349, 357, 366, 391, 402, 428, 438, 459, 462, 468
Alternate concentration limit (ACL), 30
    applications, cleanup criteria, 31
Ambient air concentrations, estimate of, 141, 142
Analytical protocols, site characterization, 83–84

Antagonism, chemicals, 468
Anthropogenic levels, corrective action assessment, 63, 64
API DSS, 504
API PRDF, 505
Aquifers, *see also* Groundwater
    contamination problems, 402
    restoration programs, 402
Archetypical chemical exposure problems
    airborne problems, 197
    contaminated soil problems, 201–202
    dermal/skin exposures, 202–204
    foods products exposure, 200–201
    human exposure, 193–195
    ingestion/oral exposures, 198
    inhalation exposures, 196–198
    water pollution problems, 199
Arithmetic mean, 182, 183, 185–188, 469, 492, 501
Asphalt batching (incorporation), 319–320, 427
ASsessment Tools for the Evaluation of Risk (ASTER), 505
Atmospheric dispersion modeling, 137
ATSDR Toxicological Profiles, 518
Attenuation, 25, 43, 94, 107
Attenuation–dilution factors, 297
Augering, 104–105
    Auger, 82, 92, 104, 105, 117, 343, 469
Average daily dose (ADD)
    carcinogenic effects, 195
    noncarcinogenic effects, 195
Averaging techniques, 182, 184–188, 209
    average concentration, 185, 188

B

Background level, 58, 66, 78, 279, 426, 434, 447, 469
Background sampling
    air samples, corrective action assessment, 65
    criteria for corrective action assessment programs, 64
    groundwater samples, corrective action assessment, 65
    sediment samples, corrective action assessment, 64, 65
    soil samples, corrective action assessment, 64, 65
    surface water samples, corrective action assessment, 65
    uses, results, 65
Background threshold, corrective action assessment, 63–64
Baseline risk assessments, 157, 158, 166, 170, 176, 177, 274–275, 278, 384, 394, 399, 445, 457, 472, 489
Benchmark risk
    carcinogens, 291
    noncarcinogens, 451, 449
Bench-scale investigations, soil contamination, 399
Bioaccessibility, 470
Bioaccumulation
    and bioconcentration factor (BCF), 122, 126, 201, 306, 470
    and octanol/water partition coefficient (K_{ow}), 124, 481